SEDUM OF NORTH AMERICA
NORTH OF THE MEXICAN PLATEAU

SEDUM OF NORTH AMERICA
NORTH OF THE MEXICAN PLATEAU

by ROBERT T. CLAUSEN

drawings by ELFRIEDE ABBE

A COMSTOCK BOOK, published by
CORNELL UNIVERSITY PRESS | Ithaca and London

Copyright © 1975 by Cornell University

All rights reserved. Except for brief quotations in a review, this book, or parts thereof, must not be reproduced in any form without permission in writing from the publisher. For information address Cornell University Press, 124 Roberts Place, Ithaca, New York 14850.

First published 1975 by Cornell University Press.
Published in the United Kingdom by Cornell University Press Ltd.,
2–4 Brook Street, London W1Y 1AA.

International Standard Book Number 0-8014-0950-0
Library of Congress Catalog Card Number 75-6084
Printed in the United States of America by Vail-Ballou Press, Inc.

Preface

The study of *Sedum* has been a major subject of my attention since 1935. *Sedum* is ideal for experimental study because of the ease with which plants reproduce vegetatively. New individuals often develop from small sections of stems or even from single leaves. These are true replicates. Since they are identical in genetic constitution, they are of great value for learning about variation and appraising differences among clones. Perennial plants generally and *Crassulaceae* in particular are useful for studying both the differences among individuals and the effects of varied environments on the same individual.

So much interesting information has accumulated concerning *Sedum* that a volume, even on part of North America, needs to be bulky in order to provide essential details. The present book, my second on *Sedum*, covers an area vastly larger than the Trans-Mexican Volcanic Belt covered by the first.

Through the years, I have gone on many trips. Financial aid for these has come from several sources. I particularly wish to express appreciation to the former Department of Botany of Cornell University, the present Division of Biological Sciences, the Office of Research of the New York State College of Agriculture and Life Sciences of Cornell University, the Faculty Research Grants Committee of Cornell University, the American Philosophical Society, and the National Science Foundation. Special grants have made possible not only travel to many parts of North America to study native populations of *Sedum*, but also trips to herbaria and the salaries of helpers. In the early years of the research, they also defrayed the expenses of a botanical artist. Two grants from the National Science Foundation, G-18011 and GB-4929X, provided major financial aid from 1961 to 1971.

Many persons who have helped on trips have increased the effectiveness of exploration for plants, as well as provided pleasant companionship and sometimes critical, independent judgment of work in progress. A list of all who have aided in this regard would be too lengthy, but I wish to mention especially my sons Eric and Tom, J. L. Edwards, L. J. Kezer, Harold Trapido, Jack Fisher, Arthur Phillips, Jr., Henry F. Dunbar, and E. M. Shields. Officials of the United States Forest Service, National Park Service, and Office of Indian Affairs have cooperated, both in making maps available and in permitting access to areas under their jurisdiction. Likewise, the Arizona Commission of Agriculture and Horticulture, the Division of Highways of the State of California, and the Oregon State Highway Committee have granted permits allowing me to collect specimens in the areas under their supervision.

Both the Plant Quarantine Division of the United States Department of Agriculture and the Plant Protection Division of the Department of Agriculture of Canada have facilitated passage of specimens, important in my studies, across the border between Canada and the United States.

The base map used for plotting the distribution of species is the Auroral Plotting Map for the International Geophysical Year, 1957–1958. P. M. Millman of the National Research Council, Ottawa, Canada, has granted permission for its use here.

Elfriede Abbe has prepared the drawings. Howard Lyon has taken many photographs, some of which are published here. Both Lillian Hollingshead and Charles Uhl contributed information about chromosomes in the early years of the research. Others who have provided helpful information about chromosomes include Charles Horn, Dorothy Niimoto, Ilse Zandstra, Barbara Joyce, Lynn and Rick Benjamin, Iris Mastrangelo, and Kathy Krathwohl. Many people have looked after the plants of *Sedum* in the greenhouse and garden. I am grateful to all, and especially to Louis Farkas who in recent time has taken care of plants in the greenhouses and cold frames. Too, I am grateful for space in the greenhouses and gardens at Ithaca to the successive chairmen of the Department of Botany, Karl M. Wiegand, Lewis Knudson, Harlan P. Banks, and George C. Kent; to the chairman of the Section of Genetics, Development and Physiology, Harry T. Stinson; and to the heads of the Department of Floriculture, Edward A. White, L. H. MacDaniels, J. G. Seeley, and J. W. Boodley. Several statisticians have advised me at different stages of the study: Walter Federer, Philip McCarthy, Douglas Robson, and Robert Steel. Curators of herbaria have been generous in making available collections under their care and occasionally lending specimens. I am grateful to all of these men for their cooperation and for their many courtesies. Several typists have worked on the manuscript. My appreciation to them, especially Patricia Thayer, is profound.

My wife, Edna, and children, Eric, Joanna, Thomas, and Heidi, through their patience and sometimes direct aid, have done what they could to further the effort on *Sedum*.

If this book is of use to others and reasonably free from error, then its fulfillment is justified.

ROBERT T. CLAUSEN

Wiegand Herbarium
Division of Biological Sciences
Cornell University

Contents

Preface		5
I Introduction		11
II Sampling		21
III History of the Study of *Sedum* in North America North of the Mexican Plateau		31
IV Geology of North America North of the Mexican Plateau		41
V The Species of *Sedum* Native in North America North of the Mexican Plateau		63

1.	*S. telephioides*	70	16.	*S. divergens*	303
2.	*S. ternatum*	92	17.	*S. debile*	312
3.	*S. nevii*	109	18.	*S. nuttallianum*	324
4.	*S. glaucophyllum*	122	19.	*S. villosum*	334
5.	*S. pulchellum*	142	20.	*S. oreganum*	345
6.	*S. pusillum*	162	21.	*S. obtusatum*	358
7.	*S. niveum*	178	22.	*S. laxum*	379
8.	*S. cockerellii*	189	23.	*S. oblanceolatum*	403
9.	*S. wrightii*	202	24.	*S. oregonense*	410
10.	*S. lanceolatum*	211	25.	*S. albomarginatum*	424
11.	*S. rupicolum*	236	26.	*S. moranii*	433
12.	*S. radiatum*	247	27.	*S. spathulifolium*	439
13.	*S. stenopetalum*	264	28.	*S. rhodanthum*	475
14.	*S. leibergii*	281	29.	*S. integrifolium*	487
15.	*S. borschii*	290	30.	*S. rosea*	517

VI The Species of *Sedum* Naturalized in North America North of the Mexican Plateau   537

1.	*S. purpureum*	537	3.	*S. acre*	547
2.	*S. sarmentosum*	543		Other Naturalized Species	552

VII The Species of *Sedum* Cultivated in North America North of the Mexican Plateau　556

1. *S. sieboldii*	557	28. *S. kamtschaticum*	566
2. *S. telephium*	558	29. *S. hybridum*	566
3. *S. purpureum*	558	30. *S. lineare*	569
4. *S. spectabile*	558	31. *S. sarmentosum*	569
5. *S. alboroseum*	558	32. *S. spurium*	569
6. *S. populifolium*	558	33. *S. stoloniferum*	569
7. *S. ewersii*	558	34. *S. magellense*	569
8. *S. anacampseros*	561	35. *S. monregalense*	571
9. *S. praealtum*	561	36. *S. album*	571
10. *S. confusum*	561	37. *S. dasyphyllum*	571
11. *S. decumbens*	562	38. *S. brevifolium*	571
12. *S. palmeri*	562	39. *S. lydium*	571
13. *S. treleasei*	562	40. *S. anglicum*	572
14. *S. pachyphyllum*	562	41. *S. gracile*	573
15. *S. lucidum*	563	42. *S. hispanicum*	574
16. *S. clavatum*	563	43. *S. multiceps*	574
17. *S. nussbaumerianum*	563	44. *S. sexangulare*	574
18. *S. adolphii*	564	45. *S. acre*	574
19. *S. rubrotinctum*	564	46. *S. caeruleum*	574
20. *S. diffusum*	564	47. *S. hirsutum*	576
21. *S. morganianum*	564	48. *S. pilosum*	576
22. *S. stahlii*	565	49. *S. chrysanthum*	576
23. *S. hintonii*	565	50. *S. sediforme*	577
24. *S. mexicanum*	565	51. *S. ochroleucum*	577
25. *S. greggii*	565	52. *S. rupestre*	577
26. *S. moranense*	565	53. *S. reflexum*	579
27. *S. aizoon*	566	Other Cultivated Species	579

VIII Species of Genera of *Crassulaceae*, Related to *Sedum*, Native in North America North of the Mexican Plateau　584

1. *Lenophyllum texanum*	585	5. *Parvisedum pentandrum*	590
2. *Parvisedum pumilum*	590	6. *Diamorpha cymosa*	600
3. *Parvisedum congdonii*	590	Other Genera and Species	606
4. *Parvisedum leiocarpum*	590		

IX Keys to Species　611

X Geography of *Sedum* in North America North of the Mexican Plateau　625

XI Relationships of Species　680

XII Conclusions　694

Gazetteer　699
Bibliography　711
Abbreviations and Symbols (and Supplementary Notes)　722
Index　725

SEDUM OF NORTH AMERICA
NORTH OF THE MEXICAN PLATEAU

CHAPTER I

Introduction

Sedum, known in English as stonecrop or orpine, is the largest genus of the *Crassulaceae*, a family of dicotyledons closely related to the *Saxifragaceae*. The number of species of *Sedum* in the world is uncertain. An estimate is about three hundred species. Although additional species will likely be described in the future, better understanding of relationships may sometimes lead to elimination of previously accepted species and to reduction of others to subspecific status. As a result, the true number of species of *Sedum* in the world may be less than three hundred.

The purpose of this book is to provide information about *Sedum* in North America for the many people who grow the plants for ornament or use them in experiments. The book answers questions such as what to call the different kinds of *Sedum*, what are their properties, and where do they occur. The book is also concerned with the origin and evolution of the species. Besides providing a basis for theoretical knowledge, the study of the evolution of *Sedum* indicates answers to practical questions about variation, relationships, and distribution.

Essential data about each species of *Sedum* include such features as distinguishing characteristics, height of stems, color and size of flowers, shape of petals, size of seeds, number of chromosomes, and ability of plants to endure freezing. These and many more items are included in the descriptions of species in the present volume. Also included are discussions of variation, relationships, geographic and ecologic distribution, and mode of reproduction, and in addition, citation of all pertinent nomenclature. The descriptive details and the interpretation of patterns of variation are prerequisite to an understanding of the evolution of the group. Likewise, any general principles about evolution must apply to *Sedum* as well as to all other genera of plants and animals.

This book provides information about all species of *Sedum* which occur naturally, are naturalized, or are commonly cultivated in North America north of the Mexican Plateau. Three kinds of data are presented—those derived from original observation, those derived from published books and journals, and those derived from labels and records in public herbaria or museums. Omitted are details available only from conversations, personal letters, or unpublished manuscripts (including theses). These last are unconfirmed and are the property of their discoverers.

Although the part of North America discussed in this book covers approximately 21,000,000 square kilometers, about two hundred sixteen times larger than the Trans-Mexican Volcanic Belt, the number of native species of *Sedum* is only thirty, one more than in the Trans-Mexican Volcanic Belt. An important objective of the present discussion is to consider the possible reasons for this situation. Another objective is to test again, using a different geographic area and other species, the ten propositions previously tested by the data for *Sedum* of the Trans-Mexican Volcanic Belt (Clausen, R. T., 1959—for references, see Bibliography). These propositions are stated here almost exactly as they appeared in the original publication. In Chapter XII, Conclusions, they are discussed briefly and evaluated from the standpoint of the data available for the species of North America north of the Mexican Plateau. Because the results from the study of *Sedum* of the Trans-Mexican Volcanic Belt seemed to sustain these propositions, they continue to be useful as working hypotheses, but they must be tested over and over again, using data from diverse groups of plants, before they can be considered as proven.

1. Evolution must be the philosophical basis for a sound classification. Evolution has occurred in the past and is occurring now.

2. Taxonomic groups at one rank are not inherently more natural than those at another. Phyletic lines are natural at all stages of evolution. Similar evolutionary processes occur at all stages.

3. The lower taxonomic ranks, as subspecies and species, are easier to define objectively than are the higher ones, but difficulties both in definition and application exist at all levels.

4. Plants which occur together, have similar characteristics, and either actually or potentially interbreed comprise a local population. Such populations may grade into each other insensibly. Sharp limits may or may not be present. Local populations are important evolutionary units.

5. Populations or series of populations which differ from each other in few or several different, but correlated characteristics (morphological, physiological, or both) and which are isolated ecologically or geographically are subspecies. These, as local populations, may intergrade insensibly, but their mean expressions may be markedly different.

6. Species may not be rigidly defined, but a threshold exists around which concepts may fluctuate. Species should differ from each other in several different, but correlated characteristics. These may be morphological or physiological or both. In addition, species should be effectively isolated biologically. Populations on the border line between species and subspecies will be difficult, perhaps impossible, to appraise one way or the other.

7. Gene mutations have been and are a primary cause of organic diversity. They provide the material on which natural selection works, and they also furnish the materials which, through hybridization, may be recombined in multiple ways. Further, they often are the basis for the development of genetic isolation.

8. Physiological changes, caused by gene mutations, have enabled plants to inhabit diverse environmental conditions. Single individuals or small numbers of individuals, fortuitously adapted to new conditions, may found new populations

(ecotypes) ecologically isolated from the ancestral ones. This is one of the most universal and important evolutionary phenomena.

9. Isolation of populations has protected them from better-adapted types and permitted them to survive. Similarly, structural changes of chromosomes and polyploidy, both leading to differences in numbers of chromosomes, have been important in producing genetic isolation within previously interbreeding populations. Chance individuals with different numbers of chromosomes may become the nucleus for the evolution of new species.

10. Hybridization, involving the recombination of characteristics of different local populations, different subspecies, different species, or even different genera, has played an important part in evolution. Sometimes new species have resulted from such hybridization.

An additional objective of the present exposition is to consider methods of taxonomic investigation. In the conduct of the study, I have given attention to both development and testing of methods which might improve the research and be useful to others. Because the work on *Sedum* of the Trans-Mexican Volcanic Belt suggested that field surveys, experimental procedures, and statistical analyses of data all are desirable, I have used these extensively in my further studies. The results of the earlier work have served to indicate how the later work should be conducted. Since sampling is an important aspect of a taxonomic study, a separate chapter is devoted to that. Other aspects of the study are considered in the present chapter.

Essential to a consideration of methods is a clear statement of the subject of discussion. *Sedum* is not an easy genus to delimit. Generic distinctions in the *Crassulaceae* are vague. For our purposes, *Sedum* is a group of plants with both stems and leaves fleshy, leaves in most cases borne spirally, flowers usually 5-merous, petals separate or almost so, stamens in two whorls, and fruits longitudinally dehiscent on the ventral sutures. Some species lack one or more of these characteristics, but none lack all. Species which appear markedly divergent from this concept still may be included if other species exist which are intermediate in the distinctive characters. The inflorescences of most species of *Sedum* terminate the floriferous stems. These generally appear terminal, though they may have originated in a previous year as axillary shoots. The apparent difference between terminal and lateral floriferous stems may result from both time and mode of development. Usually, the petals of *Sedum* are separate or connate for <1 mm. Likewise, the carpels commonly are either separate or nearly separate. Species with distinctive characteristics and not obviously connected with *Sedum* by intermediate conditions, though probably derived from the same phyletic line, have been interpreted as separate genera. Examples are *Diamorpha*, *Lenophyllum*, and *Parvisedum*. These present problems in generic segregation which are discussed in Chapter VIII. All genera of *Crassulaceae* with species native, naturalized, or cultivated in North America north of the Mexican Plateau are included in the keys in Chapter IX. With these keys, identification should be possible of all species of *Sedum* which occur in the area covered by the book, and also of genus of plants of all other *Crassulaceae* found in the area.

North America, third largest continent, is bounded on the east by the Atlantic Ocean, on the west by the Pacific Ocean, and on the north by the Arctic Ocean. Greenland is excluded from the present account. Böcher (1938), who reported on the flora of south and east Greenland, concluded that a greater floristic relationship exists between these parts of Greenland and Europe than with North America. He listed both *Sedum annuum*, unknown from North America, and *S. acre*, naturalized, but not native in North America, as occurring naturally in Greenland. The islands of the West Indies also are outside the present account. These are unimportant as far as *Sedum* is concerned, since no species are known to be native there. The portion of North America covered by this book (fig. 1) is bounded on the south successively from east to west by the Gulf of Mexico and the northern limits of the Mexican Coastal Plain, Mexican Plateau, Sonoran Desert, and Peninsular Ranges of Lower California. This southern boundary is in places arbitrary and in other places distinct, as the slope separating the Great Plains from the Mexican Plateau in Texas and New Mexico, the steep slope marking the southern limits of the Southern Rocky Mountains in New Mexico, and the escarpment at the southern border of the Colorado Plateau in Arizona. The escarpments forming these more definite parts of the boundary constitute a break in topography, climate, and vegetation. Since few species of *Sedum* occur in the arid, northern parts of the Mexican Plateau where the boundary is less distinct, the amount of overlap of species with regions to the north is slight. Thus, utilization of natural rather than political boundaries facilitates the geographical grouping of species and reduces the need for overlapping exposition concerning them.

Observation of populations in the field has been an important part of the present investigation. I have tried to study wild populations of each species and subspecies of *Sedum* known from North America. In addition, I have visited as many as possible of the physiographic regions of the continent, attempting by survey to determine what species might be present. Using random sampling of environmental conditions, as described in Chapter II, such surveys have resulted in the discovery of previously unknown populations and also provided unbiased data on the frequency of species within areas.

My early trips, though with observation of *Sedum* as an objective, were general collecting expeditions, with only brief time in any one local region. These trips were both enjoyable and instructive. They afforded experience with large numbers of species of plants, but they were not satisfactory for learning the details of variation, relationships, and distribution of species of *Sedum*. Gradually a different type of field activity has developed. This is a rigorously planned survey, with maximum attention to the subject of study. The problem of developing a suitable technique in the field has been a difficult one and is still in process.

Years of major surveys for *Sedum* in North America north of the Mexican Plateau have been 1959—Gulf of St. Lawrence; 1960—Southern Blue Ridge; 1961—Seven

Devils Mountains and Wenatchee Mountains; 1962—Southern Rocky Mountains; 1963—Pacific Mountain System; 1964—Sierra Nevada; 1965—Cascade Mountains; 1966—Central Rocky Mountains; 1968—San Bernardino Mountains; 1969—middle section of Ridge and Valley Province; 1970—local surveys in southern Blue Ridge, northern California Coast Ranges, Sierra Nevada, and on Edwards Plateau; and 1972—Cumberland Plateau. Further information about geomorphic provinces visited, time spent in each, and kinds of sampling units is available in Chapter X.

The planned surveys have been the primary source of information about localities where species occur, times of flowering, and details of ecology. In addition, herbaria have been a useful supplemental source of information. The principal value of specimens in herbaria has been as a historical record and to indicate the presence or absence of species in areas which were not part of the surveys.

In the text, abbreviations of herbaria in which types or other specimens are located are according to the suggestions of Lanjouw and Stafleu (1956):

BH	Herbarium of the Bailey Hortorium, Cornell University, Ithaca, N.Y.
BM	British Museum (Natural History), London, Great Britain
CAN	National Museum of Canada, Ottawa, Canada
CAS	Herbarium of the California Academy of Sciences, Golden Gate Park, San Francisco, Calif.
COLO	The University of Colorado Herbarium, Boulder, Colo.
CU	Wiegand Herbarium, Cornell University, Ithaca, N.Y.
DAO	Herbarium of the Department of Agriculture, Ottawa, Canada
DS	Dudley Herbarium, Natural History Museum, Stanford University, Stanford, Calif.
Duke	Herbarium of Duke University, Durham, N.C.
F	Herbarium of the Field Museum of Natural History, Chicago, Ill.
GA	Herbarium of the University of Georgia, Athens, Ga.
GH	Gray Herbarium of Harvard University, Cambridge, Mass.
JEPS	Jepson Herbarium and Library, Department of Botany, University of California, Berkeley, Calif.
K	The herbarium and library of the Royal Botanic Gardens, Kew, Richmond, Surrey, Great Britain
KY	Herbarium of the University of Kentucky, Lexington, Ky.
LAM	Herbarium of the Los Angeles County Museum, Los Angeles, Calif.
LINN	Herbarium of the Linnean Society of London, London, Great Britain
MIN	Herbarium of the Department of Botany, University of Minnesota, Minneapolis, Minn.
MONT	Herbarium of Montana State University, Bozeman, Mont.
NCU	Herbarium of the University of North Carolina, Chapel Hill, N.C.
NEBC	New England Botanical Club, Cambridge, Mass.
NY	Herbarium of the New York Botanical Garden, New York, N.Y.
NYS	Herbarium of New York State Museum, Albany, N.Y.
OAC	Herbarium of the University of Guelph, Guelph, Ont.
ORE	Herbarium of the University of Oregon, Eugene, Ore.
OSU	Herbarium of the Oregon State University, Corvallis, Ore.
Ph	Herbarium of the Academy of Natural Sciences, Philadelphia, Pa.

POM	Herbarium of Pomona College, Claremont, Calif.
RM	Rocky Mt. Herbarium, University of Wyoming, Laramie, Wyo.
RSA	Herbarium of the Rancho Santa Ana Botanic Garden, Claremont, Calif.
UC	Herbarium of the University of California, Berkeley, Calif.
US	Herbarium of the United States National Museum, Washington, D.C.
WS	Herbarium of the Department of Botany, State College of Washington, Pullman, Wash.
WILLU	Peck Herbarium of Willamette University, Salem, Ore.
WIS	Herbarium of the University of Wisconsin, Madison, Wisc.
WTU	Herbarium of the University of Washington, Seattle, Wash.
WVA	Herbarium of West Virginia University, Morgantown, W. Va.

Besides examining types or photographs of types in the herbaria listed above, I have usually studied other North American collections of *Sedum* and in addition, North American collections in other herbaria. All records of occurrence which are based solely on herbarium specimens are clearly indicated, but detailed citations of specimens in herbaria are not part of the present exposition. Part of the reason for this is that citations of specimens can give rise to erroneous concepts about distribution. For example, few specimens of *Sedum rosea* from the north shore of the Gulf of St. Lawrence are in herbaria, yet the population there is extensive and comprises millions of plants. More important than the citation of dried specimens is the status of species in an area as indicated by frequency and density. A person desiring accurate information about species in any area not surveyed in detail in the present study should go to the area himself. The indication on maps that a species occurs in such an area should be sufficient notice. Seldom will a reliable impression of the status of a species in an area result from examination of herbarium specimens alone. For that reason, I encourage readers to gain direct experience with the living plants in the field.

An important part of the present study has been the experimental culture of plants in plots outdoors and in cold frames and greenhouses at Ithaca, N.Y. For testing genetic differences among plants as well as differences among populations, randomized complete block experiments have been useful. Usually I have checked only one character at a time, but sometimes I have used linear or multiple regression to determine whether characters were varying together in groups of plants. For some species, tests of behavior have been possible in two environmental conditions in Ithaca, as well as at the original localities where the plants were obtained. For interpreting the results of experiments, I have commonly used the analysis of variance and other analytic procedures which are explained by Snedecor (1956).

Sedum is ideal material for experimental work because individuals can be subdivided into parts, each with the same genotype. Such parts grow into complete plants. Thus, identical replicates are easy to produce. In order to be sure that all are the same, not only genotypically, but with respect to physiological condition at the beginning of an experiment, my practice has been to propagate pieces of stem of equal length. This ensures that all plants have a similar start. The selected length is arbitrary for any experiment. It depends on the size of the plants from which the cuttings are made. The important point is that all plants in an experiment have been started at the same size. Likewise, all earthen pots used for plants in the greenhouse and cold frames have been of uniform size in each experiment. The usual size has been 10.5 cm. across the top. This size may be assumed unless some other size is mentioned. In the garden, spacing between plants has been a half meter.

Soil in pots in the greenhouse and cold frame has been a mixture of one-third silt loam, one-third sand, and one-third leaf mold. For drainage, gravel is placed in the bottom of each pot for about an eighth of the depth. The rest of the pot is filled with soil. The pH of soil in pots ranges from 7.6 to 7.8. My experimental plots are at the Test Garden along Fall Creek above Forest Home, in the Town of Ithaca, N.Y., and at 1421 Slaterville Road, also in Ithaca. The soil at the Test Garden is a well-drained stony silt loam with pH ranging from 6.2 to 6.6. That in the garden on the Slaterville Road is a poorly drained clay loam with pH ranging from 6.4 to 6.8.

No artificial watering has been used in the gardens. Watering in the greenhouses and cold frames has been according to the needs of the plants, not more than three times per week in winter or four times per week in summer. It has been least in cloudy weather and most when plants were actively growing. Usual time of watering is the morning.

Temperatures to which plants in the experiments are subjected vary greatly. The greenhouse is heated in the cold season by means of pipes with steam. An attempt is made to maintain a temperature of 18° C. in the day and of 7° C. at night. Since regulation of both heat and ventilation is by hand, considerable fluctuations from the desired conditions may occur. In summer, the temperature may go as high as 40° C., or even higher, but in winter the temperature has never gone as low as 0° C. In the garden, temperatures below 0° C. are the rule for about half of the year.

Light has not been controlled in any of the experiments, except that whitewash is applied to the panes of glass in the greenhouse in the growing season to reduce both temperature and light. The cold frames are partially shaded by walls to the south and west and are covered by laths in summer. Likewise, the garden on the Slaterville Road is shaded by a building to the east and trees to the west. Only the gardens at the Test Garden receive the maximum possible sunlight. Ithaca, at 42° 26′–27′ N., has potential sunlight for about fifteen hours per day in June, but for only nine hours in December.

Use of insecticides has been minimal. A solution of Black Leaf 40 and a detergent has been helpful in the greenhouse to control mealy bugs, aphids, and other pests. Weekly inspections of plants have indicated which needed treatment. In the field, neither insecticides nor fumigants have been used.

Cultivation of plants in the gardens, about once per week, has been largely to remove weeds and so to eliminate competition.

Equipment used in the field has been simple, but certain materials need special mention. Maps of the United States Army Map Service, scale 1:25,000, with 1,000 meter universal transverse Mercator grid, have been used for all areas for which they are available. For areas for which these maps are not available, topographic maps with a scale as near as possible to 1:25,000 were used. Besides maps, aerial photographs, steel measuring tape, and compasses all were helpful in locating areal sampling units. For study of plants in the field, an 8X Hensoldt hand lens, with stand and metal rule with graduations to tenths of millimeters, has made possible exact measurements of tiny structures. Likewise, a small slide rule has been a valuable part of the equipment in the field. With this, calculations of both standard deviations and necessary sizes of samples have been possible. Most estimates of pH of soil in which plants of *Sedum* are rooted have been by means of dyes: bromcresol green for range of 3.8–5; chlorphenol red for range of 5–6; bromthymol blue for range of 6–7; and phenol red for range of 7–8.2. The differences observed in determinations of the same soils in the wet and dry seasons in the Trans-Mexican Volcanic Belt (Clausen, R. T., 1959) indicate that the time of collecting the samples and conditions of collection may be important in affecting fluctuations in pH. For this reason, a quick, practical method of determination has seemed justifiable for use in the field.

For marking plants in the field, small cardboard tags have been useful for periods of a few months and copper tags for longer periods. The scheme of numbering has been simple, but definite. In local populations, each plant has received a separate number. This is the field number. When a plant is collected for special study, a permanent number is assigned. The permanent numbers comprise two parts: the first two digits indicate year of collection, then, after a hyphen, the other digits indicate the number of the collection in the year. For all areas which have been surveyed, the primary sampling units, about one kilometer square, are the localities. These are designated by the names of places included within them or by the name of some distinctive natural feature. The ultimate sampling units are clusters of plants or quadrats. Abbreviations of scientific names are usually decipherable by inspection. For example, *S* stands for *Sedum* and *r* for *rosea*. In the field, if only one species of *Sedum* is present in a cluster or quadrat, all that is necessary on tags is the number of the plant, since the number of the cluster or quadrat is marked on stakes and no confusion with other species is possible. When two or more species are in the same cluster or quadrat, the abbreviations of their names are important to separate the different series of numbers.

Observations and measurements of floral structures in the laboratory have been with the aid of a binocular stereoscopic microscope with 9X and 15X paired eyepieces and 1X, 3X, and 6X paired objectives. A binocular compound microscope with 10X, paired, wide-field, Huygenian eyepieces and 5X, 10X, 43X, and 93X, achromatic objectives also has been available. Fine measurements have been possible by means of micrometer discs inserted in the eyepieces of each microscope.

The present study is concerned primarily with gross morphology, ecology, distribution, and relationships. Techniques have included sampling of populations in the field, description of their morphologic and ecologic characteristics, and experimental culture of plants. In addition, assistants have made some studies of chromosomes.

Special anatomic or genetic studies have not been part of this investigation, though all pertinent information which has been published on these topics has been utilized in achieving interpretations about classification. Such data are cited whenever relevant. The basis for all citations from the literature has been their value to the discussion. For this reason, some publications are not mentioned. My intention has been to cite all important contributions to the knowledge of *Sedum* of North America.

Abstracting journals have been useful in obtaining references. In order to compare the relative completeness of coverage of three such bibliographical aides, Zaneta Pronsky, in 1969, listed all references on *Sedum* for the period 1960 to July 1969. The total list comprised 144 references. Of these, 87 were in *Biological Abstracts*, 82 in the *Bibliography of Agriculture*, and 15 in the parts of the *Science Citation Index* available in 1969. Further, 51 of these references were available only in *Biological Abstracts*, 30 were available only in the *Bibliography of Agriculture*, and 4 were available only in the *Science Citation Index*. The conclusion is that a researcher must check all sources if completeness is a goal.

The vocabulary of the book is simple for a reader trained in science. Others may wish to consult a good dictionary when encountering an unfamiliar word.

The methods of sampling employed in the study are discussed in Chapter II. These require careful attention to planning. Likewise, the experimental culture of plants requires both attention to planning and considerable time for execution. For these reasons, the methods may be unpopular. Too, the impression may arise that the methods are impractical if the whole plant kingdom needs to be studied in this way. Several comments are pertinent in this regard. One result of experimental studies seems to be reduction in number of species which should be recognized. Examples of this are the studies of *Potentilla* by Clausen, Keck, and Hiesey (1940) and my study of *Sedum* of the Trans-Mexican Volcanic Belt (Clausen, R. T., 1959). If the number of species becomes smaller, the total bulk of the plant kingdom is less and the magnitude of the task of study is thereby reduced. Jones (1941) estimated the total number of different species of known living plants as slightly in excess of 335,000. Though many species have been described since the figures were prepared on which Jones based his estimate, likewise many species have been reduced to either subspecific status or synonymy. Assuming that an energetic botanist could, in a lifetime, study in the field and carry on experiments involving about two hundred species, somewhat less than two thousand botanists are necessary in any generation to cover the whole plant kingdom. Considering that a society such as the Botanical Society of America has approximately one thousand nine hundred members and that probably more than two thousand taxonomists are engaged in study of parts of the plant kingdom, enough people are available to carry on planned surveys and experimental studies if they are convinced that the results are sufficiently better than those obtained by other methods.

The familiar pattern of botanists working for short periods on small problems and then publishing reports in periodicals has led to a confusing mass of literature which confronts each new student who undertakes to learn about plants. If more botanists studied groups of plants for prolonged periods of time and used experimental procedures, their publications might be fewer, but both more meaningful and more useful. In this way, the literature might be both improved and less complex. The present exposition attempts to indicate how field surveys and experimental procedures can improve our knowledge of *Sedum* in a major part of the North American continent.

CHAPTER II

Sampling

The ability to distinguish species of *Sedum*, especially in vegetative condition, from other genera of *Crassulaceae*, from each other, and from all other kinds of plants requires experience. Years of patient study are necessary to develop skill in recognition. In sampling, this skill is important if the work is to be efficient. The ideal arrangement for a comprehensive study of the genus may be for one person to do the whole study, including both field surveys and experimental culture of plants. Such conduct of the work is desirable from another standpoint, too. An understanding of relationships is possible only after experience with many details. Accordingly, methods of study are necessary that can be executed by an individual without need for a group of cooperators or assistants. This objective has been primary in organizing a scheme of sampling for the study of *Sedum* of North America.

Several kinds of data are necessary for a taxonomic study. An investigator must plan a procedure of sampling to provide them. For purposes of classification, unbiased information about variation of plants, both from the field and experiments, is essential. If the potential for hybridization is to be assessed, data on the occurrence together of species must be reliable. Likewise, unbiased information is necessary on the frequency of hybrids. Taxonomic accounts usually include information on geographic distribution, frequency, and, in addition, details about habitats. Data on density, the number of plants per unit of area, also are important and are essential for estimating the total size of populations in geographic regions. Necessary details concerning habitats include altitudes, exposure, drainage, pH, kind of soil, association with other species, and type of vegetation. All of this information, to be helpful, must be unbiased.

The problems in obtaining a large mass of reliable data for an area as enormous as North America, north of the Mexican Plateau, are many. The area's bigness makes a complete investigation of the whole area impractical. Instead, a survey is necessary. The kind of survey and the size of sampling units both are important. Whatever the type of sampling unit, the exact area of the surface may be only imperfectly known unless surveying becomes a major aspect of the work. Maps indicate horizontal extent, not area of surface.

For purposes of survey, large-scale topographic maps are most desirable. These are available for some areas, but not for others. Once a decision is reached concerning the kind and size of sampling units, maps with these units indicated are preferable to ones on which they do not appear.

The great number of plants sometimes growing in sampling units presents a serious problem in enumeration. Further, the decision concerning the delimitation of a single plant may be difficult. Most species of *Sedum* are perennial. Vegetative propagation is both easy and frequent. Enumeration of plants can be only as satisfactory as the understanding of what constitutes an individual.

Populations of plants, though sometimes remarkably distinct, may also be of vague extent. Several factors may determine the number of plants per population, among them age, the extent of suitable environment, and competition. Unless these limiting factors are the same or very similar for populations which are studied, the statistical distribution of the numbers of plants included may not be normal.

Finally, both time and money are important aspects of sampling. A person working alone and every day per year, trying to survey the area covered by this book, needs more than four thousand years to take a 1% sample, and that would be just time for the field work, without any experiments or analyses of data. Four thousand persons, all experts on *Sedum*, could do the same job in two years, assuming that they worked in the growing season of each year.

The methods of sampling must both provide necessary data and be satisfactory for the problems which are involved. In the present study, since North America is large and since the work is best carried on by one person, a survey of the entire continent is not appropriate. The possible sampling fraction would be too tiny. Further, knowledge already is available that *Sedum* is lacking from big parts of the continent. Also, some areas where it occurs need more attention than others. Accordingly, the adopted plan entailed deliberate choice of areas for study on the basis of both interest and need for attention. These selected areas, though of considerable size, could be surveyed in meaningful fashion. All data from them are kept separate from each other in the present text. They are nowhere combined in common estimates. Information on frequency and density thus applies to the several areas which were surveyed and not to the whole continent. Usually, the status of a species in a region is the information which is desired. Similarly, information on variation applies separately to each of the regions which were surveyed. Again, this separation of data is what is necessary for taxonomic appraisal. The interest in variation is from region to region.

Natural areas are the easiest kind to find in the field, even if they are unequal in size and sometimes with vague limits. Artificial units, as squares of a grid, often are difficult to locate exactly. Major emphasis on surveying may be necessary to locate them. This is both time-consuming and detractive from the main botanical activity. Further, no matter what the method of stratification, the principal expenditure of time may be to visit quadrats which are of no interest from the standpoint of the plants being studied. Unless the sampling fraction is very large, which means a larger number of quadrats than is practical for one investigator to study, rare species may be missed entirely. Just this happened when I visited the Gulf of St. Lawrence in the summer of 1959. A probability sample of quadrats in the stratum which seemed most likely to include *Sedum villosum* yielded no plants of that species.

Yet, rapid survey of natural areas, namely whole islands, in this same stratum, resulted in the discovery of five populations of *S. villosum*.

For estimates of performance, samples should be representative. This is not true when either random points or quadrats are used. The reason is that plants in areas of lesser density have a greater chance of being included in samples. A solution of this problem is to select the same fraction of plants from each quadrat or other unit, pooling the figures for units of low density, but selecting the same proportion of individuals from the group of units with low densities. A useful reminder is that the primary interest is in sampling plants, not areas.

The natural units which I have employed in sampling are topographic, as mountains, hills, ridges, valleys, and islands, or natural subdivisions of these. The primary strata in which these are grouped are the forty-three geomorphic provinces described in Chapter IV on geology. Each stratum which was surveyed was first subdivided into large, mutually exclusive, primary sampling units. Whenever practical, definition was by means of topographic features. Sometimes, however, counties, groups of counties, or even arbitrary quadrangles were more convenient than natural areas. The primary units vary considerably in size, more than I would like, but the intention was to make them about 2,500 kilometers square. Numbers assigned to these units have been the basis for choosing simple random samples, using a table of random numbers.

Because plants of *Sedum* occupy a relatively tiny part of the total environment, unrestricted randomization leads to an expected result, namely, that most selected sampling units, whether primary or listing, do not contain *Sedum*. This was the experience in 1963 when I spent six months surveying the Pacific Border region, a large stratum made by combining the Transverse Ranges of California, California Coast Ranges, Great Valley of California, Klamath Mountains, Willamette-Puget Trough, Coast Ranges of Oregon and Washington, and Olympic Mountains. Five out of nine, randomly selected, natural, primary units, and only four out of forty listing units, that is 10%, contained *Sedum*. To avoid this kind of a result in further surveys, I adopted the device of an optimum stratum deliberately selected because it was a good place for *Sedum*. This was completely sampled, a procedure worked out by Dalenius (1953). I have used optimum strata both within the primary strata and within the primary sampling units. Results have been gratifying. This design has reduced time devoted to areas lacking *Sedum*, but by permitting continuance of a random stratum, has provided a check on the data obtained as a result of restricted randomization. Each optimum stratum can count in any total estimate only as much as the fraction which it is of the total population.

The size of listing units is uneven. In most surveys, boundaries were natural features such as streams or ridges. Occasionally, boundaries were roads or even arbitrary lines. Inevitably, units with such boundaries are irregular in shape and unequal in area. Despite inequalities in size, selection of both primary and listing units has been with equal probability. The plan was to make listing units about a kilometer square. In some cases, I even used arbitrary quadrats as listing units.

My procedure was to scan completely each listing unit, noting the total number of populations or groups of plants of *Sedum* present. Frequently this has been a difficult and time-consuming job. When plants are small, the chance of missing them is great, even after prolonged and thorough search. For this reason, I have sometimes resorted to subsampling.

The pattern of distribution of plants determines the method of sampling for both density and performance. Four principal conditions of occurrence may be conceived:

A. No plants present, in which case initial scanning suffices.

B. Plants present, but so few that complete enumeration is possible.

C. Plants present in two or more discrete groups. Numbers may be assigned to these groups and a constant fraction can be chosen for study.

D. Plants generally distributed throughout the sampling unit, but either not or only partially in discrete groups. This circumstance requires some sort of arbitrary subdivision. The following paragraphs deal with this problem.

Groups of plants vary enormously in size. In addition, the limits of groups may be vague. For these reasons, use of some sort of spatial unit for subsampling within listing units has appeal. Yet spatial units may entail even greater difficulties. No small, natural sampling unit is available for plants. Trouble arises in both definition and enumeration. Further, when spatial units of arbitrary size are employed, most in any large listing unit will be unoccupied. As a result, a disproportionate amount of attention may be directed to empty units. This is a compelling argument against spatial units for subsampling within listing units. Yet I have sometimes used them and then have encountered the difficulties just mentioned.

No easy solution of the problem of sampling areas for plants is available. Each possible procedure has distinctive difficulties. Within large listing units, a balance is necessary between dispersed study and detailed investigation of groups of plants. This may be achieved by drawing a sample of groups within each listing unit. Despite the inequality of groups and the initial labor of enumeration, groups may offer the greatest promise for sampling. If variation in one group is similar to that in another, then the number of groups necessary for study may not be large. Further, proportional selection of plants within groups will correct for inequalities and yield self-weighting samples. The problem of definition may be made easier when groups are large or run together, by assigning appropriate series of numbers to such aggregations, using as the basis for assignment the estimated number of average discrete groups which probably are involved in the aggregation. If a number applying to such a cluster is randomly selected, then subdivision of that cluster is necessary. Otherwise, subdivision is unnecessary. Another way to cope with the problem of unequal groups is to cruise the whole unit quickly, tying together small groups and arbitrarily splitting large ones, to make all clusters roughly equal, at the same time excluding from consideration all zero areas which contain no plants. Jessen (1947) has provided interesting comments pertinent to this method. As a last resort in difficult cases, arbitrary lines may be used to subdivide large arrays of plants.

Simple random samples from listing units are the easiest to handle for analysis. McCarthy (1957) has explained a method for selecting a simple random sample of names from a telephone directory. This is an excellent procedure when a ready-made unit, as a page, is available, and also when about the same number of elements occurs per unit. Neither condition prevails for plants. No ready-made unit is available. Units which may be employed, whether spatial or clusters of plants themselves, are unequal in number of individuals included. Under such circumstances, if the method suggested for the telephone directory is used, the labor of enumeration will be great because most selected plants will be from different groups, requiring enumeration of many groups. Also, groups must be enumerated to find out whether selected numbers occur and many then will not yield plants.

The use of groups of plants as sampling units has the advantage of concentrating effort on a small number of groups, reducing the labor of moving around, and providing perspective for interpreting the chosen group. To understand variation, this gain in perspective may be an important part of the study. The single varying plant may be unexplained when the sample is derived from widely dispersed groups or units, but when a whole group is studied in context, explanation of variation, whether due to hybridization or environmental influence, may be obvious. In working with groups, I have normally tagged each included plant. This has provided an intimate experience with all plants of the group. Another advantage of groups as sampling units is that attention goes immediately to the plants being studied. In this respect, the use of groups results in an increase in efficiency similar to that gained by optimum stratification.

Numbering of groups of plants in listing units normally has been in serpentine fashion, beginning in the southeastern corner. Another feature has been the keeping of a record of the pattern of occurrence of *Sedum* within each unit, that is the conditions A–D listed above. The summary of this record is included in Chapter X, where the distribution of *Sedum* is discussed in detail. The data permit an indication of the number of sampling units which fit each pattern of occurrence in each geomorphic province.

For estimating the total number of plants of a species in a stratum, the formula is

$$N \left(\frac{A_1 D_1 + A_2 D_2 + \cdots A_n D_n}{n} \right)$$

if A indicates area of units, D indicates density of units, subscripts indicate the numbers of the units in the random sequence, n equals the area of all units in the sample, and N equals the total area of all units in the stratum. Measurement of areas has been by means of counting dots on acetate paper placed over maps, each dot representing an area of known size, depending on the scale of the maps. All areas were measured as plane surfaces. Time was not available to measure surface area, which often would be much greater than plane area in regions of rough topography.

Formulas for computing means and variances of density for populations are available in Cochran's (1963) *Sampling Techniques*. These are cited in the text wherever they are first used.

The usual initial sampling rate for performance has been 1%. When sampling units were enumerated completely and contained more than one hundred plants, 1% of the total was randomly selected. If the number of plants in a unit was less than one hundred, then the numbers of plants for such deficient units were placed in a single linear sequence from which plants were selected at the rate of one per hundred. When patches of plants or areal subunits had been chosen for enumeration, 1% of the plants in these was selected for study of performance. Occasionally, plants of unusual characteristics, though not part of random samples, have been selected for study. These have been handled as a distinct stratum, data for which could count only as much as the tiny fraction of the total population which such plants represent.

In studying dimensions of leaves, floral parts, or other characteristics of plants in the field, I have preferred to take random samples. This has permitted a comparison of variation within plants and among plants. Though environmental effects in the field may be profound and though replication is an essential feature for meaningful analysis, rough estimates of whether plants are the same or different have been useful in planning more precise experiments. Conceptions gained in the field have guided the work. Experiments often have confirmed them.

Sampling of plants sometimes has consisted of two stages: first, a simple random sample of stems, then, a simple random sample of structures on each stem. Stein's test, discussed by Cochran (1963: 75–81), has been useful to determine the requisite number of both stems and structures per stem. Estimates usually are with probability of 95%. Sampling in two stages greatly reduced the labor involved in taking either simple random or systematic samples from plants with many stems, leaves, and flowers.

Particular structures, as median leaves on floriferous stems and central flowers of cymes, have been useful for comparative studies. Since these sometimes vary less than leaves or flowers from diverse locations on plants, they are satisfactory too for purposes of practical identification. Whenever specific structures have been studied, they are clearly defined in comparisons and indicated in descriptions. Otherwise, descriptive data apply to all structures on plants.

Different methods of sampling may yield different estimates of parameters. Data for width of leaves of a plant of *Sedum integrifolium* in my garden, but originally from the western side of Seneca Lake, New York, demonstrate the statement. Estimates of width of leaves resulting from four different procedures of sampling are compared with the results obtained by measuring the width of every leaf on the plant. The four kinds of samples are:

1. Adequate, simple random sample. Initially, a random sample of ten leaves

was selected. The variance was calculated from the data for these. Then, using the formula for Stein's test,

$$n = \frac{t^2 s^2}{d^2}$$

the number of leaves needed for a confidence interval of 4 mm., with 95% probability, was determined.

2. A systematic sample with a random starting point. The size of this sample was arbitrarily chosen to be 2% of all leaves.

3. All median leaves of all floriferous stems. By median is meant halfway in distance from base to end of floriferous stem, the end being considered as the node where the primary branches of the cyme originate.

4. The median leaves of an adequate, simple random sample of floriferous stems, using the variance of an initial sample of ten stems to determine adequacy.

The results follow:

Sample	n	$\bar{x}$ (mm.)	s (mm.)
1. Simple random	10	9.7	2.2
2. Systematic	9	7.5	—
3. All median leaves	19	12.1	4.3
4. Median leaves of random sample of stems	19	12.1	4.3
5. All leaves	459	9.4	3.1

The estimates from the simple random sample are nearest to the parameters for all leaves. On the basis of Stein's test, for the desired confidence interval of 4 mm., with 95% probability, a sample of six leaves is adequate. The estimate of the means from the systematic sample is low. The leaves decrease upward on the stems. In the sample chosen, several upper leaves were included. When variation follows a regular pattern, systematic samples may not yield as satisfactory estimates as random samples, though they may require less time to take. Ten systematic samples, however, yielded a mean of 9.18 mm., closely similar to 9.2, the mean of ten simple random samples. Both means are close to the mean of all leaves, 9.4 mm.

Median leaves, on the stems studied, obviously are larger than most leaves. Data obtained from them are not satisfactory for estimating the condition of leaves on the whole plant. Such variability exists among median leaves from stem to stem that, according to Stein's test, all stems must be included in the sample. Measuring median leaves on all stems still is easier than drawing a simple random sample of leaves from the whole plant and may have merit if comparisons must be made rapidly.

The plant of *Sedum integrifolium* which was studied was at fruiting stage, when leaves on floriferous stems should have attained maximum expansion. Many leaves already had become yellow and dropped. Measurements were of leaves that were still green. On the majority of stems, most leaves below the middle were gone. For that reason, although the median leaves were truly median from the standpoint of height on the stems, they often were lowest or nearly lowest from the standpoint of relationship with other green leaves.

Study of structures on single stems, whether all structures or adequate samples, may not be reliable for estimating the condition on the whole plant. Data on width of leaves for each of the nineteen floriferous stems of the plant of *Sedum integrifolium* indicate highly significant differences among stems.

Stem no.	N	μ (mm.)	σ (mm.)
1	31	11	1.8
2	34	10	1.6
3	34	7	2.4
4	19	12	1.6
5	21	10	2.7
6	58	7	1.9
7	24	9	2.9
8	5	20	2.2
9	27	7	2.4
10	10	12	1.3
11	6	12	2
12	22	12	2.6
13	14	13	5
14	5	14	2.3
15	27	8	2.3
16	27	9	2.7
17	24	9	2.7
18	23	11	1.5
19	48	8	1.8

Such differences from stem to stem indicate the dangers inherent in evaluation of variation on the basis of one or few stems. Clearly, the sample should come from different stems if the whole plant is to be understood, whether for identification or for studies of variation.

Tests of independence have been useful. These have involved both environmental conditions and associated species. As with analysis of data on performance in the field, these frequently have suggested concepts, even when data were scant. Results of such tests should be more reliable than guesses not based on tests.

Studies on islands in the Gulf of St. Lawrence in the summer of 1959 have had an important influence on my ideas about sampling. The purpose of the work there was to learn about the distribution and variation of *Sedum villosum* and *S. rosea*. Sampling was multistage. Initially, two areas, each a quadrangle of one degree of latitude and longitude, were deliberately selected because of their interest in connection with *S. villosum*. Next, the area of each was subdivided into four strata: water, mainland, larger islands, and tiny islands (<100 m.^{2}) and rocks. Primary attention was devoted to the larger islands. Since available hydrographic charts have latitudinal and longitudinal lines indicated at intervals of ten minutes, these lines form a grid, the units of which seemed convenient as blocks for sampling. The plan was to survey 10% of the blocks with islands in the two one-degree quadrangles. Time was inadequate to carry out the survey as planned. Considerable bad weather, both fog and wind, further retarded the work. As a result, the sampling as originally conceived was only partially accomplished.

Primary sampling units were quadrats 100 meters square. These were chosen with probability proportional to size of islands in the blocks, ten minutes on a side, which also were selected randomly in the one-degree quadrangles. Each 100-meter square was subsampled, using either 1- or 10-meter quadrats randomly chosen within the larger units.

Nine primary units, parts of three random sequences, each on a different island, lacked *Sedum villosum*. The species occupies so little area where it occurs on islands in the Gulf of St. Lawrence that a large sampling fraction is necessary to have it included in samples. In order to find it, I deliberately visited nine islands which seemed likely to have it on the basis of accumulated information. It was on three of these. Otherwise, it was on two of the islands on which randomly selected units were located, though it was not in the units.

The eighteen islands which I visited in the Gulf of St. Lawrence provide interesting information. They range in size from about 6,000 square meters to 2,081,620 square meters. The average area is 433,681 square meters. *Sedum rosea* occurs on all islands except one. The following table shows its distribution.

Size of islands	S. rosea present	S. rosea absent	Total
<433,681 sq. m.	11	1	12
>433,681 sq. m.	6	0	6
Total	17	1	18

By the exact test of Pearson and Hartley (1954), table 38, the number of large islands with *Sedum rosea* is not significant.

The occurrences of *Sedum villosum* are too scant for the tables of Pearson and Hartley to be applicable.

Size of islands	S. villosum present	S. villosum absent	Total
<433,681 sq. m.	3	9	12
>433,681 sq. m.	2	4	6
Total	5	13	18

Corrected for continuity,

$$x^2 = \frac{(4.5 \times 3.5 - 1.5 \times 8.5)^2 \, 18}{5 \times 13 \times 12 \times 6} = \frac{162}{4,680} = .035 \text{ n.s.}$$

The results for *Sedum villosum* are in agreement with common sense, namely, that the occurrence of suitable conditions for a species of plant is not necessarily correlated with size of area. This conception may be either heeded and tested further or disregarded. Because of the expense of chartering a boat in the Gulf of St. Lawrence, I have not returned there, but I have tried to test the conception elsewhere. The results of the further studies are indicated under the appropriate species and also in the conclusions.

Information on density may be the basis for estimates concerning the sizes of populations. Data for *Sedum rosea* on Little Gull Island, a small island six kilometers west of Harrington Harbour, illustrate some of the problems involved in determining density. This island appeared larger on the hydrographic chart than it actually is. Data on charts prepared with the aid of aerial photographs may be affected by the level of the tide, whether high or low at the time of photographing, and also by the angle from which the photographs were taken.

Four discrete groups of plants of *S. rosea* occur on Little Gull Island. A randomly selected quadrat, 100 meters square, included two of these groups. Part of the quadrat was in the water, since the island appeared larger on the chart on which it was located originally. An initial sample of 10% of the area, namely ten secondary quadrats, each ten meters square, yielded no plants of *S. rosea*. After drawing this sample, I tagged every plant of *S. rosea* in the hundred meter square quadrat. This task required a day of work. The total number of plants was 375, in one patch 297 and in the other 78. With complete information available, I was able to draw repeated samples, each without replacement, and to compare the estimates with the parameters.

Sample no.	No. of quadrats with S. rosea	n (plants)	$\bar{x}$ (per 10 m.^{2})	s
1	0	0	0	0
2	3	71	7.1	33.3
3	1	50	5	15.8
4	2	24	2.4	6.9
5	1	1	.1	.3
6	2	67	6.7	17.9
7	0	0	0	0
8	2	116	11.6	26.2
9	3	121	12.1	25.9
10	2	6	.6	.06
Total, 1–10	16	456	4.6	17.5
Complete area	13	375	3.7	14

Although particular samples yielded poor estimates, the pooled results provide estimates which approximate the parameters. The big standard deviation is due to the large number of quadrats lacking *Sedum rosea*. This is a frequent problem when small artificial units are used for sampling. When standard deviations are large, confidence intervals are enormous and estimates lack precision, yet they are better than no information. Greig-Smith (1964) has provided information about density and pattern of distribution, some of which has been useful in the present study.

The experience on islands in the Gulf of St. Lawrence emphasized the need for natural sampling units which are practical in the field. It further suggested that sampling should be planned in a way that would be appropriate for the study of rare species. The methods which I have employed since 1959 are based on these considerations.

CHAPTER III

History of the Study of *Sedum* in North America North of the Mexican Plateau

History imparts perspective to thought. It indicates the growth of knowledge. From study of history may come understanding of both the present status of our information and its possible evolution in the future.

This account of the history of the study of the North American species of *Sedum* includes, first, a consideration of the discovery of the various species and subspecies, and then a review of the published information concerning their distribution, ecology, structure, chromosomes, physiological characteristics, and use in horticulture.

The period of discovery by botanists of the species of *Sedum* in North America north of the Mexican Plateau extended from 1680 to 1950, a span of 270 years. Whether the period of discovery is past cannot be stated with certainty until every mountain, slope, and cliff has been surveyed. For an area as large as North America, this degree of finality may never be attained.

Knowledge of the North American species of *Sedum* was slow to reach Europe. Perhaps the earliest printed record is that of Plukenet (1696). He listed a *Telephium Virginianum petraeum album* of John Banister. Possibly that is *Sedum telephioides*, but neither specimen nor drawing has been available to me for confirmation of the identification. J. E. Dandy, who searched in my behalf, was unable to locate Banister's specimen at the British Museum.

Gronovius (1762) listed a *Sedum annuum caule compresso, foliis obverse-ovatis*. This is *S. ternatum*. The description is based on John Clayton's collection no. 891. A photograph of that has been available to me from the British Museum through the courtesy of Dr. Dandy.

Michaux (1803) provided binomials for both *Sedum telephioides* and *S. ternatum*. He also added a third species, *S. pulchellum*, which he had collected near Knoxville, Tennessee.

The record of discovery of the North American species of *Sedum* is difficult to compile. No register is available. Instead, the history must be put together bit by bit from study of the literature and of specimens in herbaria. Names of discoverers are arranged chronologically in the following list, with indication of the species which each presumably found.

1680–1692	John Banister—*S. telephioides*
<1762	John Clayton—*S. ternatum*
1795	André Michaux—*S. pulchellum, S. pusillum*
1805	Frederick Pursh—*S. glaucophyllum*
1806	Meriwether Lewis and William Clark—*S. lanceolatum, S. stenopetalum*
<1818	Benjamin Kohlmeister—*S. rosea*
1819, 1834 (or 1835)	Thomas Nuttall—*S. nuttallianum, S. oreganum*
1820	Edwin James—*S. integrifolium*
1825	David Douglas—*S. spathulifolium*
1842	John Frémont—*S. rhodanthum*
1849	Charles Wright—*S. wrightii*
1857	Reuben Nevius—*S. nevii*
1860–1862	William Brewer—*S. obtusatum, S. radiatum*
1868	Sereno Watson—*S. debile*
1879	Joseph Howell—*S. leibergii*
1880	Wilhelm Suksdorf—*S. divergens*
1880	Joseph and Thomas Howell—*S. oregonense*
1884	Thomas Howell—*S. laxum*
1893	Elmer Wooton—*S. cockerellii*
1898	Adolf Elmer—*S. rupicolum*
1906	Joseph and Hilda Grinnell—*S. niveum*
1915	Harold St. John—*S. villosum*
1927	Kenneth Baker—*S. borschii*
1928	John and Lilla Leach—*S. moranii*
1943	George Youngs—*S. albomarginatum*
1950	Robert Whittaker—*S. oblanceolatum*

The person who first noticed each species of *Sedum* probably will never be known. These would be mostly Indians who kept no records. Even on the basis of published information, however, nonprofessional botanists discovered about half of the North American species of *Sedum*. The two Howells worked primarily as farmers and on other nonbotanical jobs; Banister and Nevius were clergymen; Lewis and Clark were explorers; the Grinnells were zoologists; Clayton was a physician; Frémont was a soldier; Kohlmeister was a missionary; Baker became a plant pathologist; and John Leach was a pharmacist. Among the thirteen botanists who discovered species of *Sedum* in North America, Douglas, Elmer, and Wright were primarily collectors. All were interested in plants in general. None specialized on either *Sedum* or the *Crassulaceae*.

At least eight North American species of *Sedum* comprise two or more subspecies. The number of recognized subspecies, not including the nomenclatural types of species, is seventeen. The presumed botanical discoverers of these are indicated in the following list.

1858	William Brewer and John Chickering—*S. integrifolium* ssp. *leedyi*
1865	John Torrey—*S. spathulifolium* ssp. *yosemitense*
1880	Thomas Howell—*S. stenopetalum* ssp. *monanthum*

1887 Drake and Dickson—*S. radiatum* ssp. *ciliosum*
1887 John Macoun—*S. spathulifolium* ssp. *pruinosum*
1888 Louis Henderson—*S. lanceolatum* ssp. *nesioticum*
~1890 Theodore Cockerell—*S. lanceolatum* ssp. *subalpinum*
1892 Louis Henderson—*S. oreganum* ssp. *tenue*
1897 H. Brown—*S. obtusatum* ssp. *boreale*
1897 N. K. Berg—*S. integrifolium* ssp. *procerum*
1901 Elmer Wooton—*S. integrifolium* ssp. *neomexicanum*
1902 Amos Heller—*S. obtusatum* ssp. *retusum*
1902 Alice Eastwood—*S. laxum* ssp. *eastwoodiae*
1902 F. W. Staunton—*S. spathulifolium* ssp. *purdyi*
1926 Doris Kildale—*S. laxum* ssp. *heckneri*
1940 Robert Clausen—*S. laxum* ssp. *latifolium*
1961 Lawrence Dempster—*S. radiatum* ssp. *depauperatum*

Six persons who discovered subspecies are or were botanists; Brewer and Torrey were scientists with special interest in botany; Cockerell and Macoun were naturalists with broad interests; and Howell was a farmer for part of his life. Information about Berg, Brown, Dempster, Drake and Dickson, and Staunton is lacking.

In order to summarize the progress of discovery of North American *Sedum*, the number of species and subspecies found in each century since 1600 is indicated in the following table. Clearly, the nineteenth century was the period of greatest discovery.

Century	No. of species discovered	No. of subspecies discovered
1601–1700	1	0
1701–1800	3	0
1801–1900	19	10
1901–	7	7

Description of species and subspecies usually is prompt. To test this impression, a comparison of the dates of discovery in North America and the dates of either description or first listing of species of *Sedum* is useful. The difference between these two dates is indicated in number of years in the third column from the left in the following table. The average period between discovery and publication is 14.8 years. The data reveal little change in habits of describers of species in the approximately four centuries in which knowledge of North American *Sedum* has been accumulating.

Year of discovery	Year of publication	Difference in years	Species
1680–1692	1696	4–16	*S. telephioides*
<1762	1762	0–?	*S. ternatum*
1795	1803	8	*S. pulchellum*
1795	1803	8	*S. pusillum*
1805	1946	141	*S. glaucophyllum*
1806	1814	8	*S. stenopetalum*
1806	1828	22	*S. lanceolatum*
<1818	1818	0–?	*S. rosea*
1819	1832	13	*S. nuttallianum*

Year of discovery	Year of publication	Difference in years	Species
1820	1832	12	S. integrifolium
1825	1840	15	S. spathulifolium
1834 (or 1835)	1840	6 (or 5)	S. oreganum
1842	1862	20	S. rhodanthum
1849	1852	3	S. wrightii
1857	1858	1	S. nevii
1860–1862	1868	6–8	S. obtusatum
1861	1883	22	S. radiatum
1868	1882	14	S. debile
1879	1882	3	S. leibergii
1880	1882	2	S. divergens
1880	1882	2	S. oregonense
1884	1903	19	S. laxum
1893	1903	10	S. cockerellii
1898	1931	33	S. rupicolum
1906	1921	15	S. niveum
1915	1922	7	S. villosum
1927	1975	48	S. borschii
1928	1930	2	S. moranii
1943	1975	32	S. albomarginatum
1950	1975	25	S. oblanceolatum

General works seldom are appraised for completeness on the basis of available published knowledge. The information concerning North American *Sedum* permits such a test. Using the species recognized in the present study as a standard, the percentage of completeness of five general works and one monograph is indicated in the following table.

Author	Publication	Year	No. of species listed from N.A.	No. of species previously collected in N.A.	Completeness (%)
Linnaeus	Species Plantarum	1753	0	1	0
De Candolle	Prodromus	1828	4	11	36
Torrey and Gray	Flora of N.A.	1838–40	9	12	75
Britton and Rose	N.A. Flora	1905	22	23	95
Berger	Nat. Pflanzenfam.	1930	21	28	75
Fröderström	The genus *Sedum*	1930–35	21	28	75

Linnaeus listed two species which later were found in North America, but at the time that he wrote he knew them only from localities in Europe. Augustin De Candolle's knowledge of the same two species was only from European sources. For that reason, those two species are not included in the numbers for Linnaeus and De Candolle in the above table. Britton and Rose, who specialized in the flora of North America and had a particular interest in the *Crassulaceae*, provided the most complete account of the species known at the time of their publication.

All together twenty-four authors have proposed fifty-eight binomials for the thirty species recognized in the present study. This is fifteen more than were proposed for the twenty-eight species of the Trans-Mexican Volcanic Belt. The summary below does not include names that anyone published as *Sedum* for species belonging to other genera or that were substituted for later homonyms. Neither does it include names published by Gray, Jones, and Nieuwland for two European species naturalized in America. Of the describers of species, only the five whose names are starred were also the discoverers of species.

Year or years	Author	No. of species proposed	Recognized in present study	Synonyms or subspecies in present study
1753	Carolus Linnaeus	2	2	0
1803	André Michaux*	3	3	0
1809	Carl Willdenow	1	0	1
1811	James Donn	1	0	1
1814	Joseph Banks	1	0	1
1814	Frederick Pursh*	2	2	0
1827	Augustin De Candolle	1	0	1
1828	John Torrey	1	1	0
1832	Constantino Rafinesque	2	2	0
1834	George Don	1	0	1
1840	William Hooker	2	1	1
1840	Thomas Nuttall*	3	1	2
1858–68	Asa Gray	4	4	0
1876–82	Sereno Watson*	5	4	1
1898	Thomas Howell*	2	0	2
1901	Per Rydberg	2	0	2
1903	Nathaniel Britton	14	2	12
1903	Joseph Rose	2	0	2
1921	Anstruther Davidson	1	1	0
1930	Louis Henderson	1	1	0
1931	George Jones	1	1	0
1936	Harald Fröderström	1	0	1
1936	Willis Jepson	1	1	0
1946	Robert Clausen	4	4	0

Knowledge about the distribution and ecology of the North American species of *Sedum* has resulted primarily from studies by botanists whose chief interest has been in either classification or synecology. Important contributions have been by Wherry (1936) on the distribution of species in the eastern part of the continent; R. T. Clausen (1942 and 1949) on subgenus *Gormania*, section *Eugormania*, and on *S. nevii*; McVaugh (1943) on *S. pusillum* on granite flatrocks of the southeastern United States; Clausen and Uhl (1943 and 1944) on *S. cockerellii* and related species, and on subgenus *Gormania*; and Heusser (1954) on *S. integrifolium* on nunataks in southern Alaska.

Several botanists have provided data on the anatomy of species of *Sedum* that are either native or naturalized in North America. Strasburger (1866–1867) described the development of the stomates of *S. spurium*. Mori (1879) published a brief discussion of the histological structure of the stems and leaves of *S. ternatum*. According to him, *S. ternatum* is structurally similar to several European species which he also studied. Henslow (1891) reported that the xylem of the ventral traces of the carpels of *S. telephium* is not inverted. Eames (1931) stated that five bundles, all originating separately from the receptacular stele, supply each carpel in *S. ternatum*. He regarded the carpels as primitive because their margins are only slightly fused and the epidermal layers of the united edges are distinct.

Mauritzon (1933) described the embryos of thirty-two species of *Sedum* and, in addition, several species of genera sometimes segregated from *Sedum*. His sample included three species—*S. oreganum*, *S. rosea*, and *S. spathulifolium*—native in North America, three commonly naturalized species, and sixteen species often cultivated in American gardens. Among these species Mauritzon found the placentas to be of the normal type, except that the carpels of *S. coerulem* each have only two ovules, one above the other. In all species studied, the nucellus is of the *Sedum*-type, with only a few cells between the usually small embryo sac and the epidermis of the elongate nucellus. The embryo sacs are of the *Polygonum*-type, monosporic and 8-nucleate, in all species of the sample except *S. populifolium* and *S. purpureum*. The embryo sacs of those two species are of the *Allium*-type, bisporic and 8-nucleate. The endosperm is cellular and, except in *S. ochroleucum* and *S. reflexum*, of the *Acre*-type. In *S. ochroleucum* and *S. reflexum*, it is of the *Rupestre*-type, which differs from the *Acre*-type only in the later stages of development. According to Mauritzon, the endosperm haustorium is either of the *Sempervivum*-type, a condition found in most species of *Sedum*, including *S. purpureum*, *S. rosea*, and *S. spathulifolium*, or of the *Echeveria*-type, round-ovoid, as in *S. acre*, *S. lydium*, *S. pachyphyllum*, and *S. sexangulare*. In summary, Mauritzon considered the members of the *Crassulaceae* as closely related from the standpoint of embryology. He concluded that the *Crassula*-type nucellus is the most reduced in the family and that *Crassula* and *Tillaea* are younger than *Sedum*.

Quimby (1940, 1971) investigated the floral anatomy of ninety species of *Crassulaceae*. He found four rather distinct types of vascular plans which form a series of increasing specialization with respect to origin and fusion of vascular bundles. The eight North American species which he studied illustrate three of the stages of specialization, including the least and most specialized.

Subramanyam (1955) described the floral anatomy of two species of *Sedum* from North America north of the Mexican Plateau, namely *S. stenopetalum* and *S. ternatum*. His results essentially agree with those of Quimby. In *S. ternatum*, for which he provided twelve figures, the carpels are connate basally, with the ventral traces partially inverted and heterocarpous, each the product of the fusion of ventral bundles of two adjacent carpels. Above the connate part of the carpels, the ventral bundles are separate as a result of division of the heterocarpous traces. Also, a little vascular tissue is present in the receptacle above the level of the heterocarpous traces. According to Subramanyam, the nectaries may be outgrowths of the carpels.

Piaget (1966) gave attention to the mode of attachment and separation from the main stem of the secondary branches of *Sedum*. He recognized seven different conditions, with some species intermediate. Since these conditions appear to be adjustments to the rigors of the environment, unrelated species may possess the same condition. Piaget had eleven North American species in his sample for study.

Jensen (1968) described the vascular anatomy of the primary stems of sixty-nine species of *Crassulaceae*, including two North American species of *Sedum*: *S. ternatum* and *S. oregonense*. He considered the phylogenetic significance of decussate versus spiral phyllotaxy and also of the relationship of open to closed vasculature.

Sherwin and Wilbur (1971) demonstrated the unique floral anatomy of *Sedum pusillum* and provided detailed information about the anatomy of *Diamorpha*.

Saint-Lager and Vivian-Morel (1876) discussed the value of glaucescence as a taxonomic character for separating species of *Sedum*. Their comments, though concerned primarily with European species, apply equally to a similar problem in North America.

J. T. Baldwin (1935, 1936, 1937, 1939, 1940, 1942 a, b, and 1943) made the earliest studies of chromosomes of North American *Sedum*. He reported numbers for eight species, discovered autoploidy in *S. ternatum* and *S. pulchellum*, and plotted the geographical distribution of plants with different numbers of chromosomes in these two species. Also, he suggested the specific distinctness between plants of *S. nevii* with six pairs of chromosomes and those with fourteen pairs of chromosomes. The gametic number of four chromosomes, originally reported by Baldwin for *S. pusillum*, is the lowest known for any species of *Sedum* or other genera of the *Crassulaceae*.

Hollingshead (1942) first reported the numbers of chromosomes of five species of the subgenus *Gormania* and indicated the similarity among them. She also described the chromosomes of *S. divergens*.

Uhl (1952, 1961, 1962, 1970, 1972, Clausen and Uhl 1943 and 1944) reported for the first time the numbers of chromosomes of eleven North American species of *Sedum* and discovered polyploidy in both *S. lanceolatum* and *S. wrightii*. He was the first to note that differences in numbers of chromosomes occur among plants sometimes identified as *S. rosea*, namely $g = 11$ and 18 (plants with the latter number being *S. integrifolium*).

No experimental data are in the literature on the genetics of North American *Sedum*. Information on genetic relationships has been derived primarily from observation of populations in the field. Some ideas about numbers of genes involved in inheritance of different characteristics are possible from comparison of replicated clones, including natural hybrids, in experiments designed in the present study to test whether differences among similar plants have a genetic basis.

Allard and Garner (1940) studied the response of plants of *Sedum glaucophyllum* (listed as *S. nevii*), *S. telephioides*, and *S. ternatum* to varying lengths of day. Also, they studied the response of three species of *Sedum* naturalized in North America and several cultivated species. They found that plants of *S. glaucophyllum* and *S. ternatum* are day-neutral or indeterminate, flowering under a wide range of lengths of day. Those of *S. ternatum* required cool temperatures, about 10° C., to produce floral buds, whereas plants under similar length of day, but at temperatures of about 22° C., did not produce floral buds. *Sedum telephioides* behaved as a typical long-day type of plant. It failed to flower when grown under lengths of day much less than fourteen hours. *Sedum telephium* also is a long-day plant. It required for flowering periods of continuous light seventeen hours or more.

Denffer (1941) studied the effects of photoperiod on the habit of growth and vegetative parts of various *Crassulaceae*, among them *Sedum telephium* and ten species of *Sedum* which are cultivated in American gardens. He found that certain species, especially those with broad flat leaves, as *S. telephium*, *S. aizoon*, and their relatives, when subjected to a short day of nine hours, developed a compact habit with small leaves in thick rosettes.

Funke (1943) investigated the influence of different wave lengths of light on flowering photoperiodicity of several species of *Sedum*. Blue supplemental light had the same effect as daylight on the flowering of some species in his experiments, but red supplemental light did not. *Sedum spectabile* was one of the species which exhibited this type of response.

Harriet Smith (1946) described the morphological and physiological differences among plants of *Sedum pulchellum* from thirty-nine localities. Twenty of her collections were diploid; sixteen were tetraploid; and three were hexaploid; but the number of collections utilized in some experiments was considerably less. Interpretation of Smith's data encounters two problems: separation of the effects of geographic variation from effects due solely to polyploidy and the probability of selection of any plant or population in her samples. Since probabilities are unknown, conditions necessary for tests of significance are not met. Despite these problems, Smith's work has suggested many useful hypotheses to test. These include the following:

1. Mature leaves of hexaploid plants are shorter, wider, and thicker than those of either diploid or tetraploid plants.

2. Primary basal leaves of hexaploid plants are thicker and longer than those of diploid and tetraploid plants.

3. The short, primary basal leaves of diploid plants are narrower than similar leaves of tetraploid and hexaploid plants.

4. Epidermal cells of primary leaves are larger in tetraploid than in diploid plants, and largest in hexaploid plants.

5. Stems of juvenile plants which are either tetraploid or hexaploid have a greater diameter than those of diploid plants.

6. Hexaploid plants have the most floriferous shoots at time of opening of the first flower; tetraploid plants the next most; and diploid plants the least.

7. Hexaploid plants have fewer flowers per inflorescence than either diploid or tetraploid plants.

8. Seeds of tetraploid and hexaploid plants are heavier, longer, and wider than those of diploid plants. Seeds of hexaploid plants are the heaviest and largest.

9. Hexaploid plants best withstand excess or deficiency of water. Diploid plants least well survive either extreme.

10. Diploid plants best withstand tannic acid. Hexaploid plants least withstand it.

11. In competition, hexaploid plants are most successful. Diploid plants are least successful.

12. Cuttings of tetraploid plants root most readily. Those of hexaploid plants develop the smallest root systems.

13. The content of growth hormone decreases progressively with increase in number of chromosomes.

14. Diploid plants flower twelve days earlier than tetraploid plants and those in turn about four days earlier than hexaploid plants.

15. The dry weight of tetraploid plants is greater than that of either diploid or hexaploid plants.

James (1958) discussed succinctly the condition of succulence in plants. His remarks are applicable to the North American species of *Sedum*. He noted that succulence occurs in many very different geographical areas and that its degree of development is expressed as the ratio of weight to surface. In addition, he wrote that the common unit of measurement is the number of grams of contained water per square centimeter of surface. Succulence varies inversely with surface. James also suggested that succulence, like the induction of flowering, is determined by a hormone produced in young leaves and conducted upward to the growing apex.

North American species of *Sedum* sometimes are cultivated as ornamentals, especially in rock gardens and borders. Some species have had a long history in cultivation. *Sedum ternatum*, for example, was illustrated in the *Botanical Magazine* as early as 1818, with a reference to an earlier listing in *Hortus Kewensis*. On the authority of Paxton (*Gardeners' Chronicle* n.s. 2:657 [1874]), *S. pulchellum*, another eastern species, was introduced in Great Britain in 1824.

For determining the number of North American species in cultivation in the recent past, four published accounts are useful. The data from these are tabulated below:

Date	Author	Publication	No. of N.A. species
1878	Masters	Hardy stonecrops: Sedums	9
1900–1902	L. H. Bailey	Cyclopedia of American Horticulture	12
1921	Praeger	Account of the genus *Sedum* as found in cultivation	14
1941	Bailey and Bailey	Hortus Second	20

All together, twenty-three species of North American *Sedum* have been in cultivation. Masters listed the species grown on the British Isles toward the end of the nineteenth century. The *Cyclopedia of American Horticulture* is an inventory of what was cultivated in the United States and Canada at the beginning of the twentieth century. Likewise, *Hortus Second* is an inventory of species cultivated in the continental United States and Canada in the decade from 1930 to 1940. According to the authors, it is based on catalogues of commercial dealers, statements in journals, exchanges between growers, correspondence, specimens in herbaria, and experience in gardens.

Only six North American species are not listed as in cultivation in the four cited works. These are *Sedum pusillum*, *S. niveum*, *S. borschii*, *S. oblanceolatum*, *S. albomarginatum*, and *S. moranii*. *Sedum glaucophyllum*, though not listed, has usually been confused with *S. nevii*.

L. H. Bailey (1900–1902 and 1914) and Bailey and Bailey (1941) have provided lists of *Sedum* from all regions in cultivation in North America. Numbers of species cited as in cultivation at different times are as follows: 1900—39; 1914—63; and 1930–1940—125.

In summary, the botanical discovery of the North American species of *Sedum* has continued over a period of four centuries. Many people have participated in the discovery, description, and investigation of these species. No single botanist has discovered more than two species. Twenty-four botanists have proposed fifty-eight binomials for the thirty species which seem worthy of recognition. Gray and Watson each described four valid species. Britton and Rose published a comprehensive taxonomic treatment in 1903. Quimby studied the floral anatomy of eight North American species; Subramanyam described the floral anatomy of *S. stenopetalum* and *S. ternatum*; Sherwin and Wilbur described the floral anatomy of *S. pusillum* and *Diamorpha*; Baldwin did pioneer cytological work; Uhl has devoted attention to chromosomes of many species; and Smith investigated the morphological and physiological variation of *S. pulchellum*. In addition, Mauritzon examined the embryos of several species and Allard and Garner studied the effects of varying photoperiods. Despite this and other research, knowledge of the variation, ecology, anatomy, chromosomes, and genetics of most North American species of *Sedum* still is incomplete.

CHAPTER IV

Geology of North America North of the Mexican Plateau

Climate and edaphic conditions, both present and past, plus method of dispersal, are important in determining the distribution of plants and animals. Since topography may have a pronounced effect on climate, is the principal explanation for edaphic conditions, and exerts a primary control on dispersal, knowledge of geology is basic to an understanding of biological distribution. The boundaries of major physiographic regions provide satisfactory limits for strata in sampling surveys for study of groups of plants or animals. Likewise, topographic distinctions in smaller regions are a practical basis for delimiting sampling units. The advantages of such units are that the boundaries are readily apparent to a trained observer and likely to persist for long periods of time. To explain the basis for sampling in the present study, as well as to interpret patterns of distribution, some discussion of geology is essential.

The definition of the part of North America (fig. 1) covered by this account of *Sedum* appears in Chapter I. Boundaries are natural and coincide with the borders of major physiographic divisions except in the cases of the Coastal Plain and the Basin and Range Province. Within these provinces, the limits are between sections, between the East Gulf and Mexican sections of the Coastal Plain, and between the Great Basin and the Sonoran Desert section of the Basin and Range Province.

North America north of the Mexican Plateau consists of six major divisions: the Canadian Shield in the northeast, the Interior Lowland in the central portion, the Appalachian Mountain System in the East, the Ouachita Mountain System south of the Interior Lowland, the Cordilleran Mountain System in the West, and the Coastal Plain along the Atlantic and Gulf Coasts. This classification, closely following King (1959) and other recent geologists, has the advantage of both simplicity and clarity.

Because rocks are the usual habitat for *Sedum*, and exposures of rocks are commonest in mountainous regions, the greatest interest in connection with these plants is in the mountain systems and the parts of the other divisions in which rocks are exposed.

The Appalachian Mountain System constitutes an elongate longitudinal belt, oriented northeastward, extending from northern Alabama and Georgia to Newfoundland. This belt often is subdivided into a northern region, the New England–

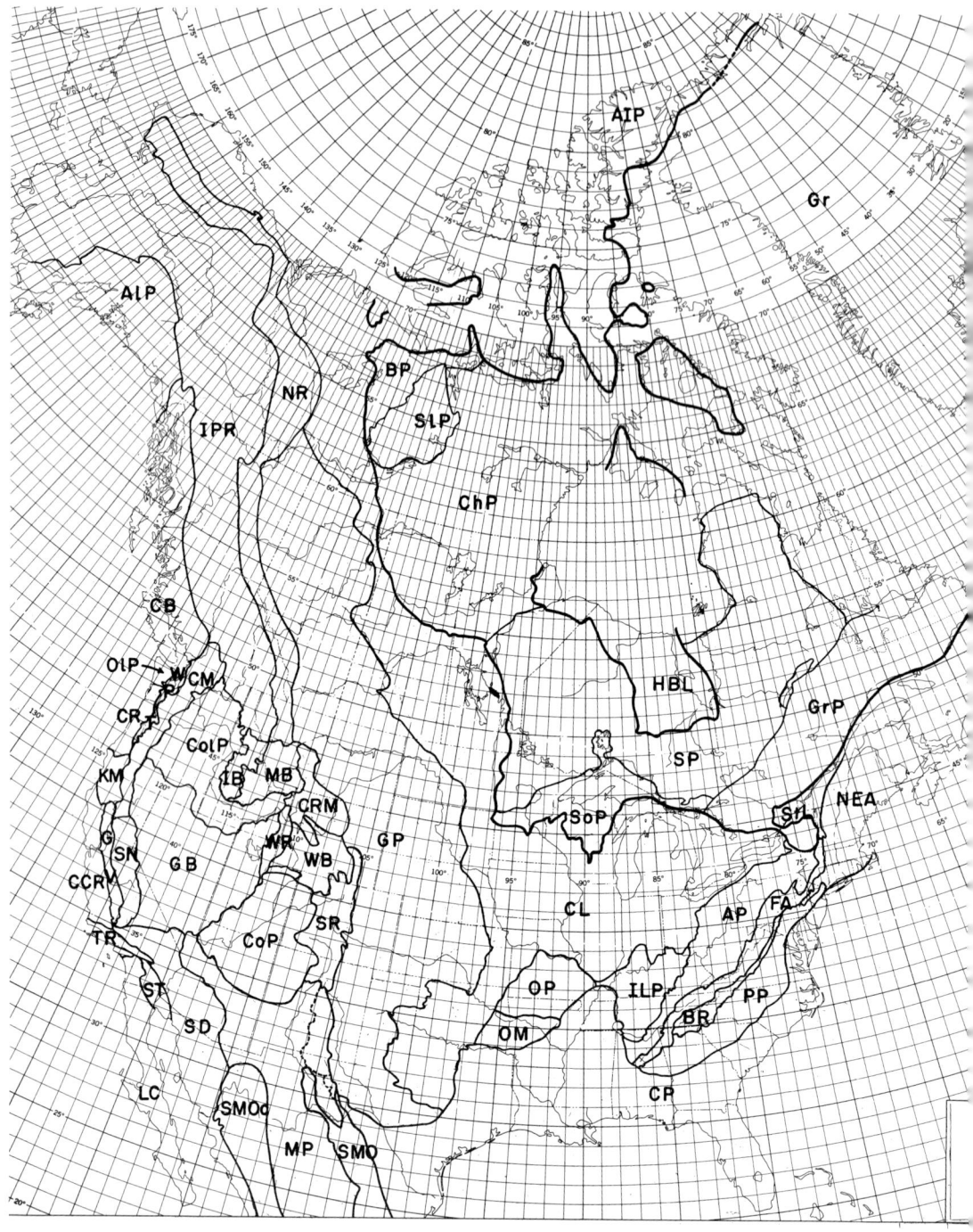

Acadian Highlands, and central and southern regions, respectively the central and southern Appalachians. The latter regions comprise three provinces: the folded Appalachian Mountains, the Blue Ridge, and the Piedmont Plateau. To the northwest of these provinces are the flat or slightly folded, Paleozoic strata of the Appalachian Plateau. This plateau belongs to the foreland of the eastern North American miogeosyncline. For understanding the distribution of plants, the structural provinces of the Appalachian Mountain System are superior to the latitudinal subdivisions because they are more homogeneous in edaphic conditions, not unduly heterogeneous in climate, and less unequal in area.

The New England–Acadian Highlands, the northernmost province of the Appalachian Mountain System, are situated primarily to the east of the valley of the Hudson River, but with a projection, known as the Reading Prong, extending southwestward across northern New Jersey and into eastern Pennsylvania. This is a region of crystalline rocks, mostly granite and gneiss. Boundaries are the St. Lawrence Valley and Gulf of St. Lawrence on the north; the Atlantic Ocean on the east; the Atlantic Ocean, Long Island Sound, and the Triassic Piedmont Lowland on the south; and the Ridge and Valley province, including the valleys of the Hudson River and Lake Champlain, on the west. The principal sections of the New England–Acadian

Fig. 1. Map of North America with the approximate boundaries of the geomorphic provinces. Abbreviations for provinces are as follows:

AIP	Arctic Island Province	KM	Klamath Mountains
Al P	Alaskan Peninsula	LC	Lower California
AP	Appalachian Plateau	MB	Mountains and basins of southwestern Montana and adjacent Idaho
BP	Bear Province		
BR	Blue Ridge	MP	Mexican Plateau
CB	Coast Batholith	NEA	New England–Acadian Highlands
CCR	California Coast Ranges	NR	Northern Rocky Mountains
Ch P	Churchill Province	Ol P	Olympic Peninsula
CL	Central Lowland	OM	Ouachita Mountain System
CM	Cascade Mountains	OP	Ozark Plateaus
Co P	Colorado Plateau	PP	Piedmont Plateau
Col P	Columbia Plateau	SD	Sonoran Desert
CP	Coastal Plain	Sl P	Slave Province
CR	Coast Ranges of Oregon and Washington	SMO	Sierra Madre Oriental
		SMOc	Sierra Madre Occidental
CRM	Central Rocky Mountains	SN	Sierra Nevada
FA	Folded Appalachian Mountains or Ridge and Valley Province	So P	Southern Province of Canadian Shield
		SP	Superior Province
GB	Great Basin	SR	Southern Rocky Mountains
GP	Great Plains	ST	Salton Trough
Gr	Greenland Province	St L	St. Lawrence Lowland
Gr P	Grenville Province	TR	Transverse Ranges of California
GV	Great Valley of California	WB	Wyoming Basin
HBL	Hudson Bay Lowland	WPT	Willamette–Puget Trough
IB	Idaho Batholith	WR	Wyomide Ranges
ILP	Interior Low Plateaus		
IPR	Interior Plateaus and Ranges of British Columbia, Yukon, and Alaska		

Highlands are the Long Range of western Newfoundland; the Shickshock Mountains of the Gaspé Peninsula; the White, Green, and Taconic Mountains; the New England Upland, including the Reading Prong; the Connecticut River Valley; and the Seaboard Lowland. The highest point is Mount Washington, in the White Mountains, with an elevation of 1,918 meters.

The Appalachian Mountains proper are folded and thrust-faulted Paleozoic sedimentary rocks. Their northwestern boundary is an erosional escarpment which forms the southeastern edge of the Appalachian Plateau. The southeastern boundary of the folded Appalachian Mountains is the more resistant mass of crystalline rocks of the New England and Blue Ridge provinces. The folded Appalachian Mountains comprise a longitudinal belt of ridges formed of sandstone and shale, and valleys underlain by limestone. This elongate region, known either as the Ridge and Valley province or the sedimentary Appalachians, consists of Southern, Middle, and Northern sections. These sections differ somewhat in topography. Boundaries are arbitrary: between the Southern and Middle sections, following Fenneman (1938), the divide between the Tennessee and New Rivers; and between the Middle and Northern sections, the southern limit of Wisconsin glaciation in eastern Pennsylvania. The Southern section is drained principally by the Tennessee River. In places, it is a series of longitudinal ridges. In other places, as in east-central Tennessee, it is largely valley. The Middle section is crossed by several large rivers, as the Susquehanna and the Potomac. It is a region of many nearly parallel mountain ridges, but with a broad valley lowland on the east. The Northern section is glaciated. It is chiefly a lowland, drained by the Hudson River and its tributaries, and also by Lake Champlain.

The Blue Ridge is a long, narrow zone of Cambrian and Precambrian igneous and metamorphic rocks. It arises in southern Pennsylvania, west of the Susquehanna River, where the limit is the contact between the older crystalline rocks of the Blue Ridge and the Triassic sedimentary rocks of the Piedmont province. This same contact forms the eastern boundary from Pennsylvania to the vicinity of the Potomac River, but southward, the base of the eastern slope of the mountains adjoins the lower crystalline rocks of the Piedmont Upland. In places, the change in relief is gradual, with the result that the distinction between the two provinces is not sharp. At the south, the Blue Ridge ends in northern Georgia, where its higher relief gives way to the lower elevations of the Piedmont Plateau. The western boundary is the base of the montane slope, where the crystalline rocks make contact with the folded sedimentary strata of the Ridge and Valley province. As thus bounded, the Blue Ridge has a latitudinal length of 1,300 kilometers, extending from Pennsylvania to Georgia. Its maximum width is 135 kilometers in North Carolina and Tennessee. The altitudinal range is from 64 meters, where the Potomac River cuts across the mountains between Maryland and Virginia, to 2,047 meters on the summit of Mount Mitchell in western North Carolina, the highest point in eastern North America. Often the Blue Ridge is divided into northern and southern sections. For this purpose, the Roanoke River serves as a practical dividing line.

The Piedmont Plateau is an area of prevailingly low relief and of complex geological structure. The predominant rocks range in age from Precambrian to Triassic. Boundaries are the Hudson River on the northeast, the fall line and the inner edge of the Coastal Plain on the east and south, and the Blue Ridge and Reading Prong on the west. Like the other Appalachian provinces, the Piedmont Plateau is long and narrow. Extending from Rockland County, New York, to Alabama, its length is about 1,500 kilometers. Its width varies from about 20 kilometers in southern New York to about 290 kilometers in southern Virginia. Elevations range from near sea level in northeastern New Jersey to 885 meters in the South Mountains of North Carolina. The major part of the Piedmont Plateau is an upland underlain by crystalline rocks. This upland reaches nearly to the Delaware River in southeastern Pennsylvania. The remaining northern part of the province is a lowland of Triassic sedimentary rocks. This consists of northern and southern sections, separated by higher land of a prong of the upland forming the divide between the Schuylkill and Susquehanna Rivers in Pennsylvania. The southern section of lowland extends from the Rappahannock River in Virginia to Pennsylvania. The northern section, though predominantly sandstone, is marked in New Jersey by extrusive ridges of igneous rock known locally as trap rock.

The Appalachian Plateau mostly is higher than the provinces to the east. Its layers of sedimentary rocks are almost horizontal or gently folded. In the north, the plateau begins at the southern edge of the Adirondack Mountains. Its eastern boundary is an escarpment, often prominent, known northward as the Allegheny Front and southward as the Cumberland Front. The Fall Line Hills of the East Gulf Coastal Plain make the boundary in the south. On the west, the surface of the plateau merges imperceptibly in many places with the adjacent lowland, but in other places, as in Kentucky and Tennessee, an escarpment about 300 meters high makes the western boundary. The length of the plateau is about 1,600 kilometers. The greatest width, about 560 kilometers is across northern Pennsylvania and Ohio. The Appalachian Plateau comprises eight sections: the Tug Hill Plateau, Mohawk Valley, Catskill Mountains, Glaciated Allegheny Plateau, Allegheny Mountains, Unglaciated Allegheny Plateau, Cumberland Mountains, and Cumberland Plateau. Elevations are lowest in the Mohawk Valley, only 70 meters at Rotterdam Junction, New York, and highest near the eastern border of the Unglaciated section, 1,431 meters on the summit of Big Spruce Knob in Pocahontas County, West Virginia.

The Ouachita Mountain System comprises nearly level-topped ridges, trending east-west. The strata are folded, most strongly in the south, and thrust-faulted westward. They consist of Paleozoic metamorphic rocks which are related to the Appalachian Mountains. The extensions of the two mountain systems toward each other are deeply buried under deposits of the Gulf Coastal Plain. Two sections of the Ouachita Mountain System are recognized: the Arkansas Valley in the north, an area of mountains and low ridges along the Arkansas River; and the Ouachita Mountains in the south, an area of anticlinal ridges, often parallel, formed of strongly folded sedimentary rocks. The present highest point, Magazine Mountain, with an elevation of 861 meters, belongs to the northern section.

The Cordillera of western North America comprises a complex system of mountains and plateaus which extends from central Mexico to Alaska. Included are the various ranges of Rocky Mountains, the intermontane plateaus, the Sierra Nevada, and the Pacific Coast Ranges. The portion of the Cordillera with which we are concerned may be considered to consist of five groups of provinces: the Eastern Ranges and Plateaus, including the Central and Southern Rocky Mountains, the Colorado Plateau, and the Wyoming Basin; a central portion, including the Idaho Batholith, Wyomide Ranges, Basin and Range Province, Sierra Nevada, Klamath Mountains, Great Valley of California, California Coast Ranges, and the Transverse Ranges of California; a northwestern volcanic region, including the Columbia Plateau, Cascade Mountains, and the Coast Ranges of Oregon and Washington; the northern mountains and plateaus, including the Northern Rocky Mountains, the interior plateaus and ranges, and the Coast Batholith; and the Alaskan Peninsula and outlying islands.

The Central Rocky Mountain region lies between the Southern and Northern Rocky Mountains and east of the main miogeosynclinal area. The ranges, highlands, and included basins sometimes distinguished by this name or the even broader designation, Middle Rocky Mountains, differ in age, structure, and orientation. In a broad classification of the Cordillera, the aggregation of these diverse elements as a geographic province has value for simplicity. For detailed purposes, including understanding the distribution of plants, subdivision into several provinces or subprovinces is desirable.

The region of the Central Rocky Mountains extends from southwestern Montana to northeastern Utah. Principal ranges are the Big Horn, Beartooth, Absaroka, Teton (fig. 2), Gros Ventre, Wind River, and Uinta Mountains. The Big Horn Basin and Jackson Hole are part of the region, although most basins between the ranges usually are grouped together under the collective name Wyoming Basin. Rocks forming the core of several ranges of the Central Rocky Mountains are Precambrian, but the Absaroka Range and Yellowstone Plateau are composed of more recent volcanic rocks, basalts and rhyolite of late Cretaceous or early Tertiary age. Sedimentary rocks, which covered the older rocks before uplift, persist on the flanks of the mountains, variously affected by the rising of the ranges.

To the south and east of the Central Rocky Mountains are the Southern Rocky Mountains. These comprise a series of ranges and intermontane high plateaus known locally as parks, extending from Casper Mountain, at the northern end of the Laramie Range in Wyoming, to the Santa Fe Range, at the southern end of the Sangre de Cristo Mountains in New Mexico. The northern boundary is the contact between the nearly level strata of the Wyoming Basin and the uplifted strata of the Laramie Range just south of Casper, Wyoming. On the east, the slopes of hogback ridges along the eastern front of the mountains mark the boundary with the Great Plains. The southern boundary, east of the Rio Grande, is the uplifted northern edge of the Glorieta Mesa. West of the Rio Grande, it is the southern base of the Jemez and Nacimiento Mountains. The western boundary in New Mexico and Colorado is the

Fig. 2. Teton Mountains: Grand Teton and Teewinot Mountain, July 29, 1966.

contact with the horizontal strata of the high Colorado Plateau. In Wyoming and northern Colorado, the border of the Wyoming Basin makes the western boundary. Altitudes range from about 1,500 meters in the valley of the Arkansas River, west of Pueblo, Colorado, to 4,399 meters above the level of the sea on the summit of Mount Elbert, southwest of Leadville, Colorado.

The principal ranges of the Southern Rocky Mountains, starting at the northeast, are the Laramie Range, Medicine Bow Range, Colorado Front Range, Sangre de Cristo Range, Park Range, Sawatch Range, and San Juan Mountains. Lesser ranges include the Rampart Range, Wet Mountains, Rabbit Ears Range, Mosquito Range, Elk Head Mountains, Flattop Mountains, Elk Range, West Elk Mountains, La Garita Hills, San Miguel Mountains, La Plata Mountains, Needle Mountains, and Jemez Mountains. The larger intermontane plateaus are North and Middle Parks, separated by the Rabbit Ears Range, South Park, the upper Arkansas Valley, and the San Luis Valley.

Basically, the structure of the Southern Rocky Mountains appears simple. The mountains have resulted from uplift of masses of Precambrian granite, pushing the overlying layers of sedimentary strata into vertical or sloping position. Although this is the general plan, the situation often is more complex. Volcanic rocks have intruded in many places. These are evident as dikes, plugs, and mountainous plateaus.

Ages of the rocks vary enormously. Likewise, uplift has occurred at many times, beginning with the Laramide orogeny at the end of the Cretaceous period. In general, the Southern Rocky Mountains are young mountains. Their present high elevation may not have been reached until middle or late Tertiary time. Much valuable information about these mountains is available in a guidebook prepared by the Rocky Mountain Association of Geologists (1948). The geologic maps of Colorado and Wyoming, prepared by the United States Geological Survey (1935 and 1955), also are helpful.

The Colorado Plateau, really a series of high plateaus drained primarily by the Colorado River, lies to the south of the Central Rocky Mountains and west of the Southern Rockies. It occupies a vast area in western Colorado, northwestern New Mexico, northern Arizona, and central and eastern Utah. A prominent escarpment separates it from the Mexican Plateau to the south and the Great Basin to the west. Principal sections are the Canyon Lands, Navajo, Datil, Grand Canyon, High Plateaus of Utah, and the Uinta Basin. Humphreys Peak in the San Francisco Mountains, part of the Grand Canyon section, with an elevation of 3,844 meters, is the highest point. Geologists believe that the present high altitudes of the Colorado Plateau resulted from uplift which began in the early Tertiary and probably continued into the Pleistocene. Rocks mostly are sedimentary in nearly horizontal strata, but volcanic areas occur, as in the San Francisco Mountains and region of Mount Taylor in the Datil section in New Mexico.

The series of high-altitude basins between the Central and Southern Rocky Mountains is known collectively as the Wyoming Basin. Included are the Laramie, Shirley, Carbon, Wind River, Great Divide (Red Desert), Washakie, and Green River Basins. Rocks are sedimentary, mostly of Eocene and Cretaceous age. Thornbury (1965) has provided a concise, modern account of this province as well as of other provinces.

The Idaho Batholith, sometimes included by physiographers in the Northern Rocky Mountains, is a great mass of quartz monzonite which forms the mountains to the north of the Snake River Plain in central Idaho and western Montana. The mountains have resulted from uplift. They lie between the miogeosynclinal sedimentary rocks of the Northern Rocky Mountains on the north and northeast, and lavas of the Columbia Plateaus on the south and west. Included mountains are the Clearwater, Bitterroot, Salmon River, and Sawtooth. According to King (1959), the Idaho Batholith is part of an eastbulging arc of the eugeosynclinal rocks of the Cordillera and had its inception during the Nevadan orogeny which produced the Sierra Nevada and Klamath Mountains. Lavas of the Columbia Plateaus now bury large areas of these eugeosynclinal rocks.

The Wyomide Ranges, geographically associated with the Central Rocky Mountains, but structurally different, occupy an area in western Wyoming, southeastern Idaho, and northeastern Utah, to the west of the Central Rockies. These, as the Northern Rocky Mountains beyond the Idaho Batholith, are formed of miogeosynclinal sedimentary rocks which are strongly folded and thrust-faulted. Principal ranges are the Wyoming, Salt, Snake River, Caribou, Bear River, and Wasatch.

The Great Basin, a section of the Basin and Range Province, is a vast area of arid intermont basins, separated by short, subparallel block ranges of mountains, of which the rocks are tilted as a result of faulting. Although centered on the state of Nevada, the Great Basin extends into western Utah, southern Oregon, southern Idaho, and western California. It lies between the Colorado Plateau and Wyomide Ranges on the east and the Sierra Nevada and Cascade Mountains on the west, and to the south of the Columbia Plateaus. The southern boundary is somewhat indefinite, roughly the Colorado River and the northern edge of the lower land of the Mohave Desert. The rocks are chiefly sedimentary, ranging in age from Paleozoic to recent, and are metamorphosed in varying degrees. Drainage is for the most part internal.

To the east of the Idaho Batholith, and between the Blackfoot River on the north and the Snake River Plain on the south, is a region of short mountain ranges and intermontane basins, similar in structure to the Basin and Range Province. Physiographers, as Fenneman (1931), have referred these mountains and basins of southwestern Montana and adjacent Idaho to the Northern Rocky Mountains. Thornbury (1965) has done likewise, but has emphasized their structural distinctness. For this reason, the region is shown as a separate unit on the map of North America.

The Sierra Nevada, in eastern California and western Nevada, is a great fault block tilted to the west. It extends for about 600 kilometers from the northeastern border of the Tehachapi Mountains, from the southern base of Breckenridge Mountain, the Walker Valley, and Kelso Valley, northward to the region north of the Feather River and south of Lassen Peak, where Tertiary lavas meet Mesozoic and Paleozoic rocks. Although the Tehachapi Mountains are structurally different according to Buwalda (1954), they are topographically continuous. The eastern boundary of the Sierra Nevada is the base of a steep escarpment which is a fault scarp, separating the range from the Great Basin. The western boundary is at the base of the foothills, where the folded metamorphic rocks and granites meet the nearly level sediments of the Great Valley of California. The altitudinal range is from about 50 meters, at the western base of the range, in the region of the lower Bear and Yuba Rivers, to 4,421 meters above the level of the sea on the summit of Mt. Whitney.

The commonest rocks of the Sierra Nevada are Mesozoic granites. Older, metamorphosed sedimentary rocks are frequent in the northern part of the range and elsewhere, as in the region of McGee Mountain in the east, and in a zone along the western border. Also, volcanic rocks, extruded for the most part in the late Tertiary, occur in many places. The ages of the rocks are various. The metamorphosed rocks of McGee Mountain and vicinity, according to Rinehart (1964), are of Paleozoic age and estimated to be four hundred million years old. At the other extreme, some lavas are comparatively recent. A description of the geology of the Sierra Nevada is available in E. H. Bailey (1966).

Northwest of the Sierra Nevada, the Klamath Mountains are situated in northwestern California and southwestern Oregon, between the folded sedimentary rocks of the California Coast Ranges on the south and the younger sedimentary and volcanic rocks of the Oregon Coast Ranges on the north, and between the recent volcanic rocks of the Cascade Mountains on the east and the Pacific Ocean on the west. Often they are called the Siskiyous. They are of complex structure. Irwin (1960) has discussed the problems of both structure and boundaries. The principal rocks are metamorphic, ranging in age from Silurian, and possibly even Ordovician, to Jurassic, with extensive granitic intrusions. The highest point is Thompson Peak, in the Trinity Alps, with an elevation of 2,744 meters.

The Great Valley of California, including the San Joaquin and Sacramento Valleys, is a structural depression between the Sierra Nevada and the Coast Ranges. The surface is a nearly level plain, with a gentle slope from both ends toward San Francisco Bay. A notable exception to the flat terrain is the group of hills known as the Sutter or Marysville Buttes. These attain a maximum elevation of 650 meters and are the eroded remains of a volcano. The Tehachapi Mountains form the southern boundary of the Great Valley and the higher land of the Klamath Mountains, north of Redding, forms the northern boundary. The surface of the Great Valley is largely alluvial material. The region probably has subsided gradually and, as it sank in relation to the adjacent mountains, alluvium filled the depression.

Bordering the Pacific Ocean and west of the Great Valley are the California Coast Ranges. These can be divided for convenience at the latitude of San Francisco Bay into a northern group of ranges (fig. 3) and a southern group. The northern boundary

Fig. 3. Red Mountain, northern California Coast Ranges, with fog on western slope, July 16, 1970. This is type locality of *Sedum laxum* ssp. *eastwoodiae* and place of occurrence of *S. spathulifolium* ssp. *spathulifolium*.

is where contact occurs with the older metamorphic rocks of the Klamath Mountains. In the south, the limit is the series of transverse ranges which are south of the Santa Maria River. The California Coast Ranges are of complex structure and history. The predominant rocks are sedimentary. These are strongly folded and faulted. Although the mountains possibly developed in the mid-Tertiary, they probably attained their present heights much latter.

The Transverse Ranges of California are distinctive in their east-west trend. Situated between the Mohave Desert and the Los Angeles Basin, their east-west extent is from the Pinto Basin to Point Arguelo. They include, among others, the Santa Ynez, Santa Monica, San Gabriel, San Bernardino, and Little San Bernardino Mountains. San Gorgonio Peak, in the San Bernardino Mountains, with an elevation of 3,500 meters, is the highest point. The rocks are of many kinds and range in age from Precambrian to recent. A detailed account of the geology of this province by Bailey and Jahns (1954) is available. Putnam (1954) has discussed the marine terraces of the Santa Monica Mountains.

Easternmost of the provinces comprising the northwestern volcanic region is the Columbia Plateau. It comprises a vast area of lava, mostly basalt, east of the Cascade Mountains in eastern Oregon and Washington, and also in adjacent western and southern Idaho and extreme northeastern Nevada. Conceived as a physiographic province, the Columbia Plateau includes two important mountainous sections, the Blue and Seven Devils Mountains. Inclusion of the Seven Devils Mountains is according to the interpretation of Anderson (1941). On the north, the province is bounded by the granitic rocks of the Okanogan Highlands, and on the east by the Idaho Batholith. The southern limit is where the high upland passes into the region of interior basins and fault-block ranges of the Great Basin. Westward, the plateaus end at the base of the eastern slope of the Cascade Mountains. The maximum width of the Columbia Plateau is about 900 kilometers. Maximum length is 720 kilometers. The altitudinal range is from about 30 meters along the Columbia River, where it cuts across the Cascade Mountains, to 2,861 meters on the summit of He-Devil Mountain in the Seven Devils Mountains of Idaho. Except in the mountains and deep canyons where older sedimentary and igneous rocks may be exposed, the prevailing rock of the plateaus is basalt of Tertiary and Quaternary age.

The Cascade Mountains (fig. 4) are west of the Columbia Plateau and north of the Sierra Nevada. They extend for about 1,200 kilometers from northeastern California to southern British Columbia. The northern boundary is where the mountains give way to the lower surface of the interior plateau of British Columbia. Specifically, the valleys of the Fraser River and its tributaries, the Thompson and Nicola Rivers, mark the northwestern limits of the range, and the valley of the Similkameen River makes the northeastern boundary. The eastern front of the Cascade Range is the valley of the Okanogan River in the north, then the western border of the Columbia Plateau, and finally the Great Basin in the south. The southern boundary is the contact of the Tertiary volcanic rocks with the Mesozoic granite of the Sierra Nevada. On the west, the boundary is in the south the contact between the volcanic rocks and the metamorphosed sedimentary and granitic rocks

Fig. 4. Mt. Stuart, habitat of *Sedum divergens*, *S. integrifolium*, and *S. rupicolum*, July 22, 1961.

of the Klamath Mountains, and farther northward, the depressions of the Willamette Valley in Oregon and the trough of Puget Sound in Washington and British Columbia. The altitudinal range is from about 150 meters in the Puget Trough to 4,392 meters above the level of the sea on the summit of Mount Rainier.

The Cascade Mountains illustrate both the complexity of a physiographic province and the difficulty of classification. As here conceived, following Williams (1957), these mountains comprise three subprovinces: the Northern Cascade Mountains, extending from near the Yakima River northward, in which intrusive granitic rocks are dominant; the Western Cascades, extending from southern Oregon to about the region of the White River in Washington, comprising for the most part old Tertiary flows of lava, tilted eastward; and the High Cascades, a line of high volcanic cones, developed in the late Tertiary and more recent time, forming the crest of the range and extending from Lassen Peak in California to Mount Baker in Washington. Present conditions are the result of a succession of extrusions of lava, uplift, explosive volcanism in the High Cascades, erosion and transport of materials by streams, glaciation, and in the north, intrusion of the older formations, both sedimentary and volcanic, by plutonic rocks. General statements about the range must be modified by detailed study and description of conditions at each site.

To the west of the Cascade Mountains is the Willamette-Puget Trough, a structural depression filled with the products of erosion from mountains on either side. Still farther west are the Coast Ranges of Oregon and Washington, and northward the Olympic Mountains. Broadly folded beds of sedimentary rocks, together with intruded lavas, form these mountains. In the north, higher elevations in the Olympic Mountains probably are the result of greater uplift. The highest point, Mount Olympus, has an elevation of 2,315 meters.

The Northern Rocky Mountains are the easternmost of the northern mountains and plateaus. Included are the mountains of the Disturbed Belt of western Montana; the Lewis and Livingstone Ranges; the Canadian Rocky, Franklin, Mackenzie, and Richardson Mountains; and at the extreme northwestern end, the Brooks Range. The Little Belt and Big Belt Mountains, although geographically close to the southern end of the Northern Rocky Mountains, are structurally allied to the Central Rockies. The Northern Rocky Mountains extend from western Montana northward to the Arctic coastal plain. For most of their length, they lie between the Great Plains on the east and the interior plateaus and ranges on the west. A prominent valley, the Rocky Mountain Trench, marks the western edge for a long distance from Montana to northern British Columbia. The predominantly sedimentary rocks are strongly folded and faulted. They are part of the Cordilleran miogeosyncline. Mount Robson, in eastern British Columbia, with an elevation of 3,948 meters, is the highest peak.

To the west of the Northern Rocky Mountains are the Interior Plateaus and Ranges of British Columbia, Yukon, and Alaska. These include, among others, the Purcell, Selkirk, and Cariboo Mountains in the south; the Interior Plateau of British Columbia; the Omineca and Cassiar Mountains, and to their west, the Stikine Plateau; the Pelly, Selwyn, and Ogilvie Mountains; and in the north, the Yukon Plateau. In general, this is a region of uplift. The rocks west of the Rocky Mountain Trench mostly are older, and also the mountains are not as high as the loftiest peaks in the ranges to the east. The highest point is Mount Dawson in the Selkirk Mountains, with an elevation of 3,390 meters.

Along the Pacific Coast, and west of the Interior Plateaus and Ranges of British Columbia, Yukon, and Alaska, is the Coast Batholith, also known as the Coast Mountains. This province extends from the valley of the Fraser River to the Yukon Territory, and includes, besides the ranges of the mainland, the adjacent coastal islands of British Columbia and southeastern Alaska. According to the summary of Eardley (1962), the rocks of the batholith are predominantly gneiss and most likely are of Lower Cretaceous age. The highest point is Mount Waddington in British Columbia, with an elevation of 4,041 meters.

The Alaska Range, Alaskan Peninsula, and adjacent islands form a geographic series which begins at the east in the vicinity of Icy Point, includes the St. Elias Range, the Wrangell Mountains, the Chugach Mountains, and the various Coast Ranges, and extends westward across the Pacific Ocean as the Aleutian Islands. The included mountains are of complex structure. Rocks of many kinds occur, especially bedded metamorphosed sedimentary and volcanic. These have been strongly folded, uplifted along great faults, and intruded by ancient crystalline rocks. Volcanism has been extensive. In the Aleutian Islands, it continues at the present time. Mount McKinley, in the Alaska Range, with an elevation of 6,187 meters, is the highest peak on the North American continent. For a detailed account of the Alaskan mountains, see Williams (1958).

The basement rocks of North America, an enormous area of Precambrian formations, mostly granite and gneiss, are exposed in the northeastern part of the continent and in Greenland. Farther south, these rocks are buried beneath the sediments of the Interior Lowland. The Canadian Shield occupies much of eastern and central Canada and extends into the United States in the Adirondack Mountains and around the western end of Lake Superior. It also includes most of Greenland, much of Baffin Island, and parts of the smaller arctic islands to the north. It is a vast area of mostly low or moderate elevation, which possibly has been in process of erosion since before Cambrian time. The highest altitudes are in the Torngat Mountains of Labrador, 1,623 meters, and on Baffin Island, about 2,592 meters, where uplift may have resulted from faulting.

The northern boundary of the Canadian Shield is either the Arctic Ocean or the sedimentary rocks of the arctic islands. The Atlantic Ocean and the Gulf of St. Lawrence form the eastern boundary. In the south, the boundary is respectively from east to west the Gulf of St. Lawrence, the St. Lawrence Valley, the Appalachian Plateau in New York, the Central Lowland and Lakes Huron and Superior in Ontario, and westward the sedimentary rocks of the Central Lowland, which also form the western boundary. For the boundary of the Canadian Shield on Ellesmere Island, the map of the principal mineral areas of Canada, prepared by the Department of Mines and Technical Surveys, Canada (1965), has been helpful.

Literature on the Canadian Shield is extensive. Publications by Wilson (1958), Stockwell (1961, 1963), and Stevenson (1962) contain valuable information and references. Geologists recognize about eight physiographic provinces on the basis of age and structure. Two provinces, the Superior and Slave, originated in the Kenoran orogeny, with an estimated age of about 2,500 million years. The oldest rocks known occur in these provinces, some even antedating the age of the Kenoran orogeny. A prevalent opinion is that the Superior and Slave provinces were the nucleus of the North American continent which then grew by accretion. The Superior Province forms a great arc around the southern end of Hudson Bay, extending from the western part of the Ungava Peninsula, around James Bay and north of the Great Lakes, to southeastern Manitoba. Granitic rocks and gneisses predominate, with included remnants of mostly eastward-trending belts of metamorphosed volcanic and sedimentary rocks. The Slave or Yellowknife Province, in western Mackenzie and eastern Yukon, differs from the Superior Province in having irregularly shaped belts of volcanic and sedimentary rocks. Younger rocks either unconformably overlie the ancient rocks at the borders of the Superior and Slave provinces or occur on their flanks as intrusions. Three provinces, the Bear, Churchill, and Southern, were formed in the Hudsonian orogeny, about 1,700 million years ago. The Bear Province in central and northern Yukon, northwest of the Slave Province, comprises rocks which are either folded or, if unfolded, then unconformably overlying the other rocks. The Churchill Province is the largest part of the Canadian Shield. It extends around the northern end of Hudson Bay, from central Manitoba northward and eastward to northern Quebec and Labrador. It includes much of Baffin Island. The boundaries are marked by unconformities, but the details still are not sufficiently worked out on the Ungava Peninsula. The rocks are mostly granitic and metamorphic. The Southern Province lies south of the Superior, on both sides of Lake Superior and to the west. Its rocks are unconformably on top of the rocks of the Superior Province. The Grenville Province, although of complex structure and with some rocks as old as those of the Superior Province, underwent its last major orogeny about 950 (800–1,100) million years ago. It is an area of granitic rocks and gneisses, trending northeastward on the northern side of the St. Lawrence River and the Gulf of St. Lawrence, and extending from central Ontario to the coast of Labrador. The Adirondack Mountains of New York are a lobe of this province. Two additional provinces, the Arctic Island and Greenland, complete the list of subdivisions of the Canadian Shield.

The Interior Lowland comprises the prairies and plains of the central stable region of the North American continent. Included are the following provinces: Central Lowland, Great Plains, Interior Low Plateaus, and Ozark Plateaus. Together they form a vast area of younger, sedimentary rocks lying on top of Precambrian rocks. The lowlands are south, southwest, and west of the Canadian Shield. Drainage is largely by the Mississippi River and its tributaries, by other streams flowing into the Gulf of Mexico, and by streams which flow into Hudson Bay and the Arctic Ocean. Although the Interior Lowland is mostly level land of low altitude, some parts have been elevated as plateaus. Certain of these plateau areas, as the Interior Low Plateaus, including the Nashville Basin, and the Ozark Plateaus, are considerably dissected through erosion. The Ozark Plateaus are built on a dome of Paleozoic limestone through which Precambrian crystalline rocks project as knobs in the east-central part. The Great Plains, lying nearest to the Cordillera, have the highest altitude of any part of the Interior Lowland. Excepting the Black Hills, which are allied to the Central Rocky Mountains and attain a maximum altitude of 2,208 meters on the summit of Harney Peak, and the Colorado Piedmont and Raton Mesa which are related to the Southern Rocky Mountains, the highest part of the Great Plains is in Colorado and New Mexico where elevations reach 1,800 meters.

The Atlantic and Gulf Coastal Plains together constitute a vast area of slight relief, extending from Cape Cod to the peninsula of Florida, and then around the Gulf of Mexico to the Yucatan Peninsula. The underlying rocks are deeply buried under sedimentary deposits which range in age from the late Jurassic Period to the present. Outcroppings of rock are scarce. Much of the area is overlain by sand. The several parts (provinces) of the Coastal Plain, although continuous, must be considered in relation to the adjacent highlands: the Atlantic section with the Appalachian Mountain System; the Gulf section with the Ouachita Mountain System; and the Mexican section with the Sierra Madre Oriental and the Trans-Mexican Volcanic Belt. The Florida Peninsula is distinctive in being the emergent part of a platform of carbonate deposits which accumulated to great depth on top of ancient crystalline rocks. Because of their great latitudinal extent, the coastal plains experience a variety of climates, ranging from humid, cool temperate to subtropical steppe and tropical. Murray (1961) has published a comprehensive account of the eastern North American coastal plains. He described them as comprising a single geologic province which, in his opinion, is a coastal geosyncline.

Continental glaciers have affected local topography and the distribution of plants in the northern part of the North American continent. In the Cordillera, local glaciers likewise have had an important role in areas south of the main continental ice sheets. Changes in climate consequent with periods of glaciation must also have greatly altered the distributional patterns of both plants and animals far away from the centers of glaciation. Details of these changes are available from study of pollen obtained from borings in bogs and swamps. The data indicate a history of climatic change, accompanied by shifts in the frequency of different species of plants.

Flint (1957) has provided a valuable review of glacial geology with emphasis on the Pleistocene epoch. Four glacial ages usually are recognized in the Pleistocene: Nebraskan, Kansan, Illinoian, and Wisconsin. Emiliani (1958), using oxygen-isotope ratios of shells obtained in cores of submarine sediments, estimated that the first glacial age, the Nebraskan, began about 300,000 years ago. Ericson *et al.* (1964), using twenty-six cores of deep-sea sediments, concluded that the Nebraskan ice age began about 1.5 million years ago and that the most recent glacial age, the Wisconsin, ended about 11,000 years ago.

The central and southern portions of the Appalachian Mountain system, together with adjacent lowlands, are unglaciated. These areas have been available for occupation by plants for millions of years. Of the areas studied for *Sedum*, the southern parts of the Appalachian provinces are among the oldest available for continuous occupation by plants.

Glaciation in the Cordillera was extensive far south of the main continental ice sheets. Ray (1940) has summarized information for the Southern Rocky Mountains. In the Wisconsin stage of Pleistocene glaciation, large glaciers existed at the higher elevations. Valley glaciers extended down the slopes to a minimum elevation of 2,440 meters. The present flora thus is a recent one which has invaded the larger part of the region in the last ten thousand years from the unglaciated lower slopes and adjacent physiographic provinces, especially from the plateau country to the south and the plains to the east.

Glaciers covered much of the Sierra Nevada in the Pleistocene epoch and had a profound effect on topography. Only the lower foothills and areas in the north were free of ice. Birman (1964) recognized seven glacial advances. Of these, the Sherwin is both the oldest and most extensive. It is pre-Wisconsin, possibly correlated with Illinoian glaciation. The two advances correlated with Wisconsin glaciation also were extensive and, according to Birman, reached a minimum altitude of about 976 meters in the valley of the San Joaquin River. Since the Wisconsin glaciation, the Sierra Nevada may have been completely free of glacial ice in the period of maximum warmth following deglaciation. Modern glacierets probably came into being since then and may be no more than four thousand years old, as suggested by Matthes (1950).

The foothills and the low northern parts of the Sierra Nevada could have been refuges for plants in the Pleistocene. Presumably, the climate at these sites was both cooler and moister than at present. Today the foothills are dry and hot in summer. Montane species can survive only in favored sites.

According to Flint (1957), glaciers reached the northern margin of the Columbia Plateau in the Pleistocene epoch. Highland glaciers existed in that epoch in the Seven Devils, Wallowa, Elkhorn, Strawberry, and Owyhee Mountains. Cook (1954) summarized data of J. V. McDonald who studied glaciation in the Seven Devils Mountains. McDonald's evidence indicates at least two periods of glaciation, and Cook suggested the possibility of a third glaciation. The high lakes in the Seven Devils Mountains are tarns. They occupy the low parts of glacial cirques. An exception is Emerald Lake which was formed by a morainic dam. In the Seven Devils Mountains, glaciers of the latest stage were small, advancing little more than 10 kilometers and reaching no lower than 1,500 meters in elevation above the sea. Their best development was on northeastern slopes. An icecap existed in the Wallowa Mountains (E. M. Baldwin, 1959). From this, valley glaciers extended down the principal streams. The evidence on glaciation suggests that much of the Columbia Plateau was available for occupation by plants in the Pleistocene. Only the highest mountains and the northern border were covered by glacial ice. If the climate then was cooler than at present, conditions may have been more favorable for growth of plants such as *Sedum*. At least, they could have survived there and repopulated the mountainous areas after deglaciation.

Major glaciers covered the northern portion of the Cascade Mountains in the Pleistocene epoch. The southernmost extent of general Cordilleran glaciation was Mount Adams, north of the Columbia River, according to Flint (1957). Southward, local glaciers covered most higher parts of the Cascade Mountains. The principal unglaciated areas were the lower slopes of the middle and southern parts of the range. Small glaciers exist on several higher peaks at the present time. Emmons Glacier, on the northeastern slope of Mount Rainier, is the largest of these. The Cascade Mountains, including Mount Rainier, perhaps were never completely free of glacial ice in the present postglacial period. Conditions now on Mount Rainier afford an opportunity to see how close populations of living plants can be to a glacier. The occurrence there of *Sedum divergens*, *S. oreganum*, and *S. integrifolium* demonstrates that these species could have survived on ridges adjacent to glaciers, and on rocks along streams below glaciers, in a time of maximum glaciation.

Flint (1957) stated that the entire Canadian Shield was covered by ice at the time of the last glacial maximum. For that reason, the antiquity of the rocks is of no consequence in explaining the present distribution of the flora. All plants must have colonized the area since deglaciation. Greenland continues to have an icecap. Plants are confined largely to its coastal regions, especially in the south. Whether plants survived along the coastal borders of the Pleistocene glaciers is speculative.

From the standpoint of the modern distribution of plants, Wisconsin glaciation is the most important to understand. Presumably the earlier ages of glaciation had similar effects on vegetation. Likewise, they must have profoundly affected the directions of evolution. We may guess that many biotypes were destroyed and that in other cases, what had been continuous, intrabreeding populations became separated into disjunct parts. Following glaciation, populations which had been separated might invade the same areas. In such cases, opportunity for hybridization became available. In this way, populations of plants could become rejuvenated, just as the soils of glaciated regions are restored to a youthful condition. Potentially, each group of plants affords an opportunity to test these ideas. *Sedum*, unfortunately, supplies us with few examples here. The conceptions are hypothetical. Changes in climate associated with glaciation may have brought *S. nevii* and *S. ternatum* together in the Appalachian region and have been involved in the origin of *S. glaucophyllum*. Similarly, changes in ranges of populations of plants, caused by glaciation, may have had a part in the origin of *S. integrifolium*. Invasion of the Seven Devils Mountains, following glaciation, by both *S. leibergii* and *S. stenopetalum*, may have led to the hybridization of these species there today. One may even guess that glaciation had a part in separating the eastern *S. ternatum* from the western *S. spathulifolium*, and separating the modern group of species known as *Parvisedum* from the ancestors of *S. nuttallianum* or *S. pusillum*. On the other hand, the change to more arid conditions in the central and western parts of the continent or the earlier development of long chains of mountains may have been responsible. To understand the true nature of the changes which have occurred in *Sedum* or any other group of organisms, one would need to have lived in the periods when the events were taking place, or else to have samples which can be dated definitely and correlated with important geological occurrences.

If the precise effects of glaciation on species of a particular genus of plants are obscure, the general impact of a condition so drastic as continental glaciation is clear. In the Wisconsin age, a great mass of ice moved southward and covered most of the northern part of the continent, south to Long Island, the northern part of the Appalachian Plateau, the northern Interior Lowland, the Missouri Plateau, and the northern part of the Cordillera, approximately down to the northern Cascade Mountains. Northern areas which escaped glaciation included much of Alaska and the driftless region in Wisconsin. The Canadian Shield was the center for Wisconsin glaciation. In the Cordillera, local glaciers covered the mountains at least as far south as the Peninsular Ranges and the mountains of Mexico. In all these places, the flora before glaciation must have been destroyed. The modern flora of these areas reinvaded the land in the period since deglaciation.

Sedum pollen has not been reported from glacial deposits or from places south of the glaciated regions. This fact is not surprising because the rocky sites in which most species of *Sedum* occur would not be conducive to preservation. Further, amounts of pollen in any one place might be small. The only report of fossil pollen of *Sedum* which has come to my attention is by Wenner (1947). His series 72 from South Stag Island, off the coast of southern Labrador, at Cape Porcupine, contained pollen of *Sedum rosea* at a depth of 15 centimeters. The site was 14 meters above sea level. Wenner interpreted the profile as indicating a postglacial change from *Salix* heath to ericaceous and *Empetrum* heath. The occurrence of *Sedum* followed this change. Presumably the *Sedum* pollen is not very old.

Important for an understanding of the distribution of plants during the glacial period is Martin's report (1958) on two cores from a marsh in the unglaciated Piedmont Plateau of southern Pennsylvania. He found evidence of a taiga-tundra vegetation at the time of the last glacial maximum. His data do not support the idea that forests characteristic of temperate climates existed close to the margin of the glacial ice. Although pollen of *Sedum* was not reported from the cores, the facts suggest two assumptions relevant to *Sedum*: first that modern Appalachian species probably would not have been able to survive near to the glacial margins, and second, that conditions as far south as the latitude of southern Pennsylvania may have been favorable for the general occurrence of species such as *S. rosea*. If the second assumption is true, then the modern occurrence of *S. rosea* along the Delaware River, south of Easton, is reasonable and represents contraction and survival from glacial times.

The soils of North America are of many kinds. In the North and in the high mountains, they are largely of glacial origin. In these regions, tundra and podzolic soils predominate. Other broad regional types are latosolic soils in the southeast, chernozemic soils in the Interior Lowland, and desertic soils in the southwest. Lithosols predominate in the mountains. The general features of the distribution of soils of the world are shown on a map published by Simonson (1957). The Soil Survey Staff of the Soil Conservation Service, United States Department of Agriculture (1960), has proposed a newer classification of soils. According to that scheme, tundra soils and many lithosols are inceptisols; podzols are spodosols; gray-brown podzolic soils are alfisols; lateritic soils in the United States are ultisols; chernozems are mollisols; and desertic soils are aridisols. Although complex, the new classification has a logical basis, but the mass of available literature employs the older terminology. For that reason, students of soils must learn two sets of nomenclature.

For detailed information about regional edaphic conditions, local reports on soil surveys are available for many counties in the United States and Canada.

Sedum is a genus of plants growing primarily on rocks. Situations inhabited by these plants often have shallow, poorly developed soils and sometimes almost no soil. Some species even grow in living mats of mosses and lichens on rocks. The majority of species occurs on lithosols. Because the niches occupied by *Sedum* must have existed before the development of the soils of forests and grasslands, this environmental condition is an old one. As far as edaphic conditions are limiting, *Sedum* could have existed in geological periods earlier than many other terrestrial angiosperms. If, on the other hand, *Sedum* is a specialized offshoot from one of the groups of angiosperms which grew in well-developed soils, then it may represent a recent adaptation to the relatively rigorous conditions of the rocky environment. Should the ideas (R. T. Clausen 1959) about *S. botterii* and phylogenetic relationships of *Sedum* be correct, namely that the primitive stonecrops were flat-leaved plants which inhabited humid forests, then we do not know when they began to invade rocky sites. From study of the distribution of modern species, we may infer that this shift in habitat occurred millions of years ago. The least specialized species, and estimates of the amounts of time that they have been separated from their nearest relatives, provide the best clues concerning this problem. *Sedum telephioides*, with distribution centered in the middle and southern Appalachian Mountains, and with nearest relatives in eastern Asia, *S. viridescens*, and Europe, *S. telephium*, may have existed already in early Tertiary times. At least, the ancestral stock may have reached the Appalachian region by the Eocene. If the fossil *S. hesperium* is correctly identified and is a relative of *S. telephioides*, it indicates that plants of this type existed as early as the Eocene and occurred in a region which might have been part of the connection between the eastern North American and eastern Asiatic floras.

That climate has fluctuated in the past is abundantly evident from the fossil record. Especially important were the shifts from warm to cold in the glacial ages. The causes of these changes are uncertain. Several authors have pondered the problem in the volume by Shapley (1953).

To explain climatic change, some authors have suggested that the axis of rotation of the earth shifts or that the crust of surface layers glides over the inner mantle. Munk and Macdonald (1960) have considered various hypotheses about the rotation of the earth. They have concluded that the empirical evidence does not require acceptance of the idea of major shifts in the axis of rotation. On the basis of dynamic considerations and rheology, they prefer to assign sufficient strength to the earth to prevent polar wandering. As we review the data for *Sedum*, we may note that none of the distributional information for this genus conflicts with the conclusion of Munk and Macdonald. Major latitudinal shifts of mountain ranges are unnecessary to explain the distributional patterns of species of *Sedum*. For other opinions about the history of the earth, see Runcorn (1962) and recent publications on plate tectonics.

Modern climates in North America, although varying with latitude, are modified by topography. They range from tropical in southern Florida to polar in the far north. Also, alpine conditions, with a short or almost no growing season, exist on the highest mountains throughout the Cordillera. A few small glaciers exist as far south as Mount Whitney in the Sierra Nevada and the Sangre de Cristo Range in the Rocky Mountains. In contrast, localities along the western coast enjoy a mild climate, even in winter. The growing season there is continuous. Inland, the intermontane plateaus and Great Plains lie in a rain shadow, shielded from the Pacific Coast by high mountains. As a result, these areas are arid or semiarid, contrasting with the humid coastal regions and moister montane slopes. The majority of species of *Sedum* appears adjusted to cool, moist conditions. These plants withstand successfully short periods of drought and are particularly common in areas where wet and dry seasons alternate.

Detailed information about the climate of North America is available from the following sources: *Climate of British Columbia* by the Department of Agriculture, British Columbia (1928——); *Arctic Summary* prepared by the Meteorological Branch of the Department of Transport, Canada (1959——); *Monthly Record* of meteorological observations in Canada, also prepared by the Meteorological Branch, Department of Transport, Canada (1916——); *The Climate of British Columbia and the Yukon Territory* by Kendrew and Kerr (1955); *Climate and Man*, U.S. Dept. of Agriculture (1941); *Climatological Data* for the various states prepared by the U.S. Weather Bureau (1885——); *Climatological Data, National Summary*, U.S. Weather Bureau (1950——); and *Monthly Climatic Data for the World*, U.S. Weather Bureau (1948——).

Because outcroppings of rocks are commoner on slopes than on level land, slopes are of particular interest in understanding the distribution of *Sedum*. In geological time, slopes may have long histories. However unstable any particular square meter of slope may appear at a time of observation, the continuity of this kind of habitat may be enormous in terms of years or even ages. Several authors have discussed this topic. Penck (1953) provided much valuable information. His ideas obviously have stimulated others. The symposium on Walther Penck's contribution to geomorphology, arranged by von Engeln (1940), dealt critically with Penck's ideas about geomorphology, especially the origin and retreat of slopes. Of particular interest is Kirk Bryan's support (von Engeln [1940]) of the opinion that slopes, once formed, persist in their inclination as they retreat.

Sharpe (1960) has given special attention to landslides, creep, and other mass movements. His comments on creep are pertinent to *Sedum* because these plants frequently are rooted in mats of mosses and thin surface soil which is subject to creep. In creep, the vegetation moves with the thin surface layer of soil. When slopes are steep, the youthful condition, creep is most active, but it persists throughout the geomorphic cycle. Creep, according to Sharpe, is the only mass movement, excepting solifluction in the subarctic, which occurs at all stages in the geomorphic cycle.

From the standpoint of populations of plants, continuity of habitat seems more important than the fate of any tiny portion of that habitat in a single season. Similarly, the survival and continuity of the general population is in the long run of greater evolutionary importance than the fate of a particular few individuals, especially if those individuals have the same or similar genotypes to those of other individuals elsewhere in the population. Sizes of populations of *Sedum* on slopes vary from small to large, as may be understood from study of the data under the several species. For the reason that the character of a slope may endure for long periods of time, the general features being preserved as the position moves backward as a result of erosion, the slope environment may be among the most stable habitats from an evolutionary standpoint. Because succulent plants such as *Sedum* may reproduce easily vegetatively, with pieces dropping from one ledge to another, a clone can occupy a large area of cliff or more gentle slope. Theoretically, such clones, all individuals of which have the same genetic constitution, could endure for thousands of years. In such circumstances, climatic change is a more important limiting factor in distribution and evolution than the durability of slopes.

In summary, North America is a continent with ranges of mountains near its eastern and western borders, on either side of a great stable interior lowland. The mountains in the West are more complex and much higher than those in the East. Present topography is of great importance in determining the modern distribution of plants, but geological changes in the past, associated with different climatic conditions, have influenced both the direction of evolution and the patterns of distribution. An understanding of the geography and evolutionary history of *Sedum*, as of any other genus of plants or animals, depends on comprehension of the geological situation, both present and past.

CHAPTER V

The Species of *Sedum* Native in North America North of the Mexican Plateau

This chapter includes all species of *Sedum* known to occur naturally in North America north of the Mexican Plateau, excluding Greenland. Naturalized species are discussed in Chapter VI. Other species, occasional escapes from cultivation and not really established in wild situations, as well as all species in general cultivation, are described in the chapter on cultivated species.

The circumscription of *Sedum* is not easy. As presently conceived, the genus includes herbs and shrubs with succulent stems and leaves, actinomorphic flowers with the sepals and petals hypogynous, the stamens usually twice as many as the petals, and the carpels either separate or somewhat connate, splitting in fruit along the ventral suture. The leaves usually are alternate; the petals chiefly are separate or only slightly connate; and the floriferous stems normally are terminal. Exceptions to each of these conditions occur. No single character separates all species of *Sedum* from the species of other genera. On the other hand, no species of *Sedum* lacks all characters just cited.

Parvisedum, a small group of annual herbs in the foothills of the Sierra Nevada, the Great Valley, and the California Coast Ranges, is distinguished by its one-seeded fruits. *Diamorpha*, a genus of restricted distribution, occurring primarily on the granite flatrocks of the southern Piedmont Upland, comprises annuals with carpels which dehisce by means of transverse flaps. *Lenophyllum*, chiefly a genus of the Mexican Gulf Coastal Plain, but extending into southern Texas, has the flowers in spikes of cymules, with the petals strongly recurved, and at least some of the leaves opposite or whorled. *Lenophyllum* probably is closely related to *Villadia*. It presents similar problems in separation from *Sedum*. *Dudleya*, predominantly a genus of the Pacific Mountain System, has the floriferous stems axillary rather than generally terminal as in *Sedum*. The rosettes of many species of *Dudleya* are distinctive and are an aid in identification.

Gormania and *Rhodiola*, though easily separable from *Sedum* in their extreme conditions, are included in *Sedum* because of the transitional series of species which makes generic status dubious.

To clarify generic relationships, the genera related to *Sedum* are included in the keys in Chapter IX. These keys make possible identification, at least to genus, of any species of *Crassulaceae* occurring in the area covered by this book.

To inform the reader of the state of my knowledge of each species of *Sedum*, a system of starring is employed. Three stars in the subhead after the name of a species or subspecies indicate that I have studied populations in the wild. Two stars indicate that I have studied living plants in cultivation, but not populations in the wild. One star indicates that I have studied herbarium specimens, but not living plants. Finally, species which I know only from the literature are unstarred.

When species are known only from a few dried specimens or from the literature, descriptions cannot be adequate. Neither can inclusion in keys be satisfactory. Taxonomic interpretations which are not based on study of living populations may be doubtful. For these reasons, effort in the present study has been to gain experience with populations of the maximum number of species.

For intelligent appraisal of a species or subspecies, a sample of at least ten flowering and fruiting plants, chosen according to the rules of probability sampling, would seem to be necessary. Only rarely have species been accepted in this treatment when the sample has been smaller. Samples must be adequate to give reliable results.

Plants growing in diverse environmental conditions may give inadequate or misleading impressions of variation. To reduce error and to minimize environmental differences, experiments are necessary. In the present study, these have mostly been of completely random design.

Each group of related species is taken up as a unit, under a special heading, usually the name of a subgenus or section. A general discussion of the group, with attention to morphology, geographical distribution, and special problems peculiar to the included species, precedes the individual accounts.

The sequence of the species is supposedly phylogenetic. The explanation for this arrangement appears in Chapter XI.

The specific accounts are according to a standard pattern. The accepted scientific name, always a Latin binomial, appears as the heading. Authors of binomials, not being part of the names themselves, are omitted from the headings. Instead, they are cited, with bibliographical references, in the section on nomenclature which is provided for each species.

The initial paragraph of each account is a brief diagnosis of the species, telling how it may be separated from other species. The purpose of the diagnosis is to provide concise, practical information which may be useful in identification.

Descriptions begin with information about the sample on which they are based, namely, number of plants and their geographical origin. The organization of descriptive details is in a series of short paragraphs, each followed by citation of statistical data to make statements more precise, indicate variability, and make possible comparisons. Headings for paragraphs are: stems, leaves, inflorescences, flowers, sepals, petals, stamens, nectaries, carpels, fruits, seeds, and chromosomes. When available, data are cited for plants in both the wild and cultivation. The place of cultivation is indicated by the following abbreviations: cf.—cold frame, Ithaca, N.Y.; gh.—greenhouse, Ithaca, N.Y.; and gd.—garden, either the Test Garden of Cornell University or the garden in my yard, 1421 Slaterville Road, Ithaca, N.Y.

Unless stated otherwise, data for plants in cultivation are from replicates in planned experiments. To save space, statistics for a stratum sometimes are combined in a single line, except in cases where differences among populations need emphasis. Headings for columns are *n-pop.*—number of populations; *n-pl.*—number of plants; *n-rep.*—number of replicates; *n-obs.*—number of observations, $\bar{x}$—the arithmetical mean; and *s*—the standard deviation among plants. The observed range, *w*, usually is omitted because *s* provides a superior measure of spread of variation. For shortcomings of the range, see the discussion by McCarthy (1957:120–121). If means of populations in a stratum differ significantly, this is shown by an asterisk after the number in the column for *n-pop.*, indicating significance at the 5% level, namely with the chances only one in twenty that the populations have a common mean, or two asterisks, indicating significance at the 1% level, with the chances one in a hundred that the difference is not real. Similarly, for plants in experiments, differences among populations are indicated in the same way and when, in addition, significant differences exist among plants within populations, this is shown by asterisks after numbers in the column for *n-pl.* In the case of two subspecies in the same stratum, data provided are on separate lines to facilitate comparisons.

Standard deviations are the square roots of variances among plants. Variation within plants was eliminated from these estimates by subtraction and division by the average number of replicates per plant, following Snedecor (1956:258–275). The same procedure, although not specially mentioned, was followed for all estimates of s^2 and *s* in *Sedum of the Trans-Mexican Volcanic Belt*.

Models for the statistical work and explanation of procedures followed are available in books by the following authors: Cochran (1963); Dixon and Massey (1951); Hansen, Hurwitz, and Madow (1953); McCarthy (1957); Snedecor (1956); and Steel and Torrie (1960). In the text, the reference to explain each statistical test is cited at the place where that test is first mentioned. Detailed demonstration is added only when essential. The purpose here is to provide information about *Sedum*, not to expound statistical theory and its application.

Measurements, unless indicated otherwise, are of mature fresh structures. Those for height apply to plants in flowering or sometimes fruiting condition. Those for floral parts are for flowers in anthesis, generally with some anthers shedding pollen and others unopened.

Colors of parts of plants, especially petals, are according to the Colour Chart of the British Colour Council (1938–1942).

In the parts of descriptions providing information about chromosomes, when the name of a cytologist is followed by a comma and the year, the data are previously unpublished. When the name is followed by a year in parentheses, the information has been published elsewhere and the reference appears in the bibliography.

A discussion of variation follows each description of a species. This provides a place for interpretation of the descriptive details. Sometimes tables are added when differences among populations in different strata are significant. Whenever data from the field fullfill the assumptions of the analysis of variance, variance ratios have been calculated. Likewise, analyses of variance and variance ratios have been calculated for data from experiments. The results of the F tests are the basis for decision about the significance of differences. In tables of comparison, asterisks indicate the levels of significance. Underlining in such tables indicates a lack of significant differences among means.

Because data from the field include an unknown environmental effect on each plant, data from experiments, where environmental effects are minimized both by means of more uniform conditions and through replication, have the greatest value in achieving taxonomic interpretations. They are the foundation for the discussions of intraspecific variation.

When significant differences exist among populations and these can be correlated with either the geographic or ecological distribution, subspecies are sometimes recognized, depending on the degree of the difference, and also on the degree of correlation. Keys to subspecies ordinarily follow the discussions of variation. They include subspecies both within and outside the part of North America north of the Mexican Plateau. Subspecies are not mentioned in the general keys in Chapter IX.

For species with subspecies, a brief diagnosis follows the heading for each subspecies and then come sections on nomenclature and distribution. For species without subspecies, the section on nomenclature follows directly the discussion of variation.

Under nomenclature are listed the various botanical names which have been applied to each species or subspecies. These are in chronological sequence, with the dates to the left, accompanied by bibliographical references and citation of both the type locality and the type collection, and including indication of the collector, the date of collection, and the herbarium in which the type is located. For transferred names, the basionym is the type. When Latin names are proposed for the first time, they appear in boldface at the place in the text which is their first formal publication.

Some authors, as Simpson (1960), have suggested other words as substitutes for type. These terms are not used here because they are unfamiliar and also may be questionable improvements. More useful might be the designation of standard populations which could serve as a living basis for interpreting names.

A section on distribution follows the discussion of nomenclature. This begins with a general statement of the known geographical range, accompanied by a map to show the distribution graphically. In order to avoid clumping of dots, the policy is to space these at intervals of one degree of latitude and longitude. Thus, a single dot may stand for one or more populations.

Tables usually follow the general statement about distribution. These contain detailed information about populations of the species. Data in the tables may include information about density, estimates of the total sizes of populations, altitudinal range, flowering time, oldest record of occurrence, pH, commonest competing species, and habitats. Nomenclature of competing species is according to monographs and revisions when available, current manuals, or even personal judgment. No single source or authority is sufficient. Further, evidence necessary for a satisfactory taxonomic interpretation often is inadequate.

Descriptions of sampling units in which populations occur sometimes are in the chapter on geography. This arrangement permits economy of space, especially when two or more species occur in the same unit. Special information about the discovery of the species and important steps in the history of knowledge about the distribution in North America come after the tables.

A section on reproduction follows distribution. This provides information on pollination and also on mode of breeding and propagation.

Relationships are the concern of the final section of each account. Topics include affinities to all other species and hypotheses concerning possible phylogenetic history. Tables of comparison sometimes show differences among related species.

References, usually to publications with original information about the species or to illustrations, may terminate accounts.

Several names in *Sedum* apply to North American species belonging to other genera. If not discussed elsewhere, these are in an alphabetical list below, with bibliographical reference and indication of type locality, type, and preferred status, with explanation if necessary. A few names of dubious status or applying to species not native in North America also appear here.

Congdonia pinetora (T. S. Brandegee) Jepson, A Manual of the flowering plants of California, p. 450 (1925). Type: *Sedum pinetorum* T. S. Brandegee. Status: same as for *Sedum pinetorum*, which see. *Congdonia* Jepson is a later homonym of *Congdonia* Jean Mueller, Flora 49:437 (1876). That applies to a genus of *Rubiaceae*.

Sedum album Rafinesque in J. D. Hooker, Index Kewensis, Suppl. 1:487 (1906). Type locality: unknown. Type: not located. Status: doubtful. Although Hooker cited First catalogues and circulars of the botanical garden of Transylvania University at Lexington in Kentucky, for the year 1824, p. 12 (1824), as the reference for this name, no evidence of *Sedum album* is in a copy of the publication at Transylvania College. The citation appears to be an error. Similarly, *S. albidum*, listed in Index Kewensis as the equivalent of Rafinesque's *S. album*, otherwise seems to be unpublished.

Sedum anoicum Praeger, Jour. Bot. 57:52 (1919). Type locality: erroneously thought to be western North America. Type: cultivated in garden of Murray Hornibrook, Abbeyleix, Queens County, Ireland. Status: synonym of *S. adenotrichum* Wall. in Edgeworth.

Sedum Blochmanae Eastwood, Proc. Calif. Acad. Sci. II, 6:422 (1896). Type locality: in hard clay soil along the road to Point Sal near Casmaila Beach, Santa Barbara County, California. Type: Alice Eastwood and Ida M. Blochman, May 13, 1896 (CAS). Status: synonym of *Hasseanthus blochmaniae* (Eastwood) Rose. Moran (1951c) indicated his preference for combining *Hasseanthus* with *Dudleya*. The main points in favor of that interpretation are the supposed occurrence of natural hybrids between *Dudleya* and *Hasseanthus* (Moran, 1951b), the hypothesized allopolyploid origin of a plant called *H. nesioticus* (Moran, 1951a), and the similarity in number of chromosomes among species of *Dudleya* and *Hasseanthus* (Uhl and Moran, 1953). Converse ideas are that the morphological distinction between the two genera is reasonably good; the supposed intergeneric hybrids have not been confirmed experimentally; the plant called *H. nesioticus* might be a derivative of the ancestral species from which *H. blochmaniae* arose; and the base number of seventeen pairs of chromosomes may have originated more than once in different, though parallel, phyletic lines. Plants from Torrey Pines, Calif., supplied by Moran and regarded by him as intermediate between *Dudleya edulis* and *Hasseanthus blochmaniae* subsp. *brevifolius*, in culture at Ithaca, N.Y., appeared intermediate between *H. blochmaniae* and *H. elongatus*, suggesting hybridization at some time between those two species. Also, a plant of the type collection of *H. nesioticus*, likewise in cultivation at Ithaca, had spreading petals and appeared more like robust *H. blochmaniae*, with wider leaves and petals, than intermediate between *Dudleya* and *Hasseanthus*. These details deserve further clarification. Meanwhile, the interpretation of *Hasseanthus* as a small genus related to *Dudleya*, but with some similarity with *Sedum*, may be justifiable.

Sedum Cotyledon Jacq., Eclogae plantarum rariorum aut minus cognitarum 1:27, pl. 17 (1811). Type locality: unknown. Type: not seen, but reported by Jacquin as having been cultivated in the University garden in Vienna since 1807. Status: synonym of *Dudleya caespitosa* (Haw.) Britton et Rose, according to Moran in Jacobsen (1960).

Sedum elongatum Fedde, Just's Bot. Jahresber. 31:828 (1904), not *S. elongatum* Ledebour (1830). Type: *Hasseanthus elongatus* Rose. Status: synonym of *H. elongatus* Rose.

Sedum Gertrudianum Eastwood, Proc. Calif. Acad. Sci. IV, 20:147 (1931). Type locality: bluffs above Morro Bay, San Luis Obispo County, California. Type: Alice Eastwood 15,112 (CAS). Status: synonym of *Hasseanthus blochmaniae* (Eastwood) Rose.

Sedum luteoviride R. T. Clausen, Cact. Succ. Jour. 18:74–77 (1946). Type locality: unknown (species described from a cultivated plant). Type: R. T. Clausen 5764c (CU). Status: species of *Sedum*, subgenus *Sedum*. Listed here because the nativity is wrongly indicated in Index Kewensis as California.

Sedum Meehani A. Gray, Proc. Am. Acad. 16:105 (1880). Type locality: cited as on City Creek, north of Salt Lake City, Utah, at the base of the mountains. Type: Thomas Meehan, June 1880 (GH), originally sent by John Reading to T. Meehan in 1874. Meehan sent the live plant to Gray in June 1880. Status: synonym of *S. hispanicum* Juslenius. Either this was introduced by a settler or some mistake has occurred. Meehan maintained the Germantown Nurseries in 1880 and cultivated there several species of *Sedum*.

Sedum multicaule Fedde, Just's Bot. Jahresber. 31:828 (1904), not *S. multicaule* Lindley (1840). Type: *Hasseanthus multicaulis* Rose. Status: synonym of *H. elongatus* Rose.

Sedum oblongorhizum Berger in Engler and Prantl, Die nat. Pflanzenfam., ed. 2, 18a:445 (1930). Type: *Hasseanthus elongatus* Rose. Status: synonym of *H. elongatus* Rose.

Sedum penthorum Crantz, Institutiones rei herbariae 2:443 (1776). Type locality: Virginia. Type: Clayton's no. 158, described by Gronovius in Flora Virginica, ed. 1, p. 51 (1739), and ed. 2, p. 71 (1762). Status: synonym of *Penthorum sedoides* L.

Sedum pinetorum T. S. Brandegee, Univ. Calif. Publ. Bot. 6:358 (1916). Type locality: uncertain, cited as the deserted Pine City above Mammoth, Mono County,

California. Type: Katherine Brandegee, July 1913, sheet no. 185,499 (JEPS). Status: uncertain, possibly the earliest name for *S. niveum* Davidson. I have studied the original collection twice, in 1940 and again in 1963. It is in packets and is a mass of soil, fragmented leaves, carpels, petals, and other debris. Reconstruction of a plant from the loose parts suggests that the roots were tuberous, 6 mm. long and 2–3 mm. wide; stems 15 mm. tall, with the peduncles 1-flowered; sepals ovate-orbicular, papillose, 2 mm. long; and petals ovate-lanceolate, white, 4 mm. long. Details of the carpels are uncertain. No sure conclusion has resulted from my study of the type. Neither could I find anything resembling it when searching at Pine City in 1940 and again in 1970. In a letter dated Jan. 28, 1941, Joseph Ewan, who has given special attention to the collecting activities of Katherine Brandegee, confirmed that she collected at Pine City in July 1913, but he had not recorded that she obtained a *Sedum* there. Both Berger (1930) and Fröderström (1930–1935) associated *S. pinetorum* with *S. compactum* and *S. minimum*. Those are species of the Sierra Madre del Sur, Trans-Mexican Volcanic Belt, and Mexican Plateau. Moran (1950) reported that a fruit in the debris of the type collection was identified as *Clethra* by A. C. Smith. Also, he suggested that a leaf fragment might be *Coldenia nuttallii*. *Clethra* occurs in Mexico, but not in California. *Coldenia nuttallii* is native on the eastern slope of the Sierra Nevada. The data with the type are "Mono County in the duffle beneath pines, near Pine City, above Mammoth." The evidence is conflicting. *Sedum pinetorum* cannot be listed with certainty as a species of the Sierra Nevada unless it is again found there. Neither can it be listed with assurance as a synonym of either *S. compactum* or *S. minimum*. The former species does not have tuberous roots. The latter species has two or more flowers per peduncle and larger sepals and petals. Another explanation of *S. pinetorum* is that it is identical with *S. niveum*. For discussion of this possibility, see the account of that species.

Sedum Sanctae Monicae Berger in Engler and Prantl, Die nat. Pflanzenfam., ed. 2, 18a:445 (1930). Type: *Hasseanthus multicaulis* Rose. Status: synonym of *H. elongatus* Rose.

Sedum subclavatum Haworth, The Philosophical Magazine n.s. 10:414 (1813). Type locality: North America. Type: possibly not preserved, a plant cultivated at Chelsea, England. Status: doubtful. See Clausen (1948).

Sedum variegatum S. Wats., Proc. Am. Acad. 11:137 (1876). Type locality: San Diego, California. Type: D. Cleveland, May 1875 (GH). Status: synonym of *Hasseanthus variegatus* (Watson) Rose.

Subgenus *Telephium*

Plants of *Sedum*, subgenus *Telephium*, are perennial. They have flat leaves and perfect flowers with separate petals and stipitate carpels. The roots usually are thick and sometimes tuberous. The rootstocks commonly are short and erect, but exceptions occur. The petals are white, pink, or green, never yellow or blue.

Quimby (1940, 1971) studied the floral anatomy of six species of subgenus *Telephium*, namely *S. sieboldii, S. spectabile, S. tatarinowii, S. telephioides, S. fabaria*, and *S. purpureum*, the last two species listed under the names *S. telephium* ssps. *fabaria* and *purpureum* respectively. These differ from the other thirty-eight species which he investigated in having six distinct whorls of vascular traces for the floral appendages. All other species of *Sedum* in Quimby's sample had fewer whorls of traces as a result of fusion. Quimby considered the anatomical condition in the flowers of subgenus *Telephium* to be the most primitive in the genus *Sedum*, an opinion consistent with information about gross morphology and geographic distribution.

Subgenus *Telephium* comprises about fifteen species. Most of these are found in Asia. Only two species occur in North America. These are *Sedum telephioides*, which is indigenous, and *S. purpureum*, the latter introduced, most likely from Europe.

Previous authors have interpreted the *Telephium* group of species either as a section of *Sedum* or as a separate genus. Because of species transitional to subgenus *Sedum*, for example *S. anacampseros*, *S. cyaneum*, and *S. ewersii*, generic status seems inappropriate. On the other hand, the anatomical situation in the flowers seems so important, especially when combined with features of gross morphology such as stipitate erect carpels, perennial habit, flat leaves, and white, pink, or greenish flowers, to make subgeneric status reasonable. Accordingly, the formal designation of this group of naturally related species may be subgenus **Telephium** (S. F. Gray) R. T. Clausen, *stat. nov.*, fundatum super *Sedum*, b, *Telephium* S. F. Gray, A natural arrangement of British plants 2:539 (1821). Type: *S. telephium* L. The usual interpretation of Gray's subgeneric name is as a section, but he did not clearly indicate a rank in his publication. Neither did he explain his usage.

Synonyms are:

1754. *Anacampseros* P. Miller, Gard. Dict., abridged ed. 4, p. 73. Type: *Anacampseros, vulgo Faba crassa* J. B. 3:681, Common Orpine. This is *S. telephium* L.
1756. *Telephium* J. Hill, British Herbal, p. 36. Type: *Telephium vulgare* C. Bauhin, which is *Sedum telephium* L. Hill's name is a later homonym of *Telephium* L., a genus of the *Caryophyllaceae*.

1. *Sedum telephioides**** (figs. 5–9)

Diagnostic features of *Sedum telephioides* (fig. 5) are the tall, tufted floriferous stems; stout roots; broad, flat, more or less elliptical leaves; dichotomous cymes; pedicellate, white or pinkish white flowers; and stipitate carpels. *Sedum telephioides* is the tallest species of *Sedum* in eastern North America, with floriferous stems attaining a length of 6.6 decimeters. Also, it is the only species of subgenus *Telephium* native in America.

Sedum purpureum, a European species now widely naturalized across cool temperate North America north of 40° N., has clusters of tuberous, carrotlike roots; leaves markedly reduced upward on the stems; and deep pink flowers. It requires

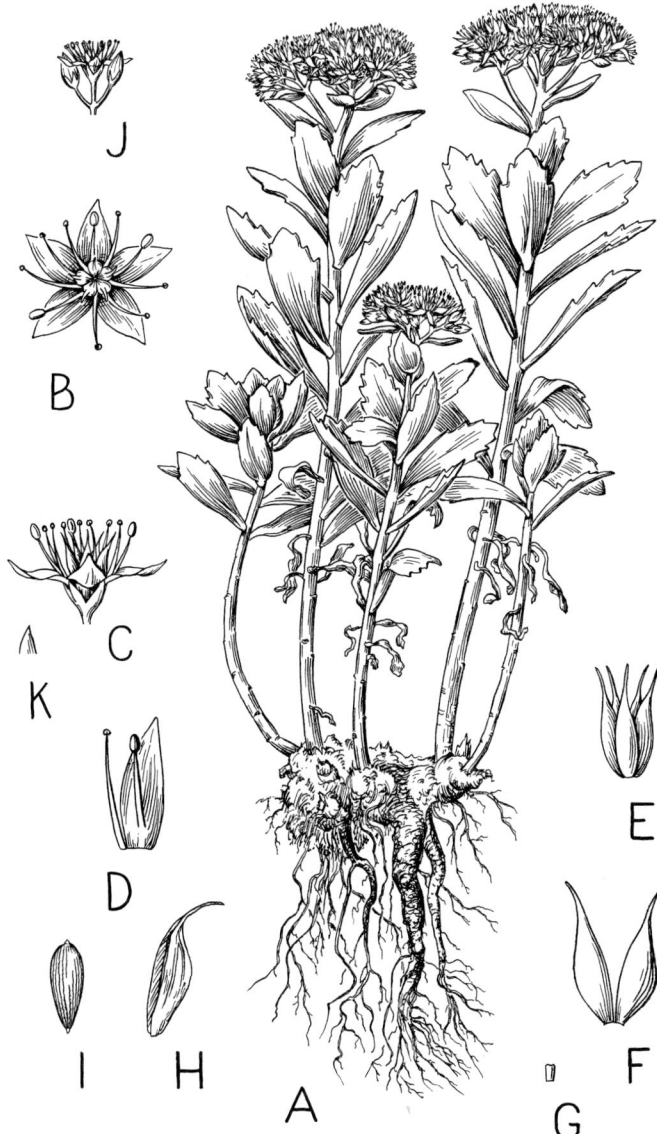

Fig. 5. Plant of *Sedum telephioides* from Rock Point in Blue Ridge, Augusta Co., Va., cultivated in garden at Ithaca, N.Y. A. Habit sketch (x .48). B. Flower from above (x 1.9). C. Flower from side (x 1.9). D. Petal and two stamens (x 2.9). E. Carpels (x 2.9). F. Two carpels (x 3.8). G. Nectary (x 4.8). H. Follicle (x 3.8). I. Seed (x 9.6). J. Cymule showing earlier development of central flower (x .96). K. Single sepal (x 1.9).

a longer light day for flowering, and has 18 pairs of chromosomes rather than 12 or 24 pairs as in *S. telephioides*.

The specific epithet, *telephioides*, alludes to the resemblance of the American plants to *Sedum telephium* of Europe. That has tuberous, carrotlike roots which often are jointed; oblong to suborbicular, usually sessile leaves; and smaller, greenish-white flowers. It sometimes is cultivated in American gardens.

Description. Sample: 69 plants. Sources and numbers of plants from each locality are: Piedmont Plateau: Pores Knob, Wilkes Co., N.C.—1; Blue Ridge: Glassy Mountain, Greenville Co., S.C.—8; Craggy Pinnacle, Buncombe Co., N.C.—9; Bluffs Mountain, Alleghany Co., N.C.—6; and Rock Point, Augusta Co., Va.—6; Ridge and Valley Province: Gilpin, Allegany Co., Md.—3; Barnes Gap, Bedford Co., Pa.—12; and Silver Mills, Bedford Co., Pa.—18; Appalachian Plateau: Cabins, Grant Co., W. Va.—1; and Interior Low Plateaus: Shawnee Hills, Garden of the Gods, Saline Co., Ill.—5. All collections are my own except the plant from Pores Knob, which E. E. Terrell obtained.

The sample from Glassy Mountain comprised accessible flowering plants. The sample from Craggy Pinnacle is simple random from a series of seventy reachable plants with floriferous stems. Likewise, the sample from Silver Mills is simple random, although, in the following tables, the data for flowers from there are pooled from the results of three kinds of sampling: simple random from all accessible plants in the population, simple random from all accessible plants with flowers, and the five largest plants in the population. Further discussion of these samples is in the section on variation, along with explanation of the reason for pooling the data. The plants from Barnes Gap were all that were reachable in seven randomly selected transects. The selections from Bluffs Mountain, Rock Point, and Gilpin were judgment samples. The sample from the Shawnee Hills comprised four plants selected randomly from a cluster of fourteen, the largest cluster found, and one large plant chosen deliberately.

Data in the description make possible comparison of the southernmost known population on Glassy Mountain with two of the northernmost populations at Barnes Gap and Silver Mills in Pennsylvania and one of the westernmost in the Shawnee Hills of Illinois. In addition, comparison is possible of populations on Craggy Pinnacle, Bluffs Mountain, and Rock Point. Data from experiments in the gardens at Ithaca, N.Y., provide a check on results obtained in the field. Plants in the gardens were in two separate experiments, both of completely random design, one (SG) in partial shade in my yard at 1421 Slaterville Road, Ithaca, N.Y., and the other (TG) in full sun at the Cornell University Test Garden. Localities from which plants in these experiments came, numbers of plants from each, and numbers of plants which yielded data are indicated in the following table.

Locality	SG, n	SG, n-data	TG, n	TG, n-data
Pores Knob	1	1	0	—
Glassy Mountain	8	5	0	—
Craggy Pinnacle	4	4	4	2
Bluffs Mountain	2	2	3	2
Rock Point	3	3	3	3
Gilpin	2	2	3	3
Barnes Gap	10	10	2	0
Silver Mills	8	4	7	3
Garden of the Gods	5	4	0	—
Totals	43	35	22	13

The experiment at 1421 Slaterville Road was organized at first as an augmented Latin-square design. The experiment at the Test Garden was planned according to the randomized complete block design. Loss of replicates forced redoing both experiments and adoption of the simpler completely random design. For explanation of this design, and also for analyses of data from experiments so planned, see Steel and Torrie (1960:99–131).

In the description, abbreviations for localities are as follows: BG—Barnes Gap, BM—Bluffs Mountain, CP—Craggy Pinnacle, G—Gilpin, GG—Garden of the Gods, GM—Glassy Mountain, PK—Pores Knob, RP—Rock Point, and SM—Silver Mills. For data from experiments, years of observation are indicated after the abbreviations for the appropriate garden. For SG, 1972, "8 pops." refers to all populations except Craggy Pinnacle, and "other pops." indicates these same populations except the one or ones specifically cited. If data are pooled, differences among means of populations are of no statistical significance.

In the tables, asterisks denote features for which differences among populations are significant on the basis of analyses of variance for wild plants. In the descriptive paragraphs preceding such features, values are indicated for D, the minimal significant difference between means. Estimation of D is according to the method of Tukey as explained by Snedecor (1956:251–253). Likewise, when differences are significant, as determined by experiment at 1421 Slaterville Road, Ithaca, these are indicated as D—SG.

Plants of *Sedum telephioides* are glabrous, perennial herbs with short rootstocks, 1 to 7 cm. long, bearing a cluster of stout, pale brown roots with white rootlets and one or many ascending or erect floriferous stems which may be green, green blotched with brown, purple, or red. All stems potentially are flower-bearing, although small stems may be flowerless. Animals which eat the ends of stems are a frequent cause of nonflowering. Height of flowering plants varies from 2 to 6.6 dm., but stems, because of their ascending habit, may be longer. The largest plant seen was at Rock Point, Va. That had a diameter of 9 dm. Values for D: diameter of plants = 24 cm., number of floriferous stems per plant = 15, and length of floriferous stems = 19 cm. D—SG = 4.5 for number of floriferous stems per plant.

	n-pl.	n-rep.	n-obs.	$\bar{x}$	s	w
Plants—diameter (cm.)*						
GM	5	—	5	42.4	30.5	15–80
BG	13	—	13	15.4	15.7	2–52
SM	5	—	5	24.6	10.8	12–40
GG	4	—	4	26	6.3	18–32
BM, PK, RP—SG, 1968	6	11	11	11	6.9	0+–14
CP, RP—TG, 1968	2	3	3	18	0	6–30
Floriferous stems (number per plant)**						
CP	8	—	8	1.6	1.2	1–4
BM	6	—	6	11.5	9	3–23
RP	6	—	6	19.8	21.9	1–58
BG	13	—	13	3	6.1	1–22

	n-pl.	n-rep.	n-obs.	$\bar{x}$	s	w
Floriferous stems (number per plant)**						
SM	5	—	5	1.8	1.3	1–4
GG	4	—	4	4	4.7	2–7
PK—SG, 1961	1	1	1	13	—	13
CP—SG, 1961	4	5	5	2	0	2
BM—SG, 1961	2	5	5	4	3.3	3–7
RP—SG, 1961	3	9	9	2.7	1.6	2–3
RP—SG, 1972	2	5	5	11	4.2	4–20
Other pops.—SG, 1972	29	36	36	5	0	2–9
BM—TG, 1961	2	5	5	2	.6	1–4
Length (cm.)*						
GM	5	—	17	44	9.8	29–62
BG	6	—	16	23	10.3	7–36
SM	4	—	7	21	14.6	7–38
GG	4	—	11	21	1.1	14–29
GG—SG, 1972	4	4	12	36	3.9	28–46
Other pops.—SG, 1972	26	36	100	25	0	11–40

Leaves are opposite or alternate and either petiolate or sessile. They vary in shape from elliptical to ovate, obovate, oblanceolate, or spatulate. The margins sometimes are entire, but often are remotely or coarsely dentate. The color varies from green to yellow-green. In some plants, the leaves are glaucous. Measurements apply to median leaves on randomly selected stems. Values for D: number of leaves per node = .5, length of leaves = 18 mm., width of leaves = 7 mm., and number of teeth per margin = 2.4. D for SG—1961 = 10.2 mm. for length of leaves and 12.8 mm. for width of leaves. Differences in length of leaves were not significant in SG in 1972.

	n-pl.	n-rep.	n-obs.	$\bar{x}$	s	w
No. per node**						
GM	8	—	24	1	0	1
CP	8	—	39	1.2	.4	1–2
BM	5	—	56	1	0	1
RP	6	—	32	1	0	1
BG	13	—	29	1.9	.3	1–2
G	3	—	5	1	0	1
SM	5	—	9	1.6	.5	1–2
BG, G—SG, 1972	12	14	37	1.5	0	1–2
Other pops.—SG, 1972	28	26	75	1	.1	1–2
Length (mm.)**						
GM	8	—	23	80	12.8	62–115
CP	8	—	40	31	8.6	12–57
BM	5	—	57	37	6.8	10–70
RP	6	—	32	31	9.4	12–52
BG	6	—	15	61	11.1	35–98
SM	4	—	7	48	14.1	31–65
GG	4	—	11	58	7.3	44–73
PK—SG, 1961	1	1	13	32	—	25–39
CP—SG, 1961	4	5	12	37	3.9	27–50
BM—SG, 1961	2	5	21	48	3.3	33–58
RP—SG, 1961	3	9	23	58	2.8	41–72

	n-pl.	n-rep.	n-obs.	$\bar{x}$	s	w
Length (mm.)**						
8 pops.—SG, 1972	30	40	106	44	8.5	22–78
CP—TG, 1961	2	5	16	26	6.2	20–40
BM—TG, 1961	2	5	10	28	4.4	18–37
RP—TG, 1961	3	8	21	34	18.9	18–74
Width (mm.)**						
GM	8	—	23	33	3.4	25–45
CP	8	—	40	16	4.4	6–27
BM	5	—	57	19	3.9	5–40
RP	6	—	32	13	2.9	5–22
BG	6	—	15	29	4.1	19–40
SM	4	—	7	24	6.7	15–37
GG	4	—	11	29	4.7	20–38
PK—SG, 1961	1	1	13	15	—	9–29
CP—SG, 1961	4	5	12	22	4.3	13–32
RP—SG, 1961	3	9	23	24	1	17–31
BM—SG, 1961	2	5	22	31	8.2	17–43
BM—SG, 1972	2	2	6	19	9.9	10–30
Other pops.—SG, 1972	28	38	100	24	0	10–47
CP—TG, 1961	2	5	16	14	5	10–25
BM—TG, 1961	2	5	10	18	13	9–26
RP—TG, 1961	3	8	21	17	11.5	8–42
Thickness (mm.)						
BG, GG, GM, SM	22	—	53	1.4	.4	.8–2.4
Teeth per left margin (n)**						
GM	8	—	23	5.5	2.1	3–11
CP	8	—	40	2.8	1.2	0–5
BM	5	—	57	3.8	.7	0–6
RP	6	—	32	1.2	.5	0–4
BG	6	—	14	1.7	1.3	0–6
SM	4	—	6	1.2	1.5	0–3
GG	4	—	11	4	.4	3–6
PK—SG, 1961	1	1	13	1.2	—	0–4
CP—SG, 1961	4	5	12	1	1.1	0–5
BM—SG, 1961	2	5	19	4	.1	0–7
RP—SG, 1961	3	9	21	2.8	.9	1–5
8 pops.—SG, 1972	30	39	100	2.4	.8	0–8
CP—TG, 1961	2	5	16	.9	.7	0–4
BM—TG, 1961	2	5	10	3.6	1.3	1–5
RP—TG, 1961	3	8	21	3.1	1.2	0–5

Inflorescences are dichotomous cymes with foliaceous bracts subtending the stalks of the major parts of the clusters. The limits of cymes are somewhat arbitrary because of axillary shoots which bear flowers. For practical purposes, an inflorescence includes all flowers which are aggregated toward the end of a primary stem. One inflorescence at Rock Point, Va., was 15 cm. across. $D = 3$ cm. for diameter of inflorescences.

	n-pl.	n-rep.	n-obs.	$\bar{x}$	s	w
Diameter (cm.)**						
GM	5	—	13	5.5	1.6	3–9
CP	8	—	13	2.7	.8	1–4
BM	6	—	20	3	1.2	0⁺–7

	n-pl.	n-rep.	n-obs.	$\bar{x}$	s	w
Diameter (cm.)**						
RP	1	—	5	2.7	—	2–4
BG	6	—	21	4.4	2.2	1–10
SM	4	—	4	6.9	2.4	4–10
Length (cm.)						
SM	12	—	28	5.4	3	1–12
Flowers (n)						
RP	5	—	22	38	14	5–99

Flowers (fig. 6) are erect on slender, green pedicels. The receptacles are convex and visible among the bases of the carpels in the centers of the flowers.

	n-pl.	n-rep.	n-obs.	$\bar{x}$ (mm.)	s (mm.)	w (mm.)
Diameter						
BM, CP, GM	11	—	43	10	2.2	5–13
BG, SM	13	—	50	10	.9	7–12
SG, 1961	7	14	33	11	1.1	8–13
TG, 1964	6	9	27	11	1.8	6–16

Sepals are lanceolate, acute or obtuse, and green. $D = .13$ mm. for width of sepals. $D\text{—}SG = .09$ for width of sepals.

	n-pl.	n-rep.	n-obs.	$\bar{x}$	s	w
Number per fl.						
BG, GM, SM	21	—	78	4.9	.2	4–5
Length (mm.)						
BG, GM, SM	21	—	78	2.4	.4	1.6–4
BM, CP, RP—SG, 1961	7	14	33	2.3	0	1.9–2.9
Width (mm.)**						
GM	6	—	25	1	.2	.8–1.6
BG	5	—	20	.8	.1	.6–1.4
SM	10	—	33	.8	.8	.6–1.1
CP—SG, 1961	2	2	3	1	.03	1.0–1.1
BM—SG, 1961	2	4	7	1	.04	.9–1
RP—SG, 1961	3	8	23	1.1	.03	.9–1.2

Petals are lanceolate or elliptic-lanceolate, obtuse and sometimes bluntly short-appendaged, white, green medianly above the middle or with dorsal midrib green. Values for D: length of petals $= 1.3$ mm. and width of petals $= .45$ mm.

	n-pl.	n-rep.	n-obs.	$\bar{x}$	s	w
Number per fl.						
GM	6	—	25	5	.2	4–5
BG	5	—	20	4.8	.3	4–5
SM	10	—	33	5	.06	4–5
Length (mm.)*						
GM	6	—	25	6.2	1.1	4.4–7.9
CP	3	—	19	5.2	1	3.5–6.7
BM	2	—	6	5.4	1.1	4.7–7
BG	5	—	20	4.9	.1	4.0–5.9
SM	11	—	38	5.6	.6	4.3–7.3
BM, CP, PK, RM—SG, 1961	8	15	37	6.3	.4	5.1–7.8

NATIVE SPECIES 77

FLOWER

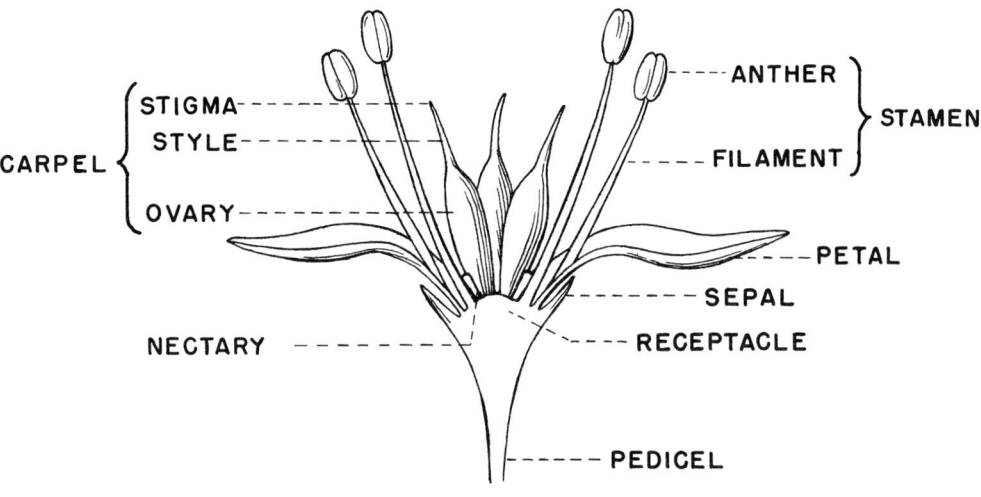

PETAL and TWO STAMENS

FRUIT

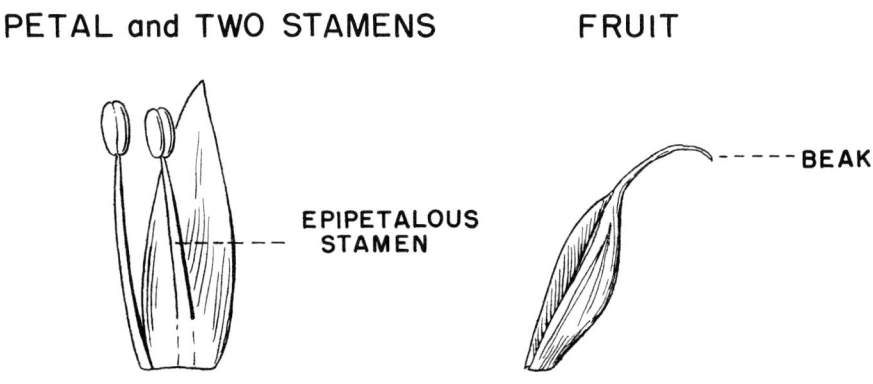

Fig. 6. Flower of *Sedum telephioides* from Rock Point in Blue Ridge, Augusta Co., Va., cultivated in garden at Ithaca, N.Y. Flower (x 6.4). Petal and two stamens (x 6.4). Fruit (x 6.4).

	n-pl.	n-rep.	n-obs.	$\bar{x}$	s	w
Width (mm.)*						
GM	6	—	25	2.2	.4	1.5–3.5
CP	3	—	19	2	.2	1.6–2.3
BM	2	—	6	2	.4	1.7–2.7
BG	5	—	20	1.7	.05	1.4–2.2
SM	10	—	33	1.9	.2	1.4–2.3
BM, CP, PK, RP—SG, 1961	8	15	37	2.2	.1	1.9–2.5

Stamens have white or pale pink filaments. The anthers are ovate-orbicular, 4-celled, and red. Sometimes they are yellow flushed with red. Rarely they are yellow. One plant on Craggy Pinnacle had flowers with 20 stamens. At the other extreme, epipetalous stamens sometimes are lacking. Values for D: length of filaments = 1.2 mm., height of epipetalous stamens on petals = .4 mm., and length of anthers = .18 mm.

	n-pl.	n-rep.	n-obs.	$\bar{x}$ (mm.)	s (mm.)	w (mm.)
Filaments—length*						
GM	6	—	25	4.4	.9	2.9–5.6
CP	3	—	19	4.1	1	2.0–5.2
BM	2	—	6	4.8	1	4.1–6
BG	5	—	20	3.5	.3	2.6–4.2
SM	10	—	33	3.8	.4	3.0–4.7
PK—SG, 1961	1	1	4	5.6	—	5.4–5.8
Epipetalous filaments— height on petals*						
GM	6	—	25	1.4	.1	.9–1.9
BG	5	—	20	.9	.1	.5–1.3
SM	10	—	33	1.1	.2	.8–1.7
Anthers—length*						
GM	5	—	14	.9	.04	.7–1.1
CP	2	—	6	.6	.2	.5–0.9
BM	2	—	2	.8	.07	.8–0.9
BG	3	—	10	.9	.06	.8–0.9
SM	7	—	25	.9	.1	.6–1.1
PK—SG, 1961	1	1	4	.9	—	.8–0.9
CP—SG, 1961	1	1	2	.9	—	.9
BM—SG, 1961	2	4	7	.9	.07	.8–1
RP—SG, 1961	3	6	14	1	.1	.8–1.1
PK—SG, 1972	1	1	3	1	—	.8–1.1
Other pops.—SG, 1972	28	37	99	.8	0	.6–1.1

Nectaries are quadrate, retuse or truncate, and white or pale yellow. D = .17 mm. for length of nectaries. D—SG = .23 mm. for length of nectaries.

	n-pl.	n-rep.	n-obs.	$\bar{x}$ (mm.)	s (mm.)	w (mm.)
Length*						
GM	6	—	25	.5	.07	.4–0.7
CP	3	—	11	.6	.06	.4–0.7
BM	2	—	6	.8	.2	.6–1.0
BG	5	—	10	.5	.1	.4–0.7
SM	10	—	33	.6	.1	.4–0.9
PK—SG, 1961	1	1	4	.9	—	.8–1.0
CP—SG, 1961	2	2	3	.5	0	.5
BM—SG, 1961	2	4	7	.8	.1	.7–0.9
RP—SG, 1961	3	8	23	.6	.1	.5–0.8
Width						
BM, CP, GM	11	—	42	.5	.09	.3–0.8
BG, SM	15	—	43	.5	.06	.3–0.6
PK—SG	1	1	4	.7	—	.6–1.0
BM, CP, RP—SG, 1961	7	14	33	.5	.05	.4–0.7

Carpels are erect, shortly stipitate, and with slender, divergent styles. The ovaries are white, pale green, or pink. The styles often are pink. The ovules are erect on submarginal placentae on the lower halves of the ovaries. Values for D: length of carpels = .5 mm. and number of ovules per ovary = 2.

	n-pl.	n-rep.	n-obs.	$\bar{x}$	s	w
Length (mm.)**						
GM	6	—	25	5	.6	3.8–6.6
BG	5	—	10	3.6	.4	2.1–4.3
SM	10	—	33	4.2	.3	3.6–4.8
Ovules per ovary (n)*						
GM	6	—	25	7	2.2	4–12
BG	5	—	10	5	.9	2–8
SM	10	—	33	6	1.3	3–10
PK—SG	1	1	4	6	—	0–9
RP—SG	2	6	15	6	2.4	4–8

Fruits are short-stipitate, erect, remotely ribbed and reticulate, pale brown, and with divergent beaks. The beaks break off readily. For that reason, the measurement for length includes only the body of the follicle. All data are for fruits collected in 1969.

	n-pl.	n-rep.	n-obs.	$\bar{x}$ (mm.)	s (mm.)	w (mm.)
Length						
GM—SG	1	1	5	2.5	—	2.0–3
BM, RP—SG	4	9	38	3.3	.3	2.6–4
BG, SM—SG	5	5	25	3.3	.2	2.2–4
CP, RP—TG	2	3	15	3.1	0	2.3–4.2
G, SM—TG	6	6	29	3.2	.7	2.0–4.9

Seeds are pyriform, finely longitudinally ribbed, and yellow-brown, pale brown, stramineous, or brown. All data are for seeds collected in 1969.

	n-pl.	n-rep.	n-obs.	$\bar{x}$ (mm.)	s (mm.)	w (mm.)
Length						
BM, GM, RP—SG	5	10	43	1.3	.06	.9–1.5
BG, SM—SG	4	4	12	1.1	.01	1.0–1.1
CP, RP—TG	2	3	15	1.2	—	1.1–1.4
G, SM—TG	5	5	21	1.1	.1	.8–1.3
Diameter						
BM, GM, RP—SG	5	10	43	.4	0	.3–.5
BG, SM—SG	4	4	12	.4	.03	.3–.4
CP, RP—TG	2	3	15	.5	—	.4–.5
G, SM—TG	5	5	21	.4	.05	.3–.5

Chromosomes are small and not as unequal as in *S. telephium*. Baldwin (1937) reported 12 pairs of chromosomes for plants from 3 localities in the Blue Ridge, 5 localities in the Ridge and Valley Province, and 1 locality on the Appalachian Plateau.

	Pl. (n)	g	sp	Cytologist
Number				
Pores Knob, N.C.	1	24	—	Charles Horn, 1965
Glassy Mt., S.C.	2	12	—	Ilse Zandstra, 1969; Barbara Joyce, 1970
Blowing Rock, N.C.	1	12	—	C. H. Uhl, 1941
Bluffs Mt., N.C.	2	12	—	Charles Horn, 1965
Rock Point, Va.	1	12	—	Charles Horn, 1965
	1	—	24	Iris Mastrangelo, 1971
Gilpin, Md.	3	12	—	Ilse Zandstra, 1969
Silver Mills, Pa.	2	—	24	L. and R. Benjamin, 1970
Barnes Gap, Pa.	2	12	—	Ilse Zandstra, 1969
	4	—	24	L. and R. Benjamin, 1970; Iris Mastrangelo, 1971
Garden of the Gods, Ill.	1	—	24	L. and R. Benjamin, 1971

	n-pl.	$\bar{x}$ (genome length; μ)	$\bar{x}$ (chromosome length; μ)	Cytologist
Length of sporophytic chromosomes				
Glassy Mt., S.C.	1	13.7	1.1	Barbara Joyce, 1970
Barnes Gap, Pa.	1	14	1.2	Iris Mastrangelo, 1971

Variation. The problem of determining whether populations of a species are different has no quick solution. Slow, careful work is necessary. Unless sampling is both representative and adequate, conclusions may be faulty. Further, data from wild plants are only suggestive. Unless differences noted in nature are confirmed by experiment, they may be environmental modifications. In the present study, the samples from Craggy Pinnacle, Barnes Gap, and Silver Mills are both representative and adequate. The sample from Glassy Mountain may be as good as can be gotten. The samples from Bluffs Mountain and Rock Point are of unknown quality. The data from the field indicate differences among these populations. Features in which significant differences occur, and the smallest difference which is significant between means for each, already have been designated in the description. Below is a table showing the number of significant differences between members of each pair of the studied populations. The first figure indicates the number of significant differences and the second figure the number of highly significant differences. The information results from analyses of variance for 22 features which were studied.

Population	BG	SM	CP	RP	BM
GM	6, 5	1, 6	1, 3	0, 3	1, 2
BM	2, 3	1, 2	2, 0	0, 1	
RP	0, 4	0, 4	0, 2		
CP	1, 3	1, 3			
SM	0, 1				

Experiments in the garden at 1421 Slaterville Road, Ithaca, N.Y., revealed that most differences observed in the field are determined by the age of the plants and the effect of the environment. The experiments do confirm, however, a few genetic distinctions among populations. Features involved are leaves per node, numbers of teeth on margins of leaves, number per plant and length of floriferous stems, and length of anthers. The following table shows the number of significant differences between populations as observed in the experiment in 1972. Tests of differences are by the method of Steel and Torrie (1960:113–114, eq. 7.17).

Population	Gilpin	Barnes Gap	Silver Mills	Pores Knob
Garden of the Gods	3	2	1	0
Glassy Mt.	2	2	1	0
Rock Point	1	2	0	0
Bluffs Mt.	1	1	0	0
Pores Knob	1	0	0	0
Silver Mills	0	0	0	0

The greatest differences are between the northeasternmost populations and the others, but evidence does not suggest the existence of any subspecies worthy of taxonomic designation. Fröderström once wrote a manuscript varietal name, taking note of the thinness of the leaves, on an herbarium specimen (NY) from Glassy Mountain, S.C. The population on Glassy Mountain is difficult to study because of the steepness of the site. A representative sample from the whole population is unattainable. Accessible plants are around the edges of the main, steep area of occurrence, mostly in shade, and slender, with thin leaves. The specimen which Fröderström studied is such a shade plant.

Variation occurs in chromosome number as well as in various details of morphology and pigmentation. Also, plants occur on a variety of rocks. Rarely, they grow as epiphytes. The most striking differences are between plants with maximum exposure to the sun and those which grow in shade. Likewise impressive is the difference between juvenile plants, with only a few flowers, and large, old plants with many floriferous stems and hundreds of flowers.

To answer the question whether large plants have bigger parts than do smaller ones, I compared data from three samples from the population at Silver Mills, Pa. The population there comprised an estimated 716 plants. Of these, 251 plants were inaccessible. The samples came from the remaining 465 plants. Sample A comprised the 5 largest plants among the 465, each 40 or more cm. in diameter. Samples B and C were selected without regard to size. Sample B was a simple random sample from the 86 accessible plants with floriferous stems. Sample C was a simple random sample from the whole series of 465 accessible plants. All figures listed in the following table are mean values. Asterisks, as usual, indicate significant differences among means. Underlining indicates that differences among means lack significance. Tests of differences are by the Wilcoxon rank sum test for comparing more than two treatments, Wilcoxon and Wilcox (1964).

Sample	A	B	C
Diameter (cm.)**	45	25	9
Floriferous stems (*n*)	3.8	1.8	1.4
Leaves per node (*n*)*	1.2	1.6	1.9

Sample	A	B	C
Median leaves—length (mm.)*	67	48	46
—width (mm.)	34	24	24
—teeth per margin (n)	2.4	1.2	.9
Flowers—diameter (mm.)	9.9	9	9.9
Sepals—length (mm.)	2.3	2.5	2.7
Petals—length (mm.)	5.6	5	5.5
—width (mm.)	2	1.8	1.8
Anthers—length (mm.)	.8	.8	.9
Nectaries—length (mm.)	.6	.6	.6
—width (mm.)	.5	.5	.5
Ovules per ovary (n)	6	7	5

Bigness is a function of both age and nutrition. Older plants are bigger. Similarly, well-nourished plants are larger than starved ones. The tests of the data from the samples at Silver Mills indicate that big plants do not have larger flowers or floral parts. In measuring floral parts, results are similar whether one studies flowers from large or small plants. On the other hand, median leaves on floriferous stems are significantly longer on large plants. Also, smaller plants more often have two leaves per node, suggesting that the condition with opposite leaves is a juvenile characteristic.

In order to check the validity of the result from Silver Mills, I made the same test on data for two samples from Barnes Gap, Pa., one comprising the 6 largest plants in the population of about 798 plants, and the other comprising all accessible plants, namely 14, in 7 randomly selected transects, each 2 meters wide, drawn from a series of 69 transects covering the entire population. The results are about the same as for the samples from Silver Mills. Except for diameter of plants, the basis on which the largest plants were chosen, number of floriferous stems per plant, and leaves per node, all other differences may be attributed to chance. Dimensions of floral parts are similar in the two samples and also in the two populations.

Sample	Largest plants	Random sample
Diameter (cm.)**	40	15
Floriferous stems (n)*	9	3
Leaves per node (n)*	1.5	1.9
Median leaves—length (mm.)	53	44
—width (mm.)	25	20
—teeth per margin (n)	2.9	1.2
Flowers—diameter (mm.)	9.4	9.2
Sepals—length (mm.)	2.4	2.2
Petals—length (mm.)	5.1	4.9
—width (mm.)	1.9	1.7
Anthers—length (mm.)	.8	.9
Nectaries—length (mm.)	.5	.5
—width (mm.)	.4	.4
Ovules per ovary (n)	5.8	6.1

The results of these two studies confirm a well-known principle, namely that floral parts are more satisfactory for use in classification than vegetative parts and are less subject to modification by the environment. Likewise, their size is less a function of the age of the plant.

Two tetraploid plants of *Sedum telephioides* have been available for study, one collected on Bluffs Mountain, Alleghany Co., N.C., by C. H. Uhl, in 1950, and the other obtained on Pores Knob, Wilkes Co., N.C., by E. E. Terrell, in 1957. The plant from Bluffs Mountain flowered in a cold frame at Ithaca, N.Y., in 1951. The plant from Pores Knob flowered in the garden in several summers. A comparison of the two plants, using the Wilcoxon rank sum test, reveals highly significant differences in three features, but since the data include an unassessed environmental effect because of the different circumstances of cultivation, conditions for the test are unsatisfactory. An appropriate test, but with too few plants, involved comparison of three diploid plants with the tetraploid plant from Pores Knob. The diploid plants included three replicates of a clone from Rock Point, Va., and a single diploid plant from Bluffs Mountain. Data for thickness of leaves also were available from an additional diploid plant from Bluffs Mountain. All plants were in an experiment of completely random design, and observations were made at the same time. Analyses of variance for each of the features studied reveal that the differences lack significance. No practical means exists for separating the tetraploid from the diploid plants. The inference is that this is an example of autotetraploidy.

Characteristics	Bluffs Mtn. tetraploid, 1951, 5 obs., Sept. 6	Pores Knob tetraploid, 1965, 4 obs., lvs. Oct. 3, fls. Sept. 18	Diploid plants, 1965, 26 obs., lvs. Oct. 3, 13 obs., fls. Sept. 18
Median leaves—thickness (mm.)	1.6**	1	1.2
Flowers—diameter (mm.)	11.7	12.2	11.1
Petals—length (mm.)	5.7**	6.8	6.1
Petals—width (mm.)	2.6	2.5	2.2
Anthers—length (mm.)	1	.9	.9
Nectaries—length (mm.)	.4**	.9	.7
Nectaries—width (mm.)	.6	.7	.4

Differences in available sunlight may have a profound effect on the appearance of plants. Shaded individuals are slenderer, with pale green stems and foliage, thinner, petiolate leaves, and more lax inflorescences. Most plants grow in partial shade. On the basis of data from random samples, the average plant at Barnes Gap, Pa., receives 37% of the potential sunlight; that at Silver Mills, Pa., receives 46%. The figure for Glassy Mountain, S.C., is higher, 53%, but most plants in the more exposed situations there are unreachable and not in my sample.

Few plants are so situated that they receive 100% of the potential sunshine. Two plants growing in shale gravel at Gilpin, Md., almost fit this condition. At the other extreme, two plants, one at Barnes Gap and the other at Silver Mills, seemed to receive only 20% of the potential sunlight. Comparison of three features of the leaves of these two pairs of plants is interesting.

Condition	Length/width	Thickness	Teeth per margin
Full sun	1.75	2.5 mm.	2.5
20% sunlight	2.2	.5 mm.	0

Seedlings (fig. 7) usually have entire leaves. Toothing appears as the plants become older. In the randomly selected transects at Barnes Gap, the leaves of plants with stems less than 5 cm. long lacked teeth. The difference between such plants and larger plants in the same transects was significant. The larger plants had an average number of 1.6 teeth per leaf.

Fig. 7. Seedlings of Sedum telephioides at Silver Mills, Pa., Sept. 4, 1969.

Differences in pigmentation of petals, anthers, and nectaries seem to have a genetic basis because plants with the petals streaked with pink or pure white, or the anthers red or yellow, or the nectaries white or pale yellow, occur side by side. Sometimes different nectaries on the same plant are white or yellow or intermediate.

Number of flowers per inflorescence varies enormously, from three per cyme on tiny plants in populations at Barnes Gap and Silver Mills, Pa., to hundreds of blossoms in cymes 15 cm. across at Rock Point, Va. Because number of flowers per floriferous stem seems to vary with age, this feature is a poor one for purposes of classification. Further, modifications occur, even on old plants. When animals browse the young shoots, as often happens, axillary shoots develop and bear small cymes. The extent of a cyme is difficult to define because lateral shoots may be either included or excluded.

Variation in number of petals per flower is slight. In a sample of 105 flowers studied in 1969, the range was from 4 to 6. A frequency distribution shows the results.

No. of petals	No. of flowers	%	Cumulative %
4	6	5.7	5.7
5	98	93.4	99.1
6	1	.9	100

The flower with 6 petals was on a plant at Silver Mills. This same plant also had a flower with 4 petals.

Reduction in the number of stamens from the usual 10 was noteworthy at Barnes Gap. Plants at that site tended to lack some or all epipetalous stamens. A random sample of 5 flowers from one large plant included 3 flowers in which only 1 petal had an epipetalous stamen, and a fourth flower had only 3 petals with stamens opposite them. Another plant had a flower which completely lacked epipetalous stamens and an additional flower in which only 1 epipetalous stamen occurred. This same plant further manifested abnormal conditions of the carpels in 4 out of a random sample of 5 flowers. In 2 flowers, a stamen had developed in place of a fifth carpel. A third flower had 2 unsealed carpels, and a fourth flower had a malformed carpel bearing a nonfunctional anther. A single flower of a plant at Silver Mills had 7 carpels.

Number of ovules per ovary is an important character because it measures potential for reproduction. In *Sedum telephioides*, it is more variable than dimensions of floral parts, as indicated in the following comparison of coefficients of variation for five features of wild and cultivated plants. Besides this greater variability, the further circumstance that only a small percentage of ovules develops into viable seeds makes reproductive potential by means of seeds both low and difficult to estimate.

Coefficients of variation for five features of *Sedum telephioides* (%)

Sample	Ovules per ovary	Carpels length	Petals per flower	Petals length	Leaves length
Glassy Mountain	29	13	4	18	16
Barnes Gap	19	10	6	3	18
Silver Mills	22	6	.4	11	29
Rock Point—S.G.	37	—	—	16	2

Finally, drying affects the dimensions of leaves, but not greatly. In order to test its effect, I measured eight fresh leaves of a plant from Glassy Mountain and then prepared herbarium specimens of each leaf. When measured three months later, the leaves had lost slightly in length and more in width, but not significantly in either dimension.

	Fresh $\bar{x}$	Dry $\bar{x}$
Length (mm.)	81	77
Width (mm.)	30	27

Nomenclature. *Sedum telephioides* Michaux, Flor. Bor. Am. 1:277 (1803). Type locality: on the highest rocks of North America, most likely in the Blue Ridge in western North Carolina. Grandfather Mountain is the probable type locality because Michaux regarded that as the highest mountain in North America. Type: a collection of Michaux in the Laboratoire de Phanérogamie, Muséum National d'Histoire Naturelle, Paris. A photograph shows that the collection comprises two floriferous stems. The leaves are coarsely dentate.

A synonym is:

1812. *Anacampseros telephioides* (Michx.) Haworth, Syn. Pl. Succ. 122. Basionym: *Sedum telephioides* Michx.

Fröderström apparently never published the Latin varietal name and description which he wrote on a sheet in the herbarium at the New York Botanical Garden. The sheet contains the specimen, already discussed, from Glassy Mountain, S.C.

Distribution. *Sedum telephioides* occurs (fig. 8) in the Appalachian provinces and adjacent Interior Low Plateaus between latitudes 35° 9′ and 39° 46′ N. and between longitudes 77° and 89° W. Geomorphic provinces in which populations occur are the Piedmont Plateau, Blue Ridge, Ridge and Valley Province, Appalachian Plateau, and Interior Low Plateaus. The species surely is native from southern Pennsylvania to western South Carolina and westward in southern Indiana and Illinois. Reports from outside this area could be based on cultivated or naturalized plants, or else on misidentifications. An old record from Sparta, New Jersey, attributed to Nuttall by Torrey and Gray (*Flora N. Am.* 1:558 [1838–1840]), lacks confirmation in the present century. Likewise, the basis of inclusion of Georgia in the range of the species by A. W. Chapman (1860) and many subsequent authors is unclear, although its occurrence in the extreme northern part of the state seems reasonable. An old specimen at the New York Botanical Garden, labeled as from Turkey Mountain, Burke County, without indication of state, could partly explain reports from Georgia. Turkey Mountain, elevation 530 meters, is not in Burke County, Georgia, but in McDowell County, North Carolina, a short distance west of the boundary with Burke County. Before 1842, when McDowell County was formed, Turkey Mountain was in Burke County. Reports from western New York are based on misidentifications. Plants from along Seneca Lake are *S. integrifolium*, and those collected by Fernald *et al.*, no. 14,298, at Fisher's Landing east of Clayton, Jefferson County,

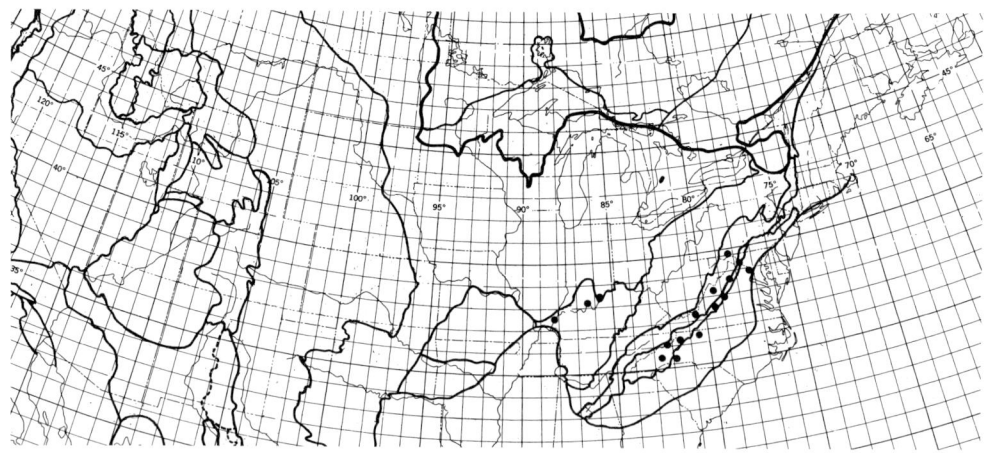

Fig. 8. Known distribution of *Sedum telephioides*.

are *S. spectabile*. Only the collection of Standley and Bollman, 12,456 (US), 1915, from Clove, Dutchess County, N.Y., appears to be *S. telephioides*, but the circumstances of occurrence there are unknown. A report of *S. telephioides* from Colchester, Vermont, Flynn (1909), probably applies to *S. purpureum*, following Knowlton (1917).

Habitats for *Sedum telephioides* include ledges and crevices of cliffs, rocky slopes, crests of bluffs, summits of sandstone ridges, gravel, and rocky woods. The plants occur on a variety of rocks: greenstone, gneiss, granite, sandstone, shale, slate, and limestone. They may grow in full sunlight or in shade where the potential sunlight is as little as 20% of the possible total. The altitudinal range is from about 46 meters on rocks below the Great Falls of the Potomac River, Md., up to 1,790 meters on Craggy Pinnacle, N.C. Altogether, I know about 78 populations of *S. telephioides*.

My principal studies of *Sedum telephioides* have been at four sites in the Blue Ridge, three sites in the Ridge and Valley Province, and one site in the Shawnee Hills of the Interior Low Plateaus. Selection of sites was deliberate in all cases. Descriptive data for the eight populations appear in the following table.

Population	Plants (n)	Area (m.^{2})	Density per m.^{2}	Exposure
Blue Ridge				
Glassy Mt., S.C.	59(–200)	80,000	.0007(–.002)	E
Craggy Pinnacle, N.C.	308	10,000	.03	S—W
Bluffs Mt., N.C.	>273	1,991	.13	S—SW
Rock Point, Va.	>360	>1,080	.33	all directions
Ridge and Valley Province				
Gilpin, Md.	>119	1,000	.12	W
Barnes Gap, Pa.	798e	1,159	.68	SSW—W
Silver Mills, Pa.	716e	6,144	.12	N—WNW
Interior Low Plateaus				
Garden of the Gods, Ill.	42	~360	~.12	SW

Population	Altitudinal range (m.)	pH range	pH tests (n)	Oldest record
Glassy Mt.	600–820	4.4–5	7	1858
Craggy Pinnacle	1,752–1,790	4.2–4.6	9	1897
Bluffs Mt.	1,120–1,130	4.2–4.6	6	1950
Rock Point	938–948	3.8–4.8	6	1960
Gilpin	250–265	—	—	1969
Barnes Gap	237–253	4.6–6.8	12	1969
Silver Mills	273–291	4.4–5.4	13	<1964
Garden of the Gods	234–240	4.6–5.2	5	<1970

The principal distribution of *Sedum telephioides* is in places where scrub pine, *Pinus virginiana*, and oaks, *Quercus*, grow. Populations also occur amid more northern associations where *Betula lutea* and *Pyrus americana*, along with *Rhododendron catawbiense* and *Kalmia latifolia*, are common. On Craggy Pinnacle, N.C., *Saxifraga michauxii* competed closely with three out of eight randomly selected flowering plants of *S. telephioides*. Other competitors, each occurring by one plant of the *Sedum*, were *Asplenium montanum*, *Acer spicatum*, a deciduous-leaved *Rhododendron*, *Heuchera villosa*, *Potentilla tridentata*, and a narrow-leaved *Carex*. Experience on Bluffs Mountain, N.C., and at Rock Point, Va., similarly revealed the miscellaneous nature of the competitors. Commonest species beside six large flowering plants of *S. telephioides* on Bluffs Mountain were *Heuchera villosa*, *Rubus flagellaris*, and *Ambrosia artemisiifolia*. A geometrid larva was feeding on the leaves of a plant on Bluffs Mountain. At Rock Point, most frequent competitors were *Heuchera* and *Festuca ovina*. Trees shading *S. telephioides* in seven randomly selected transects, each 2 meters wide, at Barnes Gap, Pa., were *Tilia americana*, *Fraxinus americana*, *Quercus alba*, *Q. borealis*, and *Ostrya virginiana*. Commonest competing species there were *Deschampsia flexuosa*, *Danthonia spicata*, *Muhlenbergia sobolifera*, *Woodsia obtusa*, *Saxifraga virginiensis*, and *Solidago caesia*. A short distance farther north, at Silver Mills (fig. 9), also along Sideling Hill Creek, trees shading plants of *S. telephioides* in my samples were *Tsuga canadensis*, *Pinus strobus*, *Acer rubrum*, *Betula lenta*, *Ostrya virginiana*, and *Amelanchier arborea*. Commonest competing species there, graded on a scale of 26 for maximum occurrence, were: *Dryopteris marginalis*—5, *Saxifraga virginiensis*—5, *Parthenocissus quinquefolia*—3, and *Asplenium trichomanes*—2.

Fig. 9. Plant of *Sedum telephioides*, on ledge of cliff, in shade of *Pinus strobus*, at Silver Mills, Pa., Sept. 6, 1969.

The variety of competitors makes futile any attempt to characterize the habitat of *Sedum telephioides* in terms of indicator species. Ability to grow in the rigorous situations where the *Sedum* occurs is a prime requisite for any competing species. After that, chance in dispersal is the controlling factor in determining what species are associated.

Reproduction. Principal reproduction of *Sedum telephioides* is by seeds. In the wet summer of 1969, many seedlings and small plants without flowers were in shaded sites in the populations along Sideling Hill Creek in southern Pennsylvania. Vegetative multiplication appears to be rare. The plants lack offsets or stolons. In order for them to spread vegetatively, a stem must break off and take root, or else the underground parts, stems and tuberous roots, must split apart. Both events do occur. Sometimes plants fall from cliffs and become established on talus or rocks below. If only part of a plant falls, then the same clone will exist in two pieces, one above and the other below.

Several species of bees, especially bumblebees, visit the flowers of *Sedum telephioides*. These achieve cross-pollination. The bees move across inflorescences and fly from plant to plant. Stigmas of flowers potentially receive pollen from many plants in a population. Unrestricted reassortment of genes seems to be the regular condition.

Sedum telephioides is a long-day plant. It flowers commonly throughout its geographic range, in which the maximum length of the light day fluctuates from 14.5 to 15 hours. Earliest flowering is southward in the range. Plants have been in flower as early as July 6, 1933, on Cathedral Mountain, Amherst Co., Va. Farther north, in southern Pennsylvania, plants do not come into anthesis until late August or early September. In 1969 the earliest flowers with fresh pollen were on September 1 at both Barnes Gap and Silver Mills.

Plants, transplanted and grown under similar conditions at Ithaca, N.Y., continued to differ in flowering time, but fluctuations occurred from year to year. In the following table, the earliest date of flowering is indicated for plants in the experimental plot at 1421 Slaterville Road, Ithaca, N.Y.

Source of plants	Latitude (°N)	Plants (n)	1961	1964	1965	1968	1969
Craggy Pinnacle	35° 42′	2	8–5	8–3	—	—	—
Pores Knob	36° 2′	1	8–15	8–17	8–24	9–7	—
Bluffs Mountain	36° 24′	2	8–15	8–3	8–3	7–30	8–9
Rock Point	37° 57′	3	8–28	8–3	8–24	8–20	<9–14
Keeper of records			Faith Miller	E. A. Abbe	E. A. Abbe	J. Freer and R. T. Clausen	R. T. Clausen

In experiments conducted by Allard and Garner (1940), *Sedum telephioides* failed to flower when grown under lengths of day much less than fourteen hours. In contrast, *S. purpureum* required continuous light periods of seventeen hours or more for initiation of flowering.

Relationships. *Sedum telephioides* lacks close relatives which are native in North America. *Sedum hesperium* Knowlton, U.S. Geol. Surv. Prof. Paper 131:163–164, pl. 37, fig. 7 (1923), described from the impression of a single leaf in a rock of Eocene age in the Green River Formation at Cathedral Bluff, 32 kilometers west of Rio Blanco, Colorado, is doubtfully a *Sedum*, but a relationship with *S. telephioides* is a remote possibility. No sure conclusions are possible unless fuller specimens become available.

In Europe, the living species, *Sedum telephium*, appears closely related to *S. telephioides*. In fact, the problem is whether to regard *S. telephioides* as a distinct species or to include it as a geographic subspecies of the polytypic *S. telephium*. Because authors have interpreted *S. telephium* in diverse ways, the usage here needs explanation. It is according to the practice of many Scandinavian botanists, for example Turesson (1938), Hylander (1945), Hultén (1950), and Jalas (1954), who considered the white-flowered plants common in Scandinavia as typifying the species. Such a definition of *S. telephium* is in agreement with both the description of Linnaeus in *Species Plantarum* and significant references which he cited in the first part of the entry, namely *Hort. Cliff.* 176, *Fl. suec.* 386, and *Mat. med.* 217. The type specimen is one of the white-flowered Scandinavian plants in Clifford's herbarium at the British Museum. This concept of *S. telephium* includes the *Telephium vulgare* K. Bauhin, *Pinax Theatri Botanici*, p. 287 (1623). It excludes the reference to the *Telephium album* of Fuchs, *De Historia Stirpium*, p. 800 (1542), adopted by Webb (1961) as the basis for typification of *S. telephium*, but listed by Linnaeus as the first-named variation, var. *album*, from the type, and not the type itself. Both *S. telephium* and *S. telephioides* include diploid as well as tetraploid populations. The two species are closely similar in morphology. Differences are small. They are indicated in the following table.

	S. telephioides $\bar{x}$	s	S. telephium $\bar{x}$	s
Roots	Stout, pale brown		Tuberous, carrotlike, often jointed, pale brown or white	
Leaves	Elliptical, sometimes ovate, obovate, or spatulate, often petiolate		Oblong to suborbicular, usually sessile, rarely short-petiolate	
Flowers	Large		Small	
Diameter (mm.)	11.4	1.7	7.5	.7
Sepals				
Length (mm.)	2.3	0	1.4	.04
Petals				
Length (mm.)	6.3	.4	4.2	.4
Nectaries				
Length (mm.)	.7	.05	.6	0
Width (mm.)	.5	.05	.35	.1
Length-width ratio	1.4	.5	1.7	.4
Chromosomes				
Length	Shorter, less unequal		Longer, unequal	

Data in the above table are for plants cultivated at the Test Garden, Ithaca, N.Y. The comparison is poor for several reasons. Only a few European plants were available; they constitute a convenience sample, and they have not all been available in one year for a satisfactory experimental study. On the other hand, the data from the samples of *S. telephioides* provide a basis for more satisfactory appraisal of relationships when randomly selected plants of European populations are available for culture and study in experiments. Meanwhile, the opinion is that *S. telephioides* and *S. telephium* are descendents from an ancient diploid species which became divided in the glacial periods into American and Eurasian populations. Perhaps *S. telephioides* is nearest to the ancestral condition. The peculiar roots of *S. telephium*, the smaller flowers, and the more northern distribution may be specializations. Information about chromosomes may help in suggesting evolutionary trends, but as yet is not good enough to be useful for this purpose.

Sedum purpureum, widely naturalized in the northern United States and southern Canada, does not interact with *S. telephioides*. South of latitude 40° N., *S. purpureum* seldom, if ever, flowers. Since the northernmost natural occurrence of *S. telephioides* is at 39° 46′ N., hybridization in nature cannot occur. Where the species are cultivated together, as in my experimental plots, their periods of flowering overlap, but no spontaneous hybrids have appeared. The difference in number of chromosomes, $g = 12$ or 24 in *S. telephioides*, and $g = 18$ in *S. purpureum*, is the probable explanation. Practical features for distinguishing *S. purpureum* are the usually deep pink flowers, carrotlike roots, and leaves markedly reduced upward on the stems.

Diploid populations of *Sedum*, subgenus *Telephium*, may occur in Asia and have relationships to *S. telephioides*. One of the ancestors of *S. viridescens*, for example, may likewise have been involved in the origin of *S. telephioides*.

A possible explanation of the occurrence of *Sedum telephioides* in eastern North America might be an origin from Eurasian populations which once ranged across the Bering Strait, but were eventually destroyed by adverse conditions in both the northern and the western part of the North American continent. Another hypothesis is that the ancestral species which gave rise to modern *S. telephioides* and *S telephium* originally had a wide distribution across the northern Atlantic region, perhaps when land areas were greater than they are at present. The ice ages would have disrupted such a distribution and left the relictual populations which exist today. Most likely the progenitor of the modern species of subgenus *Telephium* had flowers with three or more whorls of stamens.

Subgenus *Sedum*

Plants of *Sedum*, subgenus *Sedum*, are herbs or subshrubs and either perennial or annual. Roots are capillary, or sometimes tuberous, and best developed when plants are actively growing. Rootstocks usually are lacking. The leaves are various, either spirally arranged or opposite, and flat or terete. The flowers are perfect, in terminal inflorescences. Petals may be completely separate or slightly connate, and range in color from white to green, yellow, and pink. The carpels are broadest at base and not stipitate.

The species of subgenus *Sedum* are moderately specialized. Quimby (1940, 1971) studied the floral anatomy of twenty-five species. Four species in his sample manifested fusion of the lateral traces of the carpels with the staminal traces of whorl 3, and twenty-one species exhibited fusion of whorls 2 and 4. None of the species of subgenus *Sedum* possess more primitive features than species of subgenus *Telephium*, but some species may be the most advanced in the genus.

The distribution of subgenus *Sedum* is widespread in the northern hemisphere. The number of species is large, perhaps more then two hundred. Eighteen of the thirty species of *Sedum* in North America north of the Mexican Plateau belong to subgenus *Sedum*.

Previous authors have treated subgenus *Sedum* as a section, either *Seda genuina* or *Eusedum*, or else have ignored subgeneric classification. Since coordinate rank with *Telephium* is appropriate, subgeneric status is necessary. The type species is *S. album* L., the first species listed by Koch under Sect. III, *Seda genuina, Synopsis der Deutschen und Schweizer Flora*, ed. 2, p. 259 (1836). Boissier, in *Flora Orientalis* 2:775 (1872), used the designation *Eusedum*. Present rules of nomenclature require use of the generic name without a prefix or descriptive adjective for the subgenus which includes the type species of the genus. Koch established usage when he subdivided the genus and listed *S. album* first in the section which he made for the true species of *Sedum*.

For the eighteen North American species of subgenus *Sedum*, five sectional names are necessary. These are:

Section *Ternata* Berger, Die Nat. Pflanzenfam., ed. 2, vol. 18a:450 (1930). Type species: *Sedum ternatum* Michx. Other included North American species, besides *S. ternatum*, are *S. nevii*, *S. glaucophyllum*, and *S. pulchellum*.

Section **Tetrorum** (Rose), stat. nov., fundatum super *Tetrorum* Rose, N. Am. Flora 22 (1):59 (1903). Type species and only species: *Sedum pusillum* Michx.

Section *Cockerellia* Clausen et Uhl, Brittonia 5:34–35 (1943). Type species: *Sedum cockerellii* Britton. Other North American species belonging here are *S. niveum* and *S. wrightii*.

Section **Lanceolata,** sect. nov., subgeneris *Sedum*. Plantae herbae, sine rhizomis, choripetalae, cum foliis angustis et elongatis; petalis, saccis stemarum, nectariis, et ovariis plerumque luteis; cymis trifidis; et chromosomatis octo vel octamultiplicatis. Species typica *Sedum lanceolatum* Torrey, Ann. Lyc. Nat. Hist. N.Y. 2:205–206 (1828), est. Other North American species belonging in this section are *S. rupicolum*, *S. radiatum*, *S. stenopetalum*, *S. leibergii*, *S. borschii*, *S. divergens*, *S. debile*, and *S. nuttallianum*.

Section **Villosa**, sect. nov., subgeneris *Sedum*. Plantae herbae parvae cum carpis erectis, non gibbis, et stylis brevibus. Species typica *Sedum villosum* L., Sp. Pl. 1:432 (1753), est.

2. *Sedum ternatum**** (figs. 10–11)

Sedum ternatum (fig. 10) is unique among the species of *Sedum* of temperate North America in having the leaves either in whorls of three or opposite. Further, the leaves are obovate, broader than those of other white-flowered species of subgenus *Sedum* which occur in the Appalachian Highlands. Like them, it has white petals, normally

NATIVE SPECIES 93

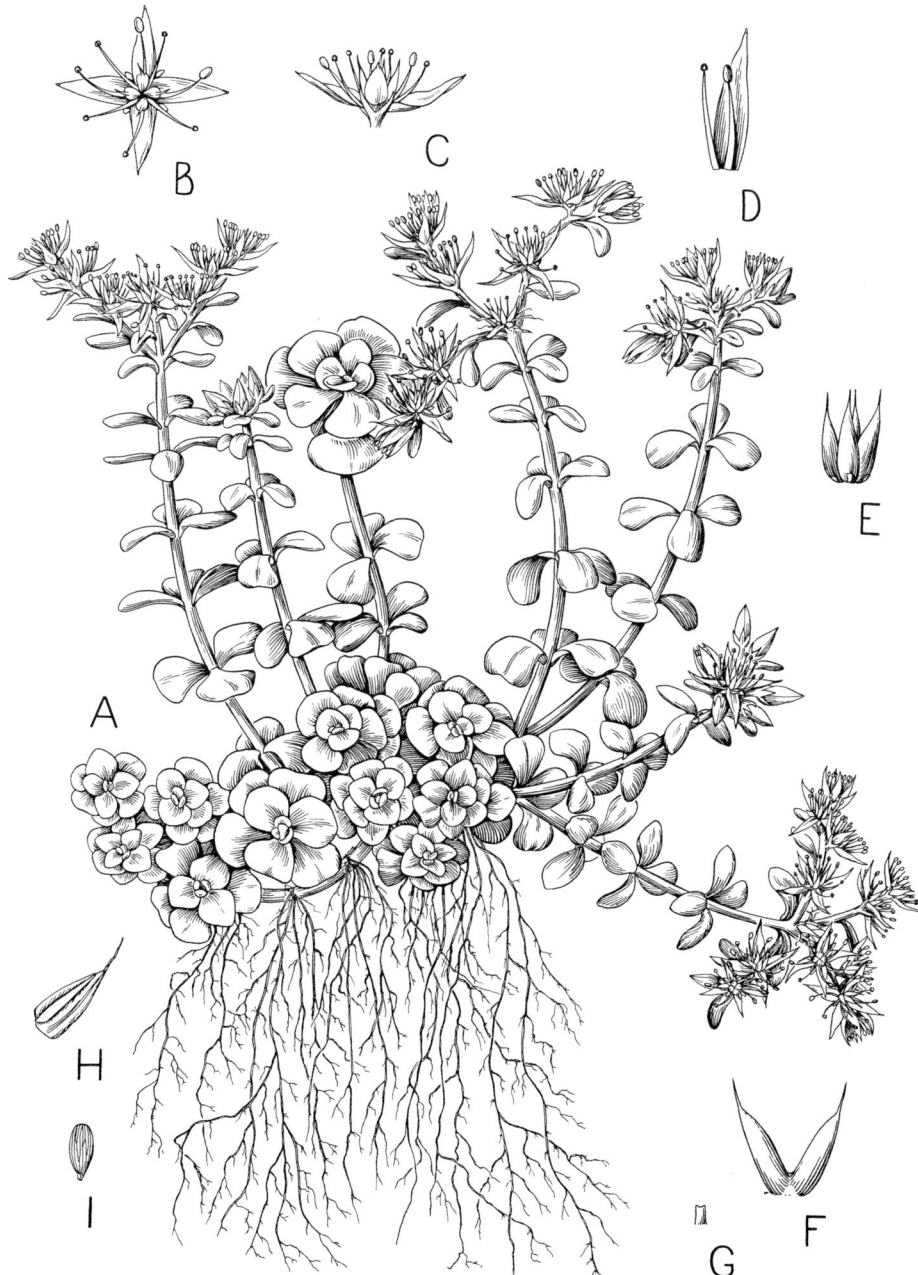

Fig. 10. Plant of *Sedum ternatum* from along Sideling Hill Creek, Silver Mills, Bedford Co., Pa., cultivated in greenhouse at Ithaca, N.Y. A. Habit sketch (x .84). B. Flower from above (x 1.7). C. Flower from side (x 1.7). D. Petal and two stamens (x 2.5). E. Carpels (x 2.5). F. Two carpels (x 3.4). G. Nectary (x 4.2). H. Follicle (x 3.4). I. Seed (x 8.4).

in whorls of four; filaments which are flattened basally and only slightly adnate to the petals; and divergent, gibbous follicles. Other eastern American species, except the taller and obviously different *S. telephioides*, have narrower, spirally arranged leaves. Western white-flowered species, besides having narrower, spirally arranged leaves, have 5-parted flowers. Also, some of them have erect, nongibbous follicles.

Of species elsewhere with leaves in whorls of three, several have yellow petals, some belonging to subgenus *Telephium* have pink or purple petals, and *S. rhodocarpum* in Mexico has angulate stems, 5-merous flowers, dark red petals, and tiny orange seeds.

Description. Sample: 42 plants from 25 localities. Six plants from the gorge of the Ocoee River in Tennessee are a random sample from there. Five plants from Allisonia, Va., are from two randomly selected quadrats chosen for sampling *Sedum glaucophyllum*. Other plants are either convenience (C) or judgment (J) samples.

The complete list of localities at which I have studied *Sedum ternatum*, organized by geomorphic provinces, follows.

Piedmont Plateau: Catawba River, Lancaster Co., S.C.; Spray, N.C.; Goode's Ferry, Va.; and Burnt Mills, N.J.

Blue Ridge: Cold Mountain, N.C.; Crawford Creek, N.C.; Buck Spring Gap, N.C.; Unaka Mountain, N.C.; Roan Mountain, N.C.; and gorge of the Ocoee River, Tenn.

Ridge and Valley Province: Pratt's Ferry, Ala.; Rutledge, Va.; Tazewell, Va.; Cleveland, Va.; Allisonia, Va.; Fincastle, Va.; Rileyville, Va.; Ingleside, W. Va.; Rawlings, Md.; Silver Mills, Pa.; and Chaneysville, Pa.

Appalachian Plateau: Holt, Ala.; Cabins, W. Va.; and Six Mile Creek and West Hill, Ithaca, N.Y.

Data in the following table are for seven populations to which I gave principal attention.

Locality	Basis of choice	Description of sample					Condition when collected	
		N_u	N_c	n	n/N_c	n/N_u	fl. (n)	fr. (n)
Piedmont Plateau								
Catawba R., S.C.	J	8	8	5	.63	.63	0	0
Blue Ridge								
Cold Mt., N.C.	C	>12	>2	2	—	—	0	2
Ocoee Gorge, Tenn.	J	192	192	6	.03	.03	6	0
Ridge and Valley Province								
Allisonia, Va.	C	>19,500	39	5	.13	.0003	5	0
Silver Mills, Pa.	C	>1	1	1	1	—	0	0
Chaneysville, Pa.	C	>5	1	1	1	—	0	0
Appalachian Plateau								
Cabins, W. Va.	C	>1	>1	1	—	—	1	0
Totals		~19,716	>241	21			12	2

(Column "Fraction studied" spans n/N_c and n/N_u.)

Experiments including plants of *Sedum ternatum* at Ithaca, N.Y., have been in the garden, cold frame, and greenhouse. Design was completely random. Plants thrive and flower in all three experimental conditions. Numbers of plants (*N*) in each experiment and numbers of plants which yielded data (*n-d*) are indicated in the following table.

Description of experiments

Source	Cold frame		Greenhouse	
	N	n-d	N	n-d
Catawba River	4	2	5	3
Cold Mountain	2	2	0	0
Ocoee Gorge	6	5	4	1
Allisonia	2	2	1	1
Silver Mills	1	1	1	1
Chaneysville	1	0	1	1
Cabins	0	0	1	1
Totals	16	12	13	8

The abbreviation for plants in the greenhouse from Silver Mills and Chaneysville is Pa.—gh.

Plants of *Sedum ternatum* are perennial herbs with decumbent stems which root at the lower nodes, then ascend. The roots are fibrous and pale brown to white. The stems are green, pink, or somewhat reddish, and finely papillose, 1–2 mm. in diameter.

	n-pl.	n-rep.	n-obs.	$\bar{x}$	s
Plants—diameter (cm.)					
Ocoee G.	6	—	6	23	16
Allisonia	4	—	4	36	18
Catawba R.—cf.	2	3	3	16	—
Cold Mt.—cf.	2	3	3	20	2
Ocoee G.—cf.	5	5	5	21	3
Allisonia—cf.	2	3	3	17	—
Silver Mills—cf.	1	1	1	29	—
Catawba R.—gh.	3	3	3	16	2
Ocoee G.—gh.	1	1	1	23	—
Allisonia—gh.	1	1	1	15	—
Pa.—gh.	2	2	2	26	1
Maximum height (cm.)					
Ocoee G.	6	—	6	8	4
Catawba R.—cf.	2	3	3	4	1
Cold Mt.—cf.	2	2	2	8	1
Ocoee G.—cf.	5	5	5	9	1
Allisonia—cf.	1	1	1	5	—
Silver Mills—cf.	1	1	1	9	—
Catawba R.—gh.	3	3	3	5	2
Ocoee G.—gh.	1	1	1	7	—
Allisonia—gh.	1	1	1	6	—
Pa.—gh.	2	2	2	9	1
Cabins—gh.	1	1	1	5	—
Primary rosettes (*n*)					
Ocoee G.	6	—	6	27	28
Allisonia	5	—	5	42	12
Catawba R.—cf.	2	3	3	55	—
Cold Mt.—cf.	2	3	3	51	18

	n-pl.	n-rep.	n-obs.	$\bar{x}$	s
Primary rosettes (n)					
Ocoee G.—cf.	5	5	5	26	16
Allisonia—cf.	2	3	3	39	14
Catawba R.—gh.	3	3	3	28	8
Ocoee G.—gh.	1	1	1	42	—
Allisonia—gh.	1	1	1	18	—
Pa.—gh.	2	2	2	54	22
Primary rosettes—diameter (cm.)					
Ocoee G.	6	—	16	2	.6
Allisonia	5	—	25	2.5	.4
Catawba R.—cf.	2	3	3	1.8	.2
Cold Mt.—cf.	2	3	18	2.4	—
Ocoee G.—cf.	5	5	15	2	.4
Allisonia—cf.	2	3	18	2.6	—
Silver Mills—cf.	1	1	3	1.8	—
Catawba R.—gh.	3	3	9	2.1	.3
Ocoee G.—gh.	1	1	1	2.2	—
Allisonia—gh.	1	1	3	2.7	—
Pa.—gh.	2	2	6	2.1	.5
Floriferous stems per plant (n)					
Catawba R.	4	—	4	1.3	1
Ocoee G.	6	—	6	3.2	3.3
Allisonia	5	—	5	2	1.3
Cold Mt.—cf.	2	3	3	7	—
Ocoee G.—cf.	4	4	4	43	1
Allisonia—cf.	2	3	3	10	—
Silver Mills—cf.	1	1	1	60	—
Pa.—gh.	2	2	2	32	17
Floriferous stems—length (cm.)					
Catawba R.	3	—	5	13	2.3
Ocoee G.	5	—	9	5	1.8
Allisonia	5	—	9	6	2.1
Cold Mt.—cf.	2	3	9	5	—
Ocoee G.—cf.	4	4	12	7	2.1
Allisonia—cf.	2	3	9	4	—
Silver Mills—cf.	1	1	3	6	—
Pa.—gh.	2	2	6	7	1.4
Cabins—gh.	1	1	1	5	—

Leaves are in whorls of three or rarely four, or occasionally are opposite. The uppermost leaves on floral stems sometimes are alternate. The bases are truncately short-spurred and either short-petiolate or sessile. Shape varies from obovate or obovate-spatulate to elliptical, rounded at apex and sometimes emarginate. Leaves are flat, from four to thirty times wider than thick. The margins sometimes are finely papillose or even crenulate. Color varies from pale yellow-green, sometimes tinged with pink, to dark green.

	n-pl.	n-rep.	n-obs.	$\bar{x}$ (mm.)	s (mm.)
Leaves of rosettes—length					
Ocoee G.	6	—	16	11	3.3
Allisonia	5	—	25	15	2.9
Catawba R.—cf.	2	3	9	11	1.4
Cold Mt.—cf.	2	3	18	15	—

NATIVE SPECIES

	n-pl.	n-rep.	n-obs.	x̄ (mm.)	s (mm.)
Leaves of rosettes—length					
Ocoee G.—cf.	5	5	15	14	2.3
Allisonia—cf.	2	3	18	17	.8
Silver Mills—cf.	1	1	3	10	—
Catawba R.—gh.	3	3	9	12	2.2
Ocoee G.—gh.	1	1	1	13	—
Allisonia—gh.	1	1	3	17	—
Pa.—gh.	2	2	6	13	2.8
Leaves of rosettes—width					
Ocoee G.	6	—	16	8	1.5
Allisonia	5	—	25	9	.9
Catawba R.—cf.	2	3	9	7	.6
Cold Mt.—cf.	2	3	18	11	—
Ocoee G.—cf.	5	5	15	10	1.3
Allisonia—cf.	2	3	18	10	—
Silver Mills—cf.	1	1	3	10	—
Catawba R.—gh.	3	3	9	8	1.6
Ocoee G.—gh.	1	1	1	7	—
Allisonia—gh.	1	1	3	11	—
Pa.—gh.	2	2	6	10	1.9
Leaves of rosettes—thickness					
Ocoee G.	6	—	16	.8	.1
Allisonia	5	—	25	1	.1
Catawba R.—cf.	2	3	9	1	.1
Cold Mt.—cf.	2	3	18	1	—
Ocoee G.—cf.	5	5	15	.9	.3
Allisonia—cf.	2	3	18	.9	—
Silver Mills—cf.	1	1	3	.8	—
Catawba R.—gh.	3	3	9	1.2	.1
Ocoee G.—gh.	1	1	1	1.2	—
Allisonia—gh.	1	1	3	1.4	—
Pa.—gh.	2	2	6	.9	.1
Median leaves of floriferous stems—length					
Catawba R.	3	—	5	15	5
Ocoee G.	6	—	10	8	2.1
Allisonia	5	—	8	12	5.4
Cold Mt.—cf.	2	3	9	13	—
Ocoee G.—cf.	4	4	12	9	.5
Allisonia—cf.	2	3	8	12	—
Silver Mills—cf.	1	1	3	6	—
Pa.—gh.	2	2	6	9	1.4
Cabins—gh.	1	1	2	16	—
Median leaves of floriferous stems—width					
Catawba R.	3	—	5	10	3
Ocoee G.	6	—	10	4	.1
Allisonia	5	—	8	4	.2
Cold Mt.—cf.	2	3	9	7	—
Ocoee G.—cf.	4	4	12	5	.8
Allisonia—cf.	2	3	8	5	—
Silver Mills—cf.	1	1	3	4	—
Pa.—gh.	2	2	6	6	.7
Cabins—gh.	1	1	2	10	—
Median leaves of floriferous stems—thickness					
Catawba R.	3	—	4	1	.1
Ocoee G.	6	—	10	.8	.2
Allisonia	5	—	8	1.1	.3

	n-pl.	n-rep.	n-obs.	$\bar{x}$ (mm.)	s (mm.)
Median leaves of floriferous stems—thickness					
Cold Mt.—cf.	2	3	9	.9	.1
Ocoee G.—cf.	4	4	12	.9	.1
Allisonia—cf.	2	3	8	.8	—
Silver Mills—cf.	1	1	3	.8	—
Pa.—gh.	2	2	6	1	.1
Cabins—gh.	1	1	1	1	—

Inflorescences are terminal cymes, usually of 3 cincinni. The axes of the cincinni may be papillose.

	n-pl.	n-rep.	n-obs.	$\bar{x}$ (n)	s (n)
Cincinni per cyme					
Ocoee G.	6	—	10	2.8	.4
Allisonia	5	—	9	2.6	.8
Cold Mt.—cf.	2	3	9	2.9	—
Ocoee G.—cf.	4	4	12	3.2	.2
Allisonia—cf.	2	3	9	2.8	—
Silver Mills—cf.	1	1	3	3	—
Pa.—gh.	2	2	6	3.5	.3
Flowers per cyme					
Ocoee G.	6	—	10	7.2	3.2
Allisonia	5	—	9	6.4	6.2
Cold Mt.—cf.	2	3	9	11.1	—
Ocoee G.—cf.	4	4	12	11.8	3.8
Allisonia—cf.	2	3	9	9	4.2
Silver Mills—cf.	1	1	3	9	—
Pa.—gh.	2	2	6	14	2.8
Cabins—gh.	1	1	2	2	—

Flowers are slightly fragrant and either sessile or on very short pedicels. The pedicels may be papillose.

	n-pl.	n-rep.	n-obs.	$\bar{x}$ (mm.)	s (mm.)
Pedicels—length					
Ocoee G.	5	—	16	.4	.2
Ocoee G.—cf.	4	4	9	.5	.3
Pa.—gh.	2	3	7	.3	—
Cabins—gh.	1	1	2	2	—
Flowers—diameter					
Ocoee G.	5	—	16	10	1.6
Allisonia	1	—	1	12	—
Cold. Mt.—cf.	2	3	8	15	1.3
Ocoee G.—cf.	4	4	9	15	1.6
Allisonia—cf.	2	3	9	14	1.3
Silver Mills—cf.	1	1	3	14	—
Pa.—gh.	2	3	7	13	1
Cabins—gh.	1	1	2	10	—
Torus—length					
Ocoee G.	5	—	15	.4	.2
Ocoee G.—cf.	4	4	8	.9	.1
Pa.—gh.	2	3	7	.4	.1

NATIVE SPECIES 99

Sepals are lanceolate-oblong or elliptical, obtuse, finely papillose, and pale yellow-green, occasionally slightly reddish at tip.

	n-pl.	n-rep.	n-obs.	$\bar{x}$	s
Number					
Allisonia	1	—	1	6	—
Cold Mt.—cf.	2	3	8	4	0
Allisonia—cf.	2	3	9	4	.2
Silver Mills—cf.	1	1	3	4.3	—
Cabins—gh.	1	1	2	4.5	—
Length (mm.)					
Ocoee G.	5	—	16	3	.6
Allisonia	1	—	1	5.4	—
Cold Mt.—cf.	2	3	8	3.8	—
Ocoee G.—cf.	4	4	9	3.8	1.2
Allisonia—cf.	2	3	9	3.7	.1
Silver Mills—cf.	1	1	3	3.4	—
Pa.—gh.	2	3	7	2.7	—
Cabins—gh.	1	1	2	3.4	—
Width (mm.)					
Ocoee G.	5	—	16	1	.2
Allisonia	1	—	1	1.7	—
Cold Mt.—cf.	2	3	8	1.5	.2
Ocoee G.—cf.	4	4	9	1.3	.4
Allisonia—cf.	2	3	9	1.2	.2
Silver Mills—cf.	1	1	3	1.3	—
Pa.—gh.	2	3	7	1.1	—
Cabins—gh.	1	1	2	.9	—

Petals are widely spreading, elliptic-lanceolate, ventrally channeled, carinate dorsally, acute, and white.

	n-pl.	n-rep.	n-obs.	$\bar{x}$	s
Number					
Ocoee G.	5	—	16	4	.2
Allisonia	1	—	1	6	—
Cold Mt.—cf.	2	3	8	4	0
Ocoee G.—cf.	4	4	9	4	0
Allisonia—cf.	2	3	9	4.1	.2
Silver Mills—cf.	1	1	3	4.3	—
Pa.—gh.	2	3	7	4	0
Cabins—gh.	1	1	2	4.5	—
Length (mm.)					
Ocoee G.	5	—	16	5.4	.9
Allisonia	1	—	1	8.9	—
Cold Mt.—cf.	2	3	8	7.3	.2
Ocoee G.—cf.	4	4	9	7	.7
Allisonia—cf.	2	3	9	7.4	.8
Silver Mills—cf.	1	1	3	6.6	—
Pa.—gh.	2	3	7	6.2	.2
Cabins—gh.	1	1	2	6.8	—
Width					
Ocoee G.	5	—	16	1.5	.2
Allisonia	1	—	1	2	—

SEDUM OF NORTH AMERICA

	n-pl.	n-rep.	n-obs.	x̄	s
Width					
Cold Mt.—cf.	2	3	8	2.4	—
Ocoee G.—cf.	4	4	9	2	.1
Allisonia—cf.	2	3	9	1.9	—
Silver Mills—cf.	1	1	3	2.3	—
Pa.—gh.	2	3	7	1.7	—
Cabins—gh.	1	1	2	1.6	—

Stamens have the filaments broad and flattened basally, but tapering upward. The filaments are white, sometimes suffused with pink. Rarely they are petaloid. The anthers are red, purple, or fuscous.

	n-pl.	n-rep.	n-obs.	x̄ (mm.)	s (mm.)
Filaments—length					
Ocoee G.	5	—	16	3.6	.5
Allisonia	1	—	1	4.9	—
Cold Mt.—cf.	2	3	8	4.8	1
Ocoee G.—cf.	4	4	9	4.7	.5
Allisonia—cf.	2	3	9	4.6	.4
Silver Mills—cf.	1	1	3	4.9	—
Pa.—gh.	2	3	7	5	—
Cabins—gh.	1	1	2	4.5	—
Epipetalous filaments—height on petals					
Ocoee G.	5	—	16	.3	.02
Allisonia	1	—	1	.05	—
Cold Mt.—cf.	2	3	8	.1	0
Ocoee G.—cf.	4	4	9	.2	.08
Allisonia—cf.	2	3	9	.1	0
Silver Mills—cf.	1	1	3	.1	—
Pa.—gh.	2	3	7	.1	0
Cabins—gh.	1	1	1	0	—
Anthers—length					
Ocoee G.	1	—	1	1.1	—
Cold Mt.—cf.	2	3	3	1.3	—
Ocoee G.—cf.	3	3	6	1.1	.3
Allisonia—cf.	2	3	3	1.4	—
Silver Mills—cf.	1	1	3	1.1	—
Pa.—gh.	1	1	3	1.1	—

Nectaries are oblong or subquadrate, truncate and sometimes submarginate, and yellow, pale yellow, or rarely white or orange.

	n-pl.	n-rep.	n-obs.	x̄ (mm.)	s (mm.)
Length					
Ocoee G.	5	—	16	.4	.08
Allisonia	1	—	1	.6	—
Cold Mt.—cf.	2	3	8	.6	.07
Ocoee G.—cf.	4	4	9	.6	.06
Allisonia—cf.	2	3	9	.5	.4
Silver Mills—cf.	1	1	3	.6	—
Pa.—gh.	2	3	7	.6	—
Cabins—gh.	1	1	1	.5	—
Width					
Ocoee G.	5	—	16	.3	.04
Allisonia	1	—	1	.6	—

	n-pl.	n-rep.	n-obs.	x̄ (mm.)	s (mm.)
Width					
Cold Mt.—cf.	2	3	8	.5	0
Ocoee G.—cf.	4	4	9	.4	.06
Allisonia—cf.	2	3	9	.4	—
Silver Mills—cf.	1	1	3	.4	—
Pa.—gh.	2	3	7	.4	—
Cabins—gh.	1	1	1	.2	—

Carpels are at first erect and white, then widely divergent, gibbous ventrally, and markedly two-lipped.

	n-pl.	n-rep.	n-obs.	x̄	s
Length (mm.)					
Ocoee G.	5	—	16	2.3	1
Allisonia	1	—	1	7.1	—
Cold Mt.—cf.	2	3	8	5.6	.7
Ocoee G.—cf.	4	4	9	5.1	1.2
Allisonia—cf.	2	3	9	4.8	—
Silver Mills—cf.	1	1	3	4.7	—
Pa.—gh.	2	3	7	4.2	.2
Cabins—gh.	1	1	1	5	—
Length of cohesion (mm.)					
Pa.—gh.	2	3	7	.6	—
Styles—length (mm.)					
Allisonia	1	—	1	1.9	—
Cold Mt.—cf.	2	3	8	1.8	0
Allisonia—cf.	2	3	9	1.8	—
Silver Mills—cf.	1	1	3	1.3	—
Ovules per ovary (n)					
Ocoee G.	5	—	16	.8	1.2
Allisonia	1	—	1	11	—
Cold Mt.—cf.	2	3	8	9.2	—
Ocoee G.—cf.	4	4	9	8	2.9
Allisonia—cf.	2	3	9	6.2	—
Silver Mills—cf.	1	1	3	8.3	—
Pa.—gh.	2	3	7	7.3	1.7

Fruits are widely divergent, gibbous ventrally, prominently two-lipped, with lips .3–.7 mm. broad, and brown, except that the lips are much paler.

	n-pl.	n-rep.	n-obs.	x̄ (mm.)	s (mm.)
Length					
Catawba R.—cf.	2	3	9	3.8	—
Cold Mt.—cf.	2	2	6	3.8	.2
Ocoee G.—cf.	4	4	12	4.2	.5
Allisonia—cf.	1	1	3	3.5	—
Catawba R.—gh.	1	1	3	5.7	—
Ocoee G.—gh.	1	1	1	3.8	—
Allisonia—gh.	1	1	3	4.9	—
Pa.—gh.	2	2	6	3.8	.6

Seeds are elliptic-pyriform, irregularly longitudinally ribbed, and brown or almost fuscous.

	n-pl.	n-rep.	n-obs.	x̄ (mm.)	s (mm.)
Length					
Catawba R.—cf.	2	3	9	.9	.3
Cold Mt.—cf.	2	2	5	.8	.03
Ocoee G.—cf.	4	4	12	.9	.06
Catawba R.—gh.	1	1	1	.9	—
Ocoee G.—gh.	1	1	1	.9	—
Pa.—gh.	2	2	2	.9	0
Diameter					
Catawba R.—cf.	2	3	9	.4	0
Cold Mt.—cf.	2	2	5	.4	.06
Ocoee G.—cf.	4	4	12	.4	.06
Catawba R.—gh.	1	1	1	.4	—
Ocoee G.—gh.	1	1	1	.5	—
Pa.—gh.	2	2	2	.4	0

Chromosomes are elongate, with median or submedian kinetochores. Diploids, triploids, tetraploids, and hexaploids occur. J. T. Baldwin, Jr. (1935, 1936, and 1942), as a result of study of 71 plants, reported tetraploids as widespread; diploids as in an area in western Virginia, West Virginia, and eastern Kentucky; triploids at two localities, one in West Virginia and the other in North Carolina; and a hexaploid from near Tuscaloosa, Ala. Uhl (1970) reported counting chromosomes of 81 additional plants: 15 diploids, 2 triploids, and 64 tetraploids. His map, fig. 17 in his article, shows sources of his plants in western Virginia and adjacent states. Citations below supplement previously published records and apply only to plants in my samples.

	n-pl.	g (n)	s (2n)	Cytologist
Number				
Catawba R., S.C.	1	—	32	D. Niimoto, 1967
Spray, N.C.	1	16	—	C. H. Uhl, 1960
Crawford Creek, N.C.	1	16	—	Charles Horn, 1966
Rutledge, Tenn.	1	8	—	C. H. Uhl, 1947
Allisonia, Va.	1	16	—	Ilse Zandstra, 1969
Lebanon, Va.	1	16	—	C. H. Uhl, 1947
Ingleside, Va.	1	8	—	C. H. Uhl, 1950
Tazewell, Va.	1	16	—	C. H. Uhl, 1950
Cabins, W. Va.	1	8	—	C. H. Uhl, 1948

Variation. Although *Sedum ternatum*, on first inspection, appears to be remarkably uniform, close scrutiny reveals variation both among plants within populations and among populations. Comparison of three populations, based on data from the wild, indicates significant differences in five out of nine features which were tested by analysis of variance. Results are shown in the following table, in which mean values are cited.

Feature	Catawba River	Ocoee Gorge	Allisonia
Median leaves of floriferous stems			
Length (mm.)	15	8	12
Width (mm.)**	10	4	4
Thickness (mm.)**	1	.8	1.1

Feature	Catawba River	Ocoee Gorge	Allisonia
Flowers—diameter (mm.)	—	10	12
Petals			
Length (mm.)*	—	5	9
Width (mm.)	—	2	2
Nectaries			
Length (mm.)	—	.4	.6
Width (mm.)**	—	.4	.6
Ovules per ovary (n)**	—	1	11

That some of these differences are environmental modifications is attested by the results of an experiment of completely random design in a cold frame in Ithaca. An interesting aspect had to do with the number of ovules per ovary. On the basis of data from the field alone, the idea could arise that *Sedum nevii* and *S. ternatum* are hybridizing in the Ocoee Gorge and, as a result, few ovules develop. Yet, the same plants that averaged one ovule per ovary in the Ocoee Gorge averaged eight ovules per ovary in cultivation. The situation in the field probably had a physiological cause, perhaps a consequence of the dry season when the observations were made. The few differences that do appear to have a genetic basis lack the kind of pattern which might suggest the formation of races.

Feature	Catawba River	Ocoee Gorge	Cold Mountain	Allisonia	Silver Mills
Median leaves of floriferous stems					
Length (mm.)	8.9	8.9	12.7	12	5.9
Width (mm.)	5	4.7	7.3	4.7	4.2
Thickness (mm.)	.8	.9	.9	.8	.8
Flowers—diameter (mm.)	16	15	15	14	14
Petals					
Length (mm.)	8	7	7	7	7
Width (mm.)*	1.9	2	2.4	1.9	2.3
Anthers—length (mm.)*	1.2	1.1	1.3	1.4	1.1
Nectaries					
Length (mm.)	.8	.6	.6	.5	.6
Width (mm.)	.4	.4	.5	.4	.4
Ovules per ovary (n)**	8	8	9	6	8

Four levels of polyploidy occur in *Sedum ternatum*. Diploid plants are less robust and do not survive as well as tetraploids in cultivation at Ithaca. As a result, comparisons are less than ideal. The following table contains data for three diploid plants, respectively from Rutledge and Ingleside, Va., and Cabins, W. Va.; and five tetraploid plants from Spray and Crawford Creek, N.C., and Allisonia, Cleveland, and Tazewell, Va. Comparisons are by means of the Wilcoxon rank sum test, except that data for diameter of flowers and length of anthers are not adequate for testing.

Feature	Diploid	Tetraploid
Median leaves of floriferous stems		
Length (mm.)	10	10
Width (mm.)	5.6	5.2
Thickness (mm.)	.6	.8

Feature	Diploid	Tetraploid
Flowers—diameter (mm.)	11	15
Petals		
Length (mm.)*	5.7	7.6
Width (mm.)	1.5	1.8
Anthers—length (mm.)	.7	1.1
Nectaries		
Length (mm.)	.5	.5
Width (mm.)*	.3	.4

The impression is that *Sedum ternatum* illustrates an autoploid series. As claimed by J. T. Baldwin, Jr. (1942), plants with the different numbers of chromosomes are not readily separable except by counting chromosomes. The occurrence of multivalents at first metaphase in the tetraploids, as reported by J. T. Baldwin, Jr. (1936), is evidence in favor of the idea of autoploidy.

Baldwin (1942) reported that diploids which he studied were dwarf, with smaller pollen, and that all grains were good. Also, the plants flowered earlier. His tetraploids had larger pollen, with some grains shriveled. They flowered later and were more resistant to fungi than the diploids. He considered that the triploids arose from hybridization of diploid and tetraploid plants. They too are more resistant to fungi than diploid plants. No comparative data are available for the one known hexaploid plant.

The common number of petals per flower is 4. A frequency distribution for a sample of 20 flowers from wild plants confirms this idea.

Petals per flower (n)	Flowers (n)	%	Cumulative %
3	1	5	5
4	18	90	95
5	0	0	95
6	1	5	100

Flowers with 5 petals occur, most often in the central position in cymes, but the 4-merous condition is more frequent.

In summary, the variation of *Sedum ternatum* does not require a subspecific classification. For those interested in number of chromosomes, indication of level of polyploidy, after the binomial, can provide the necessary information.

Nomenclature. *Sedum ternatum* Michx., Flor. Bor. Am. 1:277 (1803). Type locality: eastern North America. Michaux cited as places of occurrence rocks in the western parts of Pennsylvania, Virginia, and Carolina. Labels on the type sheet of specimens in the Muséum National d'Histoire Naturelle, Paris, indicate the origins as high mountains of Pennsylvania, Carolina, and Virginia. The precise locality of the holotype is unknown. Type: André Michaux (Paris), < 1802. The type sheet bears one shoot with leaves and three leafless floriferous stems. I select the leafy shoot and the floriferous stem below it as the holotype.

Synonyms are:

1809. *Sedum portulacoides* Willd., Enum. Hort. Berol., p. 484. Type locality: probably somewhere in Pennsylvania. Type: Muhlenberg, not seen. Willdenow cited *S. ternatum* Michx. in synonymy.

1811. *Sedum deficiens* Donn, Hort. Cantab., ed. 6, p. 126. Type locality: North America. Type: none cited. Although a description is lacking, the English name, nativity, and time of flowering suggest the correct identification.

1812. *Anacampseros ternata* (Michx.) Haw., Syn. Pl. Succ., p. 114. Based on *Sedum ternatum* Michx.

1921. *Sedum ternatum* var. *minus* Praeger, Jour. Roy. Hort. Soc. 46:161, fig. 86. Type locality: unknown. Type: a dwarf plant from Canon Ellacombe's garden, possibly not preserved. The var. *minus* may be diploid, but also it may be a depauperate tetraploid.

Gronovius, in *Flora Virginica*, ed. 2, p. 71 (1762), listed *Sedum annuum caule compresso, foliis obverse-ovatis*. The type, a collection of Clayton, no. 891, is in the herbarium of the British Museum. It is from Virginia. This is the oldest known collection of *Sedum ternatum*.

De Candolle, *Prod*. 3:403 (1828), listed *Sedum octogonum* hortul. in the synonymy of *S. ternatum*. That listing and other references to the name do not constitute valid publication, although *S. octogonum* may have been a frequent name among gardeners in the early eighteenth century. Likewise, *Sedum americanum*, listed in synonymy by Pursh, *Fl. Am. Sept.* 1:324 (1814), as a name in the herbarium of Banks, was not validly published.

Distribution. The geographic range (fig. 11) of *Sedum ternatum* is in the Appalachian Highlands from Geogia north to New Jersey, southern and western Pennsylvania, and Ohio; on the Interior Low Plateaus; and on the Central Lowland from Ohio to Missouri and Iowa. It also occurs on the Coastal Plain in Maryland and Virginia, on the Ozark Plateau in Missouri, and in the Ouachita Mountains on the eastern slope of Magazine Mountain in Arkansas. Sites on the Coastal Plain are along large rivers. Plants there may have originated from populations upstream and have been carried downward by water. Rocks on which plants occur include sandstone,

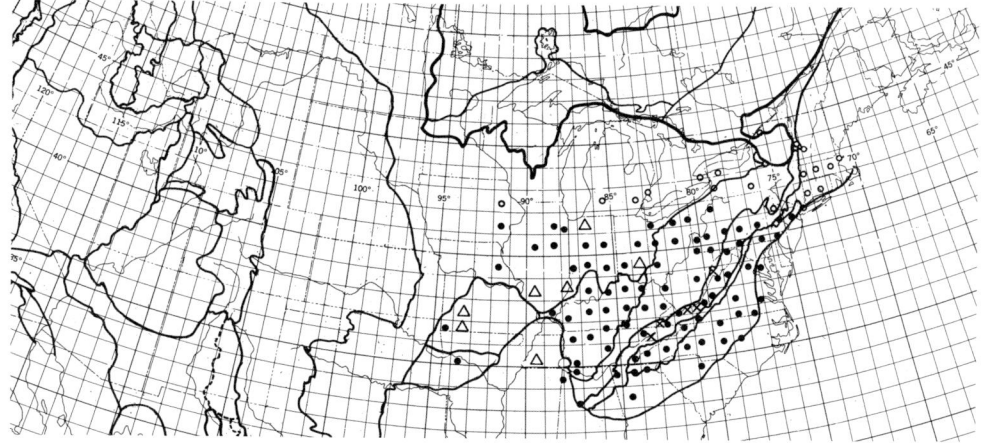

Fig. 11. Known distribution of *Sedum ternatum*: native populations (●), naturalized (○), diploid plants (x), and records from literature (△).

shale, limestone, conglomerate, quartzite, and gneiss. Habitats usually are wooded and include banks of streams, gullies, rocky slopes, ledges and crests of cliffs, and bluffs. Of all the North American species of *Sedum*, *S. ternatum* is best adapted to the deciduous forest. It thrives in shaded situations. The altitudinal range is from 50 to 1,690 meters.

Beyond its natural range northward, *Sedum ternatum* sometimes becomes established, as in New England, New York, and southern Michigan.

Data for four populations from which I obtained plants for study are in the following table. The population on Cold Mountain came to attention along the way to a randomly selected sampling unit. Selection of the population in the Ocoee Gorge was to study the interaction of *S. ternatum* and *S. nevii*. Choice of the population at Allisonia was because of work on *S. glaucophyllum* there. Attention to the population along the Catawba River was to gain information from the southern Piedmont Upland.

Population	Plants N	Area of population (m.²)	Density per m.²	Exposure	Rock
Catawba R., S.C.	7	8	.9	W	—
Cold Mt., N.C.	>12	—	—	E	gneiss
Ocoee R., Tenn.	192	1,720	.1	SW–WSW	quartzite
Allisonia, Va.	~19,500	~100,000	.2	N–ENE	sandstone, limestone

Population	Altitudinal range (m.)	Depth of soil (cm.) $\bar{x}$	s	pH range	pH tests (n)	Distance to nearest neighbor (cm.) $\bar{x}$	s	Oldest record
Catawba R., S.C.	200	—	—	6.6	1	—	—	1953
Cold Mt., N.C.	1,030–1,690	—	—	5.2–5.6	2	0	0	1960
Ocoee R., Tenn.	265–275	.8	.3	6.4–6.8	6	15	7	1947
Allisonia, Va.	573–670	—	—	5.8–6.6	13	—	—	1877

In the wooded situations in which *Sedum ternatum* grows, the shading trees and other competing species are different from site to site. The woodland condition, with shade, seems to be the important consideration, not the species of trees forming the canopy. Along the Catawba River in South Carolina, *S. ternatum* grows under elms. Cane, *Arundinaria tecta*, is an important competitor. In the gorge of the Ocoee River, the commonest shading species is *Fraxinus americana*, with *Acer nigrum* ssp. *saccharophorum*[1] and *Vitis rotundifolia* of secondary importance. At Allisonia, plants of *S. ternatum* were twice as numerous in a heavily shaded situation as at a site with small open places. *Ostrya virginiana* is the most important shading species there, with other important species being *Quercus prinus*, *Carya glabra*, and *Pinus virginiana*. Com-

---

[1] *Acer nigrum* Michx. ssp. **saccharophorum** (Koch), comb. nov., fundatum super *Acer saccharophorum* Koch, Hort. Dendr. 80 (1853). Koch based his name on plate 15 of Michaux, Hist. Arb. 2:218–237 (1812). The tree illustrated by Michaux and labeled *A. saccharinum* is the common sugar maple of eastern North America with leaves glabrous and glaucous below.

peting species are many. These include *Sedum glaucophyllum* which flowers a little later than *S. ternatum*, but overlaps in flowering time.

Reproduction. The decumbent stems of *Sedum ternatum* break readily into pieces which take root at the nodes and become new plants. This method of reproduction may be the most important means of spread. Seeds are another means of propagation. These give rise to seedlings wherever conditions of moisture and soil are suitable. In addition, vegetative apomixis sometimes occurs. For example, a plant from Crawford Creek in the southern Blue Ridge, cultivated in the cold frame, produced vegetative rosettes in the cymes in place of flowers.

Flowering time varies with latitude and altitude, April in the south, May and June northward. At Ithaca, N.Y., plants flower outdoors in late May and early June.

Pollinators are small bees. These are frequent visitors of the flowers on sunny days when there is little or no wind.

Relationships. *Sedum ternatum* is related to three other eastern North American species, namely *S. nevii*, *S. glaucophyllum*, and *S. pulchellum*. These species have in common four-parted flowers which have white or pinkish white petals; broad, flattened filaments which are scarcely adnate to the bases of the petals; and kyphocarpic carpels. *Sedum nevii* is diploid. Both *S. glaucophyllum* and *S. pulchellum* are more specialized and may be of allopolyploid origin.

The relationship of the white-flowered Appalachian species of *Sedum* with species elsewhere in the world is obscure. Three species for consideration as possibly related are: first *S. engleri* and other kyphocarpic species of eastern Asia; second, the progenitor of *S. telephioides* of subgenus *Telephium*, also with broad, flat leaves and white flowers; and third, *S. rhodocarpum* with ternate, broad leaves and slightly gibbous carpels. Perhaps the most reasonable interpretation is that the Appalachian species separated from the other species of the genus in the ancient past and have had a long period of independent evolution.

Both *Sedum ternatum* and *S. nevii* occur in the gorge of the Ocoee River. Although the two species occupy separate niches, slight overlap occurs. Separation of the two species is easy by means of arrangement and shape of leaves, especially width. The flowers of *S. ternatum* are about three weeks ahead of those of *S. nevii*. A comparison, utilizing data for fourteen features from random samples obtained in 1970, reveals twelve significant differences. The two species differ in many features, not just the obvious ones.

Feature	F	S. ternatum	S. nevii
Diameter (cm.)	18.45**	23	4
Rosettes (*n*)	6.68*	27	7
Rosettes—diam. (cm.)	33.38**	2	1
Leaves of rosettes			
Length (mm.)	18.98**	11	7
Width (mm.)	125.33**	8	3
Median leaves of floriferous stems			
Length (mm.)	.17 n.s.	8	9
Width (mm.)	48.4**	4	2
Thickness (mm.)	24.44**	.8	1.1

Feature	F	S. ternatum	S. nevii
Flowers per cyme (n)	17.83**	7	16
Flowers—diameter (mm.)	11.35**	10	8
Petals			
Length (mm.)	17.22**	5	4
Width (mm.)	.43$^{n.s.}$	1.5	1.4
Nectaries			
Length (mm.)	20.86**	.4	.3
Width (mm.)	70.69**	.4	.3

Separation of *Sedum ternatum* from *S. glaucophyllum* likewise presents no problems. The spirally arranged, oblanceolate or spatulate leaves of *S. glaucophyllum* make separation easy. Where the two species grow together, as at Allisonia, hybridization is not occurring. Even where distributional overlap exists, each species has its own niche, with the area of common occurrence restricted to a small fraction of the total habitat. Further, *S. ternatum* is ahead of *S. glaucophyllum* in time of flowering. Overlap in flowering time is small. If *S. glaucophyllum* is of allopolyploid origin, then *S. ternatum* may be one of the diploid parents. Uhl (1970) has indicated a difficulty for this explanation, namely the occurrence of distinctive, large chromosomes in *S. glaucophyllum* which he states do not occur in genomes of either *S. nevii* or *S. ternatum*. In contrast, Joyce, observing mitotic chromosomes of *S. nevii* in 1970, noted two chromosomes markedly longer than the others, 2–6 μ long, a situation similar to that in *S. glaucophyllum* and suggesting cytological as well as morphological and ecological similarity. If the two longer chromosomes of *S. glaucophyllum* came from *S. nevii*, then a third longer chromosome might have come from *S. ternatum*. This idea is compatible with the demonstration by J. T. Baldwin, Jr. (1936) of considerable variation in size among chromosomes of *S. ternatum*.

Sedum pulchellum, with pinkish white petals and linear leaves with sagittate spurs, as well as biennial growth, is easy to separate from *S. ternatum*. It is closer to *S. nevii* and *S. glaucophyllum*.

Quimby (1971) found that the vascular anatomy of the flowers of *Sedum ternatum* and *S. pulchellum* is similar, but more advanced as a result of vertical compression and consequent fusion of bundles than in *S. rhodocarpum*, a species of the Sierra Madre Oriental in Mexico. This, like *S. ternatum*, has broad leaves in whorls of three. A relationship is possible. *Sedum rhodocarpum* and *S. ternatum* may have had similar ancestries, but *S. ternatum* may be the more advanced.

Subramanyam (1955) also studied the floral anatomy of *Sedum ternatum*. He reported that receptacular stelar tissue is present at the level at which inverted ventral bundles of the carpels are distinct, but then disappears at a slightly higher level, indicating the lateral position of the carpels on the floral axis. He described the carpels as barely closed, with epidermal cells present along the line of closure. These are evidences of a primitive condition among angiosperms. Subramanyam (1963) further provided detailed descriptive information about the embryology of *S. ternatum*.

References. J. T. Baldwin, Jr. (1935, 1936, and 1942), Quimby (1971), Subramanyam (1955 and 1963), and Uhl (1970).

3. Sedum nevii*** (figs. 12–14)

Distinctive features of *Sedum nevii* (fig. 12) are basal rosettes of elliptic or oblanceolate, spirally arranged green leaves, and three-parted cymes of white, 4-merous flowers. The related *S. ternatum* has broad, obovate leaves, either in whorls of three or two per node. *Sedum pulchellum* has pinkish white petals, floriferous stems with linear leaves with sagittate bases, and larger seeds. *Sedum glaucophyllum*, which may be allopolyploid resulting from hybridization between *S. nevii* and *S. ternatum*, is most similar to *S. nevii*. Usually it is a larger plant with broader rosette leaves which sometimes are glaucous, wider leaves on the floriferous stems, longer sepals and petals, and larger seeds. Further discussion of distinctions follows in the section on relationships and also in the account of *S. glaucophyllum*.

Description. Sample: 30 plants. Sources and numbers of plants from each locality are: gorge of the Ocoee River, Polk Co., Tenn., May 1970—23, of which 14 were randomly selected, a 2% sample of accessible plants with floriferous stems in 4 randomly selected transects, 4 were early flowering, and 5 were among the largest plants seen; Pratt's Ferry on the Cahaba River, Bibb Co., Ala., March 31, 1942—1; cliffs on north side of Warrior River opposite Holt, Tuscaloosa Co., Ala., collected by H. J. Freer and R. W. Simmers, Jr., March 30, 1970—4; southern slope of Monte Sano near Huntsville, Madison Co., Ala., according to S. J. Smith and collected by him on April 11, 1949—1; and bluffs of the Chattahoochee River 183 meters below Goat Rock Dam, Harris Co., Ga., collected by C. H. Uhl, Feb. 6, 1954—1. In the description, abbreviations for localities are respectively Ocoee R., Cahaba R., Warrior R., Monte Sano, and Chattahoochee R. For comparison with *S. glaucophyllum*, the plants from the Cahaba and Chattahoochee Rivers were cultivated in an experiment of completely random design in a cold frame (cf.). Likewise, the plant from the Cahaba River was in a similar experiment in the garden (gd.).

The only adequate sample is from the gorge of the Ocoee River. For this reason, primary citation of quantitative data is from there. Unless remarkably different, data from other localities are omitted.

Growth of *Sedum nevii* at Ithaca, N.Y., is best in the greenhouse and poorest outdoors. For comparison of populations and indication of differences among plants within populations, an experiment of completely random design in the greenhouse yielded useful data in 1972. Plants included were 13 from the Ocoee Gorge, 4 from near the Warrior River, and 1 from near the Cahaba River. The plant from near the Cahaba River has survived in cultivation for thirty-one years. All plants in the experiment were started at the same time, in November 1971, from cuttings 2 centimeters long, and with two replicates of each. Unless differences among populations are significant, data are pooled. Designation of the experimental results in the following description is gh.—1972. Sometimes the variance within plants exceeded that among plants. In such cases, no values for s among plants are cited in the tables.

Plants of *Sedum nevii* are glabrous, perennial herbs with fibrous, brown roots which spread horizontally in shallow soil. The slender, pale green primary stems are decumbent and take root at the nodes. Each branch terminates in a small rosette, with the

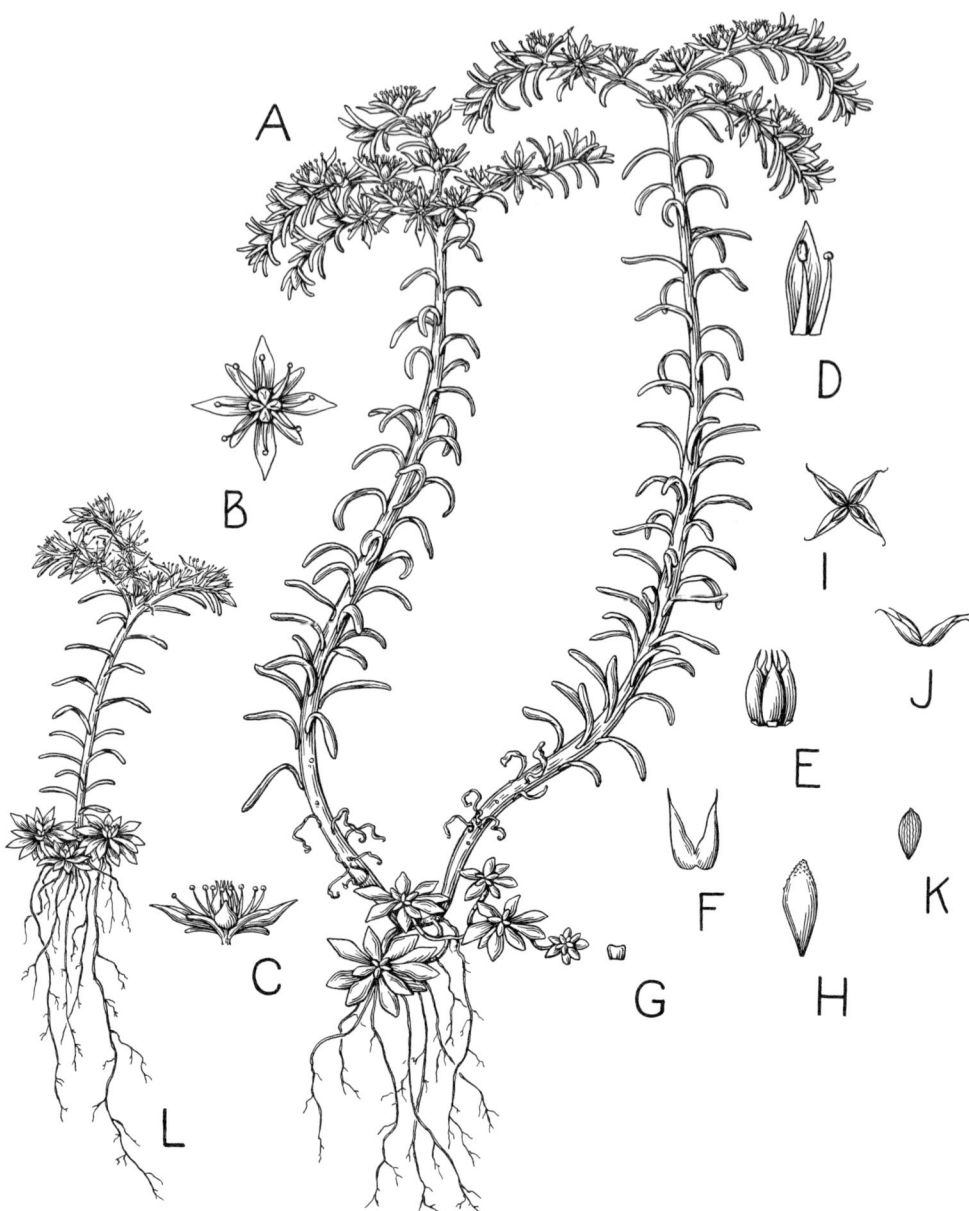

Fig. 12. Plants of *Sedum nevii* from gorge of Ocoee River, Polk Co., Tenn., cultivated in greenhouse at Ithaca, N.Y. A. Habit sketch (x 1). B. Flower from above (x 2). C. Flower from side (x 2). D. Petal and two stamens (x 3). E. Carpels (x 3). F. Two carpels (x 4). G. Nectary (x 5). H. Leaf (x 2). I. Fruits from above (x 3). J. Fruits from side (x 3). K. Seeds (x 10). L. Small plant (x 1).

result that plants usually have a tufted appearance. Rosettes in their second year may give rise to terminal, floriferous stems. If all rosettes of a plant produce floriferous stems and no new axillary shoots develop, the plant dies. In 1947 a plant in the Ocoee Gorge had a floriferous stem 21 centimeters long. Differences among samples are significant in diameter of plants and highly significant in number of floriferous stems per plant. Also, the diameter of primary rosettes in the greenhouse (gh.) is highly significantly larger.

	n-pl.	n-rep.	n-obs.	$\bar{x}$	s
Plants—diameter (cm.)					
Ocoee R.—random	13	—	13	4.1	3.6
—large	3	—	3	16.7	16.7
—gh.—1972	13	26	26	8	—
Cahaba R.—gh.—1972	1	2	2	18	—
Warrior R.—gh.—1972	4	8	8	15	.7
Primary rosettes (n)					
Ocoee R.—random	13	—	13	6.6	9
—large	4	—	4	24	15.6
—early	2	—	2	2.5	2.1
Gh.—1972	18	36	36	27	—
Primary rosettes—diameter (cm.)					
Ocoee R.—random	11	—	31	1	.2
—large	4	—	22	1.2	.2
—gh.—1970	10	18	30	5.9	.3
Floriferous stems per plant (n)					
Ocoee R.—random	11	—	11	1.7	1.3
—large	4	—	4	13.7	10.8
—early	2	—	2	7	.3
Floriferous stems—length (cm.)					
Ocoee R.—random	13	—	18	6.2	1.5
—large	4	—	22	7	1
—early	2	—	12	8.3	2

Leaves of rosettes are spirally arranged; loosely crowded; short-spurred; petiolate; elliptical, oblanceolate, spatulate, or obovate; rounded at apices; and papillose on margins. They are green or grayish green, but without a bloom. The leaves of the floriferous stems likewise are spirally arranged, short-spurred, narrowly elliptical or linear, subterete, obtuse, papillose, and yellow-green. Median leaves of floriferous stems of early-flowering plants are significantly longer than those of plants flowering later.

	n-pl.	n-rep.	n-obs.	$\bar{x}$ (mm.)	s (mm.)
Leaves of rosettes—length					
Ocoee R.—random	11	—	31	6.6	1.2
—large	4	—	22	7.3	1
—gh.—1972	13	22	63	10	.8
Cahaba R.—gh.—1972	1	2	6	11	—
Warrior R.—gh.—1972	4	7	20	6	.6
Leaves of rosettes—width					
Ocoee R.—random	11	—	31	2.5	.3
—large	4	—	22	2.5	.1
Gh.—1972	18	30	87	3.2	.5

	n-pl.	n-rep.	n-obs.	$\bar{x}$ (mm.)	s (mm.)
Leaves of rosettes—thickness					
Ocoee R.—random	11	—	31	1	.2
—large	4	—	22	.8	.1
—gh.—1970	10	18	30	2	0
Leaves of floriferous stems—length					
Ocoee R.—random	13	—	19	8.5	1.6
—large	4	—	22	9.8	2.1
—early	4	—	14	15.7	5.4
Gh.—1972	18	33	94	18	.7
Leaves of floriferous stems—width					
Ocoee R.—random	13	—	19	1.6	.3
—large	4	—	22	1.7	.3
—early	4	—	14	1.8	.3
Gh.—1972	18	33	94	3.1	—
Leaves of floriferous stems—thickness					
Ocoee R.—random	13	—	19	1.1	.1
—large	4	—	22	1	.2
—early	4	—	14	1.1	.2

Inflorescences are cymes of 2 or 3 cincinni. Rarely, a cincinnus is dichotomously branched. Floral bracts are similar to leaves of the floriferous stems, but smaller. A plant from Monte Sano, in cultivation in the greenhouse at Ithaca, produced cymes with rosettes of leaves in place of some flowers.

	n-pl.	n-rep.	n-obs.	$\bar{x}$ (n)	s (n)
Cincinni per cyme					
Ocoee R.—random	13	—	19	3	.5
—large	4	—	22	3	.5
—early	4	—	14	4.5	1.9
Flowers per cyme					
Ocoee R.—random	13	—	19	16	5
—large	4	—	22	16	4
—early	4	—	14	50	36
Cahaba R.—cf.	1	1	3	25	—
—gd.	1	2	3	11	—

Flowers are 4-merous and usually sessile. They have a short torus.

	n-pl.	n-rep.	n-obs.	$\bar{x}$ (mm.)	s (mm.)
Pedicels—length					
Ocoee R.—random	12	—	36	.02	.04
—large	2	—	6	0	0
—early	4	—	18	.01	.02
Flowers—diameter					
Ocoee R.—random	12	—	36	7.7	.7
—large	2	—	6	7.8	.7
—early	4	—	18	11	1.9
Gh.—1972	18	32	91	13	2.9
Cahaba R.—cf.	1	1	3	15.3	—
—gd.	1	2	4	8.8	—
Torus—length					
Ocoee R.—random	11	—	29	.3	.1
—large	2	—	6	.2	.1
—early	4	—	18	1	.1

Sepals are unequal, linear or linear-lanceolate, obtuse, subterete, and green.

	n-pl.	n-rep.	n-obs.	x̄ (mm.)	s (mm.)
Length					
Ocoee R.—random	12	—	36	3.8	.6
—large	2	—	6	3.9	.1
—early	4	—	18	6.6	1.9
Gh.—1972	18	32	90	6.6	—
Cahaba R.—gd.	1	2	4	4.4	—
Width					
Ocoee R.—random	12	—	36	1	.3
—large	2	—	6	.9	.1
—early	4	—	18	1.3	.2
Cahaba R.—cf.	1	1	3	1.7	—
—gd.	1	2	4	.9	—

Petals are separate, lanceolate, carinate, acute, and white.

	n-pl.	n-rep.	n-obs.	x̄	s
Number per flower					
Ocoee R.—random	12	—	36	4	0
—large	2	—	6	4	0
—early	4	—	18	4	0
—gh.—1972	13	24	67	4.04	.05
Cahaba R.—gh.—1972	1	2	6	4	0
Warrior R.—gh.—1972	4	6	18	4.28	—
Length (mm.)					
Ocoee R.—random	12	—	36	4.1	.3
—large	2	—	6	3.6	.1
—early	4	—	18	5.9	.3
Gh.—1972	18	32	92	5.3	—
Cahaba R.—cf.	1	1	3	7.1	—
—gd.	1	2	4	4.4	—
Width (mm.)					
Ocoee R.—random	12	—	36	1.4	.1
—large	2	—	6	1.3	.1
—early	4	—	18	1.8	.2

Stamens have white filaments which are flat basally. The epipetalous filaments are slightly adnate to the bases of the petals. The red anthers shed pollen as the petals expand. Anthers of stamens alternate with the petals shed pollen before those which are opposite them.

	n-pl.	n-rep.	n-obs.	x̄ (mm.)	s (mm.)
Filaments—length					
Ocoee R.—random	12	—	36	2.6	.2
—large	2	—	5	2.6	.4
—early	4	—	18	3.9	.3
Cahaba R.—cf.	1	1	3	4.6	—
Epipetalous filaments; adnation—length					
Ocoee R.—random	12	—	35	.1	0
—large	2	—	6	.1	0
—early	4	—	18	.1	.04

114   SEDUM OF NORTH AMERICA

	n-pl.	n-rep.	n-obs.	x̄ (mm.)	s (mm.)
Anthers—length					
Ocoee R.—random	12	—	31	.9	.04
—large	2	—	3	.8	0
—early	4	—	10	1	.1
Gh.—1972	17	28	79	1.1	.07
Cahaba R.—cf.	1	1	1	1.3	—
—gd.	1	1	1	1.1	—

Nectaries are subquadrate or quadrate, truncate, sometimes slightly emarginate, and white or rarely creamy white.

	n-pl.	n-rep.	n-obs.	x̄ (mm.)	s (mm.)
Length					
Ocoee R.—random	12	—	36	.3	.1
—large	2	—	6	.3	.05
—early	4	—	18	.5	.03
Width					
Ocoee R.—random	12	—	36	.4	.04
—large	2	—	6	.4	.05
—early	4	—	18	.7	.06
Cahaba R.—cf.	1	1	3	.8	—
—gd.	1	2	4	.5	—

Carpels are white and erect when anthers are in anthesis, but become divergent and gibbous ventrally as they mature. Stigmas receive many pollen grains from adjacent stamens of the same flower.

	n-pl.	n-rep.	n-obs.	x̄	s
Length (mm.)					
Ocoee R.—random	12	—	36	2.5	.2
—large	2	—	6	2.4	.2
—early	4	—	18	3.9	.3
Cahaba R.—cf.	1	1	3	4.7	—
—gd.	1	2	4	3.2	—
Length of cohesion (mm.)					
Ocoee R.—random	12	—	36	.3	.1
—large	2	—	6	.4	.1
—early	4	—	18	0	0
Ovules per ovary (n)					
Ocoee R.—random	12	—	36	9.5	1
—large	2	—	6	8.1	1.6
—early	4	—	18	16.8	3.7

Fruits are rotately spreading and brown, with prominent ventral lips.

	n-pl.	n-rep.	n-obs.	x̄ (mm.)	s (mm.)
Length					
Ocoee R.—random	8	—	24	2	.04
Warrior R.	4	—	12	2.7	.4
Cahaba R.—cf.	1	2	6	3.1	—
Gh.—1972	18	33	99	3	.1

Seeds are elliptic-pyriform, finely longitudinally ribbed, and brown.

	n-pl.	n-rep.	n-obs.	x̄ (mm.)	s (mm.)
Length					
Ocoee R.—random	4	—	12	.7	.03
Warrior R.	4	—	10	.7	.03
Cahaba R.—cf.	1	1	2	.8	—
Ocoee R.—gh.—1972	11	22	64	.8	.06
Cahaba R.—gh.—1972	1	2	6	.8	—
Warrior R.—gh.—1972	4	5	15	.7	—
Diameter					
Ocoee R.—random	4	—	12	.3	.05
Warrior R.	4	—	10	.3	.04
Cahaba R.—cf.	1	1	2	.3	—
Ocoee R.—gh.—1972	11	22	64	.3	—
Cahaba R.—gh.—1972	1	2	6	.4	—
Warrior R.—gh.—1972	4	5	15	.3	.02

Chromosomes are elongate and two-armed with median or submedian constrictions. Two chromosomes are markedly longer than the others, varying from $2-6\ \mu$ long. The standard deviation for length below is for chromosomes within a plant. Variation among samples from the same plant surpasses variation among plants. Causes of this variation include both the stage of development of the chromosomes and fluctuations in technique, especially staining.

	n-pl.	g (n)	s (2n)	Cytologist
Number				
Ocoee R., Tenn.	6	6	12	B. Joyce and L. and R. Benjamin, 1970
Cahaba R., Ala.	1	6	12	J. T. Baldwin, Jr. (1942)
	1	6	12	B. Joyce, 1970
Warrior R., Ala.	1	6	12	J. T. Baldwin, Jr. (1942)
Monte Sano, Ala.	1	—	12	C. H. Uhl, 1950
Chattahoochee R., Ga.	1	6	—	C. H. Uhl (1970)

Length of sporophytic chromosomes

	n-pl.	x̄ (genome length) (μ)	x̄ (chromosomes length) (μ)	s (μ)	Cytologist
Ocoee R.	5	28	2.4	1.1	B. Joyce and L. and R. Benjamin, 1970
Cahaba R.	1	24	2	—	B. Joyce, 1970

Variation. Modifications resulting from different environmental conditions may be great. An example of the magnitude of such modification is provided by a sample of 10 randomly selected plants which were studied first in the Ocoee Gorge and then after they had been cultivated for three months in soil in pots in the greenhouse and watered four times per week. Figures are sample means. All differences are highly significant.

	Wild	Greenhouse	Times bigger	F
Diameter (cm.)	4.8	12	2.5	151
Rosettes—diameter (cm.)	1	5.9	5.9	496
Leaves of rosettes—length (mm.)	6.8	31.5	4.6	525
—width (mm.)	2.6	6.2	2.3	89
—thickness (mm.)	1.1	2	1.8	208

Study of data in the description reveals big differences among samples from the gorge of the Ocoee River. Large plants are old. Since random samples include many young plants, averages for them are small. The first flowers to open on early-blooming plants are central or near the base of cincinni. These are the biggest flowers in a cyme. Plants selected because of large size have significantly larger diameters than the average for the population and also significantly more floriferous stems. In other respects, the large plants are not significantly different from plants in random samples. Similarly, plants that flower early have significantly longer median leaves on the floriferous stems, but their other differences may be attributed to chance.

Clusters of plants in the gorge of the Ocoee River differ significantly among themselves in details of the flower, for example length and width of the petals, width of the nectaries, and number of ovules per ovary. This circumstance reveals an advantage of subsampling, because such conditions might not be detected if a simple random sample were spread over the whole population.

The experiment in the greenhouse in Ithaca in 1971–1972 provided an objective basis for comparing populations from the valleys of the Ocoee, Warrior, and Cahaba Rivers. Of fourteen features tested for significant differences, five showed significance. These are listed in the following table, where tests of significance of individual difference are according to formula 7.17 of Steel and Torrie (1960).

	Ocoee R.	Cahaba R.	Warrior R.
Plants (n)	13	1	4
Replicates (n)	26	2	8
Diameter (cm.)**	8	18	15
Leaves of rosettes—l. (mm.)*	10	11	6
Petals (n)*	4.04	4	4.28
Seeds—l. (mm.)**	.8**	.8	.7
—diam. (mm.)*	.3	.4	.3

Plants from along the Chattahoochee River, Ga., and Monte Sano, Ala., were not remarkably different from other plants.

The disjunct populations of *Sedum nevii* probably are not subspecies. Neither are they identical. Small differences do exist among them. A related problem, namely the length of time that the populations have been separate, is unresolved.

Number of petals per flower varies little in *Sedum nevii*. Of 60 flowers in three samples observed in the field in the gorge of the Ocoee River, all had 4 petals, but 3 flowers out of 67 in the experiment in the greenhouse had 5 petals. One plant from the Ocoee Gorge, collected in 1947, had flowers with both 3 and 5 petals, but mostly 4.

NATIVE SPECIES 117

Fig. 13. Rosettes of plant of *Sedum nevii* in gorge of Ocoee River, Polk Co., Tenn., June 15, 1972 (x .4).

Most plants of *Sedum nevii* produce rosettes (fig. 13) from the bases of the old stems and by this means grow as perennials. A small number of plants, 15% in the random sample, lack rosettes and die after fruiting. Such plants are biennials or winter annuals.

In summary, *Sedum nevii* exhibits some genetic variation, but age of plants and environmental effects appear to be the principal causes of variability.

Nomenclature. *Sedum Nevii* A. Gray, Mem. Acad. Arts and Sciences, N.S. 6:373 (1858). Type locality: cliffs in the vicinity of Tuscaloosa, Ala., at the southern extremity of the Appalachian Plateau. Type: R. D. Nevius (GH), May 1858. Gray's original publication of *S. nevii*, although not a full description, is sufficient to identify the species. He confused *S. nevii* with *S. glaucophyllum* and placed specimens of the two species together on one sheet in his herbarium. The holotype is a specimen collected by Nevius in 1858 at Tuscaloosa.

Doubtful synonyms are *Sedum Beyrichianum* Masters, Gard. Chron., ser. 2, 10:376 (1878), and *S. Nevii* var. *Beyrichianum* (Masters) Praeger, Jour. of Botany 55:211 (1917). No type is available. The original description, based on a cultivated plant, could apply to *S. nevii*, but Praeger's interpretation seems to apply to a green-leaved race of *S. glaucophyllum*. Beyrich, for whom *S. beyrichianum* was named by nurserymen, traveled in the United States in 1833–1834. His journeys provided experience with *S. glaucophyllum* in Virginia, but probably were not in the area of occurrence of *S. nevii*. Yet, if a plant of *S. glaucophyllum* is designated as the type of *S. beyrichianum*, it will not be in agreement with the description by Masters. Further, Masters himself was not sure whether he was rightly naming the plant which he described. Trouble arises regardless how one interprets the name. For this reason, the best disposition is to drop *S. beyrichianum* as impossible to identify accurately, especially since Masters did not cite a definite place of origin. For further discussion, see the section on nomenclature under *S. glaucophyllum*.

Distribution. The distribution (fig. 14) of *Sedum nevii* is disjunct in the southern Appalachian Highlands. The species occurs in four regions: at two localities at the southern end of the Appalachian Plateau in Alabama, along the Warrior River near Tuscaloosa and allegedly at Monte Sano near Huntsville; along the Cahaba River in the Appalachian Ridge and Valley Province in Alabama; in the gorge of the Ocoee River in the Blue Ridge in southeastern Tennessee; and on bluffs of the Chattahoochee River at the edge of the Piedmont Plateau in Georgia. Ives (1944) cited *S. nevii* from Greenville Co., S.C., but no specimen is available to substantiate the report. Juvenile plants of *S. pulchellum* are the basis for old reports of *S. nevii* from Missouri by Robinson and Fernald (1909) and Steyermark (1934), and from Illinois by Robinson and Fernald (1908). Confusion of *S. nevii* with *S. glaucophyllum* has caused many authors to include the entire range of *S. glaucophyllum* in their statement of the distribution of *S. nevii*.

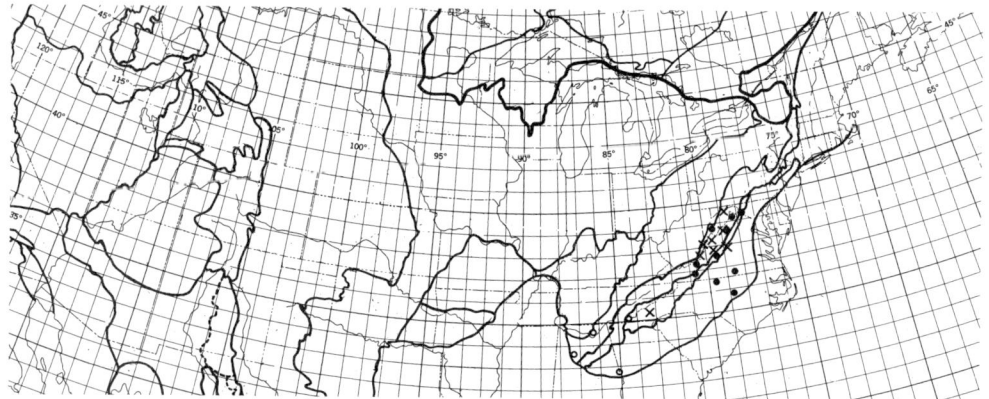

Fig. 14. Known distribution of *Sedum nevii* (○) and *S. glaucophyllum* (●) and (x for polyploid plants).

Discoverers of the various populations of *Sedum nevii* are indicated in the following table.

Locality	Year of discovery	Discoverer
Chattahoochee River, Ga.	1954	C. H. Uhl
Ocoee Gorge, Tenn.	1888	A. Gattinger
Pratt's Ferry, Ala.	1883	C. Mohr
Tuscaloosa, Ala.	1857	R. D. Nevius
Monte Sano, Ala.	1949	S. J. Smith

R. M. Harper possibly was the first to see *Sedum nevii* on Monte Sano. He thought that two species were there, but his second species may have been a perennial variation of *S. pulchellum* rather than *S. nevii*.

I have visited the populations near Tuscaloosa and at Pratt's Ferry and have made detailed studies in the gorge of the Ocoee River. Data for the population in the Ocoee Gorge are condensed in the following table.

Area (m.²)	Density per m.²	N plants, estimate	N plants, (C. I., P = .05)	Altitudinal range (m.)
84,960	.079	6,678	1,108–24,638	260–307

Exposure	pH, range	pH, n-tests	Depth of soil (cm.) (C. I., P = .05)	Distance to nearest neighbor (cm.) (C. I., P = .05)
S (SE–SW)	5.6–6.8	14	.8–1.6	3.4–14.6

Habitats of *Sedum nevii* in the Ocoee Gorge are living mats of mosses and lichens on rocks of quartzite, crevices and ledges of cliffs, and sometimes talus below cliffs. Most plants are partially shaded. Commonest shading trees are *Fraxinus americana* and *Quercus borealis*. Other shading species include *Carya glabra*, *C. ovata*, *Quercus velutina*, *Pinus virginiana*, and *Rhus toxicodendron*. Competing plants are miscellaneous. Rated on a scale of 14, the two closest competitors are a species of *Aster* with rating of 4 and a *Heuchera* with rating of 3. The two second closest competitors are a grass with rating of 4 and *Parthenocissus quinquefolia* with a rating of 3.

Along the Warrior River, *Sedum nevii* grows on sandstone and shale, and along the Cahaba River on limestone.

Fires on the slopes of the Ocoee Gorge in the spring of 1970 demonstrated that most plants of *Sedum nevii* grow on rocks where fires do not reach them, although woody plants and herbs nearby may be burned or scorched. Large plants of *S. nevii* perhaps indicate sites that are not touched by fire. Although fire may destroy some plants of *S. nevii*, surviving individuals may benefit from the destruction of competing species.

Reproduction. *Sedum nevii* reproduces both by rosettes which may become separated from the parental plants and by seeds.

Flowering time is in spring, in late April and May. My principal experience with wild plants has been when they were in bud: in late March along the Cahaba and Warrior Rivers in Alabama, and in April and May in the gorge of the Ocoee River. In 1970, when I followed closely the development of many plants in the Ocoee Gorge, the first flowers were in anthesis on May 11. No insects visited the flowers in my period of observation, but undoubtedly the blossoms are attractive to bees and flies.

At Ithaca, N.Y., plants have flowered in the greenhouse in every month from December to June, and in the garden in June and July.

Relationships and evolution. The relationships of *Sedum nevii* are with *S. glaucophyllum*, *S. ternatum*, and *S. pulchellum*. *Sedum ternatum* differs in having broader leaves either in whorls of three or opposite. *Sedum pulchellum* usually has pinkish white petals, linear leaves with sagittate, clasping bases on the floral stems, and larger seeds. Occasional perennial plants of *S. pulchellum* have rosettes which may resemble those of *S. nevii*, but the resemblance ceases when floral stems develop. Differences in base numbers of chromosomes, 6 in *S. nevii*, 8 in *S. ternatum*, and 11 in *S. pulchellum*, provide a clear genetic basis for the separation of these three species. Possibly

S. *ternatum*, with its broad leaves, weakly sealed carpels, and adjustment to mesophytic conditions, is nearest to the progenitor of this small group of species.

The relationship of *Sedum nevii* with *S. glaucophyllum* is of another sort. Some plants of *S. glaucophyllum* are so similar to those of *S. nevii* that separation is difficult. This is what is to be expected if the hypothesis is correct that *S. glaucophyllum* is the result of hybridization of *S. nevii* and *S. ternatum*, as postulated by J. T. Baldwin, Jr. (1944). *Sedum glaucophyllum* appears more variable than *S. nevii*. As a result of experiment, significant differences occur among groups of populations of *S. glaucophyllum* in diameter of primary rosettes, length of leaves of rosettes, and number of ovules per ovary, and among populations in thickness of leaves of rosettes and number of floriferous stems per plant. In contrast, populations of *S. nevii* differ significantly in at least five features. *Sedum nevii* always has 6 pairs of chromosomes. *Sedum glaucophyllum* often has 14 pairs of chromosomes, but higher levels of polyploidy occur, with sporophytic numbers ranging from 44 to about 49. Physiologically, *S. nevii* is less hardy than the polyploids. It occurs at lower altitudes and farther south. The physiological differences are important. Morphological distinctions are few. The morphological diversity of *S. glaucophyllum* is so great that few absolute differences between the two species are available. Usually, the leaves of the rosettes of *S. glaucophyllum* are wider than those of *S. nevii*. When comparisons are made at the level of populations, rather than general, pairs of populations of the two species always are separable. For example, the population of *S. glaucophyllum* geographically nearest to *S. nevii* is the one on Cedar Cliff Mountain in western Tennessee. This population differs in several features from *S. nevii* in the Ocoee Gorge, as shown in the following comparison of plants in nature. All differences are highly significant by the Wilcoxon rank sum test, Wilcoxon and Wilcox (1964).

	Ocoee Gorge	Cedar Cliff Mountain
Plants (*n*)	13	6
Diameter of plants (cm.)	4.1	29
Primary rosettes (*n*)	6.6	63
—diam. (cm.)	1	2.2
Leaves of rosettes—l. (mm.)	6.6	12.7
—w. (mm.)	2.5	3.6
—th. (mm.)	1	1.4
Fruits—l. (mm.)	2	3.5
Seeds—l. (mm.)	.68	.83
—diam. (mm.)	.28	.4
Chromosomes—sporophytic (*n*)	12	44

In addition, the plants of the *Sedum glaucophyllum* differ in having the leaves of the primary rosettes glaucous, ascending, and oblanceolate, rather than green, spreading, and elliptical.

Also on the basis of rank tests applied to samples of wild plants, the population of *Sedum glaucophyllum* at Cascades, Va., differs from *S. nevii* in the Ocoee Gorge in several features. These differences make distinction possible despite the fact that the leaves of plants at Cascades are green and those of the floriferous stems are even narrower than similar ones of *S. nevii*.

	Ocoee Gorge	Cascades
Plants (*n*)	13	3
Diameter of plants (cm.)*	4.1	19
Floriferous stems—median lvs.—w. (mm.)**	1.6	1
Flowers—diam. (mm.)**	7.7	12
Petals—l. (mm.)**	4.1	6.3
Filaments—l. (mm.)**	2.6	4
Anthers—l. (mm.)**	.9	1.1
Chromosomes—sporophytic (*n*)	12	28

Choice of Allisonia, Va., as a place for study of *Sedum glaucophyllum* was because prior information suggested that plants from there were small and resembled *S. nevii*. Investigation revealed that plants from there have fewer flowers per cyme and fewer ovules per ovary than do plants of *S. nevii*, but in other features they are larger. Significant differences occur in at least ten features, as indicated below. Data are for samples of wild plants.

	Ocoee Gorge	Allisonia
Plants (*n*)	13	11
Diameter of plants (cm.)**	4.1	17
Primary rosettes (*n*)**	6.6	21
Leaves of rosettes—w. (mm.)**	2.5	3.5
Floriferous stems—median lvs.—w./th.**	1.4	1.98
Flowers per cyme (*n*)**	15.6	8.9
Flowers—diam. (mm.)**	7.7	11.1
Petals—l. (mm.)**	4.1	6
Filaments—l. (mm.)**	2.6	3.7
Ovules per ovary (*n*)**	9.5	6.2
Seeds—l. (mm.)*	.68	.82
Chromosomes—sporophytic (*n*)	12	28

Populations of *Sedum glaucophyllum* with broad, glaucous leaves present no problems in identification. The supposed troublesome cases are the ones reviewed above. As shown, distinctions are clear in each case. Relationships are contrary to conventional prejudices about species. Once these are overcome, the situation becomes reasonable.

About 1949, Robert Lee made some crosses involving *Sedum nevii*. He produced one plant which may have been a hybrid with *S. glaucophyllum*. This had light green leaves. Otherwise, it differed from *S. nevii* in broader leaves of the floriferous stems, larger flowers, longer petals, and longer stamens. The plant had the general aspect of *S. glaucophyllum*. Both details of the cross and cytology are unclear. After flowering in June 1950, the plant died. If facts are correct, then the genomes of *S. nevii* and *S. glaucophyllum* are sufficiently similar to work together in a hybrid.

References. Baldwin, J. T., Jr. (1942a and 1944) and Clausen, R. T. (1949b).

4. Sedum glaucophyllum*** (figs. 14–17)

Sedum glaucophyllum (fig. 15) is distinguished from other North American species, except *S. nevii* and *S. cockerellii*, by its widely spreading white petals, completely separate from each other and usually in whorls of four; broad white nectaries, wider than long; only slightly epipetalous stamens with flattened filaments; and oblanceolate leaves in prominent rosettes. Its separation from *S. nevii* is difficult, but possible, especially if plants are grown in similar conditions. The leaves of the rosettes of *S. glaucophyllum* are wider, sometimes ascending rather than spreading. In addition, the median leaves of the floriferous stems tend to be two or more times wider than thick, rather than narrower, and the sepals tend to be shorter. In cultivation at Ithaca, N.Y., cymes of *S. glaucophyllum* often have more than three cincinni. The number of chromosomes is 6 pairs in *S. nevii* and either 14 pairs or some number in the twenties in *S. glaucophyllum*. Flowers of *S. cockerellii* generally are 5-merous, with the petals erect basally, then rotately spreading, and the epipetalous stamens are adnate for 1 millimeter or more. The carpels are erect in fruit, not spreading as in *S. glaucophyllum* and related eastern North American species. *Sedum ternatum*, with obovate leaves in whorls of three, and *S. pulchellum*, usually biennial, with the leaves of the floriferous stems linear and sagittate-spurred, and with pinkish white petals, are readily separable.

Description. Sample: 59 plants. Sources and numbers of plants from each are: Ridge and Valley Province, in the drainage of the New River in Virginia: Allisonia—15, Bald Knob—5, Cascades—9, Klotz—5 (1*), and Pembroke—3; and in other drainages: Rawlings, Md.—1; Clifton Forge, Va.—1*; North Mountain, Collierstown, Va.—1*; Crows, Va.—1*; Deerfield, Va.—1*; Front Royal, Va.—1*; Millboro Spring, Va.—1*; Shenandoah Mountain, Va.—1, Warm Springs, Va.—1; and Briery Branch Gap, Sugar Grove, W. Va.—1*; Blue Ridge, Cedar Cliff Mountain, N.C.—6; and in Virginia along the Blue Ridge Parkway: Rock Point—2; Rocky Knob Spring—1*; and Tye River Gap—1*; and Piedmont Plateau: Leaksville–Spray, N.C.—1; and Goode's Ferry, southwest of South Hill, Va.—1. The 10 plants marked with asterisks were collected by C. H. Uhl. Detailed information about localities in the drainage of the New River is available in the section on the Ridge and Valley Province in the chapter on geography.

Data in the description permit various comparisons. Information for plants from Cascades is separate because that is the supposed type locality. The population there serves as a standard for interpretation of the species. Data also are separate for the population at Bald Knob, Mountain Lake, Va., where plants have green leaves and grow on sandstone; the populations at Klotz and Pembroke, called in the tables New River–limestone; Allisonia; and Cedar Cliff Mountain, which is a peripheral locality. Leaksville is the designation of a plant from Leaksville–Spray, N.C. D—gh. refers to plants with 14 pairs of chromosomes, which flowered in the greenhouse at Ithaca, N.Y., from Goode's Ferry, Klotz, and the South Peak of Otter. P—gh. refers to plants with 22 or more pairs of chromosomes from Crows, Rawlings, and Warm Springs. A—gh. applies to a plant in the greenhouse, from along the Cowpasture

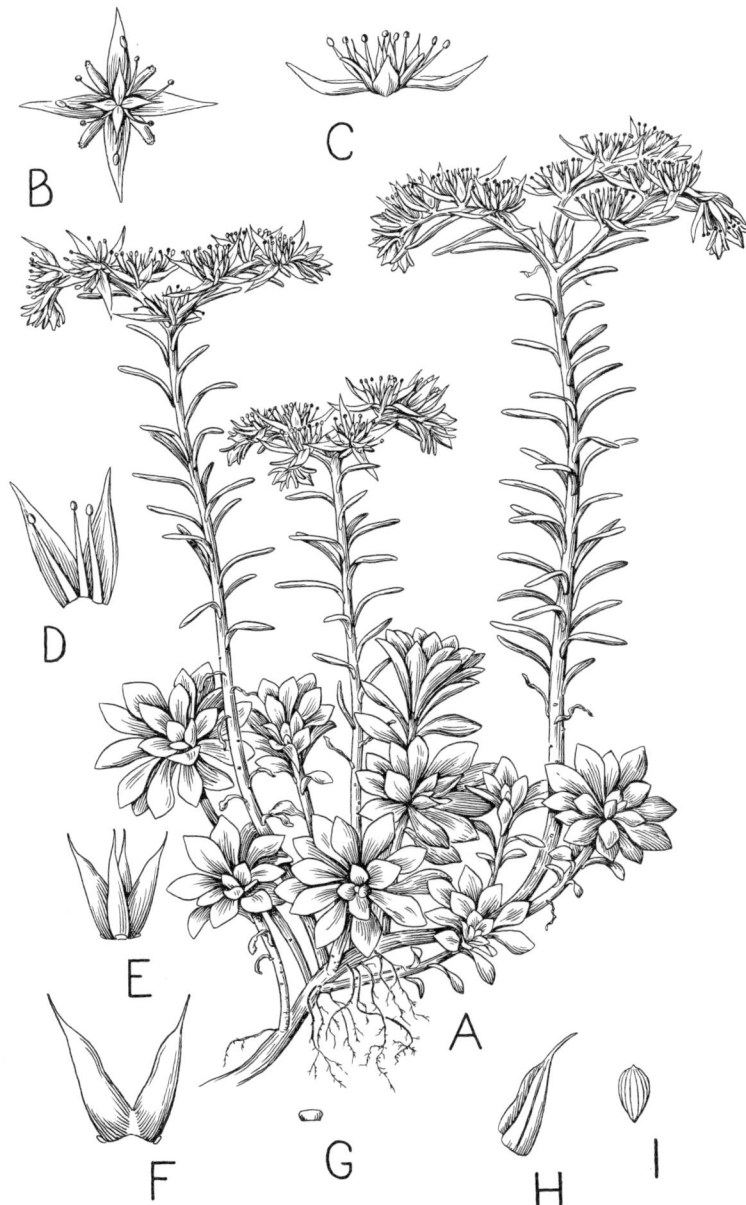

Fig. 15. Plant of *Sedum glaucophyllum* from the Cascades of Little Stony Creek, Giles Co., Va., cultivated in cold frame at Ithaca, N.Y. A. Habit sketch (x 1). B. Flower from above (x 2). C. Flower from side (x 2). D. Two petals and stamens (x 3). E. Carpels (x 3). F. Two carpels (x 4). G. Nectary (x 5). H. Follicle (x 4). I. Seed (x 10).

River northeast of Clifton Forge, which may be an autotetraploid. C—cf. applies to plants with compact rosettes, glaucous leaves, and, as far as known, 14 pairs of chromosomes, in an experiment in a cold frame at Ithaca, N.Y. L—cf. applies to plants in this same experiment with loose rosettes, green leaves, and also, as far as known, 14 pairs of chromosomes. P—cf. likewise applies to plants in the cold frame. These plants have higher numbers of chromosomes, 22, 23, or more pairs. The experiment was of completely random design. Plants included were of nearly the same age, having been started as cuttings of the same size, namely rosettes 2 cm. in diameter with stems 2 cm. long, all made within a period of ten days in June 1966. Measurements are for the growing season of 1967. Since some plants did not survive in the experiment and others grew too slowly, only forty plants yielded data in 1967. Of these, 22 plants produced flowers, 16 fruits, and 14 seeds. Although the resulting data are not ideal, they provide an objective basis for appraising variation. Outdoors, at Ithaca, N.Y., plants of *S. glaucophyllum* do poorly. Greatest success in culture was in the cold frames and greenhouse.

Stems of *Sedum glaucophyllum* are repent and much branched through the development of axillary shoots which take root and are an important means of propagation. As the old parts die, a clone becomes subdivided, but may endure for many years. The plants are perennial and cespitose, with fibrous, pale brown roots. The secondary shoots are erect, with the lower leaves withered and the upper leaves crowded and spreading, forming rosettes. The floriferous stems are annual. These develop terminally from the rosettes and die after flowering. Usually they are erect from decumbent bases. Often they are pink. Differences in the tables below among C, L, and P in diameter of rosettes, and among populations in number of floriferous stems, are significant.

	n-pl.	n-obs.	$\bar{x}$	s
Plants—diameter (cm.)				
Cascades	3	3	19	19
Bald Knob	2	2	38	16
New River–limestone	4	4	7	2
Allisonia	11	11	17	12
Cedar Cliff Mt.	6	6	29	16
C—cf.	23**	43	10.1	3.9
L—cf.	10	19	13.8	7.8
P—cf.	5	7	10.6	—
Primary rosettes (n)				
Cascades	3	3	85	112
Bald Knob	2	2	43	3
New River–limestone	4	4	10	5
Allisonia	11	11	21	33
Cedar Cliff Mt.	6	6	63	39
C—cf.	23	43	21.5	8.9
L—cf.	10	19	21.1	6.4
P—cf.	5	7	5	—
Primary rosettes—diameter (mm.)				
Cascades	3	12	12	10
Bald Knob	2	10	14	2

	n-pl.	n-obs.	x̄	s
Primary rosettes—diameter (mm.)				
New River–limestone	4	18	9	2
Allisonia	11	55	9	1
Cedar Cliff Mt.	6	29	22	4.7
C—cf.	23	221	19	3
L—cf.	10	86	27	7.4
P—cf.	5	29	16	—
Floriferous stems per plant (n)				
Cascades	3	3	6	5.7
Bald Knob	1	1	5	—
New River–limestone	3	3	3	1.5
Allisonia	11	11	2	1.7
Cedar Cliff Mt.	6	6	2	1.4
C—cf.	12	16	4.8	2.9
L—cf.	6	7	8.5	6.9
P—cf.	2	2	2	—
Floriferous stems—length (cm.)				
Cascades	3	10	6.7	4.1
Bald Knob	1	5	5	—
New River–limestone	3	8	7.5	2.6
Allisonia	11	23	5.2	1.2
Leaksville	1	2	5	—
Cedar Cliff Mt.	4	12	7.4	4.2
D—gh.	2	2	17.5	3.5
P—gh.	2	2	10.1	1.2
C—cf.	12*	30	8.1	3.1
L—cf.	6	16	8.8	3.6
P—cf.	2	3	3.7	—

Leaves are spirally arranged, both in the rosettes and on the floriferous stems. They are oblanceolate to spatulate, narrowed to petiolelike bases, broader in the rosettes and narrower on the floriferous stems. Apically they are obtuse and papillose. They are either pale green or blue-green. Rarely, both conditions occur in the same population. Sometimes the leaves of the rosettes are ascending. The leaves of the floriferous stems are short-spurred and divergent, becoming recurved. Usually they are broader on the lower part of the stem and linear upward. Differences among C, L, and P in length of leaves of rosettes, and among populations in thickness of rosulate leaves, according to the experiment in the cold frame, are significant.

	n-pl.	n-obs.	x̄ (mm.)	s (mm.)
Leaves of rosettes—length				
Cascades	3	12	9.3	4.9
Bald Knob	2	10	12	3.9
New River–limestone	4	18	7	1.9
Allisonia	11	55	6.9	1.3
Leaksville	1	2	6.5	—
Cedar Cliff Mt.	6	29	13	2.1
D—gh.	2	4	9.5	2.8
P—gh.	2	4	9.2	1.7
C—cf.	23	221	13.2	2.6
L—cf.	10	86	17.8	5.2
P—cf.	5	29	9.7	—

	n-pl.	n-obs.	x̄ (mm.)	s (mm.)
Leaves of rosettes—width				
Cascades	3	12	2.2	.8
Bald Knob	2	10	4.1	.5
New River–limestone	4	18	2.8	.4
Allisonia	11	55	3.5	.7
Leaksville	1	2	4.5	—
Cedar Cliff Mt.	6	29	3.6	.7
D—gh.	2	4	2.8	.9
P—gh.	2	4	2.2	.5
C—cf.	23	221	4.2	1.1
L—cf.	10	86	4.2	.6
P—cf.	5	29	4.3	—
Leaves of rosettes—thickness				
Cascades	3	12	.8	.2
Bald Knob	2	10	.9	.1
New River–limestone	4	18	.6	.2
Allisonia	11	55	.8	.1
Leaksville	1	1	1	—
Cedar Cliff Mt.	6	29	1.4	.2
D—gh.	2	3	.9	.4
P—gh.	2	3	.7	.2
C—cf.	23	220	1	.1
L—cf.	10	86	1	.1
P—cf.	5	29	.8	—
Leaves of floriferous stems—length				
Cascades	3	10	7.7	2.2
Bald Knob	1	2	7.5	—
New River–limestone	3	6	8.1	.8
Allisonia	11	23	7.8	1.1
Leaksville	1	2	9	—
D—gh.	3	13	15.8	5.7
P—gh.	3	14	11.7	2
A—gh.	1	9	8.8	—
C—cf.	12	34	16.6	3.3
L—cf.	6	16	17.8	4.7
P—cf.	1	1	6	—
Cedar Cliff Mt.—cf.	1	5	11	—
Leaves of floriferous stems—width				
Cascades	3	10	1	.1
Bald Knob	1	2	1.3	—
New River–limestone	3	6	1.6	.5
Allisonia	11	23	1.7	.2
Leaksville	1	3	2	—
D—gh.	3	13	2.6	.4
P—gh.	3	14	2.4	.4
A—gh.	1	9	2.9	—
C—cf.	12	34	3.6	.9
L—cf.	6	16	3.4	.8
P—cf.	1	1	2.2	—
Cedar Cliff Mt.—cf.	1	5	2.5	—
Leaves of floriferous stems—thickness				
Cascades	3	10	.8	.1
Bald Knob	1	2	.6	—
New River–limestone	3	6	.7	.4
Allisonia	11	23	.9	.1
Leaksville	1	1	1	—

	n-pl.	n-obs.	$\bar{x}$ (mm.)	s (mm.)
Leaves of floriferous stems—thickness				
D—gh.	3	12	.9	.1
P—gh.	3	13	1.1	.3
A—gh.	1	9	.5	—
C—cf.	12	33	1.3	.1
L—cf.	6	16	1.1	.3
P—cf.	1	1	.6	—
Cedar Cliff Mt.—cf.	1	5	1.2	—
Ratio, width:thickness				
Cascades	3	10	1.3	.1
Bald Knob	1	2	2.5	—
New River–limestone	3	6	2.7	.9
Allisonia	11	23	2	.3
Leaksville	1	1	2	—
D—gh.	1	9	3	—
P—gh.	1	10	3.5	—
A—gh.	1	9	6.1	—
C—cf.	12	33	2.9	.7
L—cf.	6	16	3.2	.8
P—cf.	1	1	3.7	—
Cedar Cliff Mt.—cf.	1	5	2.1	—

Inflorescences are cymes of cincinni, with the floral bracts similar to the upper leaves of the floriferous stems, but smaller.

	n-pl.	n-obs.	$\bar{x}$ (n)	s (n)
Cincinni				
Cascades	3	10	2.9	.5
Bald Knob	3	13	2.8	.6
New River–limestone	4	13	2.7	.6
Allisonia	14	28	2.4	.4
D—gh.	2	2	3.5	.2
P—gh.	2	2	3	0
C—cf.	12	35	3.7	.8
L—cf.	6	16	3.9	1.3
P—cf.	2	3	3.2	—
Cedar Cliff Mt.—cf.	1	5	2.6	—
Flowers per cyme				
Cascades	3	10	15	8.8
Bald Knob	1	5	10	—
New River–limestone	3	8	11	5.3
Allisonia	11	23	9	3.6
Leaksville	1	2	22	—
D—gh.	2	3	22	3.8
P—gh.	3	7	20	6.1
A—gh.	1	9	11	—
C—cf.	12	35	23	10.7
L—cf.	6	16	27	22.3
P—cf.	2	3	17	—
Cedar Cliff Mt.—cf.	1	5	54	—

Flowers are sessile. Central flowers in the cymes often have more parts than others in the cincinni.

	n-pl.	n-obs.	x̄ (mm.)	s (mm.)
Diameter				
Cascades	3	9	12	2.3
Bald Knob	1	3	12	—
New River–limestone	3	7	9	4.1
Allisonia	11	34	11	1.3
D—gh.	1	1	14	—
C—cf.	11	36	13.4	2.9
L—cf.	6	20	12.8	3.6
P—cf.	2	4	10	—
Cedar Cliff Mt.—cf.	1	5	12.4	—

Sepals are divergent, unequal, linear-lanceolate, subterete, obtuse and papillose at apex, and green.

	n-pl.	n-obs.	x̄	s
Number				
Cascades	3	9	4.3	0
Bald Knob	1	3	4	—
New River–limestone	3	7	4.1	.2
Allisonia	11	34	4.1	.3
Leaksville	1	1	4	—
D—gh.	2	2	4	0
P—gh.	2	2	4	0
C—cf.	11	36	4	.3
L—cf.	6	20	4.2	.2
P—cf.	2	4	4	—
Cedar Cliff Mt.—cf.	1	5	4	—
Length (mm.)				
Cascades	3	9	4.2	1.4
Bald Knob	1	3	4.6	—
New River–limestone	3	7	4	.9
Allisonia	11	34	3.6	.4
Leaksville	1	2	5	—
D—gh.	3	13	4.2	.4
P—gh.	3	13	4.2	1
A—gh.	1	9	3.7	—
C—cf.	11	36	5.5	1.3
L—cf.	6	20	5.1	.7
P—cf.	2	4	5.9	—
Cedar Cliff Mt.—cf.	1	5	6.3	—
Width (mm.)				
Cascades	3	9	1	.2
Bald Knob	1	3	1	—
New River–limestone	3	7	1	.1
Allisonia	11	34	1	.1
Leaksville	1	1	1	—
D—gh.	3	12	.9	.2
P—gh.	3	13	1	.2
A—gh.	1	9	.9	—
C—cf.	11	36	1.5	.2
L—cf.	6	20	1.2	.2
P—cf.	2	4	1.3	—
Cedar Cliff Mt.—cf.	1	5	1.6	—

Petals are widely spreading, lanceolate, acuminate, minutely hooded, and white.

	n-pl.	n-obs.	$\bar{x}$	s
Number				
Cascades	3	9	4.3	0
Bald Knob	1	3	4	—
New River–limestone	3	7	4.1	.2
Allisonia	11	34	4.1	.3
Leaksville	1	1	4	—
D—gh.	3	11	4	.1
P—gh.	3	12	4	0
A—gh.	1	9	4	—
C—cf.	11	36	4	.3
L—cf.	6	20	4.2	.1
P—cf.	2	4	4	—
Cedar Cliff Mt.—cf.	1	5	4	—
Length (mm.)				
Cascades	3	9	6.3	1.3
Bald Knob	1	3	6.7	—
New River–limestone	3	7	5.1	1.5
Allisonia	11	34	6	.5
Leaksville	1	2	5	—
D—gh.	3	13	5.8	.3
P—gh.	3	12	6.1	2
A—gh.	1	9	6.7	—
C—cf.	11	36	7.6	1.2
L—cf.	6	20	7.2	1.7
P—cf.	2	4	4.8	—
Cedar Cliff Mt.—cf.	1	5	7	—
Width (mm.)				
Cascades	3	9	2	.5
Bald Knob	1	3	1.8	—
New River–limestone	3	7	1.4	.2
Allisonia	11	34	1.7	.1
Leaksville	1	2	1.1	—
D—gh.	3	12	1.2	.2
P—gh.	3	14	1.3	.4
A—gh.	1	9	1.5	—
C—cf.	11	36	2.3	.3
L—cf.	6	20	2.1	.4
P—cf.	2	4	1.5	—
Cedar Cliff Mt.—cf.	1	5	1.2	—

Stamens have white filaments which are flattened below and broad basally. Adnation of the epipetalous filaments is slight. The anthers are basifixed, oblong, and dark red. Anthers on filaments alternate with the petals shed pollen before or as the petals spread apart, but those on the epipetalous filaments may dehisce more than a half day later.

	n-pl.	n-obs.	$\bar{x}$ (mm.)	s (mm.)
Filaments—length				
Cascades	3	9	4	.8
Bald Knob	3	11	4.8	1
New River–limestone	3	7	3.1	.9

	n-pl.	n-obs.	x̄ (mm.)	s (mm.)
Filaments—length				
Allisonia	11	34	3.7	.4
Leaksville	1	2	4	—
D—gh.	2	4	4.6	.1
P—gh.	2	2	4.5	.7
C—cf.	11	36	4.3	1
L—cf.	6	20	4.6	1
P—cf.	2	4	2.9	—
Cedar Cliff Mt.—cf.	1	5	4.5	—
Epipetalous filaments—height on petals				
Cascades	3	9	0	0
Bald Knob	1	3	0	—
New River–limestone	3	7	0	0
Allisonia	11	33	.01	.001
C—cf.	11	35	.1	.02
L—cf.	6	20	.1	.02
P—cf.	2	4	.1	—
Cedar Cliff Mt.—cf.	1	5	.1	—
Anthers—length				
Cascades	3	4	1.1	.1
Bald Knob	1	1	1.4	—
New River–limestone	3	4	1	.2
Allisonia	9	18	1	.1
D—gh.	1	1	1.1	—
P—gh.	1	3	.9	—
A—gh.	1	7	1.3	—
C—cf.	10	16	1.3	.2
L—cf.	6	10	1.3	.2
P—cf.	2	3	1.4	—
Cedar Cliff Mt.—cf.	1	5	1.2	—

Nectaries are subquadrate, truncate, and white.

	n-pl.	n-obs.	x̄ (mm.)	s (mm.)
Length				
Cascades	3	9	.3	.1
Bald Knob	1	3	.5	—
New River–limestone	3	7	.3	.1
Allisonia	11	34	.3	.04
Leaksville	1	1	.3	—
D—gh.	3	11	.4	.1
P—gh.	3	12	.4	.1
A—gh.	1	9	.2	—
C—cf.	11	36	.4	.04
L—cf.	6	20	.4	.03
P—cf.	2	4	.4	—
Cedar Cliff Mt.—cf.	1	5	.4	—
Width				
Cascades	3	9	.5	.2
Bald Knob	1	3	.6	—
New River–limestone	3	7	.4	.1
Allisonia	11	34	.5	.1
Leaksville	1	1	.5	—
D—gh.	3	11	.4	.1
P—gh.	3	12	.5	.2

NATIVE SPECIES

	n-pl.	n-obs.	x̄ (mm.)	s (mm.)
Width				
A—gh.	1	9	.4	—
C—cf.	11	36	.8	.2
L—cf.	6	20	.7	.2
P—cf.	2	4	.3	—
Cedar Cliff Mt.—cf.	1	5	.7	—

Carpels are erect, slightly connate basally, with slender styles, and white or sometimes pinkish ventrally or on styles. In age, they become divergent. Differences among C, L, and P in number of ovules per ovary are significant.

	n-pl.	n-obs.	x̄	s
Number				
Cascades	3	9	4.4	.2
Bald Knob	3	11	3.8	.2
New River–limestone	3	7	4.1	.2
Allisonia	11	34	4	.2
Leaksville	1	1	4	—
D—gh.	2	2	4	0
P—gh.	2	2	4	0
C—cf.	11	36	4	.2
L—cf.	6	20	4	.2
P—cf.	2	4	4	—
Cedar Cliff Mt.—cf.	1	5	5	—
Length (mm.)				
Cascades	3	9	4.4	1.3
Bald Knob	3	11	4.8	1.4
New River–limestone	3	7	3.4	1.6
Allisonia	11	34	3.5	.4
Leaksville	1	2	4	—
D—gh.	2	3	4.4	1.2
P—gh.	2	2	4.9	.7
C—cf.	11	36	4.7	.8
L—cf.	6	20	5.6	1.5
P—cf.	2	4	3	—
Cedar Cliff Mt.—cf.	1	5	4.5	—
Styles—length (mm.)				
Cascades	3	9	1.3	.4
Bald Knob	1	3	1.3	—
New River–limestone	3	7	1.1	.4
Allisonia	11	34	.9	.2
C—cf.	11	36	1.2	.1
L—cf.	6	20	1.7	.4
P—cf.	2	4	1	—
Cedar Cliff Mt.—cf.	1	5	.8	—
Ovules per ovary (n)				
Cascades	3	9	9	3.3
Bald Knob	1	3	12	—
New River–limestone	3	7	8	6.5
Allisonia	11	34	6	1.7
C—cf.	11	36	14	2.6
L—cf.	6	20	15	4.9
P—cf.	2	4	8	—
Cedar Cliff Mt.—cf.	1	5	11	—

132 SEDUM OF NORTH AMERICA

Fruits are widely divergent, gibbous ventrally, and brown.

	n-pl.	n-obs.	x̄ (mm.)	s (mm.)
Length				
Cascades	4	12	2.9	.6
Allisonia	1	3	2.9	—
Cedar Cliff Mt.	6	30	3.5	.6
D—gh.	1	2	3.7	—
P—gh.	1	2	3.5	—
C—cf.	10	26	3.7	.4
L—cf.	5	15	3.4	.4

Seeds are pyriform and yellow-brown, with irregular, dark brown, longitudinal ridges.

	n-pl.	n-obs.	x̄ (mm.)	s (mm.)
Length				
Cascades	4	10	.8	.08
Allisonia	1	3	.8	—
Cedar Cliff Mt.	5	25	.8	.05
D—gh.	1	1	.9	—
P—gh.	1	1	.5	—
C—cf.	9	23	.8	.1
L—cf.	4	7	.8	.1
Diameter				
Cascades	4	10	.34	.05
Allisonia	1	3	.4	—
Cedar Cliff Mt.	5	25	.4	0
D—gh.	1	1	.6	—
P—gh.	1	1	.2	—
C—cf.	9	23	.4	0
L—cf.	4	7	.4	.07

Chromosomes are of unequal size. In plants with 14 chromosomes, about 4 chromosomes of the genome are larger than the others, and one of these is much larger, with median centromere. Uhl (1970) published an extensive list of plants which he studied and also presented evidence for hybridization between plants with 14 and 22 pairs of chromosomes. Citations below are of plants which are the basis for the present description and also the discussion of variation.

	n-pl.	g	sp	Cytologist
Goode's Ferry, Va.	1	14	—	C. H. Uhl (1970)
Leaksville, N.C.	1	14	—	C. H. Uhl (1970)
Rock Point, Va.	2	—	44–49?	D. Niimoto, 1967
South Peak of Otter, Va.	1	14	—	C. H. Uhl (1970)
Tye River Gap, Va.	1	22	—	C. H. Uhl (1970)
Tye River Gap, Va.	1	—	~46	D. Niimoto, 1967
Cedar Cliff Mt., N.C.	2	—	44	I. Zandstra, 1969
Rawlings, Md.	1	22	—	C. H. Uhl (1970)
Crows, Va.	1	22	—	C. H. Uhl (1970)
Clifton Forge, Va.	1	28	—	C. H. Uhl (1970)
Warm Springs, Va.	1	22	—	C. H. Uhl (1970)

NATIVE SPECIES 133

	n-pl.	g	sp	Cytologist
Cascades, Va.	3	14	28	D. Niimoto, 1967
Bald Knob, Va.	3	14	—	D. Niimoto, 1967
Klotz, Va.	1	14	—	C. H. Uhl (1970)
Klotz, Va.	1	14	—	C. Horn, 1966
Pembroke, Va.	1	—	28	D. Niimoto, 1967
Allisonia, Va.	2	14	28	D. Niimoto, 1967
Allisonia, Va.	1	14	—	C. Horn, 1966
Boyce, Roanoke, and Mountain Lake, Va.	3	14	28	J. T. Baldwin, Jr. (1942)

Variation. Three principal races of *Sedum glaucophyllum* occur. The first race has loose rosettes of long, green leaves. It grows on sandstone and gneiss, often at elevations above 1,000 meters. The populations from Cascades, Bald Knob, and L—cf., cited in the preceding description, are of this sort. The second race has compact rosettes of glaucous leaves. It commonly grows on limestone of cliffs along rivers. The populations cited in the preceding description as New River–limestone, Allisonia (fig. 16), Leaksville, and C—cf. are of this race. The third race (fig. 17) is similar to the first, but the plants are smaller, with shorter petals, and of poorer growth in the experiment in the cold frame. In the preceding description, such plants are designated

Fig. 16. Plant of *Sedum glaucophyllum* from Allisonia, Va., cultivated in cold frame at Ithaca, N.Y., June 6, 1967 (x .4).

Fig. 17. Plants of *Sedum glaucophyllum* from Rawlings, Md., g = 22 (left), and Shenandoah Mountain, Va., g = 14 (right), cultivated in greenhouse at Ithaca, N.Y., June 3, 1947 (x .5).

as P—gh. and P—cf. Although readily distinguishable in an experiment, they may be confusing in nature.

The leaves of *Sedum glaucophyllum* vary in size, shape, and the degree to which wax is deposited on the epidermis. Not only do populations differ from each other, but plants sometimes differ in the same local population. For example, at Rock Point in the Blue Ridge, plants with green leaves and others with glaucous leaves occur side by side. In the experiment in the cold frame at Ithaca, leaves of rosettes which were glaucous tended to be directed upward, but those which were green spread widely. Leaves of rosettes vary from spatulate to linear-oblanceolate. In the experiment in the cold frame, the greatest differences in width were between plants from Bald Knob and Cascades, all with green leaves. Rosulate leaves of a plant in the greenhouse, from Goode's Ferry, developed peculiar nipplelike lobes. The cause of this unique condition is unknown.

Before experimental study, my impression was that plants with green rosettes, from sandstone above 900 meters in the drainage of the New River, had the median leaves of the floriferous stems narrower than those of plants with glaucous rosettes from cliffs of limestone in the main valley of the river. Data from the experiment in the cold frame revealed that the difference in this feature is both slight and not significant.

Floriferous stems usually are erect, but in certain plants they are prostrate. Changes noted between the orientation of floriferous stems in nature and in cultivation suggest that this feature is modified by the environment. For example, a plant with erect floriferous stems at Klotz had prostrate stems in the cold frame at Ithaca. Although such differences greatly alter the aspect of plants, they are unreliable for purposes of classification.

Fluctuations in size of flowers appear to be due to environmental differences, the vigor of the plants, and the position of a flower in the inflorescence. They are of no consequence in separating populations or groups of populations.

Flowers with different numbers of petals commonly occur on the same plant, those with higher numbers being the first to expand. A study of the frequency of different numbers of petals in 52 flowers in random samples from five populations in the drainage of the New River indicates 4 petals per flower as the commonest condition.

Petals per flower (n)	Flowers (n)	%	Cumulative %
3	1	1.9	1.9
4	42	80.8	82.7
5	9	17.3	100.0

On the basis of studies of wild plants in the drainage of the New River, filaments of plants with green leaves are longer than those with glaucous leaves, but this difference disappears when the plants are grown under similar conditions.

No unusual variations of carpels, fruits, or seeds have come to attention, except that polyploid plants have fewer ovules per ovary.

Plants with the morphological characteristics of *Sedum glaucophyllum* have different numbers of chromosomes. Fourteen seems to be the basic diploid number of chromosomes, but many plants occur with numbers in the twenties. Chromosomes of plants with the higher numbers are difficult to count. Differences among plants with different numbers of chromosomes are not marked. The impression is that plants with different genetic constitutions have been involved in independent origins of populations with more than twenty pairs of chromosomes. The strongest morphological and physiological differences are between arrays of populations with 14 pairs of chromosomes. In the drainage of the New River, plants on sandstone at the higher altitudes are larger, with longer, greener leaves than those on limestone on cliffs along the river. When grown side by side in an experiment, the plants retain their distinctive characteristics. The differences among populations with 14 pairs of chromosomes suggest a first step in the formation of ecological races. They are on the borderline for recognition as taxonomic subspecies. Plants with higher numbers of chromosomes, whether autopolyploid or allopolyploid, may have the features of either ecological race, depending on origin. This condition complicates the problem of classification and understanding.

Although *Sedum glaucophyllum* varies in many features, most of the variation results from environmental modification. The experiment of completely random design in the cold frame in Ithaca provided the best basis for appraising this variation.

The experiment included 40 clones of *S. glaucophyllum*, 3 clones of *S. nevii*, and 7 clones of *S. ternatum*, each represented by two replicates. Of the clones of *S. glaucophyllum*, 12 were of the race with green leaves, mentioned above; 25 were of the race with glaucous leaves; and 3 were polyploid, with 22 or more pairs of chromosomes. These polyploid plants grew poorly and were small. Possibly they are physiologically unbalanced. In 1967, twenty-two replicates of *S. glaucophyllum* flowered in the experiment. An analysis of variance for each of thirty-three features revealed significant differences among races in three features, among populations in two features, and among plants within populations in three features. The features in which significant differences exist among races are indicated in the following table. Column D shows the significant difference at the 5% level of confidence, as worked out by Tukey. It permits distinguishing which differences among means are significant. For further explanation, see Snedecor (1956:251–253).

Feature	Green race $\bar{x}$	Glaucous race $\bar{x}$	Polyploid race $\bar{x}$	D
Primary rosettes—diameter (mm.)	27	19	16	7.5
Leaves of primary rosettes—l. (mm.)	17.8	13.2	9.7	5.4
Ovules per ovary (*n*)	14.8	13.7	7.7	6.9

Differences in thickness of leaves and number of floriferous stems per plant are significant among populations, but other differences among populations, as width of leaves of rosettes, width and ratio of width to thickness of median leaves of floriferous stems, and length of filaments, noted as significant from study of wild plants in the drainage of the New River, are not confirmed by experiment.

Plants within populations differed significantly in diameter, width of leaves of primary rosettes, and length of floriferous stems.

The following table illustrates the method of analysis in reaching conclusions. This shows the analysis of variance for width of leaves of primary rosettes. Data are for *Sedum glaucophyllum* and *S. nevii* in the experiment in the cold frame. Numbers in parentheses are amounts before partitioning and include all figures above them.

Source of variation	Degrees of freedom	Sum of squares	Mean square	F
Between species	1	1.38	1.38	30.67*
Among races	2	.09	.045	.05 n.s.
		(1.47)		
Among populations	11	10.46	.9509	.40 n.s.
		(11.93)		
Among plants	25	59.64	2.38	5.23**
	(39)	(71.57)		
Within plants	32	14.57	.4553	
Totals	71	86.14		

Because numbers of replicates per clone yielding data were unequal, an adjustment of variances was necessary. This was done for both C—cf. and L—cf. to obtain the values of *s* cited in the description, following the method of Snedecor (1956:269).

Nomenclature. *Sedum glaucophyllum* R. T. Clausen, Cact. Succ. Jour. 18:60–61 (1946). Type locality: near Mountain Lake, Giles Co., Va., in the Ridge and Valley Province. Type: S. C. Dyal (CU), June 17, 1939. The exact site of collection of the type is uncertain, but another specimen in the Bailey Hortorium, collected by Miss Dyal on the same day as when she obtained the type, is from woods along the trail near the Cascades of Little Stony Creek, Giles Co., Va. If Miss Dyal made only one collection of *S. glaucophyllum* on June 17, 1939, the population along Little Stony Creek is the source. This locality is near Mountain Lake, only 3.8 kilometers to the west. Access is from there. On the natural routes to the Cascades from the Mountain Lake Biological Station, whether by road via Mountain Lake and Pacers Gap, or by the road down the valley of Pond Drain, the only occurrence of *S. glaucophyllum* known to me is on the bank of the road between the little and big Cascades. I now suggest that this population be regarded as the basis for interpretation of the species. The site is near the eastern base of Butt Mountain, on the western bank of the road between the little and big Cascades, just south of a point where the road rounds a rocky ridge and 450 meters north of the place where the trail curves downward to the big Cascades. The elevation is 940 meters. The populations of *S. glaucophyllum* nearest to Mountain Lake are on the slopes of Bald Knob, but a person going from Mountain Lake to the Cascades would not pass these. Fortunately, the plants near the Cascades and on Bald Knob are similar, although not the same.

The name *Sedum glaucophyllum* is inappropriate for the plants at Cascades or any other population in the highlands near Mountain Lake. They have long, green leaves. Yet, since I designated the type from there, the application of the name is fixed. Were I redoing the past, I would make the detailed study first and then, if I used the name *glaucophyllum*, designate as type a plant from a population with compact rosettes of glaucous leaves. As matters stand, that sort of plant lacks a distinctive name if a separation is made of the two diploid races. Figure 40, (*Cact. and Succ. Jour.* 18:60 [1946]), accompanying the original description of *S. glaucophyllum*, illustrates a cultivated plant of unknown origin in the wild, which is polyploid.

Although no names are available which are sure synonyms of *Sedum glaucophyllum*, the nomenclatural history of the species is intimately associated with that of *S. nevii*. Asa Gray, after describing *S. nevii* on the basis of a collection from near Tuscaloosa, Ala., included in his concept specimens sent by W. M. Canby and A. H. Curtiss from Virginia. This usage of Gray (1868) prevailed for about eighty years and continues to some extent until the present time. Because plants of *S. glaucophyllum* are commoner than those of *S. nevii* in both herbaria and gardens, the name *S. nevii* became closely associated with them. This tendency went so far that plants of the sort indentified here as *S. glaucophyllum* became the basis for interpreting *S. nevii*. In England, Masters described in 1878 a *S. beyrichianum* which he characterized as smaller than *S. nevii*, with narrower leaves, smaller flowers, and petals as long as the sepals. This could be true *S. nevii*. Otherwise it is small *S. glaucophyllum*. A specimen in the herbarium of the Missouri Botanical Garden, collected in Virginia in 1833 by Beyrich, is *S. glaucophyllum*. It does not agree with the description of *S. beyrichianum* published by

Masters. Similarly, a specimen in the herbarium at Kew, prepared by N. E. Brown in 1902 and designated by him as true *S. beyrichianum*, has petals much longer than the sepals and otherwise disagrees with the description of Masters. Neither specimen is to be regarded with certainty as the type. In the absence of a type specimen agreeing with the original description, a definite locality of origin, or a count of chromosomes, the identification of *S. beyrichianum* is doubtful. Praeger (1921) interpreted it as a variety of *S. nevii*. He distinguished it in a way which suggests that he regarded glaucous *S. glaucophyllum* to be *S. nevii* and the green-leaved race to be var. *beyrichianum*, but he based his varietal name on the specific epithet of Masters. In view of this situation, neither *S. beyrichianum* nor *S. nevii* var. *beyrichianum* can be positive synonyms of *S. glaucophyllum*. Praeger's contribution was to distinguish between the two kinds of diploid *S. glaucophyllum*.

Distribution and ecology. *Sedum glaucophyllum* (fig. 14) is endemic in the Appalachian Highlands, occurring on the Piedmont Plateau from the bank of the Neuse River four miles north of Bayleaf, Wake Co., N.C. (NCS), to the bank of the Roanoke River at Goode's Bridge, twelve miles southeast of Boydton, Mecklenburg Co., Va.; in the Blue Ridge from Cedar Cliff Mountain east of Tuckaseegee, Jackson Co., N.C. (NCS), to Harper's Ferry (G); in the Ridge and Valley Province from the drainage of the New River at Allisonia, Pulaski Co., Va., to Rawlings, Allegany Co., Md.; and on the Appalachian Plateau at Cabins, Grant Co., W. Va. (W Va).

The different races of *Sedum glaucophyllum* have distinctive distributions. Only the glaucous-leaved race, with fourteen pairs of chromosomes, occurs on the Piedmont Plateau. In the Blue Ridge, both glaucous and green-leaved races occur, and besides diploids, plants with higher numbers of chromosomes. Populations with more than 14 pairs of chromosomes occur on Cedar Cliff Mountain and from the drainage of the James River northward. In the Ridge and Valley Province, plants of the glaucous-leaved race occur on limestone in the valleys. The green-leaved race occurs on sandstone on the ridges. In the basin of the New River, only populations with 14 pairs of chromosomes are known. Northward, in the drainage of the James and Potomac Rivers, populations with higher numbers of chromosomes are also known, but other populations, especially on limestone in the Shenandoah Valley, are diploid.

Sedum glaucophyllum grows on a variety of rocks: limestone, shale, sandstone, granite, horneblende gabbro, schist, and gneiss. An idea early in the study that the distribution of *S. glaucophyllum* might be related to the occurrence of calcium in rocks stimulated interest in this subject. In 1946, Gily E. Bard, using procedures of separation and determination described by Peech (1945), determined the fraction of calcium in several samples of rock on which *S. glaucophyllum* grew. The percentages of air-dry weight of calcium were 4.025—horneblende gabbro from 3 km. north of Spray, N.C.; 8.100—granite gneiss from Peaks of Otter, Va.; 2.835—schist from 3 km. north of Fincastle, Va.; and 2.012—red oak granite from Goode's Ferry, Va. Although this calcium is not all available to plants, it can become available on weathering of the rocks.

Habitats of *Sedum glaucophyllum* are cliffs, crests of cliffs, and rocky slopes. Sites usually are shaded for half or more than half of the day.

In 1966, I studied five populations of *Sedum glaucophyllum* in the drainage of the New River. Three of these are a random selection from sixteen. The other two, Allisonia and Bald Knob, are deliberate choices and comprise an optimum stratum. Descriptions of the sampling areas containing these populations are in the chapter on geography, but data about the populations are in the following table. The population on Cedar Cliff Mountain was selected because it is the southernmost known. Study there was in 1969. Since I made complete counts of the small populations at Cascades, Pembroke, and Cedar Cliff Mountain, confidence intervals are unnecessary. The data for Klotz are not satisfactory for estimating an interval. Differences among populations in distance between nearest neighbors are significant.

Population	Sampling area	N plants, estimate	N plants, confidence interval $_{.05}$	Area of populations $(m.^2)$
Cedar Cliff Mt.	—	61	—	305
Allisonia	15–114	391	126–1,411	800
Pembroke	27–1	26	—	400
Klotz	4–48	68	—	300
Cascades	4–66	26	—	100
Bald Knob	7–93	76	9–332	1,700

Population	Altitudinal range (m.)	pH range	pH tests (n)	Depth of soil w (cm.)	Distance to nearest neighbor C.I.$_{.05}$ (cm.)
Cedar Cliff Mt.	930–935	4.8–6.8	6	1.0–3	.5–61
Allisonia	573–670	5.8–6.6	13	1.0–6	0–49
Pembroke	488–498	7.0–7.2	3	.7–4	—
Klotz	530–540	7.4–7.6	3	.4–1	0–36
Cascades	940	5.8–6.6	6	.4–2	0–15
Bald Knob	1,150–1,152	4.4–5.8	5	1.0–5	0–464

The populations at Cascades and Bald Knob, with plants with long green leaves, are at the highest elevations. The distributional data correlate with information on performance, that is morphology and physiology, and suggest racial tendencies.

In the study in the drainage of the New River, I noted the two commonest competitors of each plant of *Sedum glaucophyllum* in samples and also the species most affecting conditions of growth. The results indicate great differences in environments inhabited by the *Sedum*. Lists of both competitors and species affecting growth are miscellaneous. Competitors with the highest ratings are *Gaultheria procumbens* at Cascades, *Parthenocissus quinquefolia* at Bald Knob, *Bromus purgans* at Klotz, and *Asplenium platyneuron* at Allisonia. The more frequent species affecting conditions of growth are *Crataegus macrosperma* and *Betula lenta* at Cascades and Bald Knob, *Celtis occidentalis* at Klotz and Pembroke, and *Ostrya virginiana* at Allisonia. Study at one locality would give an erroneous idea about either competitors or species most affecting conditions of growth. The environments inhabited by *S. glaucophyllum* are diverse, as also are the other species occurring with it.

In the drainage of the New River, populations of *Sedum glaucophyllum* and *S. ternatum* commonly grow in different habitats and do not overlap. An exception is the situation at Allisonia where plants of *S. glaucophyllum* with 14 pairs of chromosomes grow within 2 centimeters of *S. ternatum* with 16 pairs of chromosomes.

Information about the distribution of *Sedum glaucophyllum* in each of the Appalachain geomorphic provinces is in the following table. Exclamation marks indicate records verified by me.

Geomorphic province	Altitudinal range (m.)	Months of flowering	Oldest record
Piedmont Plateau	60!–165!	May	?, De Chalmot (US)
Blue Ridge	152–1,067!	May–June	Btwn. 1821 and 1834, Mitchell (Ph)
Ridge and Valley	146–1,152!	May–June	1806, Pursh (Ph)
Appalachian Plateau	—	—	1939, H. A. Davis (W Va)

Reproduction. The large number of rosettes per plant and the rapidity with which they develop are evidences that vegetative reproduction must be common in *Sedum glaucophyllum*. In the description above, numbers of primary rosettes per plant are indicated for populations in the drainage of the New River and also for replicates in an experiment in Ithaca. The average number of rosettes per plant, for replicates in this experiment, is 19.7. This number indicates the rate of multiplication in a year because each replicate was started as a tiny cutting one year prior to the time of observation. In nature, whole populations could be clones which have propagated in this way. In time, the parts of such clones would become separate and appear as separate plants.

Additional to vegetative propagation, sexual reproduction and development of seeds also occurs, especially among plants at the diploid level. In cultivation, seedlings, developing from seeds dropped from follicles, are a nuisance and a potential source of contamination of cultures. After the first stages of growth, seedlings are difficult to distinguish from plantlets developed from small rosettes. For that reason, the relative importance of propagation from seeds is uncertain, but this mode of reproduction is a potential means of both perpetuating populations and providing for their spread and migration.

Although some seeds develop in follicles of polyploid plants, a high percentage of ovules were abortive in pistils of plants from Warm Springs, Va., and Rawlings, Md.

Flowering time in nature is from April to early July, depending on latitude, altitude, and exposure. In cultivation at Ithaca, N.Y., plants flower outdoors in June and early July, in the cold frame from late April to June, and in the greenhouse from April to June. The species is day-neutral or indeterminate, flowering under a wide range of lengths of day, according to Allard and Garner (1940).

Small bees, *Dialictus* and *Ceratina*, identified by R. A. Morse, appear to be important in effecting cross-pollination. At Allisonia, these were frequent visitors of flowers of both *S. glaucophyllum* and *S. ternatum*. In a quadrat ten by ten meters, I have watched a bee fly from a flower in anthesis of *S. ternatum* to a flower in anthesis of *S. glaucophyllum*. Since the two species may be in bloom at the same time, and plants sometimes are close to each other, ecological conditions for hybridization are good.

A species of beetle of the family *Alleculidae*, identified by H. Dietrich, also visits the flowers of *Sedum glaucophyllum* at Allisonia and may be another pollinator.

Relationships and evolution. The closest relatives of *Sedum glaucophyllum* are *S. nevii* and *S. ternatum*. Both species have white flowers, compressed filaments with the epipetalous ones only slightly adnate to the petals, and divergent, two-lipped follicles. *Sedum nevii* so closely resembles *S. glaucophyllum* that the two species have often been confused. Few morphological differences are available for a satisfactory separation, although a person familiar with the plants usually can distinguish them. Width of leaves of primary rosettes, 4.2 mm. in *S. glaucophyllum* and 3.5 mm. in *S. nevii* in the cold frame at Ithaca, is most satisfactory. The analysis of variance in the section on variation applies to this character. Number of chromosomes and geography are important aids in identification in this case. On the basis of higher F values in analyses of variance, *S. glaucophyllum* tends to differ from *S. nevii* in having the leaves of the floriferous stems relatively wider than thick (ratio 3 to 1.6), more cincinni per cyme (3.7 vs. 2.2), and shorter sepals (5.3 vs. 6.9 mm.).

The origin of *Sedum glaucophyllum* still is without experimental demonstration. Three hypotheses concerning its origin seem worthy of attention. The first is the possibility that the species is an amphidiploid, the result of hybridization of *S. nevii* and diploid *S. ternatum*. Such hybridization may have occurred a long time ago. Today, the ranges of *S. nevii* and *S. glaucophyllum* do not overlap. *Sedum nevii* is known only in the southern Appalachian region, and *S. glaucophyllum* is known from the middle sections. The close morphological similarity of *S. glaucophyllum* to *S. nevii* may be a point in favor of the hypothesis of amphidiploidy. Another favorable point is the occurrence of long chromosomes, similar to those of *S. nevii*, along with smaller chromosomes which may come from *S. ternatum*. The modern ranges of *S. nevii* and *S. ternatum* overlap. Diploid *S. ternatum* easily could have crossed with *S. nevii* in past geological time to produce *S. glaucophyllum* which now behaves as a functional diploid.

A second hypothesis is that a species with 7 pairs of chromosomes is the progenitor of both *S. nevii* and *S. glaucophyllum*. In that case, *S. nevii* would have resulted through loss of a chromosome and *S. glaucophyllum* would be an autotetraploid. The principal argument against this point of view is the lack of knowledge of any population of *Sedum* in eastern North America with 7 pairs of chromosomes. Another difficulty is that genomes of *S. glaucophyllum* usually have one exceptionally long chromosome. If the species were autotetraploid, two such chromosomes should occur.

The third hypothesis is that *Sedum glaucophyllum* with 14 pairs of chromosomes is ancestral to both *S. nevii* and *S. ternatum*. This view seems least tenable. *Sedum nevii* has the distribution of a relic species, known only from five widely separated populations. Its occurrences are at sites which probably have been available for enormously long periods of time. *Sedum ternatum* and *S. glaucophyllum*, on the other hand, both have more continuous distributions and partly occur in places which may have been unavailable for them as recently as the ice ages because of lower temperatures and shorter growing seasons. If 14 were a base number of chromosomes from which lower numbers were derived by means of reciprocal translocations and loss of chromosomes, then intermediate numbers between 14 and 6 might be expected to survive, but these are unknown.

Additional to plants with 14 pairs of chromosomes, populations of *Sedum glaucophyllum* occur which have as many as 22 or more pairs of chromosomes. These may be the result of hybridization with *S. ternatum* or simple doubling of the genome of *S. glaucophyllum* with consequent loss of some chromosomes. Because plants of this chromosomal constitution are not morphologically uniform, several independent origins may have occurred. Some clones from the Blue Ridge are similar. When grown in the experiment with plants with 14 pairs of chromosomes, they differ in their poorer growth, smaller size, and rosettes with few, spreading, green leaves. If such plants are hybrids of *S. glaucophyllum* and diploid *S. ternatum*, the genes of *S. glaucophyllum* are dominant over those of *S. ternatum* because the plants simulate *S. glaucophyllum*.

Although the origin of polyploid *Sedum glaucophyllum* must have been from diploid plants, the problem of which diploid race came first is not easy to resolve. Possibly the race with compact rosettes of glaucous leaves, adapted to growth on cliffs of limestone which often are hot and dry in summer, is the specialized, derived type. The greater production of wax probably is an adaptation to prevent loss of water.

In the Pleistocene epoch, when spruce forests reached their maximum extent in the southern Appalachian Mountains, survival of *Sedum glaucophyllum* could have been at low altitudes on the outer Piedmont Plateau. At that time, many present species of the Piedmont Plateau and Coastal Plain may have been out on the Atlantic continental shelf which now is submerged. For evidence in favor of such an opinion, see Whitmore, Emory, Cooke, and Swift (1967). The modern occurrence of *S. glaucophyllum* at sites along big rivers on the outer Piedmont Plateau may be residual. Survival there may be because of the presence of rocks with surfaces exposed to the north and also lower temperatures resulting from proximity to water. Possibly the plants have not washed downstream from the highlands, but instead, populations at low elevations may be relics of a once wider distribution in the low country, from which populations of the highlands have arisen. This view envisages a migration upstream and then later hybridization or autopolyploidy, involving plants with 14 pairs of chromosomes, to yield plants of higher polyploidy.

References: Atwood (1953), Shriver (1877a and b), and Uhl (1970).

5. *Sedum pulchellum**** (figs. 18–27)

Sedum pulchellum (fig. 18) belongs to a group of eastern North American species of *Sedum* with flowers predominantly white and four-parted and with filaments of the epipetalous stamens only slightly adnate to the bases of the petals. The linear leaves of the floriferous stems have sagittate spurs basally (fig. 19), a unique character not found in any other North American species. Further, the petals are pink or pinkish white, and the plants tend to be annual, but sometimes are perennial through the

NATIVE SPECIES 143

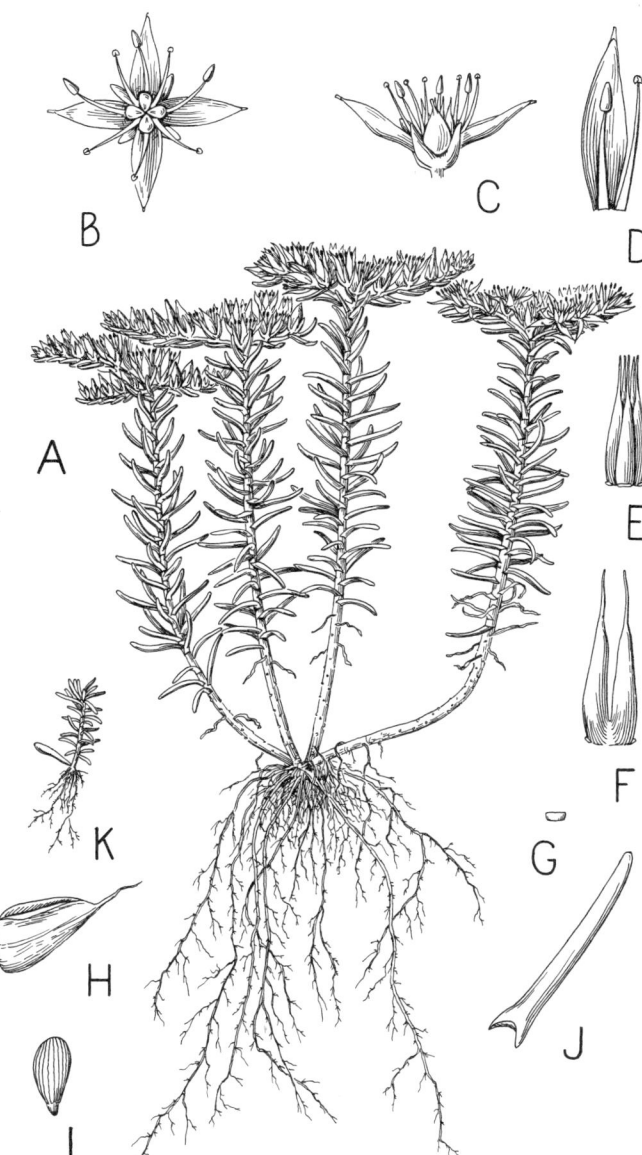

Fig. 18. Plant of tetraploid *Sedum pulchellum* from Golconda, Pope Co., Ill., cultivated in cold frame, Ithaca, N.Y. A. Habit sketch (x .44). B. Flower from above (x 1.8). C. Flower from side (x 1.8). D. Petal and two stamens (x 2.6). E. Carpels (x 2.6). F. Two carpels (x 3.5). G. Nectary (x 4.4). H. Follicle from side (x 3.5). I. Seed (x 8.8). J. Single median leaf (x 1.8). K. Plantlet developed vegetatively (x .44).

production and persistence of vegetative shoots. Also distinctive is the number of chromosomes, 11 or a multiple of 11. The distribution primarily is in the drainage of the Mississippi River, on the Interior Low Plateaus and on the Ozark Plateaus. The common name in southern Illinois is "Bluff Moss." Other English names are "Widow's Cross" (Wherry, 1950) and "Bird's-claw Sedum" (Masters, *Gard. Chron.* n.s. 2:657 [1874]), but laymen seldom use either of these names.

Fig. 19. Bases of leaves of plant of *Sedum pulchellum* from Logan Point, Monte Sano, Huntsville, Ala., cultivated in greenhouse, Ithaca, N.Y., Apr. 6, 1973 (x 15).

Description. Sample: 71 plants. Sources and numbers of plants from each locality are: Ridge and Valley Province: Pigeon Mountain, Walker Co., Ga.—1; Appalachian Plateau: Monte Sano, east of Huntsville, Ala.—49, 28 from Logan Point and 21 from along Bankhead Parkway; Interior Low Plateaus: Carthage, Tenn.—2; Lawrenceburg, Tenn., plant made available by E. T. Wherry—1; Lebanon, Tenn.—3; Nashville, Tenn.—2; Steamboat Hill, Golconda, Ill.—1; Tunnel Hill, Ill.—2; and near Makanda, Ill.—2; and Ozark Plateaus: Pennsboro, Mo.—2; Joplin, Mo.—2; Noel, Mo.—1; Yellville, Ark.—1; and Eureka Springs, Ark.—2.

My principal attention to *Sedum pulchellum* was in 1946, 1949, and 1972. In the two earlier years, my procedure at each locality was to scan whole populations and then to select plants to show extremes in dimensions. As will be shown later in the case of *S. nuttallianum*, this method provides a rapid way for estimating the range of different dimensions, but does not provide adequate data for estimating means and standard deviations. In 1972, at Monte Sano, studies were in greater detail and involved drawing random samples of plants from seven randomly selected transects which were 3 meters wide on Logan Point, a 5% sample, and 2 meters wide along the Bankhead Parkway, a 1% sample. The greater uniformity of conditions along the parkway and the smaller size of the plants there determined the smaller proportion of transects and their narrower width.

In cultivation at Ithaca, N.Y., plants of *Sedum pulchellum* which were annual or biennial in nature sometimes became perennial when watered regularly, for example, plants in the greenhouse from Pigeon Mountain, Ga., and Carthage, Tenn. Other plants from Carthage and also plants from Monte Sano, Ala., which were perennial in the wild, continued as perennials in the greenhouse. One clone from Monte Sano persisted for four years. A plant on Steamboat Hill, Golconda, Ill., had many sterile offsets there in August 1970. When transplanted to Ithaca, N.Y., it flowered in June 1971, and then produced more axillary, vegetative shoots.

Detailed data are available in the description for samples from Monte Sano. Sites there are abbreviated as LP—Logan Point and BP—Bankhead Parkway. Plants propagated from cuttings of vegetative shoots are designated as v and those grown from seeds are indicated as e. Data from other localities are cited only when they supplement the data from Monte Sano.

Plants of *Sedum pulchellum* are glabrous herbs, either annual or perennial through the development of vegetative shoots in the axils of the lower leaves. Roots are slender, fibrous, and pale brown or white. Vegetative shoots are 1—21 cm. long, and either unbranched or branched. Floriferous stems are erect, usually several-branched at or near the base, and green, pink, or even red toward base, then green upward. Rhizomes (fig. 20), sometimes developed in perennial plants, are brown. A remarkably

Fig. 20. Horizontal rootstock, 8 cm. long, with roots and sterile shoots, of plant of perennial *Sedum pulchellum* on Logan Point, Monte Sano, Huntsville, Ala., June 3, 1972 (x .6).

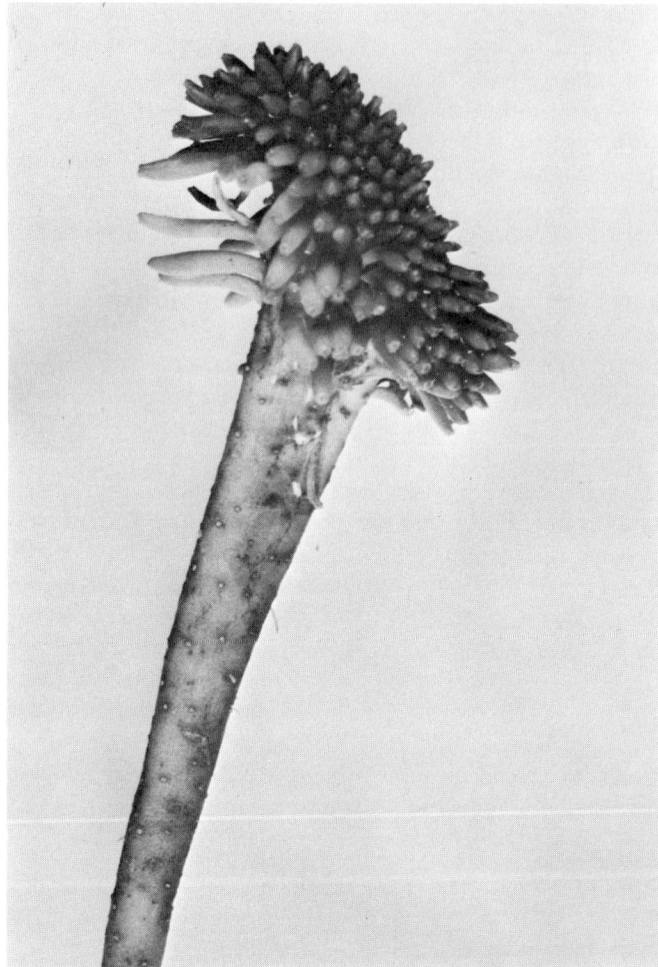

Fig. 21. Cristate stem of plant of *Sedum pulchellum* from Logan Point, Monte Sano, Huntsville, Ala., cultivated in greenhouse, Ithaca, N.Y., July 6, 1973 (x 1.5).

large plant on the crest of Logan Point at Monte Sano had 29 floriferous stems and a maximum diameter of 65 cm. A tall plant along the Harpeth River southwest of Nashville attained a height of 33 cm. Rarely, stems become cristate (fig. 21).

	n-pl.	n-rep.	n-obs.	$\bar{x}$	s
Plants—diameter (cm.)					
Monte Sano—LP	11	—	11	5	3.8
—BP	17	—	17	.4	.3
—LP—gh.—v	14	19	19	25	7.5
—BP—gh.—v	3	3	3	17	5.6
—LP—gh.—e	2	2	2	31	1.4
—BP—gh.—e	5	7	7	17	3.9

NATIVE SPECIES 147

	n-pl.	n-rep.	n-obs.	$\bar{x}$	s
Vegetative rosettes (n)					
Monte Sano—LP	11	—	11	2	3.2
—BP	17	—	17	0	0
—LP—gh.—v	14	20	20	16	24
—BP—gh.—v	3	3	3	0	0
—LP—gh.—e	2	2	2	9	4.7
—BP—gh.—e	5	7	7	9	7.9
Vegetative rosettes—diameter (cm.)					
Monte Sano—LP	9	—	16	1.2	.4
—BP	17	—	17	0	0
Maximum height (cm.)					
Monte Sano—LP	11	—	11	13	2.3
—BP	17	—	17	7	3.2
—LP—gh.—v	14	20	20	10	—
—BP—gh.—v	3	3	3	5	1.4
—LP—gh.—e	2	2	2	10	3.5
—BP—gh.—e	5	7	7	5	.5
Floriferous stems per plant (n)					
Monte Sano—LP	13	—	13	9	4.7
—BP	17	—	17	2	.8
—LP—gh.—v	11	13	13	6	24
—BP—gh.—v	3	3	3	14	12
Floriferous stems—length (cm.)					
Monte Sano—LP	13	—	52	11	1.5
—BP	21	—	27	6.4	2.3
—LP—gh.—v	11	13	29	23	7.8
—BP—gh.—v	3	3	7	7	1.2

Leaves of rosettes, when present, are spirally arranged, spreading, sometimes with two short sagittate spurs at base, occasionally petiolate, spatulate, oblanceolate, or linear, obtuse to rounded, minutely papillose at apex, and light to dark green, sometimes becoming reddish. Leaves of floriferous stems are spirally arranged, linear, subterete, prominently two-spurred at base, blunt at apex, and pale yellow-green. A plant from Pigeon Mountain, cultivated in the greenhouse, had rosettes with leaves up to 11 mm. wide. A plant near Nashville had leaves of the floriferous stems up to 36 mm. long. Measurements below for floriferous stems apply to median leaves.

	n-pl.	n-rep.	n-obs.	$\bar{x}$ (mm.)	s (mm.)
Leaves of rosettes—length					
Monte Sano—LP	9	—	16	7.1	1.7
—BP	17	—	17	0	0
—LP—gh.—v	14	20	51	9	3.5
—BP—gh.—v	3	3	3	0	0
—LP—gh.—e	2	2	6	32	2.1
—BP—gh.—e	5	7	20	11	2.6
Leaves of rosettes—width					
Monte Sano—LP	9	—	16	2.1	.7
—BP	17	—	17	0	0
—LP—gh.—v	14	20	53	2	.9
—BP—gh.—v	3	3	3	0	0
—LP—gh.—e	2	2	6	5	.6
—BP—gh.—e	5	7	20	4	1.2

	n-pl.	n-rep.	n-obs.	x̄ (mm.)	s (mm.)
Leaves of rosettes—thickness					
Monte Sano—LP	9	—	16	1	.2
—BP	17	—	17	0	0
Leaves of floriferous stems—length					
Monte Sano—LP	12	—	43	16	2.3
—BP	3	—	3	5	.4
—LP—gh.—v	11	13	27	14	1.6
—BP—gh.—v	3	3	7	9	1.9
Golconda—cf.	1	1	5	14	—
Leaves of floriferous stems—length of spurs					
Monte Sano—LP—gh.—v	11	13	27	1.4	.5
—BP—gh.—v	3	3	7	.8	.1
Leaves of floriferous stems—width					
Monte Sano—LP	12	—	43	2.1	.2
—BP	3	—	3	1.5	.3
—LP—gh.—v	11	13	27	2	.2
—BP—gh.—v	3	3	7	1.6	.1
Golconda—cf.	1	1	5	1.6	—
Leaves of floriferous stems—thickness					
Monte Sano—LP	12	—	43	1.2	.2
—BP	3	—	3	.7	.1
Golconda—cf.	1	1	5	1.3	—

Inflorescences usually are three-branched cymes, sometimes with more branches, 1–20 cm. across. A plant at Joplin, Mo., had 8 cincinni. The maximum number of flowers counted in a cyme was 148 on a plant on Logan Point at Monte Sano. Floral bracts are linear and spurred.

	n-pl.	n-rep.	n-obs.	x̄ (n)	s (n)
Cincinni per cyme (n)					
Monte Sano—LP	12	—	51	3.7	.7
—BP	21	—	29	2.7	1.2
Flowers per cyme (n)					
Monte Sano—LP	12	—	51	29	10.3
—BP	21	—	29	12	6.3
—LP—gh.—v	11	13	28	35	17.8
—BP—gh.—v	3	3	7	32	21
Golconda—cf.	1	1	5	28	—

Flowers (fig. 22) are sessile or subsessile and generally 4-merous, rarely as many as 7-merous. A short torus ranging from .1–.6 mm. long, is present. Flowers which develop in the axils of leaves, after the normal period of flowering, may be on slender pedicels to 7 mm. long, as at Joplin, Mo., in June 1949. The observed range in diameter of flowers was from 5–14 mm.

	n-pl.	n-rep.	n-obs.	x̄ (mm.)	s (mm.)
Flowers—diameter					
Monte Sano—LP	11	—	44	8	1
—BP	1	—	1	6	—
Golconda—cf.	1	1	5	12	—

Fig. 22. Flowers of plant of *Sedum pulchellum* from Logan Point, Monte Sano, Huntsville, Ala., cultivated in greenhouse, Ithaca, N.Y., Apr. 6, 1973 (x 6).

Sepals are unequal, linear-lanceolate, blunt or acute, and light green.

	n-pl.	n-rep.	n-obs.	$\bar{x}$	s
Length (mm.)					
Monte Sano—LP	11	—	44	3.4	.5
—BP	1	—	1	1.3	—
Golconda—cf.	1	1	5	5.8	—
Width (mm.)					
Monte Sano—LP	11	—	44	.9	.3
—BP	1	—	1	.5	—
—LP—gh.—v	11	12	32	1.1	—
—BP—gh.—v	3	3	9	1.1	0
Golconda—cf.	1	1	5	1.5	—

Petals are erect and subdivergent, narrowly linear-lanceolate or elliptic-lanceolate, acute or obtuse, carinate and somewhat channeled, and pale pink, cyclamen purple, pinkish white, or even pure white.

	n-pl.	n-rep.	n-obs.	$\bar{x}$	s
Number					
Monte Sano—LP	11	—	44	4.1	.2
—BP	1	—	1	4	—
Golconda—cf.	1	1	5	5	—
Length (mm.)					
Monte Sano—LP	11	—	44	6	.7
—BP	1	—	1	3.7	—
—LP—gh.—v	11	12	32	6.1	.5
—BP—gh.—v	3	3	9	5	1
Golconda—cf.	1	1	5	9.4	—
Width (mm.)					
Monte Sano—LP	11	—	44	1.5	.2
—BP	1	—	1	1.4	—
Golconda—cf.	1	1	5	2.6	—

Stamens are twice as many as the petals and have white or pinkish filaments, which are flattened basally, and red, purple, or fuscous anthers.

	n-pl.	n-rep.	n-obs.	$\bar{x}$ (mm.)	s (mm.)
Filaments—length					
Monte Sano—LP	11	—	44	3.8	.5
—BP	1	—	1	2.5	—
Golconda—cf.	1	1	5	6.9	—
Epipetalous filaments—height on petals					
Monte Sano—LP	11	—	44	.1	.2
—BP	1	—	1	.1	—
Golconda—cf.	1	—	5	.1	—
Anthers—length					
Monte Sano—LP	11	—	44	.9	.1
Golconda—cf.	1	1	2	1.6	—

Nectaries are quadrate, truncate and emarginate, and white, pinkish white, or yellow.

	n-pl.	n-rep.	n-obs.	$\bar{x}$ (mm.)	s (mm.)
Length					
Monte Sano—LP	11	—	44	.3	.07
—BP	1	—	1	.2	—
Golconda—cf.	1	1	5	.3	—
Width					
Monte Sano—LP	11	—	44	.4	.04
—BP	1	—	1	.2	—
Golconda—cf.	1	1	5	.6	—

Carpels are at first erect, later spreading and gibbous ventrally, connate basally, and white or pinkish white. The ovaries are either smooth or papillose.

	n-pl.	n-rep.	n-obs.	$\bar{x}$	s
Length (mm.)					
Monte Sano—LP	10	—	43	3.3	.4
—BP	1	—	1	.8	—
Golconda—cf.	1	1	5	6.7	—

	n-pl.	n-rep.	n-obs.	$\bar{x}$	s
Length of cohesion (mm.)					
Monte Sano—LP	10	—	43	.5	.07
—BP	1	—	1	.2	—
Golconda—cf.	1	1	5	1.1	—
Ovules per ovary (n)					
Monte Sano—LP	10	—	43	2	.4
Golconda—cf.	1	1	5	6.6	—

Follicles are spreading, tumid ventrally, either smooth or papillose, and light brown.

	n-pl.	n-obs.	$\bar{x}$ (mm.)	s (mm.)
Length				
Monte Sano—LP	13	40	3.3	.8
—BP	16	33	3.6	—
Carthage—annual	1	10	3.3	—
Lebanon	3	6	3.4	.6
Carthage—annual—cf.	1	5	3.5	—
Golconda—cf.	1	4	4.3	—
Lawrenceburg—gh.	1	2	3.2	—

Seeds are elliptic-pyriform, finely and irregularly ribbed, and brown, yellowish brown, or pale brown.

	n-pl.	n-obs.	$\bar{x}$ (mm.)	s (mm.)
Length				
Monte Sano—LP	10	30	1	.2
—BP	16	33	1	.1
Carthage—annual	1	10	1	—
Lebanon	3	6	1.1	.03
Pennsboro	1	2	.9	—
Carthage—annual—cf.	1	5	1.1	—
Golconda—cf.	1	4	1.3	—
Lawrenceburg—gh.	1	2	1	—
Diameter				
Monte Sano—LP	10	30	.5	.08
—BP	16	33	.5	.05
Carthage—annual	1	10	.5	—
Lebanon	3	6	.5	.1
Pennsboro	1	2	.4	—
Carthage—annual—cf.	1	5	.5	—
Golconda—cf.	1	4	.6	—
Lawrenceburg—gh.	1	2	.6	—

Chromosomes are somewhat smaller than those of *Sedum nevii*, unequal, and with median kinetochores. Diploids, tetraploids, and hexaploids occur. Multivalent associations are characteristic of the ployploid plants. J. T. Baldwin, Jr. (1943) reported numbers of chromosomes for 45 plants from 43 localities: 26 diploid, 16 tetraploid, and 3 hexaploid. According to his data, only diploids occur in the Interior Highlands west of the Mississippi River, tetraploids predominate on the Interior Low

Plateaus, and hexaploids are confined to the Nashville Basin. Numbers cited below are for plants collected by me which have been the basis for ideas about classification and relationships in the present study. These numbers fit the pattern described by Baldwin.

	n-pl.	g(n)	sp(2n)	Cytologist
Pigeon Mountain, Ga.	1	11	—	C. H. Uhl, 1947
Monte Sano, Ala.—Logan Pt.	1	—	~ 22	C. H. Uhl, 1947
—Logan Pt.	3	11	—	K. Krathwohl, 1973
Monte Sano, Ala.—Bankhead Pky.	2	11	—	K. Krathwohl, 1973
Midway, Ky.	1	22	—	C. H. Uhl, 1942
Carthage, Tenn.	1	22	—	C. H. Uhl, 1948
Carthage, Tenn.	1	33	—	C. H. Uhl, 1948
Nashville, Tenn.	1	22	—	C. H. Uhl, 1949
Golconda, Ill.	1	22	~ 42	I. Mastrangelo, 1971
Tunnel Hill, Ill.	1	22	—	C. H. Uhl, 1949
Makanda, Ill.	1	22	—	C. H. Uhl, 1949
Yellville, Ark.	1	~ 11	—	L. Hollingshead, 1941
Eureka Springs, Ark.	2	11	22	J. T. Baldwin (1943) and C. H. Uhl, 1941

Variation. Geographic and ecological differentiation, age of plants, tetraploidy and hexaploidy, and modifications caused by shade, moisture, and injury make *Sedum pulchellum* a variable species. From the standpoint of subspecific classification, it is a difficult species to interpret, although easy to separate from other species.

Of particular interest is a population of perennial plants on limestone ledges on Logan Point, a northern spur of Monte Sano, at the southwestern edge of the Cumberland Plateau east of Huntsville, Ala. Besides being perennial, these plants have broad, spatulate leaves on perennating sterile shoots, and they flower in late spring and early summer. Plants of this kind still were in flower on June 25, 1946, when annual *Sedum pulchellum*, only a few kilometers away, had mature fruits. Roland Harper, who found the population of perennial plants on October 15, 1939, at first regarded it as a form of *S. nevii*. The sagittate spurs of the leaves, however, are clear indication of affinity with *S. pulchellum*. Further, the number of chromosomes, $g = 11, s = 22$, is characteristic of diploid *S. pulchellum*. A plant of this kind survived for four years in the greenhouse at Ithaca, N.Y. Distinctive features were the perennial habit of growth with elongate, prostrate stems with spatulate or oblanceolate leaves (fig. 23), longer than in other diploid *S. pulchellum*, later time of flowering, and white petals.

Studies at Monte Sano from May 23 to June 12, 1972, had as the primary objective clarification of the relationships of the two kinds of *Sedum pulchellum* that grow there. Perusal of the foregoing description will reveal many differences between plants on Logan Point and those along the Bankhead Parkway. In nature, nine of these differences are highly significant: diameter of plants; diameter of vegetative rosettes; number of floriferous stems per plant; length, width, and thickness of leaves of rosettes; length and width of nectaries; and number of ovules per ovary. In addition, six differences are significant: number of vegetative rosettes per plant, maximum

Fig. 23. Stems and leaves of perennial plant of *Sedum pulchellum* from Logan Point, Monte Sano, Huntsville, Ala., cultivated in greenhouse, Ithaca, N.Y., Apr. 6, 1973 (x 1.1).

height, thickness of leaves of floriferous stems, length of torus, width of longest sepals, and length of cohesion of ovaries. When plants are grown under similar conditions at Ithaca, N.Y., few of these differences are confirmed. On the basis of ten plants from Logan Point, propagated from vegetative rosettes, compared with three plants obtained as seedlings from along the Bankhead Parkway, the populations differ highly significantly in length and width of leaves of rosettes and significantly in length of floriferous stems. Growth of plants from seeds provided another test of relationships. On the basis of two plants grown from seeds from Logan Point and seven plants grown from seeds from along Bankhead Parkway, differences are highly significant in diameter of plants and length of leaves of rosettes, and significant in maximum height. Figures to document these differences are in the description. That any differences are maintained in cultivation suggests that they have a genetic basis, but differences among plants within populations are significant and also each of the features that showed differences in cultivation is subject to great environmental modification. No sure claim of two genetic races is possible.

The plants along the Bankhead Parkway occur for a distance of 1,820 meters. The highest plants there are at an altitude of 317 meters, which is 123 meters below the lowest plants on Logan Point at an altitude of 440 meters. Differences in behavior of the plants at the two sites are consistent with the environmental conditions. The habitat on Logan Point is cooler and moister, and the plants there are more shaded. These conditions are favorable to perennial growth, whereas the warmer, drier conditions along Bankhead Parkway are conducive to early development and the annual habit. On Logan Point, late-developing seedlings (fig. 24) may produce large rosettes, and these plants do not flower until the second year. An occasional late seedling along Bankhead Parkway similarly may persist from one growing season into the next. Time of flowering varies within populations and is particularly dependent on exposure. Plants in the open on the tops of bluffs are far ahead of those on shaded ledges. Likewise, whether petals and filaments are white or pink depends on exposure, with the pink pigment most evident in plants in the more exposed situations.

Fig. 24. Seedling of *Sedum pulchellum* growing on limestone in bottom of big fissure, Logan Point, Monte Sano, Huntsville, Ala., June 1, 1972 (x .8).

Sedum pulchellum might be some kind of aneuploid derivative from *S. nevii*. The two species have many features in common and sometimes are confused. The population of perennial *S. pulchellum* at Monte Sano could be a relic, nearest to the ancestral type of the species. The circumstance may be significant that a population of *S. pulchellum* with relictual characteristics occurs in the area where the range of the species overlaps with *S. nevii*.

Of the three conditions of chromosomes in *Sedum pulchellum*, diploids have the widest distribution. Except for the population on Monte Sano, in my experience diploids always are annuals in nature. Tetraploids have a wide distribution within the area of the Interior Low Plateaus. With abundant rainfall, these plants have a tendency to produce axillary, vegetative rosettes and thus behave as perennials. I have seen plants growing in this way at Golconda and Makanda, Ill., and at Carthage, Tenn. At Carthage, tetraploid plants were perennial and still in flower on June 27, 1946, when hexaploid plants at the same locality were dry and with mature fruits.

Smith (1946) reported characteristics of twelve collections of tetraploid plants. Some features, as larger, less frequent stomates, are familiar consequences of tetraploidy. Other features, such as slightly shorter leaves and larger seeds, may be either the result of tetraploidy or the expression of the genetic constitution of plants at the localities of origin. My experience differs somewhat from that of Smith. Tetraploids may have longer leaves and also more flowers per cyme than diploids. Smith's data do not separate the geographic aspect of the problem, that is different gene frequencies in different areas, from the part due to polyploidy alone, as would be evident if tetraploids and hexaploids, experimentally produced from known diploids, were used. Other problems of the work are sampling procedure, experimental design, and components of variance. In Smith's publication, evidence of partitioning variances, with attention to variation within plants or among plants within populations, is lacking. Attention to these aspects of the analysis might yield different results from the same data.

Leaves of a plant vary considerably. The method of selection of leaves for measurement is of great importance if statistical analyses are made. Description of Smith's method of selection of leaves appears on page 501 of her paper.

A discrepancy between Smith's and my data occurs with respect to length of seeds. Smith measured 353 seeds from 5 tetraploid "selections" and obtained a mean length of 1.1 mm. with a standard error of the mean of .0081. In contrast, I measured 4 seeds from a systematic sample of 4 fructiferous stems, chosen with a random start, from a tetraploid plant from Golconda, Ill. Selection of seeds was random from central follicles on the 4 stems. The smallest seed was 1.2 mm. long and the mean length was 1.3 mm. The results are in agreement with Smith's conclusion that increase in size of seeds accompanies increase in chromosome number, but the magnitude of the increase is greater. More or fewer plants of the sort from Golconda would make a difference in the results. Emphasis must be on method of selection, not on large numbers of measurements.

Stein's test (cited with reference in Chapter II), based on a trial random sample of ten seeds from a hexaploid plant from Carthage, Tenn., reveals that measurement of two seeds per plant, selected according to a probability model, is adequate to detect a difference of .1 mm. in length.

In summary, tetraploid plants of *Sedum pulchellum*, with the exception of the perennial, diploid population on Monte Sano, exhibit the greatest tendency to be perennial. In my experience, they have larger seeds, usually more flowers per cyme, and longer leaves on the floriferous stems. In contrast, Smith reported slightly shorter, narrower leaves (averaging 10.1 mm. long and 1.6 mm. wide), fewer flowers per cyme ($\bar{x} = 49$), larger seeds ($\bar{x} = 1.1$ mm.), and larger, less frequent stomates (457μ long and 449 per square cm.). Smith further reported better survival of tetraploids than diploids in both conditions of submergence in water (35% survival) and drought (52% survival), in contrast with no survival of diploids under either condition. According to her experiment, although survival of diploid plants after 16 days of treatment with 10% tannic acid was 100%, survival of tetraploid plants was 69%. The expectation is that diploid plants would have an advantage over polyploids in acid soils.

Wherry (1950) was so impressed by the perennial habit of certain plants in cultivation, with trailing leafy stems, the tips of which become red in the autumn, that he named them var. Redtip. J. T. Baldwin, Jr., had earlier determined that a plant of this type was tetraploid; see Smith (1946). Wherry named a plant from Lawrenceburg, Tenn., which perennates by producing tiny rosettes of flattish, gray-green leaves toward the bases of the dry fructiferous stems. He called this var. Starlet. This too is tetraploid. Although Smith regarded the perennial habit to be developing from an annual condition, the reverse situation may be occurring, with the annual habit developing from the perennial one as a response to an austere environment, with a hot, dry summer. The degree to which the perennial habit persists seems to depend on both the genetic consitution of the plants and the condition of the climate in any year. Subspecific status for the tetraploid plants would produce various practical problems. That all tetraploids had the same origin is unclear. If they are derived from different diploids, then we may not properly speak of a tetraploid race, as is commonly done.

Hexaploid plants for this study came from Carthage, Tenn. These behaved as annuals in both the wild and cultivation. At Carthage, on June 27, 1946, hexaploid plants had dry fructiferous stems and fruits when tetraploid plants were still in flower. Hexaploid plants have longer leaves than diploids, but shorter than tetraploids. Similarly, they have larger seeds, but smaller than those of tetraploids. Smith (1946) reported the largest stomates for *Sedum pulchellum* on leaves of three hexaploids which she studied, 575μ long and 358 per square cm. Also, she reported shorter, wider leaves ($\bar{x} = 8.9$ mm. long and 1.8 mm. wide), fewer buds per cyme ($\bar{x} = 34$), and seeds about the same length as for tetraploids ($\bar{x} = 1.14$). Survival after 51 days submergence in water was 96% and after the same length of time without water was 93%. Under both extremes of moisture, the hexaploids have a competitive advantage. Survival was less good, only 55%, after 16 days of treatment with 10% tannic acid. In that condition, diploid plants have the advantage.

McCormick and Rushing (1964) reported differences in germination of diploid, tetraploid, and hexaploid seeds that were subjected to ionizing radiation. At the end of one week, differences in percent of germination were significant, but after two weeks, differences lacked significance. Seedlings from irradiated seeds exhibited a differential growth of roots which varied both with the radiation dose received and, at some levels of treatment, with the number of chromosomes. McCormick and Rushing reported good results in estimating numbers of chromosomes of plants by means of radiation experiments.

Environmental modification of plants of *Sedum pulchellum* is great. Plants also change in appearance with age and season. Great differences exist between seedlings which flower a few months after germination (fig. 25) and plants which flower after

Fig. 25. Flowering seedling, four months old, of *Sedum pulchellum*, grown from seed from western slope of Monte Sano, Huntsville, Ala., cultivated in greenhouse, Ithaca, N.Y., June 20, 1973 (x 1).

a full year of growth. Shade produces etiolated plants. A plant growing under a rock at Noel, Mo., had foliaceous floral bracts to 17 mm. long. Abundant moisture favors development of secondary, axillary shoots and perennial growth. Diploid plants from Pigeon Mountain, Ga., annuals in nature, became perennial in a well-watered greenhouse. Some plants on bluffs of chert along Shoal Creek, southwest of Joplin, Mo., on June 9, 1949, after flowering normally, had axillary shoots from which further flowers developed. Such plants appeared to be either much branched or with the flowers on slender pedicels in a raceme. The racemose condition is a frequent result of injury to the end of the floriferous stem. One injured plant at Noel had the leaves of the floriferous stem spatulate or the uppermost linear, 1–4.5 mm. wide. Plants of this kind have been the basis for reports of *S. nevii* from Missouri.

The number of petals per flower usually is 4, occasionally 5. A central flower of a plant in cultivation from Golconda, Ill., had 7 petals and 8 carpels. In addition, petals vary from white to pink.

In conclusion, wild plants of *Sedum pulchellum* usually are annual or biennial, but sometimes are perennial by the production of secondary shoots. In cultivation, the most extreme annual types from the wild, including diploids, tetraploids, and hexaploids, may become perennial. Annual and perennial populations occur close together at a few localities. At Monte Sano, Ala., annual plants are early-flowering, and perennial plants are late-flowering and diploid. At Carthage, Tenn., early-flowering plants are annual and hexaploid, but late-flowering plants are perennial and tetraploid. Although populations of *S. pulchellum* are different, characteristics are not enough correlated to distinguish regional subspecies or distinct chromosomal races.

Nomenclature. *Sedum pulchellum* Michx., Flor. Bor. Am. 1:277 (1803). Type locality: on rocks near Knoxville, Tenn. Type: André Michaux, <1802. Wherry (1936) thought that Knoxville was a mistake for Nashville, but since populations are known from both east and west of Knoxville, the citation by Michaux could be correct.

Synonyms are:

1821. *Aectyson sagittatum* Raf., Western Minerva, p. 41. Type locality: Kentucky, Russell Springs and on cliffs of the Dick's and Kentucky Rivers. Type: not seen.
1828. *Sedum pulchrum* DC., Prod. Syst. Nat. 3:403. Type locality: mountains of Virginia, Carolina, and Georgia, and the banks of the Ohio River. Type: not seen.
1933. *Sedum vigilmontis* Small, Manual of the Southeastern Flora, p. 587. Type locality: rocks on lower slopes of Lookout Mountain, near Rising Fawn, Ga. Type: A. H. Curtiss 6,798 (NY), May 24, 1901. Fröderström (1930–1935) published a photograph of the holotype, pl. 82.

Torrey and Gray (*Flora of North America* 1:559 [1840]) included in the synonymy of *Sedum pulchellum* an unpublished binomial of Nuttall, based on a collection from Arkansas, but that listing does not constitute valid publication.

Distribution. *Sedum pulchellum* occurs (fig. 26) in eastern North America from the Appalachian Ridge and Valley Province in eastern Tennessee and northwestern Georgia, westward through the southern Cumberland Plateau in southeastern Tennessee and northern Alabama, across the Interior Low Plateaus of central Tennessee and Kentucky and southern Illinois, the Till Plains of Missouri, the Ozark Plateaus of Missouri and Arkansas, and the Ouachita Mountains of Arkansas and Oklahoma, to the Osage Plains of Missouri, Kansas, Oklahoma, and Texas. An old collection, alleged to have come from the Till Plains at Urbana, Champaign Co., Ohio, needs confirmation. Habitats of *Sedum pulchellum* include areas of flat rocks, especially openings in cedar glades, ledges of cliffs, and bluffs. The plants grow in shallow soil or in living mats of moss on a variety of rocks, most often on limestone, sandstone, and chert. The altitudinal range is small, 91–457 meters.

My experience with *Sedum pulchellum* has been throughout its distributional area. Detailed data follow for the two populations at Monte Sano and then briefer information for twelve other populations.

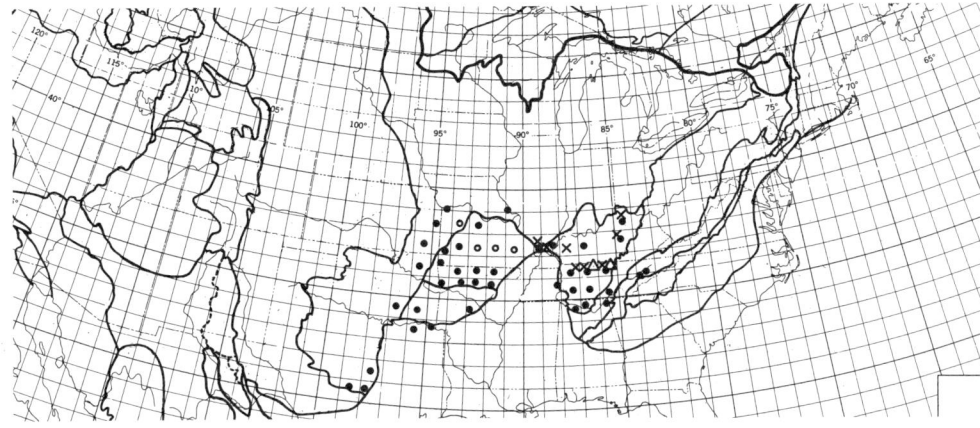

Fig. 26. Known distribution of *Sedum pulchellum* (●) and (x) for tetraploid plants, (△) for hexaploid plants, and (○) for records from literature.

Population	N plants, estimate	N plants (C.I., P = .05)	Area m.²	Exposure
Monte Sano—Logan Point	4,180	209–8,321	2,839	N, E, S, W
—Bankhead Parkway	29,666	293–70,944	8,936	N–W

Population	Altitudinal range (m.)	Depth of soil $\bar{x}$ (cm.)	s (cm.)	pH, range	pH, tests (n)	Distance to nearest neighbor $\bar{x}$ (cm.)	s (cm.)	Oldest record
Monte Sano—LP	440–470	.9	.8	7.4–8.2	11	18	17	1939
—BP	261–317	.7	.3	7.4–7.8	15	31	3	1899

Population	Altitude (m.)	Rock	Oldest record
Pigeon Mt., Ga.	305	limestone	1900
Carthage, Tenn.	~153	limestone	1946
East of Lebanon, Tenn.	255	limestone	1941
West of Nashville, Tenn.	172	limestone	1934
Golconda, Ill.	135	sandstone	1970
Tunnel Hill, Ill.	195	sandstone	1902
Makanda, Ill.	195	sandstone	1902
Pennsboro, Mo.	320	limestone	1968
Shoal Creek, Joplin, Mo.	~259	chert	1949
Noel, Mo.	~250	limestone	1908
Yellville, Ark.	213	limestone	1940
Eureka Springs, Ark.	385	limestone	1940

Other vegetation is sparse in habitats where *Sedum pulchellum* occurs. Frequent sites are open places in glades of *Juniperus virginiana*. Competing herbs and woody plants are miscellaneous. These include *Selaginella rupestris*, *Woodsia obtusa*, *Cheilanthes lanosa*, *Juniperus virginiana*, *Ulmus alata*, *Rhus toxicodendron*, *Carya glabra*, *Quercus marilandica*, and *Q. borealis*. On the western slope of Monte Sano, plants of *S. pulchellum* occur on limestone in cedar glades where *Juniperus virginiana* is the dominant species, but sometimes *Carya glabra* is important. Competitors are miscellaneous and include *Trifolium procumbens*, a species of *Cassia*, *Arenaria glabra*, *Lonicera japonica*, *Parthenocissus quinquefolia*, and grasses. On Logan Point, at Monte Sano, plants of *S. pulchellum* grow on ledges and crests of bluffs of limestone, in oak hickory forest in which *Carya glabra* and *Quercus prinus* are important species. Principal competitors are *Polymnia uvedalia* and *Parthenocissus quinquefolia*. *Sedum nuttallianum* occurs in association with *S. pulchellum* near Joplin, Mo.

Reproduction. The principal method of perpetuation of *Sedum pulchellum* is by seeds. A secondary method is by the production of axillary shoots from the lower nodes of the main stems. These develop rosettes and adventitious roots (fig. 27). When separated from the parental plants, these secondary branches become new

Fig. 27. Secondary branches and adventitious roots of *Sedum pulchellum* from Logan Point, Monte Sano, Huntsville, Ala., cultivated in greenhouse, Ithaca, N.Y., Aug. 16, 1973 (x 1.4). Note root from lower node which has reached soil.

individuals. Vegetative reproduction of this sort is common in some populations.

Caudle and Baskin (1968) discovered little germination of freshly harvested seed, but an improvement in germination over the five-month period of their experiments. An exception was that four-month old seeds germinated better at 10° C. and 20° C. than five-month old seeds. They also found that as the length of storage increased, the maximum temperature at which seeds germinated likewise increased. They concluded that high temperature dormancy is an adaptive advantage to prevent seeds of winter annuals from germinating in the hot, dry summer, when there is little chance of survival of seedlings.

J. T. Baldwin, Jr. (1943) regarded *Sedum pulchellum* to be highly self-fertile. Smith (1946) selfed plants to maintain the purity of the genetic lines which she used in experiments. In nature, sweat bees and honeybees both self-pollinate and cross-pollinate flowers of *S. pulchellum*.

Flowering time is spring and early summer, depending on the genetic constitution of the plants, latitude, altitude, and exposure. May and June are the principal months of flowering, but plants may be in bloom in April or even March in the southern part of the range.

Relationships. *Sedum pulchellum* is one of several species of *Sedum* in eastern North America that have white, 4-merous flowers with the epipetalous stamens scarcely adnate to the bases of the petals. *Sedum nevii*, *S. glaucophyllum*, and *S. ternatum* are others. *Sedum nuttallianum*, with usually 5-merous flowers, yellow petals, and epipetalous filaments more definitely adnate to the petals, may not be of this relationship. On the other hand, J. T. Baldwin, Jr. (1943) conceived *S. pulchellum* to be an amphidiploid which originated as the result of fusion of a 6-chromosome tendency, exemplified by *S. nevii*, and a 5-chromosome tendency, exemplified by *S. nuttallianum*. A similarity in shape of chromosomes of *S. pulchellum* and *S. nuttallianum* favors a relationship of the two species, but also the chromosomes of *S. nuttallianum* resemble those of *Parvisedum*. No clear conclusion is evident.

The one species to which *Sedum pulchellum* appears clearly related is *S. nevii*. Similarities are many and include both the morphology of the flowers and chromosomes. Sometimes plants of *S. nevii* fail to produce secondary vegetative shoots and then behave as annuals. *Sedum pulchellum* might be tetraploid *S. nevii* which has somehow lost one pair of chromosomes. It differs in having the leaves of the floriferous stems with sagittate spurs, the floriferous stems longer, more flowers per cyme, narrower sepals, and eleven pairs of chromosomes.

In conclusion, *Sedum pulchellum* appears to be a specialized derivative of *S. nevii*, whether through autopolyploidy and subsequent loss of chromosomes or through amphidiploidy. Its general relationships are clear, although the precise mechanisms by which it attained its present condition require further elucidation.

References. J. T. Baldwin, Jr. (1943 and 1945), Caudle and Baskin (1968), J. D. Hooker (1876), Masters (1874), McCormick and Rushing (1964), Smith (1943 and 1946), Steyermark (1942 and 1962), Wherry (1934, 1936, and 1950), and Winterringer and Vestal (1956).

6. *Sedum pusillum**** (figs. 28–33)

Michaux named this species *pusillum* (fig. 28), meaning very small. Sometimes the plants are tiny, only 7 mm. tall, with a single terminal flower. Occasional plants may be branched, to 14 cm. tall, and with as many as 118 flowers. Diagnostic features are the annual habit; short-petiolate, elliptic-oblong leaves; pedicillate, white flowers

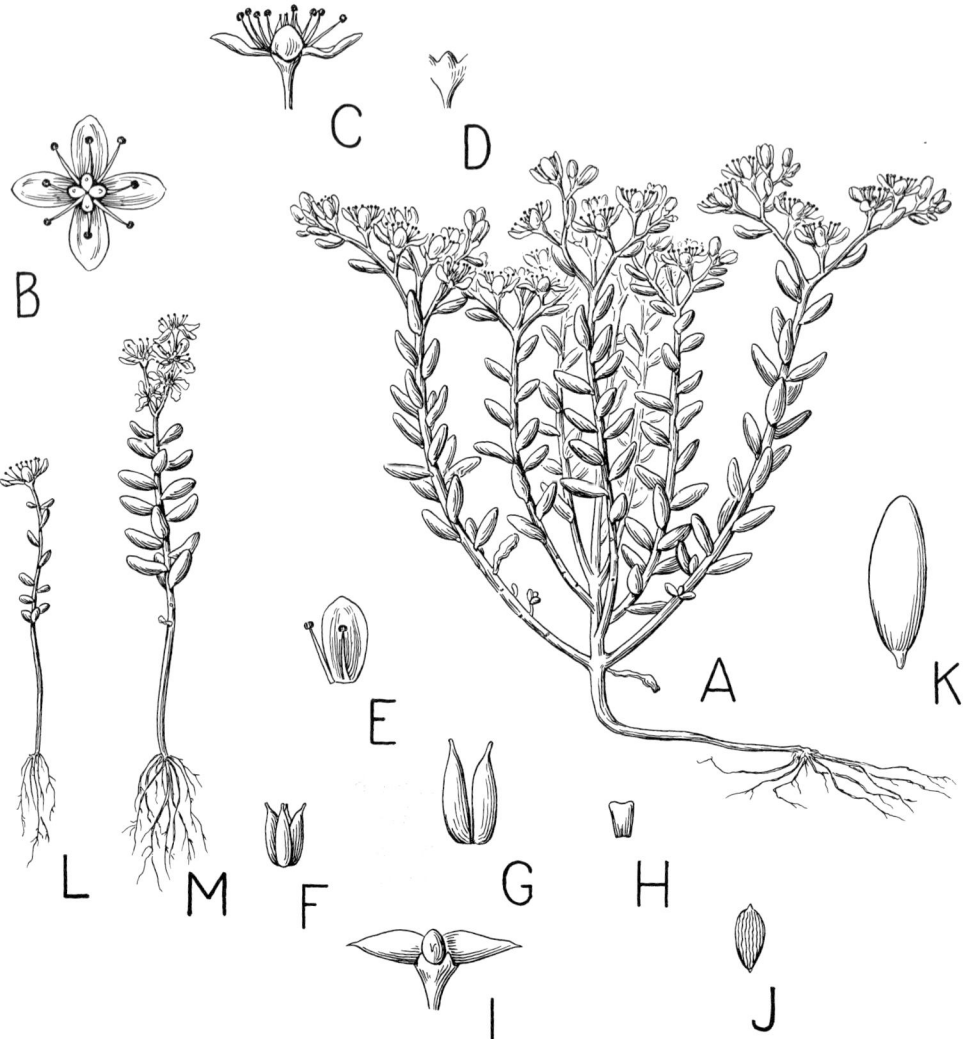

Fig. 28. Plants of *Sedum pusillum* from near Midway, Lancaster Co., S.C. A. Habit sketch of large plant (x 1.2). B. Flower from above (x 3). C. Flower from side (x 3). D. Sepal viewed from back (x 3.5). E. Petal and two stamens (x 3.5). F. Carpels (x 3.5). G. Two carpels (x 5.9). H. Nectary (x 12). I. Immature follicles (x 3.5). J. Seed (x 12). K. Median leaf viewed from top (x 3). L. and M. Habit sketches of small plants (x 1.2).

which are normally 4-merous; slightly channeled, obtuse petals; subquadrate nectaries broadest at the truncate and often emarginate apices; and widely divergent follicles which are not gibbous ventrally, but dehisce along the ventral sutures. This combination of conditions will separate *S. pusillum* from all other species of *Sedum*. In North America, *S. pulchellum* differs in having linear leaves, sessile flowers, pinkish petals, and prominently two-lipped follicles. Furthermore, its leaves and flowers usually are larger. *Diamorpha cymosa* sometimes appears similar to *S. pusillum*, but differs in the opening of the follicles along transverse sutures. In addition, its petals are distinctly hooded and the nectaries are stipitate-reniform.

Description. The sample for study comprised 142 plants: 95 in flower and 47 in fruit. Sources of plants are indicated in the following table in which C = natural clusters, E = early, Ex = extreme, and Q = quadrats one meter square. Figures indicate the number of plants in each category. Selection of both quadrats and natural clusters was random. Early plants were the first to flower at a site, whether inside or outside a chosen quadrat or cluster. Selection of extreme plants was deliberate. These included the tallest, shortest, and most branched plants, and also the one with the most petals. Detailed information about localities in South Carolina is available in the account of the Piedmont Plateau in the chapter on geography.

Locality	Flowering plants (n)				Fruiting plants (n)			Sums
	Q	C	E	Ex	Q	C	Ex	
Kelly Rock, S.C.	1	8	—	1	1	8	0	19
Midway, S.C.	21	6	9	—	5	8	1	50
Forty-acre Rock, S.C.	19	14	10	2	18	6	0	69
Heggie's Rock, Ga.	—	—	—	2	—	—	—	2
Lithonia, Ga.	—	—	—	2	—	—	—	2
Totals	41	28	19	7	24	22	1	142

The 41 flowering plants from quadrats at localities in South Carolina are random samples from 33 randomly selected quadrats within the areas of occurrence. Fractions of populations studied by means of quadrats in 1966 are indicated below.

Site	Total area of occurrence (m.²)	% of area surveyed	Plants in flower in quadrats (n)	Plants studied (n)	(%)
Kelly Rock	113	1.7	6	1	16
Midway	761	2.1	191	21	11
Forty-acre Rock	17,750	.08	294	19	6
Totals	18,624	.18	491	41	8

Fractions of populations studied by means of natural clusters likewise are indicated.

Site	Total clusters (n)	Clusters studied (n)	(%)	Plants in flower in chosen clusters (n)	Plants studied (n)	(%)
Kelly Rock	57	2	3	35	8	23
Midway	403	4	1	200	6	3
Forty-acre Rock	1,364	3	.2	796	14	2
Totals	1,824	9	.5	1,031	28	2.7

Dual methods of sampling were not carried on long enough in 1966 to get proportionate samples from each site, but work was sufficient to indicate some of the problems peculiar to the technique of using natural clusters of plants as sampling units. The definition of clusters for listing was difficult, but possible. The chance of confusion was great unless records were clear. A distinct gain was the larger number of plants in chosen clusters, more than twice as many as in quadrats. Studies of natural clusters contributed both to the description of the species and to estimates of the sizes of populations.

In the following description, sites are designated by letters: F = Forty-acre Rock, K = Kelly Rock, and M = Midway. The letter C indicates that data are derived from plants in natural clusters and Q indicates data from quadrats. Asterisks indicate significant differences among clusters or quadrats. Data from extreme plants are discussed in the appropriate parts of the description, but comparisons of plants which were the earliest to bloom with others which were in flower seventeen days later are in the section on variation.

Plants of *Sedum pusillum* (fig. 29) are annual herbs with fibrous, white roots of small extent. The roots develop horizontally because of the shallowness of the soil in which they grow. The longest roots observed had a length of 4.5 cm. Most roots are much shorter. The stems are erect, finely ribbed, and pink. The hypocotyls are remarkably long, frequently more than 40% of the height of the plants. Individuals growing closely crowded generally are unbranched. Those remote from others and several months old may be branched. The most branched plant had seven branches additional to the main stem. Maximum height observed was 14 cm. at Heggie's Rock. Before expansion of the flowers, stems are curved, with the floral buds turned to one side or downward. A collection from Forty-acre Rock, in the herbarium at Duke University, supports the idea that collectors tend to take large plants. The average height of twenty-three specimens is 7.5 cm., surpassing either of my estimates for the population at the same locality.

	n-pl.	n-obs.	$\bar{x}$	s	w
Height at time of flowering (cm.)					
K—Q	1	1	2.3	—	2.3
C	8	8	3.5	.6	2.8–4.6
M—Q**	21	21	3.6	.7	2.0–6.2
C	12	12	4.3	.8	2.8–5.5
F—Q**	19	19	5.8	1.1	3.0–8.9
C	14	14	3.1	.7	1.4–4.3
Branches (n)					
K—Q	1	1	0	—	0
C	8	8	0	0	0
M—Q	21	21	.5	1	0–3
C	12	12	.1	.3	0–1
F—Q	19	19	1	1.1	0–3
C	14	14	.1	.5	0–2

Leaves are spirally arranged, short-spurred, short-petiolate, elliptic-oblong or sometimes ovate or lanceolate, rounded, minutely papillose, often with a median

NATIVE SPECIES 165

Fig. 29. Flowering plant of *Sedum pusillum* from Midway, Lancaster Co., S.C., cultivated indoors at Ithaca, N.Y., Apr. 28, 1966 (x 3). Photo by Howard Lyon.

dorsal groove, and light green, pinkish upward. They are green in shaded situations, but pinkish in exposed sites. The lower leaves are caducous, while the upper ones persist and are either divergent or appressed. The leaf-scars are raised and have single bundle-scars. Petioles vary in length from .4 to 1.5 mm.

	n-pl.	n-obs.	$\bar{x}$ (mm.)	s (mm.)	w (mm.)
Length					
K—Q	1	1	1.6	—	1.6
C	8	8	4.2	.7	3.2–5.5
M—Q	21	21	4	1.2	2.6–7
C	8	8	3.2	.8	2.0–4.1
F—Q	18	18	4.3	1.2	2.1–5.7
C**	14	14	2.8	.7	1.4–4.8

	n-pl.	n-obs.	$\bar{x}$ (mm.)	s (mm.)	w (mm.)
Width					
K—Q	1	1	1.1	—	1.1
C	8	8	1.7	.2	1.3–2
M—Q	21	21	1.8	.3	1.2–2.3
C	8	8	1.3	.2	1.0–1.6
F—Q*	18	18	1.6	.3	.9–2.2
C**	14	14	1.4	.3	.8–2.4
Thickness					
K—Q	1	1	1	—	1
C	8	8	1.2	.2	1.0–1.4
M—Q	20	20	1.3	.2	.8–1.7
C	6	6	.8	.2	.6–1.1
F—Q	16	16	1	.3	.4–1.7
C	14	14	.6	.2	.3–1

Inflorescences either are cymes or consist of a single flower. One plant at Midway had 118 flowers. Except for being smaller, floral bracts are similar to other leaves.

	n-pl.	n-obs.	$\bar{x}$	s	w
Cymes—diameter (cm.)					
K—Q	1	1	.2	—	.2
C	8	8	.8	.2	.5–1.1
M—Q	21	21	1.2	.4	.6–2.2
C	12	12	.8	.4	.3–1.3
F—Q**	19	19	1.3	.4	.4–2.3
C*	14	14	.5	.1	.4–0.8
Cincinni (n)					
K—Q	1	1	0	—	0
C	8	8	2	0	2
M—Q	21	21	2.1	.5	1–3
C	12	12	1.7	.8	1–3
F—Q**	19	19	2.5	.4	0–3
C	14	14	1	1	0–2
Flowers per central floriferous stem (n)					
K—Q	1	1	1	—	1
C	8	8	4.5	1.5	3–7
M—Q	21	21	5.5	2.5	2–11
C	12	12	3.7	1.5	2–6
F—Q**	18	18	6.8	2.1	1–14
C	14	14	2.5	1.4	1–5

Flowers (fig. 30) are pedicellate and usually 4-merous, but range from 3 to 7-merous. The pedicels enlarge upward and are continuous with the tube of the calyx. The hypanthium is minute, about .1 mm. long.

	n-pl.	n-obs.	$\bar{x}$ (mm.)	s (mm.)	w (mm.)
Pedicels—length					
K—Q	1	1	2.3	—	2.3
C	8	8	1.6	.4	1.0–2.2
M—Q	21	21	1.6	.3	1.0–2.2
C	10	11	1.8	.2	1.3–2.3
F—Q*	15	16	1.6	.6	.6–3.1

NATIVE SPECIES 167

Fig. 30. Flowers of *Sedum pusillum*, plant from Midway, Lancaster Co., S.C., cultivated indoors at Ithaca, N.Y., Apr. 28, 1966 (x 11). Photo by Howard Lyon.

	n-pl.	n-obs.	$\bar{x}$ (mm.)	s (mm.)	w (mm.)
Flowers—diameter					
K—Q	1	1	2.4	—	2.4
C*	8	8	5.8	.7	4.2–7
M—Q	21	21	6.5	.6	5.0–7.5
C	6	7	5.4	1.2	3.7–6.5
F—Q*	15	16	4.3	.7	2.7–6.4
C**	14	16	4.6	.3	3.6–5.5

Sepals are broadly ovate to reniform and continuous with the tissue of the pedicels, connate below for .3 mm., rounded apically, and green sparsely speckled with pink. They persist at the base of the follicles.

	n-pl.	n-obs.	$\bar{x}$ (mm.)	s (mm.)	w (mm.)
Length					
K—Q	1	1	.6	—	.6
C	8	8	.7	.1	.6–1

	n-pl.	n-obs.	x̄ (mm.)	s (mm.)	w (mm.)
Length					
M—Q	21	21	.6	.1	.5–0.8
C	9	10	.6	.1	.4–0.8
F—Q	15	16	.5	.1	.4–0.6
C**	14	16	.6	.05	.4–0.7
Width					
K—Q	1	1	.8	—	.8
C	8	8	1	.1	.9–1.1
M—Q	21	21	1.1	.1	.6–1.2
C	9	10	.9	.2	.6–1.2
F—Q	15	16	1	.2	.8–1.3
C*	14	16	.9	.1	.6–1.1

Petals are separate, divergent, ovate-elliptical, slightly channeled, obtuse, and white. Sometimes they are streaked with pink.

	n-pl.	n-obs.	x̄	s	w
Number					
K—Q	1	1	4	—	4
C	8	8	4	0	4
M—Q	21	21	4	.2	4–5
C	6	7	4	0	4
F—Q	15	16	4	0	4
C	14	16	4	.1	4–5
Length (mm.)					
K—Q	1	1	1.4	—	1.4
C*	8	8	3	.2	2.5–3.3
M—Q	21	21	3.4	.5	2.2–4.2
C	6	7	2.7	.5	1.7–3.6
F—Q	15	16	2.3	.4	1.4–3.1
C*	14	16	2.2	.2	1.7–2.8
Width (mm.)					
K—Q	1	1	.8	—	.8
C	8	8	1.7	.1	1.5–1.8
M—Q	21	21	1.8	.3	.9–2.2
C	6	7	1.4	.3	.8–1.8
F—Q	15	16	1.5	.3	1.1–2
C	14	16	1.4	.2	1.1–1.8

Stamens have the filaments white, sometimes streaked with pink, and the anthers dark red.

	n-pl.	n-obs.	x̄ (mm.)	s (mm.)	w (mm.)
Epipetalous filaments—height on petals					
K—Q	1	1	.1	—	.1
C*	8	8	.1	.06	.10–.3
M—Q*	21	21	.1	.03	.05–.2
C	6	7	.1	.02	.05–.1
F—Q	15	16	.1	.04	.02–.2
C	14	16	.1	.07	.05–.3

NATIVE SPECIES   169

	n-pl.	n-obs.	x̄ (mm.)	s (mm.)	w (mm.)
Anthers—length					
K—C	8	8	.5	.05	.5–.6
M—Q	16	16	.5	.08	.4–.7
C	4	4	.5	.11	.4–.6
F—Q	6	7	.5	.06	.4–.6
C	7	8	.5	.05	.4–.6

Nectaries are subquadrate, broadest at apex, truncate, white or pale yellow, and translucent. Sometimes they are emarginate.

	n-pl.	n-obs.	x̄ (mm.)	s (mm.)	w (mm.)
Length					
K—Q	1	1	.2	—	.2
C	8	8	.4	.03	.3–.4
M—Q*	21	21	.4	.04	.3–.4
C	6	7	.3	0	.3–.4
F—Q	15	16	.3	.06	.2–.4
C	14	16	.3	.05	.3–.4
Width					
K—Q	1	1	.2	—	.2
C	8	8	.3	.03	.2–.3
M—Q	21	21	.3	.03	.2–.4
C	6	7	.2	.02	.2–.3
F—Q	15	16	.3	.04	.2–.3
C	14	16	.3	.04	.2–.4

Carpels are at first erect, but become widely divergent in age. They are separate, but may appear connate because their bases are deeply set in the tissue of the receptacle. The carpels are either white or white streaked with pink when flowers are in anthesis. As they mature, they become light green suffused with pink along the ventral suture. The margins are weakly sealed. When the stigmas are receptive to pollen, they are divergent and greenish yellow.

	n-pl.	n-obs.	x̄	s	w
Length (mm.)					
K—Q	1	1	.9	—	.9
C	8	8	1.6	.2	1.3–1.8
M—Q	21	21	1.9	.3	1.0–2.4
C	6	7	1.4	.4	.7–1.9
F—Q	15	16	1.4	.4	.5–2
C	14	16	1.4	.2	.8–1.7
Styles—length (mm.)					
K—Q	1	1	.2	—	.2
C	8	8	.3	.05	.2–.4
M—Q	21	21	.3	.06	.2–.4
C	6	7	.2	.11	.1–.4
F—Q	15	16	.2	.05	.1–.3
C	14	16	.2	.04	.1–.3

SEDUM OF NORTH AMERICA

	n-pl.	n-obs.	$\bar{x}$	s	w
Ovules per ovary (n)					
K—Q	1	1	2	—	2
C	8	8	6.5	1	5–8
M—Q	21	21	7	1.9	2–12
C	6	7	5.2	1.7	2–8
F—Q	15	16	4.7	1.1	2–6
C*	14	16	4.8	1.1	2–8

Fruits (fig. 31) are widely divergent and pale brown.

	n-pl.	n-obs.	$\bar{x}$	s	w
Length (mm.)					
K—Q	1	2	2.2	—	2.1–2.3
C	7	14	2.5	.4	2.2–3.9
M—Q	5	9	2	.4	1.5–2.6
C	6	11	2.5	.2	2.0–3.1
F—Q	11	22	3	.4	2.1–4.2
C	5	10	2.5	.4	1.3–3
Beaks—length (mm.)					
K—Q	1	2	.25	—	.2–.3
C	7	14	.24	.05	.2–.4
M—Q	5	8	.26	.04	.2–.3
C	6	11	.22	.06	.1–.3
F—Q	10	20	.25	.08	.1–.5
C	4	8	.21	.05	.1–.4
Seeds per follicle (n)					
K—Q	1	2	1	—	0–2
C	7	14	3	1.1	2–5
M—Q	5	9	1	1.6	0–4
C	6	11	4	1.7	1–9
F—Q*	11	22	5	1.8	0–10
C	5	10	4	1.2	2–7

Seeds are pyriform, finely ribbed, and brown.

	n-pl.	n-obs.	$\bar{x}$ (mm.)	s (mm.)	w (mm.)
Length					
K—Q	1	2	.7	—	.7
C	7	14	.7	.07	.6–.8
M—Q	3	6	.7	.05	.7–.8
C	6	12	.7	.1	.6–.9
F—Q**	11	22	.7	.04	.6–.9
C	5	10	.7	.04	.6–.7
Diameter					
K—Q	1	2	.3	—	.3
C	7	14	.3	.04	.2–.3
M—Q	3	6	.3	.03	.3–.4
C*	6	12	.3	.02	.3–.4
F—Q**	11	22	.3	.02	.2–.4
C	5	10	.3	.03	.2–.3

Fig. 31. Immature fruits of *Sedum pusillum*, plant from Midway, Lancaster Co., S.C., cultivated indoors at Ithaca, N.Y., Apr. 28, 1966 (x 28). Photo by Howard Lyon.

Chromosomes are unequal in size and with median or submedian centromeres.

	n-pl.	g	sp	Cytologist
Number				
M	4	4	—	Charles Horn, 1966
F	1	4	—	Charles Horn, 1966
Stone Mt., Ga.	>1	4	8	J. T. Baldwin, Jr. (1940)

Variation. Although *Sedum pusillum* occurs in a small geographic area and is present there only on exposures of granite or in their immediate vicinity, it is remarkably variable. Factors responsible for the variation appear to be age, density, depth of soil, light, moisture, and probably to a lesser degree, the genetic constitution of the plants. Seeds that germinate in the fall in deep soil, remote from competitors, have the best chance to grow into large plants. Those that germinate in the spring in dense clusters in shallow soil usually are small and slender.

A comparison of the tallest and shortest plants studied in 1966 from Forty-acre Rock shows great differences in most dimensions. Figures for flowers and floral parts of the tallest plant are averages of two observations.

	Tallest	Shortest
Date	April 9	April 9
Height (cm.)	11	1.4
Median leaves—l. (mm.)	7.8	1.7
Extra floriferous branches (n)	4	0
Flowers (n)	10	1
Flowers—diameter (mm.)	8	3.8
Petals—l. (mm.)	3.9	1.7
Anthers—l. (mm.)	.7	.5
Nectaries—l. (mm.)	.35	.3
Carpels—l. (mm.)	2.4	1.1
Ovules per ovary (n)	12	4

To compare the condition of plants collected early in the season with those gotten later, I drew two random samples from the same quadrat, one meter square, at Forty-acre Rock. Selections were two and a half weeks apart. Comparison is by means of Wilcoxon's rank sum test (Wilcoxon and Wilcox, 1964). Asterisks indicate significant differences. The results demonstrate that the time of making a study may have a profound effect on the interpretation of variation in a population.

	Early	Late
Date	March 30	April 16
n	10	9
Height (cm.)	6.1	6.5
Extra floriferous branches (n)*	2.1	.9
Median leaves—l. (mm.)	5.1	4.2
Flowers (n)**	12.1	8.2
Flowers—diameter (mm.)**	5.8	3.8
Petals—l. (mm.)**	3.1	2.1
Nectaries—l. (mm.)	.4	.3
Carpels—l. (mm.)**	2	1.3
Ovules per ovary (n)**	8.4	4

Comparison of data from quadrats and natural clusters provides an opportunity to note differences in features which might be affected by density. Expectation was that plants in dense stands might be slenderer, with fewer branches, and smaller dimensions. The following comparison is between plants in quadrat M15–5 near Midway Corners, with the highest density per square meter, and quadrat M4–7 at the same location, with lower density. Collection and study of plants was on April 14. Quadrats with lower density than M4–7, for example with 3 plants per square meter, did not yield sufficient data to make possible a meaningful comparison.

	M15–5	M4–7
N	1,143	211
n	6	5
Height (cm.)	3.9	4.4
Extra floriferous branches (n)	.2	1
Median leaves—l. (mm.)	3.8	4.7
Flowers (n)	4.6	6.8
Flowers—diameter (mm.)	6.9	6.4
Petals—l. (mm.)	3.6	3.3
Nectaries—l. (mm.)	.38	.33
Carpels—l. (mm.)	2.1	1.7
Ovules per ovary (n)	7	6.5

Another comparison involves the densest and least dense clusters studied at Forty-acre Rock. Study was on April 9.

	N.W. 264	N.W. 434
N	1,145	158
n	6	4
Density per m.^{2}	3,202	179
Height (cm.)	2.7	3.2
Extra floriferous branches (n)	0	.5
Median leaves—l. (mm.)*	1.9	3.4
Flowers (n)	2	2.5
Flowers—diameter (mm.)**	4	4.9
Petals—l. (mm.)*	2	2.4
Nectaries—l. (mm.)	.33	.36
Carpels—l. (mm.)	1.3	1.5
Ovules per ovary (n)*	3.6	6.1

Tests of differences in the preceding two tables are by means of Wilcoxon's rank sum test. The small differences between the two quadrats at Midway may be attributed to chance. Differences in sizes of floral parts and even the diameter of the flowers are contrary to expectation. The explanation may be that 95% of quadrat M15–5 is gravelly and good habitat for *S. pusillum*, whereas only 10% of quadrat M4–7 is gravelly. Since the plants are crowded in a small part of quadrat M4–7, difference in density between the two quadrats is less great than the numbers suggest. This example illustrates a problem which arises when arbitrary quadrats are employed as sampling units. The comparison of natural clusters at Forty-acre Rock results in significant differences in four features. Further, differences are according to expectation. Lack of significance in differences in height and number of floriferous branches indicates the impossibility of an infallible rule about the effect of density on stature and branching. Other factors, as depth of soil, age, and moisture may also affect variation. Depth of soil is easy to measure, but measurement must be for each plant individually. My measurements for *S. pusillum* were general for whole groups of plants or quadrats. Further, they indicate only the range in depths per plot. For these reasons, the data are inadequate for a valid test of correlation of characteristics of plants and depth of soil. Variation in depth of soil is great, even in small areas. The importance of depth of soil and the need for measurements on the basis of single plants did not impress me until the latter part of the study. Age is more difficult to appraise in the field. It requires either marking seedlings and making periodic observations, or the experimental approach, with plants started at different dates in order to detect the effect of photoperiod and temperature on rate of development. Because moisture is difficult to measure in the field, the experimental approach likewise may be most helpful in determining its effect.

The number of petals per flower varies from 3 to 7, with the average 4. The frequency of different numbers of petals in my random samples is shown in the following table. Extreme conditions did not occur in these samples.

Petals per flower (n)	Flowers (n)	Flowers %	Cumulative %
4	71	97	97
5	2	3	100
Total	73	100	

Study of populations of *Sedum pusillum* on the outcroppings of granite on the Piedmont Plateau of northern South Carolina revealed various intrapopulational differences among clusters and quadrats, as is indicated in the description. Much of this variation probably results from environmental modification.

The data in the detailed description provide some basis for comparing responses at the different sites and also for contrasting results from the use of the two different kinds of sampling units, namely quadrats and natural clusters of plants. For fifteen features, differences between samples from the same site are greater than differences among sites. Many of these differences observed in nature probably are of no genetic significance. The assumption is that the samples estimate populations with similar means. The populations do not differ from each other in any important way. Although plants vary enormously, the variation mostly is the result of environmental modification.

Nomenclature. *Sedum pusillum* André Michaux, Flora Boreali-Americana 1:276 (1803). Type locality: Flat Rock, Kershaw Co., S.C., on the Piedmont Plateau. Type: Michaux, in the herbarium of the Muséum National d'Histoire Naturelle, Laboratoire de Phanerogamie, in Paris, not seen. Two clusters of fruit, a fragment of Michaux's type, are at the Gray Herbarium. Although the type demonstrates that Michaux collected the species, doubt remains whether he also included *Diamorpha* in his concept of *S. pusillum*. Since that flowers later than the *Sedum*, he may have collected plants of only the one species. According to Gray (1889), the date of Michaux's visit to Flat Rock was April 1795. Quarrying at the type locality, with accompanying scraping of the surface of the land, has eliminated the *Sedum* from there.

A synonym is:

1903. *Tetrorum pusillum* (Michx.) Rose, N. Am. Flora 22 (1):59.

Distribution. *Sedum pusillum* (fig. 32) is a species of restricted and local range. It occurs on outcroppings of granite on the Piedmont Plateau from near Wadesboro,

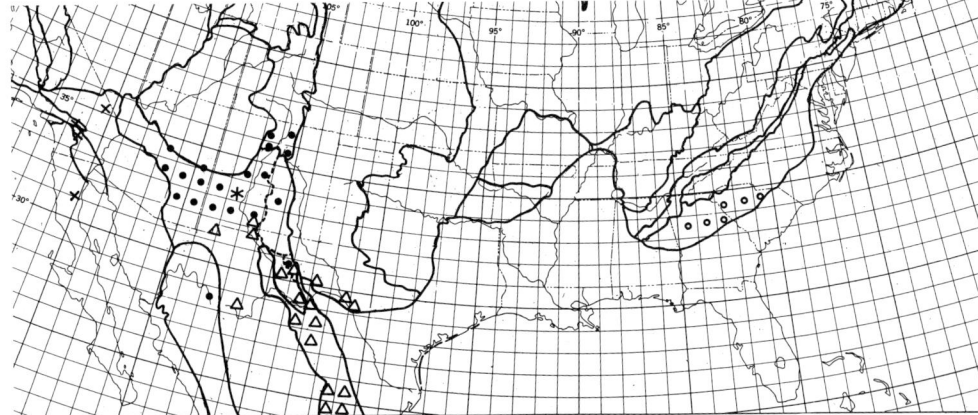

Fig. 32. Known distribution of Sedum pusillum (○), S. niveum (x), S. cockerellii (●), and S. wrightii (△). Asterisk equals record from literature for S. cockerellii.

Anson Co., N.C., to near Camak, Warren Co., Ga., and Panola Mountain, Rockdale Co., Ga., an area with a latitudinal extent of less than two degrees. A report from Henderson Co., N.C., by Radford, Ahles, and Bell (1964, 1965) is a mistake.

Seventeen disjunct populations of *Sedum pusillum* are known to me. All are on outcrops of granite. The plants characteristically grow in large numbers in shallow gravel and sand or in thin mats of moss about the edges of open areas of rock (fig. 33), either where there is little or no other vegetation or at the edge of other vegetation.

I studied three populations in northern South Carolina in March and April 1966. These were at Kelly Rock, Midway Corners, and Forty-acre Rock. They are all that are known and extant in the region between the Catawba and Lynches Rivers. Kelly Rock is the nearest site to the type locality. Data for these populations are available in the following table. Additional information about the areas in which they occur is in the chapter on geography, under the Piedmont Plateau.

Population	Sampling area	N plants, estimate	N plants (C.I., P = .05)	Area of population (m.^{2})	Density per m.^{2}
Forty-acre Rock	No. 20	968,562	—	17,750	54.57
Midway Corners	No. 21	101,880	764–207,800	375	381.50
Kelly Rock	No. 23	11,400	400–47,895	113	100.88

Population	Altitudinal range (m.)	pH, range	pH, n tests	Depth of soil (cm.)	Oldest record
Forty-acre Rock	150–168	4.8–6	7	.3–3.5	1939
Midway Corners	134–135	4.8–5.6	12	1.0–2.7	1953
Kelly Rock	165	4.6–5.6	3	.2–11	1966

The above figure for number of plants at Forty-acre Rock is the result of a ratio estimate. I did not calculate a confidence interval for that population. The other two

Fig. 33. Dense stand of *Sedum pusillum* on granite at Forty-acre Rock, Lancaster Co., S.C., Apr. 9, 1966.

estimates are expansions based on average numbers of plants per quadrat at Midway Corners and per natural cluster at Kelly Rock.

Estimate of the number of plants at Midway in 1966 was by two methods: natural clusters and quadrats.

Kind of unit	N-units	n-units	$\bar{x}$-pl.	$s_{\bar{x}}$	N-pl.	Confidence Interval .05
Natural clusters	400	3	254.7	61.55	101,880	764–207,800
Quadrats 1 m.²	725	16	180.6	87.18	130,935	2,890–265,640

The smaller standard error of the mean for natural clusters makes that estimate superior and yields a narrower confidence interval.

Sedum pusillum grows where vegetation is sparse. Other species present are diverse. Among the more frequent competitors are an *Andropogon* at Kelly Rock; a grass and *Diodia teres* at Midway Corners; and *Andropogon scoparius* and *Gelsemium sempervirens* at Forty-acre Rock. *Juniperus virginiana* often shades the *Sedum*. In this condition, the habitat is a cedar glade.

Sedum pusillum and *Diamorpha cymosa* rarely occur intermixed. Although both species occur at Kelly Rock, they grow separately there. At Midway, only the *Sedum* is present. At Forty-acre Rock, the two species are intermingled in a few places: in one randomly selected meter-quadrat out of fifteen and in one randomly selected natural cluster out of three. The *Diamorpha* usually is in depressions where water stands, but at Midway the *Sedum* was growing in water in one depression. Even where the two species appear intermixed, a sorting exists. In such situations, the *Diamorpha* is in the shallower substrate.

Reproduction. *Sedum pusillum* is an annual. Successful flowering and fruiting are essential for perpetuation of the species. Flowering occurs in early spring. The time from flowering stage to the dehiscence of fruits with ripe seeds is about five weeks. The seeds presumably germinate in the fall and winter.

In 1966 the period of flowering in northern South Carolina extended from March 30 to April 16 and possibly later. Sunlight and warmth seemed most important in causing the buds to open. At the time when the flowers expanded, small bees and flies swarmed on them. Two species of butterflies also were visitors. The flies belong to the families *Syrphidae* and *Anthomyidae*, according to L. L. Pechuman. The bees are unidentified. The butterflies are *Euchloe genutia*, the Falcate-Orangetip, and *Thecla damon*, the Olive Hair-streak. J. G. Franclemont has confirmed the identification of the *Lepidoptera*.

Where *Sedum pusillum* and *Diamorpha* occur together, the *Sedum* is ahead of the *Diamorpha*. It may have immature fruits when the *Diamorpha* still is in bud. This might have been the condition when Michaux visited Flat Rock in 1795. If so, it would explain why he collected only the *Sedum*. I saw no evidence that *S. pusillum* and *Diamorpha* hybridize at the sites I visited, although I searched for hybrids. In this respect, my experience was different from that of McCormick and Platt (1964). The difference in number of chromosomes between *S. pusillum* and *Diamorpha*, respectively 4 and 9 pairs, and the differences in time of flowering and edaphic conditions, are unfavorable for hybridization.

Relationships. The relationships of *Sedum pusillum* are clear in a general way, but obscure in a precise way. The species has the lowest number of chromosomes known in the *Crassulaceae*. It is moderately advanced in floral anatomy. Sherwin and Wilbur (1971) regarded the vascular pattern of *S. pusillum* to be unique, but derived from Quimby's (1971) group 3. According to them, three whorls of vascular traces are the usual situation. This condition could be derived from the pattern with four whorls found in certain other eastern North American species of *Sedum*, as *S. ternatum*, *S. nuttallianum*, *S. pulchellum*, and also *Diamorpha cymosa*. The carpels are widely divergent in fruit and without ventral gibbosities, as in *Crassula aquatica*, which has minimal fusion of vascular bundles supplying the floral parts.

Fröderström (1930–1935) classified *Sedum pusillum* with *S. nuttallianum* and *Diamorpha cymosa*. *Sedum nuttallianum* has yellow flowers and gibbous carpels. *Diamorpha* has follicles which dehisce transversely. *Diamorpha* may have had a parallel evolution to *S. pusillum* or it even may be derived from the same ancestral stock.

Berger (1930), in his synopsis of *Sedum*, placed *S. pusillum* near *S. nevadense* of southern Europe and northern Africa and *S. pedicellatum* of the Iberian Peninsula. On the basis of present evidence, my opinion is that *S. pusillum* may be descended from plants which were the ancestors of *S. ternatum, S. nevii,* and other 4-merous, white-flowered Appalachian species.

7. Sedum niveum*** (figs. 32 and 34)

Plants of *Sedum niveum* (fig. 34) have erect, nongibbous follicles; white flowers, 1 to 9 in a cyme; tuberous roots; and stout, prostrate, main stems which give rise to rosettes at the nodes, resulting in a dense, cespitose aspect. Both *S. cockerellii* and *S. griffithsii* resemble *S. niveum*. Those species have more flowers, 4 to 27 per cyme,

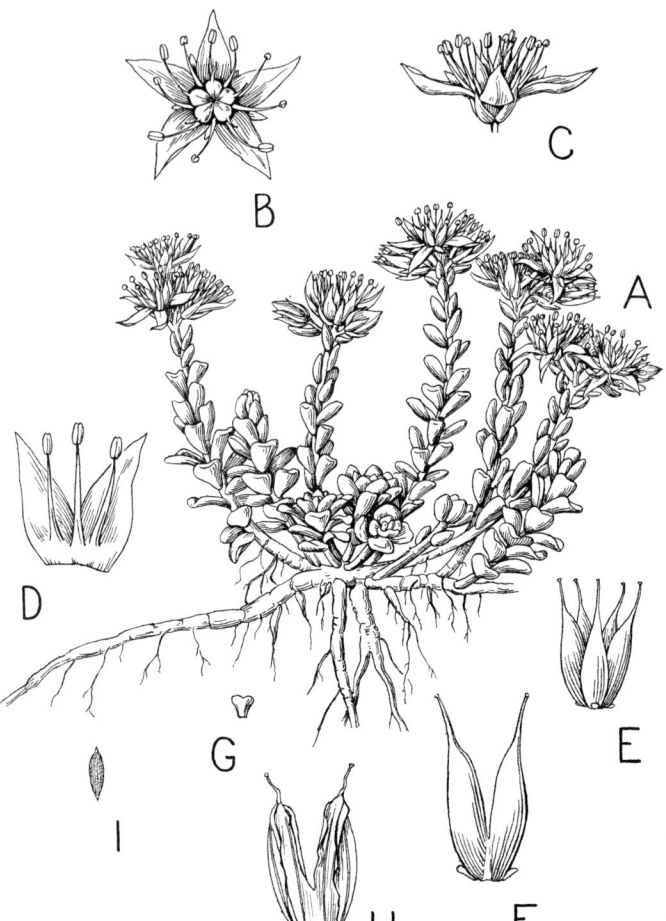

Fig. 34. Plant of *Sedum niveum* from northern slope of Sugarloaf Mountain, San Bernardino Mountains, Calif. A. Habit sketch (x .9). B. Flower from above (x 1.8). C. Flower from side (x 1.8). D. Two petals and stamens (x 2.7). E. Carpels (x 2.7). F. Two carpels (x 3.6). G. Nectary (x 4.5). H. Follicle from side (x 3.6). I. Seed (x 9).

and shorter primary stems. Also, the median leaves of their floriferous stems are longer.

In some respects, *Sedum niveum* resembles *S. minimum* of the Trans-Mexican Volcanic Belt and the Mexican Plateau. The latter differs in being biennial, with characteristic corms. The two species are similar in having erect, orthocarpic follicles; strong pigmentation of the flowers; and a tendency for loss of the stamens opposite the petals.

Description. Sample: 48 plants: 17 plants chosen randomly and proportionately, that is 5% of the plants in each of 8 randomly selected quadrats of 100 m.^{2} on the northern slope of Sugarloaf Mountain in the San Bernardino Mountains; 10 plants from two of the same quadrats, these being the only plants in flower in the two quadrats at the beginning of the flowering season; 18 plants selected randomly from a cluster of 357 plants in the upper part of the area of the quadrats; a single plant chosen randomly from a quadrat of 100 m.^{2} containing 20 plants on the lower slope of Charlton Peak east of Dollar Lake; and 2 plants collected by Reid Moran on Observatory Peak in the Sierra San Pedro Mártir, Lower California, one from 2,550 m. and the other from 2,800 m. In the following tables, C = Charlton Peak, LC = Lower California, S = randomly selected plants from random quadrats on Sugarloaf Mountain, S—e = early-flowering plants from two quadrats on Sugarloaf Mountain, and S—u = randomly selected plants from the upper cluster above the trail on Sugarloaf Mountain.

Plants of *Sedum niveum* are perennial herbs with tuberous, tufted primary roots which are white or pale brown, up to 15 cm. long and 5 mm. in diameter, and with capillary white rootlets. The primary stems are repent, branched, pale brown or light gray, to 16 cm. long and 5.5 mm. in diameter, and give rise to axillary shoots which develop into rosettes and eventually procumbent or erect floriferous stems.

	n-pl.	n-rep.	n-obs.	$\bar{x}$	s	w
Plants—diameter (cm.)						
S	17	—	17	8.6	4.9	3.5–21
S—u	18	—	18	8.9	3.1	4.0–17
C	1	—	1	6	—	6
Primary rosettes—number per plant						
S	16	—	16	14.9	11.6	2–36
S—u	18	—	18	17.2	12.4	6–49
C	1	—	1	9	—	9
Primary rosettes—diameter (cm.)						
S	16	—	45	1.1	.2	.6–1.7
S—u	13	—	39	1.1	.1	.8–1.5
C	1	—	3	1.1	—	1.0–1.2
S—gh.	9	12	36	1.5	—	.8–3.2
LC—gh.	1	2	6	.9	—	.6–1.5
Floriferous stems—number per plant						
S	17	—	17	7.2	6.9	1–19
S—u	18	—	18	8.9	6.3	0–21
C	1	—	1	2	—	2

	n-pl.	n-rep.	n-obs.	$\bar{x}$	s	w
Floriferous stems—length (cm.)						
S	17	—	40	2.3	.6	1.0–4.5
S—u	16	—	49	2.1	.4	1.0–3
S—e	10	—	12	2.1	1.3	.6–4.4
C	1	—	2	2.5	—	1.5–3.6

Leaves are alternate, sessile, and plano-convex. Those of the rosettes are obovate to oblanceolate, obtuse or rounded and bluntly submucronate, papillose apically, and dark to yellow-green sometimes speckled with red. Those of the floriferous stems are short-spurred, obovate, oblanceolate, or elliptic-ovate, obtuse to rounded, papillose, and dark to light green, variously speckled with red and sometimes appearing dark red.

	n-pl.	n-rep.	n-obs.	$\bar{x}$ (mm.)	s (mm.)	w (mm.)
Leaves of rosettes—length						
S	16	—	46	7.2	1.1	4.7–9.7
S—u	13	—	39	7.1	.9	4.8–10
C	1	—	3	6	—	5.2–7
S—gh.	10	13	39	10	—	4.8–20
LC—gh.	1	2	6	6	—	3.9–8.8
Leaves of rosettes—width						
S	16	—	46	4.2	.6	3.1–6.2
S—u	13	—	39	3.5	.3	2.8–4
C	1	—	3	4	—	3.4–4.8
S—gh.	10	13	39	4.1	—	2.5–7
LC—gh.	1	2	6	2.2	—	2.0 3
Leaves of rosettes—thickness						
S	16	—	46	1.8	.2	1.1–2.4
S—u	9	—	17	1.3	.8	1.0–1.8
C	1	—	3	1.5	—	1.3–1.8
Median leaves of floriferous stems—length						
S	17	—	37	6.5	1	4.4–9.2
S—u	16	—	47	6.2	.9	4.2–8.2
S—e	10	—	12	6	.7	5.2–7.1
C	1	—	2	6.7	—	5.4–8
S—gh.	10	13	35	9	—	4.0–25
LC—gh.	1	2	6	4.8	—	3.4–8
Median leaves of floriferous stems—width						
S	17	—	37	3.5	.6	2.1–6
S—u	16	—	47	3.2	.6	1.9–4.5
S—e	10	—	12	3.5	.6	2.2–4.4
C	1	—	2	4.4	—	3.8–5
S—gh.	10	13	35	4	—	2.1–7
LC—gh.	1	2	6	2.1	—	1.6–2.6
Median leaves of floriferous stems—thickness						
S	17	—	37	1.6	.3	.8–2.7
S—u	16	—	47	1.3	.2	.7–2
S—e	10	—	12	1.7	.3	1.3–2.4
C	1	—	2	1.7	—	1.5–2

Inflorescences are terminal cymes of 1 to 3 cincinni and 1 to 9 flowers. The floral bracts are similar to the leaves of the floriferous stems, but smaller and appressed to the flowers.

	n-pl.	n-obs.	$\bar{x}$ (n)	s (n)	w (n)
Cincinni					
S	17	37	.9	.5	0–3
S—u	16	48	1	.6	0–2
S—e	10	12	1.2	1	0–3
C	1	2	.5	—	0–1
Flowers					
S	17	37	2.2	.8	1–7
S—u	16	48	2.3	.8	1–5
S—e	10	12	2.7	1.6	1–6
C	1	2	1.5	—	1–2

Flowers are subsessile, with pedicels up to 1 mm. long, and mostly 5-merous, but variable in number of floral parts.

	n-pl.	n-obs.	$\bar{x}$ (mm.)	s (mm.)	w (mm.)
Diameter					
S	16	40	13.4	1.8	9–18
S—u	16	38	12.1	1.4	10–16
S—e	10	12	14.6	1.6	11–16
C	1	2	14	—	11–18

Sepals are divergent or suberect, very unequal, planoconvex, lanceolate, oblong-lanceolate, or oblanceolate-elliptic, acute or obtuse, sometimes papillose, green streaked with red, occasionally appearing dark red. The tissue of the sepals appears continuous with that of the pedicels.

	n-pl.	n-rep.	n-obs.	$\bar{x}$	s	w
Number						
S	16	—	42	5.1	.3	5–6
S—u	16	—	39	5.2	.6	5–8
S—e	10	—	12	5.1	.3	5–6
C	1	—	2	5	—	5
Length (mm.)						
S	16	—	42	4.8	.6	3.1–7.5
S—u	16	—	39	4.4	1.1	2.5–7
S—e	10	—	12	5	.9	3.1–6.2
C	1	—	2	5.3	—	2.7–7.9
S—gh.	8	11	17	7.2	—	2.5–19
LC—gh.	1	2	5	4.9	—	4.2–5.9
Width (mm.)						
S	16	—	42	1.4	.2	.9–2.3
S—u	16	—	39	1.2	.3	.6–2.4
S—e	10	—	12	1.7	.6	1.0–3
C	1	—	2	1.7	—	.9–2.5

Petals are erect for about one-third of their length, then widely divergent, lanceolate, acute, mucronate-appendaged, and white streaked with deep pink ventrally in region of bend and dorsally along median keel and toward apex. The petals are either slightly connate basally or held together by means of the bases of the stamens which alternate with them. Petals frequently fork. When this happens at base, an extra petal results. When there are eight petals, they may be in two whorls.

	n-pl.	n-rep.	n-obs.	$\bar{x}$	s	w
Number						
S	16	—	42	5.2	.3	5–6
S—u	16	—	39	5.2	.7	4–8
S—e	10	—	12	5.3	.5	4–6
C	1	—	2	5.0	—	5
S—gh.	8	11	17	5.6	—	5–8
LC—gh.	1	2	5	5.0	—	5
Length (mm.)						
S	16	—	42	7.1	.6	5.0–9
S—u	16	—	39	6.8	.9	5.4–9.3
S—e	10	—	12	7.3	.7	5.8–8.7
C	1	—	2	7.1	—	6.0–8.2
S—gh.	8	11	17	7.5	—	6.4–10.2
LC—gh.	1	2	5	7	—	6.2–7.8
Length of cohesion (mm.)						
S	16	—	41	.1	.1	0–0.8
S—u	16	—	38	.1	.1	0–0.5
S—e	10	—	12	.6	.4	0–1.4
C	1	—	2	.2	—	0–0.5
Width (mm.)						
S	16	—	42	2.7	.4	1.7–4
S—u	16	—	39	2.4	.5	1.5–3.2
S—e	10	—	12	2.7	.3	2.1–3.3
C	1	—	2	3.1	—	2.5–3.7

Stamens are normally twice as many as petals, but sometimes the epipetalous stamens are aborted or vestigial. The filaments are white streaked with red, appearing pink. The anthers are basifixed, four-celled, and dark red. Various irregularities occur in number of stamens. Sometimes a stamen alternating with petals is double, one on top of the other. Epipetalous stamens may be forked or double. The pollen is yellow.

	n-pl.	n-rep.	n-obs.	$\bar{x}$ (mm.)	s (mm.)	w (mm.)
Filaments—length						
S	16	—	42	5.9	.7	4.1–7.8
S—u	16	—	39	5.7	.6	4.6–7.8
S—e	10	—	12	5.7	.6	4.8–6.6
C	1	—	2	6.2	—	5.7–6.7
S—gh.	8	11	17	5.5	—	4.5–6.9
LC—gh.	1	2	5	4.8	—	3.9–5.9
Epipetalous filaments—height on petals						
S	16	—	41	1.2	.3	.5–2
S—u	16	—	39	1.3	.3	.7–2.2
S—e	10	—	12	1.2	.4	.5–1.8
C	1	—	2	1.1	—	.9–1.4

	n-pl.	n-rep.	n-obs.	$\bar{x}$ (mm.)	s (mm.)	w (mm.)
Anthers—length						
S	15	—	27	1.2	.9	.9–1.3
S—u	5	—	6	1	.2	.8–1.2
S—e	2	—	3	1.1	.2	1.0–1.4
C	1	—	1	.9	—	.9

Nectaries are reniform and stipitate or subquadrate, thickened upward, truncate or emarginate, and yellow, orange, or pink, sometimes streaked with red.

	n-pl.	n-rep.	n-obs.	$\bar{x}$ (mm.)	s (mm.)	w (mm.)
Length						
S	16	—	41	.5	.1	.3–.7
S—u	16	—	39	.5	.1	.3–.7
S—e	10	—	12	.6	.1	.4–.8
C	1	—	2	.6	—	.5–.7
S—gh.	8	11	17	.5	—	.3–.7
LC—gh.	1	2	5	.7	—	.6–.9
Width						
S	16	—	41	.7	.1	.5–1
S—u	16	—	39	.7	.1	.5–1
S—e	10	—	12	.7	.1	.5–0.9
C	1	—	2	.7	—	.6–0.8

Carpels are erect from a plane receptacle, connate basally, and white with three red veins as well as variously streaked with red. The styles are slender, short, divergent, and white. Carpels vary in size within a flower. Some are abortive.

	n-pl.	n-rep.	n-obs.	$\bar{x}$	s	w
Number						
S	16	—	42	4.9	.6	3–6
S—u	16	—	39	5.1	.4	5–7
S—e	10	—	12	5.3	.7	5–7
C	1	—	2	5	—	5
Length (mm.)						
S	16	—	41	5.9	.9	3.5–7.8
S—u	16	—	39	5.5	1	3.1–8
S—e	10	—	12	4.7	.7	3.4–5.9
C	1	—	2	5.7	—	4.8–6.6
S—gh.	8	11	17	5.9	—	4.3–7.2
LC—gh.	1	2	5	5.6	—	4.9–6.4
Styles—length (mm.)						
S	16	—	41	1.3	.2	.6–1.8
S—u	16	—	39	1.4	.3	1.0–2
S—e	10	—	12	1.2	.3	.5–1.7
C	1	—	2	1.3	—	1.1–1.5
Ovules per ovary (n)						
S	16	—	41	22	4.5	4–38
S—u	16	—	39	20	7.7	6–42
S—e	10	—	12	26	7.2	12–36
C	1	—	2	26	—	19–34

Fruits are erect, orthocarpic, three-ribbed, and pale brown.

	n-pl.	n-obs.	x̄ (mm.)	s (mm.)	w (mm.)
Length					
S	8	22	5.2	1.2	2.9–6.8
S—u	9	28	5.3	.7	2.9–6.6
C	1	3	3.9	—	3.6–4.2

Seeds are pyriform, obscurely reticulate, lustrous, short-tailed, and dark brown or reddish brown.

	n-pl.	n-obs.	x̄ (mm.)	s (mm.)	w (mm.)
Length					
S	5	11	.74	.05	.7–0.9
S—u	9	28	.8	.1	.6–1
Diameter					
S	5	11	.24	.05	.2–.3
S—u	9	28	.23	.05	.2–.3

Chromosomes are unequal in size, with three longer than the others.

	n-pl.	g	sp	Cytologist
S	1	16	—	C. H. Uhl, in Clausen and Uhl (1943)
C	1	16	—	C. H. Uhl, ibid.
S	6	16	32	Ilse Zandstra, 1969
Santa Rosa Peak	1	16	—	C. H. Uhl, in Clausen and Uhl (1943)
Sierra San Pedro Mártir	1	64	—	C. H. Uhl, in Moran (1969)

Variation. Despite its limited distribution, *Sedum niveum* is remarkably variable. The variability involves numbers of floral parts, sizes of floral parts, pigmentation, habit of growth, and size of plant, but much of this variation is the result of environmental modification.

The variability in number of floral parts is remarkable. Simple forking is common. When this happens at the base of a petal or stamen, double structures result. Coefficients of variation for the sample of plants from quadrats on Sugarloaf Mountain are 6% for number of sepals, 6% for number of petals, and 12% for number of carpels. Flowers from plants in the quadrats and also from the upper cluster on Sugarloaf Mountain varied in number of petals as shown below.

	Random quadrats			Upper cluster		
Petals (n)	Flowers (n)	%	Cumulative %	Flowers (n)	%	Cumulative %
4	0	0	0	1	2.6	2.6
5	36	85.7	85.7	30	76.9	79.5
6	6	14.3	100	6	15.3	94.8
7	0	0	100	0	0	94.8
8	0	0	100	2	5.1	99.9
Totals	42			39		

Plants in exposed situations appear to have more red pigment, especially in the leaves, sepals, and stems, but all differences in pigmentation may not be environmental because plants growing almost side by side may differ markedly in the red streaking of the petals and carpels.

In shade, plants become etiolated, with long, slender stems and leaves, and even longer floral parts. Plants in sun, on the other hand, may be densely tufted, with shorter stems and relatively broader leaves and floral parts.

The process of drying considerably alters the dimensions of plants of *Sedum niveum*. Leaves and floral parts of three plants changed significantly in eight dimensions when dried as herbarium specimens. In each instance, structures were pressed separately and measured both before drying and seven weeks afterward.

Dimension	n-obs.	Fresh	Dry
Rosettes—diam. (cm.)	3	1.1	1
Leaves—length (mm.)	3	6	5.5
—width (mm.)	3	4	3.7
Floriferous stems—length (cm.)	5	2.4	2.3
Median leaves—length (mm.)	8	6.2	5.1*
—width (mm.)	8	3.8	3.2*
Flowers—diam. (mm.)	6	12	8.7**
Sepals—length (mm.)	6	4.3	2.6**
—width (mm.)	6	1.2	.9
Petals—length (mm.)	5	6.6	4**
—width (mm.)	5	2.8	1.1**
Filaments—length (mm.)	5	5.4	3.6**
Carpels—length (mm.)	6	5.9	3.5**

An experiment of completely random design in the greenhouse at Ithaca, N.Y., provided a means of testing the differences between two replicates of a plant from the northern slope of Observatory Peak, at an altitude of 2,550 meters in the Sierra San Pedro Mártir, Lower California, and twelve replicates of eight plants from Sugarloaf Mountain, Calif. Differences indicated in the detailed description are highly significant for length and width of leaves of rosettes and for width of median leaves of floriferous stems. Differences are significant in diameter of rosettes, length of median leaves of floriferous stems, and length of nectaries. Since only one plant from Lower California provided data for comparisons in the greenhouse, the interpretation can be that the plant either is distinctive or indicates a populational difference. The second plant from Lower California, the one from the eastern face of Observatory Peak, 2,800 meters, collected in July 1968, flowered in Ithaca in September of the same year. That plant closely resembled *Sedum niveum* from the San Bernardino Mountains, except for longer leaves of the floriferous stems and longer sepals. The plants from the two sites on Observatory Peak may not be identical.

Plants of *Sedum niveum*, when cultivated at Ithaca, usually have larger rosettes, longer leaves, and particularly longer sepals. Only the longer sepals are of statistical significance, however.

The three types of samples from Sugarloaf Mountain yielded similar results. Any one of these samples is sufficient to provide satisfactory estimates of the parameters of the population.

Because few differences are significant between plants in nature and those in cultivation, data from experiments are included in the description only when they are pertinent to the discussion of variation. Despite the many differences observed among plants in the field, plants grown under similar conditions in experiments in Ithaca are remarkably alike. They exhibit no significant differences. Because variances among replicates in experiments exceed variances among plants, standard deviations among plants are not indicated in the description for data from the greenhouse.

Nomenclature. *Sedum niveum* Davidson, Bull. So. Calif. Acad. Sci. 20:53 (1921). Type locality: northern slope of Sugarloaf Mountain, San Bernardino Mountains, Calif. Type: R. Kessler no. 3430 (Los Angeles Museum), July 1920.

Many details of the description of *Sedum pinetorum* T. S. Brandegee suggest a small specimen of *S. niveum*. These include tuberous roots, basal cohesion of petals, yellow nectaries, and erect follicles. Discrepancies include arrangement of leaves—"opposite" in *S. pinetorum* and alternate in *S. niveum*; length of petals—"3.5 mm." in *S. pinetorum* versus 5–10 mm. in *S. niveum*; and color of seeds—"red" in *S. pinetorum* versus dark brown or reddish brown in *S. niveum*. The validity of these differences may be questionable. The type collection of *S. pinetorum* is so fragmentary that accurate interpretation is difficult. Evidence is conflicting. Accurate identification of *S. pinetorum* awaits rediscovery of similar plants. If *S. niveum* someday is found along the eastern slope of the Sierra Nevada and proves to be the basis for *S. pinetorum*, then Brandegee's name must be adopted for the species because it antedates Davidson's name by five years.

Distribution. The known geographical range of *Sedum niveum* (fig. 32) is in the San Bernardino, Santa Rosa, and New York Mountains of California, on northern slopes at elevations between 1,677 and 2,980 meters, and in the Sierra San Pedro Mártir, Lower California, from 2,550 to 2,835 meters. If *S. pinetorum* should turn out to be *S. niveum*, then the Sierra Nevada also is part of the range of the species.

Study in herbaria and the literature reveals that *Sedum niveum* is known from twelve sites. I have observed it at two of these locations. My most intensive studies were at the type locality on the northern slope of Sugarloaf Mountain in the San Bernardino Mountains in the summer of 1968. Arthur Phillips III assisted in the work in 1968.

Plants of *Sedum niveum* occur on Sugarloaf Mountain over an estimated area of 243,600 meters square, a tract too large for complete enumeration. After determining the apparent limits of the population, I subdivided the total area into 2,436 quadrats each of 100 meters square. Numbering of quadrats was in serpentine fashion, from east to west and south to north, on graph paper. Selection of eight quadrats for study was by means of a table of random numbers. Numbers of plants in each of the eight quadrats are shown in the following table.

Random no.	Quadrat no.	Flowering plants (n)	Sterile plants (n)	Total plants (n)	Vascular species (n)
1	853	121	36	157	5
2	841	141	19	160	7
3	2,182	38	9	47	17
4	1,261	0	0	0	4
5	2,141	0	0	0	5
6	953	41	13	54	8
7	275	0	0	0	3
8	1,547	0	0	0	3
Totals		341	77	418	

Information for the population on Sugarloaf Mountain and also for a small population on the lower slope of Charlton Peak, just east of Dollar Lake, is summarized below.

Population	N plants, estimate	N plants, (C.I., P = .05)	Density per m.², estimate	Density per m.², (C.I., P = .05)
Sugarloaf Mountain	126,672	>418–268,429	.52	0–110
Charlton Peak	26	—	.8	—

Population	Area of population (m.²)	Altitudinal range (m.)	pH, w	pH, n tests	Depth of soil— w (cm.)
Sugarloaf Mountain	243,600	2,740–2,980	5–6.6	25	1–7
Charlton Peak	300	2,867–2,883	6.6	1	1

Population	Distance to nearest neighbor— w (cm.)	Closest competitor	Species most affecting environment
Sugarloaf Mountain	1–130	Selaginella watsonii	Pinus contorta
Charlton Peak	21	Selaginella watsonii	Ribes montigenum

The lowest altitude at which I saw *Sedum niveum* in 1968 was at 2,740 meters on Sugarloaf Mountain. Information which accompanied my collections of 1940, namely 8,000 feet (=2,400 m.), may be faulty because I did not at that time have either an altimeter or satisfactory maps. Similarly, I am uncertain about data supplied by F. W. Peirson for his collection of August 16, 1920, labeled as from the creek east of Forsee Creek at an elevation of about 5,500 feet (=1,677 m.). I have visisted the slope east of Forsee Creek at the reported elevation. The vegetation there is *Quercus— Pinus ponderosa*. The situation is not reasonable for *S. niveum*. The data may be faulty or conditions may have changed. Otherwise, I have missed an interesting microhabitat.

Selaginella watsonii was the closest competitor of seventeen randomly selected plants of *Sedum niveum* in the randomly selected quadrats on Sugarloaf Mountain. Similarly, *Selaginella watsonii* was the closest competitor of sixteen out of eighteen randomly selected plants in the upper cluster on Sugarloaf Mountain. Closest competitors of the other two plants were respectively another plant of *S. niveum* and *Silene parishii*. Second closest competitors were more diverse. These were for the sample from the quadrats: other plants of *S. niveum*—13, *Penstemon caesius*—2, *Ribes montigenum*—1, and *Festuca* sp.—1. For the upper cluster they were: *Draba corrugata*—6, other plants of *S. niveum*—6, *Silene parishii*—4, and *Selaginella watsonii*—2. The species most affecting the environment of every plant of *S. niveum* in both samples was *Pinus contorta* ssp. *murrayana*.

All plants of *Sedum niveum* were on quartzite. Exposure was almost always to the north. Exceptions were a few plants exposed to the northeast or northwest, but these were shaded by cliffs, rocks, or logs which reduced light. Sunshine reaching plants was limited to less than half of what is potential at any site. Further, since clouds cover the sites of the populations on almost half of the afternoons in the summer, the total sunlight reaching the plants is small. In addition, thundershowers keep the habitat moist.

Temperatures in the sun in July occasionally reach as high as 38° C. in the area of the population of *Sedum niveum* on Sugarloaf Mountain. The average temperature in quadrats at midday was 28° C. in the sun and 17° C. in the shade. Temperatures in the sun and shade differ markedly. Since plants are alternately in both conditions, they experience frequent changes in temperature.

Reproduction. A variety of insects visit the flowers of *Sedum niveum* and appear to be effective pollinators. These are mostly *Hymenoptera*, especially bees, and *Diptera*. The larger, strong-flying insects probably are most important in carrying pollen between plants that are distantly situated from each other. The smaller insects may be more important in carrying pollen among plants that are closely associated.

Principal propagation is vegetative by means of rosettes or portions of stems which become dissociated from parental plants. Seedlings are difficult to find in summer. Seeds require four or more weeks to mature. In 1968, the first flowers were in anthesis on July 11. On August 9, four weeks later, seeds were just beginning to approach maturity.

Plants can endure considerable desiccation. Stems stored for three and a half months in a plastic bag, without moisture, sprouted when put in soil and given water.

Relationships. The relationships of *Sedum niveum* probably are with other orthocarpic species with white flowers. *Sedum cockerellii* and *S. griffithsii* are the two species of this sort which are nearest geographically. These species both have taller floriferous stems with more flowers per cyme and shorter primary stems. Another relationship might be with *S. minimum* which, like *S. niveum*, has the petals marked with red. Similarly, the petals vary in number per flower and are slightly connate basally. Even the seeds are similar in size. These similarities suggest that *S. minimum* and *S. niveum* are perhaps related and descended from a common ancestor which may have inhabited the Mexican Plateau in an earlier time. This same ancestor possibly gave rise to other modern species such as *S. napiferum*, *S. flaccidum*, and *S. vinicolor*.

Information about chromosomes still is too limited to provide a sure indication of relationships. Both *Sedum cockerellii* and *S. niveum* are $g = 16$. According to Stoutamire and Beaman (1960), *S. minimum* ssp. *minimum* is $g = 10$.

Reference. Clausen and Uhl (1943) and Moran (1969).

8. *Sedum cockerellii**** (figs. 32 and 35–38)

Plants of *Sedum cockerellii* (fig. 35) are small, perennial herbs, of tufted habit, with tiny rosettes of papillose leaves at time of flowering. The floriferous stems are erect, with the median leaves oblanceolate or spatulate. The flowers have white petals, epipetalous stamens with terete filaments adnate for >1 mm., yellow or creamy white

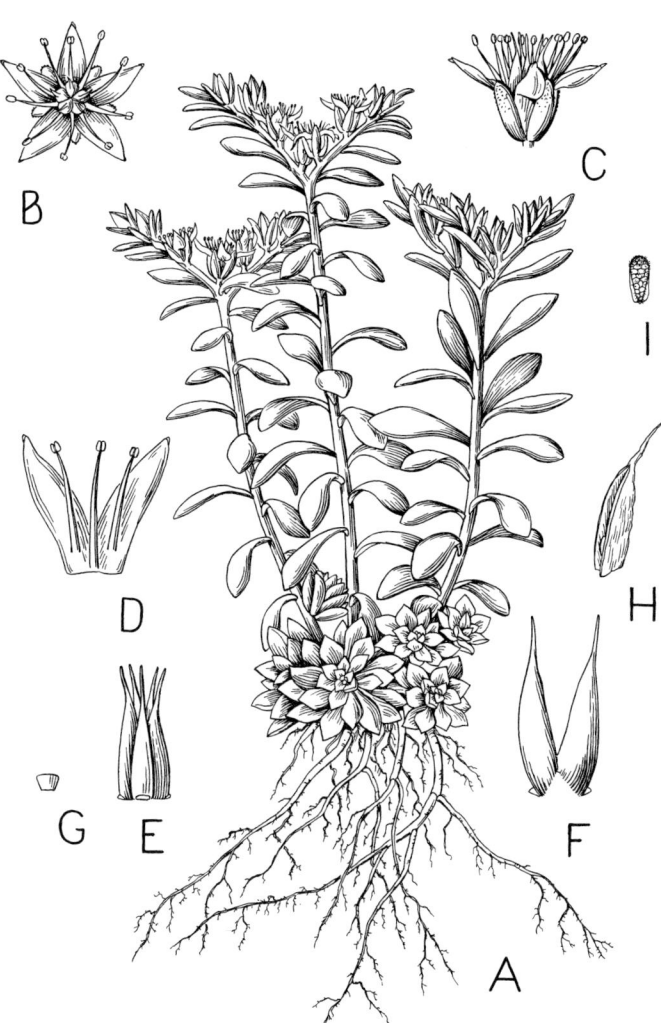

Fig.35. Plant of *Sedum cockerellii* from western slope of Tuerto Mountain, Santa Fe Co., N.M., cultivated in greenhouse, Ithaca, N.Y. A–G from Tuerto Mountain, H–I from Sandia Crest, N.M. A. Habit sketch (x .9). B. Flower from above (x 1.8). C. Flower from side (x 1.8). D. Two petals and stamens (x 2.7). E. Carpels (x 2.7). F. Two carpels (x 3.6). G. Nectary (x 4.5). H. Follicle from side (x 3.6). I. Seed (x 9).

nectaries, and orthocarpic pistils which are erect in fruit. The seeds are tiny, reddish brown, reticulate, and sparingly hairy. *Sedum nevii* and other Appalachian species with white flowers have flattened filaments scarcely adnate to the petals and kyphocarpic carpels, spreading widely in fruit and prominently two-lipped ventrally. *Sedum griffithsii*, another species of the Mexican Plateau, if separable from *S. cockerellii*, has narrower, acute leaves, narrower petals, and pink or white nectaries. *Sedum niveum* of the San Bernardino Mountains and ranges of the Lower California province has stouter, prostrate main stems, fewer flowers per cyme, broader nectaries, and dark brown seeds. *Sedum wrightii*, with distribution on the Mexican Plateau, as well as the adjacent Edwards Plateau and Sierra Madre Oriental, has thick, caducous leaves, pungently scented flowers, oblanceolate-oblong petals which are erect below and then divergent with mucronate tips, and yellow, stipitate-reniform nectaries.

Description. Sample: 27 plants. Sources and numbers of plants from each locality are: Southern Rocky Mountains: Dalton Canyon in the Santa Fe Range, Santa Fe Co., N.M., Aug. 31, 1962—4, a 20% simple random sample; Tuerto Mountain near the Santa Fe Canyon, along the old trail from Santa Fe to the Pecos River, Santa Fe Co., N.M., Aug. 1962—14, a 4.5% stratified random sample; Rio Hondo Canyon, Taos Co., N.M., June 24, 1949—1; and near Cuba in the Jemez Mountains, N.M., Delzie Demaree—1; and Colorado Plateau: Oak Creek Canyon, Coconino Co., Ariz., alt. 1,680 m., Aug. 15, 1968—3, a 31% sample. Besides these samples, I have studied four additional plants from Tuerto Mountain, including the lowermost and uppermost plants, and several plants from the Mexican Plateau: Ellis Brook and the Sandia Crest in the Sandia Mountains, Sandoval Co., N.M., 1940; and the upper canyon of Ruidoso Creek, Lincoln Co., N.M., 1949. In 1940, I collected plants along Willow Creek in the Santa Fe Range, San Miguel Co., N.M., and along Jemez Creek in the Jemez Mountains, Sandoval Co., N.M.

In the following description, abbreviations for localities are: Dalton C. for Dalton Canyon, Tuerto M. for Tuerto Mountain, Cuba for Demaree's plant from the Jemez Mountains, Rio Hondo for Rio Hondo Canyon, Oak Creek for Oak Creek Canyon, and Sandia C. for Sandia Crest. Also, gh. indicates data for plants cultivated in an experiment of completely random design in a greenhouse at Ithaca, N.Y. In the following tables, data are cited throughout for Tuerto Mountain, the type locality. These provide a basis for decision concerning the taxonomic status of *Sedum cockerellii*. Likewise, data are cited for the plant from Cuba because it differs from the others in number of chromosomes. Data for other populations from the experiment in the greenhouse are cited only when the expressions are markedly different from the condition in the wild or when differences among localities are significant.

Plants of *Sedum cockerellii* are tufted, perennial herbs (fig. 36) with fibrous, white roots, to 12 cm. long and 4 mm. in diameter, and with short, branched rootstocks, to 6 cm. long, bearing rosettes of leaves in the axils of the primary leaves. Floriferous shoots develop from the rosettes by elongation in the second year. These shoots are erect and either glabrous or papillose upward. Sometimes rosettes develop from the

NATIVE SPECIES 191

Fig. 36. Plant of *Sedum cockerellii* on northwestern slope of Tuerto Mountain, N.M., showing tufted habit, Aug. 24, 1962.

axils of the lower leaves of the floriferous stems. A plant in the Rio Hondo Canyon, in 1949, had a maximum height of 18 cm.

	n-pl.	n-rep.	n-obs.	$\bar{x}$	s
Plants—diameter (cm.)					
Tuerto M.—gh.	1	2	2	9	—
Cuba—gh.	1	2	2	12	—
Oak Creek—gh.	1	2	2	3	—
Primary rosettes (*n*)					
Tuerto M.	14	—	14	6.4	5.1
Dalton C.	4	—	4	9	5.2
Cuba—gh.	1	2	2	12.5	—
Oak Creek—gh.	1	2	2	3	—
Primary rosettes—diam. (cm.)					
Tuerto M.—gh.	1	2	6	1.2	—
Cuba—gh.	1	1	3	.5	—
Oak C.—gh.	1	2	6	.9	—

	n-pl.	n-rep.	n-obs.	$\bar{x}$	s
Floriferous stems per plant (n)					
Tuerto M.	14	—	14	1.9	1
Dalton C.	4	—	4	4.3	4.5
Tuerto M.—gh.	1	2	2	6.5	—
Cuba—gh.	1	2	2	20.5	—
Oak Creek—gh.	1	2	2	7	—
Floriferous stems—max. ht. (cm.)					
Tuerto M.	14	—	25	5.6	1.3
Dalton C.	4	—	9	9.1	.9
Tuerto M.—gh.	1	2	2	6.5	—
Cuba—gh.	1	2	2	7.5	—
Oak Creek—gh.	1	2	2	6.5	—

Leaves of rosettes are spirally arranged, sessile, spurred, plane, obovate or oblong-spatulate, broadly rounded to obtuse at apex, papillose, and green or yellow-green, occasionally glaucous and sometimes speckled with pink or red. The cotyledons are yellow-green or Scheele's green speckled with purple. They are suborbicular and broadly rounded. The first leaves after the cotyledons are ovate and prominently papillose. Leaves of floriferous stems are spirally arranged, rarely opposite or in whorls of 3 or 4, ascending or spreading, making an average angle of 77° with the stem, sessile, short-spurred, plane ventrally, convex dorsally, oblanceolate-elliptical, oblanceolate-oblong, or spatulate, obtuse to subacute, glabrous or papillose, willow-green or spinach-green, sometimes speckled with pink dorsally. One median leaf of a floriferous stem of a plant on Tuerto Mountain had a nipplelike deformity on the dorsal surface.

	n-pl.	n-rep.	n-obs.	$\bar{x}$ (mm.)	s (mm.)
Leaves of rosettes—length					
Tuerto M.	13	—	19	5.8	1.4
Dalton C.	4	—	9	4.4	.9
Cuba—gh.	1	1	3	3	—
Oak Creek—gh.	1	2	6	5.4	—
Leaves of rosettes—width					
Tuerto M.	13	—	19	3.3	.8
Dalton C.	4	—	9	2.1	.4
Cuba—gh.	1	1	3	1.6	—
Oak Creek—gh.	1	2	6	2.8	—
Leaves of rosettes—thickness					
Rio Hondo	1	—	2	1.1	—
Tuerto M.—gh.	1	2	6	1.1	—
Cuba—gh.	1	1	3	.7	—
Oak Creek—gh.	1	2	6	.8	—
Leaves of floriferous stems—length					
Tuerto M.	14	—	25	9.9	1.9
Dalton C.	4	—	9	11.7	2.6
Cuba—gh.	1	2	6	10.5	—
Oak Creek—gh.	1	2	4	14.6	—
Leaves of floriferous stems—width					
Tuerto M.	14	—	25	3.8	.6
Dalton C.	4	—	9	3.3	.7
Cuba—gh.	1	2	6	4.2	—
Oak Creek—gh.	1	2	4	6.7	—

	n-pl.	n-rep.	n-obs.	x̄ (mm.)	s (mm.)
Leaves of floriferous stems—thickness					
Rio Hondo	1	—	2	1.6	—
Tuerto M.—gh.	1	2	5	1.3	—
Cuba—gh.	1	2	6	1.9	—
Oak Creek—gh.	1	2	4	1.9	—

Inflorescences are cymes of 2 or 3, rarely 4, cincinni. Floral bracts are similar to the leaves, oblong-elliptical, but smaller. In the greenhouse, differences among plants in number of flowers per cyme are highly significant.

	n-pl.	n-rep.	n-obs.	x̄ (n)	s (n)
Flowers per cyme					
Tuerto M.	14	—	25	5.1	2.6
Dalton C.	4	—	9	3.1	.8
Tuerto M.—gh.	1	2	5	5.4	—
Cuba—gh.	1	2	6	7.7	—
Oak Creek—gh.	1	2	4	39.7	—

Flowers are short-pedicellate, usually 5-merous, and with a short torus.

	n-pl.	n-rep.	n-obs.	x̄ (mm.)	s (mm.)
Pedicels—length					
Tuerto M.—gh.	1	2	3	1.4	—
Cuba—gh.	1	2	6	1.1	—
Oak Creek—gh.	1	2	5	3.6	—
Flowers—diameter					
Tuerto M.—gh.	1	2	3	15	—
Cuba—gh.	1	2	6	16	—
Oak Creek—gh.	1	2	5	19	—
Torus—length					
Tuerto M.	11	—	19	.5	.1
Dalton C.	2	—	3	.6	.3
Tuerto M.—gh.	1	2	3	.8	—
Cuba—gh.	1	2	6	.6	—
Oak Creek—gh.	1	2	5	.8	—

Sepals are unequal, lanceolate-linear or clavate-oblong, acute or obtuse, papillose, and yellow-green speckled with pink, especially apically. Sometimes they are slightly connate basally. With age, they may become recurved.

	n-pl.	n-rep.	n-obs.	x̄ (mm.)	s (mm.)
Length					
Tuerto M.	11	—	19	4.7	1.1
Dalton C.	2	—	3	5.8	.7
Tuerto M.—gh.	1	2	3	9.3	—
Cuba—gh.	1	2	6	7.6	—
Oak Creek—gh.	1	2	5	11.9	—
Width					
Tuerto M.	11	—	19	1.4	.3
Dalton C.	2	—	3	1.4	.07
Tuerto M.—gh.	1	2	3	1.9	—
Cuba—gh.	1	2	6	2.6	—
Oak Creek—gh.	1	2	5	2.1	—

Petals are lanceolate-elliptical, obtuse, and minutely mucronate-appendaged, with the mucro .1 mm. long. They are white, faintly streaked with pink centrally or with a median patch of magnolia pink. Rarely, petals are minutely connate basally, <.5mm. long.

	n-pl.	n-rep.	n-obs.	x̄	s
Number per flower					
Tuerto M.	11	—	19	5	.1
Dalton C.	2	—	3	5	0
Cuba—gh.	1	2	6	5	—
Oak Creek—gh.	1	2	5	5	—
Length (mm.)					
Tuerto M.	11	—	19	7	.9
Dalton C.	2	—	3	6.6	1.3
Cuba—gh.	1	2	6	7.4	—
Oak Creek—gh.	1	2	5	7.4	—
Width (mm.)					
Tuerto M.	11	—	19	2.3	.5
Dalton C.	2	—	3	2	.07
Cuba—gh.	1	2	6	2.6	—
Oak Creek—gh.	1	2	5	3	—

Stamens have white filaments and chrysanthemum-crimson or brown anthers. One flower of a plant on Tuerto Mountain had a stamen alternate with the petals fused with an epipetalous stamen, giving the appearance of a forked filament.

	n-pl.	n-rep.	n-obs.	x̄ (mm.)	s (mm.)
Filaments—length					
Tuerto M.	11	—	19	5.8	.6
Dalton C.	2	—	3	5.5	.4
Tuerto M.—gh.	1	2	3	6.6	—
Cuba—gh.	1	2	6	5.3	—
Oak Creek—gh.	1	2	5	5	—
Epipetalous filaments; adnation—length					
Tuerto M.	11	—	19	1.4	.3
Dalton C.	2	—	3	1.4	.04
Tuerto M.—gh.	1	2	3	1	—
Cuba—gh.	1	2	6	1.6	—
Oak Creek—gh.	1	2	5	1.2	—
Anthers—length					
Tuerto M.	4	—	5	1	.06
Cuba—gh.	1	1	2	.6	—
Oak Creek—gh.	1	2	3	.9	—

Nectaries are quadrate, truncate and emarginate, and yellow or creamy white.

	n-pl.	n-rep.	n-obs.	x̄ (mm.)	s (mm.)
Length					
Tuerto M.	11	—	19	.5	.07
Dalton C.	2	—	3	.4	0
Tuerto M.—gh.	2	2	3	.5	—
Cuba—gh.	1	2	6	.6	—
Oak Creek—gh.	1	2	5	.4	—

	n-pl.	n-rep.	n-obs.	x̄ (mm.)	s (mm.)
Width					
Tuerto M.	11	—	19	.5	.04
Dalton C.	2	—	3	.6	.07
Tuerto M.—gh.	1	2	3	.9	—
Cuba—gh.	1	2	6	.6	—
Oak Creek—gh.	1	2	5	.9	—

Carpels are erect and white and either pale magnolia pink upward or suffused or faintly streaked with pink.

	n-pl.	n-rep.	n-obs.	x̄	s
Length (mm.)					
Tuerto M.	11	—	19	5.6	1
Dalton C.	2	—	3	4.6	2.9
Cuba—gh.	1	2	6	5.9	—
Oak Creek—gh.	1	2	5	7.4	—
Length of cohesion (mm.)					
Tuerto M.	11	—	19	.8	.2
Dalton C.	2	—	3	.8	.2
Cuba—gh.	1	2	6	1.3	—
Oak Creek—gh.	1	2	5	.7	—
Ovules per ovary (n)					
Tuerto M.	11	—	19	20	7.8
Dalton C.	2	—	3	13	18.7
Tuerto M.—gh.	1	2	3	28	—
Cuba—gh.	1	2	6	20	—
Oak Creek—gh.	1	2	5	43	—

Fruits are erect, pale brown, and with beaks somewhat divergent.

	n-pl.	n-rep.	n-obs.	x̄ (mm.)	s (mm.)
Length					
Tuerto M.	1	—	2	4.6	—
Rio Hondo	1	—	1	5	—
Ruidoso Creek	1	—	1	4	—
Tuerto M.—gh.	1	1	3	4.1	—
Cuba—gh.	1	1	3	3.7	—
Oak Creek—gh.	1	1	3	3.8	—

Seeds are elliptic-pyriform, minutely wing-margined, with testa reticulate, sparingly hairy, and reddish brown.

	n-pl.	n-rep.	n-obs.	x̄ (mm.)	s (mm.)
Length					
Sandia C.	1	—	4	.65	—
Tuerto M.—gh.	1	1	3	.7	—
Cuba—gh.	1	1	3	.7	—
Oak Creek—gh.	1	1	3	.7	—
Diameter					
Sandia C.	1	1	4	.33	—
Tuerto M.—gh.	1	1	3	.2	—
Cuba—gh.	1	1	3	.3	—
Oak Creek—gh.	1	1	3	.3	—

Chromosomes are small and elongate. Separation in smears often is poor; also, a few tiny chromosomes which may be supernumerary render counts indefinite. For these reasons, gametophytic counts for a plant from Tuerto Mountain ranged from 15 to 17 and sporophytic counts from 32 to 40. Numbers cited below are averages. Besides the information for plants from localities within the area covered by this book, three plants which I collected in the Sandia Mountains were g = 16 (Clausen and Uhl, 1943). In addition, Uhl (1972) reported gametophytic numbers of 14, 15, 16, 29, and 30 for plants with the morphological characteristics of *Sedum cockerellii*.

	n-pl.	g (n)	sp(2n)	Cytologist
Number				
S. of Terrero, Santa Fe Range	1	16	—	C. H. Uhl in Clausen and Uhl (1943)
Tuerto M.	1	17	34	L. & R. Benjamin, 1970
Along Jemez Creek, Jemez Mts.	1	16 + 4	—	C. H. Uhl in Clausen and Uhl (1943)
Cuba	1	14	28	I. Zandstra, 1969; L. & R. Benjamin, 1970
Oak Creek	1	—	32	L. & R. Benjamin, 1970

	n-pl.	$\bar{x}$ (genome length; μ)	$\bar{x}$ (chromosome length; μ)	Cytologist
Length of sporophytic chromosomes				
Tuerto M.	1	—	1	L. & R. Benjamin, 1970
Cuba	1	77.2	2.8	I. Zandstra, 1969
Oak Creek	1	—	.6	L. & R. Benjamin, 1970

Variation. Fluctuation in number of chromosomes in different populations of *Sedum cockerellii* is an important aspect of the variation of the species. The chromosomes are small and difficult to count. Many counts from a plant are necessary to gain an understanding of the true situation. Also, information from root tips is essential to confirm counts from sporocytes. Further, tiny chromosomes present in some plants appear to be supernumerary. These chromosomal fragments add to the difficulty of interpretation. As yet, the role of these small chromosomes in reproduction is unclear. Available information about chromosomes of *S. cockerellii* helps to indicate some of the problems in the classification of the species, but does not resolve them.

Information about chromosomes needs to be accompanied by descriptive data concerning the morphology and physiology of the plants that were studied. Herbarium specimens, preserved as vouchers for cytological studies, provide some gross information about plants studied, but may not reveal other features vital to an understanding of variation. For example, significant differences among plants from different populations, in an experiment, were in number of flowers per cyme and length of nectaries. Herbarium specimens show the number of flowers per cyme, provided that whole inflorescences are pressed without mutilation, but accurate interpretation of nectaries seldom is possible from dried specimens.

Replicates of a plant from the Jemez Mountains with 14 pairs of chromosomes had longer nectaries than replicates of a plant from Oak Creek Canyon with 16 pairs of chromosomes. On the other hand, the replicates of the plant from Oak Creek Canyon had more flowers per cyme than replicates of either the plant from the Jemez Mountains or a plant from Tuerto Mountain which had 17 pairs of chromosomes. This difference is highly significant. Fourteen plants in a random sample from Tuerto Mountain had an average of 5.1 flowers per cyme. Replicates of the plant from Tuerto Mountain in the experiment had an average of 5.4 flowers per cyme. In contrast, replicates of the plant from Oak Creek Canyon had an average of 39.7 flowers per cyme. The data suggest small genetic differences among the populations, but do not indicate any broad regional differences which might signify subspecies.

Responses of plants of *Sedum cockerellii* in cultivation in the greenhouse at Ithaca, N.Y., have been similar to their expressions in nature. A couple of comparisons by means of the Wilcoxon rank sum test involved a plant from Tuerto Mountain. These tests revealed no significant difference in length of sepals or width of nectaries, two features in which a difference was expected on the basis of subjective observation.

Variation in number of petals per flower is slight, $CV = 2.2\%$ for a random sample of 11 plants from Tuerto Mountain and 0% for a sample of 2 plants from Dalton Canyon. Only 1 flower in 22, or 4.5%, had 4 petals instead of the usual 5. Of the hundreds of flowers of *Sedum cockerellii* which I have seen, I have kept careful records of number of petals per flower for only 61. A frequency distribution confirms the idea that variation is slight.

Petals per flower (n)	Flowers (n)	%	Cumulative %
4	2	3.3	3.3
5	59	96.7	100
Total	61		

Morphological features for separating *Sedum cockerellii* from *S. griffithsii* include orientation, greatest width, condition of apex, and color of the leaves of the floriferous stems. Likewise, width of petals may be useful. Some of these features exhibit great variation, as indicated by coefficients of variation for plants on Tuerto Mountain. Angle of apex and width of leaves appear most satisfactory for discriminatory purposes.

Feature	CV
Angle of leaves with stems	43
Width of leaves	16
Angle of apex of leaves	8
Width of petals	20

The main problems in variation of *Sedum cockerellii* and related species probably are in the mountains of the northern part of the Mexican Plateau, for example in the Santa Rita and Huachuca Mountains of southern Arizona, where both *Sedum cockerellii* and plants of the sort which have been called *S. griffithsii* occur. Only part of the variants of this relationship have moved onto the escarpment at the southern edge of the Colorado Plateau or into the Southern Rocky Mountains.

Nomenclature. *Sedum Cockerellii* Britton, Bull. N.Y. Bot. Gard. 3 (9):41 (1903). Type locality: Tuerto Mountain east of Santa Fe, N.M., on the trail from Santa Fe to the upper Pecos River, near the Santa Fe Canyon in the Santa Fe Range, at an elevation between 2,440 and 2,600 meters. Type: T. D. A. Cockerell (NY), 1895. Cockerell, in a letter to Britton, Oct. 9, 1902, stated that the species is rare and local and that it occurs in the Canadian Zone in the Santa Fe and Las Vegas Ranges, N.M.

Synonyms are:

1888. *Sedum puberulum* S. Watson, Proc. Am. Acad. 23:273 (not *S. puberulum* DC., 1828, which is *S. hispanicum* Jusl.). Type locality: shaded cliffs of a high mountain near Guerrero, Chihuahua, probably at an altitude of about 2,865 meters. Type: C. G. Pringle 1,242 (GH), Oct. 1, 1887. Davis (1936) published information from Pringle's journal concerning the source of collection no. 1,242. Pressed specimens of this collection have the leaves blunt and broader toward the apex than below. This evidence caused a shift of the name from the synonymy of *S. griffithsii* to *S. cockerellii* (Clausen, 1946). Watson originally distinguished his species as rough-puberulent throughout, with stems from a dense cluster of fibrous and fleshy roots, and with white petals scarcely exceeding the stamens. The word papillose is more appropriate than puberulent for the vesture of the plants comprising Pringle's collection. Except for the earlier homonym, Watson's name would be the correct designation for the species now known as *S. cockerellii*.

1903. *Sedum Wootoni* Britton, Bull. N.Y. Bot. Gard. 3 (9):44 (1903). Type locality: Organ Mountains, N.M., probably from the west side of the range and possibly from near the top of the main ridge according to information in a letter from E. O. Wooton, Nov. 29, 1940. Type: E. O. Wooton (NY), Sept. 17, 1893. The original collection is a mixture of specimens of *S. cockerellii* and *S. wrightii*. Four out of twelve specimens and fragments on the lower half of the sheet are *S. cockerellii*. These are encircled. They are the primary basis for interpretation of *S. wootonii*.

1936. *Sedum anomiosepalum* Fröderström, Acta Horti Goth. 10 (App.):70, figs. 528–535, pl. 44. Type locality: Willow Creek, Santa Fe Forest, San Miguel Co., N.M., alt. 2,200 meters. Type: W. W. Eggleston 19,022 (NY). Fröderström placed this species close to *S. stenopetalum* in his classification, but the type differs from typical *S. cockerellii* only in the smaller, narrower leaves of the floriferous stems.

The oldest specimen of *Sedum cockerellii* which has come to my attention is a collection of Charles Wright, no. 233, obtained on the expedition from western Texas to El Paso, N.M., May–October 1849 (NY). This was labeled *S. wrightii*. The exact locality of origin is unknown.

Distribution and ecology. *Sedum cockerellii* occurs in the zone of pine forests in high mountains of the northern part of the Mexican Plateau, including both the Sacramento Section and the Davis Mountains, in Texas, New Mexico, and Arizona; the southern edge of the Colorado Plateau in Arizona; the southern part of the

Southern Rocky Mountains in northern New Mexico; and the northern part of the Sierra Madre Occidental in Chihuahua. Thirty-one populations, with an estimated total of 5,100 plants, are known from the Southern Rocky Mountains. There are also doubtful records from the Sierra del Carmen in Coahuila, toward the northern end of the Sierra Madre Oriental.

Because of the problem of relationships with *S. griffithsii*, some information about the distribution of *S. cockerellii* is confused. Further, herbarium specimens sometimes are difficult to distinguish from *S. wrightii*. The range, as plotted on the map, fig. 32, indicates the present state of knowledge and reveals gaps which remain in the information.

The usual habitats of *Sedum cockerellii* are cliffs (fig. 37), rocky slopes (fig. 38), and moss-covered boulders. The plants grow on a variety of rocks, including both granite and limestone. The soil usually is shallow. Sites generally are shaded. The known altitudinal range is from 1,680 meters in Oak Creek Canyon, at the edge of the Colorado Plateau, to 3,170 meters in the Sandia Mountains.

Data for three populations which had special attention in this study are in the following table.

Fig. 37. Plant of *Sedum cockerellii* on cliff in Oak Creek Canyon, Ariz., Aug. 15, 1968.

Fig. 38. Site for *Sedum cockerellii* on rocky southwestern slope of Tuerto Mountain, N.M., Aug. 23, 1962. Trees are *Pinus ponderosa*.

Population	Plants (N)	Area of population (m.²)	Density per m.²	Exposure
Tuerto Mountain	309	436	1.4	NW (–W, S, N)
Dalton Canyon	20	>1	<16	W
Oak Creek Canyon	13	—	—	W

Population	Altitudinal range (m.)	pH, range	pH, tests (n)	Oldest record
Tuerto Mountain	2,505–2,715	6.2–7.6	20	1895
Dalton Canyon	2,330	6.8	2	1962
Oak Creek Canyon	1,680	6.8	1	1961

The confidence interval, $P = .05$, for the area of the population on Tuerto Mountain is 291–1,787 square meters. Principal competing species there, rated on a scale of 14, are mosses—9, *Koeleria cristata*—6, and *Draba luteola*—4. Most important shading trees, rated on the same scale, are *Pseudotsuga menziesii*—6, *Pinus ponderosa*—4, and *Abies concolor*—4. Shading trees in Dalton Canyon are *Populus tremuloides* and *Picea pungens*. In the Oak Creek Canyon, grasses and *Holodiscus discolor* grow with the *Sedum*.

Reproduction. Principal reproduction of *Sedum cockerellii* is from seeds. Seedlings are frequent at sites where the species is well established. The stems and rootstocks are too short to be an important means of spread of the plants, although some propagation occurs from broken pieces which water or wind may move. The small rosettes take root readily, provided that moisture is adequate.

Insects pollinate the flowers. On Tuerto Mountain, I have watched two kinds of flies visiting the blossoms, one with orange-brown bases of the wings and the other with black wings.

Flowering time is from mid-August until mid-September. In the greenhouse, plants have had flowers in anthesis from July 21 until October 30.

Relationships. Species related to *Sedum cockerellii* include *S. griffithsii*, *S. niveum*, *S. lumholtzii*, *S. caducum*, and *S. wrightii*. More distantly a relationship may exist with the yellow-flowered *S. lanceolatum*. A peculiar plant from Durango surely is also related.

The status of *Sedum griffithsii* is unclear. The supernumerary chromosomes which are present in some populations and absent from others cause a special problem. The discovery that a plant in the Jemez Mountains, with the morphology of *S. cockerellii*, has 14 pairs of chromosomes and lacks the supernumerary ones, complicates the situation. A simple sorting of populations into two morphological types, correlated with number of chromosomes, is not possible. Instead, sorting must be on the basis of morphology and physiology, with the realization that the genetic story still awaits clarification. If *S. griffithsii* is separated from *S. cockerellii*, then plants assigned to it will have elliptic-linear or linear leaves, narrower petals, and pale pink or white (not yellow or creamy) nectaries. Plants with these characteristics usually grow at lower elevations, that is in a warmer environment, than do plants of *S. cockerellii*. Evolution may be going in the direction of adaptation to a warmer, drier environment. Also, since plants of this type usually lack the supernumerary chromosomes, the trend may be in the direction of reduction in number of chromosomes through elimination of fragments.

Next nearest relative to *Sedum cockerellii*, after *S. griffithsii*, is *S. niveum*. That is similar in aspect, but has elongate creeping stems, thicker leaves in the rosettes, shorter leaves on the floriferous stems, fewer flowers per cyme, slightly connate petals, and dark brown seeds. The two species appear coordinate. They replace each other geographically and could easily be descended from a common ancestor.

Sedum lumholtzii, of Sonora, with hairy leaves and inflorescences, large leaves in the rosettes, and bright red nectaries, is another relative. An interesting hypothesis to test is whether *S. griffithsii* is some sort of an interaction between *S. lumholtzii* and *S. cockerellii*.

Sedum caducum (Clausen 1950), of Tamaulipas, has caducous leaves which are channeled and prominently speckled with red. Its petals are hooded and slightly connate basally. It resembles *S. cockerellii* in many features, especially in the flowers, for example, the nectaries which are similar in size, shape, and color. The carpels are erect in both species and not gibbous ventrally. The petals are white, briefly erect at base, and then spread widely. Probably *S. caducum* and *S. cockerellii* had a similar ancestry.

Sedum wrightii, with musky-scented flowers, likewise appears related. Like *S. caducum*, it has caducous leaves, but it differs from both *S. caducum* and *S. cockerellii* in being glabrous, without papillae, and in having thicker, yellow-green leaves and petals erect to above the middle, then spreading. The origin of the distinctive chromosome numbers in *S. wrightii*, that is multiples of 12, may be the result of reduction to 6 and then development of a polyploid series. In any case, *S. wrightii* appears specialized in comparison with related species.

A remote relationship of *Sedum cockerellii* with the yellow-flowered *S. lanceolatum* is a possibility. Both species have orthocarpic carpels, a similar habit of growth, and a distribution in the Cordillera, the *S. lanceolatum* occurring northward. Eight may be the original base number of chromosomes in both species.

On November 25, 1964, Myron Kimnach collected an interesting plant, no. 556, from mossy rocks on a cliff at the top of the canyon 54 km. past El Salto, at km. 1,119 on the road to Mazatlán, Durango. Paul Hutchison previously had made a similar collection at the same place. The plant from Durango has many features in common with *Sedum cockerellii*, but differs most markedly in having the stems red upward, the petals deep pink except at the apices and bases, only 5 stamens, and longer nectaries. Studies of chromosomes by L. and R. Benjamin reveal a condition similar to that in *S. cockerellii*, but counts are indefinite. Four figures of mitotic chromosomes showed a possible range from 27 to 34 chromosomes. The relationship seems clear. The details of origin are obscure. Identification of the plant from Durango awaits further information about the population there. Were there originally several plants with only 5 stamens or was this an unusual individual among plants with other characteristics?

References. Clausen, R. T., and Uhl, C. H. (1943); Clausen, R. T. (1946 and 1950); and Uhl, C. H. (1972).

9. *Sedum wrightii**** (figs. 32 and 39–41)

Thick, fleshy, yellow-green leaves on rigidly erect stems are a distinctive feature of *Sedum wrightii* (fig. 39). On slight pressure, the leaves detach from the stems and take root from basal, callous tissue. The flowers are white, with the petals oblanceolate-oblong, erect for more than half their length, then curved outward. The blossoms have a pungent scent which distinguishes them from all other North American species, north of the Mexican Plateau. The nectaries are yellow and the anthers are red. The seeds are strongly papillose. Of other white-flowered species, *S. cockerellii* sometimes has been confused with *S. wrightii* in herbaria, but in living condition is easy to separate. It has thinner, papillose, obovate or oblong-spatulate leaves; flowers lacking a musky scent; widely spreading, lanceolate-elliptical petals; and reticulate, hairy seeds. *Sedum caducum*, with caducous leaves as in *S. wrightii*, differs in having the leaves

NATIVE SPECIES 203

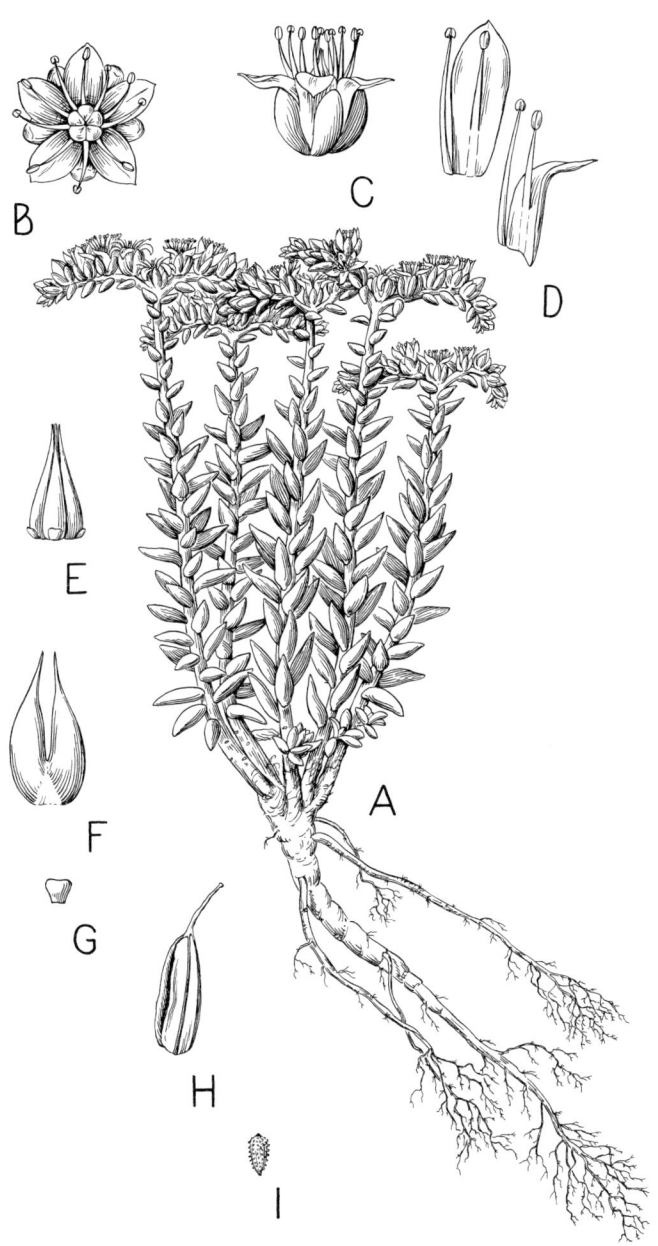

Fig. 39. Plant of *Sedum wrightii* from bluff on south side of Devils Arm of Amistad Reservoir, Val Verde Co., Tex., cultivated in greenhouse, Ithaca, N.Y. A. Habit sketch (x .41). B. Flower from above (x 1.6). C. Flower from side (x 1.6). D. Petal and two stamens (x 2.5). E. Carpels (x 2.5). F. Two carpels and nectaries (x 3.3). G. Nectary (x 4.1). H. Follicle from side (x 3.3). I. Seed (x 8.2).

ventrally channeled and prominently speckled with red; petals lanceolate and widely spreading; anthers pale yellow; nectaries creamy white; and seeds not papillose.

Description. Sample: 44 plants. Of 39 plants selected for special study on bluffs overlooking the Devils River, now the Devils Arm of the Amistad Reservoir, west of Del Rio, Val Verde Co., Texas, the basic sample comprises 16 plants chosen randomly from two randomly selected clusters, 2% from each cluster. Unless mentioned otherwise, estimates of means and standard deviations are derived from this probability sample. Choice of the other 23 plants was for diverse reasons: large size, red pigmentation, unusual environmental circumstances, availability of buds for cytological study, and also extension of the random sample, since an early idea was to study a 5% sample. In addition to my collections, a plant from the Devils River, collected in 1954 by James B. Perry, has been available for culture. Also, I have had experience with *Sedum wrightii* at La Cueva in the Organ Mountains of New Mexico, in the Chisos and Davis Mountains of Texas, and on the mountainside below the grotto of Villa García in Nuevo León. In 1940, Omer Sperry supplied a flowering specimen collected by Barton Warnock in the Davis Mountains.

In Ithaca, plants grow well in the greenhouse and survive under glass in unheated cold frames. Data for flowers, fruits, and seeds in the following description are from plants cultivated in the greenhouse (gh.) a few months after transplanting from Texas.

Plants of *Sedum wrightii* are tufted, perennial herbs with elongate, thick, primary roots and fibrous, white, secondary roots. Roots attain a length of 19 centimeters. The stems are erect or decumbent, and green, reddish, or brown, and often heavily speckled with purple. On large plants, the stems and branches are closely crowded. Animals often eat the ends of the stems. Each primary stem potentially terminates in an inflorescence. Rosettes develop in the axils of the lower leaves and become new floriferous shoots in a subsequent season. Measurement of rosettes of plants in the field was several weeks before expansion of flowers. Rosettes enlarge greatly after plants have flowered, as shown by measurements of plants in the greenhouse, after maturation of fruits.

	n-pl.	n-obs.	$\bar{x}$	s
Plants—diameter (cm.)	16	16	12.7	10.1
Rosettes (n)	16	16	3.9	2.7
Rosettes (n)—gh.	16	16	10.9	6.2
Rosettes—diameter (cm.)	13	35	1.2	.3
Rosettes—diameter (cm.)—gh.	16	47	3.3	.6
Floriferous stems per plant (n)	16	16	8.6	5.1
Floriferous stems—length (cm.)	11	32	10.5	4.3

Leaves are spirally arranged, thick, elliptical or oblanceolate-oblong, sometimes almost globular, broadly rounded, glabrous, but with margins minutely papillose, and green or yellow-green. They detach readily from the stems on slight pressure. Leaves of the floriferous stems are numerous and subdivergent. Except for being larger, they are similar to those of the rosettes.

	n-pl.	n-obs.	x̄ (mm.)	s (mm.)
Leaves of rosettes				
Length	13	36	8.9	1.6
Width	13	36	4	.5
Thickness	13	36	1.6	.3
Leaves of floriferous stems				
Length	16	46	12	2
Width	16	46	4.6	.8
Thickness	16	46	2.3	.7

Inflorescences are terminal cymes, commonly of one or two cincinni. Occasional cincinni are branched. The maximum number of flowers observed in a cyme is 33.

	n-pl.	n-obs.	x̄ (n)	s (n)
Cincinni per cyme—gh.	14	37	1.9	.5
Flowers per cyme—gh.	14	37	6.9	2.5

Flowers (fig. 40) have a pungent, musky scent. They are either sessile or short-pedicellate, and normally 5-merous. Floral bracts are similar to leaves, but smaller. A torus extends upward amid the carpels which are basally connate around it.

Fig. 40. Flowers of plant of *Sedum wrightii* from bluff on south side of Devils Arm of Amistad Reservoir, Val Verde Co., Tex., cultivated in greenhouse at Ithaca, N.Y., Oct. 16, 1970 (x 3.4). Photo by Howard Lyon.

	n-pl.	n-obs.	x̄ (mm.)	s (mm.)
Pedicels—length—gh.	14	43	.3	.3
Flowers—diameter—gh.	14	43	12.4	2.3
Torus—length—gh.	14	43	.7	.1

Sepals are very unequal, divergent or erect, elliptical or lanceolate-oblong, obtuse, and green. They are thickened upward. Measurements are of longest sepals per flower.

	n-pl.	n-obs.	x̄ (mm.)	s (mm.)
Length—gh.	14	43	7.6	1
Width—gh.	14	43	3.4	.7

Petals are erect below, then sharply curved outward, with the upper third divergent. They are oblanceolate-oblong, abruptly acute and mucronate-appendaged, and white with the midrib pale green dorsally toward the apex. They sometimes are faintly streaked with pink or with the midrib slightly pink at the apex. The observed number of petals varies from 4 to 7. The bases of the petals are separate from each other.

	n-pl.	n-obs.	x̄	s
Number—gh.	14	42	5	0
Length (mm.)—gh.	14	43	7.7	.5
Width (mm.)—gh.	14	43	3	.2

Stamens have white filaments and oblong-elliptical anthers which are red or yellow suffused with red. Anthers of the epipetalous stamens shed pollen about a day later than those alternate with the petals.

	n-pl.	n-obs.	x̄ (mm.)	s (mm.)
Filaments—length—gh.	14	43	6.4	.4
Epipetalous filaments; adnation—length—gh.	14	43	2.2	.3
Anthers—length—gh.	13	28	1.1	.1

Nectaries diverge from the bases of the carpels and are not in concavities. They are stipitate-reniform, broadly rounded or truncate, and yellow. The stipe appears like a bracket and the body of the nectary like a shelf.

	n-pl.	n-obs.	x̄ (mm.)	s (mm.)
Length—gh.	14	43	.8	.1
Width—gh.	14	43	.8	.1

Carpels are erect with the styles and stigmas close together. The carpels are white or faintly suffused with green or pink, especially basally and on the ventral walls. The ovules are pyriform and white.

	n-pl.	n-obs.	x̄	s
Length (mm.)—gh.	14	43	5.6	.5
Cohesion—length (mm.)—gh.	14	43	1.2	.1
Ovules per ovary (n)—gh.	14	43	34.2	4.3

Fruits are erect, connate basally, and pale brown. Measurements of length apply to the body of the follicle.

	n-pl.	n-obs.	x̄ (mm.)	s (mm.)
Length—gh.	13	26	4.5	.5

Seeds are elliptic- or oblong-pyriform, sometimes with persistent funicle, strongly papillose, and reddish brown.

	n-pl.	n-obs.	x̄ (mm.)	s (mm.)
Length—gh.	13	26	.6	.04
Diameter—gh.	13	26	.3	.04

Chromosomes are tiny and difficult to count. Most are $<2\mu$ long. A few exceed 2μ. One plant from the Devils River, studied by Mastrangelo, had both the triploid and hexaploid number of chromosomes in cells of the roots. Uhl (1972) has reported diploids, tetraploids, hexaploids, and octoploids.

	n-pl.	g (n)	sp(2n)	Cytologist
Number				
Devils River, Edwards Plat.	3	37, 40–41	30–33, 61–77	Iris Mastrangelo, 1971
Fern Canyon, Davis Mts.	1	12	—	C. H. Uhl (Clausen and Uhl, 1943)
La Cueva, Organ Mts.	1	12	—	C. H. Uhl (*ibid.*)
Chinese Wall, Chisos Mts.	1	36	—	C. H. Uhl (*ibid.*)

Variation. Principal variation of *Sedum wrightii* is on the Mexican Plateau and in the adjacent Sierra Madre Oriental. A possibility of geographic races exists, but is unproved. Some noteworthy variations occur. For example, a plant from the mountainside below the grotto of Villa García in Nuevo León had nectaries only half the length of those of plants at Devils River, and it had smaller petals and more flowers per cyme. Another aspect of the variation is the possibility of correlation of the level of ploidy with morphology and physiology. As yet, no clear patterns are apparent. In this connection, data for the random sample from the Devils River may be useful for comparisons with other populations.

Comparisons of subsamples studied at the Devils River reveal a couple of interesting differences. Plants on an exposed point of rocks had significantly narrower leaves in the rosettes than those on a more protected slope. Also, the difference in length of nectaries was highly significant, two plants from the point having the longest nectaries. Curiously, in cultivation in the greenhouse, plants from the point have significantly smaller rosettes.

Although there are usually 5 petals per flower, and no exception occurred in the random sample, certain plants were variable in this respect. One plant, for example, produced flowers with 5, 6, and 7 petals. This also was one of the reddest plants in the population.

Both the environment and age cause differences among plants. Diameter of plants at Devils River varied from 4 to 30 cm., and height at time of flowering varied from 2 to 15 cm. The range in number of flowers per cyme in the sample from Devils River was 1 to 19, but the plant from Villa García had 33 flowers in a cyme. Most plants were green. Others were strongly suffused with red. Sometimes, the two conditions existed side by side. In cultivation in the greenhouse at Ithaca, all plants became green, although some had minor red speckling on the stems or at the tips of the leaves. An obvious cause of modifications is difference in light. Besides the effect on pigmentation, plants in the shade of rocks or bushes are slenderer and with less turgid leaves.

The principal flowering season of *Sedum wrightii* is autumn. Time of flowering varies with latitude, moisture, and the genetic constitution of the plants. The earliest record of flowering which has come to attention is from Concepción del Oro, Zacatecas, Aug. 11, 1904. Other early records are from the southern part of the range of the species. The latest record is from hills near Chihuahua, Nov. 10, 1886. In my sample from the Devils River, the date of the first plant with flowers in anthesis in the greenhouse at Ithaca was Oct. 19, 1970. The latest plant to flower in this sample did not release pollen until Dec. 22, 1970, when other plants growing under the same conditions were long past flowering or even with fruits. In 1973 several plants from the Devils River were in flower in the greenhouse in early September, indicating a shift to an earlier time of flowering at a more northern latitude.

At present, no basis is known for recognizing subspecies in *Sedum wrightii* and no variations seem worth naming as varieties.

Nomenclature. *Sedum Wrightii* A. Gray, Plantae Wrightianae 1:76 (1852). Type locality: summit of mountain near El Paso, Texas. Gray, in the original description, cited as localities hills near the San Pedro (Devils) River, in crevices of rocks, and summits of mountains near El Paso, Texas. Wright apparently first saw the species near El Paso, then found it later near the Devils River. According to Wooton's (1906) record of localities visited by Wright, the type locality is in Texas, somewhere near El Paso. Type: Charles Wright 1,292 (GH), Oct. 2, 1849. The type sheet contains not only the holotype, but also a collection from the Devils River. Originally it also contained two collections of *S. cockerellii*, one by J. M. Bigelow from the Sandia Mountains and the other by Wright from New Mexico. The holotype is the larger specimen of *S. wrightii*, the one to the left, which I regard as no. 1,292 from near El Paso.

Distribution. *Sedum wrightii* (fig. 32) occurs in the Sierra Madre Oriental and adjacent Mexican Plateau, and also in the Davis Mountains and on the edge of the Edwards Plateau in Texas. The altitudinal range is from 335 m. along the Devils River, Val Verde Co., Texas, to 2,135 m. in the Sierra del Carmen, southwest of Piedra Blanca, Coahuila. The region inhabited by *S. wrightii* is arid. The plants grow on rocks, usually in crevices and on ledges, but sometimes out in the open, on bluffs, cliffs, and steep slopes. Shrubs or larger herbs usually provide partial shade, or else exposure is to the north. Sites are on both limestone and igneous rocks of diverse origin. At least thirty-four populations are known.

My studies of *Sedum wrightii* began in 1940 in the Organ, Chisos, and Davis Mountains. In 1949, I found it on the mountainside below the grotto of Villa García in Nuevo León. In 1970, aided by my wife Edna and son Tom, I studied the population on the north-facing bluffs along the south side of the submerged gorge of the Devils River, now the Devils Arm of the Amistad Reservoir, about 18 km. west of Del Rio, Texas. The lower part of the population there already was submerged. Plants were still alive at an altitude of 335 m. in August 1970, but below that level, down to 323 m., they were dead because of previous submergence. The normal level of the reservoir will be 341 m., with flood level at 349 m. When the reservoir is filled, more than half of the population will be eliminated. For this reason, together with the fact that the site is one of the places where Charles Wright made his original observations, study of the population is of special significance. I am grateful to the National Park Service, which administers the area, for the opportunity to make my studies. Data are for the population as it was in August 1970.

Plants (N)	Area (m.^{2})	Density per m.^{2}	Exposure
4,458	33,625	.13	N–NE (rarely SE)

Altitudinal range (m.)	pH, range	pH, tests (n)	Depth of soil $\bar{x}$ (cm.)	s (cm.)	Distance to nearest neighbor $\bar{x}$ (cm.)	s (cm.)	Oldest record
335–350	7.2–8.2	28	1.1	.7	44	56	1849

Desert shrub is the prevailing vegetation at sites where *Sedum wrightii* occurs. Competitors are miscellaneous. Using the random sample of 16 plants of *S. wrightii* at the Devils River as a measure of frequency, hence rating on a scale of 16, the commonest competitor there is *Mimosa biuncifera*, with a score of 6. Other competitors and ratings are *Jatropha dioica*—4, *Opuntia engelmannii*—1, *Echinocereus enneacanthus*—1, *Yucca torreyi*—1, a caryophyll—1, a grass—1, and an unidentified shrub—1.

Reproduction. Plants of *Sedum wrightii* produce numerous tiny seeds. The leaves fall from the stems when touched with slight pressure and take root from basal callous tissue, forming new plants. Distinction is difficult between seedlings and plantlets developed from detached leaves.

In an experiment in the greenhouse, I placed 50 loose leaves, 10 from each of 5 plants, on the surface of moist vermiculite and sand. In 7 weeks, 39 leaves, 78%, had taken root. By 11 weeks, 41 leaves, 82%, had developed into plantlets, each with a small rosette and well-developed roots, an average of 8.2 plantlets per original plant. Formation of new plants from detached leaves (fig. 41) is an effective means of propagation in *Sedum wrightii*. Furthermore, it makes the species an excellent one for experimental culture.

Some rootlets of leaves detached from a hexaploid plant, studied by Mastrangelo, had the haploid number of chromosomes, a condition also noted by Kathy Krathwohl

Fig. 41. Detached leaf of *Sedum wrightii* with small rosette on soil in greenhouse, Ithaca, N.Y., Oct. 16, 1970 (x 5.4). Photo by Howard Lyon.

in 1973 for a plant of *Sedum glaucophyllum* from Cascades, Va. The evidence for the *S. glaucophyllum* suggests a chimera, because a few cells of the root were diploid, but most cells were monoploid.

Leaves and rosettes retain vitality for a long time. Some leaves in my sprouting experiment had been detached and dry for two weeks before putting on the vermiculite. Rosettes, stored dry in plastic bags for four months, took root readily when placed on moist vermiculite and sand.

Flowering time in nature is from August to November. In the greenhouse at Ithaca, plants normally flower from September to December. Abnormally, they have flowered in March, May, June, and August. Plants which flowered in March came directly from the field two months earlier. This fact suggests the importance of moisture in breaking dormancy and stimulating growth, including flowering.

When flowers are cross-pollinated, production of seeds is good. Ripening of seeds is about two and a half months after the time of flowering.

Relationships. In general appearance, thick leaves, and pungent odor of the white flowers, *Sedum wrightii* resembles *S. ebracteatum*. It differs in details of floral structure. It lacks concavities at the bases of the carpels, a feature characteristic of *S. ebracteatum*. Further, the petals are erect for more than half their length before curving outward, and the nectaries are stipitate and reniform rather than ovate. In addition, the seeds are strongly papillose instead of glabrous and reticulate. The number of chromosomes appears to be 12 or some multiple rather than a multiple of 20. The differences may be more important than the similarities.

Sedum caducum is another possible relative of *S. wrightii*. That occurs in the Sierra Madre Oriental in the vicinity of Ciudad Victoria. It has thick, caducous leaves and white flowers, but the petals are slightly connate and spread widely, and the seeds are not papillose. A tabular comparison of the two species accompanies the original description of *S. caducum* (Clausen, 1950). At least twelve differences are available for separation.

Sedum cockerellii, also with white flowers, differs in having thinner, obovate or spatulate leaves, widely spreading lanceolate-elliptical petals, quadrate nectaries, and hairy seeds. These obvious morphological differences, plus adjustment to life in a cooler, moister environment, and a different basic number of chromosomes, suggest that *S. cockerellii* and *S. wrightii* are not close relatives.

Sedum lenophylloides, a shrubby species of the Sierra Madre Oriental, has papillose seeds, like *S. wrightii*, but slightly connate, widely spreading petals, and subquadrate, dark carmine or orange nectaries.

Sedum wrightii really is not close to any other species. It is distinctive. Nevertheless, it probably was derived long ago from the same ancestral line which produced *S. ebracteatum*, *S. caducum*, *S. cockerellii*, *S. lenophylloides*, and other species related to them. These species, despite their differences, probably are closer to each other than they are to others elsewhere in the world.

References. Clausen, R. T., and Uhl, C. H. (1943); Clausen, R. T. (1950); and Uhl, C. H. (1972).

10. *Sedum lanceolatum**** (figs. 42–49)

Among the yellow-flowered species of *Sedum* native in western North America, *S. lanceolatum* (fig. 42) is distinctive in having the pistils erect in both flower and fruit and the leaves lanceolate or elliptic-ovate, spirally arranged, and sometimes papillose. The petals are lanceolate. *Sedum stenopetalum*, which often has been confused with *S. lanceolatum*, has the pistils widely divergent in fruit and prominently two-lipped, and the petals narrowly linear-lanceolate. The leaves of *S. stenopetalum*, especially ssp. *stenopetalum*, appear subulate on drying because all except the bases wither and become bristlelike, while the bases themselves become scarious and are prominently spurred. In *S. stenopetalum*, vegetative rosettes commonly develop in place of flowers in the cymes and along the floriferous stems. *Sedum lanceolatum* somewhat resembles the European *S. reflexum* and *S. rupestre*, but differs from these in having the floriferous stems erect in bud and the vegetative shoots noncreeping. Its flowers usually are 5-merous, whereas those of the two European species often are 6-merous or with even more parts per whorl.

The closely related *Sedum rupicolum* of the Cascade Mountains in Washington, also with erect carpels, has thicker, ovate leaves which detach on the slightest pressure.

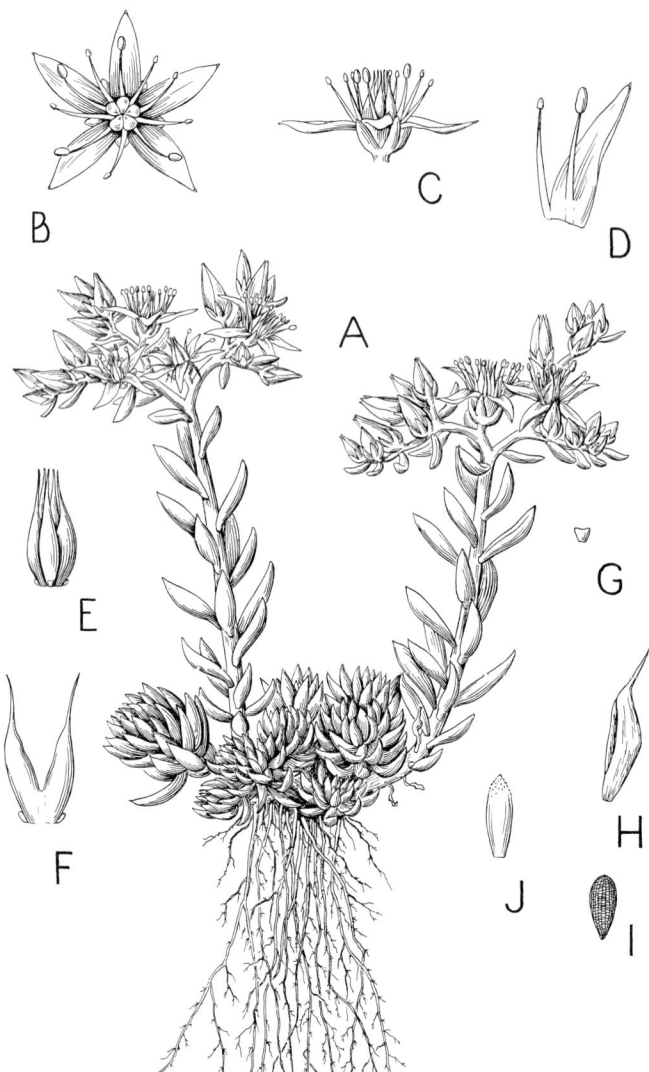

Fig. 42. Plant of *Sedum lanceolatum* ssp. *lanceolatum* from East Plum Creek about 1 km. north of Larkspur, Douglas Co., Colo., cultivated in cold frame, Ithaca, N.Y. A. Habit sketch (x .84). B. Flower from above (x 1.7). C. Flower from side (x 1.7). D. Petal and two stamens (x 2.5). E. Carpels (x 2.5). F. Two carpels and nectaries (x 3.4). G. Nectary (x 4.2). H. Follicle from side (x 3.4). I. Seed (x 8.4). J. Single leaf (x 1.7). H is follicle of plant from Macomber Peak, San Juan Co., Colo., cultivated in Test Garden at Ithaca, N.Y.

In addition, the petals and nectaries are deeper yellow. Where the two species occur together, the *S. rupicolum* is ahead of the *S. lanceolatum* in flowering.

Description. Sample: 207 plants. Principal attention to *Sedum lanceolatum* has been in the following years: 1952—Black Hills and Pine Ridge escarpment; 1961—Seven Devils Mountains, Cascade Mountains, and Northern Rocky Mountains; 1962—Southern Rocky Mountains; 1964—Sierra Nevada; 1965—Cascade Mountains and Puget Trough; and 1966—Central Rocky Mountains. In the following list of populations, the basis of choice of areas for study is indicated as C = convenience, J = judgment, and R = random selection. N_u = number of plants in a sampling unit, N_c = number of plants in a sampled cluster, and n = number of plants in samples. Selection of all plants within populations was random, regardless of the mode of choice of populations, except for populations on the Great Plains, on Summit Peak, and at Lulu and Gem Passes.

Sampling area	Basis of choice	N_u	N_c	n	Plants studied n/N_c	n/N_u	Condition when collected fl. (n)	fr. (n)
Great Plains								
W. Spanish Peak	C	—	5	3	.6	—	3	0
Monroe Canyon	J	—	—	10	—	—	10	0
Custer	C	—	—	10	—	—	10	0
Harney Peak	J	—	—	5	—	—	5	0
Little Elk Creek	J	—	—	8	—	—	8	0
Roubaix	C	—	—	10	—	—	10	0
Pluma	C	—	—	10	—	—	10	0
Deadwood	C	—	—	4	—	—	4	0
Terry Peak	J	—	—	8	—	—	8	0
NE of Buckhorn	C	—	—	6	—	—	6	0
NW of Buckhorn	C	—	—	10	—	—	10	0
So. Rocky Mts.								
Plum Creek	J	—	49	6	.12	—	6	0
Dry Gulch	R	161,200	1,240	12	.009	.00007	12	0
Pole Mt.	J	1,650	49	12	.24	.007	12	0
Platoro	C	—	>34	3	.09	—	3	0
Macomber Peak	C	—	73	4	.05	—	4	0
Tesuque Peak	J	—	25	2	.08	—	2	0
Summit Peak	R	—	—	2	—	—	2	0
Mt. Flora	J	—	19	1	.05	—	1	0
Lulu Pass	J	—	—	2	—	—	1	0
Iron Mt.	J	162	162	6	.04	.04	0	0
Central Rocky Mts.								
Teewinot Mt.	R	19,425	137	15	.11	.0008	14	14
Alaska Basin	R	111	11	1	.09	.009	0	0
N. Buffalo Fork	R	576	36	2	.06	.003	0	0
No. Rocky Mts.								
Clements Mt.	C	—	—	3	—	—	0	2
Marks Butte	J	72	72	6	.08	.08	0	6
Columbia Plateau								
Seven Devils Lake	J	—	30	8	.27	.27	0	0
Heaven's Gate	C	11	11	2	.18	.18	1	0
Sierra Nevada								
Breeze Creek	C	100	100	4	.04	.04	0	4
Gem Pass	R	12	6	2	.33	.17	0	2
Cascade Mts.								
Scotty Creek	R	204	80	10	.13	.05	10	6
Ingalls Creek	C	10	10	5	.5	.5	2	3
Chopaka Mt.	R	450	19	6	.32	.01	0	6
Puget Trough								
Lummi Rocks	J	1,020	102	9	.09	.009	0	9
Totals				207			154	52

Plants survived in Ithaca under various conditions of growth. For flowering, they did best outdoors at the Cornell Test Garden and poorest in the greenhouse. Some flowering occurred in the cold frames. Vegetative growth was good in the greenhouse. Numbers of plants (N) in three different experiments in Ithaca and numbers of plants which yielded data (n-d) are indicated in the following table.

Source	Garden N	Garden n-d	Cold frame N	Cold frame n-d	Greenhouse N	Greenhouse n-d
Great Plains	—	—	4	1	—	—
So. Rocky Mts.	10	4	41	24	—	—
Central Rocky Mts.	1	1	1	—	1	1
No. Rocky Mts.	—	—	2	1	—	—
Columbia Plateau	1	1	2	2	4	4
Cascade Mts.	2	2	5	2	3	3
Puget Trough	6	6	4	4	—	—
Totals	20	14	59	34	8	8

In the following description, geomorphic regions, elevational location, and places of cultivation are abbreviated as follows: a = alpine, cf. = cold frame, CM = Cascade Mountains, CP = Columbia Plateau, CRM = Central Rocky Mountains, gd. = garden, gh. = greenhouse, GP = Great Plains, m = lower and middle altitudes, NRM = Northern Rocky Mountains, PT = Puget Trough, SN = Sierra Nevada, and SRM = Southern Rocky Mountains.

Plants of *Sedum lanceolatum* are perennial herbs of tufted habit (fig. 43), with fibrous and capillary, white or pale brown roots and decumbent stems which root at the lower nodes and give rise to several or many shoots from the horizontal portion or rootstock. The stems are branched, green, brown, pink, or red, and sometimes papillose. In the first year, the stems terminate in rosettes of crowded leaves. In the second year, these shoots elongate and become erect, floriferous stems. Sometimes the floriferous stems are branched. After fruiting, the main stems die, but growth continues from shoots which develop in the axils of some of the lower leaves. Although the whole plant may be perennial and endure for many years, individual stems are biennial.

	n-pop.	n-pl.	n-rep.	n-obs.	$\bar{x}$	s
Plants—diameter (cm.)						
GP	1	2	—	2	10	6
CRM	2	16	—	16	16	16
SN	1	2	—	2	3	.9
CM	1	7	—	7	5	6
PT	1	9	—	9	23	27
Primary rosettes (n)						
GP	1	2	—	2	11	1.7
CRM	2	16	—	16	13	10.5
SN	2	6	—	6	3	1.4
Primary rosettes—diameter (cm.)						
GP	1	2	—	12	.8	.2
CRM	2	16	—	60	.7	.2
SN	2	6	—	20	.5	.003

NATIVE SPECIES

Fig. 43. Plant of *Sedum lanceolatum* ssp. *lanceolatum* from an altitude of 3,337 meters at Beartooth Summit, Park Co., Wyo., cultivated in greenhouse at Ithaca, N.Y., Feb. 2, 1965 (x 3). Photo by Howard Lyon.

	n-pop.	n-pl.	n-rep.	n-obs.	$\bar{x}$	s
Primary rosettes—diameter (cm.)						
CM	1	7	—	22	.9	.1
PT	1	9	—	36	1.4	.3
Leaf-bearing stems—length (mm.)						
CP—gh.	2	4	5	26	1.8	1
CM—gh.	2	2	2	8	1.3	0
ssp. *lanceolatum*—cf.	9	28	44	176	6.9	5.1
ssp. *subalpinum*—cf.	3	4	5	29	3.1	—
ssp. *nesioticum*—cf.	1	4	4	12	12.7	13.3
Floriferous stems per plant (*n*)						
SRM—m	5	29	—	29	2.9	3
—a	5	12	—	12	7	4.5
CRM	2	16	—	16	2.8	1.2
SN	2	6	—	6	1	.003
CM	1	7	—	7	2	2
SRM—m—cf.	2	4	5	5	2.4	2.3
—a—cf.	1	1	1	1	6	—
ssp. *lanceolatum*—gd.	6	6	8	8	19.5	14.7
ssp. *subalpinum*—gd.	1	1	1	1	1	—
ssp. *nesioticum*—gd.	1	6	6	6	9	6.6
Floriferous stems—length (cm.)						
SRM—m	5	29	—	64	7	1.8
—a	5	12	—	40	3.7	1.1

	n-pop.	n-pl.	n-rep.	n-obs.	$\bar{x}$	s
Floriferous stems—length (cm.)						
CRM	2	16	—	31	9.8	4.3
CP	1	2	—	2	6.5	2.6
SN	2	6	—	12	4.3	1.1
SRM—m—cf.	2	4	5	12	3.3	0
—a—cf.	1	1	1	5	3.8	—
—gd.	2	2	2	9	5.3	—
PT—gd.	1	4	4	18	9.6	—

Leaves are spirally arranged, sessile and very short-spurred, lanceolate, elliptic-lanceolate, or elliptic-ovate, obtuse or obtusely apiculate, but apparently acute, dull gray-green or bluish green, sometimes speckled or suffused with red or purple, often glaucous, and either papillose or smooth. Surfaces of leaves are finely hexagonally reticulate. Leaves of floriferous stems are spreading or subdivergent, sessile, short-spurred, subterete, elliptic-lanceolate, obtusely apiculate, and dull gray-green, often suffused with red. Bases of withered leaves may become scarious.

	n-pop.	n-pl.	n-rep.	n-obs.	$\bar{x}$ (mm.)	s (mm.)
Leaves of primary rosettes—length						
GP	1	2	—	12	9	1.2
CRM—m	2	16	—	58	5.8	1
—a	3	4	—	6	4.7	.4
NRM	2	9	—	26	5.7	.6
CP	1	6	—	21	5.4	.4
SN	2	6	—	18	5	.002
CM	2	16	—	52	6.6	1
PT	1	9	—	36	9.9	1.3
CP—gh.	2	4	5	26	4.6	.8
CM—gh.	2	2	2	8	5.3	—
CRM—a—gh.	1	1	2	12	6.7	—
ssp. *lanceolatum*—cf.	9	28	43	175	6.2	1.4
ssp. *subalpinum*—cf.	3	4	5	29	5.6	—
ssp. *nesioticum*—cf.	1	4	4	12	9.5	1
ssp. *lanceolatum*—gd.	9	9	12	72	6.2	—
ssp. *subalpinum*—gd.	2	3	4	19	4.4	.2
ssp. *nesioticum*—gd.	1	5	5	30	10.6	2.4
Leaves of primary rosettes—width						
GP	1	2	—	12	2.2	0
CRM—m	2	16	—	58	1.9	.8
—a	3	4	—	6	1.6	.1
NRM	2	9	—	26	1.5	.3
CP	1	6	—	21	1.8	.01
SN	2	6	—	18	2.2	.2
CM	2	16	—	52	2.2	.3
PT	1	9	—	36	2.8	.3
CP—gh.	2	4	5	26	1.7	.2
CM—gh.	2	2	2	8	2.1	0
CRM—a—gh.	1	1	2	12	1.5	—
ssp. *lanceolatum*—cf.	9	28	43	175	2	.2
ssp. *subalpinum*—cf.	3	4	5	29	1.8	.1
ssp. *nesioticum*—cf.	1	4	4	12	3	.3
ssp. *lanceolatum*—gd.	9	9	12	72	2.5	—

	n-pop.	n-pl.	n-rep.	n-obs.	$\bar{x}$ (mm.)	s (mm.)
Leaves of primary rosettes—width						
ssp. *subalpinum*—gd.	2	3	4	19	1.9	.1
ssp. *nesioticum*—gd.	1	5	5	30	3.2	.3
Leaves of primary rosettes—thickness						
GP	1	2	—	12	1.1	.4
CRM—m	2	16	—	58	1.2	.2
—a	3	4	—	6	.9	.06
NRM	2	9	—	26	.8	.1
CP	1	6	—	21	.9	.3
SN	2	6	—	18	.7	.14
CM	2	16	—	52	1.5	.2
PT	1	9	—	36	1.8	.2
CP—gh.	2	4	5	26	1	.1
CM—gh.	2	2	2	8	1.6	0
CRM—a—gh.	1	1	2	9	1.2	—
ssp. *lanceolatum*—cf.	9	28	43	175	1.3	.2
ssp. *subalpinum*—cf.	3	4	5	29	1.1	.2
ssp. *nesioticum*—cf.	1	4	4	12	1.9	.3
ssp. *lanceolatum*—gd.	10	10	13	74	1.7	0
ssp. *subalpinum*—gd.	2	3	4	19	1.4	—
ssp. *nesioticum*—gd.	1	5	5	30	1.8	.2
Leaves of floriferous stems—length						
GP	1	1	—	4	10.3	—
SRM—m	4	18	—	29	10.6	2.1
—a	5	5	—	10	6.9	—
CRM—m	1	1	—	1	6.2	—
CM	1	1	—	1	12	—
SRM—m—cf.	2	3	4	8	9.5	1.1
—a—cf.	1	1	1	4	10.5	—
—m—gd.	2	2	2	3	7.1	—
PT—gd.	1	4	4	14	11.4	1.8
Leaves of floriferous stems—width						
GP	1	1	—	4	2.3	—
SRM—m	4	18	—	29	3	.7
—a	5	5	—	10	2.2	—
CRM—m	1	1	—	1	2.5	—
CM	1	1	—	1	3	—
SRM—m—cf.	2	3	4	8	3.3	0
—a—cf.	1	1	1	4	4	—
—gd.	2	2	2	3	2	—
PT—gd.	1	4	4	14	3.9	.6
Leaves of floriferous stems—thickness						
SRM—m	4	17	—	28	2.2	.4
—a	5	5	—	10	1.3	—
CRM—m	1	1	—	1	.8	—
CM	1	1	—	1	2.3	—
SRM—m—cf.	2	3	4	8	1.9	.5
—a—cf.	1	1	1	4	1.8	—
—gd.	2	2	2	3	1.1	—
PT—gd.	1	4	4	14	2.3	.3

Inflorescences are terminal cymes, often of 3 cincinni, but the number varies from 1 to 6. When injured, cymes may become elongate and racemose. Floral bracts are lanceolate or elliptic-lanceolate and obtusely apiculate.

	n-pop.	n-pl.	n-rep.	n-obs.	$\bar{x}$	s
Diameter of cymes (cm.)						
GP	10	78	—	78	3.1	.6
Cincinni per cyme (n)						
GP	1	2	—	10	2.5	.9
CRM	2	15	—	30	2.5	.6
SN	2	6	—	12	1.5	.7
Flowers per cyme (n)						
GP	11	82	—	82	15.1	5.1
SRM—m	5	29	—	64	10.9	3.8
—a	5	12	—	38	4.9	2.6
CRM	2	15	—	32	8.3	3.1
CP	1	1	—	1	7	—
CM	2	15	—	30	23.8	12.9
SRM—m—cf.	2	4	5	11	19.8	0
—a—cf.	1	1	1	5	18	—
ssp. *lanceolatum*—gd.	6	6	8	43	12.2	6.7
SRM—a—gd.	1	1	1	2	4	—
PT—gd.	1	5	5	30	16.4	7.3

Flowers (fig. 44) are short-pedicellate and normally 5-merous. Young floral buds are orange, becoming yellow, with the angles yellow, green, or red. Central flowers are largest in cymes.

Fig. 44. Flower of *Sedum lanceolatum* ssp. *lanceolatum* from East Plum Creek about 1 km. north of Larkspur, Douglas Co., Colo., cultivated in cold frame, Ithaca, N.Y., May 25, 1964 (x 8). Photo by Howard Lyon.

	n-pop.	n-pl.	n-rep.	n-obs.	x̄ (mm.)	s (mm.)
Pedicels—length						
CP	1	1	—	1	.5	—
CM	2	15	—	54	.3	.2
SRM—m—cf.	2	4	5	15	.3	0
—a—cf.	1	1	1	5	.5	—
Flowers—diameter						
GP	11	81	—	81	16	1.8
SRM—m	5	12	—	17	15.3	1.9
—a	4	5	—	14	12.1	0
CRM	1	8	—	26	16	2.2
CP	1	1	—	1	11.5	—
CM	2	12	—	47	12.4	1.4
SRM—m—cf.	2	4	5	15	13.1	0
—a—cf.	1	1	1	5	14.2	—
ssp. *lanceolatum*—gd.	2	2	2	2	12	—
PT—gd.	1	4	4	20	13.6	2.9
Torus—length						
GP	1	2	—	6	1.3	.3
CM	1	10	—	42	1	.3
SRM—m—gd.	2	2	2	2	1.1	—
PT—gd.	1	4	4	20	1.3	.2

Sepals are erect, ovate or lanceolate, acute or rarely obtuse, pale green to yellow-green, sometimes speckled with red or red at apex, and often papillose.

	n-pop.	n-pl.	n-rep.	n-obs.	x̄ (mm.)	s (mm.)
Length						
GP	11	81	—	81	3.5	.5
SRM—m	5	12	—	17	3.4	.5
—a	4	5	—	14	3.2	1
CRM	1	9	—	27	3.5	.6
NRM	1	1	—	3	4.1	—
CP	1	1	—	2	2.5	—
CM	2	12	—	47	2.9	.4
SRM—m—cf.	2	4	5	15	2.8	0
—a—cf.	1	1	1	5	2.6	—
—m—gd.	5	5	6	25	2.3	.6
PT—gd.	1	4	4	15	4.5	.6
Length of cohesion						
SRM—m	5	12	—	17	.4	.1
—a	4	5	—	14	.4	.1
Width						
GP	11	81	—	81	1.2	.2
SRM—m	5	12	—	17	1.7	.3
—a	4	5	—	14	1.4	.4
CRM	1	9	—	27	1.8	.2
NRM	1	1	—	3	1.8	—
CP	1	1	—	2	1.3	—
CM	2	12	—	47	1.5	.2
SRM—m—cf.	2	4	5	15	1.3	.2
—a—cf.	1	1	1	5	1.8	—
—m—gd.	2	2	2	2	1.2	—
PT—gd.	1	4	4	20	1.7	.1

Petals are widely spreading from suberect bases, canaliculate, lanceolate, elliptic-lanceolate, or linear-lanceolate, acute or acuminate, minutely mucronate-appendaged, and some shade of yellow, ranging from canary-yellow to aureolin, rarely green medianly, sometimes with dorsal keel green with red at apex or speckled with red. A flower of a plant on Terry Peak in the Black Hills had 8 petals.

	n-pop.	n-pl.	n-rep.	n-obs.	$\bar{x}$ (mm.)	s (mm.)
Length						
GP	11	81	—	81	7.4	.7
SRM—m	5	12	—	17	8.1	.9
—a	4	5	—	14	7	.04
CRM	1	8	—	25	9.4	1.5
CP	1	1	—	2	6.7	—
CM	2	12	—	47	8.4	.8
SRM—m—cf.	2	4	5	15	6.9	.5
—a—cf.	1	1	1	5	6.9	—
—m—gd.	5	5	6	25	6.8	.7
PT—gd.	1	4	4	15	9.1	.1
Width						
GP	4	24	—	24	2.4	.3
SRM—m	5	12	—	17	2.7	.4
—a	4	5	—	14	2.5	.4
CRM	1	8	—	26	3.2	1.2
CP	1	1	—	2	1.8	—
CM	2	12	—	47	2.7	.9
SRM—m—cf.	2	4	5	15	2.2	.1
—a—cf.	1	1	1	5	2.4	—
—m—gd.	5	5	6	25	2.1	.2
PT—gd.	1	4	4	15	2.8	.5

Stamens have yellow filaments, ranging from pale canary-yellow to aureolin, and yellow anthers, ranging from lemon-yellow to aureolin and sometimes suffused with red. In the Teton Mountains, the alternate stamens sometimes are modified as carpels.

	n-pop.	n-pl.	n-rep.	n-obs.	$\bar{x}$ (mm.)	s (mm.)
Filaments—length						
GP	1	2	—	12	6	.8
CRM	1	9	—	27	5.6	1.6
Epipetalous filaments; adnation—length						
GP	1	2	—	6	1.2	.4
SRM—m	5	12	—	17	1	.2
—a	4	5	—	14	.9	.1
CRM	1	8	—	25	1.3	.4
CP	1	1	—	2	.9	—
CM	2	12	—	47	1.1	.4
SRM—m—cf.	2	4	5	15	1.1	0
—a—cf.	1	1	1	5	1.2	—
—m—gd.	2	2	2	2	.8	—
PT—gd.	1	4	4	20	1.2	.3
Anthers—length						
GP	1	2	—	4	1.3	.4
SRM—m	5	8	—	9	1.4	.1
—a	3	3	—	4	1.2	—

	n-pop.	n-pl.	n-rep.	n-obs.	$\bar{x}$ (mm.)	s (mm.)
Anthers—length						
CRM	1	4	—	5	1.3	.2
CM	1	10	—	20	1.3	.2
SRM—m—cf.	2	3	3	4	1.1	.2
—a—cf.	1	1	1	3	1	—
PT—gd.	1	3	3	6	1.3	.2

Nectaries are subquadrate, obovately quadrate, or reniform, truncate or broadly rounded and emarginate, and some shade of yellow, varying from deep yellow to lemon-yellow or yellow-green, and sometimes orange at apex.

	n-pop.	n-pl.	n-rep.	n-obs.	$\bar{x}$ (mm.)	s (mm.)
Length						
GP	1	2	—	6	.5	.1
SRM—m	5	12	—	17	.5	.1
—a	4	5	—	14	.4	.1
CRM	1	8	—	26	.4	.1
NRM	1	1	—	3	.4	—
CP	1	1	—	2	.3	—
CM	2	12	—	47	.5	.1
SRM—m—cf.	2	4	5	15	.4	.02
—a—cf.	1	1	1	5	.4	—
—m—gd.	2	2	2	2	.4	—
PT—gd.	1	4	4	20	.5	.02
Width						
GP	1	2	—	6	.8	.3
SRM—m	5	12	—	17	.6	.1
—a	4	5	—	14	.6	.1
CRM	1	8	—	26	.6	.1
NRM	1	1	—	3	.6	—
CP	1	1	—	2	.4	—
CM	2	12	—	47	.5	.1
SRM—m—cf.	2	4	5	15	.6	.04
—a—cf.	1	1	1	5	.5	—
—m—gd.	2	2	2	2	.4	—
PT—gd.	1	4	4	20	.5	.02

Carpels are erect, connate basally, smooth, and pale green or pale canary-yellow. The ovules are medianly constricted. A flower of a plant on Terry Peak, in the Black Hills, had 7 carpels. Several plants on Teewinot Mountain, in the Teton Mountains, had 10 carpels.

	n-pop.	n-pl.	n-rep.	n-obs.	$\bar{x}$	s
Length (mm.)						
GP	1	2	—	6	7	1.2
SRM—m	5	12	—	17	6.7	1
—a	4	5	—	14	5.8	.03
CRM	1	9	—	26	6.2	1.5
NRM	1	1	—	3	7.6	—
CP	1	1	—	2	4.7	—
CM	2	12	—	47	5.5	1.2

	n-pop.	n-pl.	n-rep.	n-obs.	$\bar{x}$	s
Length (mm.)						
SRM—m—cf.	2	4	5	15	5.2	.5
—a—cf.	1	1	1	5	5.8	—
—m—gd.	2	2	2	2	3.8	—
PT—gd.	1	4	4	20	5.6	.8
Length of cohesion (mm.)						
GP	1	2	—	6	1.3	.3
SRM—m	5	12	—	17	1.3	.2
—a	4	5	—	14	1.1	.1
NRM	1	1	—	3	1	—
CP	1	1	—	2	.6	—
Ovules per ovary (n)						
GP	1	2	—	6	9	2.7
SRM—m	5	12	—	17	10	2.4
—a	4	5	—	14	9	2.8
CRM	1	11	—	29	9	2.8
NRM	1	1	—	3	13	—
CP	1	1	—	2	8	—
CM	2	15	—	57	7	3
SRM—m—cf.	2	4	5	15	9	0
—a—cf.	1	1	1	5	8	—
—m—gd.	2	2	2	2	6	—
PT—gd.	1	4	4	20	11	3.2

Fruits are erect, connate basally, and brown, with divergent beaks about 1.5 mm. long and lips <.2 mm. wide.

	n-pop.	n-pl.	n-rep.	n-obs.	$\bar{x}$ (mm.)	s (mm.)
Length						
GP	4	4	—	5	5	1.2
CRM	1	14	—	28	5.3	.6
NRM	2	8	—	16	4.7	.7
SN	2	6	—	11	5.5	.7
CM	2	12	—	24	4.7	.7
PT	1	9	—	26	4.5	.6
SRM—m—cf.	2	2	—	4	4.3	.2
PT—gd.	1	3	—	6	5.1	.5

Seeds are pyriform, short-tailed, finely longitudinally ribbed, with secondary transverse ribs giving a scalariform appearance, and some shade of brown, varying from yellow-brown to dark brown.

	n-pop.	n-pl.	n-rep.	n-obs.	$\bar{x}$ (mm.)	s (mm.)
Length						
GP	1	3	—	6	1	.05
CRM	1	4	—	28	1.1	.12
NRM	2	7	—	14	1	.06
SN	2	2	—	3	1.2	0
CM	2	10	—	20	1.1	.12
PT	1	9	—	26	1.1	.1
SRM—m—cf.	3	8	9	17	1	—
PT—gd.	1	3	3	6	1.2	.13

	n-pop.	n-pl.	n-rep.	n-obs.	$\bar{x}$ (mm.)	s (mm.)
Diameter						
GP	1	3	—	6	.4	.1
CRM	1	14	—	28	.4	.07
NRM	2	7	—	14	.5	.05
SN	2	2	—	3	.4	—
CM	2	10	—	20	.4	.03
PT	1	9	—	26	.5	.06
SRM—m—cf.	3	8	9	17	.5	—
PT—gd.	1	3	3	6	.5	.06

Chromosomes (fig. 45) are elongate and large for *Sedum*. According to Mastrangelo, they vary from nearly metacentric to having the two arms unequal. Diploids, tetraploids, hexaploids, and also higher and odd ploidies occur. Diploids and tetraploids are widely distributed. Hexaploids are commonest in the Puget Trough and northern Cascade Mountains. Besides data cited below, C. H. Uhl (1962) has counted chromosomes of plants from additional localities and likewise, Margaret Powers, while a student at the University of Washington, made counts. Measurements are of chromosomes from root tips which were pretreated in a mixture of bromonaphthalene and 8-hydroxyquinoline, following the method of Mitra (1965). Chromosomes of the plant from Plum Creek were in early metaphase.

	n-pl.	g (n)	sp (2n)	Cytologist
Number				
Red Lodge, Mont.	1	16	—	Ilse Zandstra, 1969
Plum Creek, Colo.	1	8	—	Charles Horn, 1966
Plum Creek, Colo.	1	—	16	Iris Mastrangelo, 1971

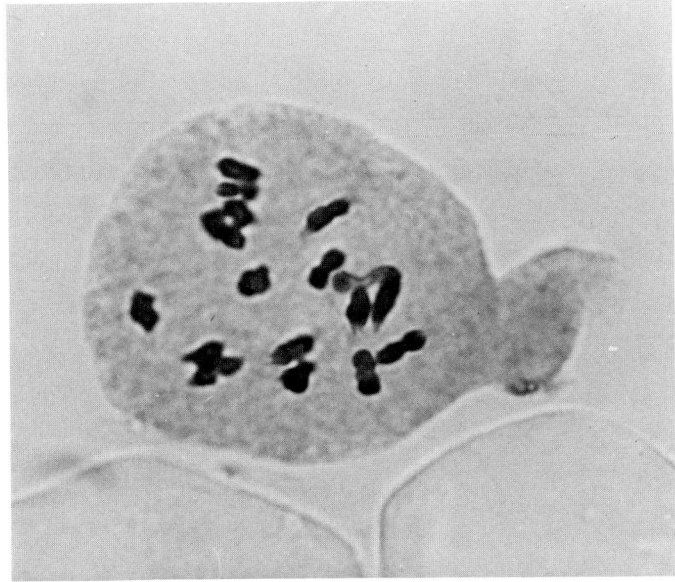

Fig. 45. Chromosomes of tetraploid *Sedum lanceolatum* ssp. *lanceolatum* from Red Lodge, Carbon Co., Mont., g = 16. Photo by Ilse Zandstra.

	n-pl.	g (n)	sp (2n)	Cytologist
Number				
Mt. Flora, Colo.	1	—	16	Iris Mastrangelo, 1971
Bergen Park, Colo.	1	8	—	C. H. Uhl, 1942
Pole Mt., Wyo.	1	16	—	Charles Horn, 1966
Macomber Peak, Colo.	1	16	—	Charles Horn, 1966
Macomber Peak, Colo.	1	—	31–38	Iris Mastrangelo, 1971
Lake Emma, Colo.	1	—	32	Iris Mastrangelo, 1971
Teton Mts., Wyo.				
Teewinot Mt.	1	—	52?	Dorothy Niimoto, 1967
Teewinot Mt.	1	—	47–50	Iris Mastrangelo, 1971
Cascade Canyon	2	—	70 ±, 54–55	Ilse Zandstra, 1969
Scotty Creek, Wash.	1	16	32	Iris Mastrangelo, 1971
Chopaka Mt., Wash.	2	24	46–50	Iris Mastrangelo, 1971
Lummi Rocks, Wash.	2	24	—	C. H. Uhl in Clausen and Uhl (1943), Ilse Zandstra, 1969
Jackson Hole, Wyo.				
(S. lanc. × sten.)	2	—	58, 59?	Dorothy Niimoto, 1967
Wenatchee Mts., Wash.				
(S. lanc. × rup.)	1	16–18	—	Charles Horn, 1966
(S. lanc. × sten.)	1	—	50–54	Dorothy Niimoto, 1967

	n-pl.	$\bar{x}$ (genome length; μ)	$\bar{x}$ (chromosome length; μ)	s (μ)	Cytologist
Length of sporophytic chromosomes					
Plum Creek, Colo.	1	46	2.8	.4	Iris Mastrangelo, 1971
Mt. Flora, Colo.	1	45	2.8	.6	Iris Mastrangelo, 1971
Lake Emma, Colo.	1	107	3.3	.4	Iris Mastrangelo, 1971

Variation. *Sedum lanceolatum* is one of the most widespread species of *Sedum* in North America, ranging from the Pine Ridge escarpment of the Great Plains in Nebraska westward to the Pacific Mountain System and from the Southern Rocky Mountains of New Mexico northward to the Yukon Plateau in Yukon Territory. Several morphological and physiological types occur in this vast area. Further, several levels of polyploidy exist. In addition, the pattern of variation is complicated by hybridization with other species, notably *S. rupicolum*, *S. stenopetalum*, and *S. divergens*.

Because chromosomes are difficult to count, especially in high polyploids, some reports lack certainty. Information about chromosomes is inadequate for purposes of classification. Other kinds of data are necessary. To sort out the major taxonomic groups within the species, I have done analyses of variance of data from both wild plants and plants cultivated in experiments in Ithaca. The hypothesis tested in each case was one of no difference among the common type of plant at low and middle altitudes, the robust type (fig. 46) characteristic of islands in Puget Sound, and the small type which occurs mostly in the alpine zone on the higher mountains. Samples for study are the ones listed at the beginning of the description. Results are indicated below in a table. As explained at the beginning of the chapter, n.s. = not significant; * = significant, P = .05; and ** = highly significant, P = .01.

Fig. 46. Rosette of *Sedum lanceolatum* ssp. *nesioticum* from Lummi Rocks, Whatcom Co., Wash., cultivated in cold frame, Ithaca, N.Y., June 22, 1966 (x 3.8). Photo by Howard Lyon.

Feature	Differences among major types			Differences among populations		
	Wild	Cf.	Gd.	Wild	Cf.	Gd.
Leaves of rosettes						
Length	**	n.s.	**	**	n.s.	**
Width	n.s.	**	*	**	n.s.	**
Thickness	n.s.	*	**	**	n.s.	**
Median leaves of flor. stems						
Length	*	n.s.	n.s.	n.s.	n.s.	n.s.
Width	**	n.s.	*	n.s.	n.s.	n.s.
Thickness	**	n.s.	n.s.	n.s.	n.s.	n.s.
Floriferous stems per plant	n.s.	n.s.	n.s.	**	n.s.	n.s.
Floriferous stems—length	**	n.s.	n.s.	*	*	n.s.
Flowers per cyme	*	n.s.	*	**	n.s.	n.s.
Flowers—diameter	**	n.s.	n.s.	n.s.	n.s.	n.s.
Sepals—length	n.s.	n.s.	*	**	n.s.	**
—width	*	n.s.	n.s.	n.s.	n.s.	n.s.
Petals—length	**	n.s.	*	n.s.	n.s.	**
—width	n.s.	n.s.	*	n.s.	n.s.	*
Nectaries—length	n.s.	n.s.	n.s.	n.s.	*	n.s.

Besides differences indicated as significant in the garden, width of sepals has a high F value and approaches significance. A difference in flowering time is marked, plants from the Puget Trough flowering much later, after the main flowering of other types is completed. A test of independence for plants in the garden confirms this conclusion. The difference is highly significant, $P = .01$.

For testing differences in dimensions of floral parts, the experiment at the Test Garden was most satisfactory. Plants there were exposed to full sun and were arranged in an experiment of completely random design. Flowering was erratic in the cold frame. Only a few plants produced flowers in any season. As a result, satisfactory tests were impossible. Failure to detect significant differences is not surprising.

In nature, plants from alpine situations have short floriferous stems. On the basis of data from wild plants, this difference is highly significant, but in cultivation, plants from lower altitudes also have short floriferous stems.

Dimensions of leaves of rosettes provide a basis for separating plants into three groups. Plants from the Puget Trough are largest in all dimensions and those from the alpine zone are smallest. The differences are most pronounced in the garden. In nature, environmental modifications may obscure them.

Leaves of rosettes often are prominently papillose, but great variation occurs in this feature. Many populations have smooth leaves, especially those in the Puget Trough. Likewise, development of anthocyanin varies among populations and even within populations, probably influenced by exposure as well as genetic constitution.

Floral size depends on several circumstances. Central flowers in cymes tend to be largest. In the Southern Rocky Mountains, plants at low and middle altitudes have flowers highly significantly larger than those in the alpine zone. Yet, in cultivation, blossoms of plants from low and middle altitudes are smaller. Floral size may not be dependable for separating subspecies.

Most flowers have 5 petals. Counts of petals in 64 flowers gave the following results:

Petals per flower (n)	Flowers (n)	%	Cumulative %
4	6	9.4	9.4
5	55	85.9	95.3
6	2	3.1	98.4
7	0	0	98.4
8	1	1.6	100.0
Totals	64	100	

The number of carpels normally is 5 per flower, with a usual range from 3 to 7. On the eastern slope of Teewinot Mountain, in the Teton Mountains, several plants had 10 carpels per flower. Flowers with the extra carpels had sepallike petals which were green medianly with yellow borders. They also had a single whorl of green filaments without anthers. The extra carpels appeared to be modified, alternate stamens. Sometimes they were intermediate between stamens and carpels. The cause of this condition is unknown. One thought, which has not been followed up or confirmed, is that radioactivity of the rocks might be involved.

Studies of *Sedum lanceolatum* revealed other interesting variations. One plant in the Black Hills had many short floral stems, each with a single pair of opposite leaves. Occasionally, axillary vegetative rosettes occur above the middle of the floriferous stems, as in *S. stenopetalum*. Vegetative apomixis is rare, but occurs with the development of leafy shoots in place of flowers. Elongate cymes with very thick fleshy leaves may develop when animals browse floriferous stems.

Correlations of gross morphology with number of chromosomes are not complete. Hexaploid plants may be either of the large type found on islands in Puget Sound or of the smaller, common type. Subspecies *nesioticum* always is hexaploid. Diploids and tetraploids occur at various altitudes. Plants of ssp. *subalpinum* sometimes are diploid. Also, a plant from along Plum Creek, near the type locality, is diploid. High polyploids may be similar to tetraploids, but with larger flowers. The population in the Teton Mountains comprises plants of this type. Information about chromosomes leads to the impression that the regional distribution of genes is more important than the level of ploidy in leading to the development of subspecies.

In summary, three subspecies merit nomenclatural designation. These are the three major types mentioned earlier. Their distinctions are indicated in the following dichotomous key. The subspecies *subalpinum* is least satisfactory as a unit for practical recognition, but its physiological distinctions indicate evolutionary importance. It is a good ecotype, but less well marked as a subspecies because experimental conditions sometimes are necessary to reveal the morphological differences. Dimensions cited in the key combine data from wild and cultivated plants. As a result, distinctions which are real when plants are grown side by side may seem obscure.

KEY TO SUBSPECIES

A. Plants of medium or small size, with leaves of rosettes 4.2–9 mm. long, 1.5–2.5 mm. wide, and .7–1.7 mm. thick, sometimes papillose; sepals 2–4 mm. long; petals 6–9 mm. long; plants flowering about two weeks earlier than ssp. *nesioticum* .... B
   B. Plants of medium size, with leaves of rosettes 4.2–9 mm. long and 1.7–2.5 mm. wide; plants adapted to growing season of more than three months .... ssp. *lanceolatum*, p. 227
   BB. Plants of small size, with leaves of rosettes 4.2–6.7 mm. long and 1.5–2 mm. wide; plants adapted to short growing season of two months or less, often in alpine habitat .... ssp. *subalpinum*, p. 231
AA. Plants of large size, with leaves of rosettes 8–13 mm. long, 3–3.5 mm. wide, and 1.6–2.2 mm. thick, smooth; sepals 4–5 mm. long; petals 9–9.2 mm. long; plants flowering later than the other two subspecies .... ssp. *nesioticum*, p. 232

Sedum lanceolatum ssp. *lanceolatum*

Nomenclature. *Sedum lanceolatum* Torrey, Annals Lyceum Nat. Hist. N.Y. 2:205–206 (1828). Type locality: near the Rocky Mountains. Type: a collection of Edwin P. James (NY), 1820, probably in early July. The type collection comprises two specimens, the larger of which I select as the holotype. The flowers are either postanthesis or in anthesis. Leaves of a secondary shoot are lanceolate and finely papillose. According to Ewan (1950), the party, of which James was a member, closely followed the course of the South Platte River. Places where James might have collected *S. lanceolatum* are where the Platte River emerges from the mountains, along the east branch of Plum Creek, along Monument Creek, and Manitou Springs. I studied plants along East Plum Creek about 1 kilometer north of Larkspur, Colo., and consider plants there as reasonable topotypes. I have also seen *S. lanceolatum* south of Larkspur and at Monument and have studied it at Manitou Springs. McKelvey (1955) provided information about the route of Stephen Long's expedition, of which James was a member.

A synonym is:

1903. *Sedum shastense* Britton, Bull. N.Y. Bot. Gard. 3:41. Type locality: unknown, but cited as north side of Mt. Shasta, alt. 5,000–9,000 ft. The type locality might be Mount Eddy. Type: a collection of H. E. Brown, no. 441 (NY), July 1–15, 1897. According to W. B. Cooke (1941), all Brown's *Seda* came from Mt. Eddy.

Many authors, probably beginning with Hooker (1829–1840), have applied the name *Sedum stenopetalum* to this species. For further discussion, see Clausen (1948). The names *S. caerulescens* Haw. and *S. subclavatum* Haw., both sometimes listed as synonyms, probably apply to other species, as indicated in the 1948 paper.

Subspecies *lanceolatum* is wide-ranging and polymorphic, with moderately large leaves which are often papillose and a distribution at low and middle altitudes. Diploids, tetraploids, hexaploids, higher ploidies, and aneuploids occur. Satisfactory classification into practical subunits is difficult. Hybridization with other species, as *S. stenopetalum* and *S. rupicolum*, adds to the problem. Sometimes hybrids and high generation back-crosses are identified as part of the aggregate subspecies.

Distribution. The geographic range of ssp. *lanceolatum* (fig. 47) includes a wide area of mountains and intermontane plateaus of western North America, extending from the western Great Plains westward to the Klamath, Olympic, and Canadian Coast Mountains, and from 35° N. in the Southern Rocky Mountains northward to the vicinity of Whitehorse, Yukon Territory, at about 61° N. In the intermontane region, it is mostly at the higher elevations in the mountain ranges. *Sedum lanceolatum* is most generally distributed in the Rocky Mountains and the ranges of the intermontane region. Its absence from big areas of the Sierra Nevada and Cascade Mountains appears to be real.

The altitudinal range is from near sea level northeast of Sydney, north of Victoria, B.C., if plants from there are correctly identified, to 3,660 m. on the Dana Plateau in the Sierra Nevada. Plants from high altitudes may be confused with ssp. *subalpinum*. Experimental culture may be necessary in difficult cases to reach decisions about subspecies. Plants grow in gravel and in crevices of rocks in open and exposed situations. Rocks include limestone, sandstone, marble, andesite, basalt, granodiorite, and granite. Flowering time is from May to September, depending on altitude and latitude.

Data for twenty-six populations surveyed in this study are organized in two tables,

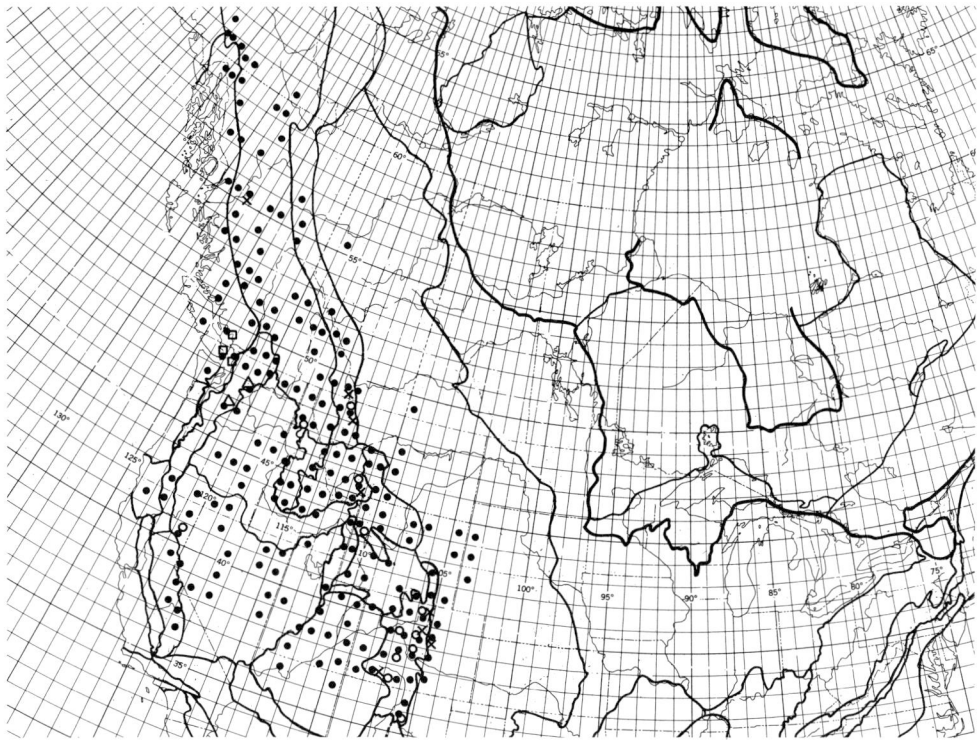

Fig. 47. Known distribution of *Sedum lanceolatum* ssp. *lanceolatum* (●), ssp. *subalpinum* (○), ssp. *nesioticum* (□), and *S. rupicolum* (△). Quadrangles in which diploid plants of *S. lanceolatum* are known to occur are indicated by (x).

the first including populations selected according to judgment or for convenience, the second devoted to populations chosen with known probability of selection.

Data for populations chosen according to judgment or for convenience:

Population	Altitudinal range (m.)	Exposure	Rock	Drainage	Oldest record
W. Spanish Peak	2,560–2,790	N	sandstone	good	1900
Monroe Canyon	1,433	N	—	moderate	1952
Custer	1,676	E	slate	good	1892
Harney Peak	2,195	N, E, S, W	granite	good	1952
Little Elk Creek	1,372	SW	granite	—	1892
Roubaix	1,676	N, E, S, W	—	moderate	1952
Pluma	1,451–1,457	NE	sandstone	excessive	1952
Deadwood	—	W	sandstone	—	1929
Terry Peak	2,149–2,155	N, E, S, W	granite	—	1927
NE of Buckhorn	1,859	NE	—	good	1952
NW of Buckhorn	1,768	—	limestone	good	1952
Plum Creek	2,049–2,050	E, W	—	good	1820
Platoro	2,998–3,000	N, E, S, W	andesite	good	1962

Population	Altitudinal range (m.)	Exposure	Rock	Drainage	Oldest record
Macomber Peak	2,935–2,945	S–SE	andesite	good	1962
Pole Mt.	2,440–2,700	ESE	granite	good–excessive	1962
Clements Mt.	2,190	SW	argillite	—	1962
Seven Devils Lake	2,120–2,140	N–NNE	andesite	good	1961
Heaven's Gate	2,380	N, E	andesite	—	1961
Breeze Creek	2,280	SW	granite	—	1964
Ingalls Creek	1,050	N, E, S, W	granodiorite	—	1961

Data for populations chosen with known probability of selection.

Population	Plants (N)	Area of population (m.^{2})	Density per m.^{2}	Exposure	Rock
Dry Gulch	161,200	13,000	12.4	SW–NW	sandstone
Teewinot Mt.	19,425	480,000	.04	NE–ESE	gneiss, granite
N. Buffalo Fork	544	1,513	.36	W	limestone
Gem Pass	12	600	.02	NNE	serpentine
Scotty Creek	204	—	—	NNE–NE	sandstone
Chopaka Mt.	450	900,000	.0005	W–NE	granite

Population	Altitudinal range (m.)	Depth of soil $\bar{x}$ (cm.)	s (cm.)	pH range	pH tests (n)	Distance to nearest neighbor $\bar{x}$ (cm.)	s (cm.)	Oldest record
Dry Gulch	2,620–2,655	—	—	7.1–8.1	11	—	—	1962
Teewinot Mt.	2,200–2,280	7.1	4.3	5.4–6.6	14	30.3	26.8	1962
N. Buffalo Fork	2,750–2,865	8.5	12	6.6–7.2	2	16.5	21.3	1966
Gem Pass	3,215	—	—	5.2–5.4	2	—	—	1964
Scotty Creek	1,175–1,185	—	—	4.6–5.2	10	—	—	1961
Chopaka Mt.	915–1,040	—	—	6.4–6.6	7	—	—	1965

In the Black Hills, ssp. *lanceolatum* occurs in rocky places in the zone of *Betula alba*, *Populus tremuloides*, *Picea glauca*, and *Pinus ponderosa*. It is absent from the warmer, drier slopes in the zone of *Juniperus scopulorum* and *Quercus*, and also from grassland. Northward on the Great Plains it grows in gravelly places with grasses and also in open groves of *Populus tremuloides*.

In the Southern Rocky Mountains, plants competing with ssp. *lanceolatum* are miscellaneous. Habitats are open areas in the zones of *Pinus ponderosa*, *P. flexilis*, and *P. contorta*. The *Sedum* grows primarily in open rocky areas. It survives for a while when grasses move in, but does not continue in dense grass or where shrubs and trees become dominant. Experience was similar in the Central and Northern Rocky Mountains, and also in the Sierra Nevada and Cascade Mountains. Where *Sedum lanceolatum* occurs in temporary clearings in the forest, populations must be of short duration because conditions are perpetually changing.

Other species of *Sedum* competing with *S. lanceolatum* ssp. *lanceolatum* include *S. stenopetalum* in Jackson Hole and the Wenatchee Mountains, *S. debile* in the Teton Mountains, *S. rupicolum* in the Wenatchee Mountains, *S. obtusatum* in the Sierra Nevada, and *S. integrifolium* in the Seven Devils Mountains and elsewhere. *Selaginella densa* is one of the commoner competitors.

Sedum lanceolatum ssp. *subalpinum*

Nomenclature. *Sedum lanceolatum* ssp. **subalpinum** (Blankinship) comb. nov., fundatum super *Sedum subalpinum* Blankinship.

Synonyms are:

1891. *Sedum stenopetalum* f. *rubrolineatum* T. D. A. Cockerell, Bull. Torr. Bot. Club 18:169. Type locality: above the Micawber Mine, vicinity of Gibbs Peak, 38° 10′ N., 105° 40′ W., eastern slope of Sangre de Cristo Range, alt. >3,050 m., Custer Co., Colo. Type: not seen; location uncertain. The original diagnosis is too brief to permit positive interpretation, but on the basis of altitude, Cockerell may have had a variation of ssp. *subalpinum* with the angles of the closed buds red.

1905. *Sedum subalpinum* Blankinship, Mont. Agr. Coll. Sci. Studies, Bot. 1:61. Type locality: rocky ledges, Sperry Glacier, alt. 2,440 m., east of Gunsight Mountain, Glacier National Park, Mont. Type: a collection of J. W. Blankinship (Mont.), Sept. 1, 1903.

1936. *Sedum stenopetalum* var. *subalpinum* Fröderström, Acta Horti Goth. 10 (App.):70, figs. 521–527, pl. 43, fig. 2. Type locality: above timber line in mountains southeast of Cameron Pass, Colo. Type: from herb. State Agric. Coll., Colo. (NY), Sept. 2, 1892. Fröderström did not mention Blankinship's earlier description of *subalpinum* as a species.

When plants of ssp. *subalpinum* are cultivated side by side with those of ssp. *lanceolatum*, their smaller size provides a means of distinction. Likewise, in nature, most plants of ssp. *subalpinum* are diminutive, with many tiny rosettes of small leaves. Occasional plants are larger and may be difficult to distinguish. Subspecies *subalpinum* appears to be a taxonomic group at an evolutionary stage somewhere between ecotype and subspecies. Its physiological distinctions include adaptations to conditions of growth on exposed summits of high mountains and ability to flower and mature seeds in a brief growing season. Plants of ssp. *subalpinum* are more difficult to maintain in cultivation at Ithaca, N.Y., than are plants of ssp. *lanceolatum*. The only available count of chromosomes is diploid.

Distribution. *Sedum lanceolatum* ssp. *subalpinum* occurs at high altitudes in the Southern, Central, and Northern Rocky Mountains, and also in the Interior Plateaus and Ranges of British Columbia, Yukon, and Alaska. Its altitudinal distribution is highest in the south and lowest in the north. In Colorado, it occurs up to 4,048 m. On Marks Butte, Idaho, it occurs at an altitude of 1,940 m. Records plotted on the map are included only if supported by data from experiments or if the identification seems positive for other reasons. This policy underestimates the range of the subspecies. Some records which apply to ssp. *subalpinum* probably are included in the data for ssp. *lanceolatum*.

Habitats of ssp. *subalpinum* are mostly exposed, alpine sites. The plants grow in gravel or stony loam. Underlying rocks include granite, andesite, and slate. Flowering time is from July to September.

Data for seven populations which I have studied are in the following table. Selection of the sites on Summit Peak and in the Alaska Basin was random. The remaining sites were judgment samples.

Population	Plants (N)	Area of population (m.²)	Density per (m.²)	Exposure	Rock
Tesuque Peak	25	24	1.04	N, E	granite
Summit Peak	—	—	—	E	andesite
Mt. Flora	19	47	.4	N, E, S, W	granite
Lulu Pass	—	—	—	—	granite
Iron Mt.	162	1,016	.16	N, E, S, W	granite
Alaska Basin	111	—	—	S	slate
Marks Butte	72	351	.2	N–NNE	granodiorite

Population	Altitudinal range (m.)	Depth of soil $\bar{x}$ (cm.)	s (cm.)	pH range	pH tests (n)	Distance to nearest neighbor $\bar{x}$ (cm.)	s (cm.)	Oldest record
Tesuque Peak	3,510	—	—	5.6	2	—	—	1962
Summit Peak	3,973–4,048	—	—	6.4–6.6	2	—	—	1962
Mt. Flora	3,870	—	—	5.6–6.4	2	—	—	1962
Lulu Pass	3,375	—	—	6.2	2	—	—	1962
Iron Mt.	3,380	—	—	5.8–6.8	6	—	—	1892
Alaska Basin	3,005–3,055	13	—	7	1	11	—	1966
Marks Butte	1,940–1,955	—	—	4.6–4.8	6	—	—	1961

Other plants competing with *Sedum lanceolatum* ssp. *subalpinum* include a variety of alpine herbs. *Trifolium nanum* was at three sites investigated and *Arenaria sajanensis* and species of *Potentilla* were at two sites. On the summit of Iron Mountain southeast of Cameron Pass, the *Sedum* was commonest in an area of dense growth of *Poa pattersonii* and next commonest in open gravelly areas. It was least common in rocky areas and among cushion plants. Densities per square meter were as follows:

Gravelly area	.11
Rocky area	.02
Area of cushion plants and sparse grass	.03
Area of dense grass	1.82

Sedum lanceolatum ssp. *nesioticum*

Nomenclature. *Sedum lanceolatum* ssp. *nesioticum* (G. N. Jones) R. T. Clausen, Cact. & Succ. Jour. 20 (10):146 (1948). Based on *S. nesioticum* G. N. Jones.
Synonyms are:
1941. *Sedum nesioticum* G. N. Jones, Madroño 6:86. Type locality: islets, Gulf of Georgia, Wash. Type: a collection of L. F. Henderson, no. 1,686 (GH).
1946. *Sedum stenopetalum* ssp. *nesioticum* (G. N. Jones) R. T. Clausen, Cact. & Succ. Jour. 18:77. Based on *S. nesioticum* G. N. Jones.
1964. *Sedum lanceolatum* var. *nesioticum* (G. N. Jones) C. L. Hitchc., Vascular plants of the Pacific Northwest, pt. 2. p. 569. Based on *S. nesioticum* G. N. Jones.

Plants of ssp. *nesioticum* are more robust than those of the other two subspecies. Leaves of rosettes are larger in all dimensions, leaves of floriferous stems are wider, flowers are more numerous per cyme, sepals are longer, and petals are both longer and wider. In addition, the leaves usually are smooth and the plants flower later. In the garden, when the subspecies are grown together, these differences are striking.

Distribution. The distributional area of *Sedum lanceolatum* ssp. *nesioticum* is on islands in Puget Sound and the Strait of Georgia, and also on adjacent Vancouver Island. On the Lummi Rocks, where I have studied the subspecies, the plants occur in crevices in graywacke. Sites include bluffs, cliffs, and flat outcroppings. Plants occur only a few meters above the level of the sea. Flowering time is from mid-May to July, depending on exposure and season.

Data for the population on Lummi Rocks, which I studied in August 1965 follow:

```
Plants—N—1,020
Exposure–N, SE–WSW
Altitudinal range (m.)—4–20
pH—range—6–6.6 (9 tests)
Oldest record—1937
```

A narrow-leaved grass is the commonest vascular plant on the Lummi Rocks. Shrubs are few: *Arctostaphylos uva-ursi*, *Holodiscus*, and a species of *Rosa*. *Sedum spathulifolium* also occurs, but in higher places and more in shade. Grasses are commonest competitors of the *Sedum lanceolatum*. Other competitors include species of *Polygonum* and *Grindelia*.

Reproduction. Vegetative reproduction is common in *Sedum lanceolatum*. Plants produce axillary shoots on the lower parts of the floriferous stems. These develop into leafy rosettes which separate easily from the primary stems, take root, and become new plants. Breaking up of clusters of rosettes appears to be a principal mode of reproduction. In addition, reproduction from seeds also occurs. Occasional clusters of seedlings below old fruits suggest that seeds sometimes simply drop to the ground and germinate when conditions are favorable. Young plants may become intermixed with parts of older plants, with the result that a patch does not necessarily consist of stems belonging to the same clone.

Seeds are light enough to travel readily if picked up by the wind. For this reason, long-distance dispersal and colonization of new sites are easy.

One plant in a population south of Custer, in the Black Hills of South Dakota, had a cyme with three leafy shoots in place of flowers. Such shoots can give rise to new individuals and are examples of vegetative apomixis.

Bees, flies, and butterflies cross-pollinate flowers of *Sedum lanceolatum*. Optimum times for pollination are sunny periods when the wind is still. On Teewinot Mountain in the Tetons, I have watched insects go from flowers of *S. debile* to *S. lanceolatum*, but I did not find hybrids there. On Pole Mountain, east of Laramie, Wyoming, a large black and white butterfly with purple spots, perhaps *Parnassius phoebus smintheus*, visits the flowers. On the same mountain, something, perhaps the larva of this butterfly, eats the stems, causing the branches of the cymes to drop. Cockerell found larvae of *Parnassius* feeding on *S. lanceolatum* ssp. *subalpinum* on Pikes Peak.

Relationships. Other species with which *Sedum lanceolatum* sometimes occurs, on the basis of studies from my surveys, are as follows. Asterisks indicate the occurrence of hybrids. In all cases except the *S. divergens*, the plants of the two species occurred intermixed at the sites which are listed.

S. debile: Teewinot Mountain
S. divergens: Ingalls Creek
S. integrifolium: Seven Devils Lake, Platoro, Summit Peak
S. obtusatum: Gem Pass
S. rupicolum: Scotty Creek, Mission Peak*
S. spathulifolium: Lummi Rocks
S. stenopetalum: Jackson Hole*, Scotty Creek*

In the Wenatchee Mountains, both *Sedum lanceolatum* and *S. rupicolum* occur, the former on sandstone and granodiorite, the latter on serpentine, basalt, and sandstone. The two species are similar morphologically and appear to be closely related. Except for the fact that they sometimes are able to coexist without intergradation, as on the Wenatchee Ridge above Scotty Creek, their status would be subspecies of a single species. The leaves of *S. rupicolum* are caducous on slight pressure, a striking feature quite different from the condition in *S. lanceolatum*. Other differences and further aspects of relationships are discussed under *S. rupicolum*.

Sedum stenopetalum is another species which appears close to *S. lanceolatum*. Many collectors have confused these two species. Even experts sometimes make mistakes. The name *stenopetalum* has been applied to both species, yet differences are marked. *Sedum lanceolatum* has erect follicles with lips not enlarged. In contrast, the follicles of *S. stenopetalum* are widely divergent with prominent, ventral lips. In addition, the leaves of ssp. *stenopetalum*, which overlaps in range with *S. lanceolatum*, are spurred, and vegetative rosettes develop in large numbers along the floriferous stems and even in place of flowers in the cymes. In common, the two species have yellow flowers, elongate leaves, and a similar habit of growth. Hybridization between species, both in Jackson Hole (fig. 48) and on the Wenatchee Ridge (fig. 49), confirms the idea of relationship. For additional information, see the section on relationships under *S. stenopetalum*.

The geographic ranges of *Sedum lanceolatum* and *S. divergens* overlap in the northern Cascade Mountains and on the Interior Plateau of British Columbia. The only place where I have found the two species growing in proximity was in the valley of Ingalls Creek in the Wenatchee Mountains. The plants never were intermixed, but populations were within less than a kilometer of each other, making cross-pollination by bees a possibility. Although I have never found in the field a hybrid between *S. divergens* and *S. lanceolatum*, Reid Moran collected a plant near Moricetown, B.C., about 55° N. and 127° 30′ W., with green, relatively plump leaves, small flowers, poorly developed pollen, and various cytological irregularities such as laggards, univalents, and variable numbers of bivalents. The evidence suggests an origin through hybridization between *S. divergens* and *S. lanceolatum*.

Sedum debile, with opposite leaves and slightly connate petals, and *S. cockerellii*, with white petals and oblanceolate leaves with flat surfaces, probably are also related to *S. lanceolatum*, although hybrids are unknown. Where *S. debile* and *S. lanceolatum* occur together, as in the Teton Mountains, the *S. debile* is diploid and the *S. lanceo-*

Fig. 48. Plant of *Sedum lanceolatum* ssp. *lanceolatum* × *S. stenopetalum* ssp. *stenopetalum* from Moose, Jackson Hole, Teton Co., Wyo., cultivated in cold frame, Ithaca, N.Y., July 11, 1967 (× 1.2). Photo by Howard Lyon.

latum is polyploid. Biological isolation is complete. Since *S. cockerellii* and *S. lanceolatum* are allopatric, opportunity for interaction is lacking.

Sedum grandipetalum, of the Trans-Mexican Volcanic Belt, is another possible relative of *S. lanceolatum*. It has divergent follicles and distinctive vegetative rosettes, with suborbicular, densely imbricate leaves.

More distantly related are *Sedum obtusatum*, *S. spathulifolium*, and *S. integrifolium*. Although these species sometimes occur in association with *S. lanceolatum*, they do not hybridize with it.

Species such as *Sedum ochroleucum*, *S. reflexum*, *S. rupestre*, and *S. sediforme*, of Europe and northern Africa, probably comprise an evolutionary line which is separate

Fig. 49. Cristate stem of *Sedum lanceolatum* ssp. *lanceolatum* × *S. stenopetalum* ssp. *stenopetalum* from Wenatchee Mountains, Wash., cultivated in greenhouse at Ithaca, N.Y., June 6, 1962 (x 2.7). Photo by Howard Lyon.

from the line to which *S. lanceolatum* belongs. Quimby (1971) found that these species have a specialized, condensed vascular anatomy, involving much fusion of bundles. Superficially, these plants have a distinctive appearance, different from *S. lanceolatum*, and their flowers more often have more than five petals.

Much of the geographic area now occupied by *Sedum lanceolatum* was under glacial ice until recent geological time. The present wide range of the species must be a modern development. Since no information is available concerning its distribution either during or before the ice ages, ideas about evolution must depend on what is known about relationships with other modern species and also on data from comparative morphology and cytology.

References. E. J. Alexander (1943) and R. T. Clausen (1948).

11. *Sedum rupicolum**** (figs. 47 and 50–52)

Sedum rupicolum (fig. 50) is similar to *S. lanceolatum*, but its leaves readily detach from the stems on slight pressure. The loose leaves, on a moist surface, sprout and produce plantlets from their bases. Otherwise, *S. rupicolum* differs from *S. lanceolatum*

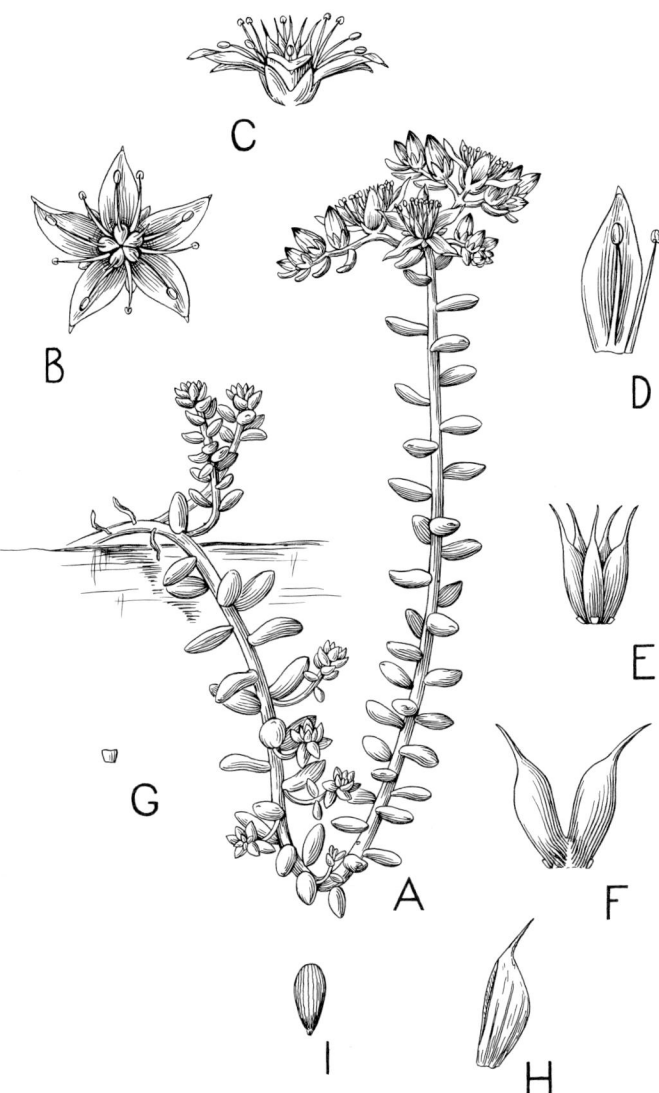

Fig. 50. Plant of *Sedum rupicolum* from slope along Peshastin Creek near Blewett Pass, Chelan Co., Wash., cultivated in cold frame, Ithaca, N.Y. A. Habit sketch (x .88). B. Flower from above (x 1.8). C. Flower from side (x 1.8). D. Petal and two stamens (x 2.6). E. Carpels (x 2.6). F. Two carpels, nectaries, and torus (x 3.5). G. Nectary (x 4.4). H. Follicle from side (x 3.5). I. Seed (x 8.8).

in having relatively broader leaves, shorter sepals and petals, orange-yellow petals, and deeper yellow nectaries. The stems tend to be red. Separations from other species are by the same distinctions as for *S. lanceolatum*.

Description. Sample: 46 plants, all except 5 from the Wenatchee Mountains and obtained in 1961. Data for the subsamples are in the following table, in which abbreviations are the same as in a similar table for *S. lanceolatum* (see p. 213). The sample can be described as follows:

Locality	Basis of choice	N_u	N_c	n	Plants studied n/N_c	n/N_u	Condition when collected fl. (n)	fr. (n)
Peshastin Creek	J	80	20	8	.4	.1	4	4
Scotty Creek	R	110	110	8	.07	.07	2	6
Earl Peak	R	108	58	6	.1	.06	4	0
Ingalls Creek	C	3	3	0	0	0	—	—
Mt. Stuart	C	>139	>139	14	.1	.1	12	0
Mission Peak	J	>117	>117	5	.04	.04	3	0
Bear Gap	J	124	124	5	.04	.04	5	0
Totals		>681	>571	46			30	10

In cultivation in Ithaca, plants of *Sedum rupicolum* do not grow as well as do plants of *S. lanceolatum*. Best survival is in the greenhouse, where a plant from Earl Peak continues in vigorous condition after eleven years of cultivation. Poorest survival is in the garden, where only one out of four plants yielded data. The others died. Two replicates of a plant from Earl Peak survived long enough in the garden to flower. The plant from Earl Peak in the greenhouse has not flowered, but a plant from along Peshastin Creek flowered in the cold frame in May 1962. Data for my three experiments follow. N = number of plants in experiments and n-d = number of plants which yielded data.

Source	Garden N	n-d	Cold frame N	n-d	Greenhouse N	n-d
Wenatchee Mts.	4	1	5	5	1	1

In the description, BG = Bear Gap, PC = Peshastin Creek, and WM = Wenatchee Mountains.

Plants of *Sedum rupicolum* are perennial herbs, of tufted habit, with capillary roots and clusters of reddish, sterile stems, with secondary shoots arising in the axils of some leaves. Floriferous stems usually are erect, but sometimes curved and decumbent. They are pink below and green upward.

	n-pop.	n-pl.	n-rep.	n-obs.	$\bar{x}$	s
Plants—diameter (cm.)						
PC—cf.	1	1	2	2	13	—
WM—cf.	2	2	3	3	8.3	.4
BG—cf.	1	1	2	2	11	—
WM—gh.	1	1	1	1	18	—
Primary rosettes (n)						
PC—gd.	1	1	1	1	1	—
WM—gd.	3	3	4	4	17	7.8
PC—cf.	1	1	2	2	26	—
WM—cf.	2	2	3	3	15	7.3
BG—cf.	1	1	2	2	11	—
WM—gh.	1	1	1	1	19	—
Leaf-bearing stems—length (mm.)						
PC—cf.	1	1	2	7	10.8	—
WM—cf.	2	3	4	19	15.9	3.7
BG—cf.	1	1	1	2	6.5	—
WM—gh.	1	1	1	1	9.8	—

	n-pop.	n-pl.	n-rep.	n-obs.	$\bar{x}$	s
Floriferous stems—length (cm.)						
PC—cf.	1	1	1	1	18	—
Maximum height (cm.)						
PC	1	4	—	4	9	3.5
WM	4	27	—	27	11.2	1.5
BG	1	4	—	5	7.6	.8

Leaves (fig. 51) are spirally arranged, divergent, and slightly upturned; ovate, elliptic-ovate, or lanceolate; obtuse or apparently acute; finely reticulate; minutely papillose; sometimes glaucous; and blue-green, purplish, or green, occasionally speckled with red. Leaves of floriferous stems are widely divergent, often incurved, elliptical, varying to suborbicular or elliptic-linear, and obtuse. All leaves are easily

Fig. 51. Leaves and stem of *Sedum rupicolum* from slope along Peshastin Creek near Blewett Pass, Chelan Co., Wash., cultivated in greenhouse, Ithaca, N.Y., Feb. 2, 1965 (x 3.8). Photo by Howard Lyon.

detachable, giving rise to new plantlets (fig. 52) from a cushion of scar tissue near the base.

	n-pop.	n-pl.	n-rep.	n-obs.	$\bar{x}$ (mm.)	s (mm.)
Leaves of primary rosettes—length						
PC	1	2	—	2	4.6	1.3
WM	4	23	—	62	5.9	1.3
BG	1	5	—	14	4.6	.2
PC—cf.	1	1	2	7	3.2	—
WM—cf.	2	3	4	19	6.2	.7
BG—cf.	1	1	1	2	4.7	—
WM—gh.	1	1	—	6	5.4	—
Leaves of primary rosettes—width						
PC	1	1	—	1	2.9	—
WM	4	24	—	62	2.7	.4
BG	1	5	—	14	2.3	.3
PC—cf.	1	1	2	7	1.9	—
WM—cf.	2	3	4	19	2.5	.1
BG—cf.	1	1	1	2	2	—
WM—gh.	1	1	—	6	2.6	—
Leaves of primary rosettes—thickness						
PC	1	1	—	1	1.8	—
WM	4	24	—	62	1.8	.3
BG	1	5	—	14	1.6	.1
PC—cf.	1	1	2	6	1.4	—

Fig. 52. Detached leaf of *Sedum rupicolum* with plantlet developing from scar tissue near base, cultivated in cold frame, Ithaca, N.Y., Nov. 24, 1965 (x 20), originally from Mission Peak, Chelan Co., Wash.

	n-pop.	n-pl.	n-rep.	n-obs.	$\bar{x}$ (mm.)	s (mm.)
Leaves of primary rosettes—thickness						
WM—cf.	2	3	4	19	2	.1
BG—cf.	1	1	1	2	1.4	—
WM—gh.	1	1	—	6	1.8	—
Leaves of floriferous stems—length						
PC	1	4	—	4	5.8	1.9
WM	2	10	—	10	6.6	.1
WM—gd.	1	1	2	4	8.8	—
PC—cf.	1	1	1	1	6.3	—
Leaves of floriferous stems—width						
PC	1	4	—	4	3.3	1.1
WM	2	10	—	10	3.1	.04
WM—gd.	1	1	2	4	4	—
PC—cf.	1	1	1	1	3	—
Leaves of floriferous stems—thickness						
WM	2	10	—	10	2	.1

Inflorescences are terminal cymes, usually three-parted, with the cincinni sometimes dichotomously branched. Floral bracts are narrowly elliptical or even linear.

	n-pop.	n-pl.	n-rep.	n-obs.	$\bar{x}$ (n)	s (n)
Flowers per cyme						
PC	1	4	—	6	8.9	2.2
WM	4	23	—	31	6.5	2.1
BG	1	5	—	8	4.6	2.6
PC—cf.	1	1	1	1	22	—

Flowers usually are 5-merous and short-pedicellate.

	n-pop.	n-pl.	n-rep.	n-obs.	$\bar{x}$ (mm.)	s (mm.)
Pedicels—length						
WM	4	23	—	47	.4	.05
BG	1	5	—	8	.5	.3
Flowers—diameter						
PC	1	4	—	12	14	1.6
WM	4	21	—	44	15.2	2.6
BG	1	4	—	7	11	1.8
WM—gd.	1	1	2	6	16	—
PC—cf.	1	1	1	5	14	—
Torus—length						
PC	1	4	—	12	.8	.3
WM	4	21	—	40	1.1	.5

Sepals normally are 5 per flower, lanceolate or lanceolate-oblong, obtuse or subacute, and green. They are slightly connate basally.

	n-pop.	n-pl.	n-rep.	n-obs.	$\bar{x}$ (mm.)	s (mm.)
Length						
PC	1	4	—	12	3.8	.5
WM	4	21	—	44	3.3	1
BG	1	5	—	8	2.6	.2
PC—cf.	1	1	1	3	3	—

	n-pop.	n-pl.	n-rep.	n-obs.	x̄ (mm.)	s (mm.)
Width						
PC	1	4	—	12	1.6	.3
WM	4	21	—	44	1.8	.6
BG	1	5	—	8	1.4	.4
PC—cf.	1	1	1	3	1.8	—

Petals are 5 per flower, shortly erect at base, then widely spreading, elliptic-lanceolate or lanceolate, obtuse or less often acute, minutely mucronate-tipped, slightly hooded, and aureolin or deep yellow.

	n-pop.	n-pl.	n-rep.	n-obs.	x̄ (mm.)	s (mm.)
Length						
PC	1	4	—	12	7.2	.7
WM	4	21	—	44	8.8	2.1
BG	1	5	—	8	6.7	.9
PC—cf.	1	1	1	3	8.1	—
Width						
PC	1	4	—	12	2.5	.2
WM	4	21	—	44	2.7	.5
BG	1	5	—	8	2.2	.2
PC—cf.	1	1	1	3	3.3	—

Stamens have yellow filaments and anthers.

	n-pop.	n-pl.	n-rep.	n-obs.	x̄ (mm.)	s (mm.)
Epipetalous filaments; adnation—length						
PC	1	4	—	12	1.6	.3
WM	4	21	—	44	1.2	.1
BG	1	5	—	8	.8	.2
PC—cf.	1	1	1	3	.5	—
Anthers—length						
PC	1	2	—	2	.9	.07
WM	4	12	—	17	1.4	.02
BG	1	4	—	5	1.2	.1
PC—cf.	1	1	1	2	1.1	—

Nectaries are subquadrate, truncate, and deep yellow.

	n-pop.	n-pl.	n-rep.	n-obs.	x̄ (mm.)	s (mm.)
Length						
PC	1	4	—	12	.4	.05
WM	4	21	—	44	.5	.07
BG	1	5	—	8	.4	.09
PC—cf.	1	1	1	3	.4	—
Width						
PC	1	4	—	12	.4	.02
WM	4	21	—	44	.4	.1
BG	1	5	—	8	.5	.1
PC—cf.	1	1	1	3	.5	—

Carpels are erect, glabrous, and greenish yellow or yellow-green. The ovules are constricted medianly.

	n-pop.	n-pl.	n-rep.	n-obs.	x̄	s
Length (mm.)						
PC	1	4	—	11	4.3	.5
WM	4	21	—	44	6	2.1
BG	1	5	—	8	4.4	.8
PC—cf.	1	1	1	3	6.3	—
Length of cohesion (mm.)						
BG	1	5	—	8	.9	.1
Ovules per ovary (n)						
PC	1	4	—	11	9	1.8
WM	4	21	—	46	9.9	3.3
BG	1	5	—	8	8.3	3
PC—cf.	1	1	1	3	10.7	—

Fruits are erect, connate basally for about a millimeter, and brown. Their ventral lips are about .2 mm. wide.

	n-pop.	n-pl.	n-obs.	x̄ (mm.)	s (mm.)
Length					
PC	1	4	12	3.8	.5
WM	1	6	22	5	.6

Seeds are pyriform, short-tailed, finely longitudinally ribbed, and brown or yellow-brown.

	n-pop.	n-pl.	n-obs.	x̄ (mm.)	s (mm.)
Length					
PC	1	4	12	1.2	.09
WM	1	6	22	1.1	.12
Diameter					
PC	1	4	12	.4	.08
WM	1	6	22	.4	.07

Chromosomes indicate that the species is polyploid.

	n-pl.	g (n)	sp (2n)	Cytologist
Number				
Peshastin Creek	2	16	—	C. H. Uhl, 1946 and 1963
Mission Peak (S. lanc.-rup.)	1	16–18	—	Charles Horn, 1966

Variation. Principal causes of variation of *Sedum rupicolum* appear to be environmental modification and hybridization with *S. lanceolatum*. As an example of modification, plants at higher altitudes have shorter floriferous stems than plants at lower altitudes. The effect of hybridization is to produce plants which surpass the usual range of variation in populations where hybridization does not occur and which are intermediate between the parental species, as on the slopes of Mission Peak and along Standup Creek on the southern slope of Earl Peak.

Polyploidy may be another cause of variation. Hexaploids may possibly occur as well as tetraploids.

Except for a double petal, with two tips, in a flower of a plant at Bear Gap, no unusual variations have come to my attention. There is no basis for naming subspecies or varieties.

Nomenclature. *Sedum rupicolum* G. N. Jones, Research Studies, State College of Washington 2 (4):125 (1930, published in 1931). Type locality: north and northwest slopes at elevations of 762–914 meters, Peshastin Creek, Wenatchee Mountains, Chelan Co., Wash. Type: Elias Nelson 1679, a plant cultivated in Yakima, collected by E. J. Newcomer in November 1928. This collection was not available in the herbarium at Washington State College in 1961. Another specimen from Nelson, possibly part of the type clone, C388, is in the Bailey Hortorium. This plant, cultivated at Ithaca, N.Y., and pressed on May 9, 1941, is in reasonable agreement with plants growing along Peshastin Creek.

A synonym is:

1964. *Sedum lanceolatum* var. *rupicolum* (G. N. Jones) C. L. Hitchcock, Vascular plants of the Pacific Northwest, pt. 2:569. Based on *S. rupicolum* G. N. Jones.

Distribution. The main area of occurrence of *Sedum rupicolum* (fig. 47) is in the Wenatchee Mountains, an eastern spur of the Cascade Mountains. There, *S. rupicolum* replaces *S. lanceolatum* at higher elevations and on serpentine. *Sedum rupicolum* also occurs along the crest of the Cascade Mountains northeast of Mount Rainier. A dried specimen from a ridge near Lookout Mountain, in the vicinity of Mount Hood, has the appearance of *S. rupicolum*, but may be *S. lanceolatum*. Likewise, a specimen collected by Thomas Stratton (US) on Mount Baker in 1868 questionably is this species.

My report of *Sedum rupicolum* from western Idaho, in the *Flora of Idaho* by Davis (1952), is a mistake. Study of the population there, in the Seven Devils Mountains, reveals that it is *S. borschii*.

T. S. Brandegee (collection no. 774, NY) discovered *Sedum rupicolum* on Mount Stuart in August 1883. Since then, botanists have found it at several places in the Wenatchee Mountains. E. J. Newcomer found it in 1930 near Bear Gap, on the crest of the Cascade Mountains, above Morse Creek (Mt. Aix Quad.).

Because information about *Sedum rupicolum* suggested that it might be a serpentine endemic, I made a deliberate search to find it on other kinds of rocks. Thirty-seven plants, to which I devoted special attention in the field, occurred on six kinds of rocks.

	In random units (n-pl.)	*In nonrandom units (n-pl.)*
Sandstone	2	7
Andesite	0	5
Basalt	0	5
Peridotite	0	6
Granodiorite	0	8
Serpentine	3	1
Totals	5	32

Distributional and ecological data for populations to which I gave special attention are in the following table.

Population	Plants (n)	Area of population (m.²)	Density per m.²	Exposure	Rock
Peshastin Creek	80	142	.6	N	sandstone
Scotty Creek	110	3,075	.04	NW–N–NE	conglomerate, sandstone
Earl Peak	108	480,000	.0002	SSE–SSW	serpentine
Ingalls Creek	3	—	—	—	serpentine
Mt. Stuart	>139	400,000	.0003	SE–SSE	granodiorite, peridotite
Mission Peak	>117	—	—	N–NNE	basalt
Bear Gap	124	2,225	.06	NE–SE	andesite

Population	Altitudinal range (m.)	pH range	pH tests (n)	Drainage	Oldest record
Peshastin Creek	880	6.2–6.4	4	moderate	1928
Scotty Creek	1,210–1,230	5.4–5.6	2	good–excessive	1961
Earl Peak	1,140–2,144	6.2–6.4	3	excessive	1961
Ingalls Creek	860	5.6	1	—	1961
Mt. Stuart	1,570–1,720	5.4	1	—	1883
Mission Peak	1,705–2,000	5.4–6.4	3	excessive	1961
Bear Gap	1,800–1,820	5.0–5.4	5	good–excessive	1930

An attempt to estimate the total number of plants of *Sedum rupicolum* in the Wenatchee Mountains resulted in a figure of 410,482. The rugged terrain makes the Wenatchee Mountains difficult to investigate, but the area surveyed included some of the best-known territory for *S. rupicolum*. The total figure is a combination of ratio estimates and expansions.

About twenty-six species of vascular plants were competitors of plants of *Sedum rupicolum* in my samples. Occurrences were miscellaneous. Thirteen different species competed with four plants of *S. rupicolum* along Peshastin Creek. Commonest were *Montia parvifolia*, a moss, and a species of *Heuchera*. On the ridge above Scotty Creek, eight species of angiosperms were competitors. None was common. Only three species competed with three plants of *S. rupicolum* on Earl Peak. *Phlox douglasii* was commonest. The others were *Arenaria nuttallii* and *Achillea lanulosa*. *Erigeron compositus* occurred with each of three plants on Mission Peak along with three other species. *Vaccinium deliciosum* grew with three of five plants of *S. rupicolum* at Bear Gap, and six other species of angiosperms also were competitors.

Trees and shrubs shading *Sedum rupicolum* include *Abies grandis*, *Pseudotsuga menziesii*, *Pinus contorta*, *Pinus ponderosa*, and *Pyrus occidentalis*.

Reproduction. The easy detachment of the leaves of *Sedum rupicolum* increases potential vegetative reproduction by as many times as there are leaves per shoot. Not only may loose leaves develop into new plants in the area where they fall, but they are more likely to be carried long distances by wind or water than are the heavier vegetative rosettes. Seeds likewise develop and these too are available for distant dispersal. Although *S. rupicolum* seems well adapted for dispersal, its present distribution is limited, suggesting a recent origin and possibly a restricted ecological amplitude.

Bees and flies visit the flowers and act as effective cross-pollinators. As an example, along Peshastin Creek, I watched a bumblebee visit the flowers of several plants in succession.

Flowering time extends from June to August. It varies with altitude and exposure. Where *Sedum rupicolum* occurs with *S. lanceolatum* and *S. stenopetalum*, it is the first to flower. On July 6, 1961, along Peshastin Creek, *S. rupicolum* had immature fruits on a northwest slope where *S. stenopetalum* was in anthesis. Elsewhere in the vicinity, *S. lanceolatum* was in flower, but generally behind the *S. stenopetalum*. In cultivation at Ithaca, N.Y., *S. rupicolum* has flowered in May in the greenhouse and cold frame and in June in the garden.

Relationships. *Sedum rupicolum* is related to *S. lanceolatum* and may be a recent derivative from that, perhaps from the ssp. *lanceolatum*. It has the attributes of an ecospecies. If it did not coexist with *S. lanceolatum* at some places, as on the ridge above Scotty Creek and on the slopes along Peshastin Creek, subspecific status might be appropriate.

Mechanisms isolating *Sedum rupicolum* from *S. lanceolatum* appear to be at least three: genetic, because at places where the two species occur together, as on the ridge above Scotty Creek and along Peshastin Creek, they do not hybridize and are distinct; temporal, the *S. rupicolum* flowering a week or more before the *S. lanceolatum*; and ecological, the *S. rupicolum* growing where other vegetation is sparser and the pH is higher.

While in the Wenatchee Mountains, to aid my recognition of *Sedum rupicolum*, I made a practical comparison of it and *S. lanceolatum*. The results may be useful for purposes of identification.

	S. rupicolum	S. lanceolatum
Height	shorter	taller
Leaves of sterile shoots		
Shape	ovate	linear-lanceolate
Thickness	thicker	thinner
Surface	finely cellular (areoles .066 mm. in diameter)	coarsely cellular (areoles .099 mm. in diameter)
Attachment	easily detachable	not easily detachable
Leaves of floriferous shoots		
Length	shorter	longer
Shape	ovate	linear-lanceolate
Sepals		
Width	narrower	broader
Apices	obtuse	acute
Petals		
Length	shorter	longer
Apices	obtuse	acute
Tips	minutely mucronate	acuminate
Length of tips (mm.)	.1	.8
Color	deep yellow	yellow
Nectaries—color	deep yellow	pale yellow

Statistical tests confirm some of the subjective impressions, but not others. Results of two tests, utilizing data from samples cited in the description, follow.

	S. rupicolum Peshastin and Scotty Creeks	S. lanceolatum Scotty Creek
Leaves of rosettes		
Width (mm.)**	3.2	2.4
Thickness (mm.)n.s.	1.7	1.6
Petals—length (mm.)*	6.8	8.4

	S. rupicolum Plant from Mission Peak	S. lanceolatum Plant from Mission Peak
Altitude (m.)	1,705	1,565
Leaves: length/width**	2.6	4.1
Sepals—length/width*	1.8	1.5
—length*	3.1	3.4

Other species of *Sedum* with which *S. rupicolum* sometimes grows, besides *S. lanceolatum*, are *S. divergens* and *S. stenopetalum*. Where *S. rupicolum* and *S. stenopetalum* occur at the same site, the *S. stenopetalum* is in the more open, drier situations. Likewise, where *S. divergens* and *S. rupicolum* occur in proximity, they are not intermingled, but usually are mutually exclusive. On the average, *S. divergens* flowers later and thrives in less well-drained sites. Hybridization with either *S. divergens* or *S. stenopetalum* is unknown.

The entire area occupied by *Sedum rupicolum* was covered by glacial ice in the late Pleistocene. Either the plants moved into their present distributional area in recent time or the species evolved recently as an offshoot of the wide-ranging *S. lanceolatum*. It could even be the result of hybridization between *S. divergens* and *S. lanceolatum*. Apparently, *S. rupicolum* is well adapted to the niche which it now inhabits. At the edges of its ecological range, it sometimes intergrades with *S. lanceolatum*. Plants which are intermediate morphologically occur on the slopes of Mission Peak and along Standup Creek, on the southern side of Earl Peak. In summary, *S. rupicolum* today occupies a distinctive niche within a small part of the geographic range of *S. lanceolatum*.

References: None, but Margaret Powers studied the relationships of *S. lanceolatum* and *S. rupicolum* in 1961 and prepared a typewritten thesis which is filed at the University of Washington.

12. *Sedum radiatum**** (figs. 53–60)

Plants of *Sedum radiatum* (fig. 53) commonly are annual herbs. Offsets or rosettes develop some distance above the ground. These may drop and take root independently from the parental plants. Leaves of the offsets are either ovate-elliptical or oblong-elliptical and papillose. Inflorescences mostly are three-parted cymes. The 5-merous

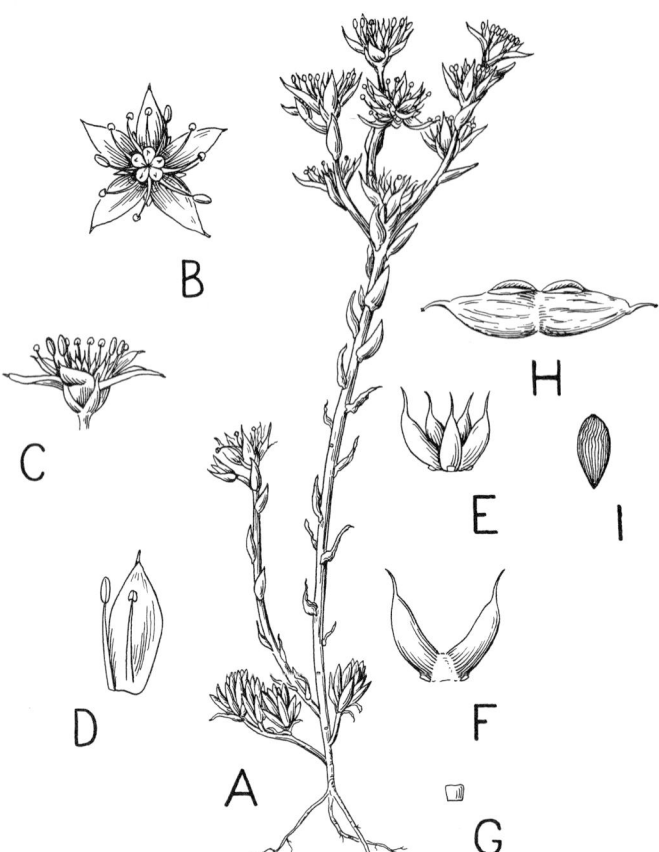

Fig. 53. Plant of *Sedum radiatum* ssp. *radiatum* from roadside bank in Manning Canyon 8 km. southwest of Lakeport, Lake Co., Calif. A. Habit sketch (x .98). B. Flower from above (x 2). C. Flower from side (x 2). D. Petal and two stamens (x 2.9). E. Carpels (x 2.9). F. Two carpels and nectaries (x 3.9). G. Nectary (x 4.9). H. Fruits (x 3.9). I. Seed (x 9.8).

flowers have white, yellow, or pale yellow petals. The carpels are connate basally and spread widely in fruit. The seeds are pyriform with fine longitudinal ribs. The offsets of the perennial *S. stenopetalum* have longer, narrower leaves with spurs which become scarious and persistent. Also, *S. stenopetalum* has larger follicles. *Sedum lanceolatum*, besides being perennial, has erect follicles and basal rosettes of oblong-lanceolate leaves. Confusion with flat-leaved species or with species with opposite leaves is unlikely.

Description. Sample: 43 plants. Sources and numbers of plants from each locality are: along the Pohono Trail in the Yosemite Gorge (YG), Mariposa Co., Calif., July 1964—10; roadside bank in canyon of Manning Creek (MC), 8 km. southwest of Lakeport, Lake Co., Calif., May and Aug. 1963—8; on south side of Rogue River (RR), .3 km. below the Hellgate Bridge, east-southeast of Galice, Josephine Co., Ore., May 1963—3; bank along road by Deadman's Point (DP), northwest of Dutchman's

Peak, Jackson Co., Ore., July and Aug. 1965—20; and bank along Beaver Creek Road (BC), below Beaver Dam, Jackson Co., Ore., Aug. 1965—2.

The 10 plants from along the Pohono Trail comprise a .5% sample of the whole population. Two randomly selected subsamples include simple random samples of 9 plants and 1 plant respectively. The 8 plants from southwest of Lakeport belong to two samples: 2 plants (.6%) chosen authoritatively because they were in flower on May 21, 1963; and 6 plants (1.8%), all with fruits and which could be reached in two randomly selected transects, each 1 m. wide, chosen from 51 transects on Aug. 13, 1963. Choice of the 3 plants from the gorge of the Rogue River on May 28, 1963, was for specimens with flowers in anthesis. The fraction which these plants are of the total population is unknown. The 20 specimens from Deadman's Point, obtained in July and August 1965, comprised two simple random samples, each of 10 plants. The procedure at Deadman's Point was first to divide the population as nearly evenly as possible, and then to select randomly one-half for study. I placed numbered tags on all flowering plants in the selected half, a total of 186 tags. Each sample was 5.3% of the chosen half of the population, or 2.6% of the whole population. Selection of the 2 plants from along the Beaver Creek Road below Beaver Dam was for large plants, as different as possible from those at Deadman's Point.

Sedum radiatum does not flourish in cultivation at Ithaca, N.Y. Although plants were included in an experiment of completely random design in the greenhouse, growth was poor and data are sparse. A few plants grew well in the cold frames. Because of inadequate experimental results, conclusions about classification depend primarily on data from wild plants.

Plants of *Sedum radiatum* mostly are annual herbs with fibrous, branched roots and simple or branched stems which are green or reddish brown and bear terminal, three-parted cymes. The stems may be horizontal below and then erect. Offsets (fig. 54) often develop along the floriferous stems, but are sometimes lacking. When present, the offsets separate easily from the parental stems and drop to the ground, where they may take root and become new plants. Old fructiferous stems, with remains of fruits, may persist for as long as a year.

	n-pl.	n-obs.	$\bar{x}$	s
Plants—diameter (cm.)				
DP	10	10	3	5.7
YG	7	7	9	4.1
Rosettes (n)				
DP	10	14	.5	.2
YG	10	20	.3	1
MC—gh.	1	1	8	—
Rosettes—diameter (mm.)				
DP	8	12	5.6	.8
MC	2	11	5	.4
YG	9	14	7.1	2.5
Rosettes—length (mm.)				
DP	4	6	.9	.4
MC	2	11	3.6	3.4

Fig. 54. Axillary shoots on floriferous stem of *Sedum radiatum* ssp. *ciliosum* from gorge of Rogue River ESE of Galice, Josephine Co., Ore., cultivated in greenhouse at Ithaca, N.Y., May 19, 1964 (x 3.4). Photo by Howard Lyon.

	n-pl.	n-obs.	$\bar{x}$	s
Floriferous stems per plant (n)				
DP	10	10	1.8	3.3
YG	10	10	2.4	2.1
Floriferous stems—length (cm.)				
RR	3	4	17	2.2
DP	10	18	5.6	3.6
MC	4	13	10	2.4
YG	7	14	9.7	1.9
RR—gh.	1	3	19	—

Leaves of the offsets are short-spurred; oblong-elliptical, oblong-lanceolate, or ovate; glabrous, papillose, or ciliate; with five green to purple veins, and also speckled with pink. The spurs are scarious, truncately obtuse, and about 1 mm. long. The leaves of the floriferous stems are divergent, lanceolate, acuminate or acute, often papillose, and yellow-green speckled with red. Their spurs are scarious and either unlobed or three-lobed. The leaves of the floriferous stems drop early. They may be dropping at the time when the flowers are in anthesis. Sometimes the lower leaves are brown and appressed.

	n-pl.	n-obs.	x̄ (mm.)	s (mm.)
Leaves of offsets—length				
DP	8	12	4.4	.7
YG	9	14	8.8	1.6
MC—gh.	1	1	14	—
Leaves of offsets—width				
DP	8	12	1.8	.2
YG	9	14	2.7	.5
MC—gh.	1	1	3	—
Leaves of offsets—thickness				
DP	8	12	1	.2
YG	9	14	.6	.3
MC—gh.	1	1	.8	—
Median leaves of floriferous stems—length				
DP	3	5	3.9	.4
MC	3	6	9.1	1
YG	1	1	11	—
RR—gh.	1	3	13	—
Median leaves of floriferous stems—width				
DP	3	5	1.9	.4
MC	3	6	3.5	1.2
YG	1	1	4	—
RR—gh.	1	3	2.7	—
Median leaves of floriferous stems—thickness				
DP	3	5	1	.4
MC	3	6	1.5	.1
YG	1	1	1.5	—
RR—gh.	1	3	1.1	—

Inflorescences are three-parted cymes. Sometimes the cincinni are dichotomously forked, especially in populations in southwestern Oregon. The axes of the cincinni are either green or red.

	n-pl.	n-obs.	x̄ (n)	s (n)
Cincinni per cyme				
RR	3	4	3.2	.3
DP	10	15	2.5	.8
MC	4	13	2.9	.2
YG	10	18	2.8	.4
RR—gh.	1	3	3	—
Flowers per cyme				
DP	10	15	9.1	6
YG	10	18	11.9	4
Flowers per cincinnus				
MC	4	38	4.9	2.4
RR—gh.	1	9	4.8	—

Flowers (fig. 55) are 5-merous and either sessile or with short pedicels. A short torus is present.

252 SEDUM OF NORTH AMERICA

Fig. 55. Flowers of *Sedum radiatum* ssp. *radiatum* from roadside bank in Manning Canyon, 8 km. southwest of Lakeport, Lake Co., Calif., cultivated in greenhouse at Ithaca, N.Y., June 8, 1965 (x). Photo by Howard Lyon.

	n-pl.	n-obs.	$\bar{x}$ (mm.)	s (mm.)
Pedicels—length				
RR	3	4	0	0
MC	2	9	1	.8
Flowers—diameter				
RR	3	4	15.3	.6
DP	7	10	10.5	1.3
MC	2	10	10.6	2.8
YG	4	25	12.4	1.9
RR—gh.	1	5	12.4	—
Torus—length				
DP	8	11	.9	.1
YG	4	14	.9	.07

Sepals are erect, lanceolate or ovate, acuminate or acute, and pale green, yellow-green, or pale empire yellow. Sometimes they are red-tipped and papillose.

	n-pl.	n-obs.	x̄ (mm.)	s (mm.)
Length				
RR	3	4	2.5	.8
DP	7	10	2.2	.4
MC	2	10	2.2	.4
YG	4	14	2.6	.2
RR—gh.	1	5	2.4	—
Width				
RR	3	4	1	.2
DP	7	10	1.6	.3
MC	2	10	1.7	.4
YG	4	14	1.6	.3
RR—gh.	1	5	1.5	—

Petals are separate, elliptic-lanceolate, carinate, acute, mucronate-appendaged, and white with pale green band on dorsal keel, creamy white, pale primrose yellow, or empire yellow streaked with red on dorsal keel. When petals are white, they may be speckled with pink. The mucronate appendages are about half a millimeter long.

	n-pl.	n-obs.	x̄	s
Number per flower				
RR	3	4	5	0
DP	7	9	5	0
MC	2	10	5	0
YG	10	22	5.1	.3
RR—gh.	1	5	5	—
Length (mm.)				
RR	3	4	9	.8
DP	7	10	6.9	.6
MC	2	10	7	1.6
YG	4	14	7.7	.8
RR—gh.	1	5	7.6	—
Width (mm.)				
RR	3	4	3	.2
DP	7	10	2.5	.2
MC	2	10	2.7	.8
YG	4	14	2.6	.3
RR—gh.	1	5	2.4	—

Stamens are twice as many as the petals, with the anthers empire yellow. The anthers of plants at Deadman's Point are suffused with red.

	n-pl.	n-obs.	x̄ (mm.)	s (mm.)
Epipetalous filaments; adnation—length				
RR	3	4	.7	.2
DP	7	10	.7	.2
MC	2	10	.5	.06
YG	4	14	.7	.08
RR—gh.	1	5	.8	—
Anthers—length				
RR	2	3	1.5	.2
DP	4	6	1.5	.1
MC	2	2	1.3	.4
YG	4	9	1.4	.1
RR—gh.	1	1	1.5	—

Nectaries are quadrate, truncate, and orange or yellow.

	n-pl.	n-obs.	x̄ (mm.)	s (mm.)
Length				
RR	3	4	.5	.06
DP	7	10	.5	.05
MC	2	10	.4	.01
YG	4	14	.4	.05
RR—gh.	1	5	.5	—
Width				
RR	3	4	.7	.2
DP	7	10	.5	.04
MC	2	10	.4	.1
YG	4	14	.4	.03
RR—gh.	1	5	.5	—

Carpels are subdivergent, somewhat connate basally, gibbous ventrally, and greenish white or empire yellow.

	n-pl.	n-obs.	x̄	s
Length (mm.)				
RR	3	4	4.4	.8
DP	7	10	4.1	.6
MC	2	10	4.3	.8
YG	4	14	4	.4
RR gh.	1	5	4.9	—
Styles—length (mm.)				
DP	7	10	1	.2
YG	4	14	1.1	.1
Ovules per ovary (n)				
RR	3	4	5.3	1.2
DP	7	10	3.9	.2
MC	2	10	4.7	2.4
YG	10	22	3.4	.2
RR—gh.	1	5	5	—

Fruits are widely divergent, connate basally for about 1.7 mm., gibbous ventrally with lips about .5 mm. wide, and straw-colored, streaked with reddish brown.

	n-pl.	n-obs.	x̄ (mm.)	s (mm.)
Length				
BC	2	4	3	.1
DP	10	19	3	.2
MC	6	20	3.1	.2
YG	2	4	3	.1

Seeds are pyriform, finely longitudinally ribbed, and brown or sometimes fuscous.

	n-pl.	n-obs.	x̄ (mm.)	s (mm.)
Length				
BC	2	4	1.1	.07
DP	10	19	1	.06

	n-pl.	n-obs.	$\bar{x}$ (mm.)	s (mm.)
Length				
MC	6	20	.9	.04
YG	2	4	1	.07
Diameter				
BC	2	4	.6	.14
DP	10	19	.55	.07
MC	6	20	.4	.04
YG	2	4	.4	.04

Chromosomes indicate that the species is diploid. Separation of chromosomes is poor in the plant, cited below, from Deadman's Point. If cells of the anther-wall are involved, the number may be 16 and sporophytic. Abbreviations in parentheses below indicate herbaria in which specimens are preserved and counts are recorded.

	n-pl.	g (n)	sp (2n)	Cytologist
Number				
Grave Creek, Ore.	1	8	—	C. H. Uhl, 1961 (in Jeps.)
RR	1	8	—	C. H. Uhl, 1961 (in CU)
DP	1	13–14 (?)	—	Charles Horn, 1965
MC	1	8	16	C. H. Uhl, 1950
Nicasio Creek, Calif.	1	8	—	C. H. Uhl, 1961 (in Jeps.)
Tuolumne Co., Calif.	1	8	—	C. H. Uhl, 1964 (in CU)

Variation. Two major regional variations of *Sedum radiatum* occur. Plants in the Klamath Mountains in southwestern Oregon and in the adjacent Coast Ranges of Oregon have white or creamy white petals, the primary branches of the cymes dichotomously forked, and the leaves of the offsets papillose-ciliate. Plants in the California Coast Ranges, the Klamath Mountains in California, and the south-central Sierra Nevada have yellow petals, the cincinni unbranched, and the leaves of the offsets glabrous or papillose, rarely papillose-ciliate. Review of means for various features in the description reveals several regional differences, but only those in length of nectaries and diameter of seeds are significant. The white-flowered plants of Oregon have relatively longer nectaries and plumper seeds.

	Yosemite Gorge	Deadman's Point
Nectaries—length (mm.)**	.37	.46
Seeds—diameter (mm.)*	.38	.55

	Manning Canyon	Deadman's Point
Seeds—diameter (mm.)**	.4	.55

Within the white-flowered variation, six of eight differences between plants in the canyon of the Rogue River and at Deadman's Point are highly significant. These differences are:

	Rogue River	Deadman's Point
Height (cm.)	17	6
Flowers per cincinnus (n)	2	9
Flowers—diameter (mm.)	15	11
Petals—length (mm.)	9	7
—width (mm.)	3	2.5
Nectaries—width (mm.)	.7	.48

In contrast, all differences between plants in the Yosemite Gorge and Manning Canyon lack significance. In cultivation, plants from Deadman's Point maintained their smaller size, but they grew so poorly that meaningful comparisons are impossible. Observations of plants along the Beaver Creek Road suggested that the plants at Deadman's Point are most similar to ssp. *ciliosum* farther down the road, but abruptly different in the six features listed above, as well as in the red pigment with which the yellow anthers are suffused. On the basis of present evidence, this population and another on Copper Butte in Siskiyou County, California, comprise a local subspecies.

The need to reach taxonomic conclusions on the basis of data from wild plants alone makes important the relative variability of features available for classification. Coefficients of variation (CVs) are a measure of this variability. They are provided in the following table for fourteen features of interest in classification along with a column for average CVs and a rating of the features on the basis of these values.

Feature	RR	DP	MC	YG	Av. CV	Rating
Height	13	70	24	33	35	14
Leaves of rosettes—length	—	16	—	18	17	11
Flowers—diameter	4	12	26	15	14	10
Sepals—length	32	18	18	8	19	12
Petals—n per flower	0	0	6	0	2	1
—length	9	9	23	10	13	9
—width	7	8	30	12	12	7.5
Anthers—length	13	7	3	13	9	4
Nectaries—length	12	10	3	13	10	5
—width	28	8	25	8	12	7.5
Ovules per ovary	23	5	51	6	21	13
Fruits—length	—	7	6	3	5	2
Seeds—length	—	6	4	7	6	3
—diameter	—	13	10	10	11	6

The table reveals that height, although used in distinguishing ssp. *depauperatum*, is more variable than features such as length and width of petals.

Variation in number of petals per flower is slight, as is further indicated by the following frequency distribution for flowers in my samples.

Petals per flower (n)	Flowers (n)	%	Cumulative %
5	53	98.1	98.1
6	1	1.9	100
Total	54		

KEY TO SUBSPECIES

A. Petals white or creamy white, seeds .5–.6 mm. in diam., primary branches of cymes often dichotomously forked .... B
    B. Plants 11–22 cm. tall, flowers 14–17 mm. in diam., petals 7–11 mm. long and 2.5–3.5 mm. wide, anthers yellow .... ssp. *ciliosum*, p. 257
    BB. Plants 3–9 cm. tall, flowers 10–11 mm. in diam., petals 6–7 mm. long and 2.4–2.6 mm. wide, anthers red .... ssp. *depauperatum*, p. 258
AA. Petals yellow, seeds .36–.44 mm. in diam., primary branches of cymes seldom dichotomously forked .... ssp. *radiatum*, p. 262

Sedum radiatum ssp. *ciliosum*

Nomenclature. *Sedum radiatum* ssp. **ciliosum** (Howell) comb. nov., fundatum super *Sedum ciliosum* Howell, A flora of northwest America 1:214 (1898).

Synonyms are:

1898. *Sedum ciliosum* Howell, A flora of northwest America 1:214. Type locality: on rocks in the Coast Mountains west of Roseburg, Ore. Type: a collection of Thomas Howell 693 (ORE), June 20, 1887. Howell described the flowers as yellow, a circumstance needing clarification.

1946. *Sedum Douglasii* ssp. *ciliosum* (Howell) Clausen, Cact. and Succ. Jour. 18:59. Based on *Sedum ciliosum* Howell.

1948. *Sedum stenopetalum* ssp. *ciliosum* (Howell) Clausen, Cact. and Succ. Jour. 20:144. Based on *Sedum ciliosum* Howell.

Distinctive features of ssp. *ciliosum* (figs. 56 and 57) are white petals, dichotomously forked primary branches of the cymes, and relatively plumper seeds than in ssp. *radiatum*. Also, the leaves of the offsets more often are ciliate.

The range of *Sedum radiatum* ssp. *ciliosum* (fig. 58) is in the northern Klamath Mountains of southwestern Oregon and in the adjacent Coast Ranges of Oregon. Sixteen populations are known. The plants occur in talus and on rocks along streams, at the base of cliffs, and on bluffs. The altitudinal range is from about 220 to 730 meters.

I have seen *Sedum radiatum* ssp. *ciliosum* at four places in the field: two sites along Grave Creek near its junction with the Rogue River (1940, 1963), in the canyon of the Rogue River east-southeast of Galice (1963), and along the Beaver Creek Road northwest of Deadman's Point (1965). The sample from the Rogue River, for which data are available in the description, came from dry talus, exposed to the north, at an altitude of 250 meters. The soil had a pH of 6.4. John and Lilla Leach discovered ssp. *ciliosum* in the canyon of the Rogue River on May 30, 1928.

In its dichotomously forked branches of the inflorescences, ssp. *ciliosum* may be less advanced than ssp. *radiatum*. Even the white petals may be less specialized. Occurring in an area which is geologically old, ssp. *ciliosum* appears central in the development of the races of *Sedum radiatum*.

The oldest collection of *Sedum radiatum* ssp. *ciliosum* which has come to attention is by Drake and Dickson (NY) from Glendale, Ore., May 27, 1887.

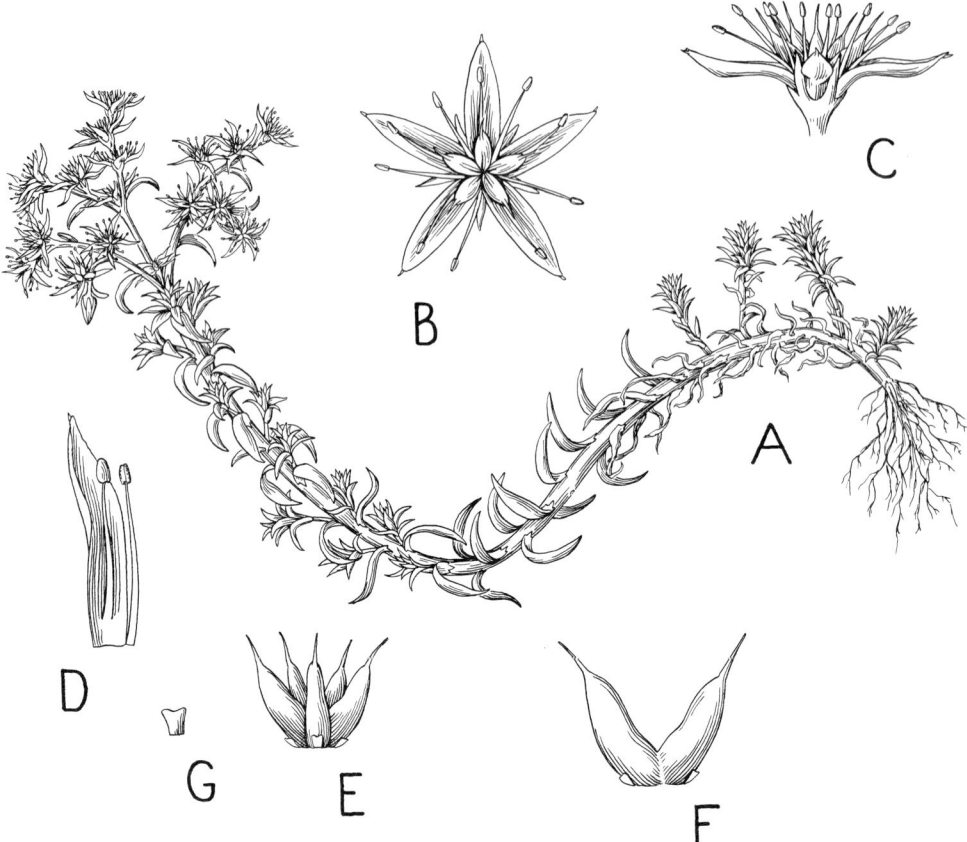

Fig. 56. Plant of *Sedum radiatum* ssp. *ciliosum* from gorge of Rogue River ESE of Galice, Josephine Co., Ore., cultivated in greenhouse, Ithaca, N.Y. A. Habit sketch (x .55). B. Flower from above (x 2.2). C. Flower from side (x 2.2). D. Petal and two stamens (x 3.3). E. Carpels (x 3.3). F. Two carpels and nectaries (x 4.4). G. Nectary (x 5.5).

Sedum radiatum ssp. **depauperatum**

Subspecies nova *Sedi radiati* cum caulibus humilioribus, 3–9 cm. proceris, cum floribus 8–10 per cincinno, 10–11 mm. in diam.; petala 6–7 mm. longa et 2.4–2.6 mm. lata; nectaria .46–.54 mm. lata. Typus in Herbario Wiegand, Universitatis Cornellianae, ab ripa viae, Deadman's Point, altitudine 1,616 m., Klamath Mountains, Jackson Co., Ore., collectio Roberti Clausenii num. 651–22, Julio 3, 1965, est.

Compared with ssp. *ciliosum*, distinguishing features of ssp. *depauperatum* (fig. 59) are the lower stature, larger number of flowers per cincinnus, smaller flowers, shorter and narrower petals, and narrower nectaries. Additionally, the anthers are strongly suffused with red.

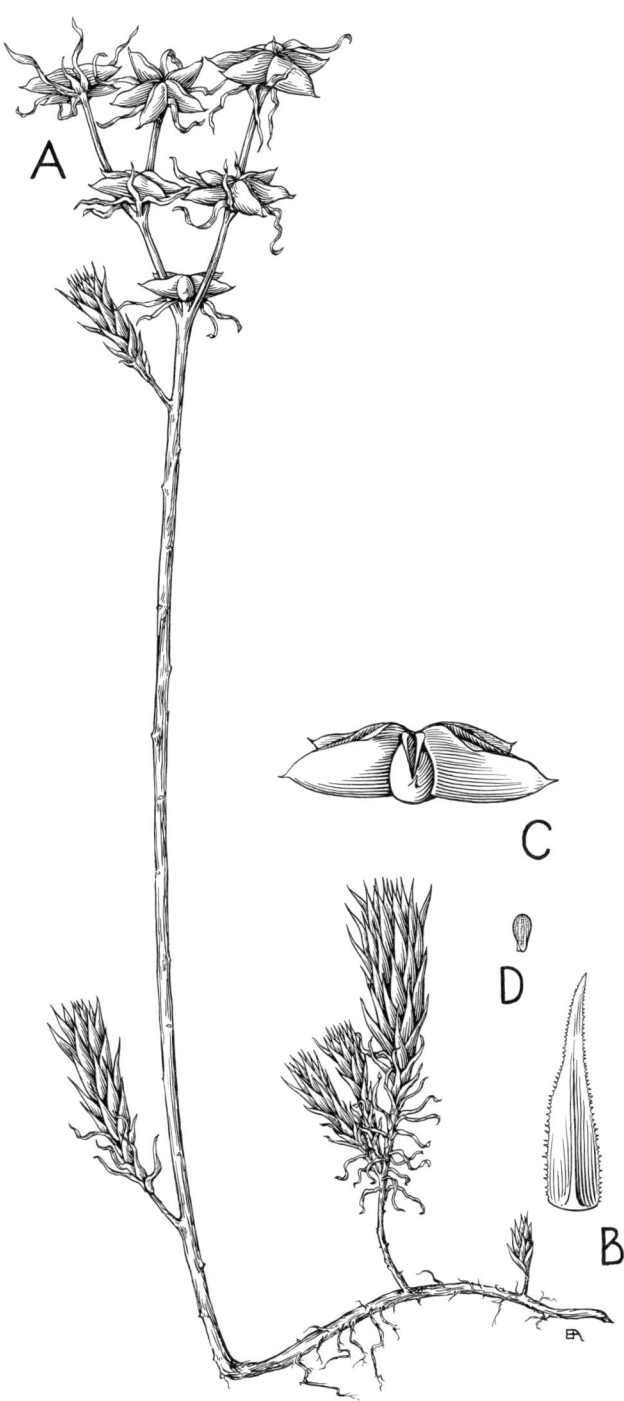

Fig. 57. Fruiting plant of *Sedum radiatum* ssp. *ciliosum* from along Grave Creek, 1.6 km. E of junction with Rogue River, Josephine Co., Ore. A. Habit sketch (x 1.2). B. Single leaf of vegetative shoot (x 2). C. Fruits (x 3.1). D. Seed (x 5.1).

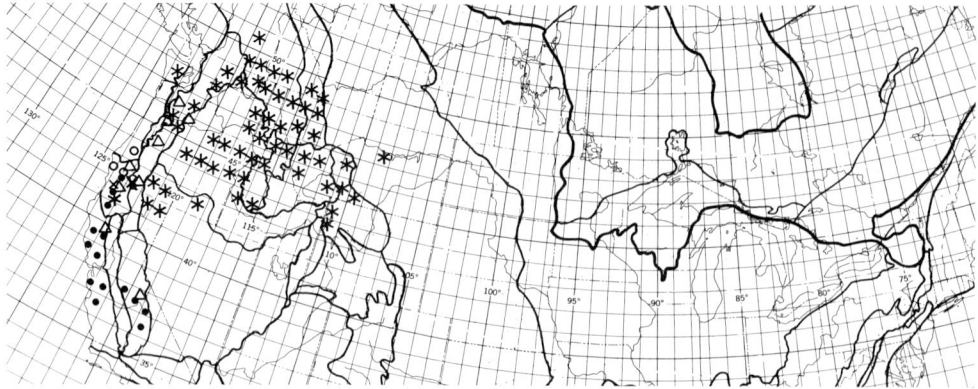

Fig. 58. Known distribution of *Sedum radiatum* ssp. *radiatum* (●), ssp. *ciliosum* (○), and ssp. *depauperatum* (x), and of *S. stenopetalum* ssp. *stenopetalum* (*) and ssp. *monanthum* (△).

Fig. 59. Plant of *Sedum radiatum* ssp. *depauperatum* on bank of road, Deadman's Point northwest of Dutchman's Peak, Jackson Co., Ore., July 3, 1973.

Besides being at Deadman's Point, ssp. *depauperatum* also occurs on Copper Butte, north of Seiad Valley, at an altitude of 1,586 m., in Siskiyou Co., Calif., where it was discovered on July 25, 1962, by Mrs. M. Williams and W. Roderick.

The site at Deadman's Point (fig. 60) is in the Rogue River National Forest. Discovery there was by Lawrence Dempster. Access is by way of the Beaver Creek Road. The plants are in a large cluster on the bank of the road. They are growing over granitic diorite. Other data about the population are condensed in the following brief table.

Flowering plants (N)	Area of population (m.²)	Density per m.²	Exposure
372	42	8.8	WNW

Altitudinal range	pH, range	pH, tests (n)	Oldest record
1,616–1,617	6–6.6	10	1961

Plants of ssp. *depauperatum* form a dense stand where they occur. Seedlings are plentiful. Few competing species occur in the population. The closest competitors of

Fig. 60. Site of type population of *Sedum radiatum* ssp. *depauperatum*, bank of road, Deadman's Point northwest of Dutchman's Peak, Jackson Co., Ore., July 3, 1973.

6 out of 10 randomly selected plants were other plants of ssp. *depauperatum*. Closest competitors of the other 4 plants were *Arctostaphylos nevadensis*, *Collinsia rattanii*, *Carex* sp., and an unidentified broad-leaved dicotyledon, each by 1 plant.

The ssp. *depauperatum* may be a dwarf derivative of ssp. *ciliosum* or, less likely, a hybrid resulting from a cross involving ssp. *ciliosum* and some other species of even smaller dimensions. Experimental work may someday help to clarify its relationships.

Sedum radiatum ssp. radiatum

Nomenclature. *Sedum radiatum* Watson, Proc. Am. Acad. Arts and Sci. 18:193 (1883). Type locality: Gavillan Peak (= Fremont Peak), Monterey Co., Calif., in dry thin soil and among limestone rocks. Type: W. H. Brewer 742 (GH), July 1, 1861. The holotype is the more branched plant in the upper right corner of the type sheet.

Synonyms are:

1946. *Sedum Douglasii* ssp. *radiatum* (Watson) Clausen, Cact. and Succ. Jour. 18:59. Based on *Sedum radiatum* Watson.

1948. *Sedum stenopetalum* ssp. *radiatum* (Watson) Clausen, Cact. and Succ. Jour. 20:144. Based on *Sedum radiatum* Watson.

Diagnostic features of ssp. *radiatum* (fig. 53) are the yellow petals, simple cincinni, and less plump seeds. The leaves of the offsets are glabrous or papillose. Less often they are ciliate.

The geographic range of *Sedum radiatum* ssp. *radiatum* (fig. 58) is at middle altitudes on the western slopes of the Sierra Nevada from Tulare to Tuolumne Counties, in California, and in the California Coast Ranges from Monterey to Humboldt Counties. It also occurs in the Klamath Mountains, especially in the southern part. The plants grow in sandy loam, gravel, or moss, or in crevices in rocks, on rocky slopes, either in open areas or in partial shade. Habitats are on a variety of rocks, including granite, sandstone, and limestone. The altitudinal range is 153–2,287 meters.

My principal attention to ssp. *radiatum* has been in the Yosemite Gorge and in the Manning Canyon southwest of Lakeport. Data for these populations follow:

Population	Plants (N)	Area of population (m.^{2})	Density per (m.^{2})	Exposure	Altitudinal range (m.)	pH, range	pH, tests (n)	Oldest record
Yosemite Gorge	1,722	1,104	1.6	N (NW–ENE)	1,380–1,635	5.2–5.8	11	1913
Manning Canyon	332	332	1	W	460–463	—	—	1950

Other vegetation is sparse where ssp. *radiatum* occurs. In the Yosemite Gorge, two plants in my random sample lacked close competitors. Other plants in the sample had miscellaneous competitors: 1—*Streptanthus tortuosus*, 1—*Mimulus moschatus*, 2—

Clarkia (*?dudleyana*), 1—*Brodiaea pulchella*, 2—*Bromus breviaristatus*, and 2— *Bromus japonicus*. No other *Crassulaceae* were with the *Sedum* in the Yosemite Gorge. A single plant of *Parvisedum pentandrum* was in one of the transects in Manning Canyon.

The ssp. *radiatum* appears closely related to ssp. *ciliosum*. With its simple cincinni, it may be more advanced. Further, it has become adapted to an environment in which the summer is both warmer and drier.

Reproduction. Vegetative offsets which become detached from the parental plants and drop to the ground may be as important as seeds in the perpetuation of *Sedum radiatum*. Small, nonflowering plants, observed in the field, could be either seedlings or rooted offsets. They were most numerous in the population of ssp. *depauperatum* at Deadman's Point.

A variety of insects visit the flowers and serve as pollinators. These include flies and small bees. Thrips frequently inhabit the blossoms.

Flowering time depends on latitude, altitude, and subspecies. It extends from May to July, and exceptionally to August.

Relationships. *Sedum radiatum* is most nearly related to *S. stenopetalum*. One interpretation is to regard these as subspecies of a single, aggregate species. The allopatric distribution and morphological similarities support this classification. Evidence in favor of specific status is of three sorts: morphological distinctness, occurrence in the same region without intergradation, and number of chromosomes. *Sedum stenopetalum* is perennial with long, narrow leaves with scarious, three-lobed spurs, and large follicles. Both *S. radiatum* and *S. stenopetalum* occur in Yosemite Park and on Dutchman's Peak in Oregon, the *S. radiatum* at lower altitudes, but not far away geographically from the *S. stenopetalum*. Populations at these sites are as distinct as those of either species anywhere. This is not the expected situation where subspecies meet. Difference in number of chromosomes explains the distinctness of populations in the field and is good evidence in favor of recognizing two species. Available evidence suggests that *S. radiatum* is diploid and *S. stenopetalum* is octoploid. Whether octoploid plants contain only genomes derived from *S. radiatum* or whether they include genomes from other species, for example *S. lanceolatum*, is unknown. If genomes of other species are involved and the octoploids are allopolyploids, then the evidence for specific status is decisive.

Several species of *Sedum* in western North America are perennial herbs with 8 or 16 pairs of chromosomes. Among them are three species with erect follicles, namely *S. lanceolatum* with yellow petals and oblong-lanceolate leaves, *S. cockerellii* with white petals and obovate or oblanceolate leaves, and *S. niveum* with tuberous roots and prostrate stems. *Sedum radiatum* ssp. *ciliosum* may be a derivative of the phyletic line which gave rise to these species. This subspecies, which has become adapted to conditions in the Klamath Mountains, appears to be the least advanced of the three subspecies of *S. radiatum*. Habit of growth, type of inflorescence, and color of flowers are in agreement with this idea.

References. Clausen, R. T. (1946 and 1948).

13. Sedum stenopetalum*** (figs. 58 and 61–64)

Plants of *Sedum stenopetalum* (fig. 61) are perennial herbs with spirally arranged, narrow leaves and widely spreading, gibbous follicles. A diagnostic feature is the production of abundant vegetative rosettes in the axils of the leaves of the floral stems and sometimes even in place of flowers. *Sedum stenopetalum* differs from *S. radiatum* in being perennial, having larger follicles, and in being octoploid rather than diploid. *Sedum lanceolatum*, another yellow-flowered species of western North America, has erect follicles and shorter, thicker leaves. *Sedum divergens* and *S. debile* have opposite leaves. *Sedum leibergii* has basal rosettes of oblanceolate or spatulate, petiolate, basal leaves, and behaves as a biennial. The most likely confusion is with *S. radiatum* or *S. lanceolatum*. The shorter, papillose leaves of the latter and the annual growth of the former usually make separation easy. In difficult cases, chromosome number can be decisive for identification.

Description. Sample: 52 plants. Collection of plants was during surveys of the Idaho Batholith, Seven Devils Mountains of the Columbia Plateau, and Wenatchee Mountains of the Cascade Mountains in 1961; the Sierra Nevada in 1964; the Klamath and Cascade Mountains in 1965; and the Central Rocky Mountains in 1966. All samples within populations were simple random, either from whole populations or from randomly selected clusters. Data about sizes of populations, number of plants in samples, and condition of plants when sampled are in the following table. N_u = number of plants in listing units, N_c = number of plants in sampled clusters, and n = number of plants in samples.

Locality	N_u	N_c	n	Plants studied n/N_c	n/N_u	Condition when collected fl. (n)	fr. (n)
Central Rocky Mts.							
Moose, Jackson Hole	214	61	2	.033	.009	0	2
Idaho Batholith							
Lolo Creek	44	44	2	.045	.045	2	0
Columbia Plateau							
Morrison Ridge	29,602	260	16	.062	.001	9	7
Kinney Point	20	20	2	.100	.100	0	2
Cascade Mts.							
Scotty Creek	478	478	11	.023	.023	6	5
O'Leary Mt.	48	41	5	.122	.104	1	0
Klamath Mts.							
Dutchman's Peak	7	7	4	.571	.571	0	0
Sierra Nevada							
Rancheria Mt.	41	41	4	.098	.098	0	0
Chilnualna Falls	440	440	6	.014	.014	0	0
Totals	30,894	1,392	52	.119	.107	18	16

Plants made good vegetative growth in the greenhouse (gh.) at Ithaca, N.Y. Experiments of completely random design, under these conditions, provided valuable comparative information about habit of growth, stems, and leaves, but many plants did not flower. Growth in the cold frames also was good, and flowering was more

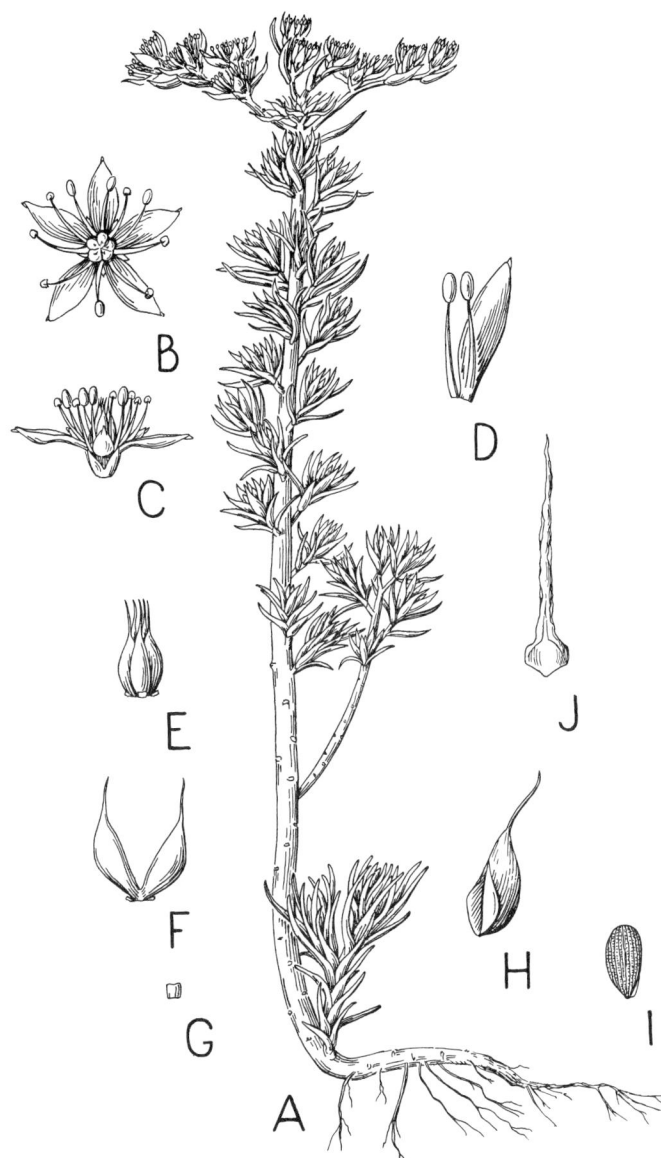

Fig. 61. Plant of *Sedum stenopetalum* ssp. *stenopetalum* from Morrison Ridge south of Shingle Creek, eastern slope of Seven Devils Mountains, Idaho Co., Ida. A. Habit sketch (x .96). B. Flower from above (x 1.9). C. Flower from side (x 1.9). D. Petal and two stamens (x 2.9). E. Carpels (x 2.9). F. Two carpels and nectaries (x 3.8). G. Nectary (x 4.8). H. Fruit (x 3.8). I. Seed (x 9.6). J. Single leaf of vegetative shoot (x 1.9).

general. Although plants of both subspecies grew outdoors in the garden, they survived for only a few seasons.

Plants of *Sedum stenopetalum* are perennial herbs with slender, fibrous roots developing from the bases of the stems and from the lower nodes. The stems are decumbent, branched, green or red, and they terminate in rosettes of leaves that take root and are the means by which the plants are perennial. Offsets also are borne in the axils of the leaves of the floriferous stems and even in the cymes in place of flowers. The production of axillary rosettes is a distinctive trait of the species. These axillary rosettes eventually break away from the stem which bears them and produce new individuals. The floriferous stems are erect except for the decumbent bases. They are green, or sometimes red, and attain a height of 3.8 dm. in nature. After the fruits mature, the floriferous stems die. Further growth is from the rosettes which become separated from the parental stems.

	n-pl.	n-rep.	n-obs.	$\bar{x}$	s
Plants—diameter (cm.)					
Moose	2	—	2	12.5	3.5
O'Leary Mt.	3	—	3	.6	.5
Chilnualna Falls	6	—	6	7.2	4.6
Moose—gh.	1	1	1	17	—
Lolo Creek—gh.	1	1	1	15	—
Morrison Ridge—gh.	1	1	1	18	—
Rancheria Mt.—gh.	4	4	4	22	2.7
Chilnualna Falls—gh.	7	7	7	19	6.8
Primary rosettes (n)					
Moose	2	—	2	5.5	.7
Lolo Creek	3	—	8	10	5
Rancheria Mt.	4	—	4	2.5	1.9
Chilnualna Falls	5	—	5	7.4	7
Moose—gh.	1	1	1	13	—
Lolo Creek—gh.	1	1	1	14	—
Morrison Ridge—gh.	1	1	1	17	—
Rancheria Mt.—gh.	4	4	4	6	4.7
Chilnualna Falls—gh.	7	7	7	14	8
Primary rosettes—diam. (cm.)					
Moose	2	—	2	.7	.2
O'Leary Mt.	3	—	6	1	.2
Dutchman's Peak	3	—	5	.9	.5
Rancheria Mt.	3	—	6	.8	.2
Chilnualna Falls	5	—	10	1	.3
Moose—gh.	1	1	3	2.7	—
Lolo Creek—gh.	1	1	2	3.7	—
Morrison Ridge—gh.	1	1	2	3.3	—
Rancheria Mt.—gh.	1	1	2	1.6	—
Rosettes on floriferous stems (n)					
Lolo Creek	3	—	8	9.8	4.7
Morrison Ridge	9	—	20	10.6	.4
O'Leary Mt.	2	—	2	.6	.02
Rancheria Mt.	2	—	2	2	1.4
Chilnualna Falls	6	—	10	11.6	5.4
Rancheria Mt.—gh.	1	1	2	37	—
Floriferous stems per plant (n)					
Moose	2	—	2	11.5	13.5
Lolo Creek	1	—	1	7	—
Morrison Ridge	9	—	9	2.3	1.7
O'Leary Mt.	4	—	4	1.5	1
Chilnualna Falls	6	—	6	1.7	1.2
Moose—gh.	1	1	1	1	—
Rancheria Mt.—gh.	1	1	1	5	—

	n-pl.	n-rep.	n-obs.	$\bar{x}$	s
Floriferous stems—length (cm.)					
Moose	2	—	4	12.3	2.5
Scotty Creek	6	—	12	12	1.4
Chilnualna Falls	6	—	9	10.1	5.1
Moose—gh.	1	1	1	25	—
Rancheria Mt.—gh.	1	1	2	43	—

Leaves are spirally arranged and either glabrous or papillose marginally. The outer, lowest leaves of the sterile rosettes may be long, lanceolate-linear, with broad, scarious bases, or they may be oblong-elliptical with scarious bases of moderate width. On drying, the leaves become brown and persistent at the bases of the stems and branches. Sometimes the blades become subulate with the bases scarious, three-lobed, and prominent. When fresh, the leaves are subterete, plane ventrally, and convex dorsally. The inner, upper leaves of rosettes are smaller than the outer ones, subterete, lanceolate, truncate, keeled laterally below the middle, papillose, and green. The leaves of the floriferous stems are linear-lanceolate, obtuse, and with scarious spurs. The mean length of the leaves of the primary rosettes of plants at Scotty Creek, cited below, is so short because at time of study the outer leaves were withered and measurements are of the inner, shorter leaves.

	n-pl.	n-rep.	n-obs.	$\bar{x}$ (mm.)	s (mm.)
Leaves of primary rosettes—length					
Moose	2	—	2	6.8	2.4
Lolo Creek	1	—	3	11	—
Scotty Creek	6	—	12	4.3	2
O'Leary Mt.	3	—	6	9.1	3
Dutchman's Peak	3	—	5	10.6	2.1
Rancheria Mt.	3	—	6	4.4	.7
Chilnualna Falls	5	—	10	7.3	1.3
Moose—gh.	1	1	3	13.3	—
Lolo Creek—gh.	2	2	8	9.6	2.7
Morrison Ridge—gh.	3	3	11	12.4	2.2
Scotty Creek—gh.	1	1	4	13.8	—
Rancheria Mt.—gh.	4	4	12	7.5	1.3
Chilnualna Falls—gh.	7	7	21	8.3	2.3
Leaves of primary rosettes—width					
Moose	2	—	2	2.2	.6
Lolo Creek	1	—	3	1.5	—
Scotty Creek	6	—	12	1.7	.1
O'Leary Mt.	3	—	6	2.2	.4
Dutchman's Peak	3	—	5	2.7	.3
Rancheria Mt.	3	—	6	2.6	.3
Chilnualna Falls	5	—	10	2.4	.3
Moose—gh.	1	1	3	2.4	—
Lolo Creek—gh.	2	2	8	1.4	.3
Morrison Ridge—gh.	3	3	11	1.4	.2
Scotty Creek—gh.	1	1	4	2.5	—
Rancheria Mt.—gh.	4	4	12	2.6	.3
Chilnualna Falls—gh.	7	7	21	2.2	.3

268 SEDUM OF NORTH AMERICA

	n-pl.	n-rep.	n-obs.	x̄ (mm.)	s (mm.)
Leaves of primary rosettes—thickness					
Moose	2	—	2	.9	.1
Scotty Creek	6	—	12	.9	.3
O'Leary Mt.	3	—	6	1	.2
Dutchman's Peak	3	—	5	1.2	.1
Rancheria Mt.	3	—	6	.3	.1
Chilnualna Falls	5	—	10	.9	.1
Moose—gh.	1	1	3	1.5	—
Lolo Creek—gh.	2	2	8	1	.3
Morrison Ridge—gh.	3	3	11	.9	.02
Scotty Creek—gh.	1	1	4	1.7	—
Rancheria Mt.—gh.	4	4	12	1.3	.2
Chilnualna Falls—gh.	7	7	21	1.3	.3
Leaves of floriferous stems—length					
Lolo Creek	2	—	5	9.2	.3
Morrison Ridge	2	—	4	10	.1
O'Leary Mt.	1	—	1	10.5	—
Dutchman's Peak	1	—	1	4.1	—
Chilnualna Falls	5	—	5	7	1.5
Moose—gh.	1	1	2	12.5	—
Leaves of floriferous stems—width					
Lolo Creek	2	—	5	1.8	.3
Morrison Ridge	2	—	4	1.9	.1
O'Leary Mt.	1	—	1	1.2	—
Dutchman's Peak	1	—	1	1.4	—
Chilnualna Falls	5	—	5	2.4	.5
Moose—gh.	1	1	2	2.7	—
Leaves of floriferous stems—thickness					
Lolo Creek	1	—	3	1.2	—
Morrison Ridge	2	—	4	.8	.02
Chilnualna Falls	4	—	4	.8	.2

Inflorescences are terminal cymes, each with a central flower, typically three-parted, but sometimes reduced to a single flower, with rosettes developing in place of the other flowers. The floral bracts are foliaceous, smaller than the leaves, linear-lanceolate, acute, and spurred.

	n-pl.	n-rep.	n-obs.	x̄ (n)	s (n)
Cincinni per cyme					
Moose	2	—	4	1.3	1.8
O'Leary Mt.	4	—	5	0	0
Dutchman's Peak	4	—	4	0	0
Rancheria Mt.	3	—	6	0	0
Chilnualna Falls	6	—	7	.5	.6
Rancheria Mt.—gh.	1	1	1	0	—
Flowers per cyme					
Moose	2	—	4	5.8	6
Morrison Ridge	3	—	8	8	1
Scotty Creek	6	—	12	2.9	1.1
O'Leary Mt.	4	—	5	.8	.5
Dutchman's Peak	4	—	4	1	0
Lolo Creek—cf.	1	1	1	44	—
Morrison Ridge—cf.	1	1	1	13	—

	n-pl.	n-rep.	n-obs.	$\bar{x}$ (n)	s (n)
Flowers per cyme					
Kinney Point—cf.	1	1	1	1	—
Scotty Creek—cf.	2	4	4	13	1.5
Moose—gh.	1	1	1	8	—
Rancheria Mt.—gh.	1	1	2	1	—

Flowers are mostly 5-merous, sessile or subsessile, and with a short torus.

	n-pl.	n-rep.	n-obs.	$\bar{x}$ (mm.)	s (mm.)
Pedicels—length					
Scotty Creek	6	—	12	.2	.2
Moose—gh.	1	1	3	.2	—
Flowers—diameter					
Lolo Creek	1	—	2	14	—
Morrison Ridge	4	—	9	12.6	1.1
Scotty Creek	1	—	1	10	—
Moose—cf.	2	4	8	15.8	1.9
Lolo Creek—cf.	1	1	3	17.3	—
Morrison Ridge—cf.	1	1	2	12	—
Kinney Point—cf.	1	1	3	16	—
Scotty Creek—cf.	2	4	12	14.1	.5
Moose—gh.	1	1	3	12.7	—
Rancheria Mt.—gh.	1	1	2	9	—
Torus—length					
Morrison Ridge	2	—	7	.9	.05
Scotty Creek	6	—	11	1.3	.05
O'Leary Mt.	1	—	1	1	—
Moose—gh.	1	1	3	.7	—
Rancheria Mt.—gh.	1	1	2	1.4	—

Sepals are erect, lanceolate or ovate, acute or long-acuminate, and pale green or yellow-green.

	n-pl.	n-rep.	n-obs.	$\bar{x}$ (mm.)	s (mm.)
Length					
Lolo Creek	1	—	3	2.9	—
Morrison Ridge	2	—	7	2.2	.3
Scotty Creek	6	—	12	2	.3
O'Leary Mt.	1	—	1	3.3	—
Moose—gh.	1	1	3	3.7	—
Rancheria Mt.—gh.	1	1	2	2.3	—
Width					
Lolo Creek	1	—	3	1.6	—
Morrison Ridge	2	—	7	.9	.04
Scotty Creek	6	—	12	1.4	.15
O'Leary Mt.	1	—	1	1.7	—
Moose—gh.	1	1	3	1.4	—
Rancheria Mt.—gh.	1	1	2	1.3	—

Petals are normally 5 per flower, lanceolate or elliptical, obtuse or acute, aristate-appendaged, and yellow with green dorsal keel, sometimes pale yellow or almost white. Exceptionally, flowers occur with 3, 4, 6, or even 8 petals.

	n-pl.	n-rep.	n-obs.	x̄ (mm.)	s (mm.)
Length					
Lolo Creek	2	—	5	7.5	1.6
Morrison Ridge	4	—	9	6.9	.4
Scotty Creek	1	—	1	5.4	—
O'Leary Mt.	1	—	1	8	—
Moose—gh.	1	1	3	7.6	—
Rancheria Mt.—gh.	1	1	2	6	—
Width					
Lolo Creek	2	—	5	2.7	.6
Morrison Ridge	4	—	9	2.4	.2
Scotty Creek	1	—	1	2	—
O'Leary Mt.	1	—	1	2.6	—
Moose—gh.	1	1	3	2.9	—
Rancheria Mt.—gh.	1	1	2	2.4	—

Stamens have pale yellow filaments and yellow anthers.

	n-pl.	n-rep.	n-obs.	x̄ (mm.)	s (mm.)
Filaments—length					
Moose—gh.	1	1	3	5	—
Epipetalous filaments; adnation—length					
Morrison Ridge	2	—	7	.5	.2
Scotty Creek	1	—	1	.5	—
O'Leary Mt.	1	—	1	.8	—
Moose—gh.	1	1	3	.4	—
Rancheria Mt.—gh.	1	1	1	.9	—
Anthers—length					
Lolo Creek	1	—	3	1.7	—
Morrison Ridge	1	—	1	1.4	—
Moose—gh.	1	1	3	1.5	—
Rancheria Mt.—gh.	1	1	1	.6	—

Nectaries are reniform-subquadrate or quadrate, truncate, greenish yellow or yellowish white, and sometimes translucent.

	n-pl.	n-rep.	n-obs.	x̄ (mm.)	s (mm.)
Length					
Lolo Creek	1	—	3	.4	—
Morrison Ridge	2	—	7	.4	.06
Scotty Creek	2	—	2	.3	.26
O'Leary Mt.	1	—	1	.4	—
Moose—gh.	1	—	3	.3	—
Rancheria Mt.—gh.	1	—	2	.3	—
Width					
Lolo Creek	1	—	3	.6	—
Morrison Ridge	2	—	7	.4	.04
Scotty Creek	2	—	2	.4	.26
O'Leary Mt.	1	—	1	.5	—
Moose—gh.	1	—	3	.4	—
Rancheria Mt.—gh.	1	—	2	.4	—

Carpels are erect and pale green or yellow-green before and when flowers are in anthesis, but soon become divergent. The ovaries are briefly connate and finely papillose.

	n-pl.	n-rep.	n-obs.	$\bar{x}$	s
Length (mm.)					
Morrison Ridge	2	—	7	4.3	.3
Scotty Creek	2	—	2	3.2	1.5
O'Leary Mt.	1	—	1	5.7	—
Moose—gh.	1	—	3	3.7	—
Rancheria Mt.—gh.	1	—	2	4.7	—
Length of styles (mm.)					
O'Leary Mt.	1	—	1	1.6	—
Ovules per ovary (n)					
Lolo Creek	1	—	3	7	—
Scotty Creek	6	—	12	5	.9
O'Leary Mt.	1	—	1	6	—
Moose—gh.	1	—	3	4	—
Rancheria Mt.—gh.	1	—	2	5	—

Fruits are divergent, gibbous ventrally with prominent lips, and brown.

	n-pl.	n-rep.	n-obs.	$\bar{x}$ (mm.)	s (mm.)
Length					
Moose	2	—	4	3.7	.3
Morrison Ridge	7	—	15	3.9	.2
Kinney Point	2	—	4	3.3	.7
Scotty Creek	5	—	10	3.4	.5
Dutchman's Peak	1	—	2	3.8	—
Morrison Ridge—gh.	1	1	2	3.2	—
Width of ventral lips					
Morrison Ridge	7	—	15	.6	.1
Kinney Point	2	—	4	.5	.07

Seeds are pyriform, finely longitudinally ribbed, and brown or pale yellow.

	n-pl.	n-rep.	n-obs.	$\bar{x}$ (mm.)	s (mm.)
Length					
Moose	2	—	4	1.1	.07
Morrison Ridge	7	—	15	.9	.06
Kinney Point	1	—	2	1.1	—
Scotty Creek	5	—	9	1	.05
Moose—gh.	2	2	4	1.3	.04
Morrison Ridge—gh.	1	1	2	1.2	—
Diameter					
Moose	2	—	4	.6	.07
Morrison Ridge	7	—	15	.4	.05
Kinney Point	1	—	2	.5	—
Scotty Creek	5	—	9	.5	.07
Moose—gh.	2	2	4	.6	.11
Morrison Ridge—gh.	1	1	2	.7	—

Chromosomes suggest that the species is octoploid in relation to *Sedum radiatum*. Natural hybrids with *S. lanceolatum* appear to be aneuploid. Data cited below apply to plants which provided descriptive details for this study.

Number	n-pl.	g (n)	sp (2n)	Cytologist
Moose	1	32	—	Ilse Zandstra, 1969
Lolo Creek	1	—	62–70	Iris Mastrangelo, 1971
Hart's Pass, Okanogan Co., Wash.	1	32	—	C. H. Uhl, 1942
Chilnualna Falls	1	—	63–64	Iris Mastrangelo, 1971
Hybrid, Moose	1	—	58	Dorothy Niimoto, 1967
Hybrid, Moose	1	—	?59	Dorothy Niimoto, 1967
Hybrid, Scotty Creek	1	—	50–54	Dorothy Niimoto, 1967

Variation. Two kinds of *Sedum stenopetalum* occur. One kind has narrow, linear leaves, cymes with several flowers, and yellow petals. The distribution of this type is from the Northern Rocky Mountains of Montana and Alberta westward to the Cascade and Olympic Mountains of Washington, and southward to mountainous parts of the northwestern Great Basin in southern Oregon and northeastern California. Pleistocene glaciers covered most of this area. Since deglaciation, the spread of the plants must have been rapid. The second kind of *S. stenopetalum* has elliptic-oblong leaves, floriferous stems with solitary flowers, and yellow or pale yellow petals. Plants of this type occur in the Klamath Mountains, the northern California Coast Ranges, the Sierra Nevada, and in the Cascade Mountains in Oregon and southern Washington. Northward in Washington and in the Seven Devils Mountains of Idaho, plants occur with solitary flowers, but with foliage resembling that of the other type. Plants of both types are octoploid.

In experiments in the greenhouse at Ithaca, N.Y., the contrast between the two kinds of *Sedum stenopetalum* is great. The plants have different habits of growth, as shown in figure 61 above and figure 64 below. They are easy to distinguish. The leaf-bearing parts of the stems are elongate in the type with elliptic-oblong leaves, and the stems are much branched. On the basis of aspect, the two kinds of *S. stenopetalum* are different species, but the occurrence of intermediate plants is evidence in favor of subspecific status. Further, variation in both dimensions of leaves and length of leaf-bearing parts of stems indicates that all differences are not as reliable as appears on first impression.

An experiment in the greenhouse in 1971 made possible a test of differences between the two types of *Sedum stenopetalum*. This experiment, of completely random design, included 3 plants of the linear-leaved type and 11 plants of the type with elliptic-oblong leaves. Analyses of variance for four features indicate that three differences are significant. The results follow.

	Linear-leaved	Oblong-leaved
Leaf-bearing part of stem—length (cm.)*	.7	2
Leaves—length (mm.)[n.s.]	11.7	8.3
—width (mm.)**	1.6	2.3
Leaves—length/width**	7.1	3.7

The reduction of the inflorescence to a single flower and eventually the suppression of all flowers with the substitution of vegetative rosettes as the principal means of reproduction seems to be a response to warmer, drier conditions in summer. A frequency distribution for number of flowers in 49 cymes of 28 wild plants in my samples reveals that 35% of the cymes have solitary flowers. Plants with this condition occur as far east as Rogers Pass, Mont., and as far north as the Wenatchee Mountains. Plants with solitary flowers usually have elliptic-oblong leaves, but the northernmost and easternmost plants have foliage resembling plants with several flowers per cyme.

Flowers per cyme (n)	Cymes (n)	Cymes %	Cumulative %
1	17	34.7	34.7
2	5	10.2	44.9
3	2	4.1	49
4	7	14.3	63.3
5	3	6.1	69.4
6	3	6.1	75.5
7	2	4.1	79.6
8	1	2	81.6
9	8	16.3	97.9
≥10	1	2.1	100
Total	49		

Color of petals varies from deep lemon-yellow, the usual condition in the linear-leaved type, to pale yellow or almost white in some plants with solitary flowers. Number of petals, normally 5, varies from 3 to 8.

Plants of the linear-leaved type of *Sedum stenopetalum* vary greatly in size and robustness. This variation seems to be a response to different conditions of growth, especially water and nutrients. However, a more slender, lower type, which has a heritable basis, occurs in the Seven Devils Mountains and in the region of the type locality in the valley of Lolo Creek, Mont. In the Cascade Mountains, taller plants have deeper yellow flowers and more anthocyanin in the stems.

A remarkable cristate variation (fig. 62) appeared spontaneously in my cultures in the cold frame at Ithaca. This produced flowers on the rim of the bizarre, coxcomblike stems.

Natural hybridization is an important cause of variation in Jackson Hole, the Seven Devils Mountains, and the Wenatchee Mountains. This is discussed below under relationships.

Where hybridization with other species is not involved, variation within populations of *Sedum stenopetalum* is slight. Study of a single plant per population is adequate to detect a difference of .2 mm. in length of seeds. Larger samples are necessary for more variable features.

KEY TO SUBSPECIES OF *SEDUM STENOPETALUM*

A. Floriferous stems bearing cymes of several flowers; leaves of principal vegetative shoots linear, 4.5–13 times longer than broad; leaf-bearing parts of principal vegetative shoots of cultivated plants, when well-watered, .1–1.5 cm. long .... ssp. *stenopetalum*, p. 274

Fig. 62. Cristate floriferous stem of plant of *Sedum stenopetalum* ssp. *stenopetalum* cultivated in cold frame, Ithaca, N.Y., May 10, 1965 (x 1).

AA. Floriferous stems terminating in a solitary flower or with a terminal vegetative rosette; leaves of principal vegetative shoots elliptic-oblong, 3–4.4 times longer than broad; leaf-bearing parts of principal vegetative shoots of cultivated plants, when well watered, 1.5–2.6 cm. long .... ssp. *monanthum*, p. 276

Sedum stenopetalum ssp. *stenopetalum*

Nomenclature. *Sedum stenopetalum* Pursh, Fl. Am. Sept. 1:324 (1814). Type locality: valley of the Clarck's (Bitterroot) River at the mouth of Traveller's Rest (Lolo) Creek in western Montana, just south of the modern town of Lolo. The vicinity of Lolo today is densely settled and managed for agriculture, but the *Sedum* still occurs up the valley of Lolo Creek. I studied it within 21 km. of the presumed type locality. Type: Lewis and Clark Expedition(Ph), July 1, 1806. The holotype comprises loose fragments in a packet, exhibiting subulate leaves and flowers with long, acuminate sepals and widely divergent pistils. Frederick Pursh described the leaves as compressed subulate and the petals as linear, both more appropriate for the species which has often been called *S. douglasii* than for *S. lanceolatum* to which the name *S. stenopetalum* has commonly been applied.

A synonym is:

1840 [1834]. *Sedum Douglasii* Hooker, Flor. Bor. Am. 1:228. Type locality: rocky banks of stream in the interior of the Columbia. Hooker stated that the species is common on rocky places of the Columbia to the mountains. Type: D. Douglas (BM), 1826.

Diagnostic for ssp. *stenopetalum* are the linear leaves which, on drying, become subulate with persistent, scarious bases. The cymes typically are composed of several flowers with deep yellow petals. The parts of the stems of the primary branches with fresh leaves are comparatively short, .1–1.5 cm. long.

The geographic range of *Sedum stenopetalum* ssp. *stenopetalum* (fig. 58) is from the Central and Northern Rocky Mountains westward through the interior plateaus and ranges to the Cascade, Klamath, and Olympic Mountains, and southward to the Modoc Plateau in northeastern California. A single collection is available from the western Great Plains in Montana. The altitudinal range is from 393 m. below the Kettle Falls of the Columbia River to 3,050 m. on Lone Mountain southwest of Bozeman, Mont. Habitats are well drained or excessively drained, open, rocky places on slopes, bluffs, banks, and even flats. The plants usually grow in rocky soil or gravel and frequently in crevices or on talus. They occur on a variety of rocks, including sandstone, granite, basalt, and andesite. Through much of its range, ssp. *stenopetalum* occurs in the altitudinal zone of *Pinus ponderosa*.

Sites where ssp. *stenopetalum* occurs often are very dry, especially in late summer. Such conditions are poor for germination of seeds and survival of seedlings. Adjustment to drought has involved the development of numerous vegetative rosettes. These even replace flowers in the cyme. Inflorescences often are few-flowered. Although the long, outer leaves of the rosettes become dry, the shorter, inner leaves remain turgid throughout the dry season. In the presence of moisture, these rosettes root easily and grow into new plants.

Of the more than 108 populations known to me, I have devoted special attention to five. Data for these are in the following table. The populations on Morrison Ridge in the Seven Devils Mountains and on the ridge north of Scotty Creek in the Wenatchee Mountains were in randomly selected sampling units in these mountain ranges. Similarly, the population at Moose was selected randomly from four known populations in Jackson Hole. Selection of the population in the valley of Lolo Creek was deliberate because it is near the type locality. Selection of the population at Kinney Point was deliberate because it was at the highest elevation studied in this investigation and the plants there had only one cluster of follicles per fructiferous stem, although they otherwise appeared similar to plants of ssp. *stenopetalum* elsewhere in the Seven Devils Mountains.

Population	Plants (N)	Area of population (m.^{2})	Density per (m.^{2})	Exposure
Moose—aspen grove	214	627	.34	all directions
Lolo Creek	44	—	—	W
Morrison Ridge	29,602	7,560	3.9	N
Kinney Point	20	—	—	NE
Scotty Creek	478	216	2.2	SW

Population	Altitudinal range (m.)	pH, range	pH, tests (n)	Oldest record
Moose—aspen grove	1,970	6.4–6.6	2	1966
Lolo Creek	1,100	—	—	1806
Morrison Ridge	1,065–1,200	5.6–6.8	6	1961
Kinney Point	2,065	5	1	1961
Scotty Creek	1,214–1,220	4.9–6.4	7	1961

The ssp. *stenopetalum* grows in open places where other species of plants are few. Sometimes it occurs in open woods of *Pinus ponderosa* and *Pseudotsuga menziesii*, as on the ridge north of Scotty Creek, or in open aspen groves, as near Moose in Jackson Hole. On the ridge above Scotty Creek, there are no plants on exposed southern slopes, but they grow on southwestern slopes where they are shaded for part of the day by the pines and Douglas firs. The openings in which the *Sedum* occurs are manmade clearings where large trees of *Pseudotsuga* have been removed. Such a population is temporary. Since the environment is continuously changing, the status of populations must be ever in state of flux. Data on exact distribution in any season are temporary.

Species competing with *Sedum stenopetalum* ssp. *stenopetalum* are miscellaneous, different in each population I studied. This result indicates the futility of trying to describe the habitat of this *Sedum* in terms of other vegetation. At Moose, the plants were in an open grove of *Populus tremuloides*, with species of *Polygonum* and *Eriogonum* prominent as competitors. On Kinney Point, two grasses, *Koeleria cristata* and a *Festuca*, and also a species of *Stellaria*, were the important competitors. On Morrison Ridge, where *Pseudotsuga menziesii* partially shaded the plants, commonest competitors were three grasses: *Bromus japonicus*, *B. brizaeformis*, and *B. mollis*, listed in order of importance. On the ridge north of Scotty Creek, where the plants were in open woods of *Pinus ponderosa* and *Pseudotsuga*, principal competitors were an unidentified annual dicotyledon, species of *Lupinus* and *Lomatium*, and *Penstemon fruticosus*.

Sedum stenopetalum ssp. *monanthum*

Nomenclature. *Sedum stenopetalum* ssp. **monanthum** (Suksdorf), comb. nov., fundatum super *Sedum monanthum* Suksdorf, Werdenda 1:19 (1927).

Synonyms are:

1898. *Sedum uniflorum* Howell, A flora of northwest America 1:213 (not *S. uniflorum* Raf., 1810, and not *S. uniflorum* Hook. et Arn., 1841). Type locality: on rocks along the Willamette River opposite Milwaukee, Ore. Type: Thomas Howell (ORE), June 1880. The type specimen is fragmentary, consisting of a single, almost naked floriferous stem, broken off at the top, without flowers, but with a single axillary rosette attached and another in an envelope. Howell specified that the stems bear numerous propagula and are terminated by a single flower.

1910. *Sedum Douglasii* var. *uniflora* (Howell) M. E. Jones, Bull. Univ. Mont., Biol. Ser. 15:30. Based on *S. uniflorum* Howell. Jones cited the variety from Upper Marias Pass.

1927. *Sedum monanthum* Suksdorf, Werdenda 1:19. Based on *S. uniflorum* Howell. Suksdorf regarded the species as very near *S. douglasii*. He regarded it as not rare along Dog Creek at "Cooks", Skamania Co., Wash., in both shady and dry, sunny places.

1936. *Sedum Douglasii* var. *monanthum* (Suksdorf) Fröderström, Acta Horti Goth. 10 (App.):105. Based on *S. monanthum* Suksdorf.

1936. *Sedum Douglasii* f. *uniflorum* (Howell) G. N. Jones, Univ. Wash. Pub. Biol. 5:164. Based on *S. uniflorum* Howell. Jones cited the form from Hurricane Ridge in the Olympic Mountains, but plants there may not be typical.

Diagnostic features of ssp. *monanthum* (fig. 63) are the solitary, terminal flowers on the floriferous stems, the elliptic-oblong leaves of the principal vegetative shoots,

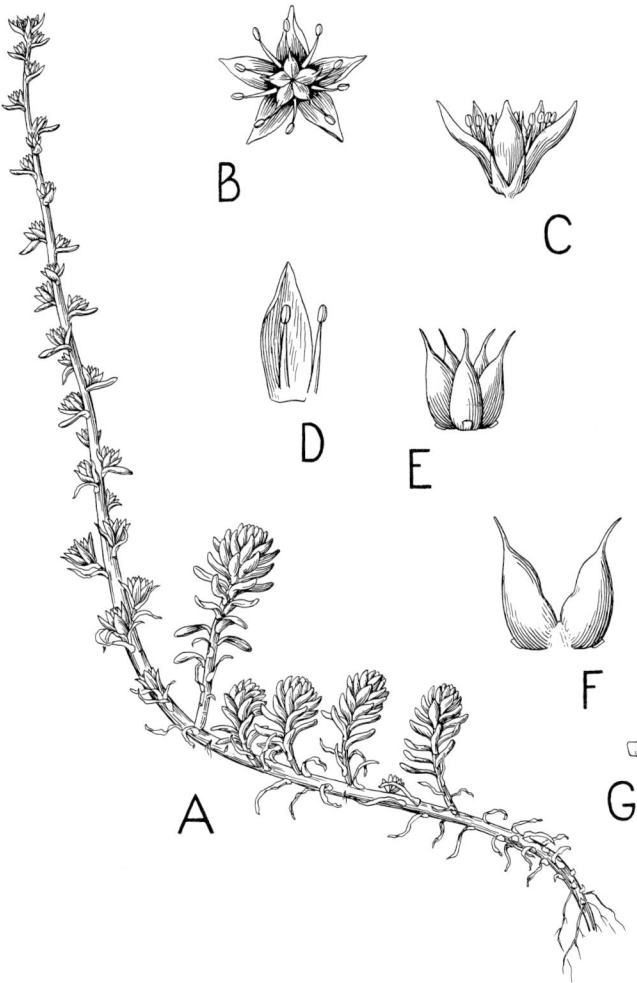

Fig. 63. Plant of *Sedum stenopetalum* ssp. *monanthum* from northwest slope of Rancheria Mountain on north side of Hetch Hetchy Reservoir, Yosemite National Park, Calif. A. Habit sketch (x .52). B. Flower from above (x 2.1). C. Flower from side (x 2.1). D. Petal and two stamens (x 3.1). E. Carpels (x 3.1). F. Two carpels and torus (x 4.2). G. Nectary (x 5.2).

and the long, leaf-bearing parts of the vegetative shoots. The last feature is best seen in cultivated plants (fig. 64) which are well watered and kept in a continual state of growth. Under such circumstances, the difference between the two subspecies is both marked and significant. To qualify as ssp. *monanthum*, plants must have the characteristic shorter, broader leaves as well as solitary flowers. Few plants north of the drainage of the Columbia River or east of the Cascade Mountains combine these two features.

Fig. 64. Plant of *Sedum stenopetalum* ssp. *monanthum* from northwest slope of Rancheria Mountain on north side of Hetch Hetchy Reservoir, Yosemite National Park, Calif., cultivated in greenhouse, Ithaca, N.Y., Dec. 23, 1964 (x .8). Note elongate vegetative shoots. Photo by Howard Lyon.

The principal geographic distribution of ssp. *monanthum* (fig. 58) is in the Cascade Mountains from Mount St. Helens southward, the Klamath Mountains, the northern California Coast Ranges, and the Sierra Nevada. Elsewhere, plants with solitary flowers occur, but they seldom have other features which characterize ssp. *monanthum*. Since some plants clearly are intermediate between the two subspecies, as in the Wenatchee and Seven Devils Mountains, subspecies status seems appropriate. Habitats include ledges and shelves, steep slopes, and cliffs. The plants grow in gravel or thin soil over a variety of rocks: andesite, granite, and schist. Drainage is good to excessive. The altitudinal range is from 487 to 1,970 meters.

Seventeen populations of ssp. *monanthum* have come to my attention. I made special studies at four sites. The resulting data are in the following table. Selection of the cluster of plants on O'Leary Mountain was random from three clusters. The population on Dutchman's Peak was the closest which could be found to *S. radiatum* ssp. *depauperatum*. This provided opportunity to note whether interaction occurs. Discovery of the population on Rancheria Mountain was in the course of the survey of the Sierra Nevada. All aspects of random selection were unrestricted. Finally, the population at Chilnualna Falls, in Yosemite National Park, is the southernmost known to me. A result of the survey of the Sierra Nevada is an estimate of the number of plants of ssp. *monanthum* there, namely 499,764.

Population	Plants (N)	Area of population (m.^{2})	Density per m.^{2}	Exposure
O'Leary Mountain	48	—	—	NE–E
Dutchman's Peak	7	4	1.8	SW
Rancheria Mountain	41	20	2	NW
Chilnualna Falls	440	372	1.2	NNW

Population	Altitudinal range (m.)	pH, range	pH, tests (n)	Oldest record
O'Leary Mountain	1,490–1,590	5.6	4	1965
Dutchman's Peak	1,790	—	—	1965
Rancheria Mountain	1,865	5–5.2	3	1964
Chilnualna Falls	1,967–1,970	5–5.2	6	1964

Other vegetation is remarkably sparse where ssp. *monanthum* occurs. On O'Leary Mountain, a species of *Poa* and *Festuca viridula* grew with the *Sedum*. On Dutchman's Peak, associated plants were *Epilobium*, a small bush of *Rhamnus*, and seedlings of another dicotyledon. On Rancheria Mountain, plants of the *Sedum* grew in little clusters by themselves, in granitic gravel on small shelves on a ridge. Although no competitors were close, remote trees of *Pinus ponderosa* provided partial shade. At Chilnualna Falls, *Gilia ciliata* was the principal competitor. *Sedum obtusatum* and a species of *Poa* were less important. An estimated sixth of the plants of ssp. *monanthum* at Chilnualna Falls lacked competitors, except other individuals of their kind.

Of the two subspecies of *Sedum stenopetalum*, ssp. *monanthum* appears to be the more advanced. Its adaptation to warmer, drier conditions in the summer and to a shorter photoperiod has involved a change from sexual to primarily vegetative reproduction. Ssp. *monanthum* appears to be relictual in the Sierra Nevada.

Reproduction. Vegetative reproduction is important in *Sedum stenopetalum*. In some populations, as at Chilnualna Falls, it may be the principal or only mode of propagation. Fruits of ssp. *monanthum* are sparse, with the result that few seeds are available to develop into new plants. Rosettes from two plants of ssp. *monanthum* at Chilnualna Falls, kept dry in a sealed can, remained alive for six months. Similarly, rosettes from plants on Dutchman's Peak and O'Leary Mountain, kept dry in plastic bags, retained life for six months. Rosettes of two plants of ssp. *stenopetalum* from Jackson Hole survived for four months without water.

Flowering time is late spring and early summer but varies with latitude and longitude. Where *Sedum radiatum* and *S. stenopetalum* occur in the same region, as in Yosemite Park and on Dutchman's Peak, the *S. radiatum* is a few weeks ahead of the *S. stenopetalum*.

Based on my experience and records from herbarium specimens, extreme dates for anthesis in the latitudinal band of 40°–45° N. are June 8 and July 20, and for the latitudinal band 45°–50° N. are May 31 and July 10.

Pollinators of *Sedum stenopetalum* are insects, especially small flies and bees. These seem to visit all yellow flowers. Hybrids between *S. lanceolatum* and *S. stenopetalum* occur near Moose in Jackson Hole and on the ridge above Scotty Creek in the Wenatchee Mountains. Likewise, a natural hybrid has developed between *S. borschii* and *S. stenopetalum* in the Seven Devils Mountains.

Relationships. *Sedum stenopetalum* is closely related to *S. radiatum*. A common practice is to interpret these as conspecific. Whether the octoploid *S. stenopetalum* is derived through autoploidy from a diploid prototype which also gave rise to *S. radiatum*, or whether it is the result of interspecific hybridization, is unclear. Likewise unclear is whether ssp. *monanthum* had an origin independent from ssp. *stenopetalum*. The differences in foliage, as well as the adaptation to warmer, drier conditions, suggest a distinctive origin.

Differences between *Sedum radiatum* and *S. stenopetalum* are few, but crucial. The main reason for treating them as separate species is that they do not intergrade where they occur together. On Dutchman's Peak, they occur within 4 km. of each other. When the *S. radiatum* is at the height of anthesis, *S. stenopetalum* still is in bud. A similar situation prevails in Yosemite Park, where populations occur within 18 km. of each other. The important distinctions are listed below. Comparisons of numbers of flowers per cyme and lengths of fruits were by analysis of variance. Although the difference in number of flowers per cyme misses significance at $P = .05$, the comparison is included because of a reasonably large F value.

	Sedum radiatum	Sedum stenopetalum
Chromosomes, gametophytic no.	8	32
Duration	annual	perennial
Flowers per cyme (n)	11	3
Fruits—length (mm.)**	3	3.6

Although *Sedum stenopetalum* has rotately spreading follicles and those of *S. lanceolatum* are erect, the ability of the two species to hybridize and backcross indicates close relationship. Hybrid swarms occur near Moose in Jackson Hole and on the Wenatchee Ridge above Scotty Creek in the Cascade Mountains. A comparison of three plants, one of either parental species, and a hybrid, all grown in 1969 in an experiment of completely random design, reveals points of distinction.

The hybrid is significantly different from typical *S. stenopetalum* in diameter of flowers and length of petals, transcending either species in both features. Tests of differences are by a Wilcoxon multiple comparison procedure. Mean expressions are cited. The sporophytic number of chromosomes of the hybrid is about 58.

	S. 1.	S. 1. × S. s.	S. s.
Rosettes—diameter (cm.)	1.1	1.1	2.4
Leaves of rosettes—length (mm.)	7	6.6	13.3
—width (mm.)	2	1.9	2.4
Flowers—diameter (mm.)*	17	18.7	12.7
Sepals—length (mm.)	2.8	2.7	3.7
Petals—length (mm.)*	8.8	10	7.6
Nectaries—length (mm.)	.4	.5	.3
Ovules per ovary (n)	8	5	4

A similar experiment in the cold frame in 1968 revealed a significantly greater width of petals of the hybrid. This experiment included two plants of the hybrid and two plants of *Sedum stenopetalum*, with two replicates of each plant. Measurements were by Jay Freer.

Hybrids between *Sedum lanceolatum* and *S. stenopetalum* have low fertility. They produce few seeds. Sporophytic chromosomes of hybrids range from 50 to 60. In Jackson Hole, hybrid plants comprised 5% of the population. On Wenatchee Ridge, they were rarer, only .2% of the total population. Both species have evolved far enough in the development of genetic incompatibilities that biological isolation is reasonably effective, even where the two species occur in the same habitat.

Only one instance of hybridization of *Sedum stenopetalum* and *S. borschii* has come to my attention. I found a single hybrid in 1961 in the Seven Devils Mountains, at an elevation of 1,745 meters on a ridge on the south side of Papoose Creek. Further consideration of this hybrid is included in the discussion of *S. borschii*.

Other species of *Sedum* with which *S. stenopetalum* occurs are *S. oreganum* and *S. divergens* on Wedge Mountain in the Cascade Mountains; *S. oregonense* and *S. oreganum* on O'Leary Mountain; and *S. obtusatum* ssp. *obtusatum*, both on Rancheria Mountain and at Chilnualna Falls.

Subramanyam (1955) studied the floral anatomy of a plant of *Sedum stenopetalum* ssp. *stenopetalum* from Hart's Pass in the Cascade Mountains, about 96 km. northwest of Okanogan, Wash. He thought that it is moderately advanced, but did not study any closely related species.

References. Clausen, R. T. (1946 and 1948).

14. *Sedum leibergii**** (figs. 65–69)

The stems of *Sedum leibergii* (fig. 65) are biennial. In the first year, they produce rosettes of oblanceolate or oblong-spatulate leaves. In the second year, these stems elongate and bear diffuse cymes of yellow flowers. All leaves are spirally arranged.

282 SEDUM OF NORTH AMERICA

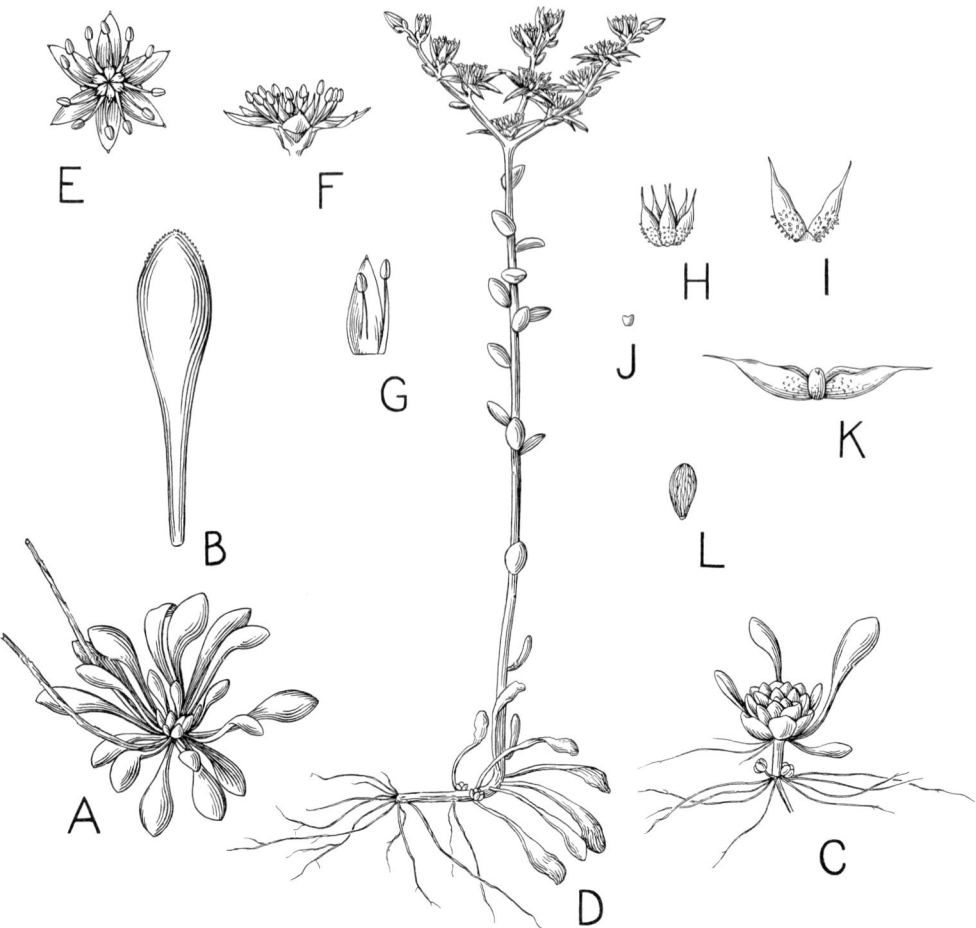

Fig. 65. Plants of *Sedum leibergii* from northern slope of ridge south of lower Papoose Creek, Seven Devils Mountains, Idaho Co., Ida. A. Rosette in autumn (x .94). B. Single leaf of rosette in autumn (x 1.9). C. Rosette in spring (x .94). D. Habit sketch of flowering plant in June 1961 (x .94). E. Flower from above (x 1.9). F. Flower from side (x 1.9). G. Petal and two stamens (x 2.8). H. Carpels (x 2.8). I. Two carpels, torus, and nectaries (x 3.8). J. Nectary (x 4.7). K. Fruits from side (x 3.8). L. Seed (x 9.4).

The elongate basal leaves may persist until time of flowering. The fruits are widely spreading and gibbous ventrally. No other western American species has this combination of characteristics.

Description. Sample: 52 plants, all collected in the Seven Devils Mountains in the eastern part of the Columbia Plateau in western Idaho, 1961 and 1970. Principal data are from 37 plants. Descriptive data about the sample are in the following brief table, where J = judgment, N_c = number of plants in sampled clusters, N_u = number of plants in sampling unit, and R = random.

Locality	Basis of choice	N_u	N_c	n	Plants studied n/N_c	n/N_u	Condition when collected fl. (n)	fr. (n)
Papoose Creek	J	864	24	16	.67	.02	6	10
Shingle Creek	R	864	144	21	.15	.02	10	11

All plants of *Sedum leibergii* in cultivation in Ithaca, N.Y., died in a few years. Five replicates of 3 plants, 2 from Papoose Creek and 1 from Shingle Creek, flowered in an experiment of random design in the cold frame in the spring of 1962. This experiment included 3 replicates of each of 6 plants of *S. leibergii* and an equal number of plants and replicates of *S. borschii*.

In the following tables, data from the wild are designated as w and from the cold frame as cf.

Plants of *Sedum leibergii* are biennial herbs with fibrous, pale brown roots and basal rosettes of leaves. The stems of axillary shoots are subterranean and white. These may attain a length of 1.5 cm. and are readily detachable. At their tips, they bear tiny rosettes which also are subterranean and have colorless leaves. In the second year, erect floriferous stems develop from the large, basal rosettes (fig. 66).

Fig. 66. Persistent basal rosette of plant of *Sedum leibergii* from Morrison Ridge south of Shingle Creek, eastern slope of Seven Devils Mountains, Ida Co., Ida., cultivated in cold frame, Ithaca, N.Y., June 6, 1962 (x 1.8). Photo by Howard Lyon.

	n-pop.	n-pl.	n-rep.	n-obs.	$\bar{x}$	s
Auxillary rosettes (n)						
w	2	16	—	16	2.9	2.3
Floriferous stems (n)						
w	2	16	—	16	1.3	.8
cf.	2	3	5	5	3.4	—
Floriferous stems—length (cm.)						
w	2	18	—	20	10.5	1.9
cf.	2	3	5	11	12	1.1

Leaves of basal rosettes are spirally arranged, oblanceolate, obovate, or narrowly spatulate, and papillose. The outermost, lowest leaves have long petioles. These are green or greenish white and may wither before or after time of flowering. Leaves of floriferous stems are divergent, ovate or elliptical, subacute, and papillose. A plant from Agatha, Ida., had basal leaves, including petioles, 55 mm. long.

	n-pop.	n-pl.	n-rep.	n-obs.	$\bar{x}$ (mm.)	s (mm.)
Leaves of rosettes—length						
w	2	10	—	10	9.5	7.4
Leaves of rosettes—width						
w	2	10	—	10	2.3	.9
Petioles—length						
w	2	10	—	10	4.3	4.3
Median leaves of floriferous stems—length						
w	2	9	—	9	3.8	.9
cf.	2	3	5	11	8.2	—
Median leaves of floriferous stems—width						
w	2	9	—	9	2.1	.7
cf.	2	3	5	11	3.3	—
Median leaves of floriferous stems—thickness						
w	2	5	—	5	1.2	.3

Inflorescences are terminal cymes of 3 to 6 cincinni. The cincinni are once or twice dichotomously forked (fig. 67).

	n-pop.	n-pl.	n-rep.	n-obs.	$\bar{x}$ (mm.)	s (mm.)
Cincinni per cyme, cf.	2	3	5	11	5.1	1.5
Flowers per cyme, cf.	2	3	5	11	59	—

Flowers (fig. 68) are short-pedicellate and often 5- or 6-merous, but sometimes 7- or more merous. Central flowers in cymes have the largest number of floral parts. A plant from Agatha, Ida., had pedicels up to 25 mm. long.

	n-pop.	n-pl.	n-rep.	n-obs.	$\bar{x}$ (mm.)	s (mm.)
Pedicels—length						
w	2	13	—	21	.2	.1
Torus—length						
w	2	13	—	22	.6	.07

NATIVE SPECIES 285

Fig. 67. Cyme with buds, view from above of *Sedum leibergii* from northern slope of ridge south of lower Papoose Creek, Seven Devils Mountains, Idaho Co., Ida., cultivated in cold frame, Ithaca, N.Y., Apr. 30, 1962 (x 1.2). Photo by Howard Lyon.

	n-pop.	n-pl.	n-rep.	n-obs.	$\bar{x}$ (mm.)	s (mm.)
Flowers—diameter						
w	2	15	—	25	10	1.3
cf.	2	3	5	12	12	1.5

Sepals are ovate, acute, and green.

	n-pop.	n-pl.	n-rep.	n-obs.	$\bar{x}$ (mm.)	s (mm.)
Length						
w	2	15	—	25	1.5	.2
cf.	2	3	5	13	1.9	.1
Width						
w	2	15	—	25	.9	.2
cf.	2	3	5	13	1.1	—

Petals are lanceolate, acute, and canary yellow with either green or dark red dorsal keels.

	n-pop.	n-pl.	n-rep.	n-obs.	$\bar{x}$ (mm.)	s (mm.)
Length						
w	2	15	—	26	4.9	.7
cf.	2	3	5	14	6.1	—
Width						
w	2	15	—	25	1.7	.2
cf.	2	3	5	14	2.1	—

Fig. 68. Flowers of *Sedum leibergii*, view from above, same cyme as in fig. 67, from northern slope of ridge south of lower Papoose Creek, Seven Devils Mountains, Idaho Co., Ida., cultivated in cold frame, Ithaca, N.Y., May 16, 1962 (x 3.1). Note central flower with eleven petals. Photo by Howard Lyon.

Stamens have yellow filaments and anthers.

	n-pop.	n-pl.	n-rep.	n-obs.	$\bar{x}$ (mm.)	s (mm.)
Epipetalous filaments; adnation—length						
w	2	13	—	22	.5	.1
Anthers—length						
w	2	3	—	4	.9	.07
cf.	1	1	2	3	1.2	—

Nectaries are subquadrate, emarginate, and deep yellow.

	n-pop.	n-pl.	n-rep.	n-obs.	$\bar{x}$ (mm.)	s (mm.)
Length						
w	2	15	—	25	.3	.04
cf.	2	3	5	14	.3	.05
Width						
w	2	15	—	25	.3	.06
cf.	2	3	5	14	.4	—

Carpels are erect at anthesis, then quickly become stellately divergent. They are connate basally and prominently glandular-papillose and yellow.

	n-pop.	n-pl.	n-rep.	n-obs.	$\bar{x}$	s
Length (mm.)						
w	2	13	—	22	3.2	.9
Ovules per ovary (n)						
w	2	13	—	22	4.5	1.3
cf.	2	3	5	14	13.3	1.8

Follicles are widely spreading, brown, and warty papillose.

	n-pop.	n-pl.	n-obs.	$\bar{x}$ (mm.)	s (mm.)
Length					
w	2	26	58	2.8	.9
Width of ventral lips					
w	2	26	56	.3	.06

Seeds are pyriform, finely longitudinally ribbed, and yellow-brown.

	n-pop.	n-pl.	n-obs.	$\bar{x}$ (mm.)	s (mm.)
Length					
w	2	20	51	.8	.09
Diameter					
w	2	20	51	.4	.05

Chromosomes are known only in diploid number.

	n-pl.	g (n)	sp (2n)	Cytologist
Number				
Celilo, Ore.	1	8	—	C. H. Uhl, 1962 (CU)

Variation. *Sedum leibergii* is noteworthy for the variation which occurs in numbers of floral parts. In a sample of 24 flowers, numbers of petals per flower were as follows: 3—1 flower, 4—0, 5—3, 6—17, 7—1, 8—1, and 11—1.

Two populations on the eastern slope of the Seven Devils Mountains, one 160 meters higher in altitude than the other, differed significantly in seven features, on the basis of data from the wild. In cultivation in an experiment in a cold frame in Ithaca, N.Y., four of these seven differences ceased to be significant. The remaining three differences were not tested. Data are included in the following tabular comparison.

	Wild		Cold frame	
Feature	Papoose C.	Shingle C.	Papoose C.	Shingle C.
Rosettes per plant (n)	1.7**	3.6**	—	—
Floriferous stems—length (cm.)	11.5*	9.8*	12$^{n.s.}$	11$^{n.s.}$
Flowers per cyme (n)	17.3*	8.8*	57$^{n.s.}$	63$^{n.s.}$
Pedicels—length (mm.)	0*	.2*	—	—
Nectaries—length (mm.)	.26*	.31*	.25$^{n.s.}$	.33$^{n.s.}$
—width (mm.)	.26*	.33*	.35$^{n.s.}$	.46$^{n.s.}$
Follicles—length (mm.)	2.3*	3.4*	—	—

One result of the comparison of the two populations in the Seven Devils Mountains was to reveal that plants of this species are subject to great environmental modification. Another comparison involved differences between plants in the wild and in the cold frame. The results indicate significant changes in seven features. In this species, although plants do not survive long in cultivation, leaves and flowers are bigger and even ovules are more numerous per ovary in the cold frame. Data about differences in responses follow.

	Wild	Cold frame
Median leaves of floriferous stems		
Length (mm.)*	3.8	8.2
Width (mm.)*	2.1	3.3
Number of floriferous stems per plant**	1.3	3.4
Flowers		
Diameter (mm.)*	10	12
Sepals		
Length (mm.)*	1.5	1.8
Width (mm.) $^{n.s.}$	.9	1.1
Petals		
Length (mm.)*	4.9	6.1
Width (mm.)*	1.7	2.1
Ovules per ovary (n)**	4.5	13.3

Except for fluctuations in number of floral parts and modifications, no unusual variations of *Sedum leibergii* have come to attention. There are no subspecies or variations which need varietal designation.

Nomenclature. *Sedum leibergii* Britton, No. Am. Flora 22 (1):73 (1905). Based on *Sedum divaricatum* Watson.

A synonym is:

1882. *Sedum divaricatum* Watson, Proc. Am. Acad. 17:372 (not *S. divaricatum* Aiton, 1789). Type locality: steep hillside, bluffs of Snake River, Ore., cited as Union County at a time when that county extended eastward to the Snake River, but the site probably is in northeastern present Baker County. Type: W. C. Cusick no. 846 (GH), June 1882. The holotype is a flowering specimen with stem bent to right. Fröderström published a photograph of part of the type sheet, but including the holotype, in Act. Hort. Goth. 10 (App.): pl. 78 (1935).

Distribution. *Sedum leibergii* (fig. 69) occurs primarily on the Columbia Plateau. Outside this geomorphic province, it is known only from the Idaho Batholith, in the drainage of the Salmon River, for example at Shoup, Lemhi Co., Ida. Other records from outside the Columbia Plateau are dubious. The plants occur on cliffs, rocky slopes, and bluffs, mostly on basalt, but sometimes on limestone. Often they are rooted in moss. The altitudinal range is from 265 to 1,190 meters. The climate of localities where *S. leibergii* occurs is wet in winter and early spring, but dry and hot in summer.

Both populations which I studied in detail in 1961 were on the eastern slope of the Seven Devils Mountains. Data for these populations are in the following table.

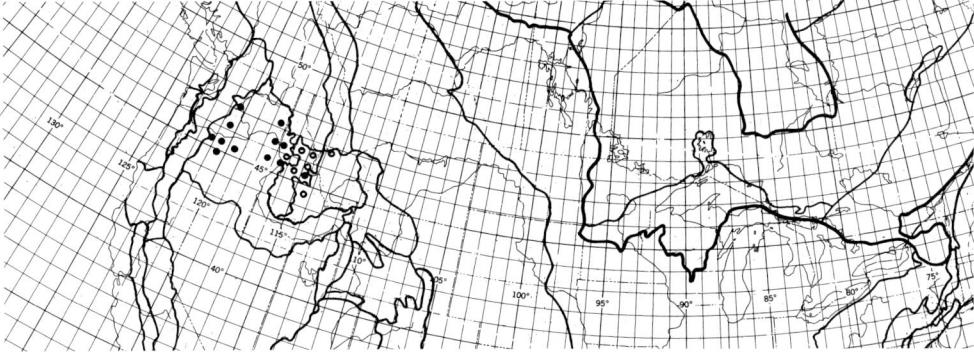

Fig. 69. Known distribution of *Sedum leibergii* (●) and *S. borschii* (○).

Population	Plants (n)	Area of population (m.²)	Density per m.²	Exposure
Papoose Creek	864	143	6	NNW–NW
Shingle Creek	864	2,213	.4	NNW–W

Population	Altitudinal range (m.)	pH, range	pH, tests (n)	Rocks	Oldest record
Papoose Creek	805–810	7.4–7.6	6	basalt	1961
Shingle Creek	970–1,005	7.2–7.8	7	limestone	1961

Sedum leibergii usually grows in open or largely bare areas where other vegetation is sparse. Miscellaneous species compete with it. In the valley of Papoose Creek, it is on a steep bank where the commonest shrubs are *Holodiscus discolor* and *Artemisia rigida*. Species of *Bromus*, *Stellaria*, and *Tonella* are associated herbs. In the drainage of Shingle Creek, *Sedum leibergii* is in openings in forest of *Pseudotsuga menziesii*. Commonest competing herbs there, rated on a scale of 12, are *Bromus japonicus*—6, *Galium aparine*—4, and a species of *Portulaceae*—3.

Specimens collected by the U.S. South Pacific Exploring Expedition under the command of Captain Charles Wilkes, no. 991 (NY), 1838–1842, are indicated as from the north fork of the Columbia River, Ore. If correctly labeled, these specimens are the oldest which have come to attention. In 1879, Joseph Howell found *Sedum leibergii* at Klickitat, Wash.

Reproduction. Plants of *Sedum leibergii*, although with biennial shoots, perpetuate themselves by tiny axillary shoots and rosettes which are subterranean in the summer. These are readily detachable and remain underground until the wet season, when they reach the surface and become the characteristic basal rosettes which eventually give rise to the floriferous stems. In addition, the species reproduces sexually and spreads by means of tiny seeds which average only .8 mm. in length. Although most seeds drop and germinate near to the plants which bear them, some probably travel long distances.

Flowering time is from April to June, depending on altitude and exposure. Fruiting time is July and August. Small bees and flies are the pollinators. These go from flower to flower and plant to plant, thoroughly mixing the genes in a population, as I have observed in the population in the valley of Papoose Creek on June 17, 1961.

Relationships. *Sedum leibergii* appears most closely related to *S. borschii*, which occurs at higher altitudes in the Seven Devils Mountains and elsewhere. Most likely that is an allopolyploid which has *S. leibergii* as one of its parents. More attention to this relationship is available in the account of *S. borschii*.

Other species of *Sedum* with yellow flowers and divergent, gibbous follicles, in the area of the Columbia Plateau and adjacent montane provinces, are *S. stenopetalum* and *S. divergens*. The *S. stenopetalum* has the habit of producing many rosettes on the floriferous stems and has long, narrow, lanceolate-linear or lanceolate leaves. *Sedum divergens* has plump, opposite leaves. Neither species is likely to be confused with *S. leibergii*.

Another relationship is with the perennial *Sedum spathulifolium* of the Pacific Mountain system. The two species are abundantly distinct, but *S. spathulifolium* may have originated long ago through hybridization, and one of the parents may have been an ancestral species, with 8 pairs of chromosomes, which may likewise have been the progenitor of the modern advanced *S. leibergii*. Besides the obvious differences in vegetative features between *S. leibergii* and *S. spathulifolium*, these species also differ in dimensions of floral parts, fruits, and seeds, as shown in the following table.

	S. leibergii	S. spathulifolium
Cold frame		
Sepals—length (mm.)**	1.9	2.7
Petals—length (mm.)n.s.	6.1	6.9
Nectaries—length (mm.)**	.3	1.1
Nectaries—width (mm.)n.s.	.4	.5
Wild		
Fruits—length (mm.)**	2.8	4.5
Seeds—length (mm.)**	.8	1.1

References: none.

15. *Sedum borschii**** (figs. 69–74)

Sedum borschii (fig. 70) is a perennial stonecrop with offsets prominent at time of flowering, yellow petals, and widely divergent, two-lipped follicles. The obovate or elliptical leaves readily separate it from *S. stenopetalum* and *S. radiatum*, and the leaves, as well as the spreading follicles, distinguish it from *S. lanceolatum*. The primary rosettes of *S. borschii*, prominent at time of flowering, with obovate or elliptical leaves, make separation from *S. leibergii* easy. Lack of stout rootstocks and smaller dimensions of leaves make confusion with *S. spathulifolium* unlikely.

Description. Sample: 33 plants from one site in each of the following three geomorphic provinces: mountains and basins of southwestern Montana and adjacent Idaho,

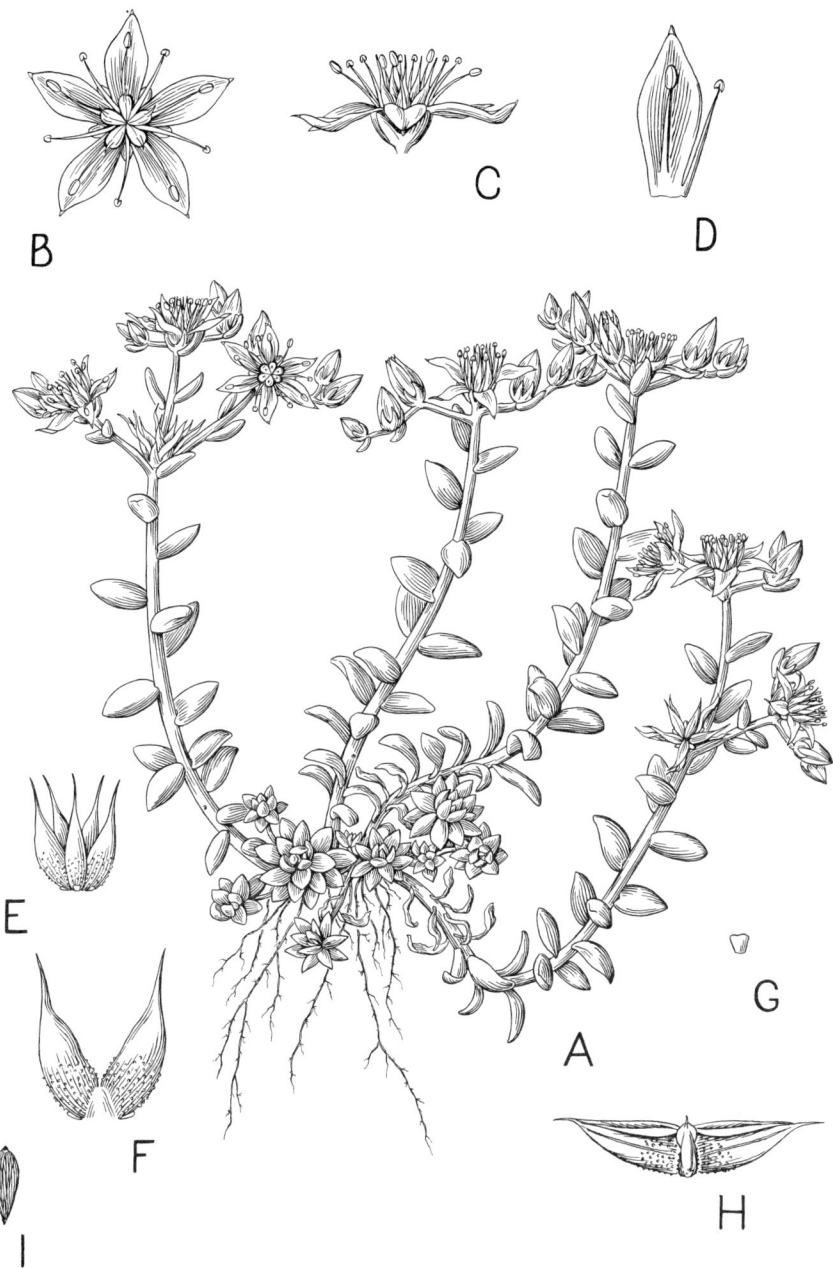

Fig. 70. Plant of *Sedum borschii* from ridge south of Papoose Creek, Seven Devils Mountains, Idaho Co., Ida. A. Habit sketch (x .88). B. Flower from above (x 1.8). C. Flower from side (x 1.8). D. Petal and two stamens (x 2.6). E. Carpels (x 2.6). F. Two carpels, torus, and nectaries (x 3.5). G. Nectary (x 4.4). H. Fruits from side (x 3.5). I. Seed (x 8.8).

Idaho Batholith, and the eastern border of the Columbia Plateau. Choice of populations was on the basis of judgment, but selection of clusters in the population in the drainage of Papoose Creek was by means of a table of random numbers, as was selection of all plants in those clusters. Information about the sample is summarized in the following table in which N_u = total number of plants in a sampling unit, and N_c = number of plants in sampled clusters.

Locality	N_u	N_c	n	Plants studied n/N_u	n/N_c	Condition when collected fl. (n)	fr. (n)
Rattlesnake Creek	?	?	3	?	?	0	3
Windy Ridge	31	31	2	.06	.06	0	2
Papoose Creek	3,096	576	28	.01	.05	20	8
Totals	>3,127	>607	33			20	13

Two experiments, both in cold frames at Ithaca, N.Y., provided valuable information about variation of *Sedum borschii*. A randomized complete block experiment, which yielded data in the spring of 1962, included 6 plants of *S. borschii*—3 from Papoose Creek, 2 from Rattlesnake Creek, and 1 from Windy Ridge, a hybrid of *S. borschii* × *stenopetalum*, and 6 plants of *S. leibergii*. Another experiment of completely random design, set up in autumn 1962, included 13 plants of *S. borschii*—9 from Papoose Creek, 2 from Rattlesnake Creek, and 2 from Windy Ridge. Data obtained in 1971 from 5 plants from Papoose Creek and 1 each from Rattlesnake Creek and Windy Ridge are cited in the description. Cited data from the cold frame (cf.) are for 1971.

In the description, localities are abbreviated as PC = Papoose Creek, RC = Rattlesnake Creek, and WR = Windy Ridge.

Plants of *Sedum borschii* (fig. 71) are perennial herbs with slender, decumbent stems and fibrous roots. Axillary shoots from the lower part of the stem bear rosettes of leaves which are a prominent feature of the plants. In their second year, rosettes give rise to erect, floriferous stems which are green or red. In dried condition, the old fructiferous stems appear angulate.

	n-pl.	n-rep.	n-obs.	$\bar{x}$	s
Plants—diameter (cm.)					
RC—cf.	1	1	1	14	—
WR—cf.	1	1	1	7	—
PC—cf.	5	6	6	9.5	5
Primary rosettes (n)					
PC	19	—	42	1.5	2.1
RC—cf.	1	1	1	38	—
WR—cf.	1	1	1	25	—
PC—cf.	5	6	6	11	2.4
Primary rosettes—diameter (cm.)					
RC—cf.	1	1	3	.8	—
WR—cf.	1	1	3	.9	—
PC—cf.	5	6	18	.8	.1

Fig. 71. Plant of *Sedum borschii* from cliff on hillside along Blackfoot River, Missoula Co., Mont., received from Frank H. Rose, June 6, 1967 (× 1.2). Photo by Howard Lyon.

	n-pl.	n-rep.	n-obs.	$\bar{x}$	s
Floriferous stems per plant (n)					
PC	19	—	19	8.6	12
RC—cf.	1	1	1	47	—
WR—cf.	1	1	1	5	—
PC—cf.	5	6	6	19	13
Floriferous stems—height (cm.)					
PC	19	—	41	7.4	1.7
Floriferous stems—length (cm.)					
RC—cf.	1	1	3	5.8	—
WR—cf.	1	1	3	4.7	—
PC—cf.	5	6	18	5.9	.9

Leaves of rosettes are divergent, sessile, obovate or elliptical, rarely lanceolate, plane ventrally and convex dorsally, obtuse, finely papillose marginally and apically, and green without a bloom. Sometimes they are speckled with red. Leaves of the floriferous stems are elliptic-oblong or oblanceolate, obtuse, papillose, and pale green, occasionally flushed with red or completely red. Measurements of leaves of floriferous stems are of median leaves.

	n-pl.	n-rep.	n-obs.	$\bar{x}$ (mm.)	s (mm.)
Leaves of primary rosettes—length					
PC	15	—	30	7.5	1.8
RC—cf.	1	1	3	4.8	—

	n-pl.	n-rep.	n-obs.	x̄ (mm.)	s (mm.)
Leaves of primary rosettes—length					
WR—cf.	1	1	3	5.2	—
PC—cf.	5	6	18	4.7	.8
Leaves of primary rosettes—width					
RC—cf.	1	1	3	3	—
WR—cf.	1	1	3	3.4	—
PC—cf.	5	6	18	2.6	.5
Leaves of primary rosettes—thickness					
RC—cf.	1	1	3	1.5	—
WR—cf.	1	1	3	1.5	—
PC—cf.	5	6	18	1.1	.2
Leaves of floriferous stems—length					
PC	10	—	28	7.1	1.7
RC—cf.	1	1	2	4	—
WR—cf.	1	1	3	4	—
PC—cf.	3	4	9	4.6	—
Leaves of floriferous stems—width					
PC	10	—	27	3	.9
RC—cf.	1	1	2	3	—
WR—cf.	1	1	3	3	—
PC—cf.	3	4	9	2.8	—
Leaves of floriferous stems—thickness					
RC—cf.	1	1	2	1.7	—
WR—cf.	1	1	3	1.6	—
PC—cf.	3	4	9	1.5	—

Inflorescences are terminal cymes, mostly two or three-parted. Floral bracts are similar to leaves of floriferous stems, but smaller.

	n-pl.	n-rep.	n-obs.	x̄ (n)	s (n)
Cincinni per cyme					
RC—cf.	1	1	3	1.3	—
WR—cf.	1	1	3	2	—
PC—cf.	5	6	18	2.4	.3
Flowers per cyme					
PC	11	—	28	7.3	5.1
RC—cf.	1	1	3	3.3	—
WR—cf.	1	1	3	4.7	—
PC—cf.	5	6	18	6.2	1.2

Flowers are either sessile or very short-pedicellate.

	n-pl.	n-rep.	n-obs.	x̄ (mm.)	s (mm.)
Pedicels—length					
RC—cf.	1	1	3	.1	—
WR—cf.	1	1	3	.1	—
PC—cf.	5	6	13	.1	.1
Flowers—diameter					
PC	19	—	45	13.2	1.9
RC—cf.	1	1	3	13.3	—
WR—cf.	1	1	3	10	—
PC—cf.	5	6	14	11.7	—

	n-pl.	n-rep.	n-obs.	$\bar{x}$ (mm.)	s (mm.)
Torus—length					
PC	2	—	12	1	—
RC—cf.	1	1	3	0	—
WR—cf.	1	1	3	0	—
PC—cf.	5	6	14	0	0

Sepals are erect, ovate, acute or blunt, and yellow-green.

	n-pl.	n-rep.	n-obs.	$\bar{x}$ (mm.)	s (mm.)
Length					
PC	2	—	12	2.3	—
RC—cf.	1	1	3	1.9	—
WR—cf.	1	1	3	1.9	—
PC—cf.	5	6	14	2	—
Width					
PC	2	—	12	1.5	—
RC—cf.	1	1	3	1.6	—
WR—cf.	1	1	3	1.5	—
PC—cf.	5	6	14	1.4	—

Petals are widely spreading, lanceolate-elliptical, acute and mucronate-appendaged, and aureolin.

	n-pl.	n-rep.	n-obs.	$\bar{x}$	s
Number per flower					
RC—cf.	1	1	3	6.6	—
WR—cf.	1	1	3	5	—
PC—cf.	5	6	14	5.1	—
Length (mm.)					
PC	19	—	45	6.5	—
RC—cf.	1	1	3	6.3	—
WR—cf.	1	1	3	5.1	—
PC—cf.	5	6	14	5.7	—
Width (mm.)					
PC	19	—	45	2.5	—
RC—cf.	1	1	3	2.9	—
WR—cf.	1	1	3	2.3	—
PC—cf.	5	6	14	2.3	—

Stamens have yellow filaments and anthers.

	n-pl.	n-rep.	n-obs.	$\bar{x}$ (mm.)	s (mm.)
Filaments—length					
RC—cf.	1	1	3	4.9	—
WR—cf.	1	1	3	3.4	—
PC—cf.	5	6	14	3.8	—
Epipetalous filaments; adnation—length					
PC	2	—	12	.9	—
RC—cf.	1	1	3	.8	—
WR—cf.	1	1	3	.4	—
PC—cf.	5	6	14	.7	.1

Anthers—length

	n-pl.	n-rep.	n-obs.	x̄ (mm.)	s (mm.)
PC	1	—	1	1.4	—
WR—cf.	1	1	1	1	—
PC—cf.	1	2	4	1.1	—

Nectaries are subquadrate or quadrate, sometimes erose at apex, and deep yellow or orange, rarely pale yellow or translucent.

	n-pl.	n-rep.	n-obs.	x̄ (mm.)	s (mm.)
Length					
PC	2	—	12	.5	—
RC—cf.	1	1	3	.4	—
WR—cf.	1	1	3	.5	—
PC—cf.	5	6	14	.4	.1
Width					
PC	2	—	12	.5	—
RC—cf.	1	1	3	.5	—
WR—cf.	1	1	3	.5	—
PC—cf.	5	6	14	.4	.05

Carpels are connate basally, divergent, and yellow or greenish yellow. The ovaries are papillose or papillose-puberulent.

	n-pl.	n-rep.	n-obs.	x̄	s
Length (mm.)					
PC	2	—	12	4.4	—
RC—cf.	1	1	3	5	—
WR—cf.	1	1	3	3.2	—
PC—cf.	5	6	14	3.8	—
Length of cohesion (mm.)					
RC—cf.	1	1	3	1.2	—
WR—cf.	1	1	3	.8	—
PC—cf.	5	6	14	.8	—
Ovules per ovary (n)					
PC	2	—	12	7.3	—
RC—cf.	1	1	3	9.3	—
WR—cf.	1	1	3	6.3	—
PC—cf.	5	6	14	5.1	—

Fruits are widely divergent and brown, with prominent ventral lips. The valves are papillose or papillose-puberulent.

	n-pl.	n-rep.	n-obs.	x̄ (mm.)	s (mm.)
Length					
RC	3	—	9	3.4	.2
WR	2	—	6	2.6	.6
PC	9	—	20	3.7	.5
RC—cf.	1	1	3	3.9	—
WR—cf.	1	1	3	2.5	—
PC—cf.	5	6	18	3.3	.3
Width of ventral lips					
PC	9	—	19	.4	.1

Seeds are pyriform, finely longitudinally ribbed, and pale or dark brown.

	n-pl.	n-rep.	n-obs.	x̄ (mm.)	s (mm.)
Length					
RC	3	—	9	.9	.1
WR	2	—	6	.8	0
PC	9	—	21	1	.08
RC—cf.	1	1	3	1	—
WR—cf.	1	1	3	1	—
PC—cf.	4	5	15	1.1	.04
Diameter					
RC	3	—	9	.4	.07
WR	2	—	6	.4	0
PC	9	—	21	.4	.03
RC—cf.	1	1	3	.5	—
WR—cf.	1	1	3	.4	—
PC—cf.	4	5	15	.4	0

Chromosomes are difficult to count. Mastrangelo, who has devoted the most attention to the problem, thinks that the plants may be segmental alloploids. Some plants had a ring of 4 chromosomes at metaphase I and also occasional trivalent and univalent configurations. Although anaphase I usually is normal, lagging chromosomes, micronuclei, and what appear to be bridges occur. A guess is that *S. borschii* usually is hexaploid with a haploid number of 24 chromosomes.

	n-pl.	g (n)	sp (2n)	Cytologist
Number				
Blackfoot River	1	—	32	I. Mastrangelo, 1971
Rattlesnake Creek	1	—	50	I. Mastrangelo, 1971
Windy Ridge	1	—	45–54	I. Mastrangelo, 1971
Papoose Creek	3	18–35?	—	Charles Horn, 1966
Papoose Creek	4	—	43–51	I. Mastrangelo, 1971

Variation. Because populations of *Sedum borschii* are disjunct, a possibility exists for differences to develop among localities. For comparisons of populations, data from experiments in which plants from all localities are grown under similar conditions are most valuable. Despite the expectation of differences, data from experiments indicate that most differences among populations lack significance. Exceptions are three features which differ highly significantly on the basis of analysis of variance of data from the experiment in 1971. The significant differences are contrasted in the following table, in which values for features are means.

	Rattlesnake Creek	Windy Ridge	Papoose Creek
n-pl.	1	1	5
n-rep.	1	1	6
Primary rosettes (n)	38	25	11
Pedicels—length (mm.)	.1	.13	.1
Petals (n)	6.6	5	5.1

A plant from along the Blackfoot River (fig. 71, above), collected by Frank Rose in 1967, is interesting because it is from the easternmost known population. Also, on the basis of one count of chromosomes, it may be a tetraploid. Since this plant has not flowered in experiments in Ithaca, comparison with plants from other localities still awaits attention.

Expressions of plants in experiments in cold frames in Ithaca suggested that great differences had occurred after cultivation for nine years. Comparison of the results, however, by means of analysis of variance, revealed that most differences may be attributed to chance. Except for number of floriferous stems, the features which changed significantly were in the direction of reduction. Data in the following table are from the same plants in both years. Asterisks in the left column, after features, indicate significant differences between years. Asterisks after figures indicate significant differences among plants.

Feature	1962	1971
Plants (n)	5	5
Floriferous stems per plant (n)	9	27**
Median leaves of floriferous stems—length (mm.)	8.8	4.4**
Flowers per cyme (n)	5.9	6.2
Flowers—diameter (mm.)	15.4	12.1
Sepals—length (mm.)*	2.3	2
—width (mm.)	1.6	1.4
Petals—length (mm.)	7.5	5.8*
—width (mm.)	3	2.4*
Anthers—length (mm.)	1.4	1.1
Nectaries—width (mm.)*	.5	.4
Ovules per ovary (n)**	11.5	6.1

Occasional flowers (fig. 72) of *Sedum borschii* have more than the usual 5 petals, as well as additional numbers of other floral parts. The following frequency distribution indicates the result of counting petals in a sample of 34 flowers.

Petals per flower (n)	Flowers (n)	Flowers %	Cumulative %
5	27	79	79
6	5	15	94
7	0	0	94
8	0	0	94
9	1	3	97
10	1	3	100
Total	34		

Available evidence, from both the field and experiments, suggests great similarity among the three disjunct populations which I have studied. Little basis exists for naming subspecies or even varieties.

Nomenclature. *Sedum* **borschii** (R. T. Clausen) comb. nov., fundatum super *Sedum leibergii* Britton var. *borschii* R. T. Clausen, Cact. Succ. Jour. 16:8 (1944).

A synonym is:

1944. *Sedum leibergii* Britton var. *borschii* R. T. Clausen, Cact. Succ. Jour. 16:8.

Fig. 72. 7 and 8-merous flowers of *Sedum borschii* on ridge south of Papoose Creek, Seven Devils Mountains, Idaho Co., Ida., June 26, 1961 (x 1.7).

Type locality: unknown, possibly in Idaho. Type: R. T. Clausen 43–52 (CU), May 28, 1943, originally from Fred J. Borsch, Maplewood, Ore.

Distribution. The distributional area (fig. 69) of *Sedum borschii* is in the northern half of the Idaho Batholith, in the western part of the mountains and basins of southwestern Montana and adjacent Idaho, and in the Seven Devils Mountains at the eastern edge of the Columbia Plateau. The plants occur on cliffs and rocky slopes, on a variety of rocks. Eleven different populations, four of them discovered by Frank H. Rose of Missoula, Mont., have come to my attention. Data for the three populations which I have studied, organized in the following table, provide an impression of the situations in which plants occur.

Population	Plants (n)	Area of population (m.^{2})	Density per m.^{2}	Exposure
Rattlesnake Creek	?	>90,000	?	W
Windy Ridge	31	?	?	N
Papoose Creek	3,096	4,987	.62	NW–NE

Population	Altitudinal range (m.)	pH, range	pH, tests (n)	Rocks	Oldest record
Rattlesnake Creek	1,270–1,570	4.8–5.6	3	sandstone	1939
Windy Ridge	2,120	4.8–5	2	schist	~1960
Papoose Creek	1,670–1,775	6.4–7.6	21	limestone	1940

Just as the rocks are different at each site which I investigated, so likewise the other vegetation is different. On the slope along Rattlesnake Creek northeast of Missoula, the plants are on a rock slide, growing in moss and lichens, including species of *Cladonia* and *Peltigera*. The only other vascular plant there is *S. stenopetalum*. On Windy Ridge, the *Sedum* is shaded by trees of *Tsuga mertensiana*. Principal competitors are a low *Vaccinium* and a species of *Synthyris*. On the ridge south of Papoose Creek, commonest competitors, rated on a scale of 21, are *Mitella stauropetala*—6, *Antennaria racemosa*—6, *Sedum stenopetalum*—6, *Heuchera* sp.—6, and *Poa secunda*—5. Fourteen plants in the samples were growing in moss. Large trees of *Pseudotsuga* shaded the *Sedum*.

Reproduction. Vegetative reproduction by means of offsets is an obvious, important means of spread of *Sedum borschii*. In addition, at least some of the seeds produced are viable and grow into seedlings.

Flowering time is May, June, and July. On the slope south of Papoose Creek, in the Seven Devils Mountains, the first plants came into flower in 1961 on June 26. In Ithaca, plants in the cold frame have flowered as early as late April. They reach the height of anthesis in late May and early June.

Small bumblebees, other small bees, and species of *Diptera* go from flower to flower and cross-pollinate *Sedum borschii*. Since *S. stenopetalum* sometimes is associated with *S. borschii* and has flowers at the same time, opportunity for hybridization is great.

Relationships. The obvious closest relatives of *Sedum borschii*, on the basis of gross morphology and ecology, are *S. leibergii* and *S. stenopetalum*. In the Seven Devils Mountains, in the drainage of Papoose Creek, *S. borschii* and *S. leibergii* occur within 12 kilometers of each other and at altitudes only 860 meters apart. Both species have papillose ovaries, divergent follicles, and pyriform, finely ribbed seeds. They differ in their mode of growth. Offsets of *S. leibergii* are tiny and subterranean at the time when the plants have flowers. The floriferous stems of *S. leibergii* have long-petiolate, oblanceolate, basal leaves. In contrast, the offsets of *S. borschii* are prominent at the time when the plants have flowers, and none of the leaves are long-petiolate. Other differences are indicated in the following table.

	S. leibergii	S. borschii
Offsets—diam. (mm.)**	5	8
Prim. ros.—lvs.—l. (mm.)n.s.	9.5	4.8
Prim. ros.—lvs.—petioles—l. (mm.)*	4.3	0
Floriferous stems—length (cm.)**	12	6
Flowers per cyme (n)*	59	5.9
Sepals—length (mm.)*	1.9	2.3
—width (mm.)n.s.	1.1	1.6
Petals—length (mm.)n.s.	6.1	7.5
—width (mm.)*	2.1	3
Nectaries—length (mm.)*	.3	.4
—width (mm.)n.s.	.4	.5

NATIVE SPECIES   301

Sedum leibergii occurs in situations which are too dry and hot in summer for the growth of trees. *Sedum borschii* grows at sites where trees such as *Pseudotsuga menziesii* and *Tsuga mertensiana* make good growth.

In the preceding, comparative table, offsets refer to the stems and rosettes which are produced at the time of flowering and afterward. Primary rosettes refer to the rosettes from which the floriferous stems arise.

Except for diameter of offsets and dimensions of leaves of rosettes, data are from the experiment in the cold frame, 1962 results.

A relationship of *Sedum borschii* with *S. stenopetalum* is indicated by at least one hybrid (figs. 73 and 74) on the ridge south of Papoose Creek. This plant had well-developed ovules, but did not produce seeds. It appeared intermediate between the

Fig. 73. Sedum borschii × stenopetalum ssp. stenopetalum from ridge south of Papoose Creek, Seven Devils Mountains, Idaho Co., Ida., cultivated in cold frame, Ithaca, N.Y., May 23, 1962 (× .75). Photo by Howard Lyon.

Fig. 74. Plants of *Sedum borschii* from ridge south of Papoose Creek, Seven Devils Mountains, Idaho Co., Ida. (left); *S. borschii* × *stenopetalum* ssp. *stenopetalum* from same site (center); and *S. stenopetalum* ssp. *stenopetalum* from Morrison Ridge south of Shingle Creek, eastern slope of Seven Devils Mountains (right); cultivated in cold frame, Ithaca, N.Y., Nov. 24, 1961 (x .7). Photo by Howard Lyon.

two species. Differences of the hybrid from *S. stenopetalum* ssp. *stenopetalum* included the following features:

1. Leaves of rosettes broader, 3 mm. wide, papillose, and pale green.
2. Leaves of floriferous stems deciduous, with the bases of those which persist less well developed and less scarious.
3. Spurs of leaves of floriferous stems shorter.
4. Buds and flowers substantially larger.

From *Sedum borschii*, the hybrid differed as follows:

1. Leaves of basal rosettes longer, 13.1 mm. long, and relatively narrower, that is, lanceolate-elliptical.
2. Taller floriferous stems.
3. More persistent leaves at the base of the floriferous stems and these with short spurs.
4. Larger flowers.
5. Vegetative rosettes in axils of leaves of floriferous stems.

Another relationship of *Sedum borschii* may be with *S. lanceolatum*. That has yellow flowers and erect follicles and occurs at higher elevations.

In the summer of 1971, Mastrangelo attempted to obtain information about the origin of *Sedum borschii*. Although her preparations from root tips did not yield precise counts, her data suggest that the species is of allopolyploid origin. Rings of chromosomes and other cytological irregularities suggest a hybrid origin. The question about the parental species remains unresolved. The best guess for one parent appears to be *S. leibergii*. Possibilities for the other parent are either *S. lanceolatum* or

NATIVE SPECIES 303

S. stenopetalum or both in the event that more than one other species is involved.

If *Sedum borschii* is of hybrid origin, its inclusion within any other potential parental species is wrong. For that reason, full specific status is necessary. Although it has a restricted range, it maintains itself in several established populations and is easy to distinguish from other, more widespread species.

Reference. Clausen, R. T. (1944).

16. *Sedum divergens**** (figs. 75 and 76)

Thick, almost globular, opposite leaves, yellow flowers with spreading petals, and divergent follicles distinguish *Sedum divergens* (fig. 75) from all other species of *Sedum*

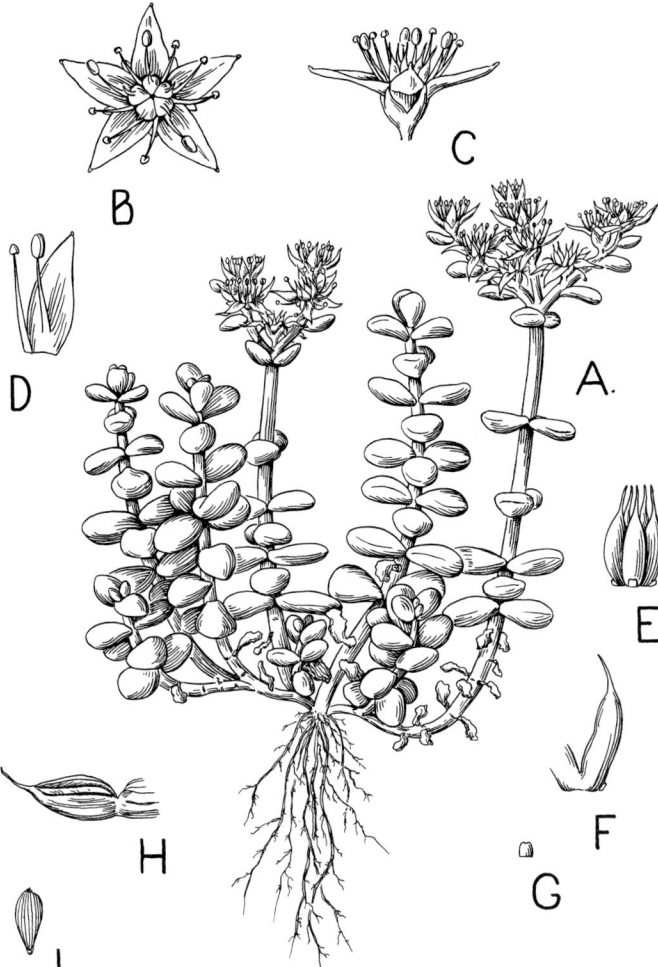

Fig. 75. Plant of *Sedum divergens* from valley of Ingalls Creek, Wenatchee Mountains, Chelan Co., Wash. A. Habit sketch (x .9). B. Flower from above (x 2.3). C. Flower from side (x 2.3). D. Petal and two stamens (x 2.7). E. Carpels (x 2.7). F. Two carpels, torus, and nectary (x 3.6). G. Nectary (x 4.5). H. Fruit from side, one whole fruit to left and base of another to right (x 3.6). I. Seed (x 9).

in western North America. The plants are glabrous throughout. The stems usually are pinkish or reddish. *Sedum debile* is similar in having opposite leaves, as well as the same number and kind of chromosomes, but differs in having the leaves pale glaucous green, the petals connate basally, and the follicles erect. *Sedum stahlii*, of the eastern Trans-Mexican Volcanic Belt and adjacent Sierra Madre del Sur, also has opposite leaves, but they are both terete and puberulent. In addition, the plants are subshrubs.

Description. Sample: 17 plants—16 from the Cascade Mountains (Wenatchee Mountains—13, Bald Mountain—3) and 1 from Groghan Hole near Trinity Summit in the Klamath Mountains. The 13 plants from the Wenatchee Mountains are from an optimum stratum. Ten plants, two random samples, each of 5 plants, an 8% sample, are from a population in the valley of Ingalls Creek. The other 3 plants were authoritatively selected, 2 from the population in the valley of Ingalls Creek and 1 from the southern base of Mount Stuart. Two of the 3 plants from Bald Mountain were randomly selected from a sampling unit also of random selection. The third was the only accessible plant in a randomly selected subunit. Means and standard deviations in the description are weighted. Abbreviations are cf. for cold frame and gh. for greenhouse.

Plants of *Sedum divergens* are perennial herbs with fibrous roots and branched, decumbent stems which root freely at the nodes and are green or reddish. The floriferous stems are either erect or decumbent and usually simple, but sometimes branched. The primary stems give rise to offsets which can produce new plants if separated from the main stem. Plants attain a maximum height of about 1 dm.

	n-pop.	n-pl.	n-obs.	$\bar{x}$ (cm.)	s (cm.)
Plants—diameter					
Wenatchee Mts.—cf.	1	3	9	13.4	2
Bald Mt.—cf.	1	3	9	14.1	0
Floriferous stems—length					
Wenatchee Mts.	2	13	29	9.4	2.5
Trinity Summit	1	1	2	9.5	—

Leaves of the sterile shoots are mostly opposite, suborbicular or spatulate-obovate, thick and turgid, but flat on ventral surface, sessile and slightly subcordate, clasping basally, rounded at apex, glabrous, and green or reddish green. The leaves are reduced in size upward on the stems. Those of the floriferous stems are similar, but usually larger. Rarely, the leaves are 3 or 4 per node.

	n-pop.	n-pl.	n-rep.	n-obs.	$\bar{x}$ (mm.)	s (mm.)
Leaves of sterile shoots—length						
Wenatchee Mts.	2	13	—	64	8.9	1.3
Bald Mt.	1	3	—	8	6.4	—
Trinity Summit	1	1	—	2	4.2	—
Wenatchee Mts.—cf.	1	2	4	16	4.1	—
Bald Mt.—cf.	1	1	2	12	4.6	—
Leaves of sterile shoots—width						
Wenatchee Mts.	2	13	—	64	6.1	.7
Bald Mt.	1	3	—	8	5.4	—
Trinity Summit	1	1	—	2	4.7	—

	n-pop.	n-pl.	n-rep.	n-obs.	x̄ (mm.)	s (mm.)
Leaves of sterile shoots—width						
Wenatchee Mts.—cf.	1	2	4	16	5.2	—
Bald Mt.—cf.	1	1	2	12	4.1	—
Leaves of sterile shoots—thickness						
Wenatchee Mts.	2	13	—	63	2.9	.4
Bald Mt.	1	3	—	8	2.4	—
Trinity Summit	1	1	—	2	2.3	—
Wenatchee Mts.—cf.	1	2	4	16	1.7	—
Bald Mt.—cf.	1	1	2	12	1.7	—
Leaves of floriferous stems—length						
Wenatchee Mts.	2	5	—	12	9.5	.7
Bald Mt.	1	1	—	2	6	—
Leaves of floriferous stems—width						
Wenatchee Mts.	2	5	—	12	7.3	.4
Bald Mt.	1	1	—	2	5.5	—
Leaves of floriferous stems—thickness						
Wenatchee Mts.	2	5	—	12	4.1	.6
Bald Mt.	1	1	—	2	2.8	—

Inflorescences are terminal cymes of either 2 or 3 cincinni, 1.5–4.5 cm. across, with elliptical, blunt floral bracts.

	n-pop.	n-pl.	n-obs.	x̄ (n)	s (n)
Flowers per cyme					
Wenatchee Mts.	2	13	29	9.7	3.2
Bald Mt.	1	2	9	7	—
Trinity Summit	1	1	1	7	—

Flowers are either pedicellate or sessile, usually 5-merous, rarely 6- or 7-merous. The pedicels are enlarged at the receptacles.

	n-pop.	n-pl.	n-obs.	x̄ (mm.)	s (mm.)
Pedicels—length					
Wenatchee Mts.	2	7	20	2.8	1.9
Trinity Summit	1	1	2	2.5	—
Flowers—diameter					
Wenatchee Mts.	2	5	24	11.4	1.4

Sepals are ovate, obtuse, and green. They are connate basally for .4–1.2 mm.

	n-pop.	n-pl.	n-obs.	x̄ (mm.)	s (mm.)
Length					
Wenatchee Mts.	2	5	25	2.2	.5
Trinity Summit	1	1	1	2.8	—
Length of connate portion					
Wenatchee Mts.	2	3	7	.7	0
Trinity Summit	1	1	1	1.2	—
Width					
Wenatchee Mts.	2	5	25	1.7	.2
Trinity Summit	1	1	1	1.6	—

Petals are elliptic-lanceolate, acute or obtuse, mucronate-appendaged, and yellow. Usually they are 5 per flower, rarely 6. Sometimes they are minutely connate basally.

	n-pop.	n-pl.	n-obs.	$\bar{x}$ (mm.)	s (mm.)
Length					
Wenatchee Mts.	2	5	24	6.1	.1
Trinity Summit	1	1	1	6	—
Width					
Wenatchee Mts.	2	5	24	2.4	.3
Trinity Summit	1	1	1	2.5	—

Stamens have yellow filaments and anthers.

	n-pop.	n-pl.	n-obs.	$\bar{x}$ (mm.)	s (mm.)
Epipetalous filaments; adnation—length					
Wenatchee Mts.	2	5	24	1	.2
Trinity Summit	1	1	1	.4	—
Anthers—length					
Wenatchee Mts.	1	3	4	1.2	0

Nectaries are subquadrate, truncate, slightly emarginate, and yellow.

	n-pop.	n-pl.	n-obs.	$\bar{x}$ (mm.)	s (mm.)
Length					
Wenatchee Mts.	2	5	24	.4	.03
Trinity Summit	1	1	1	.3	—
Width					
Wenatchee Mts.	2	5	24	.4	.1
Trinity Summit	1	1	1	.4	—

Pistils are erect at time of anthesis, gibbous ventrally, and yellow. They are connate basally and become divergent with age.

	n-pop.	n-pl.	n-obs.	$\bar{x}$	s
Length (mm.)					
Wenatchee Mts.	2	5	24	4.6	.4
Trinity Summit	1	1	1	6	—
Length of cohesion (mm.)					
Wenatchee Mts.	2	4	26	1.2	.1
Ovules per ovary (n)					
Wenatchee Mts.	2	9	32	25	3.6

Fruits are widely spreading and firmly connate below. When pulled apart, the walls break, a clear indication of cohesion.

	n-pop.	n-pl.	n-obs.	$\bar{x}$ (mm.)	s (mm.)
Length					
Wenatchee Mts.	1	7	11	4.2	.3
Bald Mt.	1	2	6	3.1	—

Seeds are pyriform, finely longitudinally ribbed, with two or three narrowly winged ridges, glabrous, and yellow-brown.

	n-pop.	n-pl.	n-obs.	$\bar{x}$ (mm.)	s (mm.)
Length					
Wenatchee Mts.	1	7	26	1	.06
Bald Mt.	1	2	6	1.1	—
Diameter					
Wenatchee Mts.	1	7	26	.4	.02
Bald Mt.	1	2	6	.4	—

Chromosomes are of unequal size. Some are elongate and markedly two-armed. Others are about as wide as long.

	n-pop.	n-pl.	g (n)	sp (2n)	Cytologist
Cascade Mts.					
Wenatchee Mts.	1	2	—	16	L. & R. Benjamin, 1970
Bald Mt.	1	2	—	16	L. & R. Benjamin, 1970
Mt. Herman, Whatcom Co., Wash.	1	1	—	16	Hollingshead (1942)
Near Puget Sound	1	1	—	16	Hollingshead (1942)
Queen Charlotte Islands	2	2	8	16	Taylor and Mulligan (1968)

	n-pl.	$\bar{x}$ (genome length) (μ)	$\bar{x}$ (chromosome length) (μ)	s (μ)	Cytologist
Length of sporophytic chromosomes					
Wenatchee Mts.	2	40	2.7	0	L. & R. Benjamin, 1970
Bald Mt.	1	44.5	3	—	L. & R. Benjamin, 1970

Variation. *Sedum divergens* appears to be a monotypic species. Variations are minor. For example, variation is so slight that random samples of 3 plants from a population are adequate to detect a difference of 1 mm. in width of leaves, P = .05.

Variation within plants includes robustness, pigmentation of the stems and leaves, number of leaves per node, and number of floral parts per whorl. Although the leaves commonly are decussately opposite, one plant from the valley of Ingalls Creek, in cultivation at Ithaca, had two stems with the leaves 3 per node and one stem with the leaves 4 per node. The petals usually are 5 per flower, rarely 6. Likewise, carpels normally are 5, but occasionally are 6 or 7 per flower.

Plants from the Wenatchee Mountains appear to be larger than those from Bald Mountain. On the basis of the Wilcoxon rank sum test, primary shoots of wild plants in the Wenatchee Mountains have highly significantly longer leaves than those on Bald Mountain, 8.9 versus 6.4 mm. This difference disappears when plants are cultivated under similar conditions. Another impression, gained in the field and still untested, is that plants in the Wenatchee Mountains have larger leaves on the floriferous stems and also larger follicles. If such differences are real, they indicate the existence of small genetic dissimilarities between populations on the eastern and western sides of the Cascade Mountains. Consistent with this idea, a plant from Mount Rainier, on the western side, in cultivation in Ithaca, is almost identical with plants from Bald Mountain.

In Ithaca, *Sedum divergens* has survived best in cold frames. In the greenhouse, it does not flower and soon dies. Likewise, its survival in the garden is poor. One experiment of completely random design, in the cold frame, included 3 plants from the Wenatchee Mountains and 3 plants from Bald Mountain, each with 3 replicates. All replicates survived for a year. Growth was sufficiently similar that any small differences in diameter of plants, either between populations or among plants, could be attributed to chance. After five years of cultivation, 3 replicates from the Wenatchee Mountains and 5 replicates from Bald Mountain still survived. Differences in diameters, whether between populations or among plants, continued to lack significance.

No clear basis is available for recognizing subspecies of *Sedum divergens*. Because the present distributional area is geologically young or was covered by glacial ice in recent time, the species either originated since the ice ages or has invaded its present range from a refuge somewhere else. Lack of subspecies may be a point in favor of a recent origin.

Nomenclature. *Sedum divergens* Watson, Proc. Am. Acad. 17:372 (1882). Type locality: Mount Adams in the Cascade Mountains, Wash. Type: W. N. Suksdorf 369 (GH), September 1880. The holotype is the specimen in the upper right part of the type sheet. This is the first collection cited by Watson. The original identification was as *Sedum debile* Watson?, but the carpels are divergent.

Lack of synonyms is in agreement with the idea of a monotypic species. *Sedum umbellatum*, listed as a synonym by Fröderström (1935), is an herbarium name which lacks valid publication. *Sedum divergens* Greene, Pittonia 1:154 (1888) is a later homonym.

The oldest known collection was by the Wilkes Expedition (NY), in the state of Washington, in 1841.

Distribution. The geographic range (fig. 76) of *Sedum divergens* is in the middle and northern Cascade Mountains north of latitude 43° N., the Olympic Mountains of Washington, the Coast Mountains of British Columbia north to the Queen Charlotte Islands, and inland on the Interior Plateaus and Ranges of British Columbia and north to about 57° N. Two relictual populations are known in the Klamath Mountains: one at Groghan Hole, near Trinity Summit, Humboldt Co., Calif., discovered by J. P. Tracy, his no. 19,294 (Jeps), Aug. 8, 1950; and the other on Lake Peak, Josephine Co., Ore., 1930, Clarice Nye (Ore.). The northeasternmost known population is at Azóuzetta Lake, along the Hart Highway, in the drainage of the Fraser River, in the Northern Rocky Mountains in eastern British Columbia, Taylor and Syczawinski (Prov. Mus., Victoria, B.C.).

The altitudinal range of *Sedum divergens* is from near sea level in the Queen Charlotte Islands to 2,288 m. on Mount Rainier in the Cascade Mountains of Washington. Habitats are rocky slopes and ledges of cliffs. Plants occur on a wide variety of rocks, including granite, andesite, and limestone.

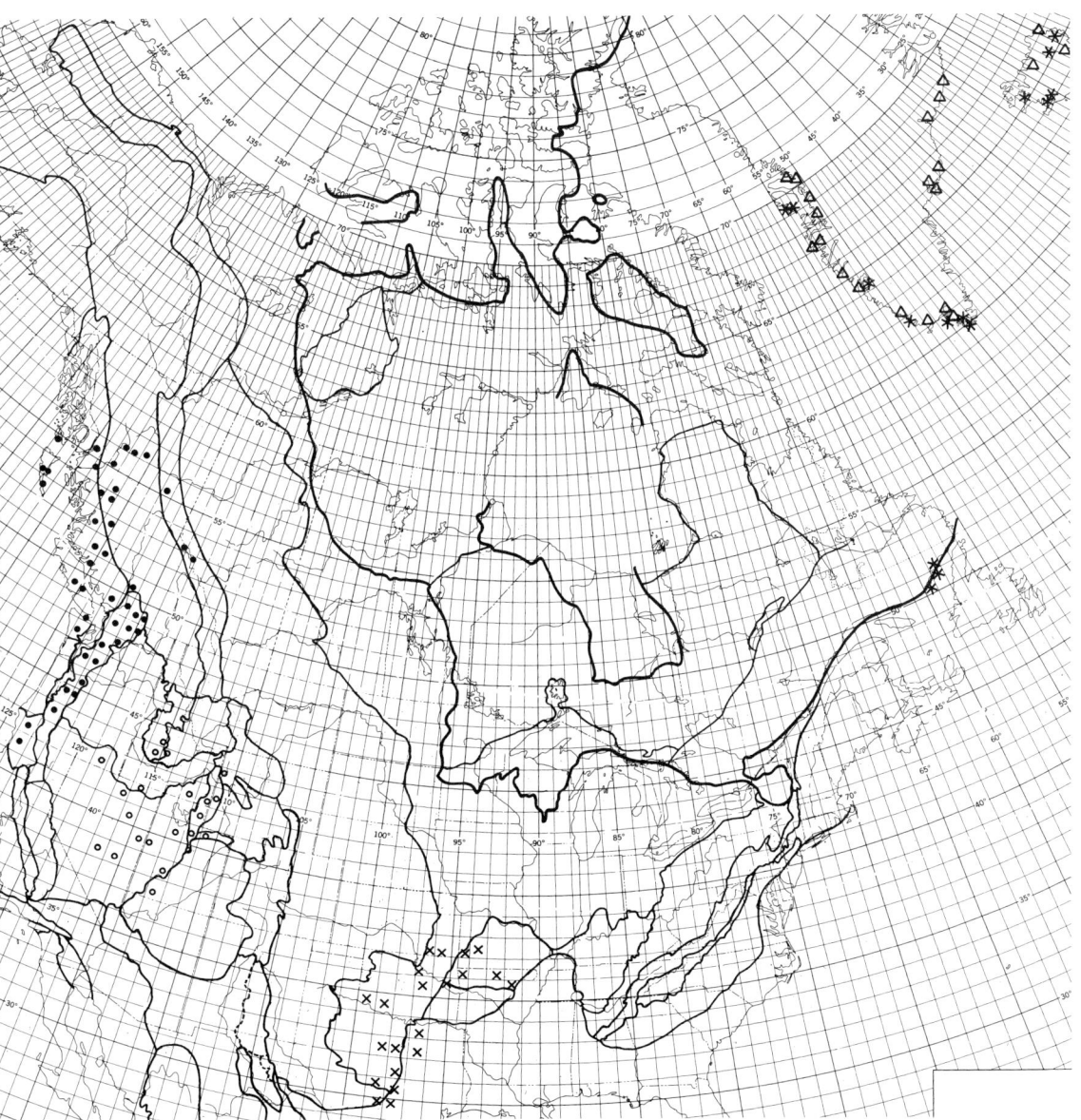

Fig. 76. Known distribution of *Sedum divergens* (●), *S. debile* (○), *S. nuttallianum* (x); and *S. villosum* (∗) and (△ for records from literature).

My principal attention to *Sedum divergens* has been to populations in the valley of Ingalls Creek in the Wenatchee Mountains, which I visited with my son Eric in 1961, and in a randomly selected area on the slope of Bald Mountain, Whatcom Co., Wash., which I surveyed in 1965. Descriptive data for these two populations follow.

Population	Plants (N)	Area of population (m.^{2})	Density per m.^{2}	Exposure
Ingalls Creek	125	7,474	.016	NNE–ESE
Bald Mt.	1,950	78,141	.025	E–ENE

Population	Altitudinal range (m.)	pH, range	pH, tests (n)	Oldest record
Ingalls Creek	1,095–1,115	5.2–6.2	10	1961
Bald Mt.	1,210–1,380	4.8–5.8	2	1965

The surveys in the Cascade Mountains permit an estimate of the total number of plants there. A ratio estimate was useful to achieve a figure for the Wenatchee Mountains.

$$\frac{266 = N \text{ plants counted}}{5.5 \text{ km.}^2 = \text{area surveyed}} = \frac{X}{4,699 \text{ km.}^2 = \text{area of Wenatchee Mts.}}$$

$$X = 227,261 \text{ plants}$$

The two randomly selected areas for survey in the Cascade Mountains were the Skagit and Okanogan primary sampling units. In the following formula, D = density per km.^{2}, A = area in km.^{2}, subscript S = Skagit p. s. u., and subscript O = Okanogan p. s. u. The formula provides an estimate of the average density per kilometer square.

$$\frac{(D_S A_S) + (D_O A_O)}{A_S A_O} = \bar{D}$$

$$\frac{2,067,000 + 0}{1,378 + 1,586} = 697.36 \text{ plants} = \text{av. density per km.}^2$$

The total area of the Cascade Mountains is about 105,000 km.^{2}. Subtracting the area of the southern part of the Cascade Mountains from this figure, and also the area of the Wenatchee Mountains, there are left 77,500 km.^{2} occupied by *Sedum divergens*. Then, multiplying the average density by the occupied area, the total number of plants is 50,768,505. To this figure must be added the estimate for the Wenatchee Mountains. The final estimate then is 50,995,766 plants. This figure may be too high. Additional surveying is necessary both to get a better estimate and to make reasonable the estimation of a confidence interval.

Competitors of *Sedum divergens* are miscellaneous. Rated on a scale of 10, only 2 species of vascular plants, growing beside 10 randomly selected plants of *S. divergens* in the valley of Ingalls Creek in the Wenatchee Mountains, had a rating of 5 or more: *Selaginella wallacei*—7 and *Lewisia columbiana*—5. Mosses of all species together had a rating of 7. Fifteen other species of vascular plants were competitors. All except one of the sample of 10 plants of *S. divergens* were in the open and received maximum potential sunlight. The exception was a plant shaded by *Pseudotsuga menziesii*. On Bald Mountain, mosses were by each of the selected plants of *S. divergens*, but *Selaginella wallacei* was by only one of them. Four other species of vascular plants competing with *S. divergens* on Bald Mountain were different from the 16 species in the valley of Ingalls Creek.

Species of *Sedum* sometimes associated with *S. divergens* are *S. rupicolum* on Earl Peak in the Wenatchee Mountains and *S. oreganum* on Wedge Mountain in the Wenatchee Mountains and on Bald Mountain.

Reproduction. Production of new plants may be either vegetative, from broken pieces of stems which root easily at the nodes and from offsets, or sexual, from seeds. Plants can endure long periods of desiccation. Two stems out of 26, of a plant collected on Bald Mountain on Aug. 21, 1965, and kept dry in a plastic bag until Feb. 24, 1966, grew readily when placed in soil in the greenhouse and provided with water.

Flowering time is primarily in July and August. Extreme dates are June 11, 1956, in the Coast Mountains of British Columbia, and Oct. 4, 1906, on Vancouver Island. June is the principal month of flowering in the garden and cold frames at Ithaca, N.Y. Extreme dates in the cold frames are May 28, 1968, and July 23, 1968, with abnormal production of flowers by one plant from Sept. 3–30, 1968.

Pollinators include bumblebees—Wedge Mountain in the Wenatchee Mountains, July 28, 1961, and Bald Mountain, Aug. 18, 1965; and small flies—valley of Ingalls Creek, Wenatchee Mountains, July 17, 1961. *Sedum divergens* and *S. oreganum* have flowers at the same time and occur together on both Wedge and Bald Mountains. Bumblebees also visit the flowers of both species. Despite optimum physical conditions for hybridization, I did not find hybrids on these mountains.

Relationships. Only two species of *Sedum* in temperate North America have the leaves always two per node and decussately opposite. These species are *S. divergens* and *S. debile*. Both species have 8 pairs of chromosomes. Further, the chromosomes are relatively large for *Sedum* and at least in part elongate and two-armed. Another interesting point is that the two species are allopatric, *S. divergens* being in the more humid Pacific Coastal and Cascade Mountains and *S. debile* in the less humid mountains of the Great Basin, Wyomide Ranges, and Central Rocky Mountains. The evidence suggests a relationship between these two species despite the kyphocarpic carpels of *S. divergens*, which develop into widely divergent, markedly two-lipped follicles, and the orthocarpic carpels of *S. debile*, which develop into erect follicles. Also, *S. debile* has glaucous leaves and the petals always connate basally.

Another Cordilleran species with 8 pairs of chromosomes is *Sedum leibergii*. That has biennial, aerial shoots with the leaves alternate. The leaves of the rosettes are oblanceolate and petiolate. Those of the axillary shoots are obovate and papillose. The flowers of *S. leibergii* are sessile or subsessile, with the sepals short and separate, 1.2–2 mm. long, and the anthers .9 mm. long. *Sedum divergens* and *S. leibergii* are so different that they are not likely to be confused. The two species have become adapted in different ways, one to the cool summers of the high mountains and coastal regions, and the other to the hot, dry summers of the interior plateaus. Both are now so specialized that the pathways of their origin from a common ancestor in the distant past are concealed.

Sedum stahlii of the Sierra Madre del Sur and Trans-Mexican Volcanic Belt, also with opposite leaves, is a subshrub with larger, puberulent, terete leaves, longer sepals which are distinct basally, broader nectaries, and smaller, papillose seeds. This is a polyploid which may be remotely related.

Other species sometimes with opposite leaves belong to the subgenus *Gormania*. Although the leaves of these species normally are spirally arranged, those of the offsets are opposite. The base number of chromosomes in these species is 15. An interesting idea is that they may be the result of ancient hybridization of species with alternate and opposite leaves, and also 8 and 7 pairs of chromosomes.

Sedum ternatum, also with 8 pairs of chromosomes, but with the chromosomes even more elongate than in *S. divergens*, has white flowers and the stamens scarcely adnate to the petals. The leaves of *S. ternatum* usually are 3 per node, but occasionally they are opposite. A relationship may be remote, but a common descent in the distant past is a possibility.

Sedum oreganum sometimes grows with *S. divergens*. Because both species have yellow flowers and spatulate-obovate leaves and also may occur together, confusion is possible. Differences between them, however, are fundamental. These include number and shape of chromosomes, 12 pairs in *S. oreganum*, arrangement of leaves, length and cohesion of petals, and orientation and gibbosity of carpels. For further discussion, see the account of *S. oreganum*. Also see the section on relationships of *S. lanceolatum* for discussion of a putative hybrid between *S. divergens* and *S. lanceolatum*.

17. *Sedum debile**** (figs. 76–80)

Pale green, almost globular rosettes, with the leaves broadly rounded and decussately opposite in four ranks, distinguish *Sedum debile* (fig. 77) from other North American species. The flowers are yellow and have erect carpels which remain upright in fruit. The only other species of *Sedum* in western North America with the leaves regularly opposite is *S. divergens*. That too has almost globular leaves, but they are greener, and the carpels both become divergent with age and spread widely in fruit. *Sedum lanceolatum*, sometimes growing with *S. debile* and also with erect follicles, has longer, narrower leaves, one per node, and separate petals. In *S. debile*, the petals are slightly connate basally. The specific epithet, meaning weak, applies to the slender, decumbent stems.

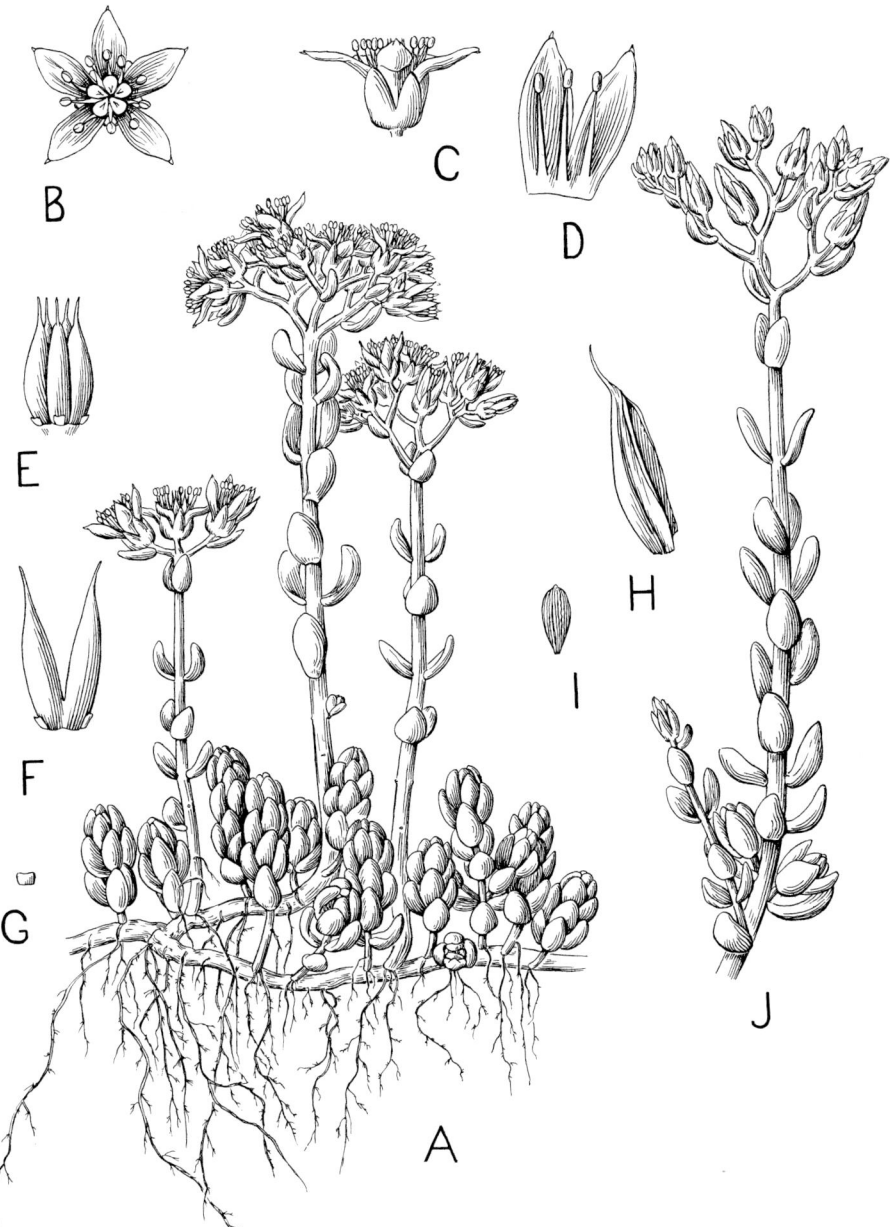

Fig. 77. Plant of *Sedum debile* from eastern slope of Teewinot Mountain, Teton Mountains, Teton Co., Wyo. A. Habit sketch (x .94). B. Flower from above (x 1.9). C. Flower from side (x 1.9). D. Petal and two stamens (x 2.8). E. Carpels (x 2.8). F. Two carpels, torus, and nectaries (x 3.8). G. Nectary (x 4.7). H. Fruit from side (x 3.8). I. Seed (x 9.4). J. Floriferous stem in bud, cultivated in cold frame, Ithaca, N.Y. (x .94).

Description. Sample: 81 plants from six areas in the Teton Mountains of Wyoming. For information about areas from which plants originated, see the discussion under Central Rocky Mountains in the chapter on geography. The principal study of *Sedum debile* in the field was in randomly selected area 3–21 on Teewinot Mountain, TM in the following account. Other studies were of several plants in an optimum stratum: area 9–141, Teton Canyon, TC; area 3–1, Cascade Canyon, CC; and area 3–19, South Fork of Cascade Canyon. Numbers of plants studied in each stratum and area follow.

Stratum	Area no.	Plants in flower (n)	Plants in fruit (n)
Random	3–21	20	33
Optimum	9–141	4	5
Optimum	3–1	1	2
Optimum	3–19	0	2

In the field, I surveyed 10% of the likely sites within areas, 10% of the clusters at sites, 3% of the plants per cluster, 6% of the vegetative rosettes, and 5% of the flowers per plant. Decisions concerning sampling fractions resulted from the application of Stein's test to initial trial samples involving length of median leaves of rosettes, length of sepals, and length of petals. In each case, the indicated sample is supposedly adequate to detect a difference of 1 mm. Initially, I oversampled clusters in area 3–21 and also surveyed one extra cluster.

With adjustment of numbers to keep the samples self-weighting, the basis for the description of vegetative parts below is a sample of 14 plants from the eastern slope of Teewinot Mountain and 4 plants from Teton Canyon. The basis of the description of fruits and seeds likewise is a sample of 14 plants from Teewinot Mountain and 3 plants from Teton Canyon. The sample for the description of flowers comprises 24 plants: 11 from site 2 and 9 from site 5 on Teewinot Mountain, and 4 from Teton Canyon. Site 2 is exposed to the northeast. Site 5 is exposed to the east-southeast. Plants at site 5 appeared larger and more vigorous. For that reason, a comparison of sites is interesting.

Besides the plants from the planned sample, I studied a few specimens from other localities, either collected by myself or donated by friends. Sources were Alaska Basin—2, Baldy Knoll—1, and Moose Basin—1. Because these do not change the description resulting from the regular sampling scheme, data are omitted, although plants were included in an experiment in Ithaca to check the impression that they are not different from those from the selected sampling areas.

Stems of *Sedum debile* are slender and break easily. They are decumbent and with fibrous roots developing from the nodes of the prostrate lower portions. Upward, they are pink and glaucous. Although single stems are biennial or even annual, the plants are perennial by means of axillary shoots which grow into compact, globular rosettes (fig. 78), forming tufts. The difference in diameter of primary rosettes between TM and TC is highly significant. The difference in length of floriferous stems is significant.

Fig. 78. Plant of *Sedum debile* from eastern slope of Teewinot Mountain, Teton Mountains, Teton Co., Wyo., cultivated in cold frame, Ithaca, N.Y., July 11, 1967 (x 1). Photo by Howard Lyon.

	n-pl.	n-obs.	$\bar{x}$	s
Plants—diameter (cm.)				
TM	14	14	13	8
TC	4	4	5	5.2
Primary rosettes (n)				
TM	14	14	26	20.4
TC	4	4	5	3.6
Primary rosettes—diameter (mm.)				
TM	14	33	7	.93
TC	4	4	5	.9
Floriferous stems—no. per plant				
TM	14	14	8	12.3
TC	4	4	2	.9
Floriferous stems—length (cm.)				
TM	14	33	7.3	1.8
TC	4	7	4	2.1

Leaves of rosettes are decussately opposite, sessile and clasping, elliptical, plano-convex, and broadly rounded, sometimes weakly emarginate, minutely papillose at the apex. The basic color is pale glaucous green speckled with pink, but in exposed situations the leaves may become pink or even red. Leaves of floriferous stems are similar to those of rosettes, but smaller and sometimes ovate-elliptical. Usually they are decussately opposite. Rarely they are alternate. Dimensions are of median leaves of both rosettes and floriferous stems.

	n-pl.	n-rep.	n-obs.	$\bar{x}$ (mm.)	s (mm.)
Leaves of rosettes—length					
TM	14	—	33	5.8	.57
TC	4	—	4	4.2	.67
TM—gh.	5	9	31	7.2	.53
TC—gh.	1	1	4	6.9	—
Leaves of rosettes—width					
TM	14	—	33	3.9	.54
TC	4	—	4	2.8	.48
TM—gh.	5	9	31	4.3	.18
TC—gh.	1	1	4	4.2	—
Leaves of rosettes—thickness					
TM	14	—	33	1.4	.24
TC	4	—	4	1.0	.34
TM—gh.	5	9	31	1.3	—
TC—gh.	1	1	4	1.2	—
Leaves of floriferous stems—length					
TM	3	—	15	4.9	.48
TC	3	—	4	4.2	.32
Leaves of floriferous stems—width					
TM	3	—	15	3	.71
TC	3	—	4	2.9	.22
Leaves of floriferous stems—thickness					
TM	2	—	12	.61	.19
TC	3	—	4	1.25	.33

Inflorescences are cymes, usually two-branched, with floral bracts similar to the leaves, but smaller. Differences between TM and TC are not significant.

	n-pl.	n-obs.	x̄ (n)	s (n)
Cincinni				
TM	14	31	2.	.76
TC	4	5	1.	.81
Flowers per cyme				
TM	14	31	6.4	2.4
TC	4	5	2.5	2.4

Flowers are sessile or on pedicels up to 1.2 mm. long, and usually 5-merous. The difference between TM and TC is not significant.

	n-pl.	n-obs.	x̄ (mm.)	s (mm.)
Diameter				
TM2	11	28	13	1.2
TM5	9	14	12.5	2.2
TC	4	4	10.7	.5

Sepals are erect, lanceolate, obtuse, pale green and glaucous, sometimes pink-tipped or faintly streaked with pink, and papillose. They are connate from one-quarter to more than half of their length. Differences among sites in length and width of sepals are not significant (n-$cl.$ = number of clusters).

	n-cl.	n-pl.	n-obs.	x̄	s
Number					
TM 2	2	11	28	5.1	.2
TM 5	3	9	14	5	0
TC	1	4	4	5	0
Length (mm.)					
TM 2	2**	11	28	3.8	.4
TM 5	3	9	14	3.4	.5
TC	1	4	4	2.7	.5
Width (mm.)					
TM 2	2	11	28	1.7	.2
TM 5	3	9	14	1.5	.2
TC	1	4	4	1.4	.5

Petals are elliptic-lanceolate, obtuse, mucronate-appendaged, spreading widely above the erect and connate bases, erose above the middle, yellow streaked with red below the centers and also on the dorsal keels. Differences among sites in length of cohesion of petals are significant.

	n-cl.	n-pl.	n-obs.	x̄	s
Number					
TM 2	2	11	28	5.1	.2
TM 5	3	9	14	5	0
TC	1	4	4	5	0
Length (mm.)					
TM 2	2	11	28	7.5	.8
TM 5	3	9	14	7.4	1.3
TC	1	4	4	7.1	1.1

	n-cl.	n-pl.	n-obs.	x̄	s
Length of cohesion (mm.)					
TM 2	2**	11	28	.62	.16
TM 5	3*	9	14	.56	.16
TC	1	3	3	.73	.35
Width (mm.)					
TM 2	2	11	28	2.4	.2
TM 5	3	9	14	2.1	.2
TC	1	4	4	1.8	.5

Stamens have yellow filaments, streaked with red, and yellow anthers. Sometimes the anthers appear compressed and winged.

	n-cl.	n-pl.	n-obs.	x̄ (mm.)	s (mm.)
Filaments—length					
TM 2	2	11	28	4.5	1.1
TM 5	3	9	14	4.1	1.6
TC	1	4	4	4.4	.6
Epipetalous filaments—height on petals					
TM 2	2	11	28	1	.2
TM 5	3	9	14	.9	.2
TC	1	4	4	1.2	.2
Anthers—length					
TM 2	1	5	12	1.1	.05
TM 5	1	1	1	.9	—

Nectaries vary from reniform to quadrate and truncate. They may be orange, orange streaked with red, red, salmon pink, or yellow.

	n-cl.	n-pl.	n-obs.	x̄ (mm.)	s (mm.)
Length					
TM 2	2	11	28	.3	.05
TM 5	3	9	14	.3	.07
TC	1	4	4	.3	.1
Width					
TM 2	2	11	28	.5	.08
TM 5	3	9	14	.4	.09
TC	1	4	4	.4	.09

Carpels are erect, vary from deep yellow to greenish yellow, and taper gradually into divergent styles. They are connate basally for .7–1.1 mm. The ovules are large and medianly constricted. Differences among sites in number of ovules per ovary are significant.

	n-cl.	n-pl.	n-obs.	x̄	s
Number					
TM 2	2	11	28	5.1	.2
TM 5	3	9	14	4.9	.2
TC	1	4	4	5	0

	n-cl.	n-pl.	n-obs.	$\bar{x}$	s
Length (mm.)					
TM 2	2	11	28	5.5	.8
TM 5	3	9	14	5	.9
TC	1	4	4	4.8	1.5
Styles—length (mm.)					
TM 2	2*	11	28	1.5	.15
TM 5	3	9	14	1.4	.4
TC	1	4	4	1.3	.5
Ovules per ovary (n)					
TM 2	2	11	28	8.4	2.4
TM 5	3	9	14	7.9	2
TC	1	4	4	5.2	2.6

Fruits are erect and straw color with purple stripes. They are red before maturity. Dehiscence begins at the upper end of the ventral suture. By the time that the follicles are dehisced, the fructiferous stems are dry, dead, and break off easily.

	n-cl.	n-pl.	n-obs.	$\bar{x}$ (mm.)	s (mm.)
Length					
TM	6	14	38	6.1	.9
TC	1	3	9	6.7	1.7

Seeds are pyriform, short-winged at the narrow end, finely ribbed vertically, and yellow to pale brown.

	n-cl.	n-pl.	n-obs.	$\bar{x}$ (mm.)	s (mm.)
Length					
TM	6	14	43	1.05	.14
TC	1	2	2	1.05	.07
Diameter					
TM	6	14	43	.39	.05
TC	1	2	2	.4	0

Chromosomes are large, about 4μ long, ranging from $3-8\mu$ (Mastrangelo in 1971), and somewhat constricted medianly, giving a two-armed appearance.

	n-pl.	g	sp	Cytologist
Number				
TM	1	8	—	Dorothy Niimoto, 1967
TM	1	8	—	Iris Mastrangelo, 1971
Teton Mts., CC	1	8	—	Charles Horn, 1966
TC	1	8	—	Charles Horn, 1966
TC	1	8	—	Dorothy Niimoto, 1967
Wasatch Range, Tony Lake, Utah, 2499 m.	1	—	14–18	L. Hollingshead, cited by Clausen (1942)

Variation. Study of plants in the Teton Mountains quickly dispelled an original notion that *Sedum debile* exhibits little variation. Plants growing on limestone in the

Teton Canyon obviously are smaller than those on granite and gneiss on Teewinot Mountain. Analysis of data from wild plants reveals highly significant differences between these populations in two features, namely diameter of primary rosettes and thickness of leaves of primary rosettes, and significant differences in four additional characters, namely length of leaves of primary rosettes, length of floriferous stems, length of cohesion of petals, and number of ovules per ovary. The plants in Teton Canyon possibly are smaller because of a drier, less suitable environment. Contrary to this idea is evidence from a small experiment which involved culture of two plants, one from Teewinot Mountain and the other from Teton Canyon. These were planted as small rosettes in a dish with vermiculite and fed with nutrient solution. At the end of four months, significant differences continued in diameter of stems and in length, width, and thickness of leaves, the plant from Teton Canyon being smaller in all features. On the other hand, another plant from Teton Canyon, compared with 9 replicates from Teewinot Mountain, after cultivation in the greenhouse at Ithaca, N.Y., for seven years, did not differ significantly in any of the characters distinguishing the plant in vermiculite. Differences noted in the field appear to be environmental modifications.

Plants in clusters exposed to the north on Teewinot Mountain usually are smaller. Significant differences exist in three characters: length of sepals, length of cohesion of petals, and length of styles. The explanation of these differences is uncertain, although they may somehow be related to available light.

The arrangement of leaves normally is opposite. Among hundreds of plants observed in the Teton Mountains in 1966, only two, both on Teewinot Mountain, had alternate leaves on the floriferous stems. Even these had opposite leaves on the secondary branches.

Number of floral parts per whorl varies from 4 to 8 (fig. 79). Three plants in three random samples totaling 24 plants had flowers with different numbers of petals. The other 21 plants were constant in number of petals. Forty-eight flowers from the sample of 24 plants had numbers of petals as follows:

Petals per flower (n)	Flowers (n)	Flowers %	Cumulative %
4	2	4	4
5	44	92	96
6	2	4	100

Color of leaves and petals varies with exposure. Plants in more exposed situations have deeper yellow petals and more red pigment in the foliage.

Most variations of *Sedum debile* appear to be environmental modifications. The conclusion is that few observed differences among plants are heritable. Studies of variation, of the sort made in the Teton Mountains in 1966, if repeated in diverse parts of the geographical area of the species, may reveal whether significant differences occur among populations isolated in other ranges of mountains. On the basis of present information, subspecies have not evolved.

NATIVE SPECIES 321

Fig. 79. Flower of *Sedum debile* with eight petals, in the Teton Canyon near the Treasure Mountain Boy Scout Camp, Teton Co., Wyo., July 26, 1966 (x 2).

Nomenclature. *Sedum debile* S. Watson, in C. King, Report of the Geological Exploration of the Fortieth Parallel 5:102 (1871). Type locality: rocky ridge, altitude of 7,000 feet, East Humboldt Mountains, northern Nevada. Type: S. Watson no. 387 (GH), July 1868. The original description, type, and type locality together form a satisfactory basis for interpretation of the species.

Synonyms are:
1903, *Gormania debilis* (S. Watson) Britton, Bull. N.Y. Bot. Gard. 3(9):30.
1904. *Cotyledon debilis* (S. Watson) Fedde, in Just, Bot. Jahresb. 31(1): 827. Based on *Sedum debile* S. Watson.
1913. *Echeveria debilis* (S. Watson) A. Nelson et J. F. Macbride, Bot. Gaz. 56:476.

Distribution and ecology. *Sedum debile* (fig. 76 above) occurs in various ranges of the Great Basin and also in the Wyomide Ranges and the Central Rocky Mountains, eastward as far as the Teton, Gros Ventre, and Uinta Mountains, and northward into the southern part of the Idaho Batholith. Southeastward, it extends onto the Colorado Plateau in Utah. A peculiar specimen, labeled as collected by S. M. Tracy (US), at

Raton, Colfax Co., N. M., in the Raton Section of the Great Plains, might be *S. debile*, but the occurrence of the species there seems so unlikely that doubt arises whether the label belongs with the specimen.

The habitat of *Sedum debile* is open rocky areas. The plants grow on a variety of rocks, including limestone, granite, and gneiss. The species does not thrive either in grassland or where growth of shrubs is dense. Although plants sometimes survive in shade of trees, *S. debile* is not a forest species.

Detailed information for populations in the Teton Mountains is available below. Descriptions of the sampling areas containing these populations are available in the chapter on geography.

Population	Sampling area	N plants, estimate	N plants, C.I. $_{.05}$	Area (m.^{2})
Teewinot Mountain				
Eastern slope	Teton 3–21	43,925	24,412–100,008	30,415e
Northeastern base	Teton 3–1	41	—	40
Cascade Canyon				
South Fork	Teton 3–19	545	99–6,037	27
Teton Canyon	Teton 9–141	1,344	—	2,787e

Population	Altitudinal range (m.)	pH (w)	Depth of soil (cm.) $\bar{x}$ (and w)	Distance to nearest neighbor, C.I.$_{.05}$; cm.)
Teewinot Mountain				
Eastern slope	2,200–2,450	5.4–6.6	3.7(.5–14)	.81–16.04
Northeastern base	2,160	5.4	2.5(1–4)	(27)
Cascade Canyon				
South Fork	2,530	5.4–5.6	2.7(1–5)	(2.3)
Teton Canyon	2,190–2,240	6.8–7.2	5.8(1–11)	7.88–47.52

On the basis of a record of the two closest competitors of each randomly selected plant of *Sedum debile* on Teewinot Mountain, the commonest competitors were *Eriogonum subalpinum*—4, *Balsamorrhiza sagittata*—4, *Selaginella densa*—3, *Sedum lanceolatum*—2, *Agropyron trachycaulum*—2, and *Stipa columbiana*—2. Figures after names indicate the number of times out of 28 that a species was either the closest or second closest competitor. Species most affecting conditions for growth, usually by shading, of 14 plants of *Sedum debile* on Teewinot Mountain, were *Balsamorrhiza sagittata*—6, *Pseudotsuga menziesii*—2, and *Artemisia tridentata*—2. In Teton Canyon, the closest competitors of 4 randomly selected plants of *Sedum debile* were *Spiraea lucida*—4 and *Pachystima myrsinites*—2. *Pseudotsuga menziesii* caused shade affecting each of these plants.

In summary, general information about the altitudinal range, months of flowering, and earliest records of *Sedum debile* is available in the following table. The data are from both herbarium specimens and original observations.

Geomorphic Province	Altitudinal range (m.)	Months of flowering	Oldest record
Teton Mountains	2,160–3,170	July, Aug.	1920, Payson (Wyo.)
Colorado Plateau	1,576	July	1905, Rydberg and Carlton (NY)
Idaho Batholith	2,440–3,086	June–Aug.	1895, Evermann (US)
Wyomide Ranges	1,617–3,508	June, July	1879, Jones (US)
Great Basin	1,800–2,958	June–Aug.	1868, S. Watson (GH)

Reproduction. Offsets are the principal means of perpetuation of *Sedum debile*. Plants may endure for many years by this mode of propagation and become broken into parts, each with the same genetic constitution. Distinction among ramets of different clones is difficult and often impossible. Only when biotypes are different is distinction possible, and even then the origin of each ramet can only be guessed.

Rosettes, kept dry (fig. 80) in plastic bags for six months, retained life and sprouted readily when put in moist vermiculite. Prolonged drought would not eliminate the species, even if the only means of propagation were vegetative.

Another common type of propagation is by means of seeds. Seedlings occur naturally, but are most successful in moist situations. These seem not to flower in the first year. Maturation of seeds requires about a month after flowering. Eventually

Fig. 80. Shoots of *Sedum debile* from Teewinot Mountain, Teton Mountains, Teton Co., Wyo., after storage in dry condition for six months in plastic bag (x 1.6).

the seeds either drop or are blown by the wind from the erect follicles. The maximum length of time that seeds retain their viability is unknown.

Plants in flower are conspicuous. They attract a variety of insects. At least three species of bees, a muscid fly, and a species of *Lepidoptera* with blue wings are pollinators. Of these, a small bee, a species of *Dialictus*, of the *Halictidae*, is most important. Individuals of *Dialictus* visit flowers of both *S. debile* and *S. lanceolatum*. Dewey Caron of Cornell University identified the *Dialictus* for me, and also another bee, *Dianthidium* of the *Megachilidae*. L. L. Pechuman identified the fly as a species of the *Tacinidae*. The pollinating insects visit the flowers in warm, sunny periods when the air is still. Few or no insects visit the flowers when the wind is strong.

In 1966, the time of flowering of *Sedum debile* in the Teton Mountains was from July 12 to August 15. Plants at the lower elevations bloomed principally in July. Those high in the mountains flowered later, mostly in August.

Relationships and evolution. The only species of *Sedum* in western North America with decussately opposite leaves and yellow petals, besides *S. debile*, is *S. divergens*. That has 8 pairs of two-armed chromosomes, as does *S. debile*. It differs in having the petals separate basally and the follicles both divergent and markedly two-lipped. Despite these differences, a relationship seems reasonable. If these two species have had a common origin, they illustrate the undesirability of grouping species solely on the basis of cohesion of petals or orientation and gibbosity of carpels. At present, the species are geographically isolated, *S. divergens* in the Pacific Mountain System and *S. debile* in the mountains of the intermontane and Rocky Mountain regions. Their origin from geographical subspecies is a possibility.

Sedum lanceolatum often occurs with *S. debile*, flowers at about the same time, but slightly earlier, and similarly has yellow flowers and erect follicles. Where the two species occur together, they do not hybridize. Differences in arrangement and shape of leaves, and in cohesion of petals easily separate them. The chromosomes of *S. lanceolatum* are smaller and not two-armed. These two species have some similar characteristics, but probably are not the most closely related.

Britton (1903) overemphasized the cohesion of the petals in placing *Sedum debile* in *Gormania*. In most other respects, *S. debile* appears distant from species of *Gormania*. Fröderström (1930–1935) reached a similar conclusion.

References. Clausen, R. T. (1942) and Hollingshead, L. (1942).

18. Sedum nuttallianum*** (figs. 76 and 81–83)

Plants of *Sedum nuttallianum* (fig. 81) are small, of annual duration, with capillary roots, lanceolate-elliptic or oblong leaves, yellow petals and stamens, and widely spreading follicles which are gibbous ventrally. The other annual species in southeastern North America, namely *S. pusillum* and *S. pulchellum*, have white or pinkish white petals and filaments. The carpels of *S. pusillum*, although widely spreading, are not gibbous ventrally. The carpels of *S. pulchellum* are both widely spreading and gibbous ventrally, but the leaves are linear with sagittate bases. The species of

NATIVE SPECIES 325

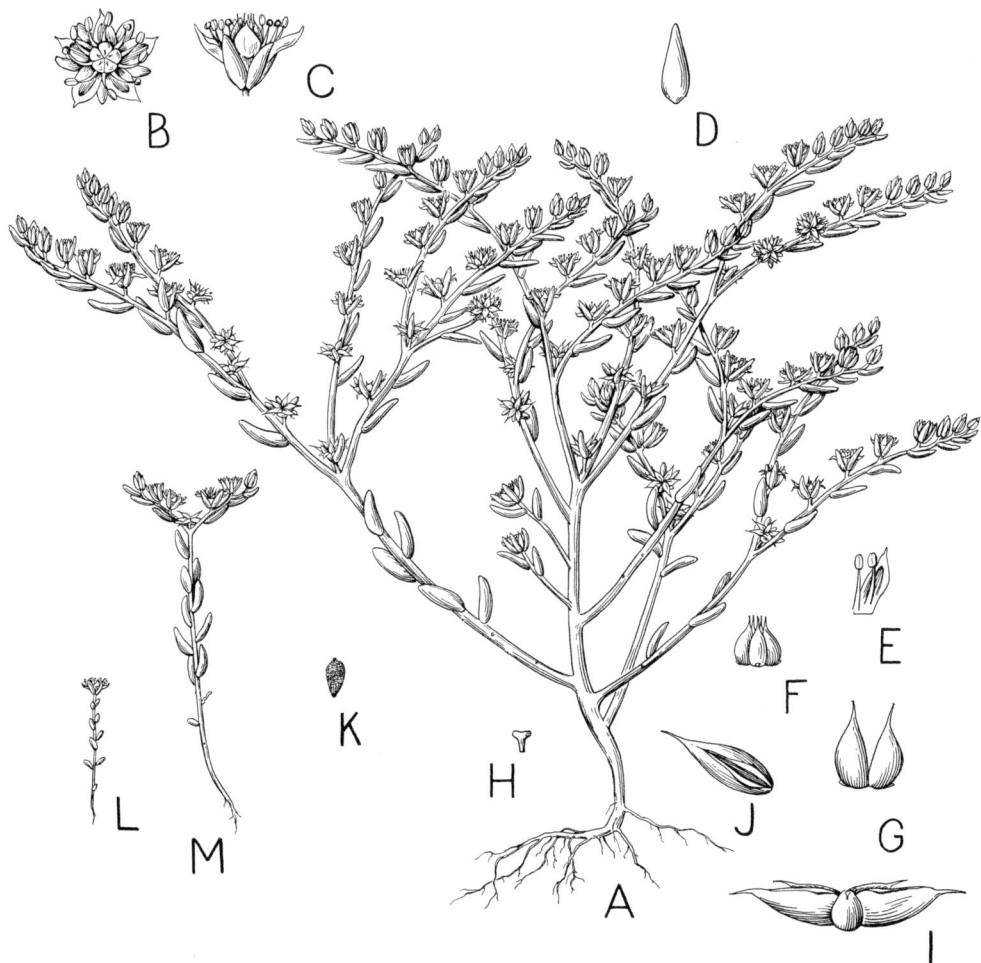

Fig. 81. Plants of *Sedum nuttallianum* from open area on flat outcrop of sandstone along west side of Jordan Creek, 4 km. west of Everton, Dade Co., Mo. A. Habit sketch of large plant (x .9). B. Flower from above (x 2.3). C. Flower from side (x 2.3). D. Sepal from back (x 2.7). E. Petal and two stamens (x 2.7). F. Carpels (x 2.7). G. Two carpels and nectaries (x 4.5). H. Nectary (x 9). I. Follicles from side (x 2.7). J. Single follicle from above (x 9). K. Seed (x 9). L. Habit sketch of small plant (x .9). M. Habit sketch of medium plant (x .9).

Parvisedum are annual and with yellow flowers, but the follicles are not gibbous and each is only one-seeded. The petals of species of *Parvisedum* are slightly connate basally. Those of *S. nuttallianum* are separate.

Description. Sample: 27 plants—25 plants chosen randomly from two randomly selected clusters, nos. 403 and 471, on exposures of sandstone in a randomly selected area along the western side of Jordan Creek, 4 km. west of Everton, Dade Co., Mo.; and 2 plants chosen deliberately to show extreme dimensions from along Shoal

Creek, about 6 km. southwest of Joplin, Newton Co., Mo. The two clusters from along Jordan Creek were selected with equal probability. They comprised respectively 294 and 753 plants. Although the clusters were studied within one day of each other, plants in the smaller cluster, exposed to the east, still had fresh flowers, while those in the larger cluster, exposed to the southeast, were mostly past flowering and with immature fruits. Data for the two clusters are kept separate in the following description when differences between the estimates of means are significant, but combined when differences are not significant. When data are pooled in the following tables, the samples are proportionate from each of the clusters, namely 2%, 6 plants from the smaller cluster and 15 plants from the larger one. Sources of data are indicated as J = Jordan Creek and S = Shoal Creek.

Selection of the plants for study from the site along Jordan Creek involved an initial survey of the sampling area. This entailed preparation of a list of all natural clusters of plants and an estimate of the number of artificial clusters of about 100 plants in each. The result of the listing was 58 natural clusters, and in these, 708 artificial clusters. From the 708 groups, I chose randomly clusters 403 and 471 for detailed study.

Plants of *Sedum nuttallianum* are annual herbs with slender primary roots and capillary rootlets which are pale brown to almost white. The stems are erect, simple or branched, glabrous, and green or pinkish. The soil in which the plants are rooted is so shallow that the roots develop horizontally. A large plant, found by Arthur Phillips III, growing in shade on the ridge west of Jordan Creek, had roots 6.5 cm. long. This same plant had the maximum number of branches observed on a primary stem, namely 13.

	n-pl.	n-obs.	$\bar{x}$	s	w
Plants—diameter (cm.)					
J	21	21	.7	.7	.2–4
Primary rosettes (n)					
J	21	21	0	0	0
S	2	2	0	0	0
Floriferous stems per plant (n)					
J	21	21	1	0	1
Floriferous stems—length (cm.)					
J	21	21	2.4	.6	1.1–4.5
S	2	2	10.7	—	2.9–18.5

Leaves are alternate, widely divergent, short-spurred, lanceolate-elliptical or oblong, rarely globular, blunt, and pale green or bluish green. They wither as the fruits mature. Measurements apply to median leaves of floriferous stems.

	n-pl.	n-obs.	$\bar{x}$ (mm.)	s (mm.)	w (mm.)
Length					
J	7	7	3	1.3	1–5
S	2	2	6	—	2–11
Width					
J	7	7	1.5	.5	.7–1.9
S	2	2	2	—	1.0–3

	n-pl.	n-obs.	x̄ (mm.)	s (mm.)	w (mm.)
Thickness					
J	6	6	1.1	.3	.8–1.7
S	2	2	2	—	1.0–3

Inflorescences are cymes of 1 to 3 cincinni or sometimes the flowers are solitary. Flowers open one by one on the cincinni, from base to tip. Floral bracts are similar to leaves.

	n-pl.	x̄ (n)	s (n)	w (n)
Cincinni				
J 403	6	1.7	1.1	0–3
J 471	15	.6	.6	0–2
S	2	1.5	—	1–2
Flowers				
J 403	6	6	5.5	1–12
J 471	15	1.4	.6	1–3
S	2	18	—	2–34

Flowers are sessile or subsessile, with pedicels up to .5 mm. long, and usually 5-merous.

	n-pl.	n-obs.	x̄ (mm.)	s (mm.)	w (mm.)
Diameter					
J 403	9	12	5	.6	4–6
J 471	1	1	3	—	3
S	2	2	7	—	4–10

Sepals are slightly spurred, very unequal, lanceolate or lanceolate-oblong, acute, and yellow-green.

	n-pl.	n-obs.	x̄	s	w
Number					
J	21	22	4.9	.3	4–5
S	2	2	5	—	5
Length (mm.)					
J 403	9	12	1.8	.6	1.1–3.1
J 471	1	1	.6	—	.6
S	2	2	3	—	2–4
Width (mm.)					
J 403	9	12	.8	.9	.4–1.4
J 471	1	1	.4	—	.4
S	2	2	1.5	—	1.3–1.7

Petals are separate, spreading, elliptic-oblong, slightly hooded and mucronate-appendaged, and yellow.

	n-pl.	n-obs.	x̄	s	w
Number					
J	21	21	4.9	.3	4–5
S	2	2	5	—	5

	n-pl.	n-obs.	$\bar{x}$	s	w
Length (mm.)					
J 403	9	12	2.8	.3	2.7–3.3
J 471	1	1	1.5	—	1.5
S	2	2	4	—	4
Width (mm.)					
J 403	9	12	.9	.1	.8–1.1
J 471	1	1	.6	—	.6
S	2	2	1.3	—	1.3

Stamens have basifixed anthers. Both filaments and anthers are yellow. Before anthesis, the anthers of the epipetalous stamens fit into hollows in the petals, with the filaments curved upward.

	n-pl.	n-obs.	$\bar{x}$ (mm.)	s (mm.)	w (mm.)
Filaments—length					
J 403	9	12	1.9	.3	1.7–2.5
J 471	1	1	.2	—	0.2
S	2	2	2.7	—	2.5–3
Epipetalous filaments—height on petals					
J 403	9	12	.3	.09	.2–.5
S	2	2	1	—	1
Anthers—length					
J 403	4	4	.65	.06	.6–.7

Nectaries are stipitate-reniform, subquadrate, or obovate, truncate at apex, and either yellow or translucent.

	n-pl.	n-obs.	$\bar{x}$ (mm.)	s (mm.)	w (mm.)
Length					
J 403	9	12	.33	.05	.3–.4
S	2	2	.4	—	.4
Width					
J 403	9	12	.27	.04	.2–.3
S	2	2	.5	—	.5

Carpels are erect, with slender styles, and greenish yellow. As they develop, they spread rotately and become gibbous ventrally. Color changes from yellow to green speckled with pink.

	n-pl.	n-obs.	$\bar{x}$	s	w
Number					
J	21	22	4.9	.3	4–5
S	2	2	5	—	5
Length (mm.)					
J 403	9	12	1.9	.3	1.4–2.5
J 471	1	1	1.3	—	1.3
S	2	2	4	—	4
Styles—length (mm.)					
J 403	9	12	.5	.2	.3–.8
J 471	1	1	.2	—	.2
S	2	2	1	—	1
Ovules per ovary (n)					
J	21	22	2.2	1.2	0–6

Fruits are widely spreading, gibbous ventrally, and stramineous.

	n-pl.	n-obs.	$\bar{x}$ (mm.)	s (mm.)	w (mm.)
Length					
J	10	16	2	.03	1.3–2.6

Seeds are ovoid, obscurely alveolate, and yellow-brown.

	n-pl.	n-obs.	$\bar{x}$ (mm.)	s (mm.)	w (mm.)
Length					
J	16	30	.6	.09	.5–.9
Diameter					
J	16	30	.3	.07	.2–.4

Chromosomes.

	n-pl.	g (n)	sp (2n)	Cytologist
Georgetown, Texas	1	10	20	J. T. Baldwin, Jr. (1940)

Variation. Variation in size of plants is great. Plants vary from tiny individuals, only 5 mm. tall, with a single small flower, up to spreading, branched ones with stems 18 cm. tall and as many as 177 flowers. On the basis of field observations, the differences in size result from dissimilar environmental conditions and unequal ages. Seeds which germinate late in adverse conditions produce the tiniest plants. The largest plants grow in shade.

In 1949, in studies along Shoal Creek, Newton Co. Mo., I sought the extremes of variation. Similarly, in 1968, when Arthur Phillips III helped me in studies along Jordan Creek, Dade Co. Mo., I had him seek plants with extreme conditions of three features. The results are indicated in the following table.

	1949, w	1968, random samples		1968, nonrandom samples, w
		w	$\bar{x} \pm t_{.05}s_{\bar{x}}$	
Height (cm.)	2.9–18.5	1.1–5	2.4 ± .3	.5–17
Flowers per plant (n)	2–34	1–12	2.7 ± 1.3	1–177
Petals per flower (n)	5	4–5	4.9 ± .01	3–5

Results of the study along Shoal Creek in 1949 are further contrasted with the results from random sampling along Jordan Creek in 1968 for four other features.

	1949	1968	
	w	w	$\bar{x} \pm t_{.05}s_{\bar{x}}$
Median leaves—length (mm.)	2.2–11	1.0–5	3.1 ± .62
Flowers—diameter (mm.)	4.0–10	4.0–6	5.2 ± .46
Petals—length (mm.)	4.0	2.3–3.3	2.8 ± .37
Nectaries—length (mm.)	.4	.3–.4	.32 ± .03

Differences among clusters along Jordan Creek are highly significant in height of plants and number of flowers per plant. They are significant in number of cincinni per inflorescence and diameter of flowers. Differences also seem to exist in length and width of both sepals and petals, length of stamens, and length of carpels and styles, but because plants in cluster 471 mostly were past flowering, confirmation with adequate data was impossible. Such differences may be attributed to different environments. A plant grown in the greenhouse from a seed from cluster 471, which contained plants with significantly smaller dimensions, surpassed in size plants in cluster 403. The cincinni were 3 per cyme, the flowers were 16 per cyme and averaged 11.7 mm. in diameter, and the petals were 3.7–4.7 mm. long and greenish yellow. The sepals became elongated, exceeding the petals (fig. 82). The phenotype in the greenhouse at Ithaca confirms that the differences among clusters observed in the field are no more than environmental modifications.

The flowers are remarkably constant in being 5-merous. Only a few are 4- or 3-merous. Of 22 flowers studied in the random samples, only 9% were 4-merous and none were 3-merous.

Fig. 82. Flower of plant of *Sedum nuttallianum* grown from seed in greenhouse, Ithaca, N.Y., Nov. 5, 1969 (x 4.1). Photo by Howard Lyon. Note that sepals are longer than petals.

Except for the profound differences resulting from inequalities in light, age, and moisture, my studies have not revealed any unusual variations.

Nomenclature. *Sedum nuttalianum* Rafinesque, Atl. Jour. 1:146 (1832). Type locality: drainage of Red River, Ark. Type: Nuttall (NY), 1819. Later authors have corrected the spelling of the specific epithet to *nuttallianum*. The locality most likely was somewhere in Oklahoma. Rafinesque based his binomial on Torrey's 171, *Sedum n. sp.* Nutt., Ann. Lyc. Nat. Hist. N.Y. 2:205 (1827). In his publication, Torrey described plants collected by Edwin James in 1820. He stated that he was not describing species which Nuttall had found in the previous year. Since Rafinesque's reference is to Torrey and since Torrey had a specimen from Nuttall, that specimen may properly be taken as the type. The brief description is appropriate, and the adjective "roundish" for the leaves is diagnostic. The discordant feature is the cited locality, near the Rocky Mountains, but that may result from Torrey's confusion of two different species. For example, he wrote on the label of the type specimen of *S. lanceolatum* that Nuttall had found the same plant on the Red River, but that seems impossible. Whether or not James ever collected *S. nuttallianum* is uncertain. Torrey evidently thought that he had, but his description could have been based solely on Nuttall's specimen. Since James, on his return trip from the Rocky Mountains, traversed country where *S. nuttallianum* grows, the possibility existed for him to see it. Even if a collection of James is available, typification by Nuttall's specimen is preferable because Torrey was trying to give credit to Nuttall for the discovery of the new species.

Synonyms are:

1833. *Sedum nuttallii* Eaton, Man. Bot. No. Am., ed. 6, p. 334. Based on Torrey's no. 171, cited above under *S. nuttallianum*.

1834. *Sedum Torreyi* G. Don, Gen. System of Gardening 3:121. Based on Torrey's *Sedum n. sp.*

1840. *Sedum sparsiflorum* Nutt. in Torrey and Gray, Flora of North America 1:559. Type locality: plains of the Red River, Ark. Type: Nuttall (NY), 1819. By the time Nuttall's manuscript name was published, several collectors had found the species and knowledge of its distribution had been extended to Texas.

Distribution. (fig. 76). The geographical range of *Sedum nuttallianum* is in the southern and western Ozark Plateaus, central and southern Osage Plains, and inner West Gulf Coastal Plain and southeastern end of the Great Plains in the vicinity of the Balcones Escarpment. It occurs northeastward to Dade Co., Mo., and Izard Co., Ark.

Plants of *Sedum nuttallianum* occur in vast numbers (fig. 83) in shallow soil on exposures of sandstone in open woods of oaks and red cedars in southwestern Missouri. Usually the sites are shaded for at least part of the day.

My studies of *Sedum nuttallianum* have been in the western part of the Ozark Plateaus, in southwestern Missouri and northwestern Arkansas. In 1968, I divided a portion of the drainage of Jordan Creek, a tributary of Sinking Creek, west of Everton, Dade Co., Mo., into nine natural units. Then I selected one of these, using

Fig. 83. Plants of *Sedum nuttallianum* in moss on sandstone on west side of Jordan Creek, 4 km. west of Everton, Dade Co., Mo., June 18, 1968.

a table of random numbers. The selected unit is an area of sandstone, .25 km.^{2}, on the western side of Jordan Creek south of the road from Everton to Antioch Church. The area is east of the road west of Jordan Creek and ranges in altitude from 286–308 m. Woodland of oaks—*Quercus alba*, *Q. velutina*, and *Q. stellata*—occupies three-quarters of the unit. Data pertinent to *S. nuttallianum* are organized in the following brief table. These are derived from study of the two clusters of which the selection is explained in the above description of the species.

Population	Sampling unit	N plants, estimate	N plants, C.I. (P = .05)	Area of population (m.^{2}) Estimate	Confidence Interval
Jordan Creek	No. 1	370,638	1,047–759,259	1,126	3.2–13,649

Population	Density per m.^{2}	Altitudinal range (m.)	pH, range	pH, n tests	Depth of soil (cm.)
Jordan Creek	329	290–308	5.2–5.8	25	.2–3

The average distance to the nearest neighbors of 21 plants chosen randomly was 6.8 mm., with s = 6.6 mm. The observed range in distance to nearest neighbor was from 1 to 25 mm. Plants growing close together usually are tiny and unbranched, with one or few flowers, as indicated in the description. Larger plants, found as a result of deliberate search, besides being in shade, appeared to be less closely spaced. Differences in distance to nearest neighbor, 5 mm. in cluster 403 and 7 mm. in cluster 471, are not significant.

Sedum nuttallianum grows in almost pure stands, where other vegetation is sparse, thus the usual competitor of any plant of *S. nuttallianum* is another plant of the same species. Other competitors, determined by listing the two species closest to each of 21 randomly selected plants at the site along Jordan Creek, and ratings for each, are: *Festuca octoflora*—16, *Rumex acetosella*—5, *Crotonopsis linearis*—4, a narrow-leaved dicotyledon which lacked flowers or fruits—3, and *Talinum* sp.—2. Ratings were achieved by counting the most important competitor of each plant of *Sedum* as 2 and the next most important one as 1, and then adding the numbers for each species. On this basis, plants of *S. nuttallianum*, competing with each other, rated as 31. Along Jordan Creek, the species most affecting the growth of *S. nuttallianum*, through shading the plants for part of the day, are *Quercus alba* and *Q. stellata*. No other *Sedum* occurs with *S. nuttallianum* along Jordan Creek. Along Shoal Creek southwest of Joplin, *S. pulchellum* was associated with *S. nuttallianum*. Other competitors there were *Talinum* sp., *Selaginella rupestris*, and *Woodsia obtusa*.

Reproduction. Plants of *Sedum nuttallianum* are annual. They die after flowering and fruiting. All reproduction is by means of seeds. Although the seeds are tiny, the majority of them must drop close to the fruiting plants, with the result that new plants develop in dense stands. Seeds blown by the wind or carried by water can give rise to new populations and spread the species to other suitable sites.

Flowering time is from April to June, depending on latitude and exposure. In cultivation, plants may flower abnormally at other times of the year, as in November.

Cross-pollination is by means of small insects, principally small bees, and also flies and at least one species of *Hemiptera*. Insects visit the flowers most commonly when many blossoms are in anthesis. Visits decrease as the period of flowering wanes. Arthur Phillips III watched a plant with 5 expanded flowers on June 19, 1968. Although this plant was in sun, no insects visited the flowers in a period of a half hour. Other plants with flowers, growing in shade, had two insect visitors in the same length of time and on the same day.

The possibility of interspecific hybridization is low. When *Sedum nuttallianum* grows where other species of *Sedum* are lacking, as at Jordan Creek, the chance of hybridization approaches zero. When *S. pulchellum* occurs nearby or even in association, it flowers earlier and is mostly past flowering when *S. nuttallianum* is at the height of bloom. A further barrier to hybridization is the difference in number of chromosomes in the two species.

Relationships. Although *Sedum nuttallianum* appears somewhat related to the other species of section *Lanceolata*, it also has affinity with *S. parvum* on the basis of morphology and geography. *Sedum parvum* occurs in the Sierra Madre Oriental and on the Mexican Plateau as far north as Brewster County, Texas. The plants have yellow petals and spreading, gibbous follicles, as in *S. nuttallianum*, but they differ in being perennial and in having ovoid leaves. The two species may be derived from the same ancestral stock. The chromosome numbers, g = 10 for *S. nuttallianum* from Georgetown, Texas, and g = 26 for *S. parvum* from Chorro Grande, Coahuila, indicate an important genetic difference, justifying distinction as species.

The species of *Parvisedum*, sp = 18, also are similar to *S. nuttallianum* in their annual habit, yellow flowers, and small leaves, but they differ in having the petals connate basally and the follicles both one-seeded and not gibbous ventrally. *Parvisedum* is perhaps a more distant and more specialized offshoot of the same phyletic line which gave rise to *S. nuttallianum* and *S. parvum*.

The Appalachian species of *Sedum*, namely the annual *S. pusillum* with g = 4, and the perennial *S. nevii*—g = 6 and *S. ternatum*—g = 8 or multiples of 8, as well as the usually biennial *S. pulchellum* with g = 11 or multiples of 11, appear to belong to another phyletic line which may have parted from the same evolutionary stem as *S. nuttallianum* in the distant past. Adjustments in numbers of chromosomes could have preceded or accompanied the important changes in physiology and morphology which distinguish the modern species.

Sedum annuum of Europe and Greenland, though superficially similar to *S. nuttallianum*, is both g = 11 and adapted physiologically to a colder climate. Small differences in morphology which distinguish *S. nuttallianum* are the slightly spurred sepals and the smaller petals, nectaries, follicles, and seeds. The sepals of *S. annuum* are not spurred and the floral parts, fruits, and seeds are larger. Despite the remarkable similarity in gross morphology, the two species may be derived from different stocks and be examples of evolutionary convergence.

Fröderström (1930–1935) included *Sedum nuttallianum* in the subgeneric group *Epeteium americanum*, along with *S. pusillum* and *S. cymosum* (= *Diamorpha cymosa*). The data of Sherwin and Wilbur (1971) do not support such a classification. The general pattern of the vascular supply of the flowers of *S. nuttallianum* is more like that of *Diamorpha* than *S. pusillum*, except that the dorsal carpellary bundle is not reduced and that the bundles are more evenly spaced in the walls of the carpels. Points of resemblance with *S. pusillum* are the ventral suture lines of the carpels which remain distinct throughout development and the unfused condition of the carpels. Unlike either *Diamorpha* or *S. pusillum*, the petals are yellow and the carpels become gibbous ventrally.

19. Sedum villosum*** (figs. 76 and 84–86)

Sedum villosum (fig. 84) is one of four North American species of *Sedum* with pink petals. The others are *S. rhodanthum*, of the Eastern Ranges and Plateaus, which is perennial with stout rootstocks; *S. pulchellum*, of the Interior Lowland, which often is annual, but with linear leaves, sagittate basally, and kyphocarpic carpels; and

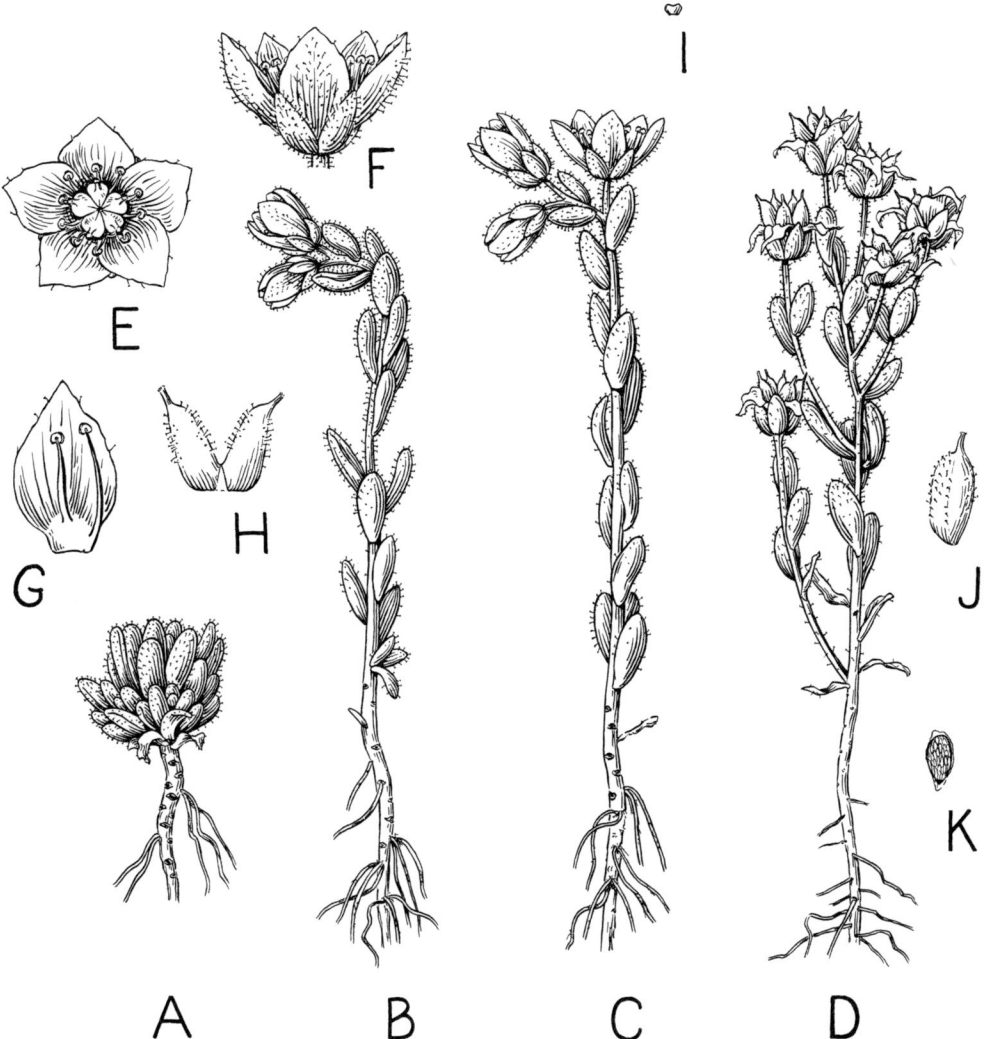

Fig. 84. Plants of *Sedum villosum* from Big Gull Island in northern part of Gulf of St. Lawrence, Que. A. Rosette of young floriferous shoot before elongation (x 2.2). B. Floriferous stem with floral buds before expansion (x 2.2). C. Floriferous stem with expanded flower (x 2.2). D. Fructiferous stem (x 2.2). E. Flower from above (x 3.2). F. Flower from side (x 3.2). G. Petal and two stamens (x 4.3). H. Two carpels and torus (x 4.3). I. Nectary (x 5.4). J. Follicle (x 4.3). K. Seed (x 11).

S. laxum, of the central Cordillera, which is perennial with thick rootstocks and rosettes of broad leaves. *Sedum villosum* differs from these species in its glandular-hairy stems, leaves, pedicels, and sepals; tiny red nectaries; and short-styled carpels which become black in fruit. The combination of characters just mentioned, plus the pink petals, will separate it from all other species of *Sedum*.

Kerner (1894) listed *Sedum villosum* as a species able to obtain some nourishment from insects which become adherent to the glands on its hairs. Others have repeated Kerner's statement, but no detailed work seems to have been done on this subject.

Description. Sample: 186 plants (94 in flower, 92 in fruit) from four islands along the northern shore of the Gulf of St. Lawrence, between the Netagamu and Kécarpoui Rivers. For description of the survey in 1959 to study *Sedum villosum*, see Chapter X. Because the species did not appear in any of the randomly selected quadrats, post-stratification was necessary. This involved creation of a special stratum which encompassed the total 194 square meters in which the species was found in 1959. Survey of this stratum included 10% of the area, 20 randomly selected quadrats, each one meter square. The intention was to study a fixed percentage of plants in each chosen quadrat, but the average number of plants in flower in a quadrat at any given time was only 10%. The probability of getting plants in flower from random samples of all plants is .1. Because I could not stay on each island until every chosen plant flowered, mandatory procedure was to take random samples of plants which were in flower at the time of visits. Another complication was that percentages of plants in flower changed as the season progressed. On Big Gull Island, for example, in the first half of July, the per cent of plants in flower in quadrats ranged from 0–14 ($\bar{x} = 2$). On Middle Kécarpoui Island, in late July, the per cent ranged from 35–52 ($\bar{x} = 45$). In August, proportions of plants in flower began to decline again. These differences required adjustment in percentages of plants selected. To get a sample from Stevenson Island in mid-August, comparable with that from Big Gull Island a month earlier, a balance seemed necessary between oversampling quadrats and undersampling the included plants in flower. Ultimate samples at flowering stage represent the following proportions:

Island	% area	% plants in flower	% total plants	n plants in flower	n plants in sample
Middle Kécarpoui	10	7	0.3	147	10
Eastern Affligée	10	0	0.0	30	0
Western Affligée	10	96	0.4	29	28
Big Gull	10	100	0.2	23	23
Stevenson	33	7	0.6	138	10

Attempts to cultivate plants in Ithaca in order to check on the validity of results obtained in the field were unsuccessful. Altogether, I brought or sent back 287 living plants for propagation. A few plants in a flat of moist cinders in the greenhouse survived longest, more than a year, and even produced floral buds. These behaved as perennials. Other plants in a pot in a cold frame quickly succumbed. Eventually all plants died before yielding meaningful results.

Because of the small size of plants and the need to take whole plants for microscopic study, measurements of fruits and seeds could not be of the same specimens examined in flower. Instead, a separate collection of fruiting plants was necessary. These came from the same quadrats as the flowering plants: 71 plants from Big Gull Island

and 19 from Stevenson Island. My departure from the other islands was too early to get fruiting material except for single specimens with fruits of the preceding year from Middle Kécarpoui Island and the Eastern Isle Affligée.

Variation in dimensions of fruits and seeds is slight. For this reason, although fruits are available from 92 plants, I measured only a few. Dimensions of one seed per plant and one plant per population are adequate to detect a difference of .2 mm. in length at the 95% level of confidence. Actually, I measured two seeds per plant and, to keep samples proportionate, 18 plants from Big Gull Island and 2 plants from Stevenson Island.

Plants of *Sedum villosum* are annual herbs, rarely becoming perennial by means of axillary shoots. The roots are fibrous and white. The stems are erect, glandular-hirtellous, and usually unbranched, though as many as 4 floriferous branches may occur. Before expansion of the flowers, the floriferous stems are bent downward.

	n-pop.	n-pl.	n-obs.	$\bar{x}$ (cm.)	s (cm.)
Height at time of flowering	4	71	71	2.7	.65

Leaves are in spiral arrangement, very short-spurred, elliptic-oblong, rounded, mostly appressed, glandular-hirtellous, and green usually streaked with dark red. Dimensions are of median leaves on floriferous stems, that is leaves at or just above the midpoints of the stems, halfway between the surface of the ground and the bases of the cymes. Data are for plants with the first flowers in anthesis.

	n-pop.	n-pl.	n-obs.	$\bar{x}$ (mm.)	s (mm.)
Length	4*	71	75	4.3	.74
Width	4	71	75	1.6	.22
Thickness	4**	71	75	.8	.09

Inflorescences are cymes of usually single, but rarely two cincinni. Auxillary branches with one or two flowers sometimes develop below the terminal cincinni.

	n-pop.	n-pl.	n-obs.	$\bar{x}$	s
Cincinni (n)	4**	71	76	1.1	.22
Flowers per floriferous stem (n)	4**	71	76	2.9	1.55

Flowers are usually 5-merous, but sometimes 4- or even 3-merous. The pedicels are glandular-hirtellous. All dimensions apply to first flowers of cymes.

	n-pop.	n-pl.	n-obs.	$\bar{x}$ (mm.)	s (mm.)
Pedicels—length	4**	70	75	2.8	.09
Flowers—diameter	4**	58	63	8.0	.73

Sepals are without spurs, lanceolate or ovate-elliptical, obtuse, glandular-hirtellous, and green speckled with red or sometimes dark purple.

	n-pop.	n-pl.	n-obs.	x̄ (mm.)	s (mm.)
Length	4*	65	70	2.4	.38
Width	4	65	70	1.2	.18

Petals are either ovate or elliptic-ovate, obtuse, and sparingly glandular-ciliate, with the dorsal keels glandular-hirtellous. They usually are some shade of pink. Petals of the first flowers in cymes sometimes are dark pink, but those of the other flowers are pale pink or pinkish white. In bud, the petals are convolute.

	n-pop.	n-pl.	n-obs.	x̄	s
Number per flower	4**	64	69	4.8	.41
Length (mm.)	4**	58	61	4.2	.40
Width (mm.)	4**	58	61	2.9	.38

Stamens have white filaments and red anthers.

	n-pop.	n-pl.	n-obs.	x̄ (mm.)	s (mm.)
Adnation of epipetalous filaments	4	52	53	.25	.19
Anthers—length	3	8	9	.41	.07
Anthers—diameter	3	8	9	.28	.09

Nectaries are reniform or subquadrate, truncate or emarginate, and bright red.

	n-pop.	n-pl.	n-obs.	x̄ (mm.)	s (mm.)
Length	4**	58	61	.22	.03
Width	4**	58	61	.42	.06

Carpels are erect, abruptly contracted to short styles, minutely glandular-hirtellous ventrally, and either yellow-green in flowers with pale pink petals or red in flowers, usually the first in cymes, with deep pink petals. As they mature, the pistils become dark purple before dehiscence, with the ventral lips slightly thickened. Also, they become slightly divergent.

	n-pop.	n-pl.	n-obs.	x̄	s
Length (mm.)	4**	58	61	3.1	.29
Ovules per ovary (n)	4**	68	69	10.3	2.43

Fruits are suberect, sparsely hirtellous ventrally, and dark purple with short divergent beaks. Many ovules do not develop into seeds. This is indicated by the smaller number of seeds per follicle than ovules per ovary.

	n-pop.	n-pl.	n-obs.	x̄	s
Length (mm.)	2	20	44	2.4	.5
Length of beaks (mm.)	2	20	43	.5	.07
Seeds per follicle (n)	2	20	79	3	1.9

Seeds are pyriform, wing-margined, amber-yellow, lustrous, irregularly ribbed, and verrucose.

	n-pop.	n-pl.	n-obs.	x̄ (mm.)	s (mm.)
Length	2*	20	57	.71	.06
Diameter	2	20	57	.29	.03

No information on chromosomes is available for North American plants. My living material died before counts were possible.

Variation. Prior to visiting the Gulf of St. Lawrence to study *Sedum villosum*, I supposed that the populations there were small and would show little diversity. Instead, I found considerable variation, not only among plants, but also among populations. In addition, the plants differed from published descriptions, such as that of Fröderström (1930–1935), in being of lower stature (maximum height 4 cm.) and in having shorter pistils, features which might result from the rigorous environment. No plants in my samples were glabrous, as var. *glabrum*, described by Hamet from Iceland, or with petals 6 mm. long, as in var. *arcticum* Fröderström, from eastern Greenland.

Flowers vary in number of petals. About 80% have 5 per corolla. The following brief chart shows the distribution for the 64 flowers cited in the description.

Petals per flower (n)	Flowers (n)	Flowers %	Cumulative %
3	1	1.562	1.562
4	12	18.750	20.312
5	51	79.687	100.000

To check on the possibility of seasonal changes in variation, I collected all 8 plants in flower in quadrat no. 29 on Big Gull Island on July 15, and again 16 plants on August 12. Measurement and comparison of these two collections revealed significant differences in three characters:

Character	July x̄	August x̄
Height (cm.)*	2.5	1.4
Floriferous stems per plant (n)**	9.6	1.1
Flowers per floriferous stem (n)*	2.6	1.6

The conclusion is that plants which flower later are smaller and have fewer blossoms, but the flowers are as large and have as big floral parts as those which bloom earlier in the season. The smaller size of the whole plant may result from a later start in growth.

Although I expected that populations of *Sedum villosum* in the Gulf of St. Lawrence would not differ significantly from each other, the studies of plants in the field indicate significant differences in sixteen characters. The failure of plants to grow satisfactorily in Ithaca precludes the possibility of checking the results from the field with experimental data. One possible measure of reliability of the data from the field is to compare differences among populations with the data from the study of seasonal variation on Big Gull Island. The sample from Stevenson Island had the smallest dimensions in eleven characters studied. Because the plants were collected late in the season, August 10–12, their smallness might result from their lateness. This does not seem to be the case, however, because nine of the characters are ones in which no significant differences exist between the July and August collections from Big Gull Island. A tenth character is length of seeds. This was not contrasted on a seasonal basis. If the differences are not genetic, they are more likely edaphic than temporal modifications. The following triangular table shows the number of significant differences among the various populations. Tukey's test of all comparisons among means is the basis for distinguishing among the means. Letters stand for populations as follows: A = Western Isle Affligée, G = Big Gull Island, K = Middle Kécarpoui Island, and S = Stevenson Island. The difference in length of seeds between plants on Big Gull and Stevenson Islands is omitted from this table because variances are lacking for two populations.

	S	G	A
K	8	3	8
A	10	5	
G	9		

To demonstrate further the differences among these populations, the sixteen characters in which significant differences occur are listed in another table, with means and standard deviations for each population, and also D, at the right, for the application of Tukey's test.

	K		A		G		S		D
	$\bar{x}$	s	$\bar{x}$	s	$\bar{x}$	s	$\bar{x}$	s	
Floriferous stems per plant (n)	1	0	1	0	4.2	5.8	1	0	3.03
Leaves—length (mm.)	4.4	.7	4.4	.8	4.3	.7	3.5	.7	.67
Leaves—width (mm.)	.81	.22	.84	.1	.81	.09	.7	.09	.09
Cincinni per floriferous stem (n)	1	0	1.25	.34	1	0	1	0	.20
Flowers per floriferous stem (n)	2.1	.6	3.9	2.3	2.8	.7	1.5	.7	1.41
Pedicels—length (mm.)	2.1	.6	3.3	1	2.7	.6	2.6	1.0	.78
Flowers—diameter (mm.)	7.8	.8	8.7	.9	8.2	.6	6.8	.5	.73
Sepals—length (mm.)	2.4	.2	2.6	.3	2.3	.5	2.2	.2	.36
Petals (n)	5	0	4.5	.6	4.8	.4	5	0	.37
Petals—length (mm.)	4.5	.2	4.1	.4	4.4	.4	3.7	.5	.41
Petals—width (mm.)	3	.4	3.2	.4	2.9	.4	2.5	.4	.37
Nectaries—length (mm.)	.2	.16	.2	0	.23	.22	.2	.09	.03
Nectaries—width (mm.)	.4	.09	.47	.23	.43	.27	.31	.22	.06
Pistils—length (mm.)	3.3	.3	3.5	.38	3.2	.18	2.7	.34	.29
Ovules per ovary (n)	13	1.4	10.2	2.9	10.1	2.3	8.1	1.7	2.28
Seeds—length (mm.)	.6	—	.7	—	.7	.06	.6	.06	—

Nomenclature. *Sedum villosum* Linnaeus, Species Plantarum 1:432 (1753). Type locality: Europe—Germany, England, and France. Type: no specimen is available with this name written by Linnaeus. The original description, though brief, is reasonably clear. This and the supporting references provide information about pubescence, color, and habitat. The older of the two references is to L'Ecluse, *Rariorum plantarum historia* 2:59 (1601). Selection of a lectotype may await better understanding of variation of the species in Europe. Preferably, the lectotype should be one of the old, but well-preserved collections from Germany.

Synonyms are:

1827. *Sedum glandulosum* Moris, Stirpium Sardoarum Elenchus, fasc. 1, p. 20. Type locality: Mt. Morgongiori in western Sardinia. Type: presumably a collection of Moris which I have not seen. The description and an illustration in the Flora Sardoa are the basis for assignment of the name to synonymy here.

1840–1843. *Sedum glandulosum* var. *minus* Moris, Flora Sardoa 2:121, pl. 73, fig. 4. Type locality: Fontana-cungiada, above Aritzo in the Gennargentu Mountains of central Sardinia. Type: a collection of Moris, not seen. The cited illustration is the basis for interpretation of the name. Distinctive features of the variety are small flowers and almost white petals.

1861. *Sedum insulare* Moris, "Enumeratio Seminum Regii Horti Taurinensis, p. 33." Type locality: unknown. Type: not seen.

Distribution. The range of *Sedum villosum* (fig. 76) is wide: the mountains of Europe, islands of the western Mediterranean Sea, the Atlas Mountains of northern Africa (var. *aristatum* Emb. et Maire, 1,300–2,000 m., Jahandiez and Maire, 1931–1941), islands in the north Atlantic Ocean, including Iceland and Greenland, and small islands along the northern shore of the Gulf of St. Lawrence in North America.

Details for the populations which I studied in the Gulf of St. Lawrence are summarized in the following tables. The figures for density apply only to the areas (stratum) in which *Sedum villosum* occurs. The sizes of these areas are indicated.

Population	Area (m.^{2})	Density per m.^{2}	N plants, estimate	N plants, C.I.
Middle Kécarpoui Island	25	149.5	3,737	299–36,863
Eastern Isle Affligée	52	228.6	11,773	1,143–33,829
Western Isle Affligée	31	202.7	6,282	608–28,882
Big Gull Island	80	147.6	11,810	1,181–35,293
Stevenson Island	6	248	1,488	496–21,381

Population	Altitudinal range (m.)	pH	Drainage	Months of flowering	Oldest record
Middle Kécarpoui Island	20	6.6–7.4	poor	July	1915
Eastern Isle Affligée	3–5	6.6–7	poor-moderate	July	1915
Western Isle Affligée	1–3	6.8–7	poor-moderate	July	1915
Big Gull Island	1–1.5	6.6–7	poor	July–Aug.	1928
Stevenson Island	2	7.0–7.2	poor	Aug.	1959

Harold St. John discovered *Sedum villosum* on a turfy hilltop on Ile Kécarpoui (= Middle Kécarpoui Island) on Aug. 11, 1915. In the same year, he also found it on the Iles Affligées. Harrison F. Lewis found it in 1928 on St. Augustin Square south of St. Augustin (see map sheet 12 N.E., Harrington–Belle Isle, National Topographic Series of Canada) and in the Mecatina Sanctuary (possibly Big Gull Island, see hydrographic chart 4469, Flat I. to Little Mecatina I.), and I found it on Stevenson Island, northernmost of the Bald Islands, in 1959. In addition, in 1959, Charles Osborne, of Harrington Harbour, reported seeing it on American Harbour Island, about 8 kilometers west of St. Augustin Square, and a Mr. Ellerd, of Toronto, reported it from Little Fishery Island south of La Tabatiére. The known extent along the coast is about 80 kilometers.

Although in 1959 I did not find *Sedum villosum* in any of the randomly selected quadrats of the prearranged survey, I did see it on five islands which either had quadrats on them or were specially visited because of its known occurrence there. For the area surveyed, the fraction known to be inhabited by *S. villosum* is very small, namely .000014. The plants occur in few, sharply delimited patches, mostly near shore, in shallow, calcareous soils.

The arrival of *Sedum villosum* on islands in the Gulf of St. Lawrence may have been by means of seeds in floating pieces of ice which originated on the shores of Greenland or even Iceland. This seems reasonable because the plants grow in moss (fig. 85) and turf in shallow depressions in gneiss gently sloping toward shore and just above the

Fig. 85. Sedum villosum growing in moss on western Isle Affligée in northern part of Gulf of St. Lawrence, Que., July 23, 1959.

usual level of high tide. Waves must occasionally wash across such sites, as is indicated by shells of *Mollusca*, pieces of *Fucus*, and driftwood where the *Sedum* occurs. The northern coast of the Gulf of St. Lawrence is rising at a rate of about 33 millimeters per year according to Lewis (1931–1932). The site on Middle Kécarpoui Island, where plants of *S. villosum* are growing in calcareous black clay at an altitude of 20 meters, may have been at sea level and an outer island less than two thousand years ago and under the sea before then. If all present habitats for *S. villosum* in the Gulf of St. Lawrence are younger than two thousand years, a recent date for the arrival of the species there is likely. Seeds may have come either on pieces of ice or in beds of *Fucus* which floated through the Straits of Belle Isle from the northern Atlantic Ocean.

Other species most often competing with *Sedum villosum*, and the number of quadrats out of 20 in which each occurred, are *Epilobium palustre*—16, *Empetrum nigrum*—13, *Rhinanthus borealis*—13, *Euphrasia arctica*—12, and *Sedum rosea*—12.

Reproduction. On islands in the Gulf of St. Lawrence, *Sedum villosum* usually is an annual. Propagation depends on the production of seeds, which in turn depends on pollination by insects. Effective pollinators appear to be *Sphaerophoria cylindrica* (identification by Stuart Neff), a syrphus fly (fig. 86) with slender abdomen with five alternating black and yellow bands, and also small black flies. I have watched these

Fig. 86. Syrphus fly on flower of *Sedum villosum*, Big Gull Island in northern part of Gulf of St. Lawrence, Que., July 15, 1959.

insects visiting the flowers on Big Gull Island in the early morning, when the air was calm. The syrphids hover over the flowers, land on them with the thoraces over the pistils, and then move their abdomens up and down. This behavior is in agreement with Günthart's (1902) observation that insects visiting flowers of *Crassulaceae* almost always land on the tips of the styles. On the other hand, the small flies, which are faster in their movements, seem to go to the nectaries. Besides visiting blossoms of *Sedum*, the syrphids visit the yellow flowers of *Potentilla anserina*.

The wind usually is strong at sites where *Sedum villosum* grows. As a result, circumstances for the activity of pollinating insects are both of short duration and infrequent. On the other hand, conditions are excellent for the dispersal of the tiny seeds by the wind.

Relationships. *Sedum villosum* appears to have no close relatives in North America. Its relationships probably are with species of Europe. Some possible related species are *S. dasyphyllum*, *S. magellense*, *S. pedicellatum*, and *S. alsinefolium*. These species have orthocarpic carpels and short styles. *Sedum dasyphyllum* is perennial and usually glandular-hairy, with the leaves either opposite or alternate and the petals pinkish white. It is native in southern Europe and northern Africa. *Sedum magellense*, also perennial, has glabrous, yellow-green leaves and elongate cymes. Its range includes the Italian and Balkan Peninsulas, as well as western Asia Minor. *Sedum pedicellatum* is annual and glabrous, except that the carpels are papillose ventrally. It occurs in the mountains of the Iberian Peninsula. *Sedum alsinefolium*, with spatulate, petiolate basal leaves may be more distantly related. It is prominently glandular-pubescent with white petals which are connate for 1 millimeter. Its range is in the Maritime Alps of southeastern France and northwestern Italy.

Subgenus *Gormania*

Plants of *Sedum*, subgenus *Gormania*, are perennial herbs with prominent rosettes of spatulate or oblanceolate leaves at time of flowering and with either horizontal rootstocks or stout decumbent stems. With the exception of one species, the chromosomes normally are in multiples of 15. The petals of all except one species are basally connate.

On the basis of gross morphology, the species of *Gormania* are moderately advanced. Quimby (1971) studied the floral anatomy of *Sedum oreganum* and two subspecies of *S. spathulifolium*. He found four whorls of vascular traces, the pattern which he designated as group 3, with the dorsal traces of the carpels arising fused with the traces of whorl 2 for the petals and stamens. Although the vascular pattern of these two species of *Gormania* is not markedly different from that of most species of subgenus *Sedum*, it is more advanced than that of the more primitive species of subgenus *Sedum*.

The known distribution of the eight species of subgenus *Gormania* is in the Pacific Mountain System from the Transverse Ranges of California northward to the Coast Batholith of British Columbia. *Gormania* is an endemic American group of species which probably originated in western North America.

Britton (1903) was the first to recognize the taxonomic distinctiveness of species of *Gormania*. He proposed *Gormania* as a genus, naming it for M. W. Gorman of Portland, Oregon. He designated as type species *Cotyledon Oregonensis* S. Watson (= *S. oregonense*). In 1942, I published reasons for reducing *Gormania* to subgeneric status and listed it as subgenus *Gormania* (Britton) R. T. Clausen, Bull. Torr. Bot. Club 69:28 (1942).

As presently conceived, subgenus *Gormania* comprises three sections, each sufficiently distinct to raise doubt whether they had a common origin.

Section *Oreganica* R. T. Clausen, Bull. Torr. Bot. Club 69:28 (1940). Type species: *Sedum oreganum* Nuttall. Characteristics of the section and only species are the narrowly lanceolate, aristate petals which are erect, but divergent upward; shiny, clavate, spatulate leaves; and erect follicles. The relationships of *S. oreganum* are unclear. Sometimes I have thought that it is too distant from the other species of *Gormania* to be included with them; see Clausen and Uhl (1944). Further experience has removed some of the doubt.

Section *Gormania* (Britton) R. T. Clausen, listed as Section *Eugormania* in Bull. Torr. Bot. Club 69:29 (1940). Type species: *Sedum oregonense* (S. Watson) M. E. Peck. Distinguishing features of the section are the petals which are connate basally and divergent upward; the vegetative rosettes with the leaves thick and leathery; and the often paniculate inflorescences. Included species are *S. obtusatum*, *S. laxum*, *S. oblanceolatum*, *S. oregonense*, *S. albomarginatum*, and *S. moranii*.

Section *Rosulata* (Berger) Clausen et Uhl, Madroño 5:166 (1944). Type species: *Sedum spathulifolium* Hooker. Distinguishing features are the petals separate to their bases, erect below for about one-tenth of their length, then widely spreading; vegetative rosettes with the leaves fleshy, but not leathery; and inflorescences which are 3-parted cymes and sometimes compound. The section had its nomenclatural inception as Group *Rosulata* Berger, in Engl. & Prantl, Nat. Pflanzenfam., ed. 2, 18a:457 (1930).

20. *Sedum oreganum**** (figs. 87–91)

The leaves of *Sedum oreganum* (fig. 87) are distinctive. They are shiny, often suffused with red, spatulate, clavate, and borne in spirals. The flowers are yellow, in leafy-bracted cymes, with erect, narrow petals, 9–11 mm. long. The follicles are erect. *Sedum divergens*, also with shiny, reddish-green foliage and yellow flowers, has the leaves opposite and often globular, the petals widely spreading and shorter, 5–7 mm. long, and the follicles both divaricate and gibbous. *Sedum spathulifolium*, another yellow-flowered species, likewise has spreading petals and divergent, gibbous follicles. *Sedum obtusatum* and *S. oregonense* have nonlustrous leaves and broader petals. Of all these species, only *S. oreganum* has the leaves clavate.

Description. Sample: 21 plants—14 in flower and 9 in fruit at time of collection. Plants came from the Cascade Mountains in 1961 and 1965 and the Coast Ranges of Oregon and Washington in 1963. Specimens from the Cascade Mountains, with

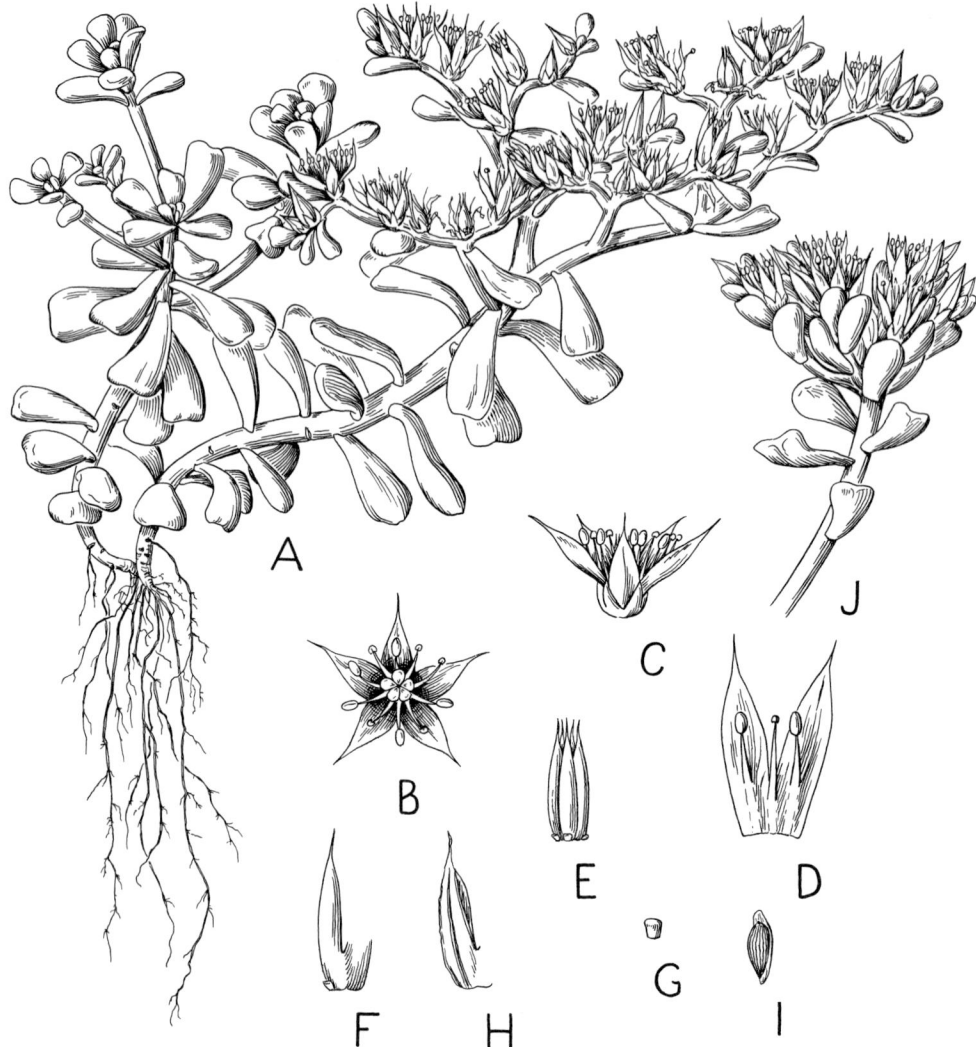

Fig. 87. Plant of *Sedum oreganum* from Seaview, Pacific Co., Wash., cultivated in greenhouse, Ithaca, N.Y. A. Habit sketch (x .9). B. Flower from above (x 1.8). C. Flower from side (x 1.8). D. Petal and two stamens (x 2.7). E. Carpels (x 2.7). F. Single carpel and nectary, and also base of a second carpel (x 3.6). G. Nectary (x 4.5). H. Follicle (x 3.6). I. Seed (x 9). J. Upper part of floriferous stem (drawn from specimen collected in 1963 at Seaview) (x .9).

the exception of Bald Mountain, are from deliberately selected primary sampling units in an optimum stratum. Those from the Coast Ranges are from optimum substrata within two randomly selected primary units. Details about numbers of plants in sampling units and sizes of samples are available in the following table.

Locality	Total plants N_u	Plants in sampled clusters N_c	Plants studied			Plants studied in	
			n	n/N_c	n/N_u	fl. (n)	fr. (n)
Pacific Coast Ranges							
Millicoma River	>15	>15	2	.13	.13	2	0
Coos Ridge	1	1	1	1	1	1	0
Beard's Hollow	2,000	2,000	2	.001	.001	2	0
Seaview	500	9	10	1	.02	3	7
Long Island	37	37	1	.27	.27	1	0
Cascade Mountains							
O'Leary Mt.	1	1	1	1	1	1	0
Wedge Mt.	53	53	3	.06	.06	3	2
Bald Mt.	130	3	1	.3	.008	1	0
Totals	2,737	2,119	21	.0099	.0077	14	9

Eight plants of *Sedum oreganum* were part of an experiment of completely random design in the greenhouse (gh.). Five of these plants flowered and produced fruits. Data for rosettes and leaves of plants in the greenhouse are from observations in the summer of 1969. In the cold frames (cf.), plants grew vigorously and flowered, but they were not in planned experiments. A plant from Wedge Mountain, which flowered in July 1969, is the source of cold frame data for the Cascade Mountains in the following description. *Sedum oreganum* grows poorly in the garden at Ithaca. It can survive the winter, but does not persist for long.

Plants of *Sedum oreganum* are glabrous, perennial herbs with fibrous roots and stout, decumbent, branched stems which are 2–5 mm. in diameter and pale yellowish green or pink. The stems are 2.5–15 cm. long, leafy toward the ends, and with slender offsets. Certain stems give rise to terminal floriferous shoots in the second year of growth.

	n-pop.	n-pl.	n-rep.	n-obs.	$\bar{x}$	s
Plants—diameter (cm.)						
Coast Ranges—gh.	4	7	14	14	22.4	2.6
Cascade Mts.—gh.	1	1	2	2	26.5	—
Primary rosettes—diameter (cm.)						
Coast Ranges	5	8	—	34	2.2	.4
—gh.	5	7	14	70	2.3	.3
Cascade Mts.	2	2	—	5	2	—
—gh.	1	1	1	10	1.2	—
Offsets per rosette (n)						
Coast Ranges—gh.	3	5	9	18	1	.9
Cascade Mts.—gh.	1	1	1	2	2	—
Offsets—length (cm.)						
Coast Ranges—gh.	3	4	5	7	.3	.2
Cascade Mts.—gh.	1	1	1	1	.3	—
Offsets—diameter of stems (mm.)						
Coast Ranges	5	8	—	34	1.5	.3
—gh.	2	3	4	6	1.2	0
Cascade Mts.	2	2	—	5	.8	—
—gh.	1	1	1	1	1.2	—
Floriferous stems per plant (n)						
Cascade Mts.—cf.	1	1	1	1	18	—

	n-pop.	n-pl.	n-rep.	n-obs.	$\bar{x}$	s
Floriferous stems—length (cm.)						
Coast Ranges	5	9	—	25	8	2.1
—gh.	1	2	2	3	10	7.4
Cascade Mts.	1	3	—	17	13	1.7
—cf.	1	1	1	2	9	—

Leaves are spirally arranged and in rosettes toward the ends of the primary stems. They are short-spurred, cuneate, spatulate, broadly rounded or truncate, thickest near the apices, lustrous, and light green suffused with red. Leaves of the floriferous stems are similar to those of the rosettes, but smaller.

	n-pop.	n-pl.	n-rep.	n-obs.	$\bar{x}$ (mm.)	s (mm.)
Leaves of rosettes—length						
Coast Ranges	5	8	—	34	9.7	2.6
—gh.	5	7	14	70	12	2.2
Cascade Mts.	3	5	—	22	11	1.4
—gh.	1	1	1	10	7	—
Leaves of rosettes—width						
Coast Ranges	5	8	—	34	9	1.5
—gh.	5	7	14	70	6.3	.9
Cascade Mts.	3	5	—	22	7.1	1.4
—gh.	1	1	1	10	4.4	—
Leaves of rosettes—thickness						
Coast Ranges	5	8	—	34	2.7	.3
—gh.	5	7	14	70	2.1	.1
Cascade Mts.	3	5	—	22	2	.2
—gh.	1	1	1	10	1.5	—
Median leaves of floriferous stems—length						
Coast Ranges	4	5	—	7	12.3	2.8
—gh.	1	1	1	1	13	—
Cascade Mts.	2	2	—	9	10	—
—gh.	1	1	1	1	10	—
Median leaves of floriferous stems—width						
Coast Ranges	4	5	—	7	8.9	3.5
—gh.	1	1	1	1	8	—
Cascade Mts.	2	2	—	9	7	—
—cf.	1	1	1	1	6	—
Median leaves of floriferous stems—thickness						
Coast Ranges	4	5	—	7	2.2	.4
—gh.	1	1	1	1	3	—
Cascade Mts.	2	2	—	9	2.5	—
—cf.	1	1	1	1	1.9	—

Inflorescences (fig. 88) are corymbose cymes, 1.5–5 cm. in diameter, with spurred, spatulate or oblanceolate floral bracts, 5–8 mm. long. The primary branches of the inflorescence are dichotomously forked.

	n-pop.	n-pl.	n-rep.	n-obs.	$\bar{x}$ (n)	s (n)
Primary branches per cyme						
Coast Ranges	5	8	—	25	3	.3
—gh.	1	2	2	3	3	0

NATIVE SPECIES 349

Fig. 88. Inflorescences of *Sedum oreganum* ssp. *oreganum*, plant on rocks north of Seaview, Pacific Co., Wash., July 26, 1963.

	n-pop.	n-pl.	n-rep.	n-obs.	$\bar{x}$ (n)	s (n)
Primary branches per cyme						
Cascade Mts.	2	2	—	4	3	—
—cf.	1	1	1	2	3	—
Flowers per cyme						
Cascade Mts.	3	5	—	21	9	.8

Flowers are either sessile or on pedicels to 3 mm. long. The receptacles are flat and disklike. The flowers are fragrant.

	n-pop.	n-pl.	n-rep.	n-obs.	$\bar{x}$ (mm.)	s (mm.)
Pedicels—length						
Coast Ranges	4	7	—	20	.6	.2
Cascade Mts.	1	3	—	17	.3	.4
Flowers—diameter						
Coast Ranges	4	7	—	28	13	1.4
—gh.	1	2	2	4	14	3.5
Cascade Mts.	3	5	—	20	12	2.7

Sepals are erect, ovate-lanceolate, acuminate or acute, and green, often suffused with red.

	n-pop.	n-pl.	n-rep.	n-obs.	$\bar{x}$ (mm.)	s (mm.)
Length						
Coast Ranges	4	7	—	28	3.4	.1
—gh.	1	2	2	4	3.6	.3
Cascade Mts.	3	5	—	21	3.7	.2
—cf.	1	1	1	2	3.3	—
Width						
Coast Ranges	4	7	—	28	2	.2
—cf.	1	2	2	4	1.8	.3
Cascade Mts.	3	5	—	21	1.8	.1
—cf.	1	1	1	2	1.6	—

Petals (fig. 89) are erect, but divergent upward and in age, connate basally, narrowly lanceolate, carinate, aristate, and primrose, empire, or sulphur yellow, sometimes green or red on dorsal keels. Rarely the petals are yellowish white.

Fig. 89. Flowers of a plant of *Sedum oreganum* ssp. *oreganum* from along the Millicoma River, Coos Co., Ore., cultivated in greenhouse, Ithaca, N.Y., June 2, 1964 (x 3.8), showing narrow petals with aristate tips. Photo by Howard Lyon.

NATIVE SPECIES

	n-pop.	n-pl.	n-rep.	n-obs.	$\bar{x}$	s
Number per flower						
Coast Ranges	4	7	—	28	5	0
—gh.	1	2	2	4	5	0
Cascade Mts.	3	5	—	21	5.2	.4
—cf.	1	1	1	2	5	—
Length (mm.)						
Coast Ranges	4	7	—	28	11.1	.3
—gh.	1	2	2	4	11	.4
Cascade Mts.	3	5	—	21	9.7	.6
—cf.	1	1	1	2	9.4	—
Length of cohesion (mm.)						
Coast Ranges	4	7	—	28	2.3	.3
—gh.	1	2	2	4	2.7	.7
Cascade Mts.	3	5	—	21	1.2	.1
—cf.	1	1	1	2	1	—
Width (mm.)						
Coast Ranges	4	7	—	28	2.7	.1
—gh.	1	2	2	4	3.1	0
Cascade Mts.	3	5	—	21	2.3	.2
—cf.	1	1	1	2	2	—

Stamens have yellowish or greenish filaments and yellow anthers.

	n-pop.	n-pl.	n-rep.	n-obs.	$\bar{x}$ (mm.)	s (mm.)
Epipetalous filaments; adnation—length						
Coast Ranges	4	7	—	28	3.3	.4
—gh.	1	2	2	4	3.2	.4
Cascade Mts.	3	5	—	21	2.2	.3
—cf.	1	1	1	2	2.6	—
Anthers—length						
Coast Ranges	2	2	—	2	1	—
—gh.	1	2	2	2	1.2	.3
Cascade Mts.	1	3	—	6	.9	.2

Nectaries are subquadrate or spatulate and yellow.

	n-pop.	n-pl.	n-rep.	n-obs.	$\bar{x}$ (mm.)	s (mm.)
Length						
Coast Ranges	4	7	—	28	.46	.02
—gh.	1	2	2	4	.55	.07
Cascade Mts.	3	5	—	21	.51	0
—gh.	1	1	1	2	.5	—
Width						
Coast Ranges	4	7	—	28	.5	.05
—gh.	1	2	2	4	.5	.07
Cascade Mts.	3	5	—	21	.5	.03
—cf.	1	1	1	2	.6	—

Carpels are erect and either yellowish or pale green.

	n-pop.	n-pl.	n-rep.	n-obs.	$\bar{x}$	s
Length (mm.)						
Coast Ranges	4	7	—	28	6.4	.5
—gh.	1	2	2	4	6.4	.07

	n-pop.	n-pl.	n-rep.	n-obs.	$\bar{x}$	s
Length (mm.)						
Cascade Mts.	3	5	—	21	6	.04
—cf.	1	1	1	2	5.3	—
Length of cohesion (mm.)						
Coast Ranges	4	7	—	28	1.4	.2
Cascade Mts.	2	2	—	4	1.4	—
Ovules per ovary (n)						
Coast Ranges	4	7	—	28	13.5	1.1
—gh.	1	2	2	4	13	1.1
Cascade Mts.	3	5	—	21	15.9	.7
—cf.	1	1	1	2	18	—

Fruits are erect and light brown or reddish.

	n-pop.	n-pl.	n-rep.	n-obs.	$\bar{x}$	s
Length						
Coast Ranges	1	7	—	14	4.9	.2
—gh.	3	4	5	10	4.7	0
Cascade Mts.	1	2	—	3	4.1	.4

Seeds are pyriform, longitudinally ridged and finely ribbed, winged, short-tailed at both ends, and brown.

	n-pop.	n-pl.	n-rep.	n-obs.	$\bar{x}$ (mm.)	s (mm.)
Length						
Coast Ranges	1	7	—	14	1	.08
—gh.	2	2	3	4	1.1	0
Cascade Mts.	1	2	—	12	1	.07
Diameter						
Coast Ranges	1	7	—	14	.4	.02
—gh.	2	2	3	4	.5	0
Cascade Mts.	1	2	—	12	.3	.02

Chromosomes are small and unequal in size. Some are elongate and two-armed. Others appear tiny and round.

	n-pl.	g (n)	sp (2n)	Total length (μ)	Average length (μ)	Cytologist
Coast Ranges						
Coos Ridge	1	—	24	17	.7	Barbara Joyce, 1970
Beard's Hollow	1	—	24	—	—	Barbara Joyce, 1970
Seaview	1	—	24	25	1.1	Barbara Joyce, 1970
Lummi Island	1	12	—	—	—	C. H. Uhl in Clausen and Uhl (1944)
Cascade Mts.						
O'Leary Mt.	1	—	24	14.5	.6	Barbara Joyce, 1970
Wedge Mt.	1	—	24	24	1	Barbara Joyce, 1970

Variation. Casual inspection of *Sedum oreganum* suggests a species with little infraspecific variation. A closer view reveals that plants along the Pacific Coast are robust, while those in the Cascade Mountains are slender. Within populations, variation occurs in pigmentation of the leaves, color of the petals—whether deep or pale yellow—and in number of petals and other floral parts per flower.

Analyses of variance of data for twenty-eight dimensions of wild plants revealed significant differences in six features between populations of the Coast Ranges and Cascade Mountains.

Feature	Coast Ranges	Cascade Mts.
Stems of offsets—diameter (mm.)**	1.5	.8
Leaves of rosettes—thickness (mm.)**	2.7	2
Leaves of floriferous stems—length (mm.)*	12.3	10
Nectaries—length (mm.)*	.46	.51
Fruits—length (mm.)**	4.9	4.1
Seeds—diameter (mm.)*	.4	.3

Plants from the Cascade Mountains do not grow as well in conditions at Ithaca, N.Y., as plants from the Coast Ranges. Neither do they flower as frequently. As a consequence, experimental testing of morphological differences has been limited. Only the difference in thickness of leaves could be tested. That continues to be significant, 2.1 versus 1.5 mm. Data for all morphological features studied are in the detailed description. The difference in growth appears to be a fundamental physiological distinction between the two regional arrays of populations.

Two distinct shades of yellow were evident in flowers of different plants at Beard's Hollow near Seaview in southwestern Washington. The prevailing color of the petals was empire yellow, but some plants had yellowish white petals. This difference in color continued in cultivation, indicating heritability rather than environmental modification.

Number of petals per flower is remarkably constant in flowers of plants in the Coast Ranges. Minor fluctuation occurs among plants in populations in the Cascade Mountains, but the regional difference is not significant. The results of counting petals of 49 flowers are indicated in the following table:

Petals per flower (n)	Flowers (n)	Flowers %	Cumulative %
5	47	96	96
6	2	4	100
Total	49		

The differences between the populations of *Sedum oreganum* in the Coast Ranges and those in the Cascade Mountains favor recognition of two subspecies. Nuttall collected the type specimen of the species from rocks at Chinook near the mouth of the Columbia River in Pacific Co., Wash. Although the herbarium specimen of the type plant is inadequate for subspecific identification, the locality indicates that Nuttall's collection belongs to the subspecies of the Coast Ranges. Further, my samples from Seaview and Beard's Hollow, only 10 km. from Chinook, are in agreement with this idea.

For purposes of identification, the reader may consult the table, several paragraphs preceding, of significant differences between populations of the Coast Ranges and Cascade Mountains, as well as the fuller data available in the description. Plants of the Coast Ranges comprise the ssp. *oreganum*. Those of the Cascade Mountains are named below as a distinctive subspecies.

Sedum oreganum ssp. oreganum

Nomenclature. *Sedum Oreganum* Nuttall in Torrey and Gray, Flora No. Am. 1:559 (1840). Type locality: on rocks at Chinook, Pacific Co., Wash., on the estuary of the Columbia River, cited in the original description as rocks near the mouth of the Oregon. Type: a collection of Nuttall (BM), probably 1835. No date appears on the label accompanying the type specimen, but Nuttall was at the estuary of the Columbia River on July 4, 1835. See Graustein (1967).

Synonyms are:

1903. *Gormania Oregana* (Nutt.) Britton, Bull. N.Y. Bot. Gard. 3:30. Based on *Sedum Oreganum* Nutt. in Torrey and Gray, Flora No. Am. 1:559 (1840).

1904. *Cotyledon oregana* (Nutt.) Fedde, Just's Bot. Jahresb. 31 (1):827. Based on *Gormania Oregana* (Nutt.) Britton, Bull. N.Y. Bot. Gard. 3:30 (1903).

1913. *Echeveria oregana* (Nutt.) Nels. et Macbr., Bot. Gaz. 56:476. Based on *Sedum Oreganum* Nutt. in Torrey and Gray, Flora No. Am. 1:559 (1840).

Plants of ssp. *oreganum* are more robust and larger than those of ssp. *tenue*. On coastal cliffs, they may have compact basal rosettes, similar in general aspect to *S. oregonense*, and petals which are erect below and then diverge above the middle.

Sedum oreganum ssp. *oreganum* occurs (fig. 90) in the Coast Ranges of Oregon and Washington, the Puget Trough, and the Coast Batholith of British Columbia and southern Alaska, from the valley of the Millicoma River in Coos Co., Ore.,

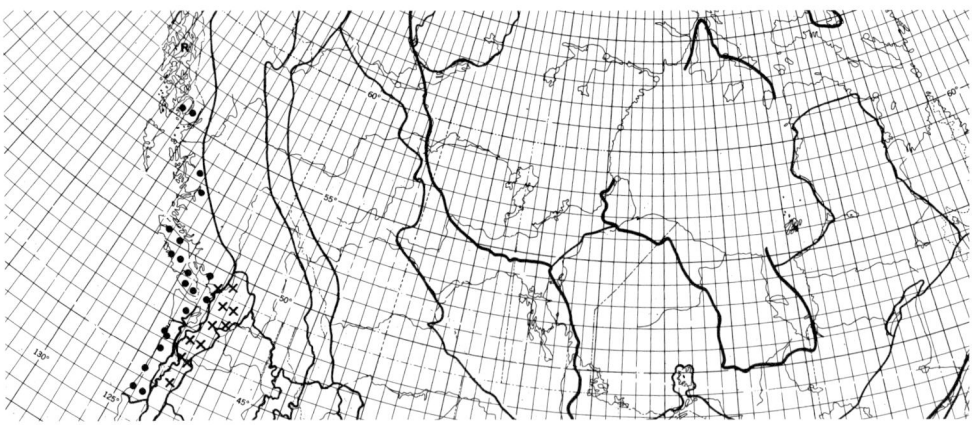

Fig. 90. Known distribution of *Sedum oreganum* ssp. *oreganum* (●) and ssp. *tenue* (x). (R = oral report.)

northward to the vicinity of Angoon, Alaska. The plants grow on rocks, especially bluffs and cliffs by the sea. Rarely, they fall from the rocks and grow in sand at the base of the cliffs. Rocks on which plants occur include basalt, sandstone, and gneiss. Usually plants are in the open, exposed to the sun. The altitudinal range is from near sea level to 460 meters.

Although ssp. *oreganum* appears common in some areas, as, for example, the cliffs by the Pacific Ocean in Oregon north of Cape Perpetua, it is disjunct in its distribution and absent from large areas. In my survey of the Pacific Border Province in 1963, I failed to find the species in any of the forty randomly selected listing units in nine primary sampling areas, although I did see it and studied it at five places outside the selected units in two of the primary areas. Despite shortcomings, the survey makes possible a tentative, minimum estimate of 17,250 plants of ssp. *oreganum* in the Coast Ranges of Oregon and Washington. Information for the five populations to which I gave some attention is in the following table.

Population	Plants (N)	Area of population (m.^{2})	Density per (m.^{2})	Exposure
Millicoma River	> 15	772	~ .02	SSE–SSW
Coos Ridge	1	.04	—	S
Beard's Hollow	2,000	—	—	—
Seaview	500	6,000	.08	SW–W
Long Island	37	—	—	SW

Population	Altitudinal range (m.)	pH, range	pH, tests (n)	Oldest record
Millicoma River	10–14	6.4	2	1963
Coos Ridge	460	6.4	1	1963
Beard's Hollow	0–20	—	—	1963
Seaview	10–20	6.6	2	1924
Long Island	5	—	—	1963

Other vegetation is sparse or lacking in the habitat occupied by ssp. *oreganum*. Competitors are few and miscellaneous. Sometimes, the *Sedum* grows by itself without competition.

Sedum oreganum ssp. **tenue**

Subspecies nova *Sedi oregani* cum caulibus tenuioribus et foliis gracilioribus quam ssp. *oreganum*; cum nectariis longioribus, .51 ± 0 mm. longis; fructibus brevioribus, 4.1 ± .4 mm. longis; et seminibus angustioribus, .3 ± .02 mm. Typus in Herbario Wiegand, Universitatis Cornellianae, ab declivite occidentale montis Wedge, altitudine 815 m., Wenatchee Mountains, Chelan Co., Wash., collectio Roberti et Erici Clausenii, num. 61·209·4, 1961, Julio 28, est.

The slenderer stems and thinner leaves of ssp. *tenue* (fig. 91), as well as its distribution in the Cascade Mountains, are distinctive. E. P. Sheldon, as early as 1903,

Fig. 91. Plant of *Sedum oreganum* ssp. *tenue* from northwestern slope of Wedge Mountain in Wenatchee Mountains, Chelan Co., Wash., cultivated in cold frame, Ithaca, N.Y., June 1962 (x .8). Photo by Howard Lyon.

thought that plants from the Cascade Mountains were different from those along the coast. He noticed differences between plants from Multnomah Falls and several coastal localities. My studies confirm his idea.

Sedum oreganum ssp. *tenue* appears to be endemic in the Cascade Mountains (fig. 90), occurring from O'Leary Mountain in Oregon to Selese Mountain in southern British Columbia. The usual habitat is on steep, rocky slopes, sometimes in partial shade. Plants grow on a variety of rocks, including granodiorite, graywacke, and andesite. The altitudinal range is from 305 to 1,550 meters.

Herbarium specimens are an inadequate basis for distinguishing subspecies of *Sedum oreganum*. Some populations need detailed study in the field before their subspecific status is clear. These include populations in the Olympic Mountains and also the population at Lake Whatcom in Washington. Tentatively, and primarily on the basis of geography, I have assigned these to ssp. *oreganum*.

Studies in the Cascade Mountains in 1961 and 1965 make possible the first estimate of the size of the total population of ssp. *tenue*, namely 3,069,637 plants. Data for the three local populations which I have studied are summarized in the following table.

Population	Plants (N)	Area of population (m.²)	Density per m.²	Exposure
O'Leary Mt.	1	.02	—	NE
Wedge Mt.	53	60,000	.0008	NW–W
Bald Mt.	130	39,071	.0033	E

Population	Altitudinal range (m.)	pH, range	pH, tests (n)	Oldest record
O'Leary Mt.	1,550	5	1	1965
Wedge Mt.	825–1,180	6.2–6.4	2	1961
Bald Mt.	1,410	5.2	1	1965

Competitors of ssp. *tenue* are miscellaneous. They include ferns, as *Cryptogramma crispa*, *Athyrium filix-femina*, and *Adiantum pedatum*, and species of *Heuchera*, *Penstemon*, and *Montia*. Trees shading the plants include *Abies grandis*, *Pseudotsuga menziesii*, and *Thuja plicata*. At each site, other species of *Sedum* were growing in close association: on O'Leary Mountain—*S. oregonense*; on Wedge Mountain—both *S. divergens* and *S. stenopetalum* ssp. *stenopetalum*; and on Bald Mountain—*S. divergens*.

Reproduction. Offsets of *Sedum oreganum*, although few per primary rosette, probably are an important means of reproduction. Multiplication of plants may come from them, breaking up of the primary stems, or seeds. Where seedlings occur, as at Seaside, they make possible the colonization of unoccupied sites as well as the production of new biotypes.

Flowering time is in July and August, mostly July for ssp. *oreganum* and August for ssp. *tenue*. Extreme dates, taken from herbarium specimens and personal experience, are June 17–Aug. 23 for ssp. *oreganum* and July 7–Sept. 7 for ssp. *tenue*. The flowers are fragrant and attract bumblebees.

The possibilities for hybridization with other species are limited. No other *Sedum* is present at the following localities: Millicoma River, Beard's Hollow, Seaview, and Long Island. *Sedum stenopetalum* ssp. *stenopetalum* is in fruit on Wedge Mountain when *S. oreganum* is in flower. When *S. spathulifolium* is at the height of bloom on Coos Ridge, *S. oreganum* is in bud. Likewise, on O'Leary Mountain in 1965, the only plant of *S. oreganum* was still in bud when most plants of *S. oregonense* were past flowering. The principal opportunity for hybridization by *S. oreganum* is with *S. divergens*. Both *S. oreganum* and *S. divergens* have yellow flowers, are visited by bumblebees, and bloom at the same time on Wedge and Bald Mountains. I found no evidence of hybrids at either place. *Sedum divergens*, although superficially similar to *S. oreganum* in general aspect, differs in having opposite, globular leaves, and gibbous, divergent follicles.

Relationships. The relationships of *Sedum oreganum* appear to be closest to species of the subgenus *Gormania*. Evidence is confusing and causes vacillation in classification. Because the species is distinct and stands by itself, I proposed in 1942 section *Oreganica* as a subdivision of *Gormania* to accommodate it. Then, in 1944, with C. H. Uhl, I suggested that the species be dropped from subgenus *Gormania* and transferred to some other subgenus of *Sedum*. Evidence for this suggestion came from the difference in habit from *Gormania*, especially the lack of prominent basal rosettes; the straight petals which do not spread above the erect bases; and principally, the different number of chromosomes. Further study has diminished the importance of these reasons. In 1963, I found many plants of ssp. *oreganum* at Beard's Hollow, Wash., with compact rosettes and with petals diverging above the middle, appearing in habit much like *S. oregonense* or *S. obtusatum*. Additional studies of chromosomes, while supporting earlier statements about number and size, reveal that some of the mitotic chromosomes appear tiny and round, similar to those of *S. obtusatum* and *S. spathulifolium*. Possibly, *S. oreganum* had an origin similar to that of *S. obtusatum*, only the resulting haploid number of chromosomes is 12 rather than 15 and some of the chromosomes are longer and two-armed. Which condition came first is unclear. Likewise, the mode of origin is a mystery. For purposes of classification, I am retaining section *Oreganica* and placing that before sections *Gormania* and *Rosulata* in subgenus *Gormania*. My concept is that neither section *Oreganica* nor *Rosulata* is closely related to the other, but that both are closer to section *Gormania* than to any other group of North American species of *Sedum*.

A few points of comparison of *Sedum oreganum* and *S. obtusatum* may be helpful. Figures in the table are ranges of averages for populations. These are taken from the descriptive tables for the two species.

	S. oreganum	S. obtusatum
Petals—length (mm.)	9.4–11.1	5.7–10.3
—length of cohesion (mm.)	1–2.7	1–3
Carpels—length (mm.)	6.4	4–7.3
—length of cohesion (mm.)	1.4	.5–1.3
Chromosomes (n)	12	15
—total length (μ)	15–25	35–36
—average length (μ)	.6–1.1	.9–1.2
Environmental adaptation	cool, moist summers	dry, warm summers

Plants of *Sedum oreganum* and *S. divergens* sometimes appear remarkably similar. They may grow together in the same habitat and have spatulate-obovate leaves, tinged with red, and yellow petals. Probably, these resemblances are superficial and result from evolutionary convergence. Differences between the two species are many and basic. *Sedum divergens* has 8 pairs of large, markedly two-armed chromosomes; opposite leaves; shorter, spreading petals, which are not connate; and gibbous carpels, which spread widely in fruit.

References. Clausen, R. T. (1942), Clausen and Uhl (1944), and Hollingshead, L. (1942).

21. Sedum obtusatum*** (figs. 92–98)

Although *Sedum obtusatum*, *S. laxum*, and *S. oregonense* are remarkably similar, *S. obtusatum* (fig. 92) is distinct wherever it occurs with or near either of the other

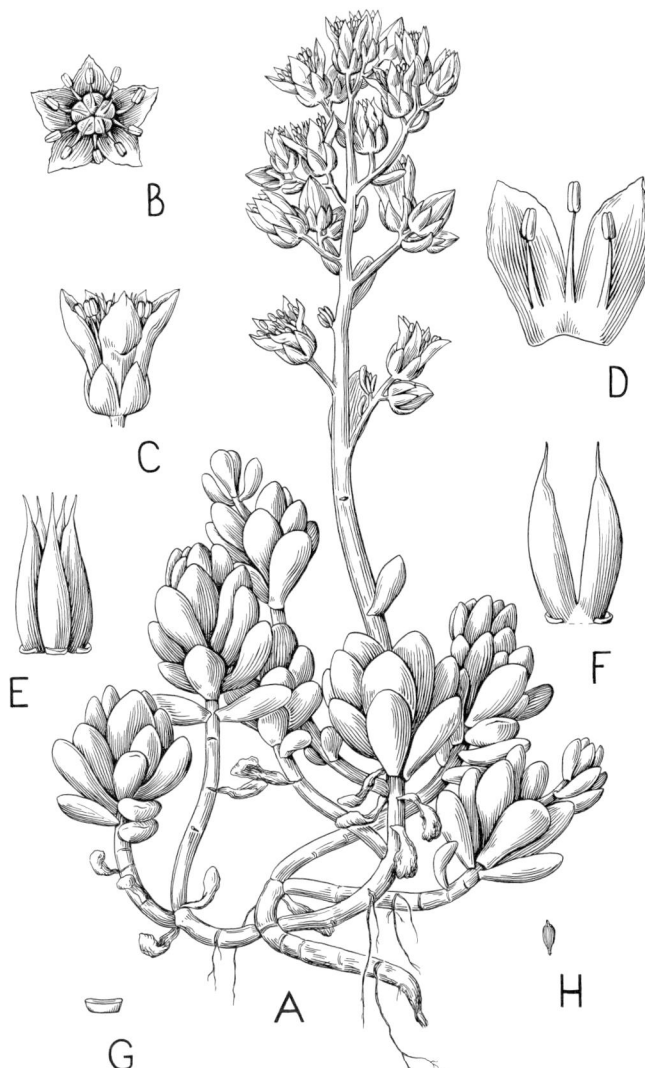

Fig. 92. Plant of *Sedum obtusatum* ssp. *obtusatum* from Yosemite Gorge, near Nevada Falls, Yosemite National Park, Calif., June 24, 1964. A. Habit sketch (x .84). B. Flower from above (x 1.7). C. Flower from side (x 1.7). D. Two petals and stamens (x 2.5). E. Carpels (x 2.5). F. Two carpels and nectaries (x 3.4). G. Nectary (x 4.2). H. Seed (x 4.2).

two species, but is difficult to distinguish by any single feature from all related species. Instead, a series of comparisons is necessary, contrasting *S. obtusatum* separately with each other species.

The elongate, glabrous inflorescences of *Sedum obtusatum* readily separate it from *S. moranii*. The two species do not occur together and present no problem in separation. Similarly, the smaller leaves which are not white-margined, as well as the longer offsets, make possible distinction from *S. albomarginatum*. Where the two species occur in the same geographic region in the northern Sierra Nevada, *S. obtusatum* is at higher altitudes and has smaller flowers. The two species do not meet. Similarly, where *S. obtusatum* occurs in the Klamath Mountains in proximity to either *S. laxum* or *S. oblanceolatum*, it is at higher altitudes and flowers later. These are ecospecies which maintain their distinctness because of differences in habitat and flowering time. Biological isolation appears effective. Practical separation of *S. obtusatum* from either *S. laxum* or *S. oblanceolatum* in the Klamath Mountains is by means of shorter petals and smaller, other floral parts. Curiously, differences in dimensions of floral parts between populations of *S. obtusatum* and these other two species are greatest where they occur nearby in the same geographic region, but least where they are remote. The situation resembles what Brown and Wilson (1956) have called character displacement. Another relationship involves difference in number of chromosomes. This produces effective isolation between the diploid *S. obtusatum* and the hexaploid *S. oregonense*. For practical recognition, the sepals of *S. obtusatum* are longer than those of *S. oregonense*.

Description. Sample: 62 plants—15 in flower and 20 in fruit at time of collection. Plants of the sample came from the Sierra Nevada in 1964 and the Klamath Mountains in 1963 and 1965. The sampling procedure was to select randomly large primary units of area and then to choose randomly listing units for survey within these. In addition, definition of optimum strata made possible survey of areas with populations of special interest. The program in each listing unit chosen for study was the same: first to enumerate the natural clusters of *Sedum obtusatum*, then to choose a random sample of clusters for attention, and finally to study samples of plants in the chosen clusters. To decide the appropriate size of a sample of plants per cluster, I appraised variation in length of leaves of rosettes in a trial sample from the Yosemite Gorge. In order to detect a difference of 5 mm., $P = .05$, an adequate sample on the basis of application of Stein's test is 4 plants per cluster. For less variable characters, a smaller sample is sufficient. Details about the sampling units surveyed, the numbers of included plants, and the sizes of the sampled clusters and of the samples of plants for study are indicated in the following table and subsequent paragraph.

Locality	Total plants N_u	Plants in sampled clusters N_c	Plants studied n	n/N_c	n/N_u	Plants studied in fl. (n)	fr. (n)
Sierra Nevada—random stratum							
Rancheria Mt.	18	18	3	.17	.17	0	0
Gem Pass	252e	42	4	.09	.016	0	3
Sierra Nevada—optimum stratum							
Little Yosemite V.	520e	24	2	.08	.004	2	0
Yosemite Gorge	2,896e	54	7	.13	.002	3	0
Chilnualna Falls	374e	374e	2	.005	.005	2	0
Sierra Buttes	407e	163	5	.03	.012	1	0
Eureka Peak	781e	334	10	.03	.013	0	6
Klamath Mountains—random stratum							
Valentine Ridge	194	194	15	.08	.08	3	8

In addition, various miscellaneous collections were part of the study. These included one plant from each of the following localities in the Sierra Nevada: near a lake tributary to Rancheria Creek, Silver Lake, Sonora Pass, and near Gold Lake; and in the Klamath Mountains: one plant from Mount Linn in the South Yolla Bolly Mountains and 4 plants from Dutchman's Peak, Jackson Co., Ore.

Experiments in Ithaca, all of completely random design, have included 43 plants. These have been in the garden, cold frame, and greenhouse. Greatest success was in a partially shaded cold frame. Least success was in the garden where the biggest loss of plants occurred in the humid, hot summers. A summary of information about the sources of plants in the experiments and the number of plants which survived and provided descriptive data (*n-d*) is available in the following table.

Locality	Garden n	Garden n-d	Cold frame n	Cold frame n-d	Greenhouse n	Greenhouse n-d
Sierra Nevada						
Chilnualna Falls	0	—	0	—	0	—
Yosemite Gorge	2	0	4	3	7	4
Little Yosemite V.	0	—	0	—	2	2
Rancheria Mt.	0	—	3	3	3	2
Rancheria Creek	0	—	0	—	1	1
Silver Lake	0	—	0	—	1	0
Gem Pass	2	0	4	3	4	1
Sonora Pass	1	0	0	—	1	1
Sierra Buttes	2	0	4	4	4	3
Gold Lake	0	—	0	—	1	1
Eureka Peak	2	0	4	3	6	4
Klamath Mountains						
Mt. Linn	0	—	1	0	1	1
Valentine Ridge	2	0	3	2	3	0
Dutchman's Peak	0	—	2	2	1	0
Totals	11	0	25	20	35	20

All data for vegetative parts of plants in the experiment in the cold frame are from observations in 1966. Data from the greenhouse are from observations in January 1969. One replicate from Gem Pass, 64–185, flowered in the cold frame in June 1967. A replicate from Sonora Pass survived in the garden for two years, but declined throughout the period. Four replicates survived for six months. The rest died in less time.

In the tables in the description, localities in the Sierra Nevada are grouped as follows: central Sierra Nevada—Chilnualna Falls, Yosemite Gorge, Little Yosemite Valley, Rancheria Mt., Rancheria Creek, Silver Lake, Gem Pass, and Sonora Pass; and northern Sierra Nevada—Sierra Buttes, Gold Lake, and Eureka Peak. Sierra Nevada is abbreviated as S.N.

Plants of *Sedum obtusatum* are perennial herbs (fig. 93) with fuscous roots and sometimes a fuscous taproot. The rootstocks are elongate and branched, brown to almost black, and stout, <5 mm. in diameter. The branches of the rootstocks, when young, are green or suffused with red and glaucous. They terminate in rosettes of blue-green leaves. Offsets arise in the axils of the leaves of the rosettes. These have pale green or pinkish glaucous stems and decussately opposite leaves. The floriferous stems are either terminal or axillary, and glaucous-green, sometimes suffused with

362   SEDUM OF NORTH AMERICA

Fig. 93. Plant of *Sedum obtusatum* ssp. *obtusatum* on granite in Yosemite Gorge near Nevada Falls, Yosemite National Park, Calif., June 24, 1964 (x .7).

red, or even purplish. One plant from Eureka Peak developed a floriferous stem 22 cm. long in cultivation in a greenhouse at Ithaca, N.Y.

	n-pop.	n-pl.	n-rep.	n-obs.	Range of $\bar{x}$	Range of s	w
Plants—diameter (cm.)							
Cent. S.N.	4	17	—	17	51–147	26–79	21–270
—cf.	3	9	16	16	6–8	2–7	1–16
—gh.	6	11	14	14	3–17	4–11	3–23
No. S.N.	2	15	—	15	26–40	16–29	6–74
—cf.	2	7	10	10	4–6	1–3	1–8
—gh.	3	8	21	21	10–11	4–7	1–26
Klam. Mts.	1	14	—	14	13	6	5–28
—cf.	2	4	6	6	1–7	1–4	1–10
—gh.	1	1	3	3	2	—	1–4
Primary rosettes—diameter (cm.)							
Cent. S.N.	5	18	—	48	1.6–3.4	.1–0.7	1.0–4.4
—gh.	6	11	14	21	1.9–3.7	.8–1.2	1.1–4.9
No. S.N.	2	15	—	28	2.1–2.6	.7–0.8	1.1–3.8
—gh.	3	8	21	34	2.3–2.6	0.7	1.5–4.4
Klam. Mts.	2	12	—	21	2.8–3.9	.6–1.4	.8–7
—gh.	1	1	3	3	1.2	—	1.1–1.5

	n-pop.	n-pl.	n-rep.	n-obs.	Range of $\bar{x}$	Range of s	w
Offsets per rosette (n)							
Cent. S.N.	5	18	—	48	1.4–2.7	.2–1.4	1–4
—cf.	3	9	16	22	1.6–3.7	1.6–6.9	0–12
—gh.	6	11	14	21	0–5	1.4–5	0–11
No. S.N.	2	15	—	28	1.9–2.4	.8–1.1	1–5
—cf.	2	7	10	15	.3–2.3	.4–4.8	0–9
—gh.	3	8	21	34	1.4–3.7	1.2–2.8	0–9
Klam. Mts.	1	4	—	6	2.2	0.9	1–3
—cf.	2	4	6	7	0–6.7	0–4.5	0–14
—gh.	1	1	3	3	0	—	0
Length of offsets (cm.)							
Cent. S.N.	5	18	—	75	.6–2.2	.2–1.5	.1–5.1
—cf.	3	8	9	14	2.2–3.4	.3–1.4	.1–6
—gh.	5	9	12	21	1.0–11	.7–5.2	.2–14
No. S.N.	2	15	—	48	1.1–1.2	.4	.2–2.5
—cf.	2	4	5	7	1.6–2.5	.5	1.2–2.5
—gh.	3	7	17	25	2.0–3.4	.3–.9	.5–7
Klam. Mts.	1	4	—	6	.8	.2	.6–1.4
—cf.	1	2	3	5	2.1	1.1	.6–3.3
Stems of offsets—diameter (mm.)							
Cent. S.N.	5	18	—	82	1.2–1.9	.1–1.5	.9–3
—gh.	5	9	12	21	1.1–1.6	.1–0.2	.8–2.2
No. S.N.	2	15	—	48	1.4–1.7	.2–0.3	1.0–2.8
—gh.	3	7	17	25	1.1–1.3	0.3	.8–2.2
Klam. Mts.	1	4	—	6	1.7	0.3	1.4–2.1
Floriferous stems per plant (n)							
Cent. S.N.	4	16	—	16	1–10	0–17	0–48
No. S.N.	2	15	—	15	3–9	2.4–16.1	1–35
Klam. Mts.	1	12	—	12	3	2.7	1–9
Floriferous stems—length (cm.)							
Cent. S.N.	4	14	—	25	5–9	1.3–2.2	3–14
—cf.	1	1	1	2	1.6	—	1–2
No. S.N.	2	15	—	15	9–11	0–2.2	7–16
Klam. Mts.	1	9	—	15	6.6	2.3	4–10

Leaves of the rosettes are oblanceolate or spatulate; rounded or truncate and obscurely mucronulate, emarginate, or retuse at apex; and green, blue-green, green suffused with red, or red. Leaves of the floriferous stems are arranged spirally and are truncately obovate or spatulate and rounded, and green, gray-green, or pale red.

	n-pop.	n-pl.	n-rep.	n-obs.	Range of $\bar{x}$ (mm.)	Range of s (mm.)	w (mm.)
Leaves of rosettes—length							
Cent. S.N.	4	16	—	44	13–20	.5–5.6	8–28
—cf.	3	9	15	23	8–13	1.5–5.2	2–19
—gh.	6	11	14	21	10–16	3.5–4.9	6–27
No. S.N.	2	15	—	28	16–19	.4–1.6	12–26
—cf.	2	7	10	15	13–14	.5–2	6–26
—gh.	3	8	21	34	13–14	1.0–1.6	5–23
Klam. Mts.	2	12	—	21	17–22	1.3–7.4	4–34
—cf.	2	4	6	7	6–14	.7–4	6–25
—gh.	1	1	3	3	6	—	6

SEDUM OF NORTH AMERICA

	n-pop.	n-pl.	n-rep.	n-obs.	Range of $\bar{x}$ (mm.)	Range of s (mm.)	w (mm.)
Leaves of rosettes—width							
Cent. S.N.	4	16	—	44	6–9	0–1.3	5–10
—cf.	3	9	15	23	4–6	.7–1.4	2–8
—gh.	6	11	14	21	5–8	.7–1.1	4–8
No. S.N.	2	15	—	28	8–9	1.5–1.7	6–13
—cf.	2	7	10	15	6–8	1.0–1.8	5–11
—gh.	3	8	21	34	7–8	.8–1.4	3–9
Klam. Mts.	2	12	—	21	8–10	.8–2.2	2–16
—cf.	2	4	6	7	4–8	.1–0.8	4–10
—gh.	1	1	3	3	5	—	4–6
Leaves of rosettes—thickness							
Cent. S.N.	4	16	—	44	2.5–4	0–.7	1.8–5
—cf.	3	9	15	23	1.8–2.5	.4–.9	.9–3.3
—gh.	6	11	14	21	1.5–2.4	.3–1	1.4–3.5
No. S.N.	2	15	—	28	2.0–2.6	.6–.8	2.0–3.9
—cf.	2	7	10	15	2.0–2.4	.3	1.7–2.9
—gh.	3	8	21	34	1.5–1.7	.1–.2	.5–2.4
Klam. Mts.	2	12	—	21	2.1–2.5	.4–.5	.8–2.9
—cf.	2	4	6	7	1.5–2.5	0.4–.3	1.3–3.2
—gh.	1	1	3	3	1	—	.8–1.3
Median leaves of floriferous stems—length							
Cent. S.N.	3	11	—	23	7.8–10.7	1.4–2.1	3–13
—cf.	1	1	1	1	6.5	—	6.5
No. S.N.	1	1	—	1	13	—	13
Klam. Mts.	3	10	—	17	10–17.9	0–4.9	10–29
Median leaves of floriferous stems—width							
Cent. S.N.	3	11	—	25	5.7–6.5	0–1.8	4–13
—cf.	1	1	1	1	4.8	—	4.8
No. S.N.	1	1	—	1	8	—	8
Klam. Mts.	3	10	—	17	4.3–8.4	.7–2.3	3.7–13
Median leaves of floriferous stems—thickness							
Cent. S.N.	3	11	—	25	2.2–2.8	.3–.6	2–4
—cf.	1	1	1	1	1.2	—	1.2
No. S.N.	1	1	—	1	2.3	—	2.3
Klam Mts.	3	10	—	17	1.6–2.2	.1–.6	1.4–3.6

Inflorescences are paniculate cymes with the floral bracts spatulate to linear-oblong. The primary branches of the inflorescences are ascending, with variable numbers of flowers, and often are further branched.

	n-pop.	n-pl.	n-rep.	n-obs.	Range of $\bar{x}$	Range of s	w
Primary axis—length (cm.)							
Klam. Mts.	2	7	—	14	9–21	13	5–45
Primary branches (n)							
Cent. S.N.	3	8	—	18	6–8	.3–3.1	4–12
—cf.	1	1	1	1	3	—	3
No. S.N.	1	1	—	2	4.5	—	4–5
Klam. Mts.	3	9	—	16	2.2–6.5	.7–2.1	0–12
Flowers per inflor. (n)							
Cent. S.N.	3	8	—	18	15–25	3.8–10.2	7–43
No. S.N.	1	1	—	2	23	—	22–25
Klam. Mts.	1	7	—	2	17	2.8	15–19

NATIVE SPECIES 365

Flowers (fig. 94) are pedicellate and erect. One flower on a plant in the Yosemite Gorge was double, with no sepals or petals between the two parts. This condition could result from either fusion or forking. A short torus is present. Distinction between torus and connate portion of carpels is difficult. Measurements of torus could apply to either or both.

Fig. 94. Flowers of *Sedum obtusatum* ssp. *obtusatum* from Gem Pass, Mono Co., Calif., cultivated in cold frame at Ithaca, N.Y., June 6, 1965 (x 3). Photo by Howard Lyon.

	n-pop.	n-pl.	n-rep.	n-obs.	Range of $\bar{x}$ (mm.)	Range of s (mm.)	w (mm.)
Pedicels—length							
Klam. Mts.	1	3	—	4	3.3	1.5	2–5
Torus—length							
Cent. S.N.	3	7	—	22	1.1–1.3	.1	1.0–1.3
No. S.N.	2	2	—	4	.9–1.2	—	.7–1.3
Klam. Mts.	3	7	—	29	.5–0.8	.3	.4–1.2
Flowers—diameter							
Cent. S.N.	2	4	—	16	10.2–11.6	.3–2	9–15
—cf.	1	1	1	2	6.7	—	6–7
No. S.N.—gh.	1	1	1	2	8.5	—	7–10
Klam. Mts.	3	6	—	23	4.7–8.2	1.1–1.3	4–9

Sepals are erect, slightly connate at base, ovate or lanceolate, acute or obtuse, somewhat convex dorsally, plane ventrally, pale green or purplish and glaucous.

366 SEDUM OF NORTH AMERICA

	n-pop.	n-pl.	n-rep.	n-obs.	Range of $\bar{x}$ (mm.)	Range of s (mm.)	w (mm.)
Length							
Cent. S.N.	3	7	—	22	4.1–5.3	.10–1.2	3.6–6.6
—cf.	1	1	1	2	4.3	—	3.9–4.8
No. S.N.	2	2	—	4	4	—	3.5–4.5
Klam. Mts.	3	6	—	23	2.9–4.4	.02–.1	2.5–5.4
Length of cohesion							
Klam. Mts.	1	2	—	3	.4	.03	.3–.6
Width							
Cent. S.N.	3	7	—	22	2.1–2.7	.10–.2	1.8–3.1
—cf.	1	1	1	2	2.1	—	2.0–2.2
No. S.N.	2	2	—	4	2.3	—	1.8–2.6
Klam. Mts.	3	6	—	23	1.5–2.2	.03–.5	1.4–2.7

Petals are mostly convolute in bud, erect and connate basally, then spreading above middle; oblanceolate-oblong, spatulate, or obovate, erose on upper margins, abruptly mucronate-appendaged; yellow, pale yellow, pale orange, creamy white, greenish white, pinkish white, pale pink, or white, with dorsal keel orange-brown or green. On withering, the petals may fade to pinkish.

	n-pop.	n-pl.	n-rep.	n-obs.	Range of $\bar{x}$	Range of s	w
Number							
Cent. S.N.	3	7	—	22	5	0–.3	4–6
—cf.	1	1	1	2	5	—	5
No. S.N.	2	2	—	4	5	—	5
Klam. Mts.	3	6	—	23	5	0–.1	5–6
Length (mm.)							
Cent. S.N.	3	7	—	22	8.4–10.3	.6–1.5	7.5–12.1
—cf.	1	1	1	2	7.4	—	6.9–7.9
No. S.N.	2	2	—	4	6.7–7.4	—	6.4–7.9
—gh.	1	1	1	2	8.8	—	8.3–9.3
Klam. Mts.	3	6	—	23	5.7–6.8	.2–.3	5.4–7.5
Length of cohesion (mm.)							
Cent. S.N.	3	7	—	22	1.7–2.1	.1–.8	0–3.3
—cf.	1	1	1	2	1.5	—	1.4–1.7
No. S.N.	2	2	—	4	2.3–3	—	2.1–3.4
—gh.	1	1	1	2	3	—	2.9–3.1
Klam. Mts.	3	6	—	23	1–2	.1–.4	.9–2.5
Width (mm.)							
Cent. S.N.	3	7	—	22	3.2–4.1	0–.4	3.0–4.4
—cf.	1	1	1	2	3.8	—	3.8–3.9
No. S.N.	2	2	—	4	2.6–3.1	—	2.5–3.2
—gh.	1	1	1	2	3.8	—	3.4–4.1
Klam. Mts.	3	6	—	23	2.4–2.9	.1–.2	2.2–3.5

Stamens have pale yellow or white filaments and yellow anthers.

	n-pop.	n-pl.	n-rep.	n-obs.	Range of $\bar{x}$ (mm.)	Range of s (mm.)	w (mm.)
Epipetalous filaments—height on petals							
Cent. S.N.	3	7	—	22	1.4–2.8	.1–.5	1.2–3.5
—cf.	1	1	1	2	1.1	—	.8–1.4

	n-pop.	n-pl.	n-rep.	n-obs.	Range of x̄ (mm.)	Range of s (mm.)	w (mm.)
Epipetalous filaments—height on petals							
No. S.N.	2	2	—	4	1.8	—	1.4–2.2
Klam. Mts.	3	7	—	29	1.2–1.8	.2–.9	.4–3.2
Anthers—length							
Cent. S.N.	2	2	—	3	1.7–1.8	—	1.7–1.8
—cf.	1	1	1	1	1.6	—	1.6
No. S.N.	2	2	—	3	1.4	—	1.4
Klam. Mts.	1	1	—	2	1.2	—	1.1–1.3

Nectaries are truncately reniform and either yellow or white.

	n-pop.	n-pl.	n-rep.	n-obs.	Range of x̄ (mm.)	Range of s (mm.)	w (mm.)
Length							
Cent. S.N.	3	7	—	22	.2–.4	0–.04	.2–.4
—cf.	1	1	1	2	.5	—	.5–.6
No. S.N.	2	2	—	4	.3–.4	—	.3–.4
—gh.	1	1	1	2	.6	—	.5–.6
Klam. Mts.	3	7	—	29	.2–.3	0–.1	.1–.3
Width							
Cent. S.N.	3	7	—	22	1.1–1.2	.03–.2	.9–1.5
—cf.	1	1	1	2	1.2	—	1.1–1.3
No. S.N.	2	2	—	4	.9–1.2	—	.9–1.2
—gh.	1	1	1	2	1.2	—	1.1–1.2
Klam. Mts.	3	7	—	29	.7–0.8	.03–.13	.4–1.1

Carpels are erect, slightly connate basally, and greenish white or pale green, sometimes pinkish upward and papillose.

	n-pop.	n-pl.	n-rep.	n-obs.	Range of x̄	Range of s	w
Length (mm.)							
Cent. S.N.	3	7	—	22	6.4–7.3	1.4–1.9	4.9–9.9
—cf.	1	1	1	2	5.1	—	5.0–5.2
No. S.N.	2	2	—	4	4.8–5.3	—	4.1–6
Klam. Mts.	3	7	—	29	4.0–5.4	.1–1.1	3.0–7
Styles—length (mm.)							
Cent. S.N.	3	7	—	22	1.3–1.5	.1–.3	1.0–2.1
No. S.N.	2	2	—	4	.7–1	—	.6–1.3
Ovules per ovary (n)							
Cent. S.N.	3	7	—	22	17–22	3.5–6.1	12–30
—cf.	1	1	1	2	16	—	16
No. S.N.	2	2	—	4	8–15	—	7–16
Klam. Mts.	3	7	—	29	15–21	4.2–7.1	5–31

Fruits are erect and brown. Following figures for length include beaks.

	n-pop.	n-pl.	n-obs.	Range of x̄ (mm.)	Range of s (mm.)	w (mm.)
Length						
Cent. S.N.	4	6	30	5.9–7.1	1.9	2.9–8.1
No. S.N.	1	6	24	7.3	.5	5.6–9.9
Klam. Mts.	1	8	32	6.5	.7	5.1–8.4

Seeds are pyriform, with short tails at both ends, finely ribbed longitudinally, and brown.

	n-pop.	n-pl.	n-obs.	Range of $\bar{x}$ (mm.)	Range of s (mm.)	w (mm.)
Length						
Cent. S.N.	3	4	22	1.1–1.5	.1	1.0–1.6
No. S.N.	1	5	17	1.3	.1	1.1–1.6
Klam. Mts.	1	7	23	1.1	.1	.8–1.3
Diameter						
Cent. S.N.	3	4	22	.4	.02	.3–.5
No. S.N.	1	5	17	.4	.03	.3–.5
Klam. Mts.	1	7	23	.4	.03	.3–.5

Chromosomes are very small and appear roughly circular or elliptical in cross-section in late prophase and at metaphase (Hollingshead). Mitotic chromosomes in a cell of a root tip of a plant from Rancheria Mountain vary in length from .7–1.1 microns, with an average length of .91 microns (Zandstra). Likewise, two cells of a plant from Eureka Peak had total complement lengths of 35.5 and 34.7 microns, and average lengths respectively of 1.18 and 1.16 microns (Zandstra). The extremely small size makes the possibility of error in measurement large.

	n-pl.	g (n)	sp (2n)	Cytologist
Number				
Yosemite Gorge, near Nevada Falls	1	—	30	L. Hollingshead, 1942
Rancheria Mt., Yosemite Nat. Park	1	—	30	I. Zandstra, 1969
Lake Vernon, Yosemite Nat. Park	1	—	30	L. Hollingshead, 1942
Mt. Hoffman, Yosemite Nat. Park	1	15	—	L. Hollingshead, 1942
Sierra Buttes	1	—	30	I. Zandstra, 1969
Eureka Peak	1	—	30	I. Zandstra, 1969
Mt. Shasta	1	—	30	L. Hollingshead, 1942

Variation. *Sedum obtusatum* exhibits dissimilar characteristics in different parts of its geographic range. Populations in the central Sierra Nevada have yellow, large flowers. Those in the northern Sierra Nevada and Klamath Mountains have white or pale yellow, smaller flowers. Although plants in the Klamath Mountains are more like those in the northern than the central Sierra Nevada, they further differ in having shorter tori in the flowers and narrower nectaries. These conclusions result from study of plants in the field and analyses of variance for 33 characters observed in wild plants. Features for which differences are significant* or highly significant** are listed in the following table. The figures in the table are estimated confidence intervals for means, $P = .05$.

Feature	Central Sierra Nevada	Northern Sierra Nevada	Klamath Mountains
Floriferous stems—length (cm.)**	5.2–9.5	8.8–11.5	6.6
Torus—length (mm.)**	1.1–1.3	.9–1.2	.5–.8
Flowers—diameter (mm.)*	10–12	—	5–8
Petals—length (mm.)**	8–10	7	6–7

Feature	Central Sierra Nevada	Northern Sierra Nevada	Klamath Mountains
Petals—width (mm.)*	3.2–4.1	2.6–3.1	2.4–2.9
Stamens—adnation (mm.)*	1.4–2.8	1.8	1.2–1.8
Anthers—length (mm.)**	1.7–1.8	1.4	1.2
Nectaries—width (mm.)**	1.1–1.2	.9–1.2	.7–.8
Pistils—length (mm.)*	6–7	5	4–5
Styles—length (mm.)*	1.3–1.5	.7–1	—

In addition, differences among populations within the three regions are significant for diameter of plants, diameter of primary rosettes, and length of nectaries. Differences are significant for ten characters among regions, but for only three characters among populations within regions.

An experiment in a cold frame in Ithaca yielded comparative data for several vegetative characters. Differences in six vegetative features among regions and among populations within regions are not significant on the basis of analyses of variance. Features for analysis were diameter of plants, number and length of offsets, and length, width, and thickness of leaves. On the other hand, on the basis of data from an experiment in the greenhouse in 1969, differences among plants from the different regions are significant for three vegetative characters. These are:

	Central Sierra Nevada	Northern Sierra Nevada	Klamath Mountains
Primary rosettes—diameter (cm.)**	1.9–3.7	2.3–2.6	1.2
Leaves of rosettes—width (mm.)*	5–8	6.8–7.7	4.8
Leaves of rosettes—thickness (mm.)**	1.5–2.4	1.5–1.7	.9

In the experiment in the greenhouse, differences in length of leaves of rosettes of plants from different populations within the same region are highly significant, but differences among regions lack significance.

A reason for the original subdivision of *Sedum obtusatum* into two subspecies was the supposed difference in length of leaves of the primary rosettes. The further study fails to support this notion. The analyses of variance of data for wild plants, as well as for the plants in experiments in the cold frame and greenhouse, demonstrate that small differences in length of leaves lack significance. Likewise, differences in color of stems of offsets and condition of apices of leaves of primary rosettes are not satisfactory for separating subspecies because too much variation in these features occurs among plants in the same population. Plants of ssp. *obtusatum* in an experiment in Ithaca had the stems of the offsets red, pale red, pink, or green, and the leaves of the primary rosettes either rounded or emarginate for .5 mm. Similarly, plants of ssp. *boreale* in the same experiment had the stems of the offsets either red or pink and the leaves rounded or emarginate for .5 mm. Other characters, less subject to modification than those previously used (Clausen 1942), are necessary to separate these two subspecies.

Plants of *Sedum obtusatum* are sensitive to environmental changes, especially in light and moisture. Plants in experiments in the cold frame and greenhouse in Ithaca appeared to have smaller, thinner leaves than those in the wild. To test this impression, I compared 7 plants for which data were available from the wild condition and also from experiments in both cold frame and greenhouse. Plants were from Yosemite Gorge—2, Rancheria Mountain—2, Gem Pass—1, and Sierra Buttes—2. Analysis was by means of Wilcoxon's test for a two-way classification, comparing all possible pairs of treatments. Except for the difference in diameter between plants in the wild and the cold frame, all other differences lack significance. The results do not support a generalization that plants in the greenhouse are smaller, with more offsets per primary rosette, and with smaller, thinner leaves. The compared characters and their average expressions in the three conditions of growth are listed below.

Character	Wild	Cold frame	Greenhouse
Diameter of plants (cm.)**	68	6	10
Offsets per primary rosette (n)	2	3	4
Leaves of rosettes—length (mm.)	14	13	14
Leaves of rosettes—width (mm.)	7	7	6
Leaves of rosettes—thickness (mm.)	3	2.6	2.3

Petals vary in number per flower from 4 to 8, although 5 is the commonest condition. Plants with 6, 7, and 8 petals per flower occur on Valentine Ridge in the Yolla Bolly Mountains. Elsewhere, I have seen flowers with 6 petals in Yosemite Gorge and near Chilnualna Falls in Yosemite National Park, and a flower with 4 petals on a plant in Yosemite Gorge.

The color of the petals varies from white to yellow. Plants in the central Sierra Nevada have yellow or pale yellow petals. Those in the northern Sierra Nevada have white, creamy white, or even pinkish white petals. Greatest variability in color is in the Klamath Mountains where petals vary from white to creamy or greenish white, and even to pale orange suffused with pink.

The anthers always are yellow. Some plants on Valentine Ridge had the filaments undeveloped or thickened and papillose. Plants in that population were unusual in several respects. Since they did poorly in cultivation in Ithaca, further attention to their variability was impossible.

Plants in my study, investigated cytologically, have been diploid. A collection from the northern California Coast Ranges, from 2.4 km. east of the Plaskett Ranger Station on the Willows-Covelo Road, C. H. Uhl 970 (Jeps.), gametic number of chromosomes 30, appears similar to ssp. *retusum*.

A reasonable subspecific classification is along geographic lines. As remarked earlier, more differences occur among populations grouped on a regional basis than among populations within the same region. The differences among subspecies, classified in this way, are greater than indicated in the key, but some differences can be detected only when plants are grown experimentally under similar conditions, as for example in my experiment in the greenhouse in Ithaca. This experiment revealed three significant differences in vegetative characters. The features contrasted in the following key should be adequate to identify most plants collected in the wild. Evidence is lacking for ecotypic differentiation within subspecies.

KEY TO SUBSPECIES

A. Petals yellow or pale yellow, averaging 8–10 mm. long and 3.2–4.1 mm. wide; anthers 1.7–1.8 mm. long; plants of the central and southern Sierra Nevada .... ssp. *obtusatum*, p. 371
AA. Petals white, greenish white, creamy white, or pale orange suffused with pink, averaging 6–9 mm. long and 2.4–3.8 mm. wide; anthers 1.2–1.5 mm. long .... B
   B. Nectaries averaging .9–1.2 mm. wide; tori 1.2 mm. long; floriferous stems averaging 9–11 cm. long; plants of the northern Sierra Nevada and southern Cascade Mountains .... ssp. *boreale*, p. 373
   BB. Nectaries averaging .7–.8 mm. wide; tori .5–.8 mm. long; floriferous stems averaging 7 cm. long; plants of the Klamath Mountains and northern California Coast Ranges .... ssp. *retusum*, p. 375

Sedum obtusatum ssp. *obtusatum*

Nomenclature. *Sedum obtusatum* A. Gray, Proc. Am. Acad. 7:342 (1868). Type locality: Sierra Nevada, Mount Hoffmann, alt. 3,292 m., Yosemite National Park, Calif. Type: W. H. Brewer's collection no. 1,678 (GH), 1860–1862.

Synonyms are:

1903. *Gormania obtusata* (A. Gray) Britton, Bull. N.Y. Bot. Gard. 3:29. Based on *Sedum obtusatum* A. Gray.

1903. *Gormania Hallii* Britton, Bull. N.Y. Bot. Gard. 3:29. Type locality: vicinity of Tuolumne Meadows, alt. 2,591–2,896 m., Yosemite National Park, Calif., perhaps on Lembert Dome. Type: collection no. 3,545 of H. M. Hall and E. B. Babcock (NY), July 1902.

1903. *Gormania Burnhami* Britton, Bull. N.Y. Bot. Gard. 3:30. Type locality: along trail between Lake Eleanor and Lake Vernon, Tuolumne County, Calif. Type: a collection of S. H. Burnham (NY), July 16, 1894.

1904. *Cotyledon obtusata* (A. Gray) Fedde, Just's Bot. Jahresb. 31 (1):827. Based on *Sedum obtusatum* A. Gray.

1904. *Cotyledon Burnhamii* (Britton) Fedde, *ibid.* Based on *Gormania Burnhami* Britton.

1904. *Cotyledon yosemitensis* Fedde, *ibid.* Based on *Gormania Hallii* Britton.

1913. *Echeveria obtusata* (A. Gray) Nels. et Macbr., Bot. Gaz. 56:476. Based on *Sedum obtusatum* A. Gray.

1913. *Echeveria Brittonii* Nels. et Macbr., *ibid.* Based on *Gormania Hallii* Britton.

1919. *Sedum rubroglaucum* Praeger, Jour. Bot. 57:51. Type locality: along the Short Trail, Yosemite Valley, Yosemite National Park, Calif. Type: a collection of H. M. Hall (place of preservation, if any, unknown), June 1915. Illustration: Jour. Roy. Hort. Soc. 46:219, fig. 125 (1921).

1921. *Sedum obtusatum* var. *Hallii* (Britton) Smiley, Univ. Calif. Pub. Bot. 9:213. Based on *Gormania Hallii* Britton.

1921. *Sedum Hallii* (Britton) Praeger, Jour. Roy. Hort. Soc. 46:241.

1930. *Sedum Burhamii* (Britton) Berger in Engl. et Prantl, Nat. Pflanzenfam., ed. 2, 18a:451. Based on *Gormania Burnhami* Britton.

Frémont (1887) collected *Sedum obtusatum* in the pass at the head of the Salmon Trout and Ebo Rivers, most likely Donner Pass, on Dec. 4, 1845. A fruiting specimen is in the herbarium at the New York Botanical Garden, Frémont's Expedition to

California, 1845–1847, no. 120. Torrey identified the plant as a species of *Echeveria* and put a manuscript name, honoring Frémont, on the label, but never published it. Rose likewise had a manuscript name for a collection of A. A. Heller, no. 705 (US), from above Donner Lake toward Donner Pass, Nevada Co., Calif. In addition, comment is necessary that Berger, in Engl. et Prantl, Nat. Pflanzenfam., ed. 2, 18a:451 (1930), wrongly listed *Echeveria Hallii* Nelson et Macbride as a synonym of *Sedum Hallii*. That is a synonym of *Dudleya lanceolata* (including *D. hallii*). Further, Fröderström (1930–1935) listed *Gormania rubroglauca* without citing either the basionym or a reference.

Wild plants of ssp. *obtusatum* have as their distinctive trait yellow petals. The only other species with which confusion could occur in nature is *Sedum spathulifolium*. That species overlaps in range on the western slopes of the Sierra Nevada. It has flatter leaves in more congested rosettes and petals which are both separate and spread widely, rather than connate below and erect for about half their length. Further, it has divergent rather than erect follicles. Other distinctive features, such as the length and width of the petals and length of the anthers, require careful measurement and even a lens for observation. For details, see the dimensions cited in the description for the populations from the central Sierra Nevada.

As the subspecies farthest from the geographic center for subgenus *Gormania* in the Klamath Mountains, ssp. *obtusatum*, with its large petals and anthers, may be closest to the ancestral condition for the species. Although its component populations are not uniform, differences among them do not seem sufficiently important to justify further taxonomic subdivision despite the several names which have been proposed for extreme plants within its geographic range.

The distribution of *Sedum obtusatum* ssp. *obtusatum* (fig. 95) is in the Sierra Nevada, on both slopes, but mostly on the western side, from the drainage of the Kaweah River northward to the drainage of the Bear River and Lake Tahoe. The altitudinal range on the western slope is from 1,370 to 3,660 meters and on the eastern slope from 2,164 to 3,150 meters. Records are available for 73 populations of ssp. *obtusatum* in

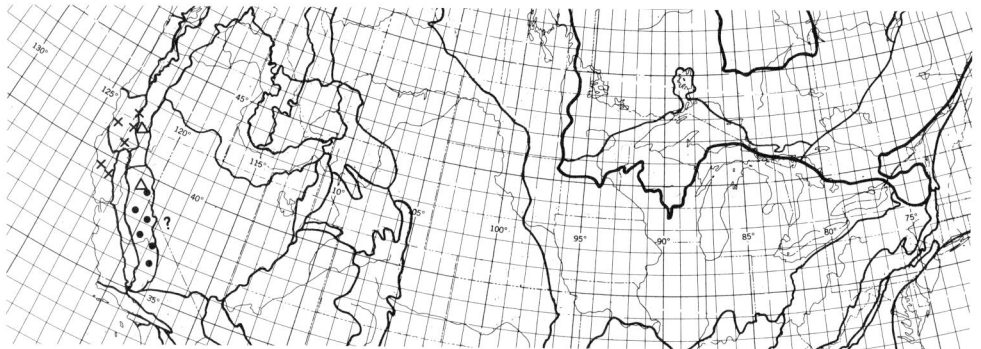

Fig. 95. Known distribution of *Sedum obtusatum* ssp. *obtusatum* (●), ssp. *boreale* (△), and ssp. *retusum* (x). (? = questionable record.)

the Sierra Nevada. The only record from outside the Sierra Nevada is from Aurora, Mineral Co., Nev., 2,286 m., Mrs. John D. Wright (SBM). A trip to Aurora for *Sedum* on Sept. 7, 1964, was unsuccessful. *Pinus cembroides* and *Artemisia tridentata* are important components of the vegetation there. The site appears too dry for *Sedum*. More likely places for *S. obtusatum* are Mount Hicks, 2,865 m., or Aurora Peak, 2,621 m. Time did not permit visits to these.

Information about the populations of ssp. *obtusatum* studied in 1964 is in the following table. Further description of the sampling units is in the account of the Sierra Nevada in the chapter on geography.

Population	Sampling Unit	Plants (N)	Area of population (m.^{2})	Density per (m.^{2})	Exposure
Rancheria Mt.	NY 2208	18	—	—	N
Gem Pass	ML 2156	252e	3,000	.084	ENE
Little Yosemite V.	SY 7807	520e	260,000	.002	N
Yosemite Gorge	SY 7802	2,896e	720,000	.004	N
Chilnualna Falls	SY 7401	374e	—	—	NNW

Population	Altitudinal range (m.)	pH, range	pH, tests (n)	Oldest record
Rancheria Mt.	1,885–1,886	4.8–5.2	3	1964
Gem Pass	3,140–3,150	5.0–5.6	3	1964
Little Yosemite V.	1,920	5	2	1964
Yosemite Gorge	1,580–1,970	4.8–5.6	7	1866
Chilnualna Falls	1,970	—	—	1964

Plants of ssp. *obtusatum* grow where other vegetation is sparse. Competing species are miscellaneous. The two closest competitors of each of 13 randomly selected plants belonged to 17 species. The only competitors which occurred more than once in the samples and the number of times that they occurred were *Poa* sp.—3, *Carex* sp.—2, *Selaginella watsonii*—2, and *Pellaea breweri*—2.

The usual habitat of ssp. *obtusatum* is in crevices or in gravel. Most populations are on granite, but some are on andesite or even serpentine.

Sedum obtusatum ssp. *boreale*

Nomenclature. *Sedum obtusatum* ssp. *boreale* R. T. Clausen, Bull. Torr. Bot. Club 69:32 (1942). Type locality: rocky slope, east side of Mud Creek Canyon, Mount Shasta, Calif., alt. 1,707 m. Type: collection no. 4,952 of R. T. Clausen and H. Trapido (CU), July 26, 1940. W. B. Cooke discovered ssp. *boreale* in the Mud Creek Canyon and collected it there on Aug. 10, 1939. He guided me to the population in 1940. Specimens in herbaria, labeled as from the north side of Mount Shasta, most likely are from Mount Eddy, according to A. A. Heller as quoted by Cooke (1941). Probably they are ssp. *retusum*.

The features which separate ssp. *boreale* (fig. 96) from ssp. *obtusatum* are the color of the corollas—white, creamy white, or pinkish white; the shorter, narrower petals,

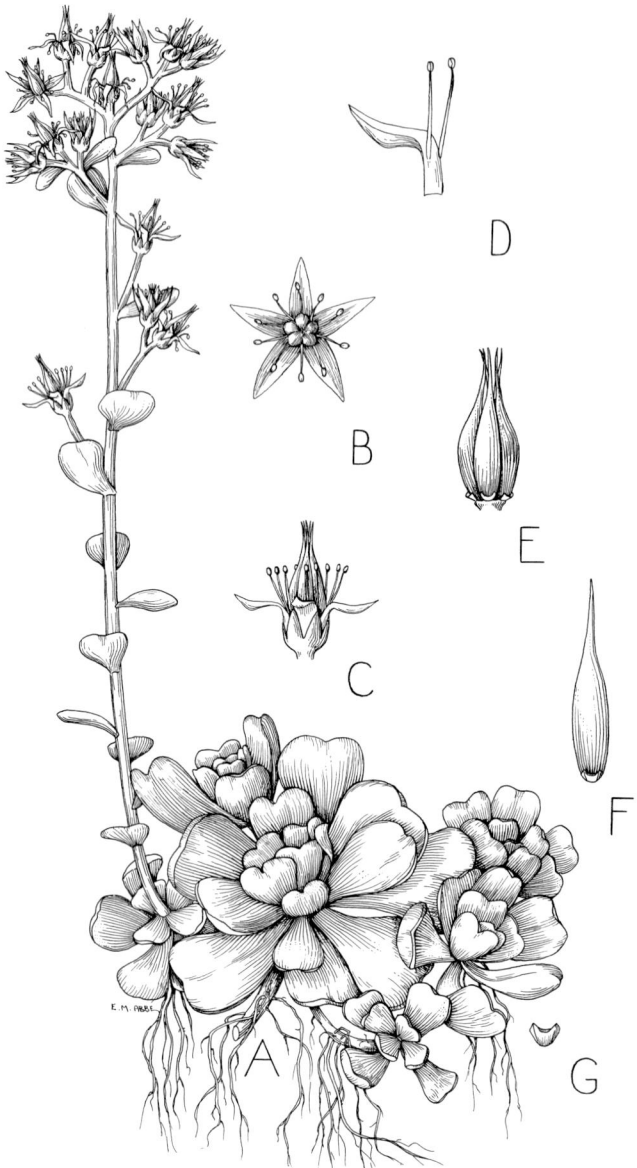

Fig. 96. Type plant of *Sedum obtusatum* ssp. *boreale* from Mud Creek Canyon, Mt. Shasta, Siskiyou Co., Calif., cultivated in greenhouse, Ithaca, N.Y., May 1943. A. Habit sketch (x .86). B. Flower from above (x 1.7). C. Flower from side (x 1.7). D. Petal and two stamens (x 2.6). E. Carpels (x 2.6). F. Single carpel and nectary (x 3.4). G. Nectary (x 4.3).

averaging less than 7.5 mm. long and 3.2 mm. wide; and the shorter anthers, 1.4 mm. long. From ssp. *retusum*, it differs in its broader nectaries, averaging .9–1.2 mm. wide, and longer tori, averaging .9–1.2 mm. long. Also, the floriferous stems are taller, averaging 9–11 cm. long.

The range of ssp. *boreale* (fig. 95) is in the region of volcanic rocks in the northern

Sierra Nevada and southern Cascade Mountains. Knowledge of its extent still is imperfect, but better than at the time of description in 1942. The known geographic extent is from the Sierra Buttes, Sierra Co., Calif., to Mount Shasta. The known altitudinal range is from 1,570 to 2,033 meters. Besides the localities already cited, ssp. *boreale* occurs northeast of Gold Lake, Sierra Co., above Lake Center Camp, Plumas Co., and probably elsewhere.

Data for two populations studied in 1964 are in the following table. Further description of the sampling units is in the account of the Sierra Nevada in the Chapter on geography.

Population	Sampling Unit	N plants, estimate	N plants, C. I. $_{.05}$	Area of population (m.²)	Density per m.²
Sierra Buttes	FR 3,328	407e	225–5,282	2,505	.16
Eureka Peak	FR 16,801	781e	504–4,775	14,447	.05

Population	Altitudinal range (m.)	pH, range	pH, tests (n)	Oldest record
Sierra Buttes	1,970–2,033	4.8–5.2	5	1964
Eureka Peak	1,570–1,695	4.6–5.2	10	1951

Species competing with *Sedum obtusatum* ssp. *boreale* are miscellaneous. Fifteen species of vascular plants were growing in association with 15 randomly selected plants of the *Sedum* in 4 randomly selected clusters on the Sierra Buttes and Eureka Peak. Only 3 of these species occurred more than once as associates. These were *Pellaea densa*, *Cheilanthes gracillima*, and *Quercus vaccinifolia*.

Plants of *Sedum obtusatum* ssp. *boreale* grow in open areas of siliceous argillite, quartz porphyry, andesite, or granite. They occur on rock slides, in crevices, and on ledges. They are rooted in shallow soil, which may consist of gravel, sand, or even moss.

Sedum obtusatum ssp. *retusum*

Nomenclature. *Sedum obtusatum* ssp. **retusum** (Rose) comb. nov., fundatum super *Gormania retusa* Rose, Bull. N.Y. Bot. Gard. 3:31 (1903).

Synonyms are:

1903. *Gormania retusa* Rose, Bull. N.Y. Bot. Gard. 3:31. Type locality: Sanhedrin Mountains, Lake Co., Calif., alt. 1,500 m. Type: collection of A. A. Heller (US), Aug. 6, 1902.

1904. *Cotyledon retusa* (Rose) Fedde, Just's Bot. Jahresb. 31 (1):827. Based on *Gormania retusa* Rose.

1930. *Sedum sanhedrinum* Berger in Engler et Prantl, Nat. Pflanzenfam., ed. 2, 18a:451. This is a renaming of *Gormania retusa* Rose, but without bibliographical citation.

1942. *Sedum laxum* ssp. *retusum* (Rose) Clausen, Bull. Torr. Bot. Club 69:39, as to name bringing synonymy, not as to principal concept. Based on *Gormania retusa* Rose.

Ssp. *retusum* is similar to ssp. *boreale* (fig. 97), but differs in the narrower nectaries, averaging .7–.8 mm. wide, and the shorter tori of the flowers, .5–.8 mm. long. The petals vary from pale orange to white and may fade to pink. The leaves of the rosettes vary from broadly rounded to emarginate. The floriferous stems are on the average shorter than in ssp. *boreale*.

Ssp. *retusum* occurs in disjunct populations in the northern California Coast Ranges and Klamath Mountains. It ranges (fig. 95) from the Sanhedrin Mountains in northwestern Lake Co., Calif., to Dutchman's Peak, Jackson Co., Ore. The altitudinal range observed by me is from 1,480 to 2,130 meters, but specimens are available in herbaria with altitudes indicated as from 465 to 2,286 meters. Records are available for 33 populations in the California Coast Ranges and 30 in the Klamath Mountains.

Data for a population on Valentine Ridge (fig. 98), the only one studied in any detail, follow.

Population	Sampling Unit	Plants (N)	Area of population (m.^{2})	Density per (m.^{2})	Exposure
Valentine Ridge	YB 16–68	194	547	.35	W

Population	Altitudinal range (m.^{2})	pH, range	pH, tests (n)	Oldest record
Valentine Ridge	1,480–1,575	5.4–5.6	4	1963

Plants of ssp. *retusum* occur on rocky slopes and ridges, in crevices, and in gravel. They grow on a variety of rocks: metamorphosed sandstone, granite, quartz diorite, and serpentine. Commonest conifers in the altitudinal zone in which this subspecies occurs are *Pseudotsuga menziesii*, *Abies magnifica*, and *Pinus lambertiana*.

Reproduction. The common method of propagation which is evident in every population of *Sedum obtusatum* is by means of vegetative offsets. These take root and become new plants with the same genetic constitution as the original rosette from which each developed. Likewise, older parts of rhizomes die, breaking a large plant into separate parts. Because of these means of multiplication, individual clones may survive for a long time. Stems with rosettes of leaves can endure long periods of desiccation. Five plants collected in the summer of 1964 and stored in dry condition in plastic bags in a cardboard box in Ithaca, N.Y., remained alive for eighteen months.

Seedlings are seldom seen in the summer. Small plants of *Sedum obtusatum* usually are rooted offsets rather than seedlings. Production of seeds is good, but mortality of seedlings must be enormous, though seeds make possible the principal spread of the species.

Pollination is by insects, usually *Diptera*, especially bombyliids and various flies. Time of flowering is late spring and early summer: June 14–Aug. 5 for ssp. *obtusatum*; June 27–July 31 for ssp. *boreale*; and June 1–Aug. 8 for ssp. *retusum*.

Fig. 97. Plant of *Sedum obtusatum* ssp. *retusum* from Valentine Ridge, Tehama Co., Calif., cultivated in cold frame, Ithaca, N.Y., June 8, 1965 (x 2.2). Photo by Howard Lyon.

Fig. 98. Sedum obtusatum ssp. *retusum* on metamorphosed sandstone, Valentine Ridge, Tehama Co., Calif., June 9, 1963.

Relationships. Sedum obtusatum is most closely related to *S. laxum*, *S. oregonense*, and *S. albomarginatum*. It is a generalized, diploid ecospecies which is easily separable from other species where it occurs with or near them, but is difficult to characterize in general terms if the idea is to separate it in absolute fashion from all other species. *Sedum laxum* and *S. albomarginatum* have longer petals and other floral parts in areas where their ranges overlap with *S. obtusatum*, and also they occur at lower altitudes, flower earlier, and grow on serpentine. *Sedum oregonense* is hexaploid and has shorter sepals.

Sedum obtusatum is less advanced than related species. Hexaploidy in *S. oregonense* and adaptation by *S. laxum* and *S. albomarginatum* to serpentine and drier conditions in the summer are specializations. If peripheral populations are most like ancestral types, but those nearest centers of origin are most differentiated from each other, then ssp. *obtusatum* with its large yellow flowers is nearest the biological type of the species. In opposition to this, white may have been the original color of the petals and yellow may be the derived condition.

Present populations of *Sedum obtusatum* probably have moved into the mountains in recent geological time from lower elevations where they survived Pleistocene glaciation. During the glacial period, the modern species or its antecedent probably had a more restricted range, and distinctions between *S. obtusatum* and related species may not have progressed to their present state.

Plants ancestral to *Sedum obtusatum* may have had large flowers in elongate inflorescences with the primary branches dichotomously forked. *Crassulaceae* with these characters may likewise have given rise to such derived genera as *Dudleya* and *Echeveria*.

Sedum obtusatum does not occur in the same habitats with species which are most nearly related. Where it does occur intermixed with other species of *Sedum* or other genera of *Crassulaceae*, distinctions are sharp and no interaction through hybridization occurs. Sites where *S. obtusatum* occurs in close association with other *Crassulaceae* are: Yosemite Gorge, with *S. spathulifolium* ssp. *yosemitense*; Chilnualna Falls, with *S. stenopetalum* ssp. *monanthum*; Rancheria Mountain, with *S. stenopetalum* ssp. *monanthum* and *Dudleya cymosa* ssp. *cymosa*; and Gem Pass, near *S. lanceolatum*. At most sites, for example Little Yosemite Valley, Sierra Buttes, Eureka Peak, Valentine Ridge, and Dutchman's Peak, *Sedum obtusatum* is the only species of *Crassulaceae* in the habitat where it occurs.

References. Clausen, R. T. (1942) and Hollingshead, L. (1942).

22. Sedum laxum*** (figs. 99–112)

Pinkish white or white petals are a distinctive feature of *Sedum laxum* (fig. 99). These easily separate it from *S. moranii*, *S. albomarginatum*, and *S. obtusatum* ssp. *obtusatum*, which have yellow petals, and from *S. oregonense* with creamy white petals. Rarely the petals are yellowish white, for example in certain populations in the drainage of the South Fork of the Trinity River. Plants with yellowish petals must be distinguished by other features, especially the thicker leaves of the rosettes. *Sedum laxum* has the thickest leaves of any species of the subgenus *Gormania*. Like *S. moranii* and *S. albomarginatum*, it occurs on serpentine and is at lower altitudes than either *S. obtusatum* or *S. oregonense*.

The anthers of *Sedum laxum* are red in ssps. *latifolium* and *laxum*, red or yellow suffused with red in ssp. *eastwoodiae*, but yellow in ssp. *heckneri*. The yellow-flowered species have yellow anthers, as does *S. oregonense*.

Sedum oblanceolatum has white petals; more elongate, thinner leaves in the rosettes; longer sepals; and yellow anthers. It is more glaucous or even pruinose. It occurs in the drainage of the Rogue River upstream from the area of occurence of *S. laxum*.

Sedum obtusatum ssp. *boreale*, likewise with white petals, besides having shorter, thinner leaves in the rosettes, has shorter petals. Its occurrence at higher altitudes on granitic rocks removes the possibility of confusion in the field.

Description. The sample for study comprised 27 plants: 10 chosen randomly and proportionately from the two clusters on the northeastern slope of the southeastern spur of Red Mountain, Mendocino Co., Calif.; 4 chosen randomly from a cluster

380 SEDUM OF NORTH AMERICA

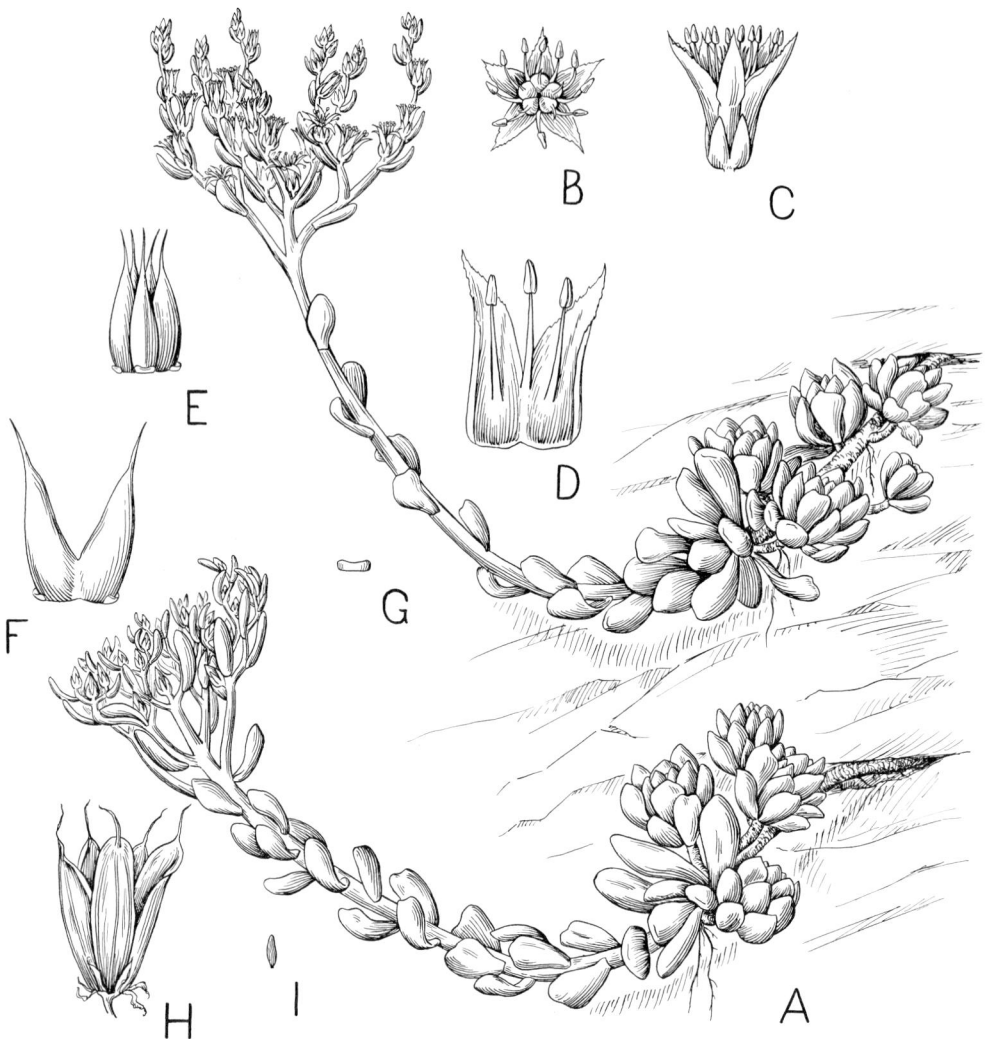

Fig. 99. Plants of *Sedum laxum* ssp. *laxum* from canyon of Rogue River, Josephine Co., Ore. A. Habit sketches (x .45). B. Flower from above (x 1.8). C. Flower from side (x 1.8). D. Two petals and stamens (x 2.7). E. Carpels (x 2.7). F. Two carpels and nectaries (x 3.6). G. Nectary (x 4.5). H. Follicles (x 3.6). I. Seed (x 4.5).

of 24 west of Wildwood, Trinity Co., Calif.; 4 chosen randomly from a cluster of 10 along Post Creek northeast of Forest Glen, Trinity Co., Calif.; 4 chosen deliberately to show extreme conditions of foliage from the canyon of the Smith River 3.7 km. west of Patrick Creek, Del Norte Co., Calif.; and 5 chosen randomly from a cluster of 10 in a larger group of 34 in the canyon of the Rogue River 10 km. south of Galice, Josephine Co., Ore. Data for plants cultivated in Ithaca, N.Y., are indicated in the following tables as gh. = greenhouse or cf. = cold frame. Data without special explanatory abbreviation are for plants in the wild.

Selection of sites had to be deliberate because no randomly selected unit in any primary stratum contained *Sedum laxum*. Although I visited 40 sampling units in nine major regions of the Pacific Borderland in 1963, all were negative for this species. This experience attests to the comparative rarity of the species. Its distribution is localized in the Klamath Mountains and adjacent parts of the northern California Coast Ranges. Although it occurs at widely scattered localities in this area, it is nowhere common.

Plants of *Sedum laxum* are perennial herbs with fibrous roots from stout, branched, fuscous, gray, or brown rootstocks which are 3 to 15 mm. in diameter. The observed maximum length of roots in nature is 16 cm. Rosettes of thick, fleshy leaves terminate the branches of the rootstocks. Each rosette may give rise to one or more axillary offsets with opposite leaves and eventually a single, terminal, erect, floriferous stem. The floriferous stems are glabrous and green, pink, or red. They may attain a maximum height of 4 decimeters. After bearing fruits, they die. Further growth is from offsets which may take root and become separate plants. In addition, although the older parts of the rootstocks die, the younger parts may continue to grow as separate plants. Diameters of plants in cultivation are for one year after propagation from cuttings 2 cm. long.

	n-pl.	n-rep.	n-obs.	$\bar{x}$	s	w
Plants—diameter (cm.)						
Wildwood—gh.	4	9	9	20	5.1	13–25
—cf.	2	4	4	7.7	4.4	7–13
Post Creek—gh.	4	9	9	9.8	8.8	1–22
Smith River—gh.	3	5	5	12.4	8.2	6–19
—cf.	2	3	3	10	0	10
Rogue River—gh.	5	10	10	18.1	2.3	16–22
—cf.	2	3	3	15	4.9	8–19
Red Mt.	10	—	10	19	13.9	5–50
—cf.	9	11	11	10	2.8	7–15
Primary rosettes—diameter (cm.)						
Wildwood	4	—	8	5.5	1.9	4.0–8
—gh.	4	8	15	3.5	.9	2.0–5
Post Creek	4	—	12	3.9	.7	3.0–5
—gh.	4	8	14	2.6	.7	1.0–3
Smith River	4	—	7	3.7	1.3	2.4–4
—gh.	3	5	8	4.8	1.1	3.4–6
Rogue River	5	—	12	3.5	1.1	2.5–5
—gh.	5	10	18	3.6	1.2	1.9–6
Red Mt.	10	—	29	2.6	.4	1.6–3.5
—cf.	4	4	9	4	1.3	2.5–6
Offsets (n)						
Wildwood—cf.	2	2	5	6.6	2.7	3–12
Post Creek—gh.	2	2	3	6.2	3.9	3–9
Smith River—gh.	2	2	3	.5	.7	0–1
—cf.	2	3	3	6	4	4–8
Rogue River—gh.	2	2	3	6.2	3.8	1–9
—cf.	2	3	4	7	7	1–12
Red Mt.	10	—	10	27	37	2–122
—cf.	4	4	4	2	1.6	0–4

	n-pl.	n-rep.	n-obs.	$\bar{x}$	s	w
Offsets—length (cm.)						
Wildwood—cf.	2	4	5	1	.1	.7–1.2
Post Creek—gh.	2	2	4	1.8	1.3	.5–3
Smith River—gh.	2	2	2	1.1	1.6	0–2.3
—cf.	2	2	3	3	0	3
Rogue River—gh.	2	2	4	3.8	1.7	1.9–5
—cf.	2	3	4	3.2	2.5	1.8–8
Stems of offsets—diameter (mm.)						
Wildwood	4	—	8	2.9	.5	2.2–3.5
Post Creek	2	—	6	2.3	.1	2.0–2.6
—gh.	2	2	4	1.9	.5	1.5–2.5
Smith River	1	—	2	3.2	—	3.2–3.3
—gh.	1	1	1	1.4	—	1.4
Rogue River	5	—	12	2.6	.4	1.8–3.5
—gh.	2	2	4	2.2	.6	1.2–2.9
Floriferous stems per plant (n)						
Wildwood—gh.	2	4	4	2.2	2	1–4
Post Creek—gh.	3	4	4	1.2	.6	1–2
Smith River—gh.	1	1	1	8	—	8
Rogue River—gh.	2	2	2	1.5	.7	1–2
Red Mt.	10	—	10	1.1	.3	1–2
—cf.	9	11	11	1.6	—	1–2
Floriferous stems—length (cm.)						
Wildwood	3	—	8	18	3.6	12–22
—gh.	2	2	6	17.8	4.4	11–21
Post Creek	4	—	10	14.6	1.4	11–16
Smith River	2	—	3	22.7	6	16–27
—cf.	2	2	3	26	2.5	25–29
Rogue River	5	—	10	18.5	2.5	15–24
—cf.	2	3	4	20.2	4	17–26
Red Mt.	10	—	12	7	2.1	4–11
—cf.	5	5	8	9.8	3.1	5–13

Leaves are spirally arranged in primary rosettes and on floriferous stems, but opposite on offsets. Rosulate leaves are divergent, sessile, and oblanceolate, spatulate, or obovate, with the apices rounded, emarginate, or retuse. They are either glaucous or green, usually the former. The leaves of the floriferous stems are sessile, truncate or cordate at base, and oblanceolate-oblong, oblong, obovate, or suborbicular, with the apices acute or rounded. They vary from glaucous-green to bright green or pink.

	n-pl.	n-rep.	n-obs.	$\bar{x}$ (mm.)	s (mm.)	w (mm.)
Leaves of rosettes—length						
Wildwood	4	—	8	29.8	11.2	22–46
—gh.	4	8	15	28	7.7	20–54
—cf.	2	4	5	22.7	2.5	17–26
Post Creek	4	—	12	23.8	4.4	19–32
—gh.	4*	8	14	14.5	9.1	7–21
Smith River	4	—	7	22.2	4.8	16–29
—gh.	3	5	8	22.2	1.7	17–28
—cf.	2	3	4	30.8	11	20–40
Rogue River	5	—	12	19.6	1.3	11–26
—gh.	5	10	18	18.2	5.5	10–29
—cf.	2	3	4	36.2	13	23–49

NATIVE SPECIES

	n-pl.	n-rep.	n-obs.	$\bar{x}$ (mm.)	s (mm.)	w (mm.)
Leaves of rosettes—length						
Red Mt.	10	—	29	16	1.9	12–23
—cf.	4	4	9	23.1	4.3	17–29
Leaves of rosettes—width						
Wildwood	4	—	8	12.4	3.4	9–17
—gh.	4	8	15	8.7	1.6	6–13
—cf.	2	4	5	10.1	2.7	6–12
Post Creek	4	—	12	11.1	1.3	10–14
—gh.	4	8	14	7.5	2.4	5–11
Smith River	4	—	7	17.2	2.6	12–21
—gh.	3*	5	8	16.3	2.6	13–20
—cf.	2	3	4	19.1	2.2	12–24
Rogue River	5	—	12	10.1	1.3	6–13
—gh.	5	10	18	9	1.3	6–14
—cf.	2	3	4	12.5	0	12–13
Red Mt.	10	—	29	9.2	1.5	5–13
—cf.	9	11	30	10.1	1.3	8–14
Leaves of rosettes—thickness						
Wildwood	4	—	8	3.5	.6	2.8–4.1
—gh.	4	8	15	3.7	.7	2.6–4.9
—cf.	2	4	5	3.1	1.7	2.5–3.7
Post Creek	4	—	12	3.6	.3	2.8–4.1
—gh.	4	8	14	2.8	.7	1.9–3.5
Smith River	4	—	7	3.2	.2	2.5–3.6
—gh.	3*	5	8	3.4	1.5	2.6–5.3
—cf.	2	3	4	3.4	.5	2.5–3.8
Rogue River	5	—	12	3.4	.9	2.0–4.7
—gh.	5	10	18	3.6	.3	2.7–4.7
—cf.	2	3	4	3.6	.1	3.3–3.9
Red Mt.	10	—	29	3.8	.5	2.7–4.8
—cf.	9	11	30	3.7	.4	3.2–4.6
Median leaves of floriferous stems—length						
Wildwood	3	—	6	11.3	4.1	8–16
—gh.	2	4	5	10.7	.3	7–12
Post Creek	2	—	6	8.9	.6	8–10
—gh.	3	4	4	8.7	2.5	6–11
Smith River—gh.	1	1	2	9.5	—	9–10
—cf.	2	2	3	19.5	2.1	17–21
Rogue River	4	—	9	17.5	1.6	14–19
—gh.	2	2	2	14	0	14
—cf.	2	3	4	16.1	2.2	14–18
Median leaves of floriferous stems—width						
Wildwood	3	—	6	8.3	1.5	7–10
—gh.	2	4	5	9	1.1	6–10
Post Creek	2	—	6	10.2	1.1	8–11
—gh.	3	4	4	8.7	3.5	6–13
Smith River—gh.	1	1	2	8	—	8
—cf.	2	2	3	13.2	1.1	12–14
Rogue River	4	—	9	7.8	1.3	6–11
—gh.	2	2	2	7.5	.7	7–8
—cf.	2	3	4	6.8	2.3	5–9
Median leaves of floriferous stems—thickness						
Wildwood	3	—	6	2	.7	1.3–2.8
—gh.	2	2	6	3.5	.5	2.8–3.9
Post Creek	2	—	6	2.4	.4	2.0–3.3

	n-pl.	n-rep.	n-obs.	x̄ (mm.)	s (mm.)	w (mm.)
Median leaves of floriferous stems—thickness						
Smith River—cf.	2	2	3	2.9	.7	2.4–3.7
Rogue River	4	—	9	3.9	.5	2.7–4.6
—cf.	2	3	4	2.8	.9	2.4–3.6

Inflorescences are elongate, with three or more primary branches which may be dichotomously forked. In bud, the ends of the cincinni are incurved. In the early stages of development, the floriferous stems of ssp. *heckneri* and *latifolium* are bent downward.

	n-pl.	n-rep.	n-obs.	x̄	s	w
Primary branches (n)						
Wildwood	3	—	8	5.7	1.4	2–12
—gh.	1	1	1	10	—	10
Post Creek	4	—	10	3.4	.5	3–4
Smith River	2	—	3	3.7	.3	3–4
—cf.	2	2	3	10.7	4.5	6–14
Rogue River	5	—	10	5.2	.6	4–6
—cf.	2	3	4	5.3	.8	4–6
Red Mt.	9	—	9	4	1.2	3–8
—cf.	5	5	8	4	.8	3–5
Primary axis—length (mm.)						
Wildwood	3	—	10	6.5	6.9	0–18
Post Creek	4	—	10	3.5	4.4	0–16
Smith River	2	—	3	2.1	1.2	.8–3
Rogue River	4	—	9	14.3	2.3	6.0–30

Flowers (fig. 100) are pedicellate and erect, with a short torus.

	n-pl.	n-rep.	n-obs.	x̄ (mm.)	s (mm.)	w (mm.)
Basal pedicels—length						
Wildwood	3	—	6	6.3	.6	5–7
Post Creek	4	—	10	5.2	.9	4–7
Smith River	2	—	3	1.5	.7	1–2
Rogue River	5	—	10	.6	.8	0–2
Red Mt.	10	—	30	1.8	.7	1–5
—cf.	9	11	17	2.3	1.3	1–5
Torus—length						
Wildwood	3	—	10	.8	.2	.6–1.2
—gh.	2	2	7	1.3	.1	1.2–1.5
Post Creek	4	—	17	1.1	.2	.8–1.4
Rogue River	1	—	5	.9	—	.6–1.3
Flowers—diameter						
Wildwood	3	—	11	7.5	1.1	6–11
—gh.	2	2	7	13.4	.1	12–15
Post Creek	4	—	17	9	1.6	6–11
Smith River—cf.	2	2	4	8.4	.2	8–9
Rogue River	1	—	5	11.6	—	8–16
—cf.	2	3	5	9.6	1.6	8–13

Sepals are erect, appressed to tube of corolla, connate basally, ovate or lanceolate, acute or sometimes obtuse, and pale green.

NATIVE SPECIES 385

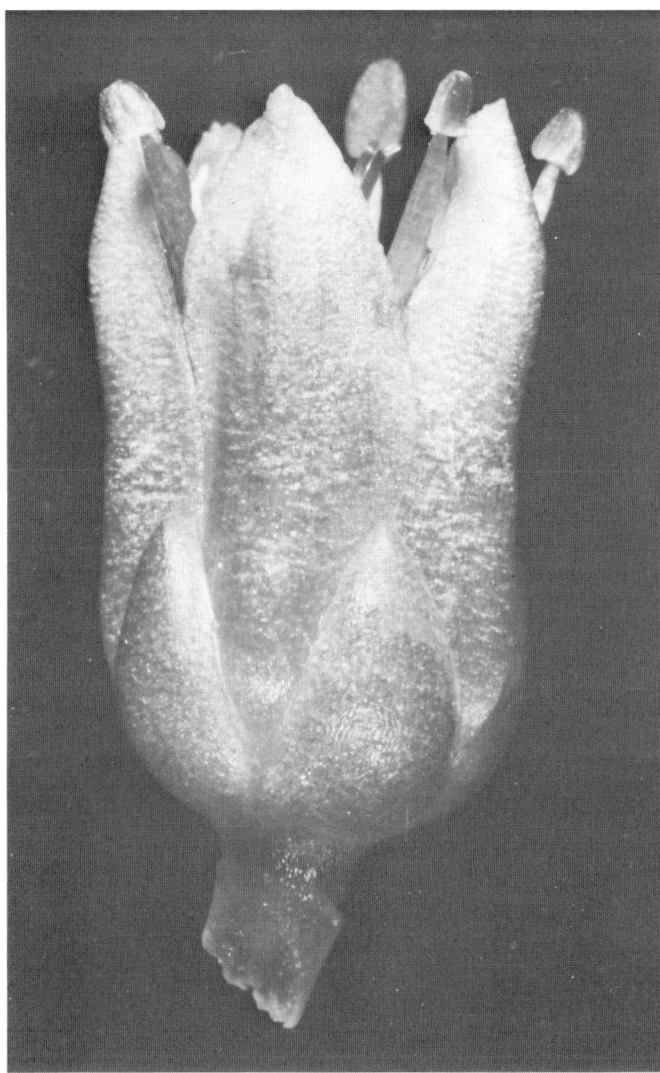

Fig. 100. Flower of *Sedum laxum* ssp. *eastwoodiae* from Red Mountain, Mendocino Co., Calif., cultivated in cold frame, Ithaca, N.Y., June 21, 1971 (× 10). Photo by Howard Lyon.

	n-pl.	n-rep.	n-obs.	x̄ (mm.)	s (mm.)	w (mm.)
Length						
Wildwood	3	—	11	3.7	.3	2.5–5.6
—gh.	2	2	7	3.4	.4	3.0–3.8
Post Creek	4	—	17	3.9	.4	3.1–5
Smith River—cf.	2	2	4	4.1	.7	3.4–5.5
Rogue River	1	—	5	3.8	—	3.5–4.4
—cf.	2	3	5	4.4	.7	3.5–5.1
Red Mt.	10	—	29	2.8	.7	2.0–4.1
—cf.	9	11	20	2.6	.4	2.1–3.3

	n-pl.	n-rep.	n-obs.	$\bar{x}$ (mm.)	s (mm.)	w (mm.)
Width						
Wildwood	3	—	11	2	.1	1.5–2.7
—gh.	2	2	7	2.7	.2	2.2–2.9
Post Creek	4	—	17	2.1	.3	1.7–2.9
Smith River—cf.	2	2	4	1.9	.4	1.4–2.6
Rogue River	1	—	5	2.1	—	1.9–2.2
—cf.	2	3	5	2.1	.4	1.6–2.6

Petals are erect basally, divergent above middle; lanceolate-oblong, oblanceolate-oblong, or elliptic-oblong; obtuse and short mucronate-appendaged or aristate-appendaged; pink, white, or yellowish white and pinkish dorsally. Sometimes the margins are erose.

	n-pl.	n-rep.	n-obs.	$\bar{x}$	s	w
Number						
Wildwood	3	—	11	5	0	5
—gh.	2	2	7	5	0	5
Post Creek	4	—	17	5	0	5
Smith River—cf.	2	2	4	5	0	5
Rogue River	1	—	5	5	—	5
—cf.	2	3	5	5	0	5
Length (mm.)						
Wildwood	3	—	11	9.3	.6	7.4–10.7
—gh.	2	2	7	10	.5	9.5–10.5
Post Creek	4	—	17	10.5	1.2	8.1–12.9
Smith River—cf.	2	2	4	8.6	1	7.8–9.5
Rogue River	1	—	5	9.2	—	8.6–9.8
—cf.	2	3	5	9.7	1.2	8.6–11.2
Red Mt.	10	—	29	6.5	1	4.0–8
—cf.	8	10	19	6.2	—	5.7–7.1
Length of cohesion (mm.)						
Wildwood	3	—	11	3	.4	2.2–3.9
—gh.	2	2	7	2.8	.3	2.0–3.2
Post Creek	4	—	17	3.5	.6	2.8–4.8
Smith River—cf.	2	2	3	1.4	.9	0–2.1
Rogue River	1	—	5	1.7	—	1.1–2.5
—cf.	2	3	5	2.2	.9	1.6–3.4
Red Mt.	1	—	3	2.9	—	2.7–3.4
—cf.	8	10	19	1.8	.1	1.3–2.1
Width (mm.)						
Wildwood	3	—	11	2.7	.2	2.0–3.6
—gh.	2	2	7	3.7	.2	3.1–4.1
Post Creek	4	—	17	3.5	.8	2.9–4.1
Smith River—cf.	2	2	4	2.9	.6	2.3–3.5
Rogue River	1	—	5	3	—	2.9–3.1
—cf.	2	3	5	3.6	.6	3.4–3.9

Stamens have pink, white, or greenish white filaments and either red or yellow anthers. The filaments alternating with the lobes of the corolla hold the petals together.

	n-pl.	n-rep.	n-obs.	x̄ (mm.)	s (mm.)	w (mm.)
Epipetalous filaments, height on petals						
Wildwood	3	—	11	3.7	.1	3.0–4.2
—gh.	2	2	7	4.2	.7	3.4–4.8
Post Creek	4	—	17	3.9	.2	3.0–5.8
Smith River—cf.	2	—	4	1.9	.5	1.4–2.5
Rogue River	1	—	5	2	—	1.4–2.4
—cf.	2	3	5	2.7	.7	2.4–3.6
Anthers—length						
Wildwood	3	—	6	1.5	.1	1.1–1.8
Post Creek	3	—	6	1.4	.7	1.2–1.6
Smith River—cf.	2	2	2	1.7	.3	1.5–2
Rogue River	1	—	1	1.7	—	1.7
—cf.	2	3	3	2.3	.1	1.9–2.7

Nectaries are reniform or transversely oblong; deeply concave to truncate at apex; yellow, pink, or white; and often translucent.

	n-pl.	n-rep.	n-obs.	x̄ (mm.)	s (mm.)	w (mm.)
Length						
Wildwood	3	—	11	.4	.05	.3–.5
—gh.	2	2	7	.4	.06	.3–.5
Post Creek	4	—	17	.4	.4	.3–.5
Smith River—cf.	2	2	4	.4	.07	.3–.5
Rogue River	1	—	5	.3	—	.3–.4
—cf.	2	3	5	.5	.4	.4–.5
Width						
Wildwood	3	—	11	1.1	.2	.7–2.1
—gh.	2	2	7	1.4	.01	1.2–1.5
Post Creek	4	—	17	.9	.1	.7–1.2
Smith River—cf.	2	2	4	.8	.1	.8–0.9
Rogue River	1	—	5	1	—	.9–1.1
—cf.	2	3	5	1.2	.1	1.1–1.3
Red Mt.	1	—	3	.8	—	.6–1
—cf.	9	11	20	.8	.1	.6–0.9

Carpels are erect, connate basally, and green, very pale green, pink, or white, with pink styles.

	n-pl.	n-rep.	n-obs.	x	s	w
Length (mm.)						
Wildwood	3	—	11	6.6	1.3	4.1–9
—gh.	2	2	7	7.9	.9	6.9–8.7
Post Creek	4	—	17	8.1	.6	6.8–9.7
Smith River—cf.	2	2	4	5.8	.2	5.5–6.4
Rogue River	1	—	5	7.1	—	6.6–7.3
—cf.	2	3	5	6.6	.1	6.4–7.3
Ovules per ovary (n)						
Wildwood	3	—	11	17	5.4	8–32
—gh.	2	2	7	17	6.3	12–24
Post Creek	4	—	17	17.5	4.6	12–26
Smith River—cf.	2	2	4	18.2	.2	16–20
Rogue River	1	—	5	14.4	—	12–16
—cf.	2	3	5	18.3	2	14–24
Red Mt.	1	—	3	11	—	7–18

Fruits are erect and brown.

	n-pl.	n-rep.	n-obs.	x̄ (mm.)	s (mm.)	w (mm.)
Length						
Wildwood	1	—	1	7.2	—	7.2
Post Creek	2	—	2	6.6	.6	6.2–7.1
Smith River	2	—	3	6.5	.7	6–7
—cf.	1	1	2	4	—	3.7–4.3
Rogue River	1	—	1	5.8	—	5.8
—cf.	2	3	6	4.3	.02	3.8–5
Red Mt.	3	—	9	5.8	.12	4.9–6.4
Beak—length						
Wildwood	1	—	1	1.7	—	1.7
Post Creek	2	—	2	1.6	.3	1.4–1.8

Seeds are pyriform, short-tailed at both ends, finely ribbed, and yellow-brown to dark brown.

	n-pl.	n-rep.	n-obs.	x̄ (mm.)	s (mm.)	w (mm.)
Length						
Post Creek	2	—	4	1.3	.07	1.3–1.4
Smith River	2	—	12	1.3	.01	1.2–1.4
—cf.	1	1	1	1.2	—	1.2
Rogue River—cf.	1	1	2	1	—	1.0–1.1
Red Mt.	3	—	9	1.2	.16	1.1–1.5
Diameter						
Post Creek	2	—	4	.4	.04	.4–.5
Smith River	2	—	12	.5	.01	.5–.6
—cf.	1	1	1	.4	—	.4
Rogue River—cf.	1	1	2	.3	—	.3
Red Mt.	3	—	9	.4	.07	.4–.5

Chromosomes are very small and appear either roughly circular or elliptical in cross section in late prophase and at metaphase.

	n-pl.	g (n)	sp (2n)	Cytologist
Number				
Mendocino Co., Calif.	1	—	30	L. Hollingshead (1942)
Mendocino Co., Calif.	1	15	—	C. H. Uhl, in Clausen and Uhl (1944)
Smith River Canyon, Calif.	1	—	30	L. Hollingshead (1942)
Waldo, Ore.	2	—	30	L. Hollingshead (1942)
Waldo, Ore.	1	15	—	C. H. Uhl, in Clausen and Uhl (1944)
Galice, Ore.	2	—	30	L. Hollingshead (1942)
Locality unknown	1	—	30	L. Hollingshead (1942)

Paul Hutchison, in a letter of July 31, 1962, reported a tetraploid from Horse Mountain, Humboldt Co., Calif. Charles Uhl, in a seminar at Cornell University in 1962, mentioned a plant which was g = 14 and sp = 29. The plant from Horse Mountain would go under ssp. *heckneri* as interpreted here. The relationship of Uhl's plant is unclear.

Variation. *Sedum laxum* is the most variable species of the section *Gormania*. In some respects, it is central in the section. Since it occupies an area which both is geologically old and had only local glaciers in the Pleistocene, the possibility exists that populations in the Klamath Mountains have had a long history there and have had time to adjust to a variety of conditions. The resulting pattern of variation has led to different interpretations and classifications, as is indicated by the synonymies under the nomenclature of the subspecies. Significant differences exist among plants within populations, among populations, and among groups of populations. Differences may justify recognition of four subspecies. Of these, the ssp. *heckneri* is the most variable and perhaps the least specialized. It may be nearest to the ancestral type from which the species is descended. Its broad leaves on the floriferous stems, with cordate-clasping bases, are distinctive. Close to ssp. *heckneri* is ssp. *latifolium*. That has large, broad leaves both in the rosettes and on the floriferous stems (fig. 101). Northward, in the drainage of the Rogue River, is a subspecies with narrower leaves and pinkish flowers. It occurs on serpentine. It is the nomenclatural type, ssp. *laxum*. Southwest of the Klamath Mountains, in the Coast Ranges of northern California, populations occur with shorter sepals, petals, and floriferous stems. These seem to represent a fourth subspecies. Still other subspecies may be recognized. Opinions

Fig. 101. Leaf of floriferous stem of *Sedum laxum* ssp. *latifolium* from canyon of Smith River west of Patrick Creek, Del Norte Co., Calif., cultivated in greenhouse, Ithaca, N.Y., Mar. 15, 1965 (x 7). Photo by Howard Lyon.

vary as to how many differences are necessary to distinguish a subspecies and how strong must be the geographic or ecological separation. At one extreme is naming each geographically disjunct population which differs significantly from others. At the other end of the spectrum is grouping widely different populations into one or a few broad regional subspecies. The treatment here is a compromise between these extreme points of view.

Paul Hutchison similarly has given attention to the problem of variation of *Sedum laxum*. His views, expressed in a letter of July 31, 1962, have been useful to me, although my studies have not always led to the same conclusions.

Because of the poor growth of *Sedum laxum* in Ithaca, results of experiments have been disappointing, but have yielded some interesting information. The experiments indicate few differences among plants within populations. Most plants in the small populations could be parts of the same clone. Exceptions are the population of ssp. *heckneri* along Post Creek, in which significant differences exist among plants in length of leaves, and the population of ssp. *latifolium* from the Smith River at Patrick Creek, in which significant differences occur among plants in both width and thickness of leaves.

Significant differences among subspecies are few, yet they seem to be correlated with distribution. Part of the trouble in classification here is a considerable within-plant variation which eliminates many features for discriminatory purposes. Another problem has been that some potentially good discriminatory features were not realized until after studies had been made. The difference in length of pedicels between ssp. *heckneri* and the other subspecies is highly significant on the basis of comparison of populations in the field. Using Tukey's test of all comparisons among means, a difference of 1.67 mm. in length of pedicels is significant. Compare with the data for pedicels in the above description. Similarly, on the basis of experiment in Ithaca, the leaves of ssp. *latifolium* are significantly wider than those of the other subspecies. The minimal significant difference among populations in an experiment in the greenhouse is 4.1 mm. Likewise, plants of ssp. *latifolium* have more floriferous stems, but this difference could be primarily an indication of greater vigor under the conditions of the experiment. Data in the description provide the basis for comparisons of populations and subspecies, and also of plants in nature and in cultivation.

KEY TO SUBSPECIES

A. Leaves of floriferous stems with cordate bases, clasping stems; pedicels of basal flowers 4–7 mm. long; petals pink, white, or yellowish white; anthers yellow .... ssp. *heckneri*, p. 391

AA. Leaves of floriferous stems with truncate bases, not clasping stems; pedicels of basal flowers 0–4 mm. long; petals pink or white; anthers red or yellow suffused with red .... B

    B. Leaves of primary rosettes obovate, 17–30 mm. broad, green or glaucous; nectaries .8–.9 mm. wide .... ssp. *latifolium*, p. 394

    BB. Leaves of primary rosettes 9–17 mm. broad, usually glaucous; nectaries .8–1.4 mm. wide C

        C. Sepals 3.5–5.1 mm. long; petals 9–11 mm. long; floriferous stems 15–26 cm. long .... ssp. *laxum*, p. 394

        CC. Sepals 2–4.1 mm. long; petals 4–8 mm. long; floriferous stems 4–13 cm. long .... ssp. *eastwoodiae*, p. 398

Sedum laxum ssp. *heckneri*

Nomenclature. *Sedum laxum* ssp. *heckneri* (M. E. Peck) R. T. Clausen, Bull. Torr. Bot. Club 69:39 (1942). Based on *S. heckneri* M. E. Peck.

A synonym is:

1937. *Sedum Heckneri* M. E. Peck, Proc. Biol. Soc. Wash. 50:121. Type locality: dry cliff along the Middle Fork of the Applegate River 5.9 km. above the mouth of Carberry Creek, Siskiyou Co., Calif. Type: M. E. Peck's collection no. 16421 (WILLU), June 26, 1931. Paul Hutchison has brought to my attention that the type locality is in California and not in Jackson Co., Ore.

The distinctive features of ssp. *heckneri* (fig. 102) are the nearly suborbicular leaves of the floriferous stems, which have cordate, clasping bases (fig. 103); the relatively

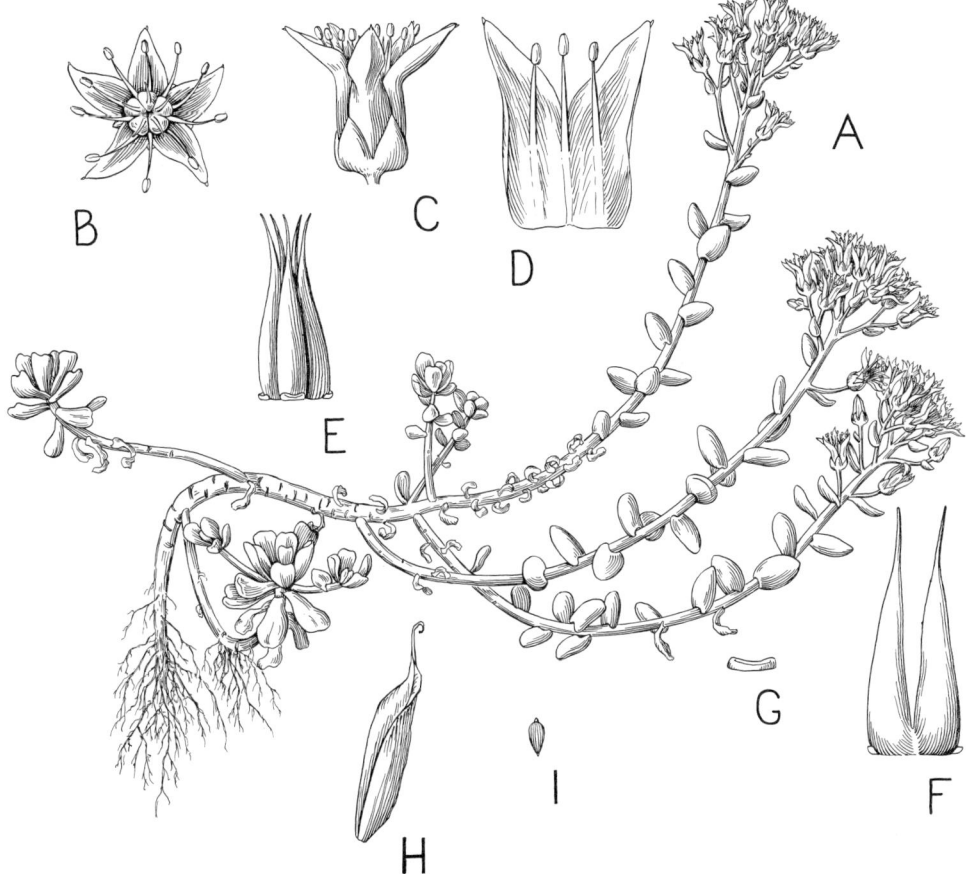

Fig. 102. Plant of *Sedum laxum* ssp. *heckneri* from west of Wildwood in the Klamath Mountains, Trinity Co., Calif., cultivated in greenhouse, Ithaca, N.Y., May 1964. A. Habit sketch (x .46). B. Flower from above (x 1.8). C. Flower from side (x 1.8). D. Two petals and stamens (x 2.8). E. Carpels (x 2.8). F. Two carpels and nectaries (x 3.7). G. Nectary (x 4.6). H. Follicle (x 3.7). I. Seed (x 4.6).

Fig. 103. Leaf of floriferous stem of *Sedum laxum* ssp. *heckneri* from along Post Creek northeast of Forest Glen in the Klamath Mountains, Trinity Co., Calif., cultivated in greenhouse, Ithaca, N.Y., Mar. 15, 1965 (x 5.4). Photo by Howard Lyon.

large, thick leaves of the primary rosettes; and the long pedicels of the flowers (fig. 104). Before expansion of the flowers, the floriferous stems are bent downward. The petals usually are pink, but they are white or even yellowish white in some populations in Trinity Co., Calif. The anthers are yellow.

The question about the proper taxonomic status of *Sedum laxum* ssp. *heckneri* is not easy to answer. No single item of evidence resolves the problem. Perhaps the most important considerations are that the populations are not uniform and that some are transitional to either ssp. *latifolium* or ssp. *laxum*. Such a condition is an argument against specific status. The ssp. *heckneri* is the most variable of the subspecies of *S. laxum*. Further, the number of features which are diagnostic is small. Another point of importance is the lack of occurrence of ssp. *heckneri* together with any other subspecies of *S. laxum*. A population near the mouth of the Rogue River at Gold Beach, Oregon, is variable with respect to the shape and bases of the leaves of the floriferous stems, but instead of appearing as two species occurring together, suggests a single variable subspecies which is best identified as ssp. *laxum*. The possibility exists that at some time ssp. *heckneri* washed downstream and, through hybridization, contributed to the variability.

Fig. 104. Flowers of *Sedum laxum* ssp. *heckneri* from west of Wildwood in the Klamath Mountains, Trinity Co., Calif., cultivated in greenhouse, Ithaca, N.Y., May 1964 (x 1.5). Photo by Howard Lyon.

Distribution. *Sedum laxum* ssp. *heckneri* has a wide distribution (fig. 105) in the Klamath Mountains in the drainages of the Trinity and Klamath Rivers and adjacent portions of the drainage of the Rogue River in Oregon, at altitudes from 152 meters in the valley of the Klamath River to 1,769 meters at Salmon Summit. Populations are known to me from 22 localities. This is the most widely distributed subspecies of *S. laxum*. The plants occur on serpentine and flower in May or June. Ecological data for the two populations which I have studied in the field are summarized in the following table.

Population	N plants	Altitude (m.)	pH	Oldest record
West of Wildwood	24	923–925	6.6	1937
Post Creek	10	790	6.2–6.4	1951

Plants in shade at the site west of Wildwood had large, green, retuse leaves, resembling ssp. *latifolium*.

The oldest collection which has come to my attention is by Harvey P. Chandler (UC), June 1901, from along the Klamath River at an altitude of 305 meters in Humboldt Co., Calif.

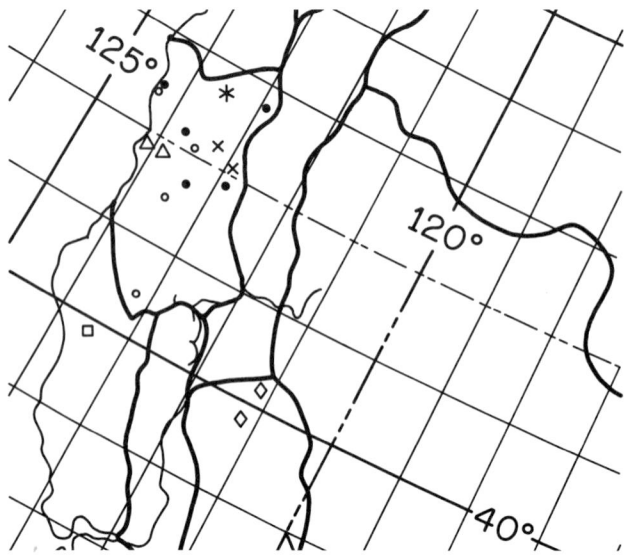

Fig. 105. Known distribution of Sedum laxum ssp. heckneri (○), ssp. latifolium (△), ssp. laxum (●), and ssp. eastwoodiae (□), and also of S. oblanceolatum (x), S. albomarginatum (◇), and S. moranii (·).

Sedum laxum ssp. *latifolium*

Nomenclature. *Sedum laxum* ssp. *latifolium* R. T. Clausen, Bull. Torr. Bot. Club 69:38 (1942). Type locality: rocky slope along Smith River 38 km. northeast of Crescent City, Del Norte Co., Calif. Type: R. T. Clausen's collection no. 4941 (BH), July 24, 1940.

Plants of ssp. *latifolium* (fig. 106) are the most robust of any of the subspecies of *Sedum laxum*. The rootstocks may be stout and the leaves are both broad and thick, with the apices deeply emarginate and green, without a bloom. Because the distinctions observed in the field (fig. 107) are maintained in cultivation and continue to distinguish plants from the drainage of the Smith River, they are the basis for status as a subspecies. Against this idea is the evidence that some plants from the Smith River are less robust than others and approach either ssp. *heckneri* or ssp. *laxum*. Populations somewhat intermediate between ssps. *heckneri* and *latifolium* occur in the drainage of the Trinity River, for example near Peanut.

Distribution. *Sedum laxum* ssp. *latifolium* (fig. 105) occurs only in the drainage of the Smith River, Del Norte Co., Calif. W. L. Jepson discovered it along the South Fork of the Smith River on July 16, 1907. The known altitudinal range is from 61 to 549 meters. Flowering time is June and July. A total of nine different localities for this subspecies have come to my attention. At 3.7 km. west of Patrick Creek, the source of my sample in 1963, the plants grew on serpentine, exposed to the northeast, at an altitude of 260 meters. Competing species included *Polystichum munitum*, *Holodiscus discolor*, a species of *Aster*, and grasses. The pH of the soil was 6.6.

Sedum laxum ssp. *laxum*

Nomenclature. *Sedum laxum* (Britton) Berger, in Engler's Nat. Pflanzenfam., ed. 2, 18a:451 (1930). Based on *Gormania laxa* Britton.

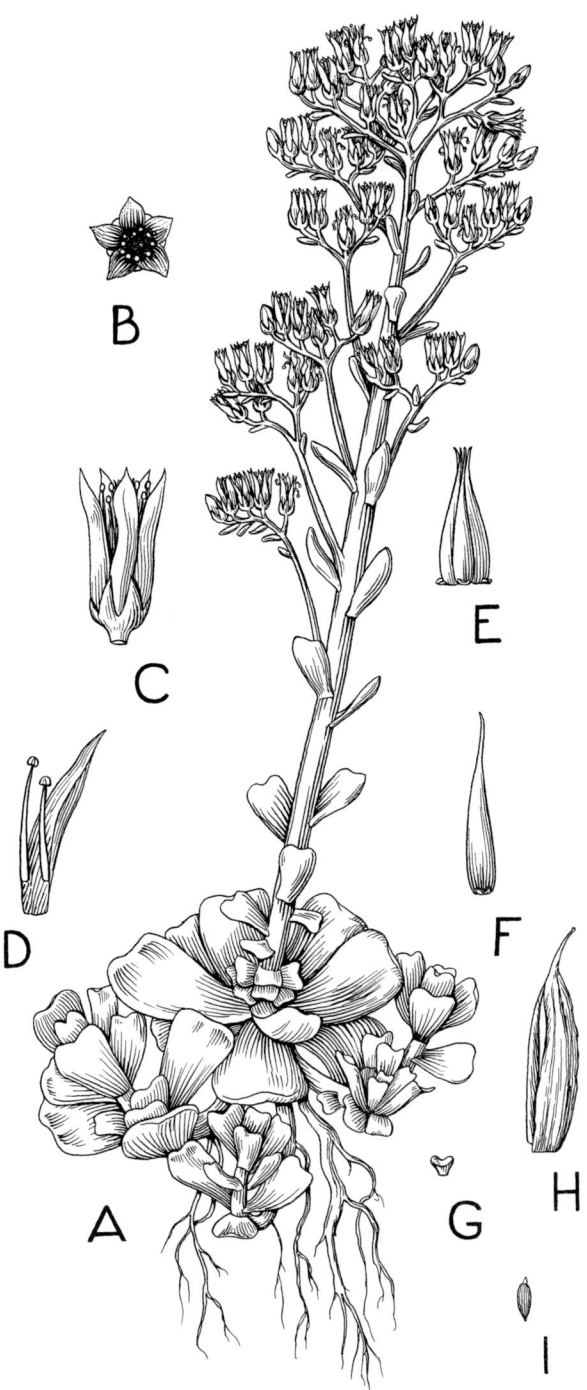

Fig. 106. Type plant of *Sedum laxum* ssp. *latifolium* from slope along Smith River, 38 km. northeast of Crescent City, Del Norte Co., Calif., cultivated in greenhouse, Ithaca, N.Y., June 1944. A. Habit sketch (x .5). B. Flower from above (x 2). C. Flower from side (x 2). D. Petal and two stamens (x 3). E. Carpels (x 3). F. Single carpel and nectary (x 4). G. Nectary (x 5). H. Follicle (x 4). I. Seed (x 5).

Fig. 107. Plant of *Sedum laxum* ssp. *latifolium* at type locality, on rocks above Smith River, 38 km. northeast of Crescent City, Del Norte Co., Calif., July 24, 1940. Photo by Harold Trapido.

Synonyms are:

1903. *Gormania laxa* Britton, Bull. N.Y. Bot. Gard. 3:29. Type locality: Waldo, Ore. Type: Thomas Howell's collection of June 4, 1884 (NY).

1904. *Cotyledon Brittoniana* Fedde, Just's Bot. Jahresb. 31 (1):827. Based on *Gormania laxa* Britton.

1913. *Echeveria Gormania* Nels. & Macbr., Bot. Gaz. 56:476. Based on *Gormania laxa* Britton.

1942. *Sedum laxum* ssp. *perplexum* R. T. Clausen, Bull. Torr. Bot. Club 69:36–37. Type locality: cliff near mouth of Rogue River, Ore. Type: M. E. Peck's collection no. 8703 (BH), July 8, 1919.

The name *Sedum Jepsonii* H. M. Butterfield, Desert Plant Life 8:7 (1936) is not a valid name since a Latin diagnosis is lacking. Further, there is no indication of a type. Application of the name has been to *S. laxum* ssp. *laxum*.

Spatulate or oblanceolate leaves of the rosettes and floriferous stems, narrower than in either of the two preceding subspecies, distinguish ssp. *laxum*. Some plants at Waldo produce the tallest floriferous stems known for the species, but others at the same site are as short as the smallest specimens described as ssp. *perplexum*. Howell, on whose collection *S. laxum* is based, was aware of this situation. An interesting series of specimens in the herbarium of the University of Oregon illustrates the variation in size. Howell even had a varietal name in manuscript for the largest plants. Size is not a satisfactory basis for separating ssp. *perplexum* from ssp. *laxum*.

Distribution. *Sedum laxum* ssp. *laxum* (fig. 105) occurs primarily in the lower portion of the drainage of the Rogue River in southwestern Oregon, but also in adjacent drainages in northern California where it may be transitional to other subspecies. The altitudinal range is from near sea level to 1,220 meters. Flowering time is June. Twenty localities are known to me for this subspecies. In addition, plants at several other localities may be intermediate between ssp. *laxum* and other subspecies. The plants usually grow on serpentine. In the canyon of the Rogue River 10 km. south of Galice, Ore., ssp. *laxum* occurs on bluffs of serpentine (fig. 108), exposed to the southwest, at an altitude of 280 meters. I studied it there on May 28, 1963. The population comprised about 300 plants in discrete groups on about eight sections of bluff. Other vegetation consisted of moss and a species of *Eriogonum*.

Fig. 108. Plants of *Sedum laxum* ssp. *laxum* on serpentine in canyon of Rogue River, 10 km. south of Galice, Josephine Co., Ore., May 28, 1963.

Drainage was excessive, and the range of pH of the substrate in which the *Sedum* was rooted was 6.4–6.6. Plants were in bud, almost ready to come into flower, at the time of study.

Sedum laxum ssp. *eastwoodiae*

Nomenclature. *Sedum laxum* ssp. **eastwoodiae** (Britton) R. T. Clausen, comb. nov., fundatum super *Gormania Eastwoodiae* Britton, Bull. N.Y. Bot. Gard. 3:31 (1903). Synonyms are:

1903. *Gormania Eastwoodiae* Britton, Bull. N.Y. Bot. Gard. 3:31. Type locality: Red Mountain, northern Mendocino Co., Calif. Type: a collection of Alice Eastwood (CAS34297), August 1902.
1904. *Cotyledon mendocinoana* Fedde, Just's Bot. Jahresb. 31 (1):827. Based on *Gormania Eastwoodiae*.
1930. *Sedum Eastwoodiae* (Britton) Berger, in Engler's Nat. Pflanzenfam., ed. 2, 18a:451.

Shorter sepals, petals, and floriferous stems, and also a disjunct distributional area, are distinctive features of ssp. *eastwoodiae* (fig. 109). Study of the type population on Red Mountain has revealed that the leaves of the rosettes actually are as thick as in the other subspecies. The degree of congestion of the inflorescence has not proved to be a satisfactory character for distinction. The new name is necessary for this subspecies because the epithet *retusum*, used in my publication of 1940, now is transferred to status under *S. obtusatum*. Some populations may be intermediate between ssp. *eastwoodiae* and other subspecies.

Distribution. *Sedum laxum* ssp. *eastwoodiae* (fig. 105) occurs in the northern California Coast Ranges where it grows on serpentine. The elevational range is from 1,065 to 1,173 meters. On Red Mountain, the type locality, in Mendocino Co., Calif., plants occur on the rocky northeastern slope of the southeastern spur of the mountain, in an area where both logging and fires have occurred between 1950 and 1970. The vegetation in 1970 included occasional trees of *Libocedrus decurrens*, *Pseudotsuga menziesii*, *Pinus ponderosa*, and *P. lambertiana*, and shrubs of *Arctostaphylos*. Plants receive about 60–80% of potential sunlight. The rock is serpentine. Data for the population on Red Mountain are organized in the following table.

Population	Plants (n)	Area (m.^{2})	Density per m.^{2}	Exposure
Red Mt.	885	9,219	.096	N–NE

Population	Altitudinal range (m.)	pH, range	pH, tests (n)	Depth of soil (C.I.$_{.05}$; cm.)	Distance to nearest neighbor (C.I.$_{.05}$; cm.)
Red Mt.	1,065–1,100	5.6–6.8	18	1.4–3.2	.6–69.4

Plants of *Sedum laxum* on Red Mountain did not appear to be in good condition. Some were dying. Others had spots on the leaves (fig. 110) and appeared to be

Fig. 109. Plant of *Sedum laxum* ssp. *eastwoodiae* from Red Mountain, Mendocino Co., Calif., cultivated in cold frame, Ithaca, N.Y., June 21, 1971 (x .9). Photo by Howard Lyon.

diseased. Closest competitors of 10 randomly selected plants were *Iris purdyi*—3, a grass—3, *Libocedrus decurrens*—2, and *Hieracium bolanderi*—2.

Reproduction. Plants of *Sedum laxum* reproduce either vegetatively by means of rootstocks and rosettes which become separated from the parental parts or by seeds. Rosettes of ssp. *heckneri* and ssp. *laxum*, stored in dry condition in plastic bags in

Fig. 110. Plant of *Sedum laxum* ssp. *eastwoodiae* on Red Mountain, Mendocino Co., Calif., July 16, 1970. Note spots on leaves and dead shoots.

1963, remained in a good state for two months, indicating ability to withstand lengthy periods without water.

The period of flowering of the various subspecies of *Sedum laxum* is from May to July, principally in June. Insects visit the flowers and effect cross-pollination. At the site west of Wildwood and along Post Creek, both in the Klamath Mountains of California, a large black fly appears to be a pollinator of ssp. *heckneri*.

Hybrids with other species are rare. In the southern Klamath Mountains, *S. laxum* ssp. *heckneri* and *S. spathulifolium* ssp. *purdyi* occur near to each other and flower at the same time without crossing. Similarly, in the canyon of the Rogue River four species of *Sedum* occur in adjacent sites but do not hybridize, although times of flowering overlap. *Sedum moranii* and *S. radiatum* ssp. *ciliosum* are in anthesis by late May. *Sedum laxum* ssp. *laxum* follows in early June. In addition, *S. spathulifolium* ssp. *spathulifolium* sheds pollen at the same time. Genetic isolation of some kind keeps these species distinct.

Relationships. The closest relationships of *Sedum laxum* appear to be with *S. oregonense* and *S. oblanceolatum*. *Sedum oregonense*, as far as known, is hexaploid.

It has white or yellowish white flowers, and the leaves of the floriferous stems are sessile. If ideas about an affinity to *S. laxum* are correct, then the relationship is with ssp. *heckneri*. That has leaves which clasp the floriferous stems and either pink or white flowers. Plants in some populations in the drainage of the Trinity River in California can be confused with *S. oregonense*. The physiological differences, indicated by different habitat and altitudinal range, are fundamental, but the morphological distinctions are weak.

The relationship of *Sedum laxum* with *S. oblanceolatum* may be through ssp. *laxum*. Populations of both kinds occur in the drainage of the Rogue River, with the *S. oblanceolatum* upstream and peculiar to part of the drainage of the Applegate River. It is diploid, with the leaves of the rosettes and upper parts of the floriferous stems pruinose. Both the sepals and seeds are longer than in *S. laxum*. The two kinds of populations are close enough morphologically to raise a question whether *S. oblanceolatum* is better interpreted as a subspecies of *S. laxum* or as a separate species.

A more distant relationship is with *Sedum obtusatum* which has smaller leaves and petals. The flowers of that are yellow in two subspecies, but white in the ssp. *boreale*.

Wide crosses are not unusual in the *Crassulaceae*. Three natural intergeneric hybrids came to attention in the study of *Sedum* in the Trans-Mexican Volcanic Belt (Clausen 1959). Two plants, found by Helen Payne of Dallas, Ore., on cliffs along the Trinity River in Humboldt Co., Calif., appear to be hybrids of *S. laxum* ssp. *heckneri* and *S. spathulifolium* ssp. *spathulifolium*. One plant (fig. 111), 71–3, received on May 14, 1971, as *Sedum* Silvermoon, was regarded at first as an unusual variation

Fig. 111. Plant of *Sedum laxum* ssp. *heckneri* × *spathulifolium* ssp. *spathulifolium*, nothomorph Silvermoon, from cliff above Trinity River, Klamath Mountains, Humboldt Co., Calif., sent by Helen E. Payne, photographed on May 14, 1971 (x 1.2). Photo by Howard Lyon.

of *S. spathulifolium*. The second (fig. 112), 71–6, received on June 7, 1971, clearly was intermediate between *S. laxum* and *S. spathulifolium*. According to Mrs. Payne, both parental species occur in the vicinity of the site of discovery of the hybrids. A comparison of the two hybrids with the parental species follows. Data for each of the hybrids are means of 5 observations per plant. Data for the two species are ranges of means for wild plants. Note that 71–3 exhibits 6 features of *S. laxum*, 7 features of *S. spathulifolium*, and 4 which are intermediate. 71–6 has 6 features of *S. laxum*, 10 features of *S. spathulifolium*, and 1 intermediate feature. Petals of both plants were

Fig. 112. Plant with flowers of *Sedum laxum* ssp. *heckneri* × *spathulifolium* ssp. *spathulifolium* from cliff above Trinity River, Klamath Mountains, Humboldt Co., Calif., sent by Helen E. Payne, photographed on June 8, 1971 (x 1). Photo by Howard Lyon.

yellow. These two hybrid clones are different. Possibly one is the result of backcrossing with one of the parental species.

Feature	S. laxum	71–3	71–6	S. spathulifolium
Primary ros.—diam. (cm.)	2.6–5.5	2.9	4	1.5–3.7
Offsets per ros. (n)	.5–7	.8	5	1.3–4
Offsets—diam. of stems (mm.)	.1–.6	1.6	1.8	1.2–1.9
Lvs. of ros.—width (mm.)	8–19	7.6	7	5–9
—thickness (mm.)	2.8–3.7	3.2	3.3	1.4–2.5
Med. lvs. of flor. st.—w. (mm.)	6.8–13.2	6.3	7	3–5.7
—th. (mm.)	2–3.9	3.1	3.1	1.4–2.1
Cincinni per cyme (n)	3–11	3	8	2–3.3
Flowers—diam. (mm.)	8–13	15	12	10–16
Sepals—length (mm.)	3.4–4.4	4.3	2.4	2.1–3.6
Petals—length (mm.)	8.6–10.5	8	5.6	5.8–8.7
—length of cohesion (mm.)	1.7–3.5	.8	.5	0
—width (mm.)	2.7–3.7	2.6	1.9	1.7–2.7
Stamens—adn. epipet. fil. (mm.)	1.9–4.2	1.9	1	.8–1.5
Nectaries—width (mm.)	.8–1.4	.7	.5	.3–.6
Carpels—length (mm.)	5.8–8.1	6.8	4.5	4.6–6.2
Ovules per ovary (n)	14–18	10.8	4.2	4–11

Fruits collected in the greenhouse in Ithaca, N.Y. on Aug. 9, 1971, from 71–3, nothomorph Silvermoon, are subdivergent, less divergent than in *Sedum spathulifolium*, but more so than in *S. laxum*. They are somewhat gibbous ventrally and 4.7–5.7 ($\bar{x}$ = 5.1) mm. long. Of hundreds of ovules, only 5 are developed as seeds. These are 1.1 mm. long and .5 mm. in diameter.

In summary, *Sedum laxum* is the most variable species of subgenus *Gormania*. In some respects it is central in relationships. Perhaps it is nearest to the ancestral type from which the other modern species are derived. Its greater variability, elongate inflorescences, and large leaves all suggest a generalized condition. Further, its occurrence in an area which has been available for occupancy by plants for a long period is a point of importance. Other features, as the greater cohesion of petals compared with the condition in related species and the occurrence on serpentine, are specializations to be expected if the species has had a long history and has had time to adapt to special environmental situations.

References. Clausen, R. T. (1942) and Hollingshead, L. (1942).

23. *Sedum oblanceolatum**** (figs. 105 and 113–114)

Plants of *Sedum oblanceolatum* (fig. 113) are similar to *S. oregonense*, but slenderer, with the leaves of the rosettes and the upper parts of the floriferous stems pruinose; the leaves of the floriferous stems oblanceolate-oblong, averaging in nature 22 mm. long and 6 mm. wide; and the sepals longer, averaging 5.4 mm. long. The petals are white, with the lobes erect, except slightly divergent at apex. *Sedum laxum* usually has pink flowers; *S. laxum* ssp. *heckneri* has suborbicular, clasping leaves

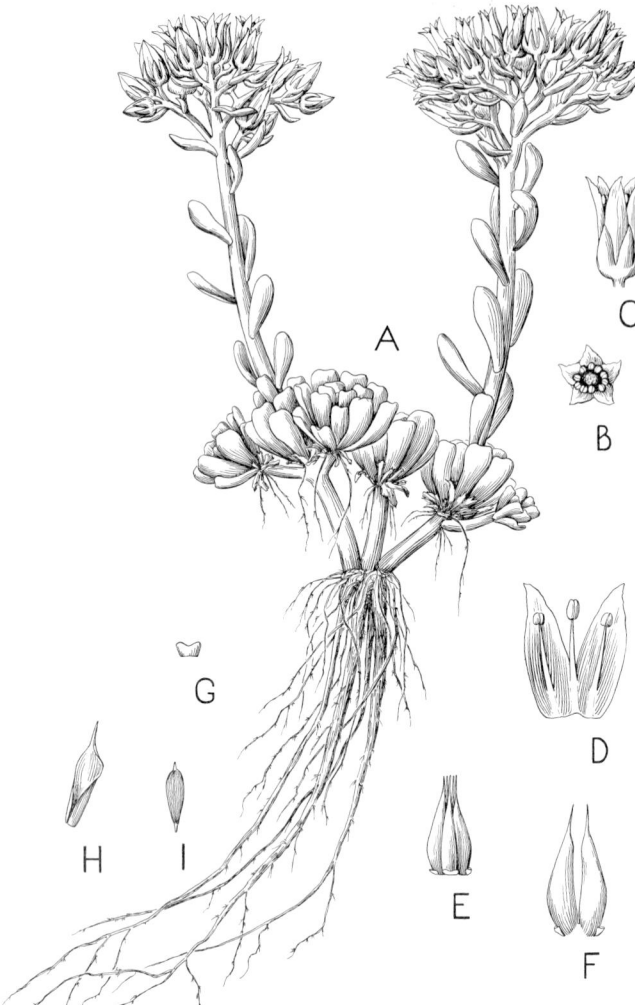

Fig. 113. Plant of *Sedum oblanceolatum* from slope by road, .8 km. below Deadman's Point, on ridge northwest of Dutchman's Peak, Jackson Co., Ore. A. Habit sketch (x .68). B. Flower from above (x 1.4). C. Flower from side (x 1.4). D. Two petals and stamens (x 2). E. Carpels (x 2). F. Two carpels and nectaries (x 2.7). G. Nectary (x 3.4). H. Follicle (x 2). I. Seed (x 3.4).

on the floriferous stems; and *S. obtusatum* ssp. *boreale* has the leaves of the rosettes merely glaucous and the sepals and petals shorter.

Because *Sedum oblanceolatum* is described here for the first time, a brief description in Latin is provided.

Descriptio originalis. *Sedum* **oblanceolatum,** sp. nov., affine *Sedi oregonensis*, sed tenuiore, foliis rosularum et caulibus floriferibus pruinosis, foliis caulium floriferum oblanceolate-oblongis, fere 22 mm. longis, et fere 6 mm. latis, et sepalis longioribus, fere 5.4 mm. longis. Petala alba sunt, cum lobis erectis praeter exigue divaricatis ad apicem. Typus in herbario Wiegand, Universitatis Cornellianae, ab declivitate saxoso prope viam .8 km. subter Deadman's Point, monte dorso inter septentriones et occasum solis spectante ex Dutchman's Peak, Jackson Co., Ore., alt. 1,582 m., collectio Roberti Clausenii, num. 651,034, 1965, Julio 8, est.

Description. Sample: 14 plants—6 in flower and 8 in fruit, all from slope by road .8 km. below Deadman's Point, on ridge northwest of Dutchman's Peak, Jackson Co., Ore. This sample comprised half of the plants which were found with flowers in 1965. The procedure was to tag all plants in flower and then to draw a simple random sample. Plants with the first 6 randomly chosen numbers, collected on July 3, 1965, comprised the sample for study of flowers. Plants with the other 8 randomly chosen numbers, collected on August 7, 1965, comprised the sample for study of fruits and seeds. A detailed description of the site is in the chapter on geography, under Klamath Mountains.

In the following tables, w = data for plants in the wild and cf. = cold frame in Ithaca, N.Y.

Plants of *Sedum oblanceolatum* are perennial herbs with long, coarse, fibrous roots from short, slender, branched rootstocks bearing rosettes in dense clusters. Upper parts of floriferous stems are pruinose. Before anthesis the floriferous stems are recurved.

	n	n-rep.	n-obs.	$\bar{x}$	s	w
Diameter (cm.)						
w	8	—	8	11.7	4.9	6–20
cf.	2	4	4	6	1.8	4–8
Primary rosettes—diameter (mm.)						
w	6	—	10	35.3	11	16–54
cf.	2	4	7	16.8	2.9	11–20
Offsets (n)						
w	6	—	10	2.7	1.6	1–6
cf.	2	4	7	2.4	1.6	0–5
Offsets—length (cm.)						
w	6	—	11	1.4	.7	.6–2.7
cf.	2	4	5	1.4	.9	.5–2
Offsets—diameter of stems (mm.)						
w	6	—	11	1.8	.2	1.5–2.2
Floriferous stems (n)						
w	8	—	8	1.6	.9	1–3
cf.	1	1	1	1	—	1
Floriferous stems—length (cm.)						
w	8	—	8	10.1	3.1	6–15
cf.	1	1	1	6	—	6

Leaves of rosettes are erect, oblanceolate-oblong, obtuse to truncate and sometimes emarginate, and pruinose. Leaves of floriferous stems are oblanceolate-oblong and sessile.

	n	n-rep.	n-obs.	$\bar{x}$ (mm.)	s (mm.)	w (mm.)
Leaves of rosettes—length						
w	6	—	10	26.7	7.4	10–38
cf.	2	4	8	8.7	.9	7–11

	n	n-rep.	n-obs.	$\bar{x}$ (mm.)	s (mm.)	w (mm.)
Leaves of rosettes—width						
w	6	—	10	7.1	1.2	5–9
cf.	2	4	7	5.2	.5	5–6
Leaves of rosettes—thickness						
w	6	—	10	2.3	.3	2.0–2.8
cf.	2	4	8	1.1	.1	.8–1.3
Median leaves of floriferous stems—length						
w	6	—	9	21.6	3.3	16–27
cf.	1	1	1	15	—	15
Median leaves of floriferous stems—width						
w	6	—	9	6.2	.5	5–7
cf.	1	1	1	6.2	—	6.2
Median leaves of floriferous stems—thickness						
w	6	—	9	2	.3	1.5–2.4
cf.	1	1	1	3.1	—	3.1

Inflorescences are dense, paniculate cymes.

	n	n-rep.	n-obs.	$\bar{x}$ (n)	s (n)	w (n)
Primary branches						
w	6	—	9	7.4	.5	6–9
cf.	1	1	1	5	—	5
Flowers and buds						
w	6	—	9	41	8.8	31–57

Flowers are pedicellate and 5-merous.

	n	n-rep.	n-obs.	$\bar{x}$ (mm.)	s (mm.)	w (mm.)
Flowers—diameter						
w	6	—	10	6	1	5–8
cf.	1	1	2	9.3	—	9–9.7
Torus—length						
w	6	—	11	1	.2	.7–1.4

Sepals are erect, connate basally, lanceolate-oblong, acute, papillose, and glaucous.

	n	n-rep.	n-obs.	$\bar{x}$ (mm.)	s (mm.)	w (mm.)
Length						
w	6	—	12	5.4	.7	4.1–7.1
cf.	1	1	2	5	—	4.9–5.1
Width						
w	6	—	12	2	.4	1.5–3.2
cf.	1	1	2	1.6	—	1.6–1.7

Petals are erect, connate basally for about a quarter of their length, oblong, with erose margins, acute and minutely mucronate-appendaged at apex, and creamy white.

	n	n-rep.	n-obs.	$\bar{x}$	s	w
Number						
w	6	—	12	5	0	5
cf.	1	1	2	5	—	5

	n	n-rep.	n-obs.	x̄	s	w
Length (mm.)						
w	6	—	12	10.2	.8	8.8–11.3
cf.	1	1	2	9	—	8.8–9.2
Length of cohesion (mm.)						
w	6	—	12	2.6	.5	1.8–3.7
cf.	1	1	2	2.2	—	1.8–2.6
Width (mm.)						
w	6	—	12	2.9	.3	2.5–4
cf.	1	1	2	2.7	—	2.5–3

Stamens have yellow anthers.

	n	n-rep.	n-obs.	x̄ (mm.)	s (mm.)	w (mm.)
Epipetalous filaments, height on petals						
w	6	—	12	2.7	.4	2.2–3.9
cf.	1	1	2	2.1	—	2.0–2.3
Anthers—length						
w	2	—	3	1.4	.03	1.4–1.5
cf.	1	1	1	1.4	—	1.4

Nectaries are transversely oblong, deeply retuse, and yellow, varying from pale yellow to orange.

	n	n-rep.	n-obs.	x̄ (mm.)	s (mm.)	w (mm.)
Length						
w	6	—	12	.5	.14	.3–.8
cf.	1	1	2	.3	—	.3–.4
Width						
w	6	—	12	1.2	.14	.9–1.4
cf.	1	1	2	1	—	1.0–1.1

Carpels are erect, with pale green ovaries and pink styles.

	n	n-rep.	n-obs.	x̄	s	w
Length (mm.)						
w	6	1	12	6.9	.7	5.3–7.9
cf.	1	—	2	5.9	—	5.8–6
Styles—length (mm.)						
w	6	—	12	1.6	.3	1.1–2.5
Ovules per ovary (n)						
w	6	—	12	20.8	5.8	12–28
cf.	1	1	2	17	—	16–18

Fruits are erect, connate basally, and brown.

	n	n-rep.	n-obs.	x̄	s	w
Length (mm.)						
w	8	—	24	5.3	.7	4.1–7.3
cf.	1	1	2	5.4	—	5.1–5.8

Seeds are pyriform, short-winged at both ends, finely longitudinally ribbed, and brown.

	n	n-rep.	n-obs.	$\bar{x}$ (mm.)	s (mm.)	w (mm.)
Length						
w	8	—	24	1.1	.2	.9–1.5
cf.	1	1	2	1.1	—	1.1
Diameter						
w	8	—	24	.34	.05	.3–.4
cf.	1	1	2	.4	—	.4

Chromosomes are small.

	n-pl.	g (n)	sp (2n)	Cytologist
Deadman's Point	1	15	—	C. Horn

Variation. The most noteworthy observed variation of *Sedum oblanceolatum* is the differential performance of plants on the slope at Deadman's Point, Ore., and in cultivation at Ithaca, N.Y. The cultivated plants are smaller, with the leaves of the rosettes only about a third of the length of those in the wild. The modification of vegetative parts is not matched by similar differences in flowers and fruits. The floral parts, fruits, and seeds of a plant in cultivation in a cold frame were similar to those of plants in the wild, but the petals were more divergent.

Except for a flower in which two sepals had grown together, no aberrations have come to attention. All flowers studied were 5-merous.

Nomenclature. *Sedum oblanceolatum* Clausen, proposed as a species on p. 404. The type is my collection number 651,034, of July 8, 1965, from the rocky bank of the road .8 km. below Deadman's Point, on the ridge northwest of Dutchman's Peak, Jackson Co., Ore. Specimens in herbaria have been labeled as near *S. oregonense* or *S. laxum*, or with a subspecific name under *S. laxum*.

Distribution. *Sedum oblanceolatum* (fig. 105) is known from three localities, all in the drainage of the Applegate River in the Klamath Mountains of Jackson Co., Ore. R. H. Whittaker discovered the species at Deadman's Point on July 13, 1950, his collection no. 201 (WS). Paul Hutchison found the other two populations.

The known altitudinal range of *Sedum oblanceolatum* is from 457 meters along the Applegate River below the Star Ranger Station to 1,582 meters below Deadman's Point. The complete known range of the species has a diameter of only 20 kilometers.

I studied the population at Deadman's Point in 1965. Plants (fig. 114) there grow on a rocky slope, exposed to the southwest and south-southwest, in crevices and about the edges of diorite, at an altitude of 1,580 to 1,582 meters. I counted 28 plants with floriferous stems in an area of 408 square meters. The range of pH of soil about the roots of 9 randomly selected plants was from 5 to 5.8. Drainage was good to excessive. Scattered large trees of *Pinus ponderosa* occurred at the site for the *Sedum*. Smaller competing plants were miscellaneous. Of competing species by 9 randomly selected plants of *S. oblanceolatum*, only *Arctostaphylos nevadensis* occurred by as many as 2 plants.

Reproduction. *Sedum oblanceolatum* was in flower at the site below Deadman's Point in early July 1965. The only other *Sedum* nearby was *S. radiatum* ssp.

Fig. 114. Sedum oblanceolatum on rocky slope, .8 km. below Deadmans Point, on ridge northwest of Dutchman's Peak, Jackson Co., Ore., July 3, 1965.

depauperatum. That too was in bloom, but the plants were not intermixed. The chances of cross-pollination appeared negligible. The only pollinators observed visiting flowers were small bees.

Propagation can be either vegetative, by means of rosettes, or sexual, by means of seeds. The small size of the population near Deadman's Point suggests that neither method of spread is effective under present conditions.

Rosettes, kept dry for seven months, grew satisfactorily when put in soil and provided with water.

Relationships. The relationships of *Sedum oblanceolatum* are with *S. laxum*, *S. obtusatum* ssp. *boreale*, and *S. oregonense*.

The separation of *S. oregonense* is easy. When plants are in flower, the sepals provide a good means of distinction. A sample of *S. oregonense* from Deer Butte had a mean length of sepals of 3.2 mm. The sample of *S. oblanceolatum* from Deadman's Point had a mean length of sepals of 5.4 mm. According to the Wilcoxon rank sum test, the difference is highly significant. In cultivation in an experiment of completely random design, clones of *S. oregonense* from four localities had significantly longer leaves of the primary rosettes than *S. oblanceolatum*, but other dimensions of leaves were not significantly different. Tests of significance of differences in dimensions of leaves were by means of analysis of variance.

Sedum oblanceolatum has longer, narrower leaves on the floriferous stems than *S. laxum* ssp. *heckneri*. On the basis of the Wilcoxon rank sum test, plants from below Deadman's Point differ significantly from plants from between Wildwood and Peanut, Calif., in both length and width of median leaves of floriferous stems.

Both *Sedum obtusatum* ssp. *boreale* and *S. oblanceolatum* occur on Dutchman's Peak, Lane Co., Ore. The former is still in bud when the latter is in flower. The *S. obtusatum* occurs at a higher altitude, is on the northern side of the ridge, and grows on gneiss. Its petals are much shorter than those of either *S. oblanceolatum* or *S. oregonense*.

Sedum oblanceolatum, *S. oregonense*, *S. laxum*, and *S. obtusatum* all may have had a common evolutionary background. Physiologically, *S. oblanceolatum* appears close to *S. laxum*, which also occurs at low altitudes and flowers early in the summer. It replaces that in the drainage of the Applegate River. If it is not a separate species, then it is a subspecies of *S. laxum*. Absence of intermediate populations is the reason for adoption of specific status here.

24. *Sedum oregonense**** (figs. 115–119)

Distinctive features of *Sedum oregonense* (fig. 115) are stout rhizomes; prominent rosettes of obovate or oblanceolate, emarginate leaves; and erect petals which are connate for about one-quarter of their length and white, yellowish white, or greenish white. Plants often are large, with many rosettes. The glabrous leaves and stems readily separate *S. oregonense* from such pubescent-leaved species as *S. chrysanthum* which is sometimes cultivated. Its nearest relatives appear to be *S. laxum* ssp. *heckneri* and *S. obtusatum* ssp. *boreale*. The former has the leaves of the floriferous stems partially amplexicaul. It occurs on serpentine at lower elevations than *S. oregonense*. The ssp. *boreale* of *S. obtusatum* has longer sepals and shorter, white, creamy white, or pinkish petals. It is diploid, whereas *S. oregonense* is hexaploid. *Sedum laxum* ssp. *laxum*, another diploid, has pink petals and occurs on serpentine at low elevations.

Description. The sample for study comprised 32 plants: 23 of these studied in flower and 21 studied in fruit. Selection of plants was random from randomly chosen clusters in an optimum stratum, in Lane Co., Ore., comprising the summit of O'Leary Mountain, the rocks along the old trail up Deer Butte, and the portion of the slope south of Glacier Creek known as Glacier Way; and in a random stratum comprising the western slope of the Cascade Mountains between the Middle Fork of the Willamette River and the Mackenzie River, except the part included in the optimum stratum. Plants occurred in only one randomly selected unit out of five surveyed in the random stratum, namely on a ridge south of Mud Lake and north of Mink Lake

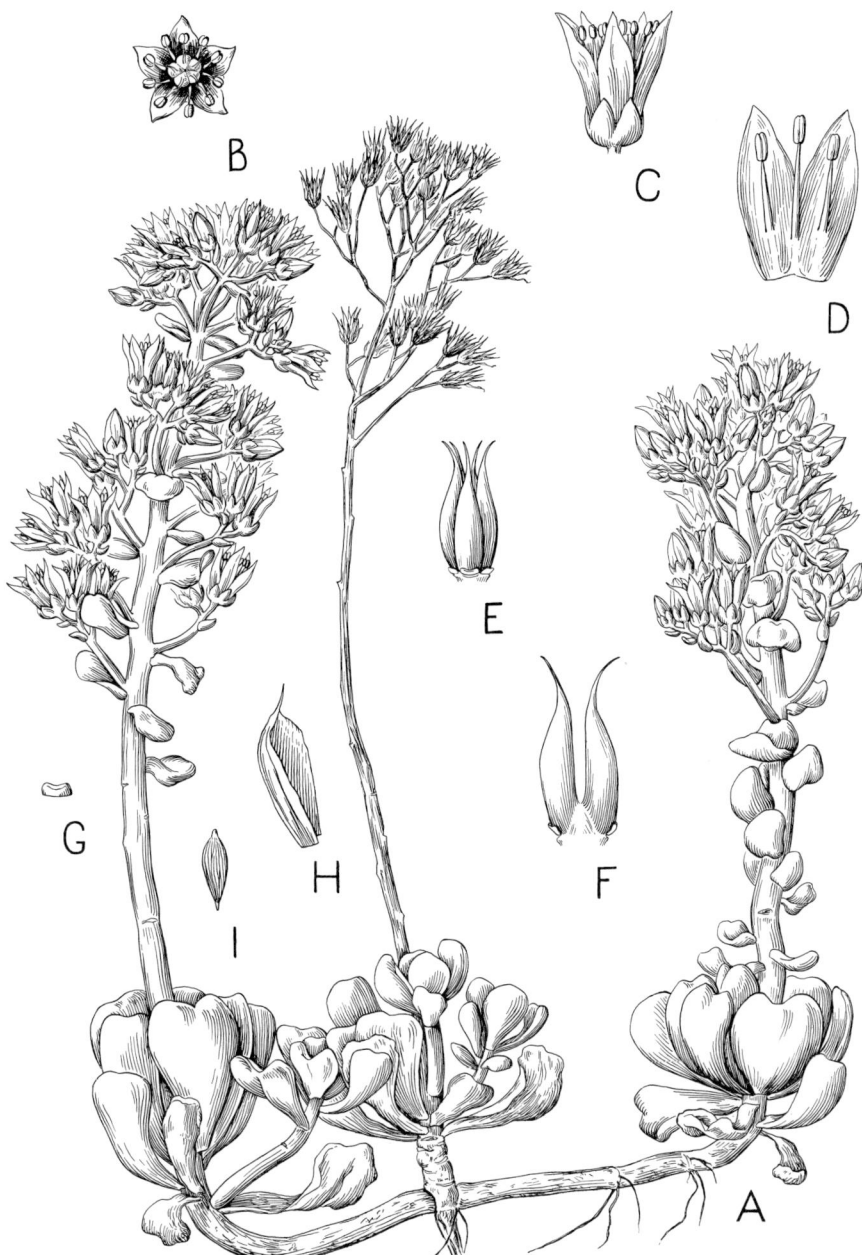

Fig. 115. Plant of *Sedum oregonense* from O'Leary Mountain, Cascade Mountains, Lane Co., Ore. A. Habit sketch (x .84). B. Flower from above (x 1.7). C. Flower from side (x 1.7). D. Two petals and stamens (x 2.5). E. Carpels (x 2.5). F. Two carpels and nectaries (x 3.4). G. Nectary (x 4.2). H. Follicle (x 3.4). I. Seed (x 8.4).

in Lane Co., Ore. The procedure in the field was first to list the natural clusters of plants in each sampling unit, then to select one-fifth of these for study. Numbered tags were attached to all plants with floriferous stems in chosen clusters. The details of the sample are indicated in the following table.

Locality	Clusters			Flowering plants			Fruiting plants	
	N	n	%(n/N)	N	n	%(n/N)	n	%(n/N)
O'Leary Mt.	9	2	22	279	14	5	9	3
Deer Butte Trail	17	3	17	184	9	5	9	5
Glacier Way	0	0	—	0	0	—	0	—
Mud Lake	16	3	18	16	2	12.5	2	12.5
Totals	42	8	(19)	479	25	(5)	20	(4)

A second table shows the estimated number of flowering plants at each locality, the per cent of the total of the flowering plants in the chosen clusters, and the per cent of the total of the plants in the samples for detailed study.

Locality	Plants (N)	Plants in chosen clusters (%)	Plants in samples (%)
O'Leary Mt.	2,142	13	.6
Deer Butte Trail	1,043	17	.9
Mud Lake	565	2.8	.3

In addition to the sample described above, I have also studied and cultivated two plants from the lava south of Clear Lake, Linn Co., Ore. Data for plants from the vicinity of Crater Lake are not repeated here. They were the basis for the description in an earlier publication (Clausen 1942).

In the following description, data are kept separate for the principal populations studied in the field, both to permit comparisons and because percentages of plants in samples are unequal. Names of localities are abbreviated as D = Deer Butte Trail, M = Mud Lake, and O = O'Leary Mountain. In addition, cf. applies to plants cultivated in a cold frame at Ithaca, N.Y., and cl. = cluster. The experiment in the cold frame included two clones from each of the populations investigated in Oregon and two replicates of each of these.

Plants of *Sedum oregonense* (fig. 116) are perennial herbs with stout fibrous roots arising from thick rootstocks with diameter of 5 to 7 mm. The bark of the rootstocks is smooth and gray to fuscous. The rootstocks are much branched, with the branches giving rise to rosettes of fleshy leaves. Eventually, older portions die, with the result that a clone may consist of separate pieces. Unbroken sections of rootstock along the old Deer Butte Trail were 21 cm. long and the maximum extent of what appeared to be one plant there, with rootstocks all interconnected, was 5.4 m. The floriferous stems are erect, either terminal on branches of the rootstock or arising in the axils of the leaves of the primary rosettes. They are glabrous and either green or suffused with red.

Fig. 116. Plant of *Sedum oregonense* from rocks below Deer Butte, Cascade Mountains, Lane Co., Ore., cultivated in cold frame, Ithaca, N.Y., June 22, 1966 (x .9). Photo by Howard Lyon.

	n-cl.	n-pl.	n-obs.	$\bar{x}$	s	w
Plants—diameter (cm.)						
M	3	4	4	31	4	17–50
O	2	14	14	29	11	13–65
D	3	9	9	96	46	19–230
cf.	4	8	15	15	6	4–26
Primary rosettes—diameter (cm.)						
M	3	4	7	3.6	.9	2.3–5.1
O	2	14	26	3.6	.7	1.8–4.8
D	3	9	17	4.2	.6	2.3–6.1
cf.	4	8	25	3.3	.9	1.6–4.7
Offsets (n)						
M	3	4	7	2	1.1	1–3
O	2	14	26	1.2	.8	0–4
D	3	9	17	1.7	1	0–4
cf.	4	8	25	2.2	1.7	0–6
Offsets—length (cm.)						
M	3	4	8	2.1	.4	.3–6.5
O	2	14	26	.9	.8	0–2.7
D	3	9	16	2	1.2	0–4.8
cf.	4	8	19	1.3	1.3	0–3.5
Stems of offsets—diameter (mm.)						
M	3	4	8	1.9	.1	1.6–2.2
O	2	11	21	1.8	.3	1.1–2.5
D	3	8	15	2.3	.5	1.5–3.3
cf.	4	8	18	1.6	.8	.8–2.2
Floriferous stems per plant (n)						
M	2	2	2	1	0	1
O	2	14	14	1.3	.7	1–3
D	3	9	9	2.5	3	1–10
cf.	2	2	2	1.5	—	1–2
Floriferous stems—length (cm.)						
M	2	2	2	13.5	3.5	11–16
O	2	14	14	13.3	4.7	9–28
D	3	9	9	18.4	5.3	11–28
cf.	2	2	3	6	—	6

Leaves are spirally arranged in rosettes and on floriferous stems, but opposite on offsets. The rosulate leaves are divergent, sessile or sometimes subpetiolate, obovate or oblanceolate, emarginate to retuse, plane dorsally, and usually glaucous and green suffused with red dorsally. The leaves of the floriferous stems are sessile, suborbicular or obovate, and either rounded or bluntly apiculate.

	n-cl.	n-pl.	n-obs.	$\bar{x}$ (mm.)	s (mm.)	w (mm.)
Leaves of rosettes—length						
M	3	4	7	23.7	2.1	20–28
O	2	14	26	21.4	2.8	10–29
D	3	9	17	26.7	4.1	21–36
cf.	4	8	25	17.1	4.6	10–23
Leaves of rosettes—width						
M	3	4	7	9.6	1.8	7–11
O	2	14	26	10.8	2.2	5–16
D	3	9	17	11.5	1.5	9–16
cf.	4	8	25	10.4	3.4	6–16

	n-cl.	n-pl.	n-obs.	$\bar{x}$ (mm.)	s (mm.)	w (mm.)
Leaves of rosettes—thickness						
M	3	4	7	2.9	0	2.3–3.6
O	2	14	26	2.6	.4	2.1–3.8
D	3	9	17	3	.5	2.2–4.1
cf.	4	8	25	2.3	.5	1.4–3.5
Median leaves of floriferous stems—length						
M	2	2	2	9	—	6–12
O	2	11	14	10.2	1.8	7–12
D	2	2	3	11.5	—	9–14
cf.	2	2	3	8.9	—	7–11
Median leaves of floriferous stems—width						
M	2	2	2	7	—	5–9
O	2	11	14	8.6	1.8	6–11
D	2	2	3	9.5	—	8–11
cf.	2	2	3	6.9	—	6–8
Median leaves of floriferous stems—thickness						
M	2	2	2	2.4	—	1.6–3.2
O	2	11	14	2.4	.6	.9–3.5
D	2	2	3	2.1	—	1.6–2.6
cf.	2	2	3	2.6	—	2.1–2.8

Inflorescences are paniculate cymes with the floral bracts similar to the uppermost leaves of the floriferous stems, but smaller. The primary branches of the cymes are dichotomously branched.

	n-cl.	n-pl.	n-obs.	$\bar{x}$ (n)	s (n)	w (n)
Primary branches						
M	2	2	2	6.5	—	6–7
O	2	14	17	8.3	2.7	5–12
D	3	9	12	10.9	3.2	6–16
cf.	2	2	3	4.5	—	3–6
Flowers per cyme						
M	2	2	2	28.5	—	22–35
O	2	14	17	38.3	14.4	15–67
D	3	9	12	58.8	32.2	24–121

Flowers are pedicellate and erect or nearly so. An apparent torus may consist largely of the connate basal parts of the carpels. On dissection, it breaks readily from the receptacle.

	n-cl.	n-pl.	n-obs.	$\bar{x}$ (mm.)	s (mm.)	w (mm.)
Diameter						
M	2	2	4	9.7	—	9–11
O	2	12	24	8.4	1.2	6–11
D	3	7	13	9.1	1.3	7–12
cf.	2	2	4	10.7	—	8–12

Sepals are erect, appressed to tube of corolla, connate basally, with lobes ovate and subacute, and pale green or greenish yellow.

SEDUM OF NORTH AMERICA

	n-cl.	n-pl.	n-obs.	x̄ (mm.)	s (mm.)	w (mm.)
Length						
M	2	2	4	3.2	—	2.8–3.8
O	2	12	24	3.4	.3	2.4–3.9
D	3	8	14	3.2	.5	2.5–3.8
cf.	2	2	4	3.5	—	2.9–4.1
Length of cohesion						
M	2	2	4	.6	—	.6
O	2	12	24	.5	.3	.2–1.4
D	3	8	14	.6	.2	.3–0.9
Width						
M	2	2	4	2.1	—	1.9–2.5
O	2	12	24	2.1	.1	1.7–2.7
D	3	8	14	2	.2	1.4–2.3
cf.	2	2	4	2.6	—	1.7–3.5

Petals are erect below and then divergent apically, elliptic-oblong or oblanceolate-oblong, erose marginally, hooded, and white varying to yellowish or greenish white, but becoming pinkish with age.

	n-cl.	n-pl.	n-obs.	x̄	s	w
Number						
M	2	2	4	5	—	5
O	2	12	24	5	.3	4–6
D	3	8	14	5	0	5
cf.	2	2	4	5	—	5
Length (mm.)						
M	2	2	4	9.1	—	8.7–9.6
O	2	12	24	9	1	7.2–11
D	3	7	13	9.6	1	8.5–11.5
cf.	2	2	4	9.2	—	8.7–9.9
Length of cohesion (mm.)						
M	2	2	4	2.1	—	2.0–2.2
O	2	12	24	2	.4	1.1–3.1
D	3	7	13	2.3	.7	0–3
cf.	2	2	3	1.8	—	1.4–2.4
Width (mm.)						
M	2	2	4	3.4	—	3.1–3.6
O	2	12	24	3.3	.2	2.8–4.2
D	3	7	13	3	.2	2.1–3.5
cf.	2	2	4	2.6	—	1.7–3.5

Stamens have pale yellow filaments and yellow anthers. The bases of the filaments alternating with the lobes of the corolla seem to hold the petals together.

	n-cl.	n-pl.	n-obs.	x̄ (mm.)	s (mm.)	w (mm.)
Adnation of epipetalous filaments						
M	2	2	4	2.8	—	2.6–3
O	2	12	24	2.6	.4	1.8–4.3
D	3	7	13	3.2	.6	2.2–4.2
cf.	2	2	4	2.5	—	1.9–2.9

NATIVE SPECIES 417

	n-cl.	n-pl.	n-obs.	x̄ (mm.)	s (mm.)	w (mm.)
Anthers—length						
M	2	2	3	1.6	—	1.4–2
O	2	10	16	1.7	.1	1.4–2
D	2	2	3	1.6	—	1.6–1.7
cf.	2	2	3	1.6	—	1.4–1.7

Nectaries are subreniform, truncate, and white or yellow.

	n-cl.	n-pl.	n-obs.	x̄ (mm.)	s (mm.)	w (mm.)
Length						
M	2	2	4	.5	—	.4–.5
O	2	12	24	.4	.03	.3–.5
D	3	7	13	.4	.03	.3–.5
cf.	2	2	4	.5	—	.4–.6
Width						
M	2	2	4	1	—	.9–1.3
O	2	12	24	.9	.3	.7–1.4
D	3	7	13	.8	.4	.5–1.2
cf.	2	2	4	.9	—	.7–1.1

Carpels are erect, slightly connate basally for .8 to 1.7 mm., without clear distinction between ovary and style, and pale green or greenish white. The stigmas are white and expanded when receptive to pollen. Sometimes the carpels are not completely closed.

	n-cl.	n-pl.	n-obs.	x̄	s	w
Length (mm.)						
M	2	2	4	7	—	6.4–8
O	2	12	24	5.9	.9	3.7–8
D	3	7	13	6.9	.8	5.5–8.6
cf.	2	2	4	6.6	—	4.5–8.1
Styles—length (mm.)						
M	2	2	4	1.7	—	1.4–2.2
O	2	12	24	1.3	.3	.5–1.9
D	3	7	13	1.4	.2	1.0–1.8
Ovules per ovary (n)						
M	2	2	4	13.7	—	10–19
O	2	12	24	17.7	2	12–22
D	3	8	14	12.8	3.5	6–20
cf.	2	2	4	10.2	—	7–12

Fruits are erect, strongly 5-veined, with distinct beaks about 1 mm. long, and brown. The sepals and petals persist in withered, brown condition around the follicles.

	n-cl.	n-pl.	n-obs.	x̄	s	w
Length (mm.)						
M	2	2	4	5.1	—	4.5–6.1
O—cl. 5	1	7	14	4.7	.9	3.4–6
O—cl. 7	1	2	4	6.3	.4	6.1–6.6
D	3	9	18	5.9	.5	4.8–7.3
cf.	2	2	4	5.5	—	5.2–5.8

Seeds are pyriform, short-tailed at both ends, finely ribbed longitudinally, and pale brown.

	n-cl.	n-pl.	n-obs.	$\bar{x}$	s	w
Length						
M	1	1	2	1.45	—	1.4–1.5
O	2	7	14	1.4	.06	1.2–1.7
D	3	9	18	1.5	.2	1.2–1.7
cf.	2	2	4	1.4	—	1.3–1.4
Diameter						
M	1	1	2	.3	—	.3
O	2	7	14	.4	.08	.3–.5
D	3	9	18	.4	.5	.3–.5
cf.	2	2	4	.35	—	.3–.4

Chromosomes are tiny.

	n-pl.	g	sp	Cytologist
Number				
Crater Lake	3	—	90±	L. Hollingshead (1942)
M	1	45	—	D. Niimoto, 1967
O	1	—	90	I. Zandstra, 1969
D	1	—	90	I. Zandstra, 1969

Variation. Populations of *Sedum oregonense* are remarkably similar. Comparison of plants from rocks along the old trail to Deer Butte and the summit of O'Leary Mountain indicated only small differences between sites, of no statistical significance. Greater differences exist among clusters at a site than between sites. Analyses of variance for 21 features revealed no significant differences between sites but significant differences among clusters within sites for 5 features, namely diameter of plants, diameter of primary rosettes, width of petals, length of follicles, and length of seeds. In addition, since means for sites were either the same or closely similar for 12 other features, tests of significance seemed unnecessary. Age, exposure, and moisture may be factors involved in the intrapopulational variation.

In the experiment in the cold frame in Ithaca, variation among plants from within sites was greater than variation among sites. Analyses of variance for 8 features revealed significant differences among plants in 2 features: diameter of offsets and width of leaves of primary rosettes. The implication is that real genetic differences exist among plants within populations.

Although differences among sites are not significant, plants on the summit of O'Leary Mountain tended to be smaller, with smaller leaves, fewer floriferous stems, and smaller flowers, with smaller floral parts, than those at Deer Butte. Also, the variation among plants tended to be less in some features, as is indicated by the following comparison of coefficients of variation for the two populations.

Feature	D—C.V.	O—C.V.
Rosulate leaves—length	15	13
—width	13	20
—thickness	17	15
Sepals—length	16	9
Petals—number	0	6
—length	10	11
—length of cohesion	30	20
Nectaries—length	7	7
—width	50	33
Seeds—length	13	4

Two points are important to remember in considering the above results. The population on O'Leary Mountain is nearly twice the size of the population along the trail to Deer Butte. Further, the climate on the summit of O'Leary Mountain is more rigorous. The results are consistent with these circumstances. Likewise, the general results of no significant differences among populations are consistent with the hexaploid status of the species. Study of one population gives about the same information as may be obtained from any other.

Length of sepals seems important in distinguishing *Sedum oregonense*. On the basis of variation in this dimension in a simple random sample of 10 plants in a cluster on O'Leary Mountain, 1 plant is adequate to detect a difference of 1 mm. The fractions of plants which I studied from clusters are more than adequate to detect differences among features of diagnostic importance.

Age and edaphic factors may affect the shape of plants. Young plants, actively growing, may be nearly square or circular, growing out evenly in all directions from a center. Older plants (fig. 117) may assume irregular shapes or become elongate. On steep slopes, gravel and other debris may accumulate on the upper side of plants, retarding or preventing growth in that direction. Because there are two ends but only one front, lateral growth may be twice as great as anterior expansion. The differential may be even greater if the slope downward is inhospitable for rooting. As a result, the plant may become a horizontal band on the slope. Many plants had this aspect at Deer Butte.

A plant from the ridge south of Mud Lake, in flower in the greenhouse on Jan. 1, 1968, was somewhat abnormal. The cyme had only two primary branches and the petals spread more widely than usual, with the result that a flower had a diameter of 13 mm. In other respects, dimensions of floral parts and leaves were within ranges observed for the species in the field. This same plant produced another abnormal flower (fig. 118) in January 1969, with four epipetalous and one alternate stamen modified and petallike.

Nomenclature. *Sedum oregonense* (S. Watson) M. E. Peck, Man. Higher Plants of Oregon, p. 361 (1941) and Madroño 6:134 (1941). Basionym: *Cotyledon oregonensis* S. Watson, Proc, Am. Acad. 17:373 (1882).

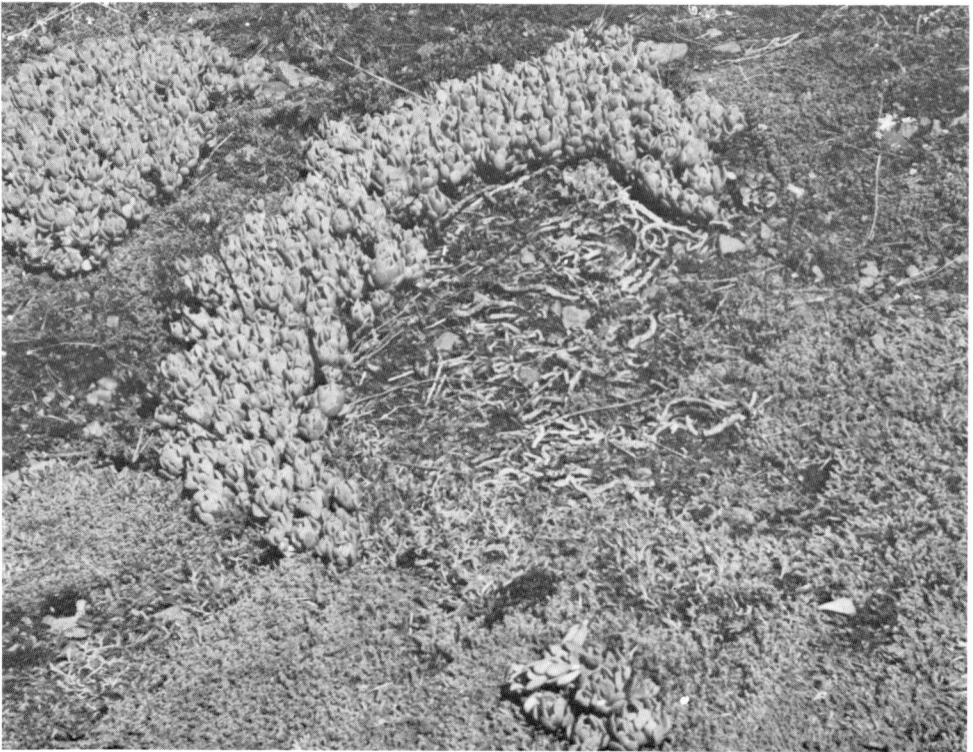

Fig. 117. Old plant of *Sedum oregonense*, with elongate, semicircular shape and big, dead part, on rock below Deer Butte, Cascade Mountains, Lane Co., Ore., July 24, 1965.

Synonyms are:
1882. *Cotyledon oregonensis* S. Watson, Proc. Am. Acad. 17:373. Type locality: Cascade Mountains, Ore. Type: Joseph and Thomas J. Howell no. 322 (GH), July 1880. Watson mentioned northern Oregon as the region of occurrence and described the petals as pale yellow. Mount Hood might be the type locality.
1903. *Gormania Watsoni* Britton, Bull. N.Y. Bot. Gard. 3:29. Basionym: *Cotyledon oregonensis* S. Watson.
1913. *Echeveria Watsonii* (Britton) A. Nelson et J. F. Macbride, Bot. Gaz. 56:476.
1927. *Sedum Watsoni* (Britton) Tidestrom, Proc. Biol. Soc. Wash. 40:119.

Distribution. The known distribution (fig. 119) of *Sedum oregonense* is in the Western and High Cascade Mountains of Oregon, north to Mount Hood, and the Klamath Mountains of southwestern Oregon and northern California. A record from Seaview near Ilwaco, Wash., is doubtful. It is based on a collection of L. F. Henderson (DS, OSU). *Sedum oreganum* is common at Seaview, but I was unable to find *S. oregonense* there in 1965. The trouble might be confusion of data.

Fig. 118. Abnormal flower of plant of *Sedum oregonense* from ridge south of Mud Lake, Cascade Mountains, Lane Co., Ore., cultivated in greenhouse, Ithaca, N.Y., Jan. 17, 1969 (x 6). Photo by Howard Lyon.

On the basis of herbarium collections, *Sedum oregonense* is known from 19 localities in the Cascade Mountains and 12 localities in the Klamath Mountains. I have seen it at 6 localities in the Cascade Mountains, but have yet to see it in the Klamath Mountains. Plants which I have seen there were either *S. laxum* ssp. *heckneri* or *S. obtusatum* ssp. *retusum*. Because these species are easier to separate when in fresh, flowering condition, I believe that each population in the Klamath Mountains needs careful checking. Meanwhile, their status is uncertain.

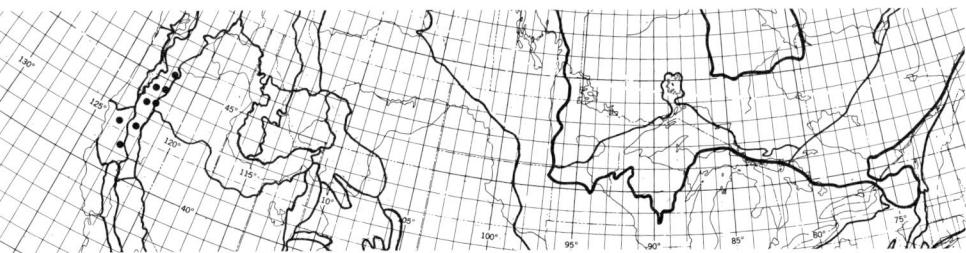

Fig. 119. Known distribution of *Sedum oregonense* (●).

Data for the three populations which I have studied in detail are in the following table. The plants occur in discrete groups, pattern C of the chapter on sampling.

Population	Sampling unit	N plants, estimate	N plants, C.I.$_{.05}$	Area of population (m.^{2})	Density per m.^{2}
Mud Lake	2,469	565	106–1,221	7,000	.08
Deer Butte	4,421	1,043	274–1,811	6,000	.17
O'Leary Mountain	2,850	2,142	1,456–2,828	7,347	.29

Population	Altitudinal range (m.)	pH	Oldest record
Mud Lake	1,580–1,590	5.4–6.2	1967
Deer Butte	1,190–1,226	5.4–6	1942
O'Leary Mountain	1,530–1,685	5.2–5.6	1938

Sites where *Sedum oregonense* occurs have sparse vegetation. Competing species are diverse. A record of the two principal competitors of each of 27 randomly selected plants of *S. oregonense* in the three populations listed in the foregoing table includes 25 species of vascular plants, among them *S. stenopetalum* ssp. *monanthum*. Only 2 competitors appeared in samples from two populations, namely a *Lomatium* from Deer Butte and Mud Lake, and *Arctostaphylos nevadensis* from O'Leary Mountain and Mud Lake, but their ratings were low: *Lomatium*—twice at Deer Butte and once at Mud Lake; and the *Arctostaphylos*—once on O'Leary Mountain and once at Mud Lake. The commonest competitor on Deer Butte was *Selaginella wallacei*, with a score of 7 out of 9 possibilities. The two commonest competitors on O'Leary Mountain were *Lupinus laxiflorus* and *Festuca viridula*, each with a score of 3 out of 14 possibilities. At Mud Lake, no competing species appeared more than once in a sample. On O'Leary Mountain, other species of *Sedum* present, besides *S. stenopetalum* ssp. *monanthum*, were *S. oreganum* and *S. spathulifolium*.

Sedum oregonense occurs in gravel or in mats of *Selaginella* or moss on rocky slopes and on ledges and in crevices of cliffs, usually on andesite or basalt. The plants do best in full sun. Those in shade may be small and nonflowering. The following table summarizes certain aspects of the distributional status.

Geomorphic province	Altitudinal range (m.)	Months of flowering	Oldest record
Klamath Mountains	1,800–2,681	June–Aug.	1903, E. B. Copeland (US)
Cascade Mountains	1,190–2,400	June–Aug.	1880, J. and T. J. Howell (GH)

Reproduction. The relative importance of vegetative reproduction versus development from seeds is difficult to assess. Small rosettes, broken from a larger plant and rooted, may appear similar to seedlings. My studies have not resolved the problem, beyond demonstrating that both types of propagation occur. Vegetative reproduction involves not only breaking away of rosettes, but also the death of portions of rhizomes connecting clusters of rosettes. In this way, a large plant becomes divided into separate parts. Clones may come to be widely distributed. Vegetative growth may be rapid. In two years, 15 plants in an experiment in a cold frame at Ithaca increased from a diameter of 2 cm. to an average diameter of 15 cm.

Stems of *Sedum oregonense* can endure long periods of desiccation. To learn the length of time that rosettes might remain alive in dry condition, I put samples in plastic bags and stored these in cartons in a cabinet in the basement of my house at Ithaca, N.Y. The results are indicated in the following table, in which G stands for good condition and A for alive but not good. Inspections of plants were six months and two and a half years after collection. All plants were in good condition at the beginning of the experiment.

		After 6 months		After 2.5 years	
Population	N	G	A	G	A
M	3	2	1	0	0
O	9	0	9	0	0
D	5	4	1	0	0
Clear Lake	1	0	1	0	0
Totals	18	6	12	0	0

Rosettes of a clone from Deer Butte, put in soil in two pots in the greenhouse after six months in dry storage, grew and multiplied. After two years these plants had attained diameters of 15 and 12 cm.

A low percentage of ovules developed into seeds in cultivation in cold frames at Ithaca: 1.9 to 18.8% according to Avanelle Morgan, based on study in 1967 of 15 follicles of 5 plants. Some seeds germinated within three weeks of collection, but seedlings were discarded because of mould.

Flowering time varies with altitude, exposure, and latitude. Extreme dates for plants in anthesis are May 27 and Aug. 10. Principal month of flowering is July.

Several species of insects are pollinators of *Sedum oregonense*. These include two kinds of bees, a fly, and two species of butterfly, one large and black and white, the other small and blue. In addition, my daughter Heidi saw a hummingbird visiting flowers at Deer Butte on July 25, 1965. Three times at the same locality, I have watched bees which were working the yellow flowers of *S. spathulifolium* switch to the white blossoms of *S. oregonense*. Were these species cross-compatible, one might expect many hybrids. Once, when a bee was switching from the yellow to the white flowers, it appeared uncertain, as though it wanted further yellow ones, but it finally made a decisive switch.

Although both *Sedum spathulifolium* and *S. oregonense* may occur together and be in flower at the same time, the *S. spathulifolium* is substantially ahead in development. The two species do not occur with equal frequency. Ratios of *S. oregonense* to *S. spathulifolium* are respectively 82 to 151, 47 to 75, and 55 to 4 for three sites at Deer Butte; and 232 to 3 and 244 to 0 on O'Leary Mountain. No *S. spathulifolium* was present at Mud Lake. Hybrids are either lacking or very rare. The conclusion from study in the field is that hybrids do not occur at the studied sites.

Relationships. The closest relatives of *Sedum oregonense* appear to be *S. obtusatum* ssp. *boreale*, which has longer sepals, but sometimes white petals, and *S. laxum* ssp. *heckneri*, with amplexicaul leaves on the floriferous stems. Although *S. laxum* ssp. *heckneri* differs in important ways such as lower number of chromosomes, habitat on serpentine, and occurrence at lower altitudes, it resembles *S. oregonense* in many respects. Statistical tests reveal that various supposed differences lack significance. The most useful diagnostic feature is the vertical distance from the median point of attachment of leaves of floriferous stems to the bases of the basal lobes. The leaves of *S. laxum* ssp. *heckneri* clasp the floriferous stems, but those of *S. oregonense* simply are sessile. When the petals are pink, *S. laxum* ssp. *heckneri* is distinctive, but some populations have creamy white corollas, in which case the leaves are an essential aid in identification.

The population of *Sedum oregonense* on the slope at the southern base of Deer Butte most nearly resembles *S. laxum* ssp. *heckneri* in dimensions of floral parts. Similarly, plants of *S. laxum* ssp. *heckneri* on bluffs of serpentine between Wildwood and Peanut, Calif., approach *S. oregonense*. Small differences between these populations in measurements of floral parts are not significant, $P = .05$, using the Wilcoxon rank sum test, for samples from the wild, except that the epipetalous stamens are adnate to the petals for a greater length and the nectaries are wider in *S. laxum* ssp. *heckneri*, $P = .1$.

The precise origin of *Sedum oregonense* is obscure, yet certain details seem clear. The species is hexaploid and must be derived from plants with lower numbers of chromosomes, either through simple doubling or hybridization. Presently, it behaves as a functional diploid. Its modern range in the Cascade Mountains must be of recent origin because part of the area was covered by alpine glaciers in the Pleistocene. At that time, the species or its progenitors may have existed on the lower slopes of the mountains or on outlying ridges. The species appears youthful. Its evolution may be related to that of *S. laxum* ssp. *heckneri* and *S. obtusatum* ssp. *boreale*.

References. Clausen, R. T. (1942) and Hollingshead, L. (1942).

25. Sedum albomarginatum*** (figs. 105 and 120–122)

Descriptio originalis. Sedum **albomarginatum**, sp. nov., affine *Sedi laxi* et *S. moranii*; foliis rosularum magnis, 1.4–6.7 cm. longis et .5–2.3 cm. latis, albomarginatis, glaucis; caulibus floriferis papillosis vel glabris, non pubescentibus; foliis caulerum floriferorum longioribus, 1.5–3.5 cm. longis; cymis paniculatis ramis saepe dichotomis; floribus petalis gilvis, erectis ad basim, divaricatis sursum, 7.9–11.8 mm. longis. Typus in herbario Wiegand, Universitatis Cornellianae, ab angustiis fluvii Feather, Plumas Co., Calif., alt. 855 m., collectio Roberti Clausenii, num. 63,169, 1963, Junio 6, est.

The distinctive features of *Sedum albomarginatum* (fig. 120) are the large leaves of the primary rosettes, with whitish margins; the floriferous stems which are not glandular-pubescent; the paniculate cymes of which the branches are sometimes two-

NATIVE SPECIES 425

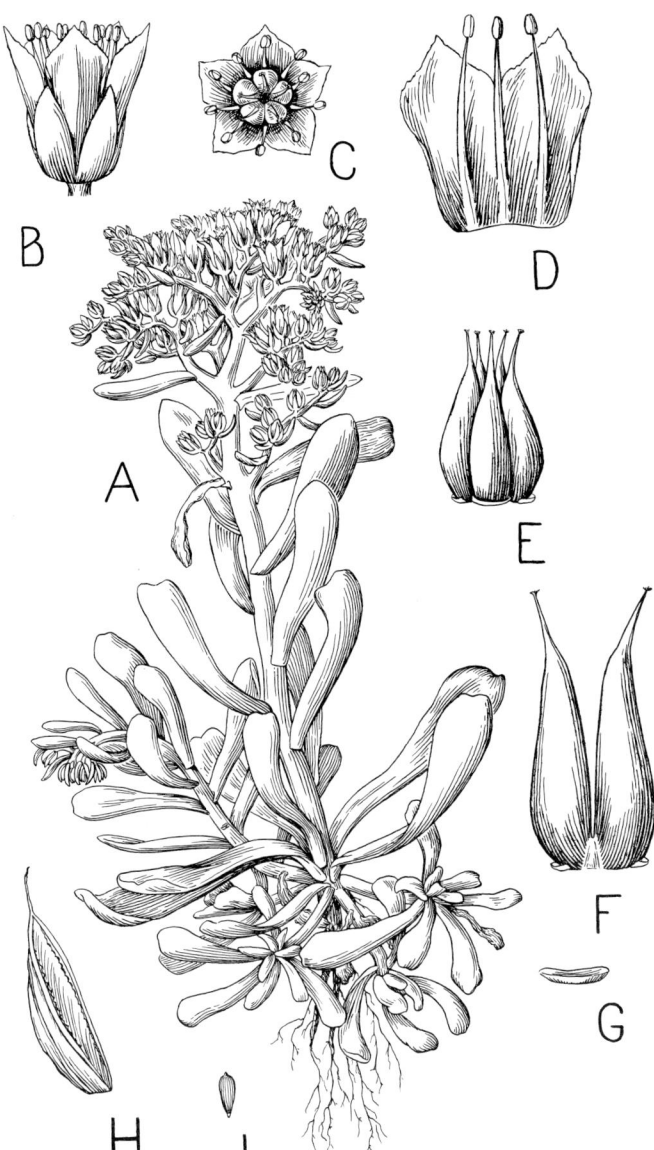

Fig. 120. Plant of *Sedum albomarginatum* from Serpentine Canyon of East Branch of North Fork of Feather River, Plumas Co., Calif., cultivated in greenhouse, Ithaca, N.Y., Apr. 5, 1965. A. Habit sketch (x .52). B. Flower from side (x 2.1). C. Flower from above (x 2.1). D. Two petals and stamens (x 3.1). E. Carpels (x 3.1). F. Two carpels and nectaries (x 4.2). G. Nectary (x 5.2). H. Follicle (x 4.2). I. Seed (x 5.2).

or three-parted; and the flowers with erect pale yellow petals. The species most resembles *S. laxum*, but it differs in the white margins of the leaves of the rosettes, the longer leaves of the floriferous stems, and the yellow petals. *Sedum albomarginatum* also resembles *S. moranii*, but it differs in the larger leaves of the rosettes, elongate inflorescences with usually more than three branches, pale yellow flowers, and lack of glandular-pubescence. The separation from *S. obtusatum*, the other yellow-flowered species, is even easier. Both leaves and floral parts are bigger, and the plants occur only on serpentine.

Description. Sample: 30 plants from the northern side of the Serpentine Canyon of the East Branch of the North Fork of the Feather River between Rich bar and Rich Gulch, Plumas Co., Calif., and from a short distance up the valley of the North Fork. Descriptive data apply to 4 plants chosen randomly from a cluster of 10 on June 6, 1963, and studied in flower; 10 plants chosen randomly from 48 which could be reached in a larger cluster of 68 in a randomly selected area studied on July 18, 1964, for details of leaves and ecology and for seeds when available; a single plant obtained on Aug. 2, 1964, from along the North Fork of the Feather River 2 kilometers above the mouth of the East Branch; and 15 plants chosen deliberately for preserving, pressing, or supplementary study. Data for the 15 plants last-mentioned are cited only when they add to the information from the other samples.

Plants of *Sedum albomarginatum* have been part of three experiments devoted to species of *Gormania* in Ithaca, N.Y. Plants have done poorly. Greatest survival has been of plants not in the experiments. Nine plants collected in 1963 were in the first experiment in the greenhouse, started in the same year. These yielded data on leaves; 2 of them flowered. Only 3 plants were included in a second experiment in the greenhouse, started in December 1964, and 2 plants were in an experiment in the cold frame, begun in April 1965. Plants in the two later experiments were begun as cuttings 2 cm. in length and also in diameter.

In the following tables, 1963 refers to the random sample of 4 flowering plants studied in that year; 1964 refers to the random sample of 10 fruiting plants studied in that year; N. F. refers to the single plant from the North Fork of the Feather River; gh. 1 refers to the experiment begun in the greenhouse in 1963; gh. 2 refers to the experiment begun in the greenhouse in 1964; and cf. refers to the experiment in the cold frame.

Plants of *Sedum albomarginatum* are perennial with short, branched rootstocks bearing densely compacted rosettes of large, thick leaves. Old plants may have a woody primary root > 10 cm. long. Offsets arise in the axils of the leaves of the primary rosettes and generally are short-stemmed. One plant, which in the wild had only 1 primary rosette, developed 8 in cultivation. The floriferous stems are terminal from the center of the rosettes. They are pink and, before the opening of the buds, recurved.

	n-pl.	n-obs.	$\bar{x}$	s	w
Diameter (cm.)					
1964	10	10	11	7	6–30
N. F.	1	1	9	—	9
gh. 1	8	12	6	2.9	.3–11
cf.	1	1	4	—	4
Primary rosettes (n)					
1964	10	10	1.7	.9	1–3
N. F.	1	1	1	—	1

	n-pl.	n-obs.	$\bar{x}$	s	w
Primary rosettes—diameter (cm.)					
1963	4	8	5	2	4–9
1964	10	17	7	1.4	3–9
N. F.	1	1	8	—	8
gh. 1	4	7	4	.8	3–5
gh. 2	1	3	3	—	3
Offsets (n)					
1964	10	18	.9	.8	0–3
N. F.	1	1	2	—	2
Offsets—length (mm.)					
1964	8	16	7	4.4	0–17
N. F.	1	2	12	—	8–17
Offsets—diameter of stems (mm.)					
1963	3	6	3.3	.3	2.5–4
1964	6	11	3.5	.7	2.1–4.6
N. F.	1	2	3.1	—	3.0–3.2
Floriferous stems (n)					
1964	10	10	1.2	.4	1–2
N. F.	1	1	1	—	1
gh. 1	2	2	1.5	.7	1–2
Floriferous stems—length (cm.)					
1963	4	5	22	3.8	16–25
1964	9	10	20	3.2	16–25
N. F.	1	1	23	—	23
gh. 1	1	1	20	—	20

Leaves of primary rosettes (fig. 121) are stiff, oblanceolate, oblanceolate-oblong, or spatulate, broadly rounded or truncate, sometimes emarginate, glaucous, white- or pink-margined, and somewhat erose. Those of the floriferous stems are oblanceolate-oblong or oblanceolate-elliptic and incurved. One plant in cultivation had a median leaf on a floriferous stem 45 mm. long.

	n-pl.	n-obs.	$\bar{x}$ (mm.)	s (mm.)	w (mm.)
Leaves of rosettes—length					
1963	4	8	31	1.8	22–42
1964	10	17	42	8.7	21–67
N. F.	1	1	45	—	45
gh. 1	3	7	20	4.4	15–25
gh. 2	1	1	14	—	14
cf.	1	2	17	—	16–18
Leaves of rosettes—width					
1963	4	8	10	2.1	5–15
1964	10	17	12	2.7	7–18
N. F.	1	1	23	—	23
gh. 1	3	7	10	1.8	8–12
gh. 2	1	1	7	—	7
cf.	1	2	7	—	6–8
Leaves of rosettes—thickness					
1963	4	8	3.4	.1	2.9–3.8
1964	10	17	3.4	.4	2.5–4.6
N. F.	1	1	3.1	—	3.1
gh. 1	3	7	2.2	.5	1.8–2.8
gh. 2	1	1	1	—	1
cf.	1	2	2.4	—	2.3–2.5

Fig. 121. Rosettes of plant of *Sedum albomarginatum* from Serpentine Canyon of East Branch of North Fork of Feather River, Plumas Co., Calif., cultivated in greenhouse, Ithaca, N.Y., Dec. 23, 1964 (x .8). Photo by Howard Lyon.

	n-pl.	n-obs.	$\bar{x}$ (mm.)	s (mm.)	w (mm.)
Median leaves of floriferous stems—length					
1963	4	5	26	4.7	20–31
gh. 1	2	3	20	7.1	15–25
Median leaves of floriferous stems—width					
1963	4	5	10	1.6	8–13
gh. 1	2	3	9	1.4	8–10
Median leaves of floriferous stems—thickness					
1963	4	5	3	.6	2.2–3.8

Inflorescences are elongate with 3 to 14 branches which are sometimes two- or three-parted.

	n-pl.	n-obs.	$\bar{x}$	s	w
Primary axis—length (mm.)					
1963	4	5	30	17	8–55
Primary branches (n)					
1963	4	5	9	6.8	5–14
1964	8	10	5	1.1	3–7
Flowers (n)					
1964	8	10	37	11.6	10–53

Flowers (fig. 122) are pedicellate and have a pungent smell.

	n-pl.	n-obs.	x̄ (mm.)	s (mm.)	w (mm.)
Basal pedicels—length					
1963	4	10	3.9	.8	2–5
Torus—length					
1963	4	18	.8	.1	.6–1.2
gh. 1	2	6	.7	.2	.6–1.1
Diameter					
1963	4	18	6.7	.8	5.1–8
gh. 1	2	6	6.7	.4	6.0–8

Fig. 122. Flower of plant of *Sedum albomarginatum* from Serpentine Canyon of East Branch of North Fork of Feather River, Plumas Co., Calif., cultivated in greenhouse, Ithaca, N.Y., Mar. 31, 1964 (x 7.6). Photo by Howard Lyon.

Sepals are erect, connate basally for about a quarter of their length, lanceolate, acute or subacute, glabrous and green or slightly glaucous, and sometimes minutely papillose-ciliolate.

	n-pl.	n-obs.	x̄ (mm.)	s (mm.)	w (mm.)
Length					
1963	4	18	4.1	.2	3.5–5.1
gh. 1	2	6	5.1	1.1	4.3–7.2
Width					
1963	4	18	2.5	.2	1.9–3.2
gh. 1	2	6	2.7	.3	2.5–3.1

Petals are erect, connate below, then slightly divergent upward, imbricate, oblanceolate-oblong, with the upper margins erose, abruptly acute or obtuse and slightly mucronate-appendaged, and pale primrose yellow suffused with green medianly. One plant had a flower with 6 petals.

	n-pl.	n-obs.	x̄	s	w
Number					
1963	4	18	5	0	5
gh. 1	2	6	5	.1	4–5
Length (mm.)					
1963	4	18	9.2	.6	8.0–10.2
gh. 1	2	6	9.8	.4	9.1–10.8
Cohesion—length (mm.)					
1963	4	18	2.6	.2	1.6–3.6
gh. 1	2	6	3.9	1	2.3–4.6
Width (mm.)					
1963	4	18	3.3	.3	1.4–4.6
gh. 1	2	6	3.6	.1	3.5–4

Stamens are epipetalous with pale green filaments and primrose yellow anthers.

	n-pl.	n-obs.	x̄	s	w
Number					
1963	4	18	10	.1	10–11
gh. 1	2	6	10	.1	8–11
Adnation—length (mm.)					
1963	4	18	1.6	.3	1.0–2.6
gh. 1	2	6	2.1	.4	1.9–3
Anthers—length (mm.)					
1963	3	7	1.4	.2	1.1–1.8
gh. 1	2	3	1.7	.3	1.5–2.1

Nectaries are transversely oblong, thickened and concavely truncate apically, white, and translucent.

	n-pl.	n-obs.	x̄ (mm.)	s (mm.)	w (mm.)
Length					
1963	4	18	.4	.1	.4–.6
gh. 1	2	6	.7	.1	.6–.8

	n-pl.	n-obs.	x̄ (mm.)	s (mm.)	w (mm.)
Width					
1963	4	18	1.4	.1	1.1–1.7
gh. 1	2	6	1.5	.1	1.4–1.8

Carpels are erect, inrolled to the tips of the styles, and green.

	n-pl.	n-obs.	x̄	s	w
Length (mm.)					
1963	4	18	7.4	.8	5.8–9.1
gh. 1	2	6	7.5	.4	6.2–9
Styles—length (mm.)					
1963	1	1	1.4	—	1.4
Ovules per ovary (n)					
1963	4	18	28	7.1	16–42
gh. 1	2	6	30	14.4	20–48

Fruits are suberect and brown.

	n-pl.	n-obs.	x̄ (mm.)	s (mm.)	w (mm.)
Length					
1963	5	5	7	.34	7–8
1964	6	13	6	.89	5–7
N. F.	1	1	7	—	7

Seeds are pyriform, finely ribbed, short-tailed, and brown.

	n-pl.	n-obs.	x̄ (mm.)	s (mm.)	w (mm.)
Length					
1963	6	6	1.1	.07	1–1.2
1964	6	14	1.1	.06	1–1.2
N. F.	1	1	1.1	—	1.1
Diameter					
1963	6	6	.4	.04	.3–.4
1964	6	14	.4	.04	.3–.4
N. F.	1	1	.4	—	.4

Chromosomes.

	n	g (n)	sp (2n)	Cytologist
Number				
Feather River Canyon, 4.6 km. E. of Rich Bar	1	15	—	C. H. Uhl (on label in Jepson Herbarium)

Variation. Sedum albomarginatum, although of limited distribution, varies in the dimensions of the leaves of the rosettes, but differences among plants are not confirmed by experiment. The species may comprise only a few biotypes.

Variation within plants is small. Adequate samples are as follows when $P = .05$: 2 leaves to detect a difference of 3 mm. in width, 2 sepals to detect a difference of

.8 mm. in length, 2 petals to detect a difference of 1 mm. in length, and 1 seed to detect differences of .1 mm. in both length and diameter. Similarly, 8 plants per population are adequate, $P = .05$, to detect a difference of 7 mm. in length of leaves of primary rosettes and 4 plants are adequate to detect a difference of 3 mm. in width of leaves of rosettes.

In cultivation in the greenhouse at Ithaca, N.Y., plants have become infested with a mould. This caused the leaves, pedicels, and floral parts to appear downy.

Nomenclature. *Sedum albomarginatum* R. T. Clausen, p. 424. Type locality: on ledges of serpentine beside brook, northern side of Serpentine Canyon of East Branch of North Fork of Feather River, alt. 855 m., 4.5 km. by road east of Rich Bar, Plumas Co., Calif. Type: R. T. Clausen no. 63,169 (CU), June 6, 1963.

Paul Hutchison has used a manuscript name for this species on herbarium labels and in a mimeographed list. A previous listing (Clausen and Uhl 1944) was under *S. obtusatum* and was based on a collection of George B. Youngs, made in 1943, from about 4 km. up the North Fork of the Feather River from Belden.

Distribution. The known range (fig. 105) of *Sedum albomarginatum* is on serpentine at disjunct sites on the northern side of the canyon of the Feather River in the northern Sierra Nevada of California. Its maximum extent up and down the canyon is 40 kilometers, from the bridge below Pulga to Rich Gulch east of Rich Bar. The known altitudinal range is 493 meters, from 365 to 858 meters.

Data for a randomly selected sampling area which was surveyed in 1964 are in the following table. Further details about the site are included in the chapter on geography. See there the description of Serpentine Canyon unit no. 20,408 in the section on the Sierra Nevada. Data obtained in 1964 are fuller than details noted at four selected sites in a briefer survey in 1963.

Area (km.^{2})	Plants (n)	pH, range	pH, tests (n)	Altitude (m.)	Exposure
1.3	68	6.4–7.2	10	850–858	SSE–SW

Plants of *Sedum albomarginatum* were contagiously distributed in five main groups in unit no. 20,408. Few or no plants of other kinds occurred in the crevices and on the ledges of serpentine where the *Sedum* grew. Of 14 randomly selected plants, 10 lacked close competitors. In data with an herbarium specimen, Hutchison has noted *Orobanche fasciculata* and *Phacelia heterophylla* ssp. *virgata* as associated species. All plants in my samples were shaded for part of the day. They were rooted in thin soil or organic matter and gravel. Drainage was good.

Assuming similar frequency for *Sedum albomarginatum* in each of the seven sampling areas which I delimited along the northern side of the Serpentine Canyon, the total size of the population there is 476 plants. Unfortunately, a confidence interval cannot be indicated for the size of this population because I surveyed only one randomly selected unit and have no estimate of variance for plants per unit. However, the evidence is clear that *S. albomarginatum* is a narrow endemic which should not be collected both because of its rarity and its inability to survive long in cultivation.

Reproduction. Vegetative reproduction is one means by which *Sedum albomarginatum* spreads. Seedlings noted in crevices indicate that the species also multiplies by means of seeds. The relative importance of the two modes of reproduction is unassessed, but the short rootstocks suggest that vegetative spread may be less easy than in species in which they are longer.

A rosette collected in July 1964 and kept in dry condition was still alive in February 1965 and had produced etiolated shoots. Similarly, rosettes collected on June 6, 1963, and kept dry in plastic bags were still in good condition on July 20, 1963. Evidently the plants can endure long periods of drought.

Paul Hutchison has told me about a moth which visits the flowers at night. Since I have not observed the plants in the Serpentine Canyon in periods of darkness, I am unable to confirm this interesting observation. The relative importance of diurnal and nocturnal pollinators is a problem. In cultivation in Ithaca, N.Y., flowers of *Sedum albomarginatum* remain widely open in daylight. In nature, the petals appeared to be more erect and less widely spread.

Plants of *Sedum albomarginatum* are in flower in June. In cultivation in the greenhouse in Ithaca, they have flowered in April.

Relationship. *Sedum albomarginatum* resembles *S. laxum* enough to sugest a close relationship. The problem is whether to separate it as a species or to regard it as a distinctive subspecies. Diagnostic features include shorter offsets, longer leaves on floriferous stems, smaller diameter of flowers at anthesis, and a larger number of ovules per ovary. The white or pinkish margins of the leaves of the rosettes also are distinctive. The populations in the canyon of the Feather River together comprise a morphological and ecological unit which appears to be on the threshold between subspecies and species.

Another relationship of *Sedum albomarginatum* is with *S. moranii* from which it differs in its elongate inflorescences with dichotomously forked branches, paler yellow petals, and absence of hairs on the upper portions of the floriferous stems.

Lack of intermediate populations between *Sedum albomarginatum* and either *S. laxum* or *S. moranii* may be a point in favor of specific status. From an evolutionary standpoint, available evidence favors an origin either from *S. laxum* or from ancestral populations which gave rise to the modern species of *Gormania*.

Reference. Clausen and Uhl (1944).

26. *Sedum moranii**** (figs. 105 and 123–124)

A unique feature of *Sedum moranii* (fig. 123) is the glandular-pubescence of the cymes and upper parts of the floriferous stems. The cymes are mostly three-parted, also an unusual condition for subgenus *Gormania*. The petals are yellow, the leaves of the rosettes are papillose-crenulate at the apices, and the floriferous stems are recurved before the time of flowering. Perhaps the closest resemblance is to *S. albomarginatum*; but that has cymes with several branches, is not glandular-pubescent, has shorter and

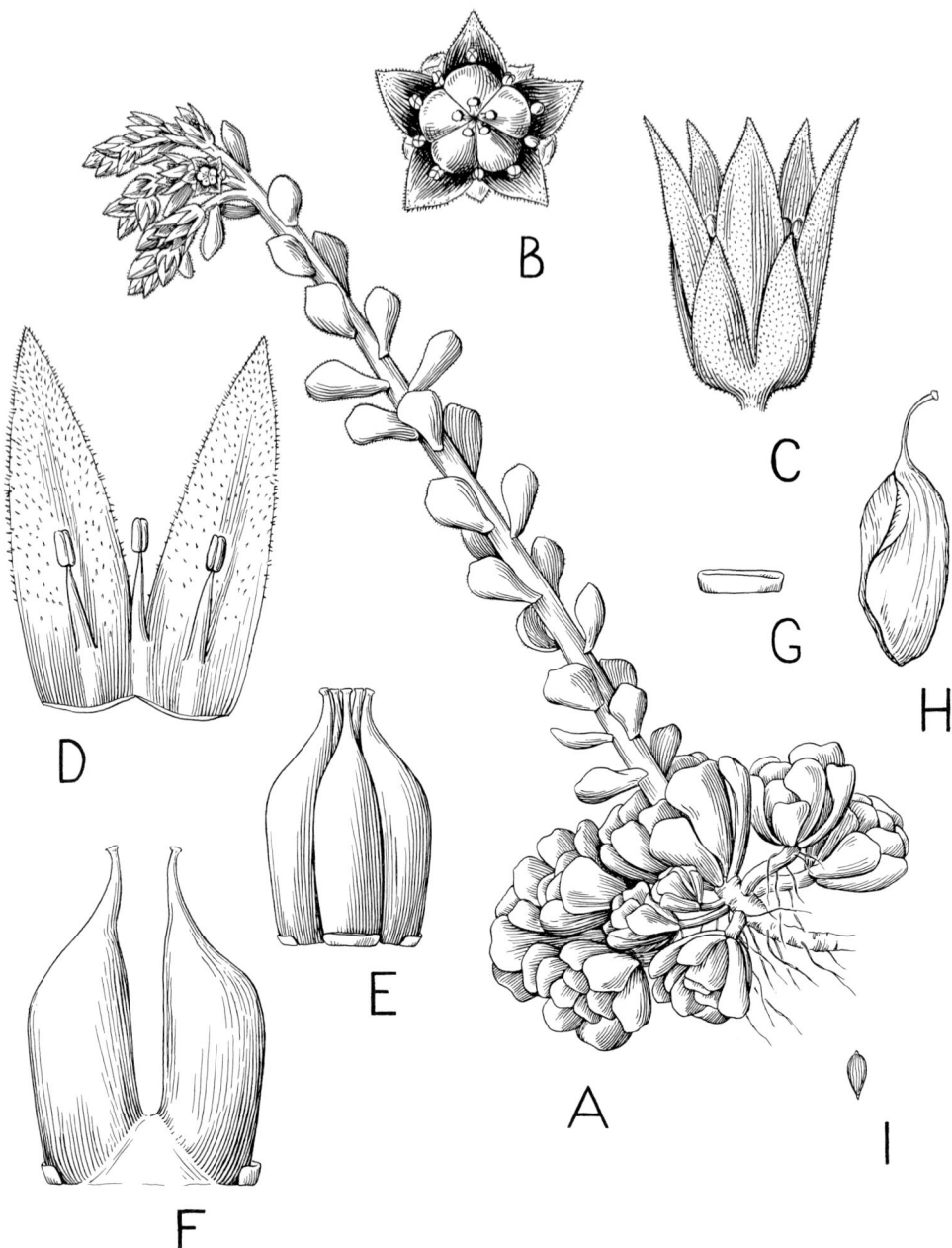

Fig. 123. Plant of *Sedum moranii* from canyon of Rogue River, Josephine Co., Ore. A. Habit sketch (x .57). B. Flower from above (x 2.3). C. Flower from side (x 2.3). D. Two petals and stamens (x 3.4). E. Carpels (x 3.4). F. Two carpels and nectaries (x 4.6). G. Nectary (x 5.7). H. Follicle (x 4.6). I. Seed (x 5.7).

paler yellow petals, and has the leaves of the primary rosettes both larger and white-margined.

Description. Sample: 5 plants chosen randomly from a cluster of 10 on the northern side of the gorge of the Rogue River just below the bridge at the mouth of Grave Creek in Josephine Co., Ore. These 5 plants became part of experiments on *Gormania* in the greenhouse and cold frame in Ithaca. Four of them flowered in the greenhouse. Data for plants cultivated in Ithaca are indicated in the following tables by cf. = cold frame and gh. = greenhouse. All other data apply to plants in the wild.

Plants of *Sedum moranii* are perennial with rootstocks up to 1 cm. thick, bearing at the ends of the branches compact rosettes which eventually give rise to terminal floriferous stems which are either erect or decumbent, but recurved in development. The upper parts of the floriferous stems are glandular-puberulent.

	n-pl.	n-obs.	$\bar{x}$	s	w
Plants diameter (cm.)—gh.	5	9	8.1	1.3	4–10
—cf.	1	1	1	—	1
Primary rosettes—diameter (cm.)	5	10	4	1.1	2.2–6
—gh.	5	15	3.2	.7	2.1–4.4
Offsets (n)—gh.	1	2	5	—	1–9
—cf.	1	1	0	—	0
Offsets—length (cm.)—gh.	1	2	2.5	—	1–4
Offsets—diameter of stems (mm.)	5	10	3.3	.4	2.8–4.5
—gh.	1	2	1.6	—	1.2–2.1
Floriferous stems per plant (n)	5	5	1.8	2.6	1–4
—gh.	5	8	.9	.8	0–2
Floriferous stems—length (cm.)	5	9	21	5.2	13–30
—gh.	1	1	27	—	27

The leaves of the primary rosettes are stiff and ascending, flat and oblong-spatulate, papillose-crenulate at the apices, emarginate, broadly rounded, or obtuse, and gray-green and glaucous when young, but becoming lustrous green and sometimes red toward apices with age. The leaves of the floriferous stems are sessile, oblong-oblanceolate, tuberculate-mucronate, and green suffused with red or mostly red. The upper ones are glandular-puberulent.

	n-pl.	n-obs.	$\bar{x}$ (mm.)	s (mm.)	w (mm.)
Leaves of rosettes—length	5	10	23	5.4	14–32
—gh.	5	15	17	3.6	10–24
—cf.	1	1	9	—	9
Leaves of rosettes—width	5	10	12	1.7	9–14
—gh.	5*	15	8	1.6	6–10
—cf.	1	1	5	—	5
Leaves of rosettes—thickness	5	10	3.4	.4	2.8–4.2
—gh.	5*	15	2.8	.7	1.5–3.3
—cf.	1	1	2.5	—	2.5
Median leaves of floriferous stems—length	5	9	15	3.6	11–21
—gh.	4	7	15.8	3.2	11–21
Median leaves of floriferous stems—width	5	9	9	2.8	7–14
—gh.	4	7	9	.6	7–11
Median leaves of floriferous stems—thickness	5	9	3.2	.5	3–4
—gh.	1	1	2.1	—	2.1

Inflorescences are three- or sometimes two-parted cymes. The floral bracts are oblong-spatulate and glandular-ciliate, largest by the first flowers on the cincinni and much reduced by the distal flowers.

	n-pl.	n-obs.	x̄ (n)	s (n)	w (n)
Cincinni	5	9	2.4	.4	2–3
—gh.	1	1	3	—	3
Flowers per cincinnus	5	26	4.7	2	1–9

Flowers (fig. 124) are short-pedicellate and erect. Each has a short torus.

Fig. 124. Flower and buds of plant of *Sedum moranii* from canyon of Rogue River, Josephine Co., Ore., cultivated in greenhouse, Ithaca, N.Y., May 10, 1965 (x 3.8). Photo by Howard Lyon.

	n-pl.	n-obs.	x̄ (mm.)	s (mm.)	w (mm.)
Basal pedicels—length	5	9	1.8	.8	1.0–3.6
Torus—length	3	4	1	.4	.5–1.4
Diameter	3	4	9	1	8–10

Sepals are erect, ovate or lanceolate, obtuse or acute, united basally for 1–2 mm., and glandular-puberulent.

	n-pl.	n-obs.	x̄ (mm.)	s (mm.)	w (mm.)
Length	3	4	7.4	.6	6.8–8.3
—gh.	1	2	7.9	—	7.5–8.4
Width	3	4	4.1	.5	3.3–4.6
—gh.	1	2	3.8	—	3.8

Petals are erect, convolute, oblong-lanceolate, aristate-tipped, united basally, glandular-puberulent and ciliate, and sulphur yellow.

	n-pl.	n-obs.	x̄	s	w
Number	3	4	5	0	5
—gh.	1	2	5	—	5
Length (mm.)	3	4	14.7	.9	13.4–15.8
—gh.	1	2	13.1	—	12.5–13.8
Length of cohesion (mm.)	3	4	1.9	.4	1.4–2.4
—gh.	1	2	2	—	1.6–2.5
Width (mm.)	3	4	4.4	.6	3.9–5
—gh.	1	2	4	—	4

Stamens all are epipetalous with greenish yellow filaments and yellow anthers which are truncate basally and rounded apically. The tips of the anthers are at the same level as the stigmas.

	n-pl.	n-obs.	x̄ (mm.)	s (mm.)	w (mm.)
Adnation of filaments	3	4	2.2	.5	1.4–2.7
—gh.	1	2	2.1	—	1.8–2.4
Anthers—length	1	1	1.8	—	1.8

Nectaries are narrowly reniform, white, and translucent.

	n-pl.	n-obs.	x̄ (mm.)	s (mm.)	w (mm.)
Length	3	4	.5	.1	.4–.6
—gh.	1	2	.5	—	.5
Width	3	4	1.5	.4	1.1–1.8
—gh.	1	1	1.8	—	1.7–1.9

Carpels are erect, connate basally, glabrous, and pale green, with short, stout styles. The carpels are inrolled all the way to the tips of the styles.

	n-pl.	n-obs.	x̄	s	w
Length (mm.)	3	4	7.3	.7	6.5–8
—gh.	1	2	7.2	—	6.5–8
Ovules per ovary (n)	3	4	51	13.1	42–66
—gh.	1	2	74	—	70–78

Follicles are subdivergent and brown.

	n-pl.	n-obs.	x̄ (mm.)	s (mm.)	w (mm.)
Length	1	1	5.8	—	5.8

Seeds are oblong-pyriform, finely striate, and brown.

	n-pl.	n-obs.	x̄ (mm.)	s (mm.)	w (mm.)
Length	1	1	1.1	—	1.1
Diameter	1	1	.4	—	.4

Chromosomes are small.

Number	n-pl.	g	sp	Cytologist
Rogue River Canyon	2	—	30	L. Hollingshead (1942), reported as *S. glanduliferum*

Variation. Plants of *Sedum moranii* appear to be similar. The growth of clones cultivated in experiments in Ithaca, N.Y., has not been good enough to allow satisfactory comparisons involving flowers and seeds, but only comparisons on the basis of leaves. One clone had a markedly wider median leaf of the floriferous stem in the field. This clone has not flowered in cultivation, but, in an experiment which ran from 1963 to 1965 in a greenhouse in Ithaca, differed significantly from two other clones in width and from three other clones in thickness of leaves of primary rosettes. On the basis of present evidence, all plants of *S. moranii* could belong to as few as four clones.

Nomenclature. *Sedum moranii* R. T. Clausen, Bull. Torr. Bot. Club 69:40 (1942). Basionym: *Cotyledon glandulifera* L. F. Henderson, Rhodora 32:26 (1930).
Synonyms are:
1930. *Cotyledon glandulifera* L. F. Henderson, Rhodora 32:26. Type locality: along the trail down the Rogue River, 3 miles below "Alameda," Josephine Co., Ore. Type: Mr. and Mrs. J. R. Leach no. 1599 (ORE), June 1, 1928.
1941. *Sedum glanduliferum* (L. F. Henderson) M. E. Peck, Man. Higher Plants Ore., p. 361, and Madroño 6:134 (1941). Basionym: *Cotyledon glandulifera* L. F. Henderson. Not *Sedum glanduliferum* Gussone (1827).
1944. *Gormania glandulifera* (Henderson) Abrams, Ill. Fl. Pacific States 2:343. Basionym: *Cotyledon glandulifera* L. F. Henderson. Peck's citation of *Gormania glandulifera* in the synonymy of *S. glanduliferum*, Manual of the Higher Plants of Oregon, p. 361 (1941), did not constitute valid publication.

Distribution. *Sedum moranii* (fig. 105) grows on serpentine in the canyon of the Rogue River north of the Almeda Mine, Josephine Co., Ore. See the Galice 15' topographic map. The known distribution is very limited, with a diameter of less than 2 kilometers and an altitudinal range of about 10 meters, from 210 to 220 meters.

The site of the type population is just below the bridge across the Rogue River at the mouth of Grave Creek. This bridge is 4.5 kilometers in a straight line north of the Almeda Mine. In 1963, the type population comprised 83 plants. Of these, I studied environmental conditions of 5 plants selected randomly from a cluster of 10 in a major group of 27 on either side of a small rivulet at an altitude of 220 meters. Exposure was to the southwest or west. The range of pH, on the basis of five tests, was 6.4 to 7.2. The plants of *Sedum* grew in moss on serpentine. *Selaginella wallacei* was closely associated with 3 of the plants. Other species nearby were *Rhus diversiloba*, *Achillea lanulosa*, and *Pityrogramma triangularis*.

Reproduction. *Sedum moranii* multiplies both by means of vegetative offsets and by seeds. The species seems to be self-incompatible. Ovules of a plant in cultivation, without chance for cross-pollination, aborted.

Flowering time in nature is late May and early June. Plants in cultivation in the greenhouse in Ithaca have flowered in April and May.

Relationships. *Sedum moranii* is one of the most specialized species of the subgenus *Gormania*. Points in favor of this idea are the inflorescence, a two- or three-parted cyme rather than a paniculate one with the branches dichotomous, glandular-pubescence, and the occurrence on serpentine. Perhaps the most closely related species is the almost equally specialized *S. albomarginatum*. These two species may be offshoots from the ancestral stock which gave rise to *S. obtusatum* and *S. laxum* ssp. *heckneri*.

References. Clausen, R. T. (1942), and Hollingshead, L. (1942).

27. *Sedum spathulifolium**** (figs. 125–135)

Distinctive features of *Sedum spathulifolium* (fig. 125) are prominent rosettes of spatulate leaves, offsets, bright yellow flowers, widely spreading petals, and gibbous follicles. The stems are creeping and much branched, producing a matted appearance. *Sedum spathulifolium* is the commonest stonecrop in the Pacific Coast Ranges of North America.

The flowers of *Sedum spathulifolium* are in cincinni in three-parted cymes, unlike the paniculate inflorescences of *S. obtusatum*, *S. laxum*, and all other species of subgenus *Gormania* except *S. moranii*. The separate, spreading petals differentiate it from *S. moranii* as well as from the other Gormanias. Some eastern North American species, as *S. nevii*, *S. glaucophyllum*, and even *S. ternatum*, have a similar habit of growth, flat leaves in rosettes, and kyphocarpic carpels, but differ in their usually 4-merous flowers, white petals, and flattened filaments with those opposite the petals scarcely adnate.

Description. Sample: 130 plants. Collection of most of the plants was during surveys of the Pacific Coast Ranges in 1963, the Sierra Nevada in 1964, the Cascade Mountains in 1965, and the San Bernardino Mountains in 1968. The usual procedure in sampling was first to scan selected sampling units, enumerating either plants or natural clusters of plants in the unit. Then, depending on the size of the population, samples were either simple random of plants from the whole population or of clusters. When clusters were chosen, sampling was two-stage, with proportionate numbers of plants selected randomly from each chosen cluster. An exception to plotless sampling was in the San Bernardino Mountains where samples were random from clusters

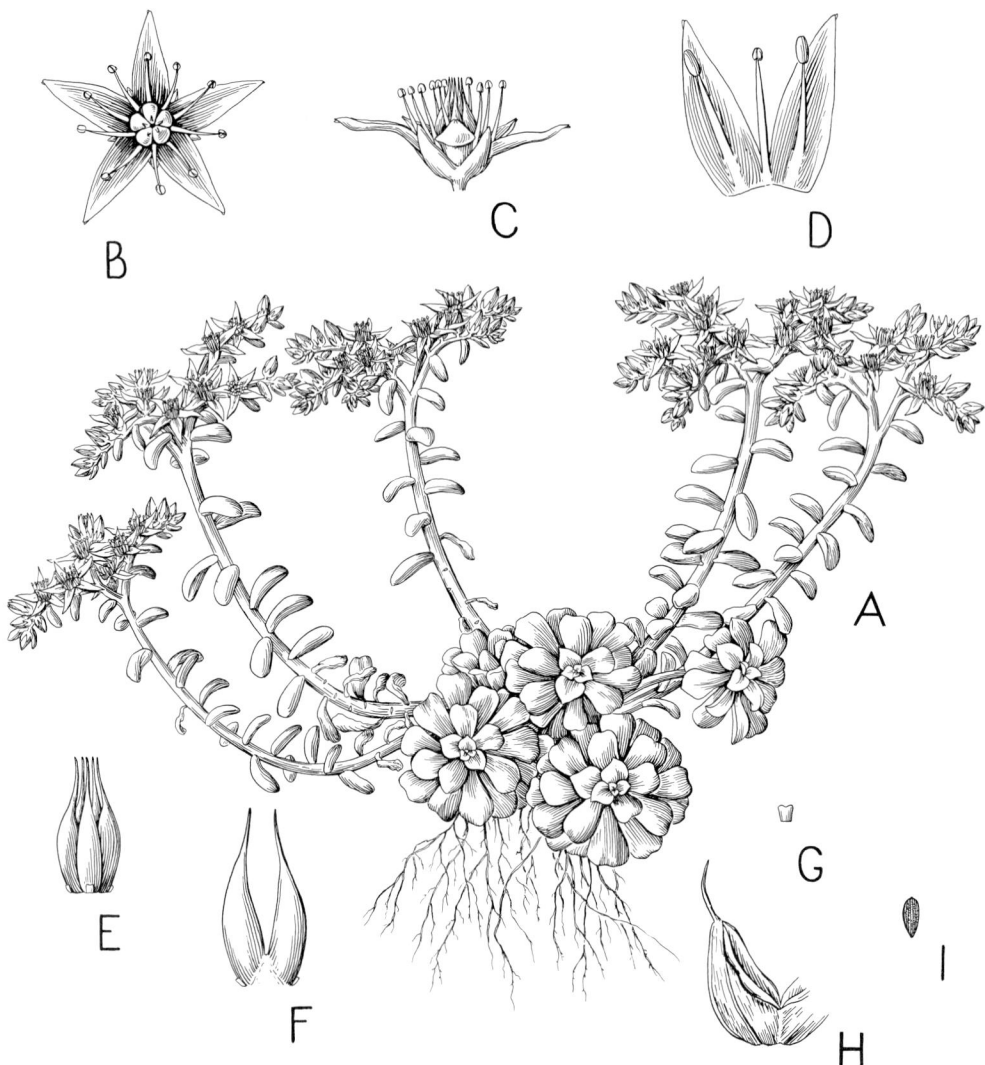

Fig. 125. Plant of *Sedum spathulifolium* ssp. *spathulifolium* from Bear Mountain in the Mayacamas Mountains, Sonoma Co., Calif., cultivated in greenhouse, Ithaca, N.Y., May 13, 1965. A. Habit sketch (x .48). B. Flower from above (x 1.9). C. Flower from side (x 1.9). D. Two petals and stamens (x 2.9). E. Carpels (x 2.9). F. Two carpels and nectaries (x 3.8). G. Nectary (x 4.8). H. Follicle (x 3.8). I. Seed (x 4.8).

in quadrats of 100 square meters. Information about sizes of populations, number of plants in samples, and condition of plants when collected is available in the following table. N_u = number of plants in listing units, N_c = number of plants in sampled clusters, and n = number of plants in samples. Question marks result from lack of enumeration of populations at sites outside regular sampling units. Selection of plants from such places was because of interesting variations or unusual environmental conditions.

Locality	N_u	N_c	n	n/N_c	n/N_u	Condition when collected fl. (n)	fr. (n)
Pacific Coast Ranges							
Burnt Mountain	100	4	4	1	.04	4	0
Middle Brummit Creek	186	21	7	.3	.04	3	0
Bear Mountain	454	74	16	.2	.03	6	10
Coos Ridge	>2	>2	2	?	?	2	0
Pinnacles	>92	25	4	.2	<.04	3	3
Bear Mt. (Mayacamas Mts.)	520	520	2	.004	.004	0	2
Hopland Grade	2,249	26	12	.5	.005	1	11
Klamath	>2	>2	2	?	?	0	0
Willamette–Puget Trough							
Lummi Island	>2	>2	2	?	?	0	1
Lummi Rocks	47	5	1	.2	.02	0	1
Cascade Mountains							
Dartmouth Creek	101	51	3	.06	.03	0	3
Deer Butte Trail	13,022	230	14	.06	.001	2	11
O'Leary Mountain	48	4	2	.5	.04	2	0
Oregon Coast							
Bandon	33	22	5	.2	.15	1	0
Transverse Ranges							
Mountain Home Creek	592	148	6	.04	.01	0	6
W. of Forsee Creek	72	4	3	.75	.04	0	3
Sierra Nevada							
Vernal Falls	24,233	340	13	.04	.0005	12	0
El Portal	>1	>1	1	?	?	0	1
Hetch Hetchy	>2	>2	2	?	?	0	2
Stringtown	345	345	1	.003	.003	0	1
Rock Creek	197	197	9	.04	.04	0	9
Feather R. Canyon	>7	>7	7	?	?	1	5
Keddie	1,252	1,252	2	.002	.002	0	2
Spanish Creek	462	185	4	.02	.01	0	4
Klamath Mountains							
Cottonwood Creek	151	28	5	.17	.03	0	1
Wildwood	>1	>1	1	?	?	0	0
Totals	>44,133	>3,498	130	<.04	<.003	37	76

My intention was to select samples proportional to the sizes of the populations of *Sedum spathulifolium*. Great inequality of populations and lack of time prevented attainment of this goal. As a result, weights (W) are necessary in computations in order that small populations may not be overrepresented and large populations underrepresented in estimates. The various strata likewise require weights based on estimates of included total numbers of plants. Sizes of populations and weights are available in the following table. Some estimates, as for the Sierra Nevada, may be too small. These resulted when plants were lacking in randomly selected units. The absence of *S. spathulifolium* from vast areas of the Sierra Nevada is certain, but the total number of plants there is uncertain.

Areal unit	Listing units N	Listing units W	Tertiary strata N	Tertiary strata W	sampling units N	sampling units W	Secondary strata N	Secondary strata W	Primary strata N	Primary strata W
Pacific Coast R.									813,552	.801
Random stratum							813,550	1		
Oregon Coast R.					61,244	.955				
Random stratum			60,788	.993						
Burnt Mt.	100	.35								
Mid. Brum. C.	186	.65								
Optimum stratum			456	.007						
Bear Mt.	454	.996								
Coos Ridge	2	.004								
Gabilan Range					92	.002				
Pinnacles	92	1								
Mayacamas Mts.					2,769	.043				
Bear Mt.	520	.188								
Hopland Grade	2,249	.812								
Optimum stratum							2	—		
Klamath	2	1								
Puget Trough									517	.001
Lummi Island	2	.041								
Lummi Rocks	47	.959								
Cascade Mts.									200,982	.198
Three Sist. area					185,955	1				
Random stratum			172,885	.93						
Dartmouth Cr.	101	1								
Optimum stratum			13,070	.07						
Deer Butte Tr.	13,022	.996								
O'Leary Mt.	48	.004								
Pacific Coast									891	1
Bandon	33	1								
Transverse Ranges									4,980	.17
San Bern. Mts.					664	1				
Mtn. Home Cr.	592	.892								
W. of Forsee Cr.	72	.108								
C. Sierra Nevada									24,236	.83
S. Yosemite area					24,234	1				
Vernal Falls	24,233	1								
El Portal	1	—								
N. Yosemite area					2	—				
Hetch Hetchy	2	—								
N. Sierra Nevada									2,263	.139
Feather R. area					2,263	1				
Stringtown	345	.153								
Rock Creek	197	.087								
Feather R. Can.	7	.003								
Keddie	1,252	.553								
Spanish Cr.	462	.204								
Klamath Mts.									13,961	.861
Yolla Bolly Mts.					13,961	1				
Random stratum			13,960	1						
Cottonwood Cr.	151	1								
Optimum stratum			1	—						
Wildwood	1	1								

Data are grouped in the description under the following abbreviated designations: CR, WP, C = Pacific Coast Ranges, Willamette–Puget Trough, and Cascade Mountains. Data for all populations at localities in these three geomorphic provinces, as indicated in the preceding table, are pooled under this heading (= ssp. *spathulifolium*).

PC = Pacific Coast. This applies to the sea cliffs. Only the population at Bandon falls in this category (= ssp. *pruinosum*).

CSN, T = central Sierra Nevada and Transverse Ranges of California (= ssp. *yosemitense*).

NSN, K = northern Sierra Nevada and Klamath Mountains (= ssp. *purdyi*).

The grouping of localities adopted here was achieved after first making comparisons at the level of populations, without assumptions concerning regional groups. Information from this approach is available in the section on variation. The scheme for pooling data condenses a long series of descriptive tables and permits comparison of subspecies for each feature which was studied.

Sedum spathulifolium grows poorly outdoors at Ithaca, N.Y. Experiments in the garden primarily were tests of survival. Greatest loss was in late spring and summer. Plants grew well and flowered in the cold frames and greenhouses. A measure of success in experiments is provided by the following table in which figures in the *n-d* columns indicate the number of plants which provided descriptive data in each situation.

	Garden		Cold frame		Greenhouse	
Source	N	n-d	N	n-d	N	n-d
CR, WP, C	6	0	13	12	13	13
PC	2	0	2	2	2	2
CSN, T	3	0	4	4	7	7
NSN, K	9	0	6	5	11	11
Totals	20	0	25	23	33	33

All means and standard deviations in the following tables are weighted as indicated in the table of population sizes and weights. Standard deviations resulting from the data of the experiments are derived from variances among plants with the variance component for replicates removed, following the method of Snedecor (1956:268–270).

Plants of *Sedum spathulifolium* are perennial herbs with stout, branched rhizomes and fibrous, brown roots. Primary stems are procumbent or creeping and pale green, white, pink, or red. They vary in diameter from 1 to 3 mm., take root at the nodes, and are much branched, with each branch terminating in a rosette which in turn gives rise to offsets. These offsets eventually take root. In this way, a single plant may form a complex patch. Floriferous stems are either terminal or axillary and either erect or decumbent. At time of flowering they may be green, pink, or red, and often glaucous. After dehiscence of fruits, the fructiferous stems wither and die.

	n-pop.	n-pl.	n-rep.	n-obs.	$\bar{x}$	s
Plants—diameter (cm.)						
CR, WP, C	7	39	—	39	17	14
PC	1	4	—	4	20	9
CSN, T	4	282	—	282	112	124

	n-pop.	n-pl.	n-rep.	n-obs.	$\bar{x}$	s
Plants—diameter (cm.)						
NSN, K	5	21	—	21	65	39
CR, WP, C—cf.	8	10	17	17	12	0
PC—cf.	1	2	4	4	18	2
CSN, T—cf.	2	2	3	3	7	0
NSN, K—cf.	2	3	6	6	12	4
CR, WP, C—gh.	7	13	23	23	13	4
PC—gh.	1	2	4	4	17	2
CSN, T—gh.	4	8	14	14	10	0
NSN, K—gh.	7	11	27	27	11	0
Primary rosettes—diameter (cm.)						
CR, WP, C	12	50	—	169	3	.4
PC	1	5	—	23	1.5	.1
CSN, T	5	24	—	53	3.2	.8
NSN, K	7	25	—	79	3.7	.8
CR, WP, C—cf.	8	10	15	26	2.5	.7
PC—cf.	1	2	3	6	2.1	.5
CSN, T—cf.	2	2	3	5	2.6	0
NSN, K—cf.	2	3	5	10	2.8	1.6
CR, WP, C—gh.	7	13	23	43	3.3	0
PC—gh.	1	2	4	8	1.7	.1
CSN, T—gh.	4	8	14	26	2.4	.2
NSN, K—gh.	7	10	25	47	3	0
Offsets—number						
CR, WP, C	5	22	—	41	1.6	.6
CSN, T	5	23	—	49	2.3	.8
NSN, K	4	16	—	62	1.9	.2
CR, WP, C—cf.	8	10	15	27	1.3	.9
PC—cf.	1	2	4	8	2.4	.7
CSN, T—cf.	2	2	3	5	4	0
NSN, K—cf.	2	3	6	11	3.8	2.7
CR, WP, C—gh.	7	13	23	37	2.4	0
PC—gh.	1	2	4	8	2.1	1.2
CSN, T—gh.	4	8	14	22	2.9	1.2
NSN, K—gh.	7	10	25	47	3.8	1.5
Offsets—length (cm.)						
CR, WP, C	5	22	—	41	1.7	.5
CSN, T	5	21	—	68	1.9	.5
NSN, K	4	14	—	102	1.6	1
CR, WP, C—cf.	7	9	13	22	1.4	.2
PC—cf.	1	2	4	8	1.4	.6
CSN, T—cf.	2	2	3	5	2.8	0
NSN, K—cf.	2	3	6	12	2.2	0
CR, WP, C—gh.	7	13	21	39	2.2	0
PC—gh.	1	2	4	7	2.2	.8
CSN, T—gh.	4	8	13	24	2.3	0
NSN, K—gh.	7	10	25	46	2.6	1.6
Offsets—diameter of stems (mm.)						
CR, WP, C	13	49	—	167	1.8	.2
PC	1	5	—	23	1.6	.2
CSN, T	5	21	—	68	1.4	.3
NSN, K	7	22	—	118	1.5	.3
CR, WP, C—cf.	7	9	13	22	1.3	.8
PC—cf.	1	2	4	8	1.3	.1
CSN, T—cf.	2	2	3	5	1.4	0
NSN, K—cf.	2	3	5	10	1.2	0

	n-pop.	n-pl.	n-rep.	n-obs.	$\bar{x}$	s
Offsets—diameter of stems (mm.)						
CR, WP, C—gh.	7	13	18	31	1.9	0
PC—gh.	1	2	4	8	1.6	.1
CSN, T—gh.	4	8	13	24	1.4	0
NSN, K—gh.	7	10	25	43	1.3	0
Floriferous stems per plant (n)						
CR, WP, C	8	38	—	38	3.5	2.6
PC	1	1	—	1	2	—
CSN, T	4	282	—	282	19.2	42.1
NSN, K	6	20	—	20	2.5	1.5
Floriferous stems—length (cm.)						
CR, WP, C	9	41	—	59	11.6	3.2
PC	1	2	—	2	3.2	1.8
CSN, T	2	271	—	271	11.8	2.2
NSN, K	7	22	—	29	14.4	1.7
CR, WP, C—cf.	8	11	18	34	7.1	0
PC—cf.	1	2	3	6	4.7	1.2
CSN, T—cf.	3	4	6	11	5.8	3.1
NSN, K—cf.	3	4	7	13	7.1	0
CR, WP, C—gh.	8	12	17	31	11.9	.8
PC—gh.	1	2	4	8	7.4	.2
CSN, T—gh.	3	4	7	11	3.7	0
NSN, K—gh.	7	10	20	31	7.8	2.4

Leaves of rosettes are spirally arranged, spatulate, papillose marginally, rounded or truncate and often submucronate apically, and green, sometimes suffused with red and often glaucous or pruinose. Surfaces are paler dorsally than ventrally. Old leaves may persist in withered condition below living ones. Leaves of offsets, below the rosettes, are opposite or whorled. Those of floriferous stems are alternate, divergent, spatulate-oblong or elliptic-oblong, rounded apically, and green or red, and also sometimes glaucous. A single leaf of one rosette was three-lobed.

	n-pop.	n-pl.	n-rep.	n-obs.	$\bar{x}$ (mm.)	s (mm.)
Leaves of rosettes—length						
CR, WP, C	13	50	—	169	17.2	1.7
PC	1	5	—	23	7.1	1.3
CSN, T	5	24	—	53	16.7	3.1
NSN, K	7	25	—	79	19.2	6.2
CR, WP, C—cf.	8	10	15	27	15.4	2.4
PC—cf.	1	2	4	8	10.1	2.2
CSN, T—cf.	2	2	3	5	15	—
NSN, K—cf.	2	3	6	12	15.2	4.5
CR, WP, C—gh.	8	13	23	43	17.5	2.5
PC—gh.	1	2	4	8	9.1	.7
CSN, T—gh.	4	8	14	26	13	1.9
NSN, K—gh.	7	11	26	49	14.8	0
Leaves of rosettes—width						
CR, WP, C	13	50	—	169	8.5	.7
PC	1	5	—	23	4.6	.4
CSN, T	5	24	—	53	7.5	1
NSN, K	7	25	—	79	8.4	1.3
CR, WP, C—cf.	8	10	15	27	8.1	1.5
PC—cf.	1	2	4	8	6	.5

	n-pop.	n-pl.	n-rep.	n-obs.	$\bar{x}$ (mm.)	s (mm.)
Leaves of rosettes—width						
CSN, T—cf.	2	2	3	5	7	—
NSN, K—cf.	2	3	6	12	7.8	.7
CR, WP, C—gh.	8	13	23	43	9.3	0
PC—gh.	1	2	4	8	5.6	.2
CSN, T—gh.	4	8	14	26	7	.7
NSN, K—gh.	7	11	26	49	7.3	1
Leaves of rosettes—thickness						
CR, WP, C	13	50	—	169	1.9	.2
PC	1	5	—	23	2.3	.2
CSN, T	5	24	—	53	2.1	.3
NSN, K	7	25	—	79	1.9	.4
CR, WP, C—cf.	8	10	15	26	1.8	.6
PC—cf.	1	2	4	8	2.5	.2
CSN, T—cf.	2	2	3	5	2	—
NSN, K—cf.	2	3	6	12	1.7	0
CR, WP, C—gh.	8	13	23	43	1.6	0
PC—gh.	1	2	4	8	2.1	.6
CSN, T—gh.	4	8	14	26	1.4	.2
NSN, K—gh.	7	11	26	49	1.5	.1
Leaves of floriferous stems—length						
CR, WP, C	8	19	—	35	11.3	1
PC	1	1	—	1	4	—
CSN, T	1	13	—	28	9.2	1.7
CR, WP, C—cf.	8	11	18	32	8.8	.8
PC—cf.	1	2	3	6	7.5	.7
CSN, T—cf.	3	4	6	9	10.5	0
NSN, K—cf.	3	4	7	11	8.5	0
CR, WP, C—gh.	8	12	18	31	11.6	0
PC—gh.	1	2	4	8	10	1.5
CSN, T—gh.	3	4	7	11	9.5	0
NSN, K—gh.	7	10	20	29	11.5	.7
Leaves of floriferous stems—width						
CR, WP, C	8	19	—	35	5.7	.8
PC	1	1	—	1	3	—
CSN, T	1	13	—	29	4.7	.5
CR, WP, C—cf.	8	11	18	34	4.7	.6
PC—cf.	1	2	3	6	3.7	.2
CSN, T—cf.	3	4	6	9	4.2	0
NSN, K—cf.	3	4	7	11	3.9	0
CR, WP, C—gh.	8	12	18	31	6	0
PC—gh.	1	2	4	8	5	.2
CSN, T—gh.	3	4	7	11	3.7	0
NSN, K—gh.	7	10	20	29	5.1	0
Leaves of floriferous stems—thickness						
CR, WP, C	8	19	—	34	1.8	.4
PC	1	1	—	1	1.5	—
CSN, T	1	12	—	27	2.1	.2
CR, WP, C—cf.	8	11	18	34	1.6	0
PC—cf.	1	2	3	6	2	.8
CSN, T—cf.	3	4	6	9	1.7	.3
NSN, K—cf.	3	4	7	11	1.7	.5
CR, WP, C—gh.	8	12	18	31	1.6	0
PC—gh.	1	2	4	8	1.8	.5
CSN, T—gh.	3	4	7	11	1.4	.3
NSN, K—gh.	7	10	20	29	1.5	0

Inflorescences are cymes of usually three cincinni which sometimes are once or twice dichotomously forked. Floral bracts are oblong-spatulate or linear and smaller than the leaves of the floriferous stems.

	n-pop.	n-pl.	n-rep.	n-obs.	$\bar{x}$ (n)	s (n)
Cincinni per cyme						
CR, WP, C	11	42	—	71	3.1	.4
PC	1	2	—	2	2	0
CSN, T	2	4	—	7	2.8	.6
NSN, K	2	3	—	5	2.7	.7
CR, WP, C—cf.	8	11	18	34	2.9	.1
PC—cf.	1	2	3	5	3	0
CSN, T—cf.	3	4	6	11	3	.3
NSN, K—cf.	3	4	7	13	3	0
CR, WP, C—gh.	8	12	18	32	3.3	0
PC—gh.	1	2	4	8	3	0
CSN, T—gh.	3	4	7	11	3	0
NSN, K—gh.	7	10	20	32	3	0
Flowers per cyme						
CR, WP, C	4	5	—	20	28	20
CSN, T	2	3	—	13	22	7

The flowers usually are 5-merous and either sessile or short-stalked on pedicels up to 8 mm. long, and may be sweetly fragrant. A short torus is present.

	n-pop.	n-pl.	n-rep.	n-obs.	$\bar{x}$ (mm.)	s (mm.)
Flowers—diameter						
CR, WP, C	8	21	—	65	13	.6
PC	1	1	—	1	10	—
CSN, T	1	12	—	30	12	1.6
NSN, K	1	1	—	5	12	—
CR, WP, C—cf.	8	11	18	36	14	0
PC—cf.	1	2	3	5	11	.6
CSN, T—cf.	3	4	6	9	13	0
NSN, K—cf.	3	4	7	14	12	.9
CR, WP, C—gh.	8	12	18	36	16	0
PC—gh.	1	2	4	8	12	1.7
CSN, T—gh.	3	4	7	14	16	.8
NSN, K—gh.	7	10	20	40	15	0
Torus—length						
CR, WP, C	8	21	—	65	.8	.06
CSN, T	1	12	—	28	.9	.1
NSN, K	1	1	—	5	.6	—

Sepals are lanceolate, linear-lanceolate, oblong-ovate, or ovate, connate basally, acute or obtuse, green or yellow-green, and either glaucous or pruinose.

	n-pop.	n-pl.	n-rep.	n-obs.	$\bar{x}$ (mm.)	s (mm.)
Length						
CR, WP, C	8	21	—	65	2.6	.4
PC	1	1	—	1	2.1	—
CSN, T	1	12	—	28	2.1	.3

	n-pop.	n-pl.	n-rep.	n-obs.	x̄ (mm.)	s (mm.)
Length						
NSN, K	1	1	—	5	2.5	—
CR, WP, C—cf.	8	11	18	36	2.7	0
PC—cf.	1	2	3	5	2.8	1.3
CSN, T—cf.	3	4	6	9	2.7	.5
NSN, K—cf.	3	4	7	14	2.6	0
CR, WP, C—gh.	8	12	18	36	3.6	0
PC—gh.	1	2	4	8	2.9	.7
CSN, T—gh.	3	4	7	14	2.9	.9
NSN, K—gh.	7	10	20	40	3.3	0
Width						
CR, WP, C	8	21	—	65	1.1	.3
PC	1	1	—	1	1.4	—
CSN, T	1	12	—	28	1.4	.2
NSN, K	1	1	—	5	1.2	—
CR, WP, C—cf.	8	11	18	36	1.5	0
PC—cf.	1	2	3	5	1.4	.3
CSN, T—cf.	3	4	6	9	1.5	.4
NSN, K—cf.	3	4	7	14	1.6	0
CR, WP, C—gh.	8	12	18	36	1.6	0
PC—gh.	1	2	4	8	1.4	.2
CSN, T—gh.	3	4	7	14	1.7	0
NSN, K—gh.	7	10	21	42	1.8	—

Petals are separate and erect basally, then widely spreading. They usually are lanceolate, varying from linear to oblanceolate, acute, and some shade of yellow, ranging from almost white to lemon-yellow, canary-yellow, or aureolin. The number of petals usually is 5, but varies from 4 to 7.

	n-pop.	n-pl.	n-rep.	n-obs.	x̄	s
Number per flower						
CR, WP, C	8	21	—	65	5.2	.4
CSN, T	1	12	—	28	5	.1
NSN, K	1	1	—	5	5	—
CR, WP, C—cf.	8	11	18	36	5	.04
PC—cf.	1	2	3	5	5	0
CSN, T—cf.	3	4	6	9	5	0
NSN, K—cf.	3	4	7	14	4.9	.3
CR, WP, C—gh.	8	12	18	36	5.2	.4
PC—gh.	1	2	4	8	5	0
CSN, T—gh.	3	4	7	14	5	0
NSN, K—gh.	7	10	20	40	5.1	.3
Length (mm.)						
CR, WP, C	8	21	—	65	6.8	.3
PC	1	1	—	1	6	—
CSN, T	1	12	—	28	5.8	.7
NSN, K	1	1	—	5	5.8	—
CR, WP, C—cf.	8	11	18	36	7.3	.1
PC—cf.	1	2	3	5	7	2.2
CSN, T—cf.	3	4	6	9	6.4	.6
NSN, K—cf.	3	4	7	14	6.1	0
CR, WP, C—gh.	8	12	18	36	8.7	.3
PC—gh.	1	2	4	8	7.6	.3
CSN, T—gh.	3	4	7	14	7.9	.2
NSN, K—gh.	7	10	20	40	7.9	0

	n-pop.	n-pl.	n-rep.	n-obs.	$\bar{x}$	s
Width (mm.)						
CR, WP, C	8	21	—	65	1.8	.3
PC	1	1	—	1	1.7	—
CSN, T	1	12	—	28	1.8	.3
NSN, K	1	1	—	5	1.8	—
CR, WP, C—cf.	8	11	18	36	2.2	0
PC—cf.	1	2	3	5	2	.6
CSN, T—cf.	3	4	6	9	1.8	.3
NSN, K—cf.	3	4	7	14	2.1	.3
CR, WP, C—gh.	8	12	18	36	2.4	0
PC—gh.	1	2	4	8	2.3	.1
CSN, T—gh.	3	4	7	14	2.5	.3
NSN, K—gh.	7	10	20	40	2.7	0

Stamens have yellow filaments and basifixed anthers which vary from pale yellow to aureolin. Those alternating with the petals shed pollen before the epipetalous ones. Fused stamens, staminodia additional to functional stamens, and structures which are half stamen and half petal rarely occur.

	n-pop.	n-pl.	n-rep.	n-obs.	$\bar{x}$ (mm.)	s (mm.)
Epipetalous filaments; adnation—length						
CR, WP, C	8	21	—	67	1.1	.2
CSN, T	1	12	—	28	1	.2
NSN, K	1	1	—	5	.8	—
CR, WP, C—cf.	8	11	18	36	1.3	.2
PC—cf.	1	2	3	5	1.2	.6
CSN, T—cf.	3	4	6	9	1	0
NSN, K—cf.	3	4	7	14	.8	.07
CR, WP, C—gh.	8	12	18	36	1.5	.5
PC—gh.	1	2	4	8	1.4	.1
CSN, T—gh.	3	4	7	14	1.1	0
NSN, K—gh.	7	10	20	40	1.2	.1
Anthers—length						
CR, WP, C	8	17	—	38	.9	.2
CSN, T	1	7	—	11	1.2	.1
NSN, K	1	1	—	3	.9	—
CR, WP, C—cf.	6	7	8	8	1.2	0
PC—cf.	1	1	1	1	1	—
CSN, T—cf.	3	4	5	5	1.2	.3
NSN, K—cf.	3	4	6	8	1.3	0
CR, WP, C—gh.	8	11	14	15	1.3	0
PC—gh.	1	1	1	1	.8	—
CSN, T—gh.	1	1	1	1	1.1	—
NSN, K—gh.	6	9	15	17	1.4	.1

Nectaries are reniform or nearly quadrate, truncate, sometimes erose, and yellow, varying from almost white to orange. Occasionally they are translucent.

	n-pop.	n-pl.	n-rep.	n-obs.	$\bar{x}$ (mm.)	s (mm.)
Length						
CR, WP, C	8	21	—	65	.4	.05
CSN, T	1	12	—	27	.5	.1
NSN, K	1	1	—	5	.4	—
CR, WP, C—cf.	8	11	18	36	.5	0

	n-pop.	n-pl.	n-rep.	n-obs.	$\bar{x}$ (mm.)	s (mm.)
Length						
PC—cf.	1	2	3	5	.5	.1
CSN, T—cf.	3	4	6	9	.6	.1
NSN, K—cf.	3	4	7	14	.5	.1
CR, WP, C—gh.	8	12	18	36	.5	.06
PC—gh.	1	2	4	8	.5	.02
CSN, T—gh.	3	4	7	14	.6	.2
NSN, K—gh.	7	10	20	40	.5	.06
Width						
CR, WP, C	8	21	—	65	.4	.06
CSN, T	1	12	—	27	.4	.04
NSN, K	1	1	—	5	.3	—
CR, WP, C—cf.	8	11	18	36	.5	.01
PC—cf.	1	2	3	5	.5	.1
CSN, T—cf.	3	4	6	9	.5	0
NSN, K—cf.	3	4	7	14	.5	0
CR, WP, C—gh.	8	12	18	36	.6	.05
PC—gh.	1	2	4	8	.6	.05
CSN, T—gh.	3	4	7	14	.5	.2
NSN, K—gh.	7	10	20	40	.6	.04

Carpels are erect when flowers are in anthesis, but in maturing become divergent, gibbous ventrally, and two-lipped. They vary from 3 to 6 per flower and from canary yellow to green. Basally, they are connate for .6 to 1.2 mm. The styles are slender, rolled, and hollow. Carpels rarely are unsealed.

	n-pop.	n-pl.	n-rep.	n-obs.	$\bar{x}$	s
Length (mm.)						
CR, WP, C	8	21	—	65	4.7	.4
CSN, T	1	12	—	28	4.6	.6
NSN, K	1	1	—	5	4.8	—
CR, WP, C—cf.	8	11	18	36	5.4	.7
PC—cf.	1	2	3	5	4.8	2.2
CSN, T—cf.	3	4	6	9	5.3	0
NSN, K—cf.	3	4	7	14	5.4	.9
CR, WP, C—gh.	8	12	18	36	6.2	0
PC—gh.	1	2	4	8	5.1	.3
CSN, T—gh.	3	4	7	14	6.2	0
NSN, K—gh.	7	10	20	40	6	0
Styles—length (mm.)						
CR, WP, C	2	4	—	8	1.4	.2
CSN, T	1	12	—	28	1.7	.1
Ovules per ovary (n)						
CR, WP, C	8	24	—	68	6.3	3.5
PC	1	1	—	2	4	—
CSN, T	1	12	—	28	5.6	2.3
NSN, K	1	1	—	5	5.2	—
CR, WP, C—cf.	8	11	18	36	10	1.6
PC—cf.	1	2	3	5	7	6.5
CSN, T—cf.	3	4	6	9	8	0
NSN, K—cf.	3	4	7	14	8	2.6
CR, WP, C—gh.	8	12	18	36	11	0
PC—gh.	1	2	4	8	7	1
CSN, T—gh.	3	4	7	14	8	.5
NSN, K—gh.	7	10	20	40	11	0

Fruits (fig. 126) are divergent or subdivergent, 5-ribbed, brown, and with two prominent ventral lips to .8 mm. wide.

	n-pop.	n-pl.	n-rep.	n-obs.	x̄ (mm.)	s (mm.)
Length						
CR, WP, C	8	43	—	84	4.1	.5
CSN, T	4	11	—	22	4	.5
NSN, K	6	22	—	43	3.7	.6
CR, WP, C—cf.	8	11	19	38	3.9	0
PC—cf.	1	2	4	8	3.8	0
CSN, T—cf.	2	3	4	8	3.5	0
NSN, K—cf.	2	2	5	10	3.9	0
CR, WP, C—gh.	8	10	15	30	4.1	0
PC—gh.	1	2	3	6	5.2	.1
CSN, T—gh.	2	3	3	6	3.3	1.1
NSN, K—gh.	4	5	12	26	4.3	0

Fig. 126. Fruits of *Sedum spathulifolium*, on left—ssp. *spathulifolium* from Burnt Mountain, Coos Co., Ore., and on right—ssp. *purdyi* from Stringtown, Butte Co., Calif., cultivated in greenhouse, Ithaca, N.Y., Nov. 23, 1966 (x 1.5). Photo by Howard Lyon.

Seeds (fig. 127) are pyriform, short-tailed at the broad end, finely longitudinally ribbed, and yellow-brown.

	n-pop.	n-pl.	n-rep.	n-obs.	x̄ (mm.)	s (mm.)
Length						
CR, WP, C	8	43	—	86	1.1	.13
CSN, T	4	11	—	22	1	.12
NSN, K	6	21	—	41	1.1	.02
CR, WP, C—cf.	8	11	19	32	1	.04
PC—cf.	1	2	4	8	.9	.04
CSN, T—cf.	1	1	1	1	1	—
NSN, K—cf.	2	2	4	6	1.1	0
CR, WP, C—gh.	8	10	14	27	1.1	.06
PC—gh.	1	1	1	1	1	—
CSN, T—gh.	1	1	1	2	1.1	—
NSN, K—gh.	5	7	9	17	1.1	.14
Diameter						
CR, WP, C	8	43	—	86	.4	.03
CSN, T	4	11	—	22	.4	.05
NSN, K	6	21	—	41	.4	.01
CR, WP, C—cf.	8	11	19	34	.36	0
PC—cf.	1	2	4	8	.4	0
CSN, T—cf.	1	1	1	1	.4	—
NSN, K—cf.	2	2	4	6	.4	0
CR, WP, C—gh.	8	10	14	27	.44	0
PC—gh.	1	1	1	1	.3	—
CSN, T—gh.	1	1	1	2	.5	—
NSN, K—gh.	5	7	9	17	.44	.09

Fig. 127. Seeds of *Sedum spathulifolium* ssp. *purdyi* from Stringtown, Butte Co., Calif., cultivated in greenhouse, Ithaca, N.Y., Jan. 4, 1967 (x 35). Photo by Howard Lyon.

Chromosomes are small and round, varying in size within the genome. Approximate average lengths of mitotic chromosomes measured by Barbara Joyce in 1970 are 1μ for a plant from Bandon and $.93\mu$ for a plant from Cottonwood Creek.

Number	n-pop.	n-pl.	g	sp	Cytologist
CR, WP, C	10	13	15	30	L. Hollingshead (1942); C. Horn, 1966; C. H. Uhl in Clausen and Uhl (1944); and I. Zandstra, 1969
PC	1	1	—	30	B. Joyce, 1970
CSN, T	5	11	15	—	C. Horn, 1966; C. H. Uhl in Clausen and Uhl (1944); and I. Zandstra, 1969
NSN, K	9	13	15	30	L. Hollingshead (1942); C. Horn, 1966; C. H. Uhl in Clausen and Uhl (1944); I. Zandstra, 1969; and B. Joyce, 1970

Uhl (1972) reported a plant which was $g = 14$ and another which was triploid. The latter (UC) occurred east of Mount Hamilton, Santa Clara Co., Calif.

Variation. *Sedum spathulifolium* ranges from 33° to 54° N. and in altitude from near sea level to 1,970 meters. It has adapted to a variety of habitats, including slopes of mountains along the Pacific Coast where fogs are frequent, sea cliffs within reach of spray of salt water, and the slopes of the inner Coast Ranges, Transverse Ranges, and Sierra Nevada, which are dry and hot in summer.

Study of plants in 25 populations in the wild and also in experiments at Ithaca, N.Y., has indicated features with significance for distinguishing populations or groups of populations on a regional basis. Of 33 morphological features which were studied, 27 exhibit significant differences either among regions or among populations within regions, but half of these features are unimportant because the differences disappear when plants are grown under similar conditions in experiments. Tests of significance are by means of analysis of variance. Regional groupings of populations are by geomorphic provinces, except that populations in the northern and central Sierra Nevada are grouped separately, and Klamath is treated by itself.

	Differences among regions			Differences among populations		
Feature	Wild plants	Cold frame	Greenhouse	Wild plants	Cold frame	Greenhouse
Primary rosettes						
Diameter	*	**	*	**	n.s.	n.s.
Offsets						
No. per rosette	n.s.	**	n.s.	*	n.s.	n.s.
Length	n.s.	*	n.s.	n.s.	n.s.	n.s.
Diameter of stems	**	n.s.	**	**	n.s.	n.s.
Floriferous stems						
No. per plant	**	n.s.	n.s.	n.s.	n.s.	n.s.
Length	n.s.	n.s.	n.s.	**	n.s.	n.s.
Leaves of rosettes						
Length	**	n.s.	n.s.	*	n.s.	n.s.
Width	*	n.s.	n.s.	**	n.s.	**
Thickness	n.s.	n.s.	*	**	n.s.	n.s.

	Differences among regions			Differences among populations		
Feature	Wild plants	Cold frame	Greenhouse	Wild plants	Cold frame	Greenhouse
Median lvs.—flor. stems						
Length	n.s.	n.s.	n.s.	*	n.s.	**
Thickness	n.s.	n.s.	n.s.	*	n.s.	n.s.
Cincinni						
No. per cyme	n.s.	n.s.	n.s.	**	n.s.	n.s.
Flowers						
Diameter	n.s.	*	*	*	n.s.	n.s.
Torus						
Length	n.s.	n.s.	n.s.	**	n.s.	n.s.
Sepals						
Length	n.s.	n.s.	*	**	n.s.	n.s.
Width	n.s.	n.s.	n.s.	**	n.s.	n.s.
Petals						
Number	n.s.	n.s.	n.s.	n.s.	n.s.	*
Length	n.s.	n.s.	n.s.	**	n.s.	n.s.
Width	n.s.	n.s.	n.s.	**	n.s.	n.s.
Stamens						
Length of adnation of epipetalous filaments	n.s.	n.s.	*	**	n.s.	n.s.
Nectaries						
Length	n.s.	n.s.	n.s.	**	n.s.	n.s.
Width	n.s.	n.s.	n.s.	**	n.s.	n.s.
Carpels						
Length	n.s.	n.s.	n.s.	**	n.s.	n.s.
Ovules per ovary (n)	n.s.	n.s.	*	*	n.s.	n.s.
Follicles						
Length	n.s.	n.s.	n.s.	**	n.s.	n.s.
Seeds						
Length	*	n.s.	n.s.	**	n.s.	*
Diameter	n.s.	*	n.s.	*	n.s.	n.s.

Examination of the preceding table reveals features for which differences among regions or among populations are significant, but does not make clear which regions or populations are different. Although multiple range tests, such as Tukey's, are useful for providing information about particular differences, unequal subsamples make their use here inappropriate. Instead, individual comparisons are necessary. Steel and Torrie (1960:113–114, eq. 7.17) have described an appropriate test, utilizing the error mean square and error degrees of freedom. Data in the following table, resulting from such tests, involved all populations of *Sedum spathulifolium* which were studied. Names of populations appear in the left margin and their initials in the headings of the vertical columns, for example, C.R. = Coos Ridge. Figures indicate the number of significant differences between each pair of populations. Despite horizontal abridgement of the table, all combinations of populations showing significant differences are included. The only exceptions to the preceding statement are the following: Lummi Rocks differs significantly from 5 other populations, Wildwood from 4, Dartmouth Creek from 3, and Feather River Canyon from 1. Abbreviations for headings of columns may be deciphered by noting that localities are in reverse order from the left column, with the addition of Coos Ridge (CR) for the first column of figures.

Population	CR	P	Be M CA	BM	DB	MH C	YG	CC	Be M O	W FC	K	S	B	LI	MB C	HH	HG	Σ r
Wildwood	3	1	0	0	0	0	0	0	0	1	0	1	2	0	0	1	0	9
Lummi Rocks	5	5	4	3	3	4	1	2	2	2	1	1	1	3	1	0	1	39
Dartmouth Creek	5	4	4	3	1	4	3	1	3	1	1	1	5	3	0	1	1	41
El Portal	5	5	4	4	1	3	2	3	3	2	3	1	2	1	2	0	1	42
Feather R. Can.	11	15	3	3	2	2	1	1	4	4	2	1	4	1	1	2	1	58
Spanish Creek	6	7	6	4	3	5	3	4	3	1	3	2	5	2	2	2	0	58
Keddie	6	8	5	4	1	5	3	4	3	3	3	2	4	1	4	2	0	58
Rock Creek	5	7	5	5	4	5	3	5	5	2	4	1	5	2	3	0	1	62
O'Leary Mt.	13	14	5	6	3	7	3	4	4	1	2	1	5	3	2	0	4	77
Hopland Grade	6	5	2	2	2	5	5	1	5	5	1	0	6	1	5	2	—	53
Hetch Hetchy	4	5	2	2	2	4	2	2	1	1	1	1	2	2	1	—	—	32
Mid Brummit Cr.	14	12	1	5	5	3	3	1	2	3	0	1	6	1	—	—	—	57
Lummi Island	6	2	3	2	0	3	1	3	1	1	2	2	2	—	—	—	—	28
Bandon	10	12	6	7	5	6	5	6	5	5	4	5	—	—	—	—	—	76
Stringtown	3	4	2	0	2	3	2	1	0	2	0	—	—	—	—	—	—	19
Klamath	3	2	0	1	3	2	3	1	1	4	—	—	—	—	—	—	—	20
W. of Forsee Cr.	11	18	7	7	5	3	3	7	4	—	—	—	—	—	—	—	—	65
Bear Mt., Ore.	13	15	1	3	5	6	4	2	—	—	—	—	—	—	—	—	—	49
Cottonwood Cr.	2	5	1	2	2	4	1	—	—	—	—	—	—	—	—	—	—	17
Yosemite Gorge	12	13	3	6	4	8	—	—	—	—	—	—	—	—	—	—	—	46
Mountain Home Cr.	9	13	4	4	4	—	—	—	—	—	—	—	—	—	—	—	—	34
Deer Butte Tr.	7	14	4	5	—	—	—	—	—	—	—	—	—	—	—	—	—	30
Burnt Mt.	5	8	1	—	—	—	—	—	—	—	—	—	—	—	—	—	—	14
Bear Mt., Cal.	1	4	—	—	—	—	—	—	—	—	—	—	—	—	—	—	—	5
Pinnacles	7	—	—	—	—	—	—	—	—	—	—	—	—	—	—	—	—	7
Σ(columns)	172	198	73	78	57	82	48	48	46	38	27	20	49	20	21	10	9	—
Totals (r c)	172	205	78	92	87	116	94	65	95	103	47	39	125	48	78	42	62	—

Using the total number of significant differences between a population and other studied populations as a measure of distinctness, that is the addition of the sums of rows and columns in the above table, the four most distinctive populations are Pinnacles, Coos Ridge, Bandon, and Mountain Home Creek. These four populations belong to three different regional groups: the first two to the subspecies of the Coast Ranges, Bandon to the coastal subspecies, and Mountain Home Creek to the subspecies of the central and southern Sierra Nevada and Transverse Ranges. Numbers of differences shown in the table are minimal figures. Since some populations could not be studied at all stages of development, more differences may exist than are shown. The pattern of variation among populations is complex. Since many differences observed among wild plants result from environmental modifications, experiments were necessary to distinguish between the genetic and environmental differences.

A result of the experimental culture of plants in Ithaca has been the discovery that many differences observed in the field are environmental modifications without genetic basis. The few differences which are heritable permit distinction of populations on a regional basis as shown in the following two tables, the first based on results from the experiment in the cold frame and the second based on data from the greenhouse. Figures indicate the number of significant differences between regions.

Cold frame	Cent. S.N.	N.S.N.	Puget Tr.	Oreg. C.	P.C.R.
Cascade Mts.	2	2	0	1	0
Pacific Coast Ranges	2	3	1	2	
Oregon Coast	2	1	1		
Puget Trough	1	1			
Northern Sierra Nevada	0				

Greenhouse	Klam. Mts.	Oreg. C.R.	Cent. S.N.	N.S.N.	Calif. C.R.	Calif. C.
Oregon Coast	3	3	2	1	1	0
Calif. Coast	1	1	2	1	0	
Calif. Coast Ranges	1	1	3	1		
N. Sierra Nevada	0	0	0			
Cent. Sierra Nevada	0	0				
Oregon Coast Ranges	1					

Fig. 128. Plant of *Sedum spathulifolium* ssp. *pruinosum* from Bandon, Coos Co., Ore., cultivated in greenhouse, Ithaca, N.Y., June 2, 1964 (x .6). Photo by Howard Lyon.

NATIVE SPECIES 457

Plants from Bandon on the Oregon Coast (fig. 128) appear distinctive in all evaluations. They most resemble plants from the California Coast, and those in turn seem intermediate between the Oregon plants and populations elsewhere in the Coast Ranges and Cascade Mountains. Another distinction appears between the populations of the Sierra Nevada and those of the Pacific Coast Ranges, Puget Trough, and Cascade Mountains, but separations are not uniformly good. Although plants from the Klamath Mountains most resemble those of the Sierra Nevada, they are somewhat intermediate between them and populations of the Coast Ranges.

Information from the general regional comparisons favored putting together all populations from the Sierra Nevada and Klamath Mountains, interpreting them as a single subspecies. Yet they looked and behaved differently in experiments. A crucial test was possible from the experiment in the greenhouse. If plants of the central Sierra Nevada, *S. yosemitense* (fig. 129), and those of the northern Sierra Nevada and Klamath Mountains, *S. purdyi* (fig. 130), are the same, then measurements of the stem of the longest offset on each replicate of plants of the two kinds ought to reveal no difference. Instead, the difference in diameter is highly significant and in length significant. The experiment was of completely random design and with plants of the

Fig. 129. Plant of *Sedum spathulifolium* ssp. *yosemitense* from Panorama Point, Yosemite Gorge, Mariposa Co., Calif., cultivated in greenhouse, Ithaca, N.Y., Dec. 23, 1964 (x .8). Photo by Howard Lyon. Note relatively short, stout stems of offsets and also green leaves.

Fig. 130. Plant of *Sedum spathulifolium* ssp. *purdyi* from along Rock Creek, Plumas Co., Calif., cultivated in greenhouse, Ithaca, N.Y., Dec. 23, 1964 (x .8). Photo by Howard Lyon. Note long, slender stems of offsets.

same age. Date of observations was Feb. 19, 1970. Testing was by means of the Wilcoxon rank sum test.

	yosemitense	*purdyi*
Plants (*n*)	7	11
Replicates (*n*)	13	20
Stems of offsets—length (cm.)*	3.4	4.6
—diameter (mm.)**	2	1.6

Since other students have noted differences between these two kinds of plants and names are available, their continuation as taxonomic subspecies is reasonable.

Physiological features probably are most important in distinguishing the regional groups of populations of *Sedum spathulifolium*. Differences in vigor and survival in experiments at Ithaca, N.Y., have seemed impressive. Yet, objective tests of independence have yielded less impressive results. Chi-square is 5.85 with probability of .22 for independence of survival in the experiment in the greenhouse. For the experiment in the cold frame, $\chi^2 = 14.79$ with P $= < .025$, suggesting real differences. Testing procedure is according to Snedecor (1956:227–228).

Source	Greenhouse N	Survival 1 yr. %	4 yrs. %	Cold frame N	Survival 3 yrs. %	Garden N	Survival 1 yr. %
Sea cliffs	4	100	100	4	100	2	50
Coast Ranges	26	84	69	17	82	6	0
Cascade Mts.	—	—	—	8	37	—	—
Klamath Mts.	9	100	55	6	50	4	0
C. Sierra Nevada	13	92	77	8	37	4	0
N. Sierra Nevada	24	75	50	12	25	6	0
Totals	76			55		22	

Another physiological difference involves time of flowering. Plants from the Sierra Nevada and Klamath Mountains flower earlier than those from the Pacific Coast Ranges and Cascade Mountains. This condition is confirmed by a test of independence for plants in the experiment in the greenhouse in 1969, $\chi^2 = 5.45$, $P = < .025$. Both flowering time and ability to survive under different conditions of moisture appear to be important regional differences.

Still another regional difference has to do with the waxiness of the leaves. The young leaves of many plants are glaucous, but the wax wears off with age, with the result that older leaves appear green. In some populations, especially along the coast, the wax is thick and persists through the life of the leaf. Experiments in Ithaca confirm the idea that the presence or absence of a bloom has a genetic basis and also is associated with geographic distribution. For the experiment in the greenhouse, testing the hypothesis of no association with distribution in the Coast Ranges or Sierra Nevada, $\chi^2 = 52.28**$, and for the experiment in the cold frame, $\chi^2 = 18.85**$. Plants in the Sierra Nevada and San Bernardino Mountains have green leaves. The only exception in my experience is the population along Spanish Creek in the northern Sierra Nevada. Leaves of plants in the Klamath Mountains are either green or glaucous. Those of plants in the Pacific Coast Ranges and Cascade Mountains usually are glaucous. Plants with the maximum wax on leaves are on cliffs by the sea. For purposes of identification, degree of waxiness is a practical feature, but not always dependable because the wax wears off and exceptions occur.

Most plants in cultivation in Ithaca continue either to have a bloom or not, depending on their tendency in the wild. Exceptions include plants from Burnt Mountain along Middle Brummit Creek in Oregon. These plants were shaded in nature and had either green or glaucous leaves. In cultivation, their leaves are glaucous. Occasional leaves are green on plants which otherwise are strongly glaucous or pruinose.

Petals per flower vary from 4 to 7 (fig. 131). Five is the prevailing number. Only a small fraction of flowers on certain plants depart from this pattern. Further, experience of several years with plants in experiments reveals that the tendency to produce flowers with other than 5 petals varies from season to season. The frequency of different numbers of petals per flower in a sample of 320 flowers examined in both the wild and experiments is indicated in the following table.

Fig. 131. Flower with seven petals of *Sedum spathulifolium* ssp. *spathulifolium* from along Middle Brummit Creek, Coos Co., Ore., cultivated in greenhouse, Ithaca, N.Y., June 6, 1969 (x 4.5). Photo by Howard Lyon.

Petals per flower (n)	Flowers (n)	Flowers %	Cumulative %
4	8	2.5	2.5
5	294	91.9	94.4
6	17	5.3	99.7
7	1	.3	100.0
Total	320		

Petals of *Sedum spathulifolium* vary from pale canary yellow to lemon yellow. Part of the variation is genetic because plants growing under similar conditions of light may have flowers of different hues, for example on Coos Ridge in western Oregon. Other variation, especially with regard to tint, may be influenced by the intensity and duration of the light. Plants with aureolin petals at the Pinnacles National Monument, Calif., had empire yellow petals in the greenhouse at Ithaca. The yellowest flowers observed in the wild were on plants of ssp. *pruinosum* growing on cliffs by the Pacific Ocean at Bandon, Ore. The palest flowers—very pale canary yellow almost white—were on a plant on Coos Ridge. Aureolin is the commonest color. Buds may be yellower than expanded petals. Likewise, petals of plants in exposed situations may be yellower than those in shade. On withering or drying, petals may lose pigment and become white. For this reason, persons working with herbarium specimens may think that petals are white, for example, Jepson in describing *S. purdyi* and Britton in describing *S. californicum*. No truly white petals on living plants have come to my attention.

Of the 130 plants which have been the principal basis for study of variation of *Sedum spathulifolium*, counts of chromosomes are available for 30 plants from 16 localities. All plants investigated have either a gametophytic number of 15 or a sporophytic number of 30 chromosomes. Variation in morphology and physiology within this species results from genic changes and not from differences in whole chromosomes or sets of chromosomes. Natural selection, working upon the products of gene mutation, is the explanation of the evolution of the intraspecific variation. Since plants with other than 15 pairs of chromosomes have not appeared in my random samples, the inference is that they must be rare in nature. Because of the small size of the chromosomes and the difficulty of separation, the chance for error in counting is great.

Important objectives of the study of variation of *Sedum spathulifolium* have been to separate environmental modifications from features which have a genetic basis and also to note the extent to which plants are modified. Slender plants, with thin green leaves, occur in shaded situations, as in deep canyons in the southern part of the distributional area, where as little as a quarter of the potential sunlight is available. Such sites are cooler and moister than adjacent, exposed slopes. In contrast, plants on rocks by the shore of the Pacific Ocean are robust, with thick, pruinose leaves. Although environmental modifications account for part of the observed differences in nature, the principal differences may be heritable. An experiment with 5 plants, each from a different locality, namely Bandon, Burnt Mountain, Coos Ridge, Rock Creek, and Yosemite Gorge, revealed no significant differences in responses of vegetative features in three circumstances of growth—the wild, a cold frame in Ithaca, and a greenhouse in Ithaca. Data in the following table are average responses of the 5 plants in each situation. Interpretation of results is by means of the Wilcoxon (1964) rank-test for more than two treatments and a two-way classification. All differences may be attributed to chance. If conditions of cultivation are intermediate between environmental extremes in nature, the result is reasonable. A plant from a wet situation, as Coos Ridge, has the largest leaves in nature. One from a dry site, with smaller leaves in nature, has larger leaves in cultivation. The different plants become most alike in cultivation.

Feature	Wild	Cold frame	Greenhouse
Diameter of plants (cm.)	47	14	18
Leaves of rosettes			
Length (mm.)	17	16	15
Width (mm.)	8	8	8
Thickness (mm.)	2.5	2	1.7
Offsets			
Diameter of stems (mm.)	1.8	1.4	1.6

A highly significant difference in average diameter of plants in the Coast Ranges as compared with the Sierra Nevada may have several explanations. The humid Coast Ranges are a favorable place for growth of seedlings. With more seedlings, competition increases. Likewise, higher humidity favors fungi which cause rot and death of *Sedum*. A result may be smaller plants. In contrast, the drier circumstances in summer of the interior mountains are poor for the survival of seedlings and also for pathogens. Consequences may be that plants of *Sedum* live longer, have less competition, and become larger in the Sierra Nevada, Klamath Mountains, and Transverse Ranges.

In summary, a series of populations of *Sedum spathulifolium*, with certain common characteristics, occurs in the Pacific Coast Ranges. These plants usually have glaucous, moderately thick leaves. The type specimen of the species probably belongs to this series which may be called ssp. *spathulifolium*. On cliffs by the Pacific Ocean, distinctive plants occur with pruinose leaves, compact rosettes, and small flowers. The type specimen of *S. pruinosum* is close enough to this concept to include in the coastal subspecies, which then may be named ssp. *pruinosum*. Another distinctive series of populations occurs at low and middle elevations in the central Sierra Nevada and Transverse Ranges. These plants have green leaves and offsets with robust stems. They constitute ssp. *yosemitense*. Finally, in the northern Sierra Nevada and Klamath Mountains, populations occur with the leaves of the rosettes closely compacted and the offsets with long, slender stems. These comprise ssp. *purdyi*.

A key is a sort of synoptic summary of data on variation. For *Sedum spathulifolium*, it is an oversimplification. The quantitative data of the whole description provide a detailed basis for distinguishing subspecies and subspecific identification. What follows is a condensation of the data of the description and the information on variation. Dimensions in the key are averages for the expressions in nature and the experiments in Ithaca. For fuller exposition, see the figures in the description.

KEY TO SUBSPECIES

A. Primary rosettes 2.5–3.7 cm. in diameter, with leaves glaucous or green and 1.4–2.1 mm. thick; flowers 12–16 mm. in diam. Plants of diverse habitats .... B
    B. Leaves of rosettes usually glaucous; offsets per rosette 1–2; stems of offsets 1.3–1.9 mm. in diameter. Plants of the Pacific Coast Ranges, Willamette–Puget Trough, and Cascade Mountains .... ssp. *spathulifolium*, p. 462
    BB. Leaves of rosettes normally green, without waxy bloom; offsets per rosette 2–4; stems of offsets 1.2–2 mm. in diameter. Plants of the Sierra Nevada, Klamath Mountains, and Transverse Ranges C
        C. Offsets with stems 1.4–2 mm. in diameter and 1.9–3.4 cm. long; leaves of rosettes not closely compacted. Plants of the central and southern Sierra Nevada and Transverse Ranges .... ssp. *yosemitense*, p. 467
        CC. Offsets with stems slenderer, 1.2–1.6 mm. in diameter and 1.6–4.6 cm. long; leaves of rosettes closely compacted. Plants of the northern Sierra Nevada and Klamath Mountains .... ssp. *purdyi*, p. 468
AA. Primary rosettes 1.5–2.1 cm. in diameter with leaves pruinose, 2.1–2.5 mm. thick; flowers 10–12 mm. in diameter. Plants of cliffs by the sea .... ssp. *pruinosum*, p. 465

Sedum spathulifolium ssp. *spathulifolium*

Nomenclature. *Sedum spathulifolium* Hooker, Flor. Bor. Am. 1:227 (1834). Type locality: dry rocks of the Columbia River somewhere between Vancouver and the Pacific Ocean. Type: a collection of Douglas (BM), May 10, 1825. Hooker's description of the leaves and petals, as well as a photograph of the type specimen, kindly provided by the Keeper of Botany of the British Museum, make identification certain. The date of collection is from the journal of Douglas. Data on the sheet with the type specimen are "rocky banks of the Columbia near the ocean, 1825."

Synonyms are:

1903. *Gormania anomala* Britton, Bull. N.Y. Bot. Gard. 3:30. Type locality: sandy hills in path of strong daily sea winds, San Luis Obispo Co., Calif. Type: a collection of Mrs. R. W. Summers (NY), June 1883. Evidence favoring the listing of *G. anomala* in the synonymy of ssp. *spathulifolium* is as follows: 1—Britton described it as light green, slightly pinkish pruinose; 2—the type specimen has a whitish aspect on the lower part of a floriferous stem and at the base of a cyme, possible evidence of a glaucous condition; 3—Rose identified as *Sedum anomalum* plants with glaucous leaves which are ssp. *spathulifolium* from Los Gatos Canyon, Santa Clara Co., Calif.; 4—Britton included in his concept of *S. anomalum* all plants of *S. spathulifolium* from San Luis Obispo Co. to Santa Clara Co., Calif.; and 5—*G. anomala* more likely belongs to the ecotype of *S. spathulifolium* in the Coast Ranges than to an ecotype of the foothills and lower slopes of the Sierra Nevada.

1904. *Cotyledon anomala* (Britton) Fedde, Just's Bot. Jahresb. 31 (1):827. Based on *Gormania anomala* Britton.

1905. *Sedum anomalum* Britton, No. Am. Flora 22:72. Based on *Gormania anomala* Britton.

1905. *Sedum Woodii* Britton, No. Am. Flora 22:73. Type locality: Oregon City, Ore. Type: a collection of A. Wood (NY), 1866.

1941. *Sedum spathulifolium* var. *minus* Henderson in M. E. Peck, Man. High. Plants Oregon, p. 341. Type locality: rocks, Snow Camp Lookout, alt. 1,525 m., Curry Co., Ore. Type: a collection of Mr. and Mrs. J. R. Leach, no. 2,292 (ORE), June 24, 1929.

1944. *Sedum spathulifolium* ssp. *anomalum* (Britton) Clausen et Uhl, Madroño 7:174. Based on *Gormania anomala* Britton as to type and name, but not as to principal concept.

Doubtful synonyms are:

1903. *Sedum californicum* Britton, Bull. N.Y. Gard. 3:44. Type locality: unknown, although published as the north side of Mount Shasta, which certainly is wrong. Type: H. E. Brown 336 (NY). Fröderström (1930–1935) published a photograph of the type specimen, plate 74. The shape of the rosulate leaves suggests ssp. *spathulifolium*, but the color of the petals, which have faded to white, suggests ssp. *purdyi*. The robustness of the specimens and the pruinose condition suggest ssp. *pruinosum*. The precise identity of these specimens must remain a mystery since we are unable to restore them to life and to study them satisfactorily. A. A. Heller, who knew Brown, wrote to W. B. Cooke in 1941 that Brown was at one time at Mendocino City on the coast and also that he labeled as from Mount Shasta plants which he had collected on Mount Eddy.

1921. *Sedum spathulifolium* var. *purpureum* Praeger, Jour. Roy. Hort. Soc. 46:239. Type locality: unknown. Type: a plant cultivated by R. L. Praeger (whether or not preserved is uncertain, but not at Dublin or Glasnevin according to A. Brady at the National Botanic Gardens, Glasnevin). Distinctive features, according to Praeger, are the deep purple leaves, which are white and mealy when young, and the large inflorescences.

Distribution. *Sedum spathulifolium* ssp. *spathulifolium* (fig. 132) occurs in the Pacific Coast Ranges from near San Luis Obispo, Calif., northward to the valley of the Skeena River in British Columbia; in the Willamette–Puget Trough of Oregon and Washington; and on the western slope of the Cascade Mountains in Oregon, Washington, and southern British Columbia. It occurs up the Columbia River to Hood River on the southern side and to the Big White Salmon River on the northern side. A specimen collected by Albert R. Sweetser (ORE) at Castle Rock, Morrow Co., Ore., if data are correct, may indicate that the species extends farther up the river into the Columbia Plateau. It also occurs up the valley of the Fraser River, on the Interior Plateau of British Columbia, as far as Alexandria Bridge.

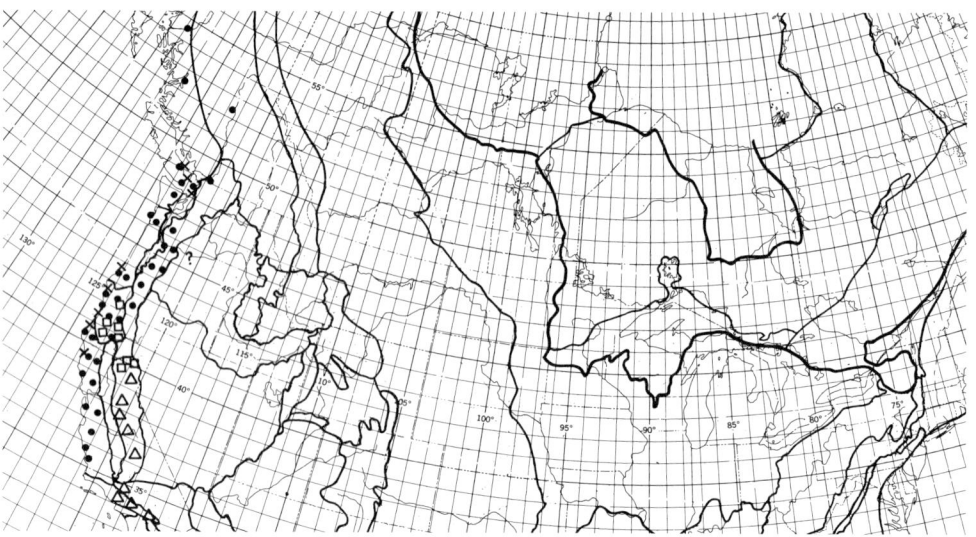

Fig. 132. Known distribution of *Sedum spathulifolium* ssp. *spathulifolium* (●), ssp. *pruinosum* (x), ssp. *yosemitense* (△), and ssp. *purdyi* (□).

Plants of ssp. *spathulifolium* grow in shaded situations on cliffs and rocky slopes at elevations from near sea level to 1,590 meters. Usual habitats are in mats of moss and in shallow soil in crevices on sandstone, andesite, and rhyolite breccia. Flowering time is from April 1 to August 1. More than 156 populations are known.

I have studied ssp. *spathulifolium* at eight localities and in addition have made brief observations on cliffs along the Hopland Road in the Mayacamas Mts.,

Mendocino Co., Calif.; cliffs 11 km. south of Klamath, Calif.; Coos Ridge, Coos Co., Ore.; and Lummi Island and the Lummi Rocks in Puget Sound, Whatcom Co., Wash. Data follow for the eight populations studied in detail.

Population	Sampling unit	Plants (N)	Area of population (m.^{2})	Density per (m.^{2})	Exposure
Pinnacles	—	>92	—	—	E–W
Bear Mountain (Calif.)	—	520	800	.02	N
Burnt Mountain	43–32	100	1,300	.08	NE–ENE
Middle Brummit Creek	43–30	186	75	2.48	SE
Bear Mountain (Ore.)	—	454	315	1.44	N–NE
Dartmouth Creek	437	101	100	1.11	S–WSW
O'Leary Mountain	2,850	48e	3,715	.01	NE
Deer Butte Trail	4,421	13,022e	6,000	2.17	ESE–SW

Population	Altitudinal range (m.)	pH, range	pH, tests (n)	Oldest record
Pinnacles	405–450	5.2–5.4	4	1921
Bear Mountain (Calif.)	420	6.4	1	1963
Burnt Mountain	680	5.6	2	1963
Middle Brummit Creek	630	5.6–6.4	4	1963
Bear Mountain (Ore.)	952–965	5.4–6.4	4	1963
Dartmouth Creek	630–670	5.4–5.6	4	1965
O'Leary Mountain	1,590	5.6	1	1965
Deer Butte Trail	1,140–1,229	5.2–6	14	1942

Other vegetation is sparse where *Sedum spathulifolium* occurs. Often the plants are rooted in moss, as at the Pinnacles, on Bear Mountain (Ore.), and along Dartmouth Creek. *Montia parvifolia* grew by all plants studied in two randomly selected tracts along Middle Brummit Creek. Otherwise, competitors were miscellaneous: a composite at the Pinnacles; *Whipplea modesta* and *Gaultheria shallon* along Middle Brummit Creek; *Berberis aquifolium*, *Heuchera micrantha*, and *Symphoricarpos* on the summit of Bear Mountain (Ore.); *Sedum oregonense* and *Achillea lanulosa* on O'Leary Mountain; and *Selaginella wallacei* and *Melica harfordii* along Dartmouth Creek. Along the old Deer Butte Trail, closest competitors of 14 randomly selected plants were moss by 5 plants, *Selaginella wallacei* by 4, *Gilia capitata* by 3, *Amelanchier florida* by 1, and *Agrostis exarata* by 1.

Sedum spathulifolium ssp. *pruinosum*

Nomenclature. *Sedum spathulifolium* ssp. *pruinosum* (Britton) Clausen et Uhl, Madroño 7:172 (1944). Based on *S. pruinosum* Britton.
Synonyms are:
1905. *Sedum pruinosum* Britton, No. Am. Flora 22:72. Type locality: Crescent City, Calif. Type: a collection of Alice Eastwood (NY), 1903, which flowered at the New York Botanical Garden on May 8, 1904. Fröderström (1930–1935, pl. 77, fig. 2) published a photograph of the type specimen.

1966. *Sedum spathulifolium* var. *pruinosum* (Britton) Boivin, Nat. Canad. 93:646. Based on *S. pruinosum* Britton.

Two features, namely the pruinose condition and compactness, distinguish ssp. *pruinosum* (fig. 133).

Distribution. *Sedum spathulifolium* ssp. *pruinosum* (fig. 132) occurs only by the seaside from the Mattole River south of Cape Mendocino, Calif., northward to Vancouver Island. The plants grow from just above the level of high tide up to a

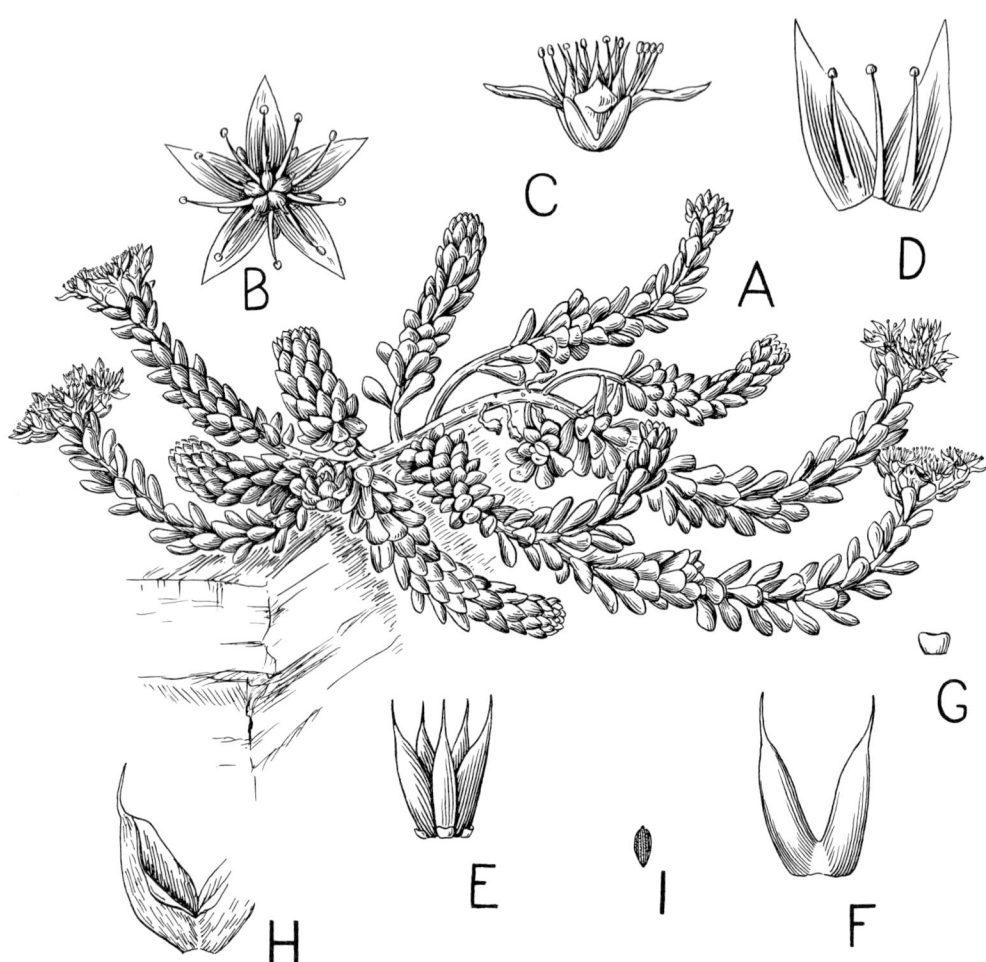

Fig. 133. Plant of *Sedum spathulifolium* ssp. *pruinosum* from Bandon, Coos Co., Ore., cultivated in greenhouse, Ithaca, N.Y., June 22, 1966. A. Habit sketch (x .57). B. Flower from above (x 2.3). C. Flower from side (x 2.3). D. Two petals and stamens (x 3.4). E. Carpels (x 3.4). F. Two carpels and nectaries (x 4.6). G. Nectary (x 5.7). H. Follicle (x 4.6). I. Seed (x 5.7).

maximum altitude of 30 meters. Twenty-nine populations are known. Flowering time is from mid-May until late July. I studied ssp. *pruinosum* on cliffs south of Crescent City, Calif., on July 24, 1940, and on the bluffs at Bandon, Ore., on July 14, 1963. Data for the population at Bandon are summarized in the following brief table. Plants there were rooted in sand over gray sandstone. Competing species included *Plantago maritima* ssp. *juncoides* and *Fragaria chiloensis*.

Population	Plants (N)	Exposure	Altitudinal range (m.)	pH, range	pH, tests (n)	Oldest record
Bandon	33e	W	3–5	6.4–6.6	2	1965

The earliest collection of ssp. *pruinosum* which has come to my attention is by John Macoun (NY) from Cedar Hill, Vancouver Island, in May 1887.

Sedum spathulifolium ssp. *yosemitense*

Nomenclature. *Sedum spathulifolium* ssp. **yosemitense** (Britton) comb. nov., fundatum super *Sedum Yosemitense* Britton.

A synonym is:

1903. *Sedum Yosemitense* Britton, Bull. N.Y. Bot. Gard. 3:44. Type locality: between Vernal and Nevada Falls, Yosemite National Park, Calif., alt. 1,676 m. Type: a collection of H. M. Hall and E. B. Babcock no. 3425 (NY), July 1902. The specimens of the type collection are in late-flowering condition.

A doubtful synonym is:

1921. *Sedum spathulifolium* var. *majus* Praeger, Jour. Roy. Hort. Soc. 46:238, fig. 138, *a*. Type locality: unknown. Type: a plant cultivated by R. L. Praeger (whether or not preserved is unknown). Described as with rosettes twice as large as in typical *S. spathulifolium* and with green leaves which are scarcely glaucous.

Since the type of *Sedum anomalum* belongs to ssp. *spathulifolium*, that name cannot apply to this subspecies.

Distribution. *Sedum spathulifolium* ssp. *yosemitense* (fig. 132) occurs on the eastern slope of the Sierra Nevada, northward to the drainage of the American River, and in canyons in the Transverse Ranges of California. The plants grow on rocky slopes and cliffs at elevations from 430 to 2,286 meters. Usual habitats are open situations where other vegetation is sparse and which are shaded for part of the day. The plants grow in crevices, gravel, soil, or mats of moss on granite, quartz monzonite, or slate. Drainage varies from excessive to moderate. Flowering time is from April 30 to July 4, with the principal flowering in May and June.

Twenty-seven populations of ssp. *yosemitense* are known. I have studied the type population in the Yosemite Gorge between Vernal and Nevada Falls, and also two populations in the San Bernardino Mountains, one in the canyon of the East Fork of Mountain Home Creek and the other on the slope west of Forsee Creek. Details concerning these populations are in the following table. Further information about the sampling area in Yosemite Gorge is in the chapter on geography.

Population	Plants (N)	Area (m.²)	Density per m.²	Exposure
Yosemite Gorge	24,233e	790,000	.031	N–WNW
Mountain Home Creek	592e	80	7.4	NW–SE
West of Forsee Creek	72e	18	4	N

Population	Altitudinal range (m.)	pH, range	pH, tests (n)	Oldest record
Yosemite Gorge	1,480–1,970	5.0–6.4	12	1865?
Mountain Home Creek	1,560–1,580	6.6–6.8	6	1968
West of Forsee Creek	1,605–1,615	6.4–6.6	3	1968

Additional to the populations for special attention, I have studied plants at the following localities: Cascade Canyon, San Gabriel Mountains; El Portal, along the Merced River; and near Mather on the southern slope of the Hetch Hetchy Valley.

The commonest close competitors of plants of ssp. *yosemitense* in Yosemite Gorge, out of 27 species of vascular plants occurring by plants in the sample, are *Heuchera micrantha*, a species of *Poa*, and *Ribes roezlii*. *Orobanche uniflora* sometimes is a root parasite. Commonest competing species in the San Bernardino Mountains are *Quercus wislizenii*, *Q. chrysolepis*, *Festuca californica*, *Bromus* sp., *Ribes nevadense*, and *Polystichum munitum*. Species shading plants of ssp. *yosemitense* in the Yosemite Gorge include *Pseudotsuga menziesii*, *Libocedrus decurrens*, and *Abies concolor*, and in the San Bernardino Mountains include *Quercus wislizenii*, *Q. chrysolepis*, *Pseudotsuga macrocarpa*, and *Libocedrus decurrens*.

Sedum spathulifolium ssp. *purdyi*

Nomenclature. *Sedum spathulifolium* ssp. **purdyi** (Jepson) comb. nov., fundatum super *Sedum Purdyi* Jepson.

A synonym is:

1936. *Sedum Purdyi* Jepson, Fl. Calif. 2:110. Type locality: Etna Mills, Siskiyou Co., Calif., in the Klamath Mountains. Type: a flowering specimen sent to Jepson by Carl Purdy, no. 2,848 (JEPS), June 14, 1920. The type, consisting of a floriferous shoot and a single rosette with four offsets, had been cultivated at the Terraces, Ukiah, Calif. Jepson called the species Ray Sedum. Two earlier publications of *S. Purdyi*, without adequate description, were by Butterfield, Desert Plant Life 8:7 (1936), and Purdy, in his printed catalogs of perennial plants, beginning with the issue for 1920–1921.

The problem with ssp. *purdyi* (fig. 134) has been whether to give it some taxonomic status or to include it in the synonymy of ssp. *yosemitense*. The demonstration of significant differences in diameter and length of stems of offsets strengthens the case for taxonomic recognition. The ability to cross with ssp. *spathulifolium* and to produce some good seeds, a circumstance to be discussed under reproduction, favors subspecific status within *Sedum spathulifolium*.

Distribution. The distributional area of ssp. *purdyi* (fig. 132) is in the Klamath Mountains of northern California and extreme southern Oregon and in the drainage

NATIVE SPECIES 469

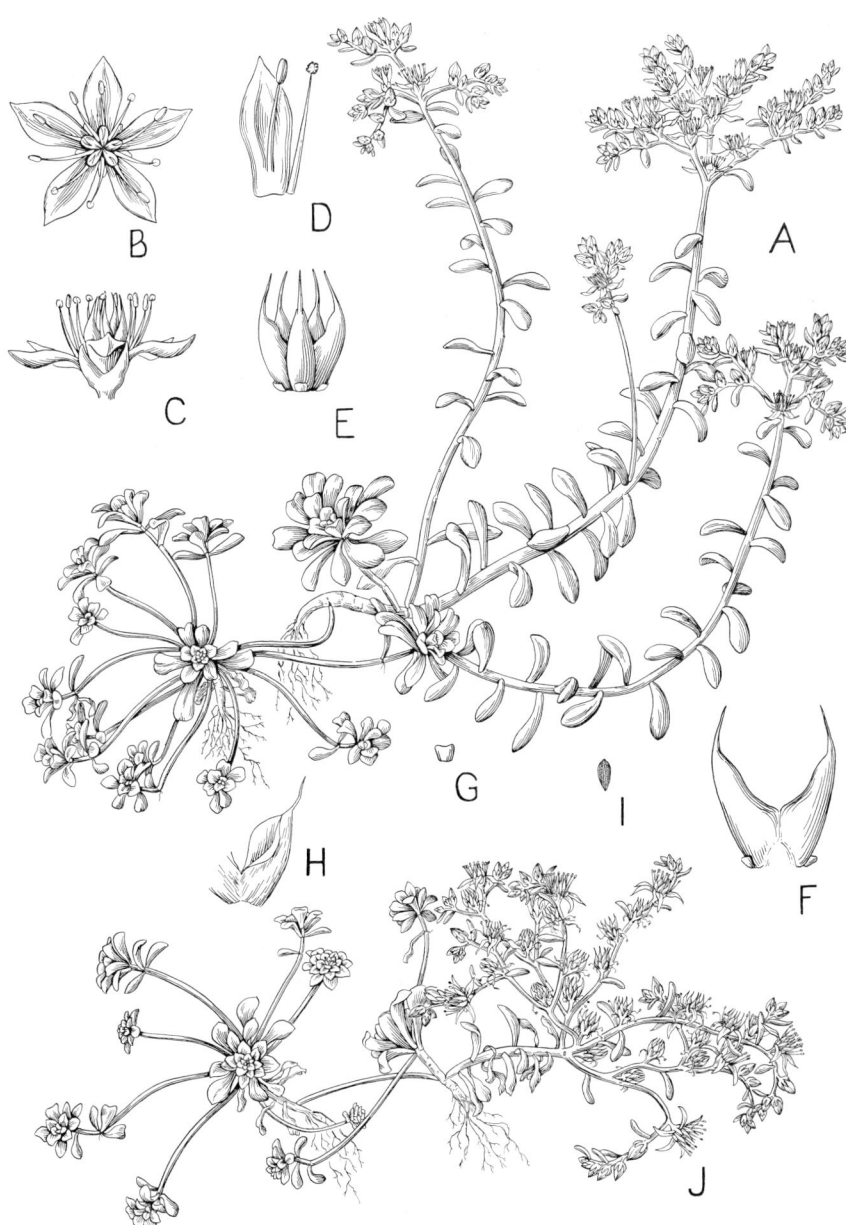

Fig. 134. Plants of *Sedum spathulifolium* ssp. *purdyi*, A–I from along Cottonwood Creek, Shasta Co., Calif., and J from Wildwood, Trinity Co., Calif., cultivated in greenhouse, Ithaca, N.Y., Jan. 1964. A. Habit sketch (x .39). B. Flower from above (x 1.6). C. Flower from side (x 1.6). D. Petal and two stamens (x 2.3). E. Carpels (x 2.3). F. Two carpels and nectaries (x 3.1). G. Nectary (x 3.9). H. Follicle (x 3.1). I. Seed (x 3.9). J. Habit sketch (x .39).

of the Feather River in the northern Sierra Nevada. The plants grow in partial shade in thin mats of moss and soil or in gravel on rocky slopes and cliffs of granite or slate. Drainage in these situations is moderate to good. Flowering time is from April 20 to June 9.

Twenty-five populations of ssp. *purdyi* have come to my attention. I have studied plants at five localities and in addition have made brief observations northwest of Wildwood and along Canyon Creek, 1.6 km. from the Scott River and 19 km. south of Hamburg, in the Klamath Mountains, and in the canyon of the Feather River near the Rock Creek Power Dam. Further information about sites for study in the Sierra Nevada is in the chapter on geography. Data concerning the populations for special attention are in the following table.

Population	Plants (N)	Area (m.²)	Density per m.²	Exposure
Stringtown	345	116e	2.97	N
Rock Creek	197	5,810e	.03	N–NE
Keddie	1,252	—	—	N–NNE
Spanish Creek	462e	2,020e	.23	N–NNW
Cottonwood Creek	151e	27	5.59	N–NNW

Population	Altitudinal range (m.)	pH, range	pH, tests (n)	Oldest record
Stringtown	390	5.2	1	1964
Rock Creek	580–650	5.0–5.6	9	1961
Keddie	1,050–1,080	5.2–5.4	2	1964
Spanish Creek	1,065–1,080	5.2–6.2	4	1961
Cottonwood Creek	445–460	6.6	1	1963

Few other vascular plants grow at the sites inhabited by ssp. *purdyi*. Competing plants are miscellaneous, with no one species a consistent competitor in all populations. Plants occurring in association with ssp. *purdyi* include *Polypodium vulgare*, *Calycanthus occidentalis*, *Rhus diversiloba*, and species of *Selaginella* and *Diplacus*. Occasionally, the *Sedum* is without close competition.

The oldest collection of ssp. *purdyi* which I have seen dates from 1902, F. W. Staunton 15,157 (NY). It consists of two sheets, one with drawings and the other with specimens, and is labeled as from Redding, Calif.

Reproduction. The offsets of *Sedum spathulifolium* are an easy and effective means of vegetative reproduction. These may take root while still attached to the parental plant or break loose and then become rooted. In either case, a large patch of *S. spathulifolium* may consist of many shoots which, whether attached or not, have the same genetic constitution. A further way by which plants become broken into separate parts is by the death of old parts of rhizomes, leaving the branches separate. Sometimes vegetative apomixis (fig. 135) occurs, with rosettes of leaves developing in place of flowers in the inflorescences.

NATIVE SPECIES

Fig. 135. Plant of *Sedum spathulifolium* ssp. *purdyi* from Keddie, Plumas Co., Calif., cultivated in greenhouse, Ithaca, N.Y., May 10, 1965 (x 1.5). Photo by Howard Lyon. Note rosettes of leaves in inflorescence, an example of vegetative apomixis.

Reproduction by seeds is common. The critical factor in the survival of seedlings is sufficient moisture in the driest part of the year, the summer, when mortality must be high. Where moisture is adequate, seedlings may be plentiful. Whenever seedlings persist, they are a means of introducing new genotypes into a population or of colonizing new sites.

Flowering time is the spring and early summer, from April until August. It varies with subspecies, latitude, and altitude. Following is a list of dates when I observed plants in flower in different regions which I surveyed.

Subspecies	Year	Region	Dates of anthesia
spathulifolium	1963	California Coast Ranges	May 6–20
spathulifolium	1963	Oregon Coast Ranges	June 26–Aug. 17
spathulifolium	1965	Cascade Mountains	July 9–Aug. 1
pruinosum	1963	Oregon Coast	Postanthesis July 14
yosemitense	1964	Central Sierra Nevada	June 18–27
purdyi	1963	Sierra Nevada	June 6

Plants of the different subspecies have distinctive times of flowering in Ithaca, as indicated in the following table. Quoted records for 1966 are by Miss Abbe.

Experiment	1966	1967	1968	1969
Greenhouse				
ssp. *spathulifolium*	2/15–"6/6"	3/25–5/10	3/20–7/2	4/14–7/6
ssp. *pruinosum*	"5/23"–6/8	6/29–7/2	—	6/25–7/6
ssp. *yosemitense*	2/15–"5/18"	3/3–25	1/16–5/7	6/7–8
ssp. *purdyi*	1/27–"5/27"	2/18–5/15	11/28/67–7/2	1/20–6/11
Cold frame				
ssp. *spathulifolium*	"5/16–6/6"	5/22–6/12	5/7–6/17	5/24–6/15
ssp. *pruinosum*	—	5/22–6/12	6/4–11	6/2–8
ssp. *yosemitense*	"4/3–5/23"	5/1–6/3	4/30–6/11	5/8–25
ssp. *purdyi*	3/7–"5/18"	3/7–6/4	4/16–5/28	5/8–25

Pollinators of *Sedum spathulifolium* are *Hymenoptera* and *Diptera*. In July 1965, along the old Deer Butte Trail in Oregon, two kinds of bees and a species of fly visited the blossoms of ssp. *spathulifolium*. Three times I observed bees switching from the yellow flowers of *S. spathulifolium* to the white blossoms of *S. oregonense*. Once the switching bee behaved uncertainly, as though it wanted more yellow and not white flowers. In Yosemite Gorge, in June 1964, bumblebees, small bees, and two kinds of bombyliids visited flowers of ssp. *yosemitense*.

Plants of *Sedum spathulifolium* appear to be nearly self-incompatible. Of 12 replicates checked for seeds in 1967 in the greenhouse, where flying insects are scarce, 9 (75%) yielded seeds. Of 29 replicates in the cold frame, where bees are frequent, 25 (86%) yielded seeds in 1967.

In 1969, several plants were selfed, crossed, or open-pollinated. Results of the experiments are available in the following table. The last column, % success, indicates the percentage of seeds, out of the estimated total number of ovules involved, which developed.

Species or subspecies	Procedure	Plants (n)	Seeds (n)	% success
ssp. *spathulifolium*	selfed	2	15	2
ssp. *spathulifolium*	open-pollinated	1	3	.2
ssp. *spathulifolium* ♀ × *purdyi* ♂	crossed	2	15	2.7
S. sp. ssp. *spathulifolium* ♀ × *S. moranii* ♂	crossed	1	0	0
ssp. *yosemitense*	open-pollinated	3	20	.6
ssp. *yosemitense* ♀ × *purdyi* ♂	crossed	2	0	0
ssp. *purdyi*	selfed	1	0	0
ssp. *purdyi* ♀ × *purdyi* ♂	crossed	2	12	1.1
ssp. *purdyi* ♀ × *spathulifolium* ♂	crossed	2	4	.7
ssp. *purdyi* ♀ × *yosemitense* ♂	crossed	1	0	0
S. moranii ♀ × *S. sp.* ssp. *spathulifolium* ♂	crossed	1	0	0

The above data suggest that *Sedum spathulifolium* may not be completely self-incompatible and also that some exchange of genes is possible between the subspecies *spathulifolium* and *purdyi*. Unfavorable conditions of growth probably are responsible for the low production of seeds.

Apomixis is unknown in *Sedum spathulifolium*. All development of seeds presumably occurs after sexual reproduction following pollination.

Relationships. Of the species of *Sedum* in western North America, *S. moranii* is the one which seems closest to *S. spathulifolium*. The two species have in common similar types of inflorescences, yellow petals, and 15 pairs of chromosomes. In other respects, *S. moranii* differs in its glandular-hairy inflorescences, larger flowers with elongate sepals, erect and connate petals, and basally connate, erect carpels with short, stout styles. Both species are specialized, but in different ways. Possibly they are derived from the same ancestral stock, but diverged long ago.

A less apparent relationship is indicated by two natural hybrids, *Sedum laxum* ssp. *heckneri* × *S. spathulifolium* ssp. *spathulifolium*, discovered along the Trinity River in Humboldt Co., Calif., by Helen Payne of Dallas, Ore. Principal discussion of these plants is in the section on relationships, under *S. laxum*. One of the plants, 71–3, sufficiently resembles *S. spathulifolium* in vegetative characters that it could be taken to be a minor variant. Neither hybrid grew well in cultivation in a greenhouse at Ithaca, N.Y., and 71–3 died 2.5 years after receipt.

Sedum leibergii is another specialized western North American species which may be derived from the same stock which gave rise to *S. spathulifolium*. That has developed the biennial habit of growth and has become adapted to the semiarid Columbia Plateau where it endures high summer temperatures and drought. Like *S. spathulifolium*, it has rosettes of flat leaves, yellow petals, and divergent, gibbous follicles.

A more distant relationship may be with species of eastern North America with flat leaves, gibbous carpels, and inflorescences of 3 cincinni. These species include *Sedum ternatum*, *S. nevii*, and *S. glaucophyllum*. Diploid numbers of chromosomes in this group are respectively 8, 6, and 14. The filaments of the species of this relationship are distinctive in being flattened and scarcely adnate to the petals. These features suggest a less specialized status than that of *S. spathulifolium*. Further, the 15 pairs of chromosomes of *S. spathulifolium* and other species of *Gormania* may be of allopolyploid origin, a consideration compatible with the idea that *S. spathulifolium* is more advanced than the eastern North American species and perhaps of different lineage.

Search for possible relatives of *Sedum spathulifolium* among species of the Mexican Plateau or Trans-Mexican Volcanic Belt is unsuccessful unless an obscure relationship exists with *S. palmeri*. Elsewhere, *S. ambiflorum* of the Philippine Islands and *S. formosanum* of Formosa may be distantly related, but they have a different habit of growth, spurred sepals, slightly connate petals, papillose seeds, and other distinctive features. *Sedum engleri* of western China likewise may be remotely related, although its unbranched caudex and petiolate leaves easily distinguish it.

References. Brown, R. M. (1970); Clausen, R. T., and Uhl, C. H. (1944); Gartenflora 21: pl. 741 b (1872); and Hollingshead, L. (1942).

Subgenus *Rhodiola*

Plants of *Sedum*, subgenus *Rhodiola*, are perennial herbs with rootstocks with usually scalelike, but sometimes well-developed leaves, and axillary, annual, floriferous stems. The pistils are erect, orthocarpic, and nonstipitate. Most species have perfect, 5-merous flowers, but a few species have the flowers imperfect and 4-merous.

The species of subgenus *Rhodiola* exhibiting the most primitive characteristics occur in Asia. Their anatomy still remains to be studied. In America, Quimby (1971) investigated the floral anatomy of a plant of *Sedum integrifolium*, cited as *S. roseum*, from Watkins Glen, N.Y. He found maximum fusion of traces in a vertical direction, with only three whorls of traces as compared with the six whorls in subgenus *Telephium*. According to his data, the floral anatomy of *S. integrifolium* and of four European species of the relationship of *S. sediforme* is the most advanced in the *Crassulaceae*. Whether other species of *Rhodiola*, especially the ones in Asia with primitive features, resemble *S. integrifolium* in floral anatomy is unknown. Their status ultimately will determine the optimum phylogenetic position of subgenus *Rhodiola* among the subgenera of *Sedum*. Embryological data, such as that of Mauritzon (1933) who studied four species of *Rhodiola*, is indecisive. Likewise, information about chromosomes does not clearly indicate the relationships of *Rhodiola* with respect to the other subgenera.

The number of species comprising the subgenus *Rhodiola* must remain uncertain until the details of relationships of species in the mountains of Asia are satisfactorily investigated. My present knowledge favors recognition of about 35 species. These species are widely distributed in the northern parts of Eurasia and North America and extend southward at high elevations in the mountains. Southern occurrences in North America are on Roan Mountain in the Blue Ridge, the Sierra Blanca of the Sacramento Ranges in New Mexico, and near Mineral King in the Sierra Nevada of California. The range in Asia is southward to the Himalayan Mountains, the northern Burmese Ranges, and the Great Snowy Mountains.

Rhodiola has been interpreted at different times as a genus or as a section of *Sedum*. The interpretation depends on the point of view, whether one wishes to emphasize the distinctions which separate the most advanced species from *Sedum* or whether one takes note of the similarities between the two groups of species and also of the species which form a continuum between the two groups. The preference here is for an intermediate course. The suggestion is to elevate *Rhodiola* to subgeneric status.

Sedum, subgenus **Rhodiola** (L.) R. T. Clausen, stat. nov., fundatum super *Rhodiola* L., Sp. Pl., ed. 1, 2:1035 (1753). Species typicum *Sedum rosea* (L.) Scop. est.

A synonym is:

1763. *Rodia* [Dioscorides] Adanson, Fam. Pl. 2:248. Based on *Telephium* Morison, Hist. Pl. Univ., sect. 12, p. 468, pl. 10, fig. 8, and *Rhodiola* L.

Two sections are represented in North America.

Section *Clementsia* (Rose) A. Bor., Flora URSS 9:28 (1939). Based on *Clementsia* Rose, Bull. N.Y. Bot. Gard. 3:3 (1903). Type species: *Sedum rhodanthum* A. Gray. Distinguishing features of the section are the elongate, racemiform cymes and the bisexual, mostly 5- to 6-merous flowers. *Sedum rhodanthum* is the only North American species.

Section *Rhodiola* (L.) Scop., Introd. ad Hist. Nat., p. 255. Based on *Rhodiola* L.

Type species: *Sedum rosea* (L.) Scop. Characteristics of the section are scalelike primary leaves, axillary annual floriferous stems, and corymbose cymes of small 4- or 5-merous flowers which often are unisexual. Two included species occur in North America, namely *S. integrifolium* and *S. rosea*.

28. Sedum rhodanthum*** (figs. 136–139)

Elongate, dense, racemelike inflorescences are a distinctive feature of *Sedum rhodanthum* (fig. 136). Like other species of subgenus *Rhodiola*, it has stout rootstocks with small, brown, scalelike leaves in the axils of which the floriferous stems arise. The petals are erect, separate, and pink. Few other North American species have pink flowers: *S. pulchellum*, *S. villosum*, and *S. laxum*. Of these, *S. pulchellum* and *S. villosum* usually do not have rootstocks. *Sedum pulchellum* has linear leaves and divergent, gibbous follicles. *Sedum villosum* generally is an annual with glandular-hairy stems, leaves, pedicels, and sepals. It has tiny red nectaries and short-styled carpels which become black in fruit. *Sedum laxum* has thick, oblanceolate or spatulate basal leaves, paniculate cymes, and pyriform seeds. The follicles of *S. rhodanthum* are erect and brown and the seeds are fusiform-oblong. The leaves are elliptic-oblanceolate, with midribs depressed, and glabrous. The pink-flowered North American species are vastly different from each other. They are not likely to be confused.

Sedum semenovii of central Asia most resembles *S. rhodanthum*. It has longer, narrower leaves which always are entire. It is rare in cultivation in American gardens.

Description. Sample: 29 plants from 9 localities in 3 geomorphic provinces.

Population	Geomorphic province	Kind of sample	Plants (n)
Berthoud Pass	So. Rocky Mts.	random	9
W. Fork of Clear Creek	So. Rocky Mts.	judgment	1
Joe Wright Creek	So. Rocky Mts.	random	2
Below Lake Emma	So. Rocky Mts.	judgment	4
Elwood Pass	So. Rocky Mts.	judgment	1
Meadow E. of Beartooth Lake	Central Rocky Mts.	convenience	7
Frozen Lake	Central Rocky Mts.	judgment	1
Teton Mts.	Central Rocky Mts.	convenience	3
Fremont Peak	Colorado Plateau	judgment	1
Total			29

Plants in cultivation in Ithaca do poorly. They barely survive and rarely flower. Longest survival in the garden was 2 years, in the cold frame 10 years, and in the greenhouse 2 years. Plants which survived 10 years in the cold frame came from Berthoud Pass and the valley of Joe Wright Creek. Rot of the rootstock appears to be the usual cause of death. At the end, the tissue of the rootstock becomes reddish brown and soft.

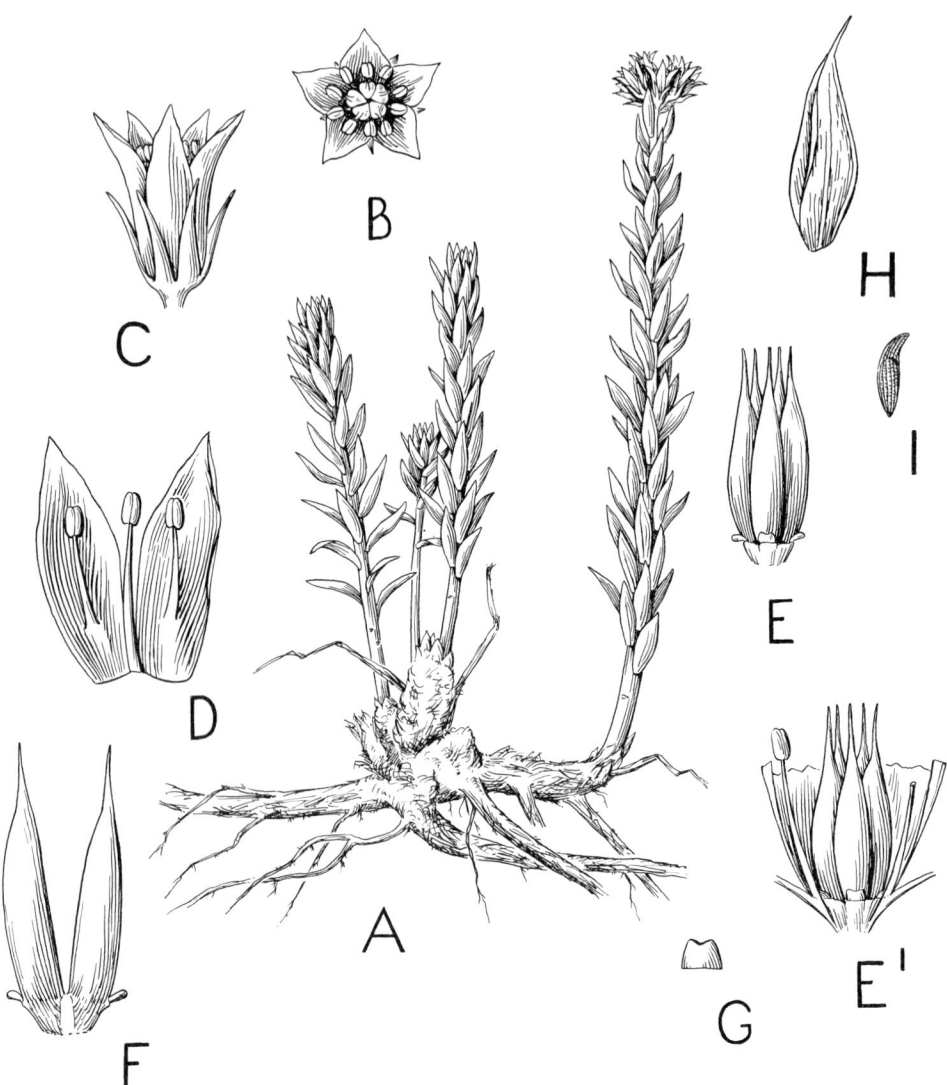

Fig. 136. Plant of *Sedum rhodanthum* from Berthoud Pass, Southern Rocky Mountains, Clear Creek Co., Colo. A. Habit sketch (x .59). B. Flower from above (x 2.4). C. Flower from side (x 2.4). D. Two petals and stamens (x 3.5). E. Carpels (x 3.5). E′. Carpels, and also short hypanthium and bases of other floral parts (x 3.5). F. Two carpels and nectaries (x 4.7). G. Nectary (x 5.9). H. Follicle (x 4.7). I. Seed (x 5.9). H and I are drawn from specimens collected in a meadow east of Beartooth Lake in the Beartooth Mountains, Park Co., Wyo.

Attempts at experimental culture of plants are indicated in the following table.

	Garden		Cold frame		Greenhouse	
Population	n	n-data	n	n-data	n	n-data
Berthoud Pass	3	0	11	0	0	0
W. Fork Clear Creek	0	0	1	0	0	0
Joe Wright Creek	2	0	1	0	0	0
Below Lake Emma	1	0	1	0	0	0
Elwood Pass	1	0	1	0	0	0
Fremont Peak	0	0	0	0	1	0
Totals	7	0	15	0	1	0

Abbreviations for names of localities in the following description are as follows:

BL = swamp 2.5 km. E. of Beartooth Lake, Park Co., Wyo.
BP = Berthoud Pass, Colo.
CC = valley of W. Fork of Clear Creek, S. of Berthoud Pass, Colo.
FL = along brook flowing into Frozen Lake, Beartooth Mts., Park Co., Wyo.
JW = valley of Joe Wright Creek, near Chambers Lake, Larimer Co., Colo.
LE = swampy area below Lake Emma, N.E. of Silverton, Colo.
TM = canyon of South Fork of Cascade Creek, Teton Mts., Wyo.

All data are for plants in nature.

Plants of *Sedum rhodanthum* are perennial herbs (fig. 137) with stout, branched rootstocks, 1 to 7 cm. in diameter, and fibrous, pale brown roots. Rootstocks are erect or decumbent and brown. The floriferous stems are erect, up to 4 dm. tall, and arise in the axils of scalelike primary leaves. Sometimes erect leafy stems, non-flowerbearing, arise from the rootstocks. Old fructiferous stems may persist in dry condition in the second season of growth.

	n-pl.	n-obs.	$\bar{x}$	s
Plants—diameter (cm.)				
BL	1	1	14	—
FL	1	1	>9	—
Maximum height (cm.)				
BP	9	9	20	4.4
CC	1	1	17	—
JW	2	2	31	1.3
LE	4	4	21	3.8
BL	1	1	13	—
FL	1	1	13	—
Floriferous stems per plant (n)				
BP	9	9	2.3	2
CC	1	1	2	—
JW	2	2	8	2.1
LE	4	4	4	2.4
BL	1	1	2	—
FL	1	1	>2	—
Floriferous stems—length (cm.)				
BL	1	2	13	—
FL	1	2	13	—

Fig. 137. Plant of *Sedum rhodanthum* on bank in Berthoud Pass in Southern Rocky Mountains, Clear Creek Co., Colo., July 20, 1962.

Leaves of rootstocks are scalelike, sessile, ovate or oblong-lanceolate, acute to acuminate-attenuate, and brown, 6–10 mm. long and 4–6 mm. wide. Leaves of floriferous stems are elliptic-oblanceolate, oblanceolate-oblong, or oblong, acute or obtuse, entire or sparsely toothed, and lettuce green, RH 861, or spinach green, RH 960/1. Midveins of leaves are prominently impressed, appearing as grooves. Late in summer, leaves become red or brown.

	n-pl.	n-obs.	$\bar{x}$	s
Length (mm.)				
BP	9	14	18	2.9
CC	1	2	17	—
JW	1	3	28	—
LE	4	9	20	1.9
BL	1	2	14	—
FL	1	1	19	—
Width (mm.)				
BP	9	14	5	.9
CC	1	2	4	—
JW	1	3	6	—
LE	4	9	6	.4
BL	1	2	5	—
FL	1	1	7	—

	n-pl.	n-obs.	$\bar{x}$	s
Length/width				
BP	9	14	3.5	.3
CC	1	2	4.2	—
JW	1	3	4.3	—
LE	4	9	3.5	.2
BL	1	2	3.1	—
FL	1	1	2.7	—
Thickness (mm.)				
BL	1	2	.6	—
Teeth per margin (n)				
BP	9	14	.2	.5
CC	1	2	0	—
JW	1	3	.7	—
LE	4	9	.8	.7
BL	1	2	0	—
FL	1	1	0	—

Inflorescences are elongate, racemelike cymes, or are sometimes congested and appear capitate (fig. 138). Each cyme has a terminal central flower. Floral bracts are elliptic-oblanceolate or linear.

	n-pl.	n-obs.	$\bar{x}$ (cm.)	s (cm.)
Length				
BP	5	9	1.8	.3
CC	1	2	1.2	—
JW	2	6	2.4	.1
LE	4	8	2.5	.5
FL	1	2	1.6	—

Flowers are perfect and unusual in having a short hypanthium, .9–1.1 mm. long, adnate to the bases of the ovaries. Buds are peony purple, RH 729.

	n-pl.	n-obs.	$\bar{x}$ (mm.)	s (mm.)
Pedicels—length				
BP	5	9	.7	.5
CC	1	2	.8	—
JW	2	6	.4	.2
LE	4	8	1	.5
Flowers—diameter				
BP	5	9	9	1.5
CC	1	2	8	—
JW	2	6	7	1.8
LE	4	8	10	.8
Torus—length				
BP	1	1	1	—

Sepals are unequal, divergent, lanceolate, acute, and pale lettuce green tipped with peony purple.

480 SEDUM OF NORTH AMERICA

Fig. 138. Headlike inflorescence on plant of *Sedum rhodanthum* from valley of Joe Wright Creek near Chambers Lake in the Southern Rocky Mountains, Larimer Co., Colo., cultivated in cold frame, Ithaca, N.Y., June 9, 1964 (x 2.5). Photo by Howard Lyon.

	n-pl.	n-obs.	$\bar{x}$ (mm.)	s (mm.)
Length				
BP	5	9	6.9	.8
CC	1	2	6.7	—
JW	2	6	6.8	1
LE	4	8	8	.5
Width				
BP	5	9	1.5	.2
CC	1	2	1.5	—
JW	2	6	1.4	0
LE	4	8	1.6	.3

Petals are separate, erect, except divergent at apex, oblanceolate-oblong, carinate, and acute or abruptly acuminate. They are imbricate in bud. Color usually is fuchsine pink, RH 627/2, or pale rose pink, RH 427/3, or rarely white suffused with pale pink.

	n-pl.	n-obs.	$\bar{x}$	s
Number				
BP	3	4	5.8	.3
CC	1	2	7	—
JW	2	6	5.2	.3
LE	4	8	5.3	1.1
Length (mm.)				
BP	5	9	9.7	.9
CC	1	2	9.9	—
JW	2	6	9.2	1.2
LE	4	8	12.3	.8
Width (mm.)				
BP	5	8	3.3	.3
CC	1	2	2.6	—
JW	2	6	2.5	.7
LE	4	8	3.4	.6

Stamens are twice as many as the petals or rarely one less than that number. The filaments are fuchsine pink, sometimes white below, and linear-subulate, broadest basally. The anthers are Indian-lake, RH 826, chrysanthemum crimson, RH 824, or fuscous.

	n-pl.	n-obs.	$\bar{x}$ (mm.)	s (mm.)
Filaments—length				
BP	5	8	5.7	.9
CC	1	2	4.9	—
JW	2	6	6.3	.9
LE	4	8	7.1	.6
Epipetalous filaments—height on petals				
BP	5	9	1.9	.5
CC	1	2	2	—
JW	2	6	2.2	.3
LE	4	7	2.6	.6
Anthers—length				
BP	1	1	1.2	—
CC	1	1	.9	—
LE	1	1	1.4	—

Nectaries are reniform or reniform-ovoid, truncately rounded, sometimes emarginate, and aureolin or pale yellow.

	n-pl.	n-obs.	$\bar{x}$ (mm.)	s (mm.)
Length				
BP	5	9	.7	.05
CC	1	2	.6	—
JW	2	6	.6	.02
LE	4	8	.8	.16

	n-pl.	n-obs.	x̄ (mm.)	s (mm.)
Width				
BP	5	9	.9	.15
CC	1	2	1	—
JW	2	6	.7	0
LE	4	8	.9	.08

Carpels are either equal to number of petals or one or two less, with slender styles, and fuchsine pink or white. The ovules are caudate and white.

	n-pl.	n-obs.	x̄	s
Length (mm.)				
BP	5	9	7.9	.9
CC	1	2	7.4	—
JW	2	6	8.4	.5
LE	4	8	10.6	2.2
Length of cohesion (mm.)				
BP	5	9	.9	.1
CC	1	2	.8	—
JW	2	6	1	0
LE	4	8	1	.06
Ovules per ovary (n)				
BP	5	9	11	.8
CC	1	2	10	—
JW	2	6	8	.8
LE	4	8	11	.7

Fruits are erect, with slender beaks, and pale brown.

	n-pl.	n-obs.	x̄ (mm.)	s (mm.)
Length				
BL	7	13	5.9	.8
TM	3	14	5.7	.3

Seeds are fusiform-oblong or elliptic-oblong, short-tailed, winged, obscurely longitudinally ribbed and also with scalariform riblets, and pale brown. Seeds remain in fruits until follicles are tilted, when they drop out. Seedlings sometimes develop in large numbers at ends or below old fruiting stalks.

	n-pl.	n-obs.	x̄ (mm.)	s (mm.)
Length				
BL	6	13	1.9	.1
TM	3	14	1.9	.1
Diameter				
BL	6	13	.5	.06
TM	3	14	.5	.02

Chromosomes are small, but probably larger than in *Sedum rosea* (See Uhl 1952: figs. 1–2).

	n-pl.	g (n)	sp (2n)	Cytologist
Number				
Rocky Mt. Nat. Park	1	7	—	Wiens & Halleck (1962)
Fremont Peak	1	7	—	C. H. Uhl (1952)

Variation. On July 5, 1966, I drew a small sample of fruits and seeds from a population of *Sedum rhodanthum* in a meadow 2.5 km. east of Beartooth Lake in the Beartooth Mountains of Wyoming. The fruits and seeds were of the crop of 1965. On Sept. 3, 1966, I drew another sample from the same population. Meanwhile, on Aug. 28, 1966, I drew a sample of fruits and seeds from a population in the Canyon of the South Fork of Cascade Creek in the Teton Mountains. Here are the results. Dimensions are arithmetical means.

Population	BL	BL	TM
Date	7–5	9–3	8–28
Plants (*n*)	3	4	3
Fruits—length (mm.)	5.8	5.9	5.7
Seeds—length (mm.)	1.8	2	1.9
—diameter (mm.)	.5	.5	.5

The remarkable thing about the data is the small difference among the three samples. Variation is slight. Although this is true of most characteristics of *Sedum rhodanthum*, populations are not the same. On the basis of study of wild plants, significant differences occur in at least three features. When tests are made, using the method of Steel and Torrie (1960: 113–114), to see which means are significantly different, no pattern emerges, as is indicated below, where means which are not significantly different are underlined.

Feature	LE	JW	BP	FL	CC
Cymes—length (cm.)*	2.5	2.4	1.8	1.6	1.2
Petals—length (mm.)**	12.3	9.2	9.7	—	9.9
Ovules per ovary (*n*)**	11	8	11	—	10

Other features, height for example, vary enormously from population to population, but variation within populations is so great that the differences among populations lack significance.

Because of the poor growth of plants in cultivation, no check on conclusions from studies in nature could be obtained from experiments. However, the lack of experimental results may not be serious because the data for wild plants do not indicate regional tendencies.

Variation within plants is sufficiently limited that only small samples are necessary, provided that they are selected according to a probability model. For example, measurement of 3 leaves is adequate to detect a difference of 2 mm. in width, $P = .05$, and measurement of 2 seeds is adequate to detect a difference of .5 mm. in length, $P = .05$.

Color of petals is relatively stable, either fuchsine pink or pale rose pink, but sometimes almost white flushed with pink. Petals tend to vary in number. Commonest numbers in a small sample of 18 flowers from 9 plants were 5 and 6.

No. of petals	No. of flowers	%	Cumulative %
4	1	5.6	5.6
5	8	44.4	50
6	6	33.3	83.3
7	1	5.6	88.9
8	2	11.1	100
Total	18	100	

To check on the value of herbarium specimens as a source of descriptive data, I compared measurements from a series of dried specimens with data from my samples of living plants. Analyses of variance were for three features, two of them, length of cymes and length of petals, in which significant differences occur among populations on the basis of data from the field. The results indicate in each case a highly significant difference. Probably as a result of shrinkage in drying, leaves of herbarium specimens are narrower and petals are shorter. None of the wild populations had a mean petal length as short as the mean for the herbarium specimens. On the other hand, collectors tend to take the larger plants at a site, as is indicated by the longer inflorescences of the herbarium specimens. The conclusion is that dried specimens are not a reliable source of descriptive data.

Feature	Live plants	Herbarium specimens
Leaves—width (mm.)**	5.2 (n = 16)	4.2 (n = 36)
Cymes—length (cm.)**	2.1 (n = 13)	3.3 (n = 37)
Petals—length (mm.)**	10.5 (n = 12)	8.5 (n = 37)

The evidence on variation suggests that *Sedum rhodanthum* is a monotypic species, lacking races of evolutionary significance at the present time.

Nomenclature. *Sedum rhodanthum* A. Gray, Am. Jour. Sci., ser. 2, 33:405 (1862). Type locality: moist places in high alpine region, headwaters of Clear Creek and alpine ridges east of Middle Park. Type: C. C. Parry (GH), 1861. The holotype is on the right side of the type sheet. Study of Parry's account of his explorations in 1861 suggests that the type locality may be Berthoud Pass. See Am. Jour. Sci., ser. 2, 33:231–237 (1862).

A synonym is:

1903. *Clementsia rhodantha* (A. Gray) Rose, Bull. N.Y. Bot. Gard. 3:3. Type: *Sedum rhodanthum* A. Gray. Rose regarded the elongate inflorescence to be so distinctive that he described the new genus to accommodate the species. He believed that *S. semenovii* also belongs in *Clementsia*.

Distribution. Areas of principal occurrence of *Sedum rhodanthum* (fig. 139) are the Southern and Central Rocky Mountains. The species also occurs in disjunct fashion in the higher parts of the Colorado Plateau, for example on the Aquarius

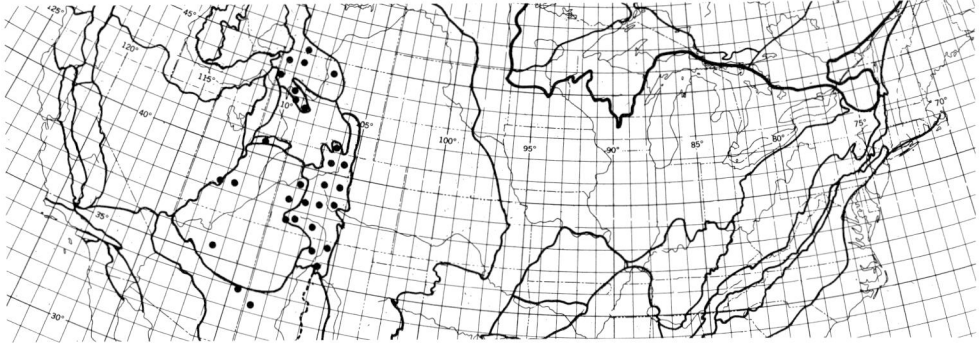

Fig. 139. Known distribution of *Sedum rhodanthum* (●).

Plateau of Utah and in the San Francisco and White Mountains of Arizona. In addition, Hess has found *S. rhodanthum* in the Mogollon Mountains, which are part of the Mexican Plateau in Catron Co., N.M. Habitats include willow swamps, meadows, banks of streams, and gravelly places and turf in alpine and subalpine situations. The plants usually are deeply rooted in organic matter, alluvium, or gravelly loam, generally poorly drained, over granite or andesite. The altitudinal range is from 1,981 to 4,115 meters (to 4,267 m. according to Riedel and Wilson, 1967).

Data for seven populations which I have studied are available in the following table.

Population	Plants (N)	Exposure	Rock
Berthoud Pass	~642	E–S, W–NW	granite
Joe Wright Creek	~25	W–N	granite
Lake Emma	>4	E–S	—
Elwood Pass	>1	SSE	andesite
Md. E. of Beartooth Lake	~100	NW	granite
Frozen Lake	~100	—	granite
Teton Mts.	15	E	granite

Population	Alt. range (m.)	pH, range	pH, tests (n)	Oldest record
Berthoud Pass	3,380–3,390	4.8–5.6	12	1868
Joe Wright Creek	2,785	6.2–6.4	2	1894
Lake Emma	3,645	—	—	1962
Elwood Pass	3,490	5.6	1	1962
Md. E. of Beartooth Lake	2,850	5	1	1937
Frozen Lake	3,050	—	—	1966
Teton Mts.	2,650	—	—	1899

In Berthoud Pass, in the area of occurrence in 1962, density was .32 plant per square meter.

Competing species are miscellaneous. *Salix phylicifolia* ssp. *planifolia* is the most important competitor in Berthoud Pass, where the habitat is a swamp in which this willow is dominant. Two other common competitors there are *Caltha leptosepala* and a grass with glaucous leaves. Plants of *Sedum rhodanthum* grow in alluvium, on the swampy bank of Joe Wright Creek, in the lower valley near Chambers Lake. The dominant tree there is *Picea engelmannii* and commonest competitors are *Equisetum arvense*, *Erigeron peregrinus* ssp. *callianthemus*, *Taraxacum officinale*, a species of *Senecio*, and a species of the *Apiaceae*. Associated species in the swampy area below Lake Emma are *Salix phylicifolia* ssp. *planifolia*, *Caltha leptosepala*, and *Pedicularis groenlandica*. Both *Salix phylicifolia* ssp. *planifolia* and *Caltha leptosepala* likewise are present in the meadow south of Elwood Pass. In the Teton Mountains, the plants grow in moss on ledges in the canyon of the South Fork of Cascade Creek. In the Beartooth Mountains, they grow in swampy meadows along with willows, sedges, and a variety of herbaceous dicotyledons.

Reproduction. *Sedum rhodanthum* reproduces abundantly from seeds. The moist situations in which it occurs are ideal for germination and development of seedlings. When old fruiting stems bend down or fall down, numerous seedlings may develop where the follicles touch ground. Sometimes seeds germinate in the follicles of an erect stem. Another method of reproduction, this one vegetative, is by division of the rootstock.

Bees, especially large bumblebees, are effective cross-pollinators of the flowers. Riedel and Wilson (1967) found that bees at pollen traps at an elevation of 4,267 meters on Mount Evans, in the Southern Rocky Mountains, did not collect pollen of *Sedum rhodanthum* at that altitude, although plants were present there, but gathered pollen from flowers of the *Sedum* at lower altitudes. Besides bees, small *Diptera* visit the flowers and also may be pollinators.

Flowering time is from late June to early September. Plants may have mature fruits with seeds by late August, but fruiting stalks persist through the winter and seeds sometimes remain in the follicles until the following spring or summer.

Relationships. The North American species most closely related to *Sedum rhodanthum* are *S. rosea* and *S. integrifolium*. These species also are members of the subgenus *Rhodiola* and have stout rootstocks with scalelike primary leaves and axillary floriferous stems. They differ from *Sedum rhodanthum* in having smaller, usually imperfect flowers, with greenish yellow or deep purple petals, in dichotomous cymes. Also, the midribs of their leaves are not depressed. *Sedum rhodanthum*, with 7 pairs of chromosomes, may have had a role in the evolution of *S. integrifolium*, with 18 pairs of chromosomes, but this is unproved.

Another relationship of *Sedum rhodanthum* is with *S. semenovii* of central Asia. J. N. Rose, in a letter of July 5, 1902, to N. L. Britton, suggested placing the two species together in a separate genus, *Clementsia*. *Sedum semenovii* occurs in the western Tien Shan and Alai Mountains, 38°–43° N., 68°–78° E. It is remarkably similar to *S. rhodanthum*, but differs in having longer, narrower leaves and carpels as long as the petals at time of anthesis. For comparison with my samples of *S. rhodanthum* from western North America, I have used data for a plant of *S. semenovii* cultivated at Kew Gardens in 1925.

	S. rhodanthum	S. semenovii
n-plants	16	1
Leaves		
Length (mm.)	18.3 (12–27)	31 (29–35)
Width (mm.)**	5.2 (4–7)	1.6 (1.1–2)
Length/width**	3.5 (2.7–4.3)	19.8 (15.5–17.2)
Margins	entire or sparsely toothed	entire
Petals		
Length (mm.)	10.5 (8.1–13.3)	8.2 (8–8.5)
Carpels		
Length (mm.)	8.9 (5.1–13.8)	8.2 (8–8.5)

Sedum dumulosum, another Asiatic species sometimes placed close to *S. rhodanthum* in classification, differs in having elliptic-linear leaves on the floriferous stems, corymbose cymes, and white, erose, acuminate-aristate petals. Uhl (1952) reported a gametophytic number of 7 chromosomes for a cultivated plant of *S. dumulosum*, the same number as in *S. rhodanthum*.

Illustration. Jour. Roy. Hort. Soc., London 46:68, fig. 28 (1921).

29. *Sedum integrifolium**** (figs. 140–150)

Sedum integrifolium (fig. 140) and *S. rosea* are similar. Like *S. rhodanthum*, they have thick rhizomes with scalelike leaves, and axillary, annual floral stems. Unlike *S. rhodanthum*, they usually are dioecious, with smaller flowers in corymbose cymes and petals either deep red or yellow, and the midribs of the leaves are not depressed. Leaves commonly are toothed, but sometimes are entire. Although *S. integrifolium* differs from *S. rosea* in many ways, few absolute distinctions are available. Wider petals of *S. integrifolium*, 1.3–1.7 mm. broad in staminate flowers, are most dependable for recognition. Several other tendencies of *S. integrifolium* aid identification. These are green rather than glaucous leaves, deep red rather than yellow petals, and a higher percentage of 5-merous flowers. Longer, narrower leaves are a further distinction in some populations. Occasional plants, difficult to determine by morphology alone, may be identified by considering a combination of the average expressions of all characteristics of the population of origin.

English names for *Sedum integrifolium* include Western Roseroot or just Roseroot, and also King's Crown and Stonecrop.

Description. Sample: 160 plants from 23 populations. Information about the subsamples is available in the following table (C = convenience, J = judgment, and R = random).

Population	Year of study	N	n	n/N	Mode of selection	Stage when studied
Glenora, N.Y.	1947–69	10,422	46	.004	R & J	Fl., Fr.
Watkins Glen, N.Y.	1946–72	1	1	1	C	Fl.
Root River, Minn.	1949	535	13	.024	C	Fl.
W. Spanish Peak, Colo.	1962	511	14	.027	R & J	Fl.
Sierra Blanca, N.M.	1949	>35	11	.314	J	Bud, Fr.

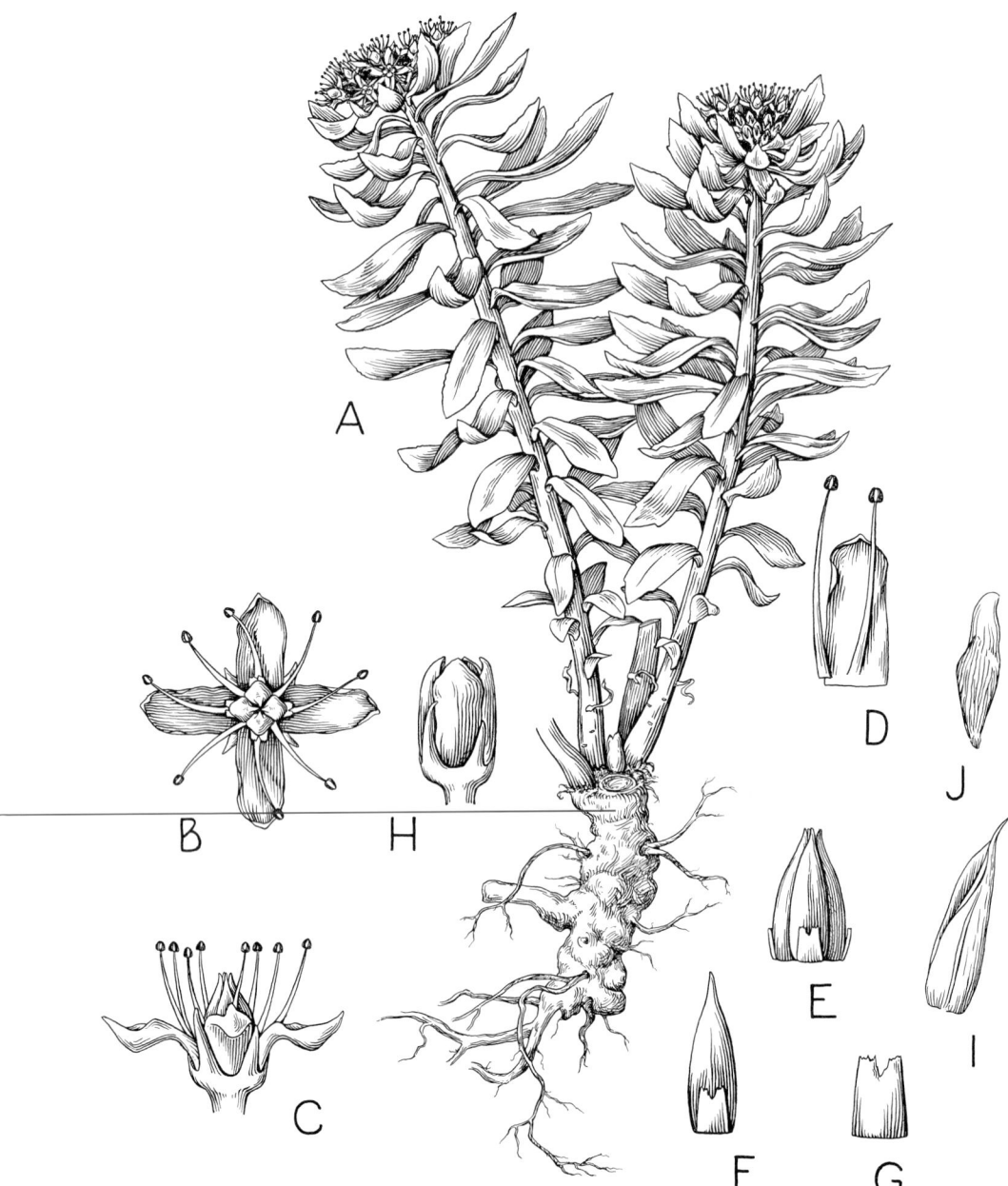

Fig. 140. Plant of *Sedum integrifolium* ssp. *leedyi* from Watkins Glen, Schuyler Co., N.Y., cultivated in garden, Ithaca, N.Y., June 3, 1947. A. Habit sketch (x .96). B. Flower from above (x 3.8). C. Flower from side (x 3.8). D. Petal and two stamens (x 4.8). E. Carpels (x 4.8). F. Carpel and nectary (x 5.8). G. Nectary (x 9.6). H. Bud (x 3.8). I. Follicle (x 4.8). J. Seed (x 9.6).

NATIVE SPECIES 489

Population	Year of study	N	n	n/N	Mode of selection	Stage when studied
Sandia Mts., N.M.	1940	?	1	—	C	Fl.
Lake Peak, N.M.	1962	168	4	.024	R & J	Fr.
Rio Hondo Canyon, N.M.	1949	200	10	.050	J	Fl.
Pikes Peak, Colo. (NW spur)	1962	221	9	.041	R	Fl.
S.W. of Mt. Flora, Colo.	1962	2	1	.500	J	Fl.
Lulu Pass, Colo.	1962	?	3	—	C	Fl.
Platoro, Colo.	1962	754	4	.005	R	Fl.
Elwood Pass, Colo.	1962	?	3	—	C	Fl.
Summit Peak, Colo.	1962	5,000	4	.001	J	Fl.
Bonita Peak, Colo.	1962	2,000	7	.004	J	Fl.
Eureka Gulch, Colo.	1962	?	2	—	C	Fl.
Beartooth Summit, Wyo.	1963–64	?	4	—	C	Fr.
Alaska Basin, Wyo.	1966	34	2	.059	R	Fl.
Clements Mountain, Mont.	1961	?	1	—	C	Fr.
Seven Devils Lake, Ida.	1961	88	11	.125	R	Fl.
Hilton Lake 4, Calif.	1964	130	3	.023	J	Fr.
Gem Pass, Calif.	1964	104	4	.038	R	Fr.
Piute Peak, Calif.	1964	?	2	—	C	Fr.
Totals		>20,205	160			

Combinations of data in the following descriptive tables are as follows: So. Rocky Mts.—Lake Peak, Rio Hondo Canyon, Pikes Peak, Mount Flora, Lulu Pass, Platoro, Elwood Pass, Summit Peak, Bonita Peak, and Eureka Gulch; and Sierra Nevada—Hilton Lake 4, Gem Pass, and Piute Peak.

Sedum integrifolium does poorly in cultivation at Ithaca, N.Y. Best results came from an experiment in a shaded cold frame (cf.) and in a partially shaded plot at 1421 Slaterville Road (SR). In the following table indicating plants in experiments, n = number of plants in each experiment, n-d = number of plants which yielded data, and S-$yrs.$ = maximum number of years of survival of any clone from the population cited.

	TG			SR			Cf.		
Source of sample	n	n-d	S-yrs. (n)	n	n-d	S-yrs. (n)	n	n-d	S-yrs. (n)
Glenora	1	1	5	19	14	19	0	—	—
Watkins Glen	1	1	35	1	1	6	0	—	—
Root River	1	1	4	0	—	—	0	—	—
W. Spanish Peak	10	0	2	1	0	0	15	8	11
Sierra Blanca	3	3	7	0	—	—	0	—	—
Lake Peak	1	0	2	1	1	2	4	4	11
Rio Hondo Canyon	1	1	7	0	—	—	0	—	—
Pikes Peak	4	0	2	1	0	1	6	0	11
Mt. Flora	0	—	—	0	—	—	1	0	3
Lulu Pass	1	0	0	0	—	—	2	0	7
Platoro	2	1	2	1	1	2	5	3	11
Elwood Pass	0	—	—	2	0	0	3	0	7
Summit Peak	3	0	2	2	1	5	3	1	11
Bonita Peak	3	0	2	1	0	0	5	4	11
Eureka Gulch	2	0	2	0	—	—	2	1	11
Beartooth Summit	0	—	—	0	—	—	5	2	11
Clements Mountain	0	—	—	0	—	—	1	0	0
Seven Devils Lake	0	—	—	0	—	—	3	1	8

	TG			SR			Cf.		
Source of sample	n	n-d	S-yrs. (n)	n	n-d	S-yrs. (n)	n	n-d	S-yrs. (n)
Hilton Lake 4	1	0	0	0	—	—	2	1	5
Gem Pass	0	—	—	0	—	—	4	1	8
Piute Peak	0	—	—	0	—	—	1	1	8
Lake Tahoe	0	—	—	0	—	—	1	1	11
Totals	34	8		29	18		63	28	

Plants of *Sedum integrifolium* are dioecious, perennial herbs with stout rootstocks and pale brown, fibrous roots to 13 cm. long. The rootstocks are either erect or horizontal, branched, brown, 1–5 cm. in diameter, and up to 20 cm. long. The floriferous stems arise in the axils of scalelike primary leaves and may not be completely elongated when plants are in flower. They are green and sometimes finely papillose. Old floral stems may persist on rhizomes into a second growing season. Plants sometimes become quite large. One plant on Bonita Peak, for example, covered about 9 square meters. Occasionally, as at Gem Pass, slender stolons develop from the rootstocks. Differences among populations in height are significant on the basis of wild plants and highly significant in the experiment in the cold frame.

	n-pop.	n-pl.	n-rep.	n-obs.	$\bar{x}$	s
Maximum diam. (cm.)	3	9	—	9	33	18
Floriferous stems—*n* per plant	14	77	—	77	12	12.8
Height (cm.)						
Glenora	1	9	—	9	23	4.1
Root River	1	4	—	4	16	12.4
Sierra Blanca	1	2	—	2	17	2.8
W. Spanish Peak	1	11	—	11	8	9.8
Pikes Peak	1	7	—	7	5	1.3
Bonita Peak	1	6	—	6	28	12.6
So. Rocky Mts.	9	31	—	31	19	9.3
Seven Devils Lake	1	11	—	11	6	1.7
Sierra Nevada	3	9	—	9	7	2.8
Sierra Blanca—TG	1	3	3	3	9	1.7
Glenora—SR	1	6	8	8	13	—
W. Spanish Peak—cf.	1	8	12	12	4	.7
So. Rocky Mts.—cf.	5	13	18	18	15	5.9
Seven Devils Lake—cf.	1	2	2	2	4	.7
Sierra Nevada—cf.	4	4	5	5	6	—

Leaves of rhizomes are scalelike, lanceolate to ovate, acute or obtuse, 1–9 mm. long, 1.5–8 mm. wide, and reddish brown. Leaves of floriferous stems are spirally arranged; short-spurred; sessile; elliptic-oblong, oblong-oblanceolate, elliptical, or ovate; acute or obtuse; and either entire or coarsely dentate. They usually are spinach green, but sometimes are pale green and glaucous. Leaves are longest toward the middle of floral stems. Dimensions are for median leaves. Differences among populations are significant in length and highly significant in width and number of teeth per margin on the basis of wild plants. In the experiment in the cold frame, differences in length and width lack significance, but in number of teeth per margin and length-width ratios are highly significant.

NATIVE SPECIES   491

	n-pop.	n-pl.	n-rep.	n-obs.	$\bar{x}$	s
Length (mm.)						
Glenora	1	9	—	21	31	5.9
W. Spanish Peak	1	10	—	18	13	2.4
Pikes Peak	1	8	—	14	9	1.7
Bonita Peak	1	6	—	13	18	14.2
So. Rocky Mts.	8	57	—	81	16	9
Alaska Basin	1	2	—	5	10	.5
Seven Devils Lake	1	11	—	18	12	4.4
Sierra Nevada	3	4	—	9	10	2.5
Sierra Blanca—TG	1	3	3	7	15	1.5
Glenora—SR	1	6	8	17	29	6.6
W. Spanish Peak—cf.	1	6	9	15	12	8.4
So. Rocky Mts.—cf.	4	11	13	23	19	6
Seven Devils Lake—cf.	1	1	1	2	14	—
Sierra Nevada—cf.	4	4	5	8	15	—
Width (mm.)						
Glenora	1	9	—	21	9	2.8
Root River	1	13	—	13	13	3.8
Sierra Blanca	1	10	—	10	5	1.2
W. Spanish Peak	1	10	—	18	7	1.7
Pikes Peak	1	8	—	14	5	1.6
Bonita Peak	1	6	—	13	11	3
So. Rocky Mts.	9	67	—	91	9	3.2
Alaska Basin	1	2	—	5	6	.4
Seven Devils Lake	1	11	—	18	6	.7
Sierra Nevada	3	4	—	9	5	1.4
Sierra Blanca—TG	1	3	3	7	4	1
Glenora—SR	1	6	8	17	10	—
W. Spanish Peak—cf.	1	6	9	15	6	3.2
So. Rocky Mts.—cf.	4	11	13	23	8	—
Seven Devils Lake—cf.	1	1	1	2	7	—
Sierra Nevada—cf.	4	4	5	8	7	—
Length/width						
Glenora	1	9	—	21	3.6	1.1
W. Spanish Peak	1	10	—	18	2	.3
Pikes Peak	1	8	—	14	1.8	.3
Bonita Peak	1	6	—	13	3	.8
So. Rocky Mts.	8	57	—	81	2.4	.4
Alaska Basin	1	2	—	5	1.6	.1
Seven Devils Lake	1	11	—	18	1.7	.3
Sierra Nevada	3	4	—	9	2.1	.1
Sierra Blanca—TG	1	3	3	7	3.6	.5
Glenora—SR	1	6	8	17	2.7	.4
W. Spanish Peak—cf.	1	6	9	15	2	.9
So. Rocky Mts.—cf.	4	11	13	23	2.5	—
Seven Devils Lake—cf.	1	1	1	2	2.1	—
Sierra Nevada—cf.	4	4	5	8	2.1	—
Teeth per margin (n)						
Glenora	1	9	—	21	3	1.8
W. Spanish Peak	1	10	—	18	2.9	2.3
Pikes Peak	1	6	—	10	3.9	3
Bonita Peak	1	6	—	13	3.1	4.4
So. Rocky Mts.	9	38	—	67	2.7	2.7
Seven Devils Lake	1	11	—	18	.1	.3
Sierra Nevada	3	4	—	9	0	0
Glenora—SR	1	6	8	17	4	—

	n-pop.	n-pl.	n-rep.	n-obs.	$\bar{x}$	s
Teeth per margin (n)						
W. Spanish Peak—cf.	1	3	3	5	1	1
Lake Peak—cf.	1	1	1	2	5	—
Bonita Peak—cf.	1	3	3	5	4.2	.3
Sierra Nevada—cf.	2	2	2	3	0	—

Inflorescences are terminal, corymbose cymes, often dense, 1–8 cm. across. A cyme of a staminate plant at Glenora had 283 flowers in 1948. Floral bracts are elliptical, smaller, and narrower than foliage leaves. The difference in number of flowers per cyme between the population in the Alaska Basin and two populations in the Sierra Nevada is significant.

	n-pop.	n-pl.	n-obs.	$\bar{x}$	s
Length of axis of cyme (cm.)					
W. Spanish Peak	1	6	10	1.5	.8
Bonita Peak	1	6	11	.9	.5
So. Rocky Mts.	7	20	34	1.1	.7
Flowers (n)					
Alaska Basin	1	2	3	14.8	.5
Sierra Nevada	2	7	21	8.1	2.9

Flowers usually are functionally either male or female, but sometimes perfect. They most often are 5- or 4-merous, rarely 3-merous. The receptacle sometimes appears enlarged and makes a short torus, $\leq$ 1 mm. The sepals, petals, and stamens form a very short hypanthium, $\leq$ .7 mm., around the carpels. Rarely, flowers are asymmetrical, with parts smaller on one side, as on a plant on Summit Peak. Differences among populations in length of pedicels are highly significant. Likewise, differences between sexes in diameter of flowers are highly significant.

	n-pop.	n-pl.	n-rep.	n-obs.	$\bar{x}$ (mm.)	s (mm.)
Pedicels—length						
W. Spanish Peak	1	6	—	10	1.6	.7
Bonita Peak	1	5	—	17	2	.9
So. Rocky Mts.	6	15	—	54	2.8	.6
Flowers—diameter						
Glenora ♂ 1948	1	23	—	39	10.2	1.2
— ♀ 1948	1	5	—	5	6	1.2
Root River ♂	1	12	—	23	6.8	1.5
— ♀	1	1	—	2	3.3	—
W. Spanish Peak ♂	1	5	—	16	5.1	.6
— ♀	1	5	—	9	3.6	.7
Bonita Peak ♂	1	4	—	14	6.9	1.9
So. Rocky Mts. ♂	7	29	—	77	7.4	1
— ♀	1	2	—	4	3.8	1
Seven Devils Lake ♂	1	6	—	21	4	.8
— ♀	1	4	—	19	3.3	.5
Bonita Peak—cf. ♂	1	1	1	2	5.8	—
Hilton Lake 4—cf. ♂	1	1	1	2	6	—

Sepals are lanceolate, linear-lanceolate, or ovate, obtuse or acute, erect or divergent at an angle of 45°, unequal in length, and green, oxblood red, ruby red, or Indian lake. Differences between staminate and pistillate plants in dimensions of sepals lack significance.

	n-pop.	n-pl.	n-rep.	n-obs.	$\bar{x}$ (mm.)	s (mm.)
Length						
Glenora	1	1	—	2	2	—
W. Spanish Peak	1	11	—	26	1.8	.5
Bonita Peak	1	5	—	15	2.5	.8
So. Rocky Mts.	6	14	—	50	2.4	.6
Alaska Basin	1	2	—	6	2.5	0
Glenora—SR	1	9	11	31	2.8	.5
Width						
Glenora	1	1	—	2	1.1	—
W. Spanish Peak	1	6	—	10	.7	.2
Bonita Peak	1	5	—	15	.8	.2
So. Rocky Mts.	6	14	—	50	.9	.1
Alaska Basin	1	2	—	6	1	.2
Glenora—SR	1	9	11	31	1.4	—

Petals are elliptic-oblong or oblanceolate-oblong, obtuse, hooded and spreading in ♂ flowers, plane or hooded and erect in ♀ flowers, chrysanthemum crimson or ruby red, rarely yellow streaked or flushed with crimson or red just at apex. Differences among populations in dimensions lack significance except differences among pistillate plants in length, which are highly significant. Differences between sexes are significant in length and highly significant in width, but only the difference in width is confirmed by experiment.

	n-pop.	n-pl.	n-rep.	n-obs.	$\bar{x}$	s
Number						
Glenora	1	29	—	47	4.1	.5
Root River	1	12	—	20	4.2	.2
W. Spanish Peak	1	11	—	26	4.3	.6
Alaska Basin	1	2	—	6	4	0
Length (mm.)						
Glenora ♂	1	14	—	18	4.3	.7
— ♀	1	5	—	7	3.8	.7
Root River ♂	1	12	—	12	3.7	.6
— ♀	1	1	—	2	2.3	—
W. Spanish Peak ♂	1	5	—	16	3.4	.2
— ♀	1	6	—	9	2.5	1.2
Bonita Peak ♂	1	5	—	19	4.3	.3
So. Rocky Mts. ♂	6	19	—	68	4	.4
— ♀	2	2	—	5	3.5	—
Seven Devils Lake ♂	1	6	—	21	3.3	.2
— ♀	1	4	—	19	1.9	.4
Alaska Basin ♀	1	2	—	6	2.2	.1
Glenora—SR ♂	1	6	8	22	4.2	.4
— ♀	1	2	2	6	2.4	1.1
W. Spanish Peak—cf. ♂	1	6	9	18	3.2	—
— ♀	1	1	1	2	3.8	—

	n-pop.	n-pl.	n-rep.	n-obs.	$\bar{x}$	s
Length (mm.)						
Bonita Peak—cf. ♂	1	4	4	8	4.2	.4
So. Rocky Mts.—cf. ♂	5	11	14	29	3.9	—
— ♀	1	1	1	2	4.3	—
Beartooth Summit—cf. ♀	1	1	1	2	2.1	—
Seven Devils Lake—cf. ♂	1	1	1	1	3.7	—
— ♀	1	1	1	1	2.8	—
Sierra Nevada—cf. ♂	3	3	4	8	3	—
— ♀	1	1	1	2	2.7	—
Width (mm.)						
Glenora ♂	1	14	—	19	1.5	.3
— ♀	1	5	—	8	.9	.1
Root River ♂	1	12	—	14	1.4	.3
— ♀	1	1	—	1	1	—
W. Spanish Peak ♂	1	5	—	16	1.5	.3
— ♀	1	5	—	9	1.4	.2
Bonita Peak ♂	1	5	—	19	1.3	.3
So. Rocky Mts. ♂	6	19	—	68	1.4	.2
— ♀	2	2	—	5	1.4	—
Seven Devils Lake ♂	1	6	—	21	1.3	.2
— ♀	1	4	—	19	.9	.1
Alaska Basin ♀	1	2	—	6	.6	0
Glenora—SR ♂	1	6	8	22	1.7	.3
— ♀	1	2	2	6	.6	.5
W. Spanish Peak—cf. ♂	1	6	9	18	1.4	—
— ♀	1	1	1	2	1.8	—
Bonita Peak—cf. ♂	1	4	4	8	1.5	.2
So. Rocky Mts.—cf. ♂	5	11	14	29	1.5	—
— ♀	1	1	1	2	1.8	—
Beartooth Summit—cf. ♀	1	1	1	2	.6	—
Seven Devils Lake—cf. ♂	1	1	1	1	1.5	—
— ♀	1	1	1	1	.9	—
Sierra Nevada—cf. ♂	3	3	4	8	1.5	—
— ♀	1	1	1	2	.8	—

Stamens have ruby red or chrysanthemum crimson filaments and crimson or rhodonite red anthers. Rarely the anthers are yellow. Stamens opposite petals are adnate to their bases. One plant from West Spanish Peak had a second superposed stamen above an epipetalous one. Differences among populations in length of filaments are highly significant and in adnation of filaments are significant. Differences among populations in length of anthers are highly significant on the basis of data from the experiment in the cold frame.

	n-pop.	n-pl.	n-rep.	n-obs.	$\bar{x}$ (mm.)	s (mm.)
Filaments—length						
Glenora	1	1	—	2	3	—
Bonita Peak	1	5	—	15	4.6	.3
So. Rocky Mts.	5	12	—	47	4.5	.5
Glenora—SR	1	7	9	25	4.3	.4
Epipetalous filaments—length of adnation						
Glenora	1	1	—	2	.3	—
Bonita Peak	1	5	—	15	1.1	.2
So. Rocky Mts.	5	12	—	47	1	.1

	n-pop.	n-pl.	n-rep.	n-obs.	x̄ (mm.)	s (mm.)
Epipetalous filaments—length of adnation						
Seven Devils Lake	1	4	—	4	.7	.2
Glenora—SR	1	7	9	25	.6	.2
Anthers—length						
Glenora	1	1	—	2	.7	—
W. Spanish Peak	1	3	—	10	.7	.1
Bonita Peak	1	3	—	12	1	.2
So. Rocky Mts.	5	10	—	36	.8	.1
Seven Devils Lake	1	2	—	5	.6	.03
Root River—TG	1	1	1	2	.8	—
Platoro—TG	1	1	1	2	.8	—
Glenora—SR	1	4	5	7	.8	.1
W. Spanish Peak—cf.	1	6	8	11	.8	.2
Bonita Peak—cf.	1	4	4	6	.8	.1
So. Rocky Mts.—cf.	5	10	12	19	.8	—
Hilton Lake 4—cf.	1	1	2	3	.6	.03

Nectaries are quadrate or oblong; truncate, emarginate, and sometimes erose at apex; and yellow, yellow flushed with crimson, maroon, pale ruby red, or orange. Differences among populations in length are highly siginficant for wild plants. Differences between sexes in dimensions lack significance.

	n-pop.	n-pl.	n-rep.	n-obs.	x̄ (mm.)	s (mm.)
Length						
Glenora	1	28	—	29	1.1	.2
Root River	1	13	—	15	.7	.1
W. Spanish Peak	1	10	—	24	.7	.1
Bonita Peak	1	5	—	15	.6	.2
So. Rocky Mts.	8	29	—	76	.6	.1
Alaska Basin	1	2	—	6	.9	.1
Seven Devils Lake	1	11	—	40	.9	.3
Glenora—SR	1	9	11	31	1	.2
W. Spanish Peak—cf.	1	3	4	8	.8	—
Bonita Peak—cf.	1	2	2	4	.7	.2
Sierra Nevada—cf.	2	2	2	4	1	—
Width						
Glenora	1	1	—	2	.5	—
Root River	1	1	—	1	.7	—
W. Spanish Peak	1	10	—	24	.5	.1
Bonita Peak	1	5	—	15	.7	.1
So. Rocky Mts.	8	20	—	68	.7	.2
Alaska Basin	1	2	—	6	.6	.1
Seven Devils Lake	1	11	—	40	.5	.2
Glenora—SR	1	9	11	31	.8	—
W. Spanish Peak—cf.	1	4	5	10	.7	.1
Lake Peak—cf.	1	1	1	2	.9	—
Bonita Peak—cf.	1	2	2	4	.8	.1
Beartooth Summit—cf.	1	1	1	2	.6	—
Sierra Nevada—cf.	3	3	3	6	.6	—

Carpels are erect, with short styles. The ovaries are Indian lake, ruby red, or green. In staminate flowers, carpels are rudimentary. Differences between sexes in length of carpels are significant. Likewise, differences among populations in number of ovules per ovary are significant.

	n-pop.	n-pl.	n-rep.	n-obs.	$\bar{x}$	s
Length (mm.)						
Glenora ♂	1	14	—	14	3.2	.9
— ♀	1	5	—	5	7.7	1.5
W. Spanish Peak ♂	1	5	—	16	2.1	.5
— ♀	1	6	—	9	4.6	1.7
Bonita Peak ♂	1	5	—	15	3.4	1
So. Rocky Mts. ♂	4	11	—	45	3.3	.8
— ♀	2	2	—	3	9.1	—
Seven Devils Lake ♀	1	4	—	19	3.7	.4
Ovules per ovary						
W. Spanish Peak	1	6	—	8	7	1.5
So. Rocky Mts.	2	2	—	3	8	—
Alaska Basin	1	2	—	6	9	.8
Seven Devils Lake	1	1	—	1	10	—

Fruits are erect, brown follicles with short, divergent beaks. Measurements of length exclude beaks. On the basis of data for wild plants, differences among populations in length are significant, but on the basis of results of the experiment in the cold frame, these differences lack significance.

	n-pop.	n-pl.	n-rep.	n-obs.	$\bar{x}$ (mm.)	s (mm.)
Length						
Glenora	1	4	—	20	6.5	1.2
Sandia Mts.	1	1	—	2	5.4	—
Beartooth Mts.	1	3	—	6	4.8	.4
Seven Devils Lake	1	5	—	25	3.7	.7
Sierra Nevada	3	4	—	11	5.2	.2
Sierra Blanca—TG	1	2	2	4	4.3	.4
Glenora—SR, 1970	1	2	2	10	7.8	.4
W. Spanish Peak—cf.	1	1	1	5	4.8	—
Bonita Peak—cf.	1	1	1	5	4.6	—
So. Rocky Mts.—cf.	3	3	3	12	4.8	—
Beartooth Mts.—cf.	1	1	1	2	6.7	—
Seven Devils Lake—cf.	1	1	1	2	4.7	—

Seeds are pyriform or oblanceolate, finely longitudinally ribbed, winged, and reddish brown or yellowish brown. Very fine riblets occur transversely between the longitudinal ribs. Differences among populations in length are highly significant on the basis of data for wild plants, but experimental confirmation is lacking because of the small number of plants which produced seeds in cultivation.

	n-pop.	n-pl.	n-obs.	$\bar{x}$ (mm.)	s (mm.)
Length					
Glenora	1	9	89	2.4	.4
Sandia Mts.	1	1	2	1.4	—
Beartooth Mts.	1	3	5	1.9	.1
Seven Devils Lake	1	5	26	1.9	.1

	n-pop.	n-pl.	n-obs.	x̄ (mm.)	s (mm.)
Length					
Sierra Nevada	3	3	11	1.4	—
Sierra Blanca—TG	1	2	7	1.5	.04
Glenora—SR	1	1	5	2.6	—
W. Spanish Peak—cf.	1	1	5	1.9	—
Platoro—cf.	1	1	5	1.7	—
Bonita Peak—cf.	1	1	5	1.9	—
Beartooth Mts.—cf.	1	1	3	1.9	—
Seven Devils Lake—cf.	1	1	3	2	—
Diameter					
Glenora	1	4	17	.6	.1
Sandia Mts.	1	1	2	.4	—
Beartooth Mts.	1	3	5	.7	.1
Seven Devils Lake	1	5	26	.5	.1
Sierra Nevada	3	3	11	.4	—
Sierra Blanca—TG	1	2	7	.6	.04
Glenora—SR	1	1	5	.5	—
W. Spanish Peak—cf.	1	1	5	.6	—
Platoro—cf.	1	1	5	.6	—
Bonita Peak—cf.	1	1	4	.6	—
Beartooth Mts.—cf.	1	1	3	.5	—
Seven Devils Lake—cf.	1	1	3	.5	—

Chromosomes are unequal in length. Sporophytic chromosomes vary from globular to elliptical. Sometimes they show obscure constrictions. Counts cited below are for plants used in preparing the foregoing description. Information is available for all four subspecies.

	n-pl.	g (n)	sp (2n)	Cytologist
Number				
Glenora	1	18	36	C. H. Uhl (1952)
Glenora	2	18	36	L. & R. Benjamin, 1970
Watkins Glen	2	18	36	C. H. Uhl (1952)
Root River	1	18	—	C. H. Uhl (1952)
W. Spanish Peak	1	18	—	Ilse Zandstra, 1969
Sierra Blanca	1	18	—	C. H. Uhl (1952)
Sandia Mts.	1	18	—	C. H. Uhl (1952)
Lake Peak	1	18	—	Ilse Zandstra, 1969
Rio Hondo Canyon	1	18	—	C. H. Uhl (1952)
Eureka Gulch	1	—	36	Ilse Zandstra, 1969
Hilton Lake 4	1	18	—	Ilse Zandstra, 1969

	n-pl.	x̄ (μ)	s (μ)	Cytologist
Length of sporophytic chromosomes				
Glenora	2	.83	.2	L. & R. Benjamin, 1970

Variation. *Sedum integrifolium* is a variable species. Five specific epithets and two distinctive varietal names proposed for various populations attest to the variability. Physiological adaptation to different environmental conditions may be an important reason for the variability. Geographic variation may be another reason. Hybridization less likely is important because the most distinctive populations are isolated and in places where other species do not occur. However, hybridization may have figured in the origin of the species, although details are unknown.

Differences among populations are indicated in the following table. Although the list is short and some differences were not confirmed by experiment, those which are real make for distinctive plants which are easy to recognize.

Feature	Wild plants	Cold frame
Height	*	**
Median leaves		
Length	*	n.s.
Width	**	n.s.
Length/width	n.s.	**
Teeth per margin	**	**
Petals		
Length (♀ plants)	**	—
Width (♀ plants)	**	—
Stamens		
Filaments—length	**	—
Epipetalous filaments—length of adnation	*	—
Anthers—length	n.s.	**
Nectaries		
Length (♂ plants)	**	n.s.
(♀ plants)	*	n.s.
Width (♂ plants)	n.s.	*
Pistils		
Length	**	—
Ovules per ovary	*	—
Follicles		
Length	*	n.s.
Seeds		
Length	**	—
Totals	8**	4**
	6*	1*

Plants from most populations of *Sedum integrifolium* grow poorly in cultivation at Ithaca, N.Y. This circumstance makes study of variation difficult and reduces the chance for satisfactory results. Some plants grew with sufficient vigor in experiments to permit comparison between their condition in nature and in cultivation. The results, shown in the following tables, reveal diverse modifications: tall plants becoming both shorter and taller, and short plants becoming shorter. Some plants evidently are capable of more modification than others. The idea of Clements *et al.* (1950:173 and 201), that short plants from alpine situations grow taller at low altitudes when moisture is adequate, is not sustained. Plants from West Spanish Peak and Seven Devils Lake, for example, made good growth in the experiment in the cold frame, but after several years of cultivation, were even shorter than they were when collected in nature. Some plants do grow taller when given increased moisture and shade, but others retain a dwarf, compact habit under these same conditions. Difference in response to conditions of growth is an important reason for recognizing two subspecies in the Southern Rocky Mountains. Changes in cultivation are indicated for several features.

NATIVE SPECIES 499

Height

Source of plant and no.	Ht. in nature (cm.)	Yrs. in cult. (n)	Ht. in cult., 270 m. (cm.)
Glenora 69–20 ♂, 138 m.	30	3	15
69–23 ♂, 137 m.	15	3	17
West Spanish Peak 62–79 ♂, 3,300 m.	15	9	3
62–100 ♂, 3,305 m.	7	6	2
Lake Peak 62–242 ♂, 3,590 m.	30	3	17
62–243 ♂, 3,520 m.	16	3	27
Bonita Peak 62–165 ♂, 4,050 m.	20	9	15
62–166 ♂, 4,032 m.	15	6	24
Seven Devils Lake 61–182–11 ♂, 2,260 m.	8	4	4

Leaves—length

Source of plant and no.	Lgth. in nature (cm.)	Yrs. in cult. (n)	Lgth. in cult. (cm.)
Glenora 69–20	3.5	3	3.1
62–23	1.6	3	4
West Spanish Peak 62–79	1.2	9	.7
62–100	1.4	6	1
Lake Peak 62–242	3	3	2.2
62–243	2.2	3	3.6
Bonita Peak 62–165	2.5	9	1.5
62–166	1.5	6	2.3
Seven Devils Lake 61–182–11	1.3	4	1.4

Leaves—length/width

Source of plant and no.	Lgth./wdth. in nature	Yrs. in cult. (n)	Lgth./wdth. in cult.
Glenora 69–20	2.9	3	3.4
69–23	2.4	3	4.1
West Spanish Peak 62–79	1.7	9	1.6
62–100	1.7	6	2.5
Lake Peak 62–242	4.6	3	3.9
62–243	2.4	3	3.8
Bonita Peak 62–165	3	9	2.1
62–166	1.9	6	2.5
Seven Devils Lake 61–182–11	2.4	4	2

Color of leaves

Source of plant and no.	Color in nature	Yrs. in cult. (n)	Color in cult.
Glenora 69–20	blue-green	3	glaucous
69–23	glaucous	3	glaucous
West Spanish Peak 62–79	spinach green	9	spinach green
62–100	spinach green	6	spinach green
Lake Peak 62–242	lettuce green	3	spinach green
62–243	spinach green	3	spinach green
Bonita Peak 62–165	pale spinach green	9	spinach green
62–166	pale spinach green	6	spinach green
Seven Devils Lake 61–182–11	pale green	4	spinach green

	Color of petals		
Source of plant and no.	Color in nature	Yrs. in cult. (n)	Color in cult.
West Spanish Peak 62–79	ruby red	9	chrysanthemum crimson
62–100	Naples yellow tipped with red	6	yellow-green tipped with red
Bonita Peak 62–165	ruby red	9	chrysanthemum crimson
62–166	ruby red	6	chrysanthemum crimson
Seven Devils Lake 61–182–11	deep purple	4	chrysanthemum crimson

Age, vigor, and freedom from disease are factors which determine the morphological expressions of plants. All plants investigated in this study were old enough to flower and appeared free from disease.

Proportions of male and female plants in populations fluctuate. In some populations, as along the western side of Seneca Lake, pistillate plants outnumber staminate plants. At the other extreme, male plants are six times as numerous as female plants at Seven Devils Lake. Although plants with flowers which are functionally hermaphrodite are sometimes lacking in populations, they are occasional at other sites, 16% on West Spanish Peak and 22% on Pikes Peak. Differences in sex ratios and a higher incidence of perfect flowers distinguish *S. integrifolium* from *S. rosea*. In the following table, showing sex ratios, all data are from random samples.

Population	Plants studied (n)	♂(n)	♀(n)	Hermaph. (n)	Sterile (n)	Ratio ♂:♀
Glenora	965	347	451	0	167	.77:1
W. Spanish Peak	37	15	16	6	0	.94:1
Lake Peak	28	16	12	0	0	1.3:1
Pikes Peak	27	14	7	6	0	2:1
Platoro	58	41	10	2	5	4.1:1
Seven Devils Lake	90	77	13	0	0	5.9:1
Totals	1,205	510	509	14	172	1:1

Petals of staminate plants tend to be longer and wider than those of pistillate plants. Experiments confirm the value of width of petal as a secondary sexual characteristic. Some data for demonstration of differences in secondary sexual characteristics observed in nature follow.

Feature	n-pop.	n-pl.	♂	♀
Flowers—diam. (mm.)**	12	63	5.9	3.4
Petals—length (mm.)*	12	76	3.7	2.4
—width (mm.)**	12	75	1.4	1
Pistils—length (mm.)*	10	49	3.1	5.6

The following two differences in secondary sexual characteristics were observed in experiments.

Feature	n-pop.	n-pl.	n-rep.	♂	♀
Petals—length (mm.)n.s.	15	26	33	3.5	3.1
—width (mm.)**	15	26	33	1.5	1.1

A greater, general variability of *Sedum integrifolium* is illustrated by fluctuations in number of petals per flower. The frequency of flowers with 3 petals is almost four times as great as in *S. rosea*. Likewise, flowers occur with more petals than usual, for example 10 petals (fig. 141) on a plant from Glenora. The 5-merous condition is a little commoner than the 4-merous, unlike *S. rosea* in which 4-mery is usual. The following frequency distribution shows the condition in flowers from random samples of plants.

No. of petals	No. of flowers	%	Cumulative %
3	9	14.3	14.3
4	26	41.3	55.6
5	28	44.4	100
Totals	63	100	

The taxonomic intepretation of the variational pattern of *Sedum integrifolium*, although beset with problems, is not impossible. To ignore group differences and to

Fig. 141. Flower of *Sedum integrifolium* ssp. *leedyi* with 10 petals. Plant is from west side of Seneca Lake, north of Glenora, Yates Co., N.Y., cultivated in garden, Ithaca, N.Y., June 11, 1956 (x 7). Photo by Howard Lyon.

place together all populations under the binomial, *S. integrifolium*, appending the locality for precise designation, misses the opportunity for classification to show relationships. A better way is to distinguish the isolated populations which exhibit distinctive traits and also the major evolutionary trends in the species. Important because of geographic isolation are the populations by Seneca Lake in New York and along the north branch of the Root River in Minnesota. Plants in these populations are tall, with long, glaucous leaves and elongate seeds. Another distinctive population occurs on the Sierra Blanca in New Mexico. Plants there have narrower leaves than in any of the Rocky Mountain populations. This difference is real. Plants in the garden in Ithaca, after five years in cultivation, continued to have the narrowest leaves of any population of *S. integrifolium*. Also, the petals are yellow, red only at the apex. In the Rocky Mountains and farther west, two types of *S. integrifolium* occur, one with the leaves elongate, mostly 2.5 or more times longer than broad, and the stems tall, and the other with the leaves relatively shorter, < 2.5 times longer than broad, and the stems short. These differences exist even at the highest altitudes in the alpine zone.

KEY TO SUBSPECIES

A. Leaves long and narrow, ≤ 5 mm. wide, and 3.6 times longer than wide; petals yellow, red only at the tips .... ssp. *neomexicanum*, p. 509
AA. Leaves of various shapes, but ≥ 5 mm. in width; petals usually some shade of dark red or crimson .... B
    B. Leaves glaucous, appearing blue-green, averaging 30 mm. long; seeds elongate, averaging 2.4 mm. long .... ssp. *leedyi*, p. 502
    BB. Leaves normally spinach green, sometimes glaucous; seeds averaging less than 2 mm. long C
        C. Leaves elongate, mostly more than 2.5 times longer than broad; stems usually more than 10 cm. tall; .... ssp. *procerum*, p. 506
        CC. Leaves seldom more than 2.5 times longer than broad; stems short, usually less than 10 cm. long; .... ssp. *integrifolium*, p. 509

Sedum integrifolium ssp. *leedyi*

Nomenclature. *Sedum integrifolium* ssp. **leedyi** (Rosendahl et Moore), comb. nov., fundatum super *Sedum Rosea* (L.) Scop. var. *Leedyi* Rosendahl et Moore, Rhodora 49:198 (1947).

A synonym is:

1947. *Sedum Rosea* var. *Leedyi* Rosendahl and Moore, Rhodora 49:198, pl. 1,086.
Type locality: northeast-facing cliffs of limestone, North Branch of Root River opposite Herman McDaniel farm about 5 km. south of Simpson, Olmstead Co., Minn. Type: a collection of John L. Leedy (MIN), July 12, 1942 (pl. 1,086).

Plants of ssp. *leedyi* (fig. 140) produce tall floral stems with glaucous, oblong leaves. Except for some populations of ssp. *procerum* in the Sangre de Cristo and San Juan Mountains of New Mexico and Colorado, ssp. *leedyi* surpasses the other subspecies in robustness of stems and leaves. Likewise, the seeds are the longest known for the species. Relationships appear to be with ssp. *procerum*. Plants from Lake Peak, N.M., are similar in some dimensions, but differ in their response to conditions for growth in Ithaca, their spinach green leaves, fewer teeth on margins of leaves, and smaller follicles. Color of the petals provides another distinction. Petals are dark red, usually greenish yellow basally, in ssp. *leedyi*, but solidly chrysanthemum crimson in ssp. *procerum*. Two populations by Seneca Lake have the most yellow in the petals. Some plants in the population in Minnesota have petals dark red to the base. Others have the petals greenish white basally.

Ssp. *leedyi* (fig. 142) occurs in two disjunct areas, by Seneca Lake, N.Y., and along the North Branch of the Root River in Minnesota. The populations on the western side of Seneca Lake and in Minnesota undoubtedly are natural. The one at Watkins Glen might have been introduced from farther up the lake. In any case, discovery was not until 1918, a surprisingly late date for a place so much visited as Watkins Glen. The population is tiny and has been steadily declining in number of plants: 1961—7, but 1972—only 1. The population on the western side of Seneca Lake (fig. 143) extends northward along the lake from Glenora Point for 2.7 kilometers. The plants (fig. 144) there grow on ledges and in crevices of cliffs and on talus, in partial shade. Commonest competitors of randomly selected plants in the population by Seneca Lake, rated on a scale of 24, are: *Poa compressa* 7, *Geranium robertianum* 3, *Parthenocissus vitacea* 3, *Parietaria pennsylvanica* 2, and a species of *Bromus* 2. Other plants, rare in New York, occurring in association with *Sedum integrifolium* on the western side of Seneca Lake north of Glenora are *Draba arabisans* (1969) and *Achillea borealis* (1958). The population along the North Branch of the Root River is .5 kilometers long. Competing species there are *Cystopteris bulbifera*, *Aquilegia canadensis*, *Draba arabisans*, and *Rhus toxicodendron*. Further ecological data are in the following table.

Population	Plants (n)	Area (m.^{2})	Density per m.^{2}	Exposure
Glac. Allegheny Plat.				
Glenora, Seneca Lake, N.Y.	10,422	33,489	.3	E
Watkins Glen, N.Y.	1	.04	—	S
Driftless Area of Central Lowland				
Root River, Minn.	535	~18,500	.03	NNE–NW

Population	Altitudinal range (m.)	pH, range	pH, tests (n)	Oldest record
Glenora	136–141	6.8–7.6	11	<1842
Watkins Glen	157	—	—	1918
Root River	347–384	—	—	1936

Population	Latitude (° N.)	Rock	Drainage	Vitality
Glenora	42° 30–31'	shale, limestone	good-poor	vigorous
Watkins Glen	42° 22' 30"	shale	poor	mod. vig.
Root River	43° 53'	limestone	good	vigorous

The population along the North Branch of the Root River is in the western part of the Driftless Area and may have survived there during Wisconsin glaciation. Since

504 SEDUM OF NORTH AMERICA

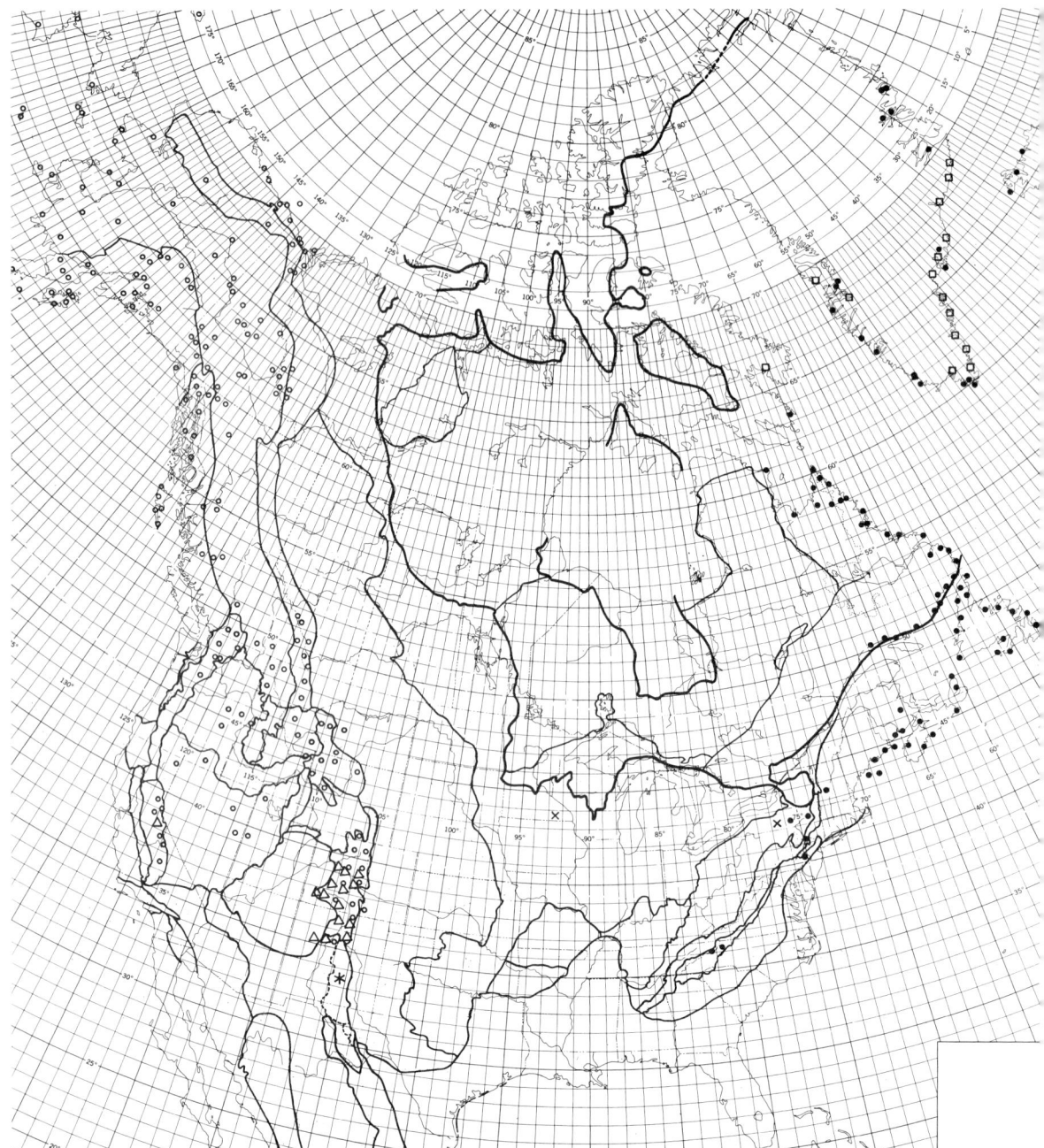

Fig. 142. Known distribution of Sedum integrifolium ssp. integrifolium (○), ssp. leedyi (x), ssp. procerum (△), and ssp. neomexicanum (∗); and S. rosea (●) and (□ for literature records).

NATIVE SPECIES 505

Fig. 143. Habitat of *Sedum integrifolium* ssp. *leedyi* on cliffs of shale on western side of Seneca Lake about 1 km. north of Glenora Point, Yates Co., N.Y., July 15, 1969.

Fig. 144. Plant of *Sedum integrifolium* ssp. *leedyi* on ledge of shale on western side of Seneca Lake about .5 km north of Glenora Point, Yates Co., N.Y., July 2, 1969.

the two populations by Seneca Lake are in glaciated territory, the plants must have spread to the present sites in the last ten thousand years, probably from an unglaciated area farther west. The present occurrence in the Finger Lakes Region of New York is another example of a western species which supports the idea of an eastern extension of the steppe in postglacial time (Gleason, 1922; Guilday, Martin, and McCrady, 1964; Schmidt, 1938; and Transeau, 1935).

Sedum integrifolium ssp. **procerum**

Subspecies nova *Sedi integrifoliae* cum caulibus procerioribus, 11–49 cm., et foliis semper fere viridis, aliquando glaucis. Flores rubri sunt. Typus in Herbario Wiegand, Universitatis Cornellianae, ab Bonita Peak, declivitate occidentale, inter rupibus, altitudine 3,932 m., Southern Rocky Mountains, borealis ab Silverton, San Juan Co., Colo., collectio Roberti Clausenii num. 62–69, Augusto 2, 1962, est.

Plants of ssp. *procerum* (figs. 145 and 146) are tall, to 49 cm., with usually green, or sometimes glaucous, leaves, and petals which are ruby red from base to apex. Even at high altitudes, plants retain their tall stature, whereas ssp. *integrifolium* becomes dwarf and compact in the alpine zone. Plants of ssp. *procerum* resemble plants of ssp. *leedyi*, but grow less well in the garden at Ithaca and normally have green leaves. In nature, plants with glaucous leaves most often are in shaded situations. Ssp. *leedyi* always has glaucous leaves and grows better in cultivation. Ssp. *neomexicanum* has narrower leaves and yellow petals which are red only at the apex. *Rhodiola alaskana* Rose is similar in its tall stems, but has glaucous leaves. It may be ecotypically different, but possibly is only a modification of ssp. *integrifolium*. In any case, its survival at Ithaca is poorer than the survival of ssp. *integrifolium*.

The principal area of occurrence of ssp. *procerum* (fig. 142) is in the Southern Rocky Mountains, especially in the Sangre de Cristo and San Juan Mountains. Populations elsewhere, as on Piute Peak in the Sierra Nevada, are not entirely typical, but are somewhat intermediate between ssp. *procerum* and ssp. *integrifolium*. Ssp. *procerum* is a high-altitude subspecies. Plants occur on rocky, montane slopes, among rocks and in gravel or stony loam, from the zone of *Picea pungens* and *P. engelmannii* up into the alpine region. Competing plants are miscellaneous and include species of grasses, especially *Festuca ovina* (?), *F. brachyphylla*, and *Trisetum spicatum*, as well as *Saxifraga bronchialis* ssp. *austromontana*, *Geum rossii*, and others. A summary of data for eight populations follows.

Population	Plants (n)	Area (m.^{2})	Density per m.^{2}	Exposure
So. Rocky Mts.				
Lake Peak, N.M.	168	—	—	N,E,S,W
Rio Hondo Canyon, N.M.	200	~4,000	~.05	S
Platoro, Colo.	754	~2,080	~.36	NW–NE
Summit Peak, Colo.	5,000	~2,000,000	~.0025	E,S
Elwood Pass, Colo.	?	—	—	SW
Bonita Peak, Colo.	2,000	~2,000,000	~.001	W,E
Eureka Gulch, Colo.	?	—	—	S,SW
Sierra Nevada				
Piute Peak, Calif.	?	—	—	N

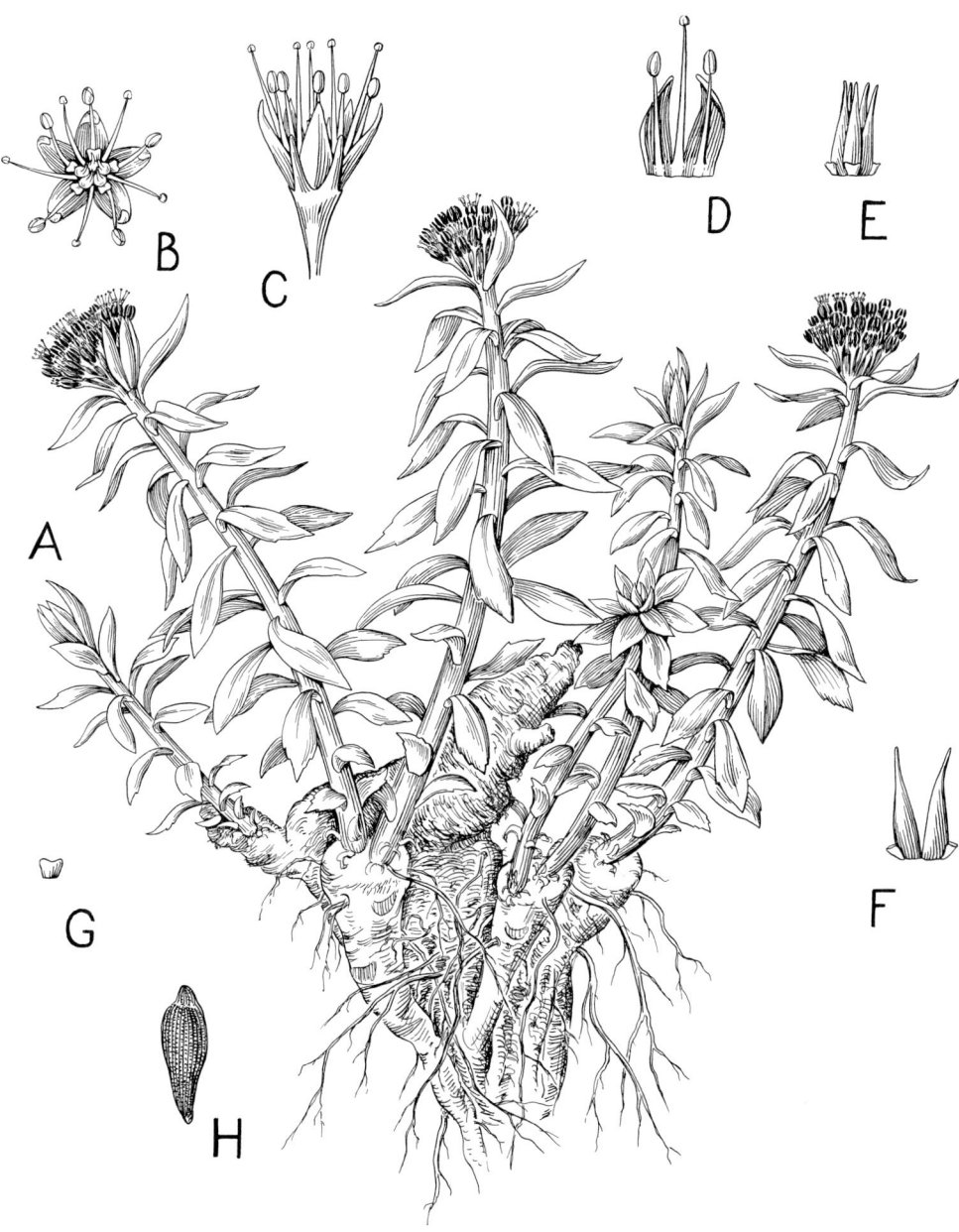

Fig. 145. Plant of *Sedum integrifolium* ssp. *procerum* from Lake Peak, Santa Fe Co., N.M., cultivated in cold frame, Ithaca, N.Y., April 24, 1964. A. Habit sketch (x 1). B. Flower from above (x 3.1). C. Flower from side (x 3.1). D. Two petals and stamens (x 3.1). E. Carpels (x 4.1). F. Two carpels and nectaries (x 5.1). G. Nectary (x 5.1). H. Seed from pistillate plant (x 10).

Fig. 146. Plants of *Sedum integrifolium* ssp. *procerum* cultivated in cold frame, Ithaca, N.Y., June 2, 1964 (x .4). Plant on left is from Lake Peak, Santa Fe Co., N.M., and has green leaves. Plant on right is from Eureka Gulch, near Lake Emma, in the San Juan Mountains north of Silverton, San Juan Co., Colo., and has glaucous leaves. Photo by Howard Lyon.

Population	Altitudinal range (m.)	pH, range	pH, tests (n)	Oldest record
Lake Peak	3,450–3,590	5.4	1	1962
Rio Hondo Canyon	2,715	—	—	1949
Platoro	2,993–2,999	6.2–6.6	5	1962
Summit Peak	3,378–3,958	5.6–6.6	3	1962
Elwood Pass	3,490	5.2–5.4	2	1962
Bonita Peak	3,412–4,050	6.4–6.6	4	1962
Eureka Gulch	3,580–3,625	6.4–6.6	2	1962
Piute Peak	2,776	—	—	1964

Population	Latitude (°N.)	Rock	Drainage	Vitality
Lake Peak	35° 48′	granite	good	vigorous
Rio Hondo Canyon	36° 35′	—	moderate	vigorous
Platoro	37° 21′ 30″	andesite	moderate-excessive	vigorous
Summit Peak	37° 21′ 30″	andesite	moderate-good	vigorous
Elwood Pass	37° 28′ 30″	andesite	good	vigorous
Bonita Peak	37° 54′	andesite	good	vigorous
Eureka Gulch	37° 54′ 30″	andesite	good	vigorous
Piute Peak	38° 1′	granite	moderate	moderate

Sedum integrifolium ssp. *neomexicanum*

Nomenclature. *Sedum integrifolium* ssp. **neomexicanum** (Britton) comb. nov., fundatum super *Rhodiola Neomexicana* Britton, Bull. N.Y. Bot. Gard. 3(9):38 (1903). Synonyms are:

1903. *Rhodiola Neomexicana* Britton, Bull. N.Y. Bot. Gard. 3(9):38. Type locality: White Mountain Peak, Lincoln Co., N.M., alt. 3,500 meters. Type: a collection of E. O. Wooton (NY), Aug. 1, 1901.

1930. *Sedum roseum* var. *neo-mexicanum* (Britton) Berger in Engler and Prantl, Nat. Pflanzenfam., ed. 2, 18a:440. Based on *Rhodiola neomexicana* Britton.

Elongate, narrow leaves, averaging 5 mm. wide or less, and 3.6 times longer than wide, are distinctive features of ssp. *neomexicanum*. No other North American population of *Sedum integrifolium* is so extreme in the narrowness of the leaves. The basic color of the petals is yellow, with red only at the apex and on the dorsal keel, or less frequently above the middle of the petal. Other populations of the species only exceptionally have plants with yellow petals.

Plants with characteristics of ssp. *neomexicanum* are restricted (fig. 142) to the Sierra Blanca, a peak on the Mescalero Indian Reservation northwest of Ruidoso, N.M., and northeast of the White Sands. The Sierra Blanca is a dry mountain in summer, with few springs. The situation where the *Sedum* occurs is exposed, on rocks on the southern slope of the Middle Ridge, northeast of the highest peak, and also on the eastern slope of the main peak. Two ferns, *Woodsia oregana* and *Cystopteris fragilis*, are the principal competitors. Plants of the *Sedum* flower on the Sierra Blanca in July and August. Other pertinent data are summarized in the following table.

Population	Plants (n)	Area (m.^{2})	Density per m.^{2}	Exposure
Sierra Blanca, N.M.	>35	10,000	>.0035	E,S

Population	Altitudinal range (m.)	pH, range	pH, tests (n)	Oldest record
Sierra Blanca	3,300–3,595	—	—	1901

Population	Latitude (°N.)	Rock	Drainage	Vitality
Sierra Blanca	33° 22′ 30″	porphyry	good	moderate

A population intermediate between ssps. *neomexicanum* and *procerum* occurs below Winsor's Ranch in the northwestern corner of San Miguel Co., N.M.

Sedum integrifolium ssp. *integrifolium*

Nomenclature. *Sedum integrifolium* (Raf.) Nels. ssp. *integrifolium* is based on *Rhodiola integrifolia* Raf., Atlantic Jour. and Friend of Knowledge 1:146 (1832). The basionym is species no. 173, *Sedum Rhodiola* DC., in Torrey, Ann. Lyc. Nat.

Hist. N.Y. 2:206 (1828). Type locality: Rocky Mountains, possibly near the sources of the Platte River. Type: a collection of Edwin James (NY), 1820. The holotype, on the right side of the sheet, is 13 cm. tall to the base of the cyme. The leaves are elliptical and crenate-serrate, 10–11 mm. long and 4–7 mm. wide. Both specimens of the type collection are staminate, with dark red petals. Torrey had written on the label: Rhodiola rosea "*fol. integerrimis*" Hook.

Synonyms are:

1901. *Sedum frigidum* Rydb., Bull. Torr. Bot. Club 28:282. Type locality: Old Hollowtop, near Pony, Mont., alt. 2,745 m. Type: a collection of P. A. Rydberg and E. A. Bessey 4,248 (NY), July 7, 1897. Plants of this sort, with stems often less than a decimeter tall, are common in the Rocky Mountains and adjacent provinces, as well as in Alaska. Plants of the type collection are good matches for plants in populations which I have studied at Seven Devils Lake and on Mount Clements in Glacier National Park. Rydberg's citation of a collection of James suggests that he included the type of *Rhodiola integrifolia* in his species.

1901. *Sedum polygamum* Rydb., Bull. Torrey Bot. Club 28:283. Type locality: West Spanish Peak, Colo., alt. 3,000–3,800 m. Type: a collection of P. A. Rydberg and F. K. Vreeland 6,113 (NY), July 9, 1900. Rydberg emphasized as distinctions stoutness and the presence of stamens in the pistillate flowers. Study of plants on West Spanish Peak (fig. 147) in 1962 revealed that the prevailing plants there above 3,000 m. are slender, as on Pikes Peak, and similar to the type specimen of *S. frigidum*. Only occasional plants are stout. The longest floriferous stem of the holotype is merely 1.5 dm. tall. Of a sample of 37 plants, 6, 16%, had perfect flowers, a lower percentage than on Pikes Peak.

1903. *Rhodiola Alaskana* Rose, Bull. N.Y. Bot. Gard. 3(9):39. Type locality: Misty Harbor, Nagai Island, Alaska. Type: a collection of C. H. Townsend (US), July 22, 1893. In its taller stature, this approaches ssp. *procerum*, but it is slenderer, with strongly toothed, oblanceolate leaves. This may be a coastal modification of a northern ecotype of ssp. *integrifolium*. Plants from Alaska have not remained alive long enough in experiments in Ithaca to yield satisfactory results. Their inability to survive at 42° 30′ N. suggests a physiological difference.

1903. *Rhodiola polygama* (Rydb.) Britton and Rose, Bull. N.Y. Bot. Gard. 3(9):39. Based on *Sedum polygamum* Rydb.

1909. *Sedum integrifolium* (Raf.) Nelson, in Coulter and Nelson, New Manual of Botany of the Central Rocky Mountains, p. 233. Based on *Rhodiola integrifolia* Raf.

1915. *Sedum alaskanum* (Rose) Henry, Fl. South. Brit. Columbia, p. 157. Based on *Rhodiola alaskana* Rose, but without appropriate citation of basionym.

1925. *Rhodiola rosea* var. *integrifolia* (Raf.) Jepson, Manual of the Flowering Plants of California, p. 450. Based on *Rhodiola integrifolia* Raf.

1930. *Sedum roseum* var. *polygamum* (Rydb.) Fröderström, Act. Hort. Goth. 5, App.: 40. Based on *Sedum polygamum* Rydb.

1930. *Sedum roseum* var. *alaskanum* (Rose) Berger, in Engler and Prantl., Nat. Pflanzenfam., ed. 2, 18a:440. Based on *Rhodiola alaskana* Rose.

Fig. 147. Plant of *Sedum integrifolium* ssp. *integrifolium*, topotype of *S. polygamum*, from West Spanish Peak, Huerfano Co., Colo., cultivated in cold frame, Ithaca, N.Y., May 19, 1971 (x 1). Photo by Howard Lyon.

1930. *Sedum roseum* var. *integrifolium* (Raf.) Berger, in Engler and Prantl, Nat. Pflanzenfam, ed. 2, 18a:440. Based on *Rhodiola integrifolia* Raf.

1937. *Sedum roseum* var. *aleuticum* Fröderström ex Hultén, Fl. Aleutian Is., p. 205. Type locality: False Pass, Unimak Island, Aleutian Islands. Type: a collection of W. J. Eyerdam, no. 1,896 ("LD"), Aug. 1, 1932.

1945. *Sedum roseum* ssp. *integrifolium* (Raf.) Hultén, Årsskr. Lunds Univ., N. F. Avd. 2, 41(1):895. Based on *Rhodiola integrifolia* Raf.

1945. *Sedum roseum* ssp. *integrifolium* var. *frigidum* (Rydb.) Hultén, Årsskr. Lunds Univ., N. F. Avd. 2, 41(1):897. Based on *Sedum frigidum* Rydb.

1952. *Rhodiola Rosea* subsp. *integrifolia* (Raf.) Hara, Jour. Fac. Sci., Univ. Tokyo, sect. 3, Bot. 6:62. Based on *Rhodiola integrifolia* Raf.

Although Hara (1952) made *Sedum atropurpureum* Turcz. a synonym of Rafinesque's species, this and other named Asiatic populations may not be the same as North American populations. Data for a satisfactory comparison are lacking.

Ssp. *integrifolium* (fig. 148) is the common, widely distributed type of the species at high altitudes in the western United States and Canada and at high altitudes in Alaska

Fig. 148. Pistillate plant of Sedum integrifolium ssp. integrifolium on northwestern slope of Pikes Peak, Colo., July 6, 1942.

and the Yukon. The plants frequently are of low stature, often under a decimeter in height, with short, elliptical or ovate, spinach green leaves, and usually red or crimson petals. Typical plants are easy to separate from other subspecies, but populations occur which are intermediate between this and ssp. *procerum*. Plants in Alaska and Yukon may be tall, as reported by Porsild (Nat. Mus. Canada Bull. 121:204. 1951). The plants in Alaska, Yukon, and Mackenzie constitute a distinctive ecotype, but satisfactory morphological distinctions are lacking.

The distributional area of ssp. *integrifolium* (fig. 142) is in the mountains of western North America from the Sandia Mountains in New Mexico, 35° 10′ N., and the Sierra Nevada in California, 36° 25′ N., northward to the Arctic coastal plain, north of the Brooks Range, in Alaska. Usual habitats are on rocky slopes or in alpine meadows or tundra. Plants root in crevices in rocks or in shallow soil or gravel. Soils often are lithosols or organic matter. Flowering time is July and August. Data for eleven populations, the principal basis for ideas about the subspecies, follow.

Population	Plants (n)	Area (m.^{2})	Density per m.^{2}	Exposure
Great Plains—Raton sect.				
W. Spanish Peak, Colo.	511	42,000	.012	WNW–NW
Mexican Plateau				
Sandia Mts., N.M.	?	—	—	W
So. Rocky Mts.				
Pikes Peak, Colo. (N.W. spur)	221	4,784	.046	S
Mt. Flora, Colo.	2	.24	—	S
Lulu Pass, Colo.	?	—	—	S,E,W

Population	Plants (n)	Area (m.²)	Density per m.²	Exposure
Central Rocky Mts.				
Beartooth Summit, Wyo.	?	—	—	—
Alaska Basin, Wyo.	34	~16	~2.125	S,W
Northern Rocky Mts.				
Clements Mt., Mont.	?	—	—	SW
Columbia Plateaus—Seven Devils Mts.				
Seven Devils Lake, Ida.	88	3,640	.024	N
Sierra Nevada				
Hilton Lake 4, Calif.	130	—	—	N
Gem Pass, Calif.	104	200	.520	NE

Population	Altitudinal range (m.)	pH, range	pH, tests (n)	Oldest record
W. Spanish Peak	3,240–3,570	5.6–6.6	13	1900
Sandia Mts.	3,170	—	—	<1940
Pikes Peak	3,590	4.8–5.4	5	<1878
Mt. Flora	3,955	5.4	1	1962
Lulu Pass	3,370	5.4–6.2	2	1962
Beartooth Summit	3,250	—	—	1963
Alaska Basin	3,050	6.6	1	1966
Clements Mt.	2,190	—	—	<1937
Seven Devils Lake	2,250–2,290	4.8–5.6	10	1961
Hilton Lake 4	3,140	5.2	1	1964
Gem Pass	3,150–3,151	6.4	2	1964

Population	Latitude (° N.)	Rock	Drainage	Vitality
W. Spanish Peak	37° 23'	andesite	moderate-good	good
Sandia Mts.	35° 11'	limestone	—	good
Pikes Peak	38° 51'	granite	good	good
Mt. Flora	39° 48' 15"	granite	good	good
Lulu Pass	40° 29'	granite	good	good
Beartooth Summit	44° 58'	granite	—	good
Alaska Basin	43° 42'	gneiss	moderate	good
Clements Mt.	48° 42'	argillite	—	good
Seven Devils Lake	45° 20' 40"	andesite	excessive-moderate	medium
Hilton Lake 4	37° 29'	granite	good	good
Gem Pass	37° 47'	serpentine	good	good

Associated species are miscellaneous. Commonest competitors on West Spanish Peak, rated on a scale of 11, are *Saxifraga bronchialis* ssp. *austromontana* 6, *Festuca brachyphylla* 6, *Senecio werneriaefolius* 4, *S. crassulus* 4, and *Carex atrata* 4; on Pikes Peak, rated on a scale of 4, are *Geum rossii* 4, *Carex* sp. 4, and *Trifolium dasyphyllum* 3; by Seven Devils Lake, rated on a scale of 11, are *Lewisia columbiana* 8, *Sedum lanceolatum* ssp. *lanceolatum* 5, and *Arnica gracilis* 5; and by Hilton Lake 4, rated on a scale of 2, *Selaginella watsonii* 2.

Pathology. Most populations of *Sedum integrifolium* appear free from disease. An exception is West Spanish Peak where a few plants had transversely wrinkled stems and pustulate leaves. Likewise, several plants on Bonita Peak showed evidence of disease. The pathogen is unknown.

Reproduction. Perpetuation of *Sedum integrifolium* is by means of both seeds and growth and breaking up of rootstocks. The winged seeds are adapted for dispersal by wind. This mode of dissemination may explain the wide distribution of the species and also its occurrence on many high peaks. The rootstocks endure for a long time. Since the older parts die, the younger parts become severed and as a result a single old plant eventually breaks into many separate plants, each with the same genotype. In cultivation, one clone, now in two parts, has survived for 36 years.

Seeds sometimes germinate in position in the follicles (fig. 149). When these fall to the ground, the seedlings may take root.

Flowering time varies with latitude and altitude. July is the principal month for flowers in the western mountains, but June is the month in the two eastern populations.

Pollinators are bees and syrphus flies.

Relationships. The relationships of *Sedum integrifolium* (fig. 150) appear to be with *S. rosea*. Some authors combine the two species, treating them as subspecies. Different

Fig. 149. Seedlings on old follicles of *Sedum integrifolium* ssp. *leedyi* on western side of Seneca Lake, Yates Co., N.Y., June 16, 1958 (x 2.5). Photo by Howard Lyon.

Fig. 150. Inflorescences of *Sedum integrifolium* ssp. *procerum* (left) from Lake Peak, Santa Fe Co., N.M., and *S. rosea* (right) from Gull Island on the north side of the Gulf of St. Lawrence near Harrington Harbour, Que., cultivated in cold frame, Ithaca, N.Y., June 2, 1964 (x 1.5). Photo by Howard Lyon.

ecological and geographic distributions suggest profound physiological differences, but morphological distinctions are minor. Although few absolute differences are available for separating all populations of one species from all populations of the other, comparison of any two local populations provides many differences. In the following table, data for wild plants of three populations of *S. integrifolium* are contrasted with data for a population of *S. rosea*. Asterisks indicate significant differences among populations, as determined by analysis of variance; (sp.*) indicates significant differences between species; and (sp.**) indicates highly significant differences between species.

Feature	*Sedum integrifolium*			*S. rosea*
	Glenora	West Spanish Peak	Pikes Peak	Mt. Horrid
Staminate plants (*n*)	5	5	8	40
Pistillate plants (*n*)	4	6	1	43
Height (cm.) ♂**	23	9	5	18
—♀**	23	7	10	20
Median leaves of floriferous stems				
Length (mm.) ♂**	29	14	9	19
—♀**	33	12	12	20

	Sedum integrifolium			S. rosea
Feature	Glenora	West Spanish Peak	Pikes Peak	Mt. Horrid
Median leaves of floriferous stems				
Width (mm.)♂**	9	7	5	9
—♀*	9	6	5	8
Length/width ♂**	3.3	2	1.8	2.3
—♀**	4.1	2.1	2.5	2.2
Teeth per left margin (n)♂	3.6	2.9	3.4	—
— ♀	2.3	2.9	2	—
Petals				
Length (mm.)♂** (sp.*)	4.3	3.4	2.7	2.3
Width (mm.)♂** (sp.**)	1.5	1.5	1.1	.7
—♀**	.9	1.4	—	.6
Nectaries				
Length (mm.)♀	.9	.7	—	—
Pistils				
Length (mm.)♀*	8	5	—	6
Follicles				
Length (mm.)♀	7	—	—	6
Seeds				
Length (mm.)♀ (sp.**)	2.4	—	—	1.9
Diameter (mm.)♀	.6	—	—	.4

A test of differences between species and among three populations, based on samples in an experiment of completely random design in a cold frame at Ithaca, N.Y., confirmed the difference in width of petals of staminate plants between species and indicated significant differences in length and width of median leaves of floriferous stems of pistillate plants. Indications of significance are as for wild plants.

	Sedum integrifolium		S. rosea
Feature	West Spanish Peak	Bonita Peak	Gulf of St. Lawrence
Staminate plants (n)	6	4	5
Height (cm.)	4	18	4
Median leaves of floriferous stems			
Length (mm.)	12	19	7
Width (mm.)	6	8	6
Length/width	2	2.4	1.2
Teeth per left margin (n)	1	4.2	—
Petals			
Length (mm.)**	2.1	4.2	2.5
Width (mm.) (sp.**)	1.4	1.5	1.1

	West Spanish Peak	Tesuque Peak	Gulf of St. Lawrence
Pistillate plants (n)	2 (2 reps.)	1 (2 reps.)	11 (19 reps.)
Height (cm.)	5	15	5
Median leaves of floriferous stems			
Length (mm.) (sp.*)	16	22	8
Width (mm.) (sp.*)	8	9	5
Length/width	2.1	2.4	1.7
Teeth per left margin (n)	—	5	1
Petals			
Width (mm.) (sp.**)	1.8	1.7	.8

	West Spanish Peak	Tesuque Peak	Gulf of St. Lawrence
Nectaries			
Length (mm.)	1.1	.7	.9
Follicles			
Length (mm.)	4.8	4.3	4.3
Seeds			
Length (mm.)	1.9	—	1.9
Diameter (mm.)	.6	—	.6

Petals of *Sedum integrifolium* usually are dark red. Petals of most populations of *S. rosea* are greenish yellow, except sometimes suffused with red at the apex in exposed situations. The differences in pigmentation and width of petals, plus a difference in number of chromosomes, 11 versus 18, makes impossible any claim that the plants of the two species are the same. They are different and deserve specific status. *Sedum integrifolium* is the more variable of the two species. This condition is in harmony with the idea that it contains a genome of *S. rosea*, together with a genome of some other species, but more information about chromosomes is necessary. An alternate hypothesis, namely that it is an older species which has become adapted to a greater variety of environmental situations, is equally reasonable.

Usually the leaves of *Sedum integrifolium* are spinach green and not glaucous, and the petals are chrysanthemum crimson. Conversely, the leaves of *S. rosea* are glaucous and the petals are yellow. Data from an experiment of completely random design, including plants of both species, 28 of *S. integrifolium* and 12 of *S. rosea*, make possible a test of independence of these two features. The resulting χ^2, corrected for continuity, 7.94, is highly significant and causes rejection of the hypothesis that glaucosity of the leaves and yellow color of the petals are independent. A similar test, based on 28 plants—13 of *S. integrifolium* and 15 of *S. rosea*, cultivated in an experiment of completely random design in a partially shaded garden at 1421 Slaterville Road,—involved toothing of leaves and length-width relationships. Chi square, 1.16, is not significant, suggesting that shape and toothing of leaves are independent.

Ideas about relationships of *Sedum integrifolium* with various Asiatic species await further information about populations there. Some names of Asian plants may belong in the synonymy of American subspecies, but intelligent grouping requires data similar to what is available for North American populations. Names for consideration are *S. stephanii* Chamisso, with elliptic-linear or lanceolate-oblong leaves, 1–7 mm. wide, and yellow petals; *S. algidum* Led., with leaves 2–4 mm. wide and petals 9–10 mm. long; *S. kirilowii* Regel, which is close to *S. stephanii*; and *S. atropurpureum* Turcz., which may be an Asiatic subspecies of *S. integrifolium* or even a synonym of ssp. *integrifolium*.

Reference: Kennedy (1912).

30. Sedum rosea*** (figs. 142 and 151–155)

Sedum rosea (fig. 151) is characterized by stout rootstocks with scalelike leaves; annual floriferous stems with glaucous, ovate or oblong leaves; and corymbose

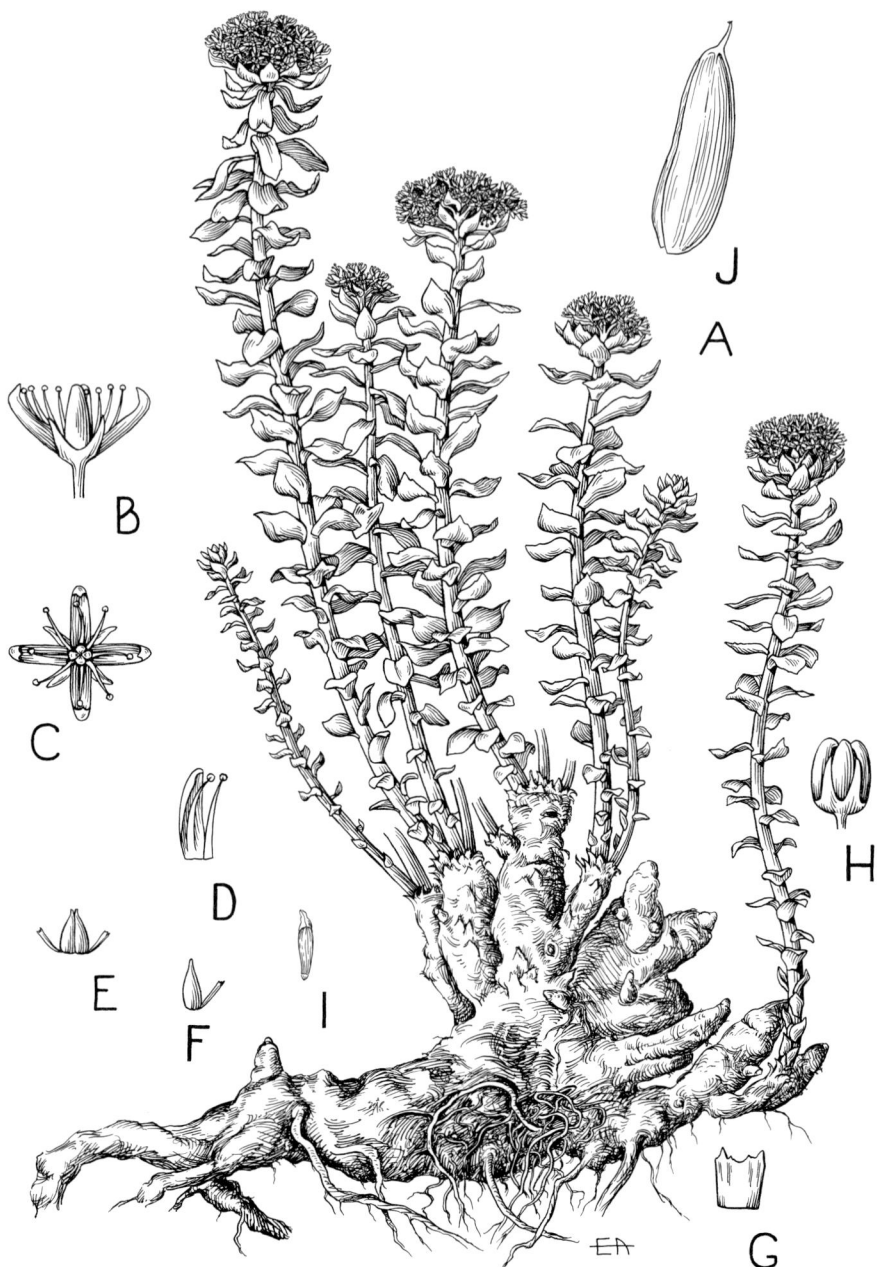

Fig. 151. Plant of *Sedum rosea* from Mount Horrid, Rochester Township, Windsor Co., Vt. A. Habit sketch (x .48). B. Flower from side (x 3.8). C. Flower from above (x 3.8). D. Petal and two stamens (x 4.8). E. Carpels (x 4.8). F. Single carpel and nectary (x 5.8). G. Nectary (x 9.6). H. Bud (x 3.8). I. Seed (x 4.8). J. Follicle (x 4.8).

inflorescences with imperfect flowers with narrow, yellow petals, 1.1 mm. or less in width, and sometimes red-tipped. The similar *S. integrifolium* commonly has the leaves green, not glaucous, and the petals wider, $\geq$ 1.1 mm. wide, and usually red. *Sedum rhodanthum* has racemose inflorescences with larger, perfect flowers with pink petals. Except for populations of *S. integrifolium* along the western side of Seneca Lake and at Watkins Glen, *S. rosea* is the only species of subgenus *Rhodiola* in northeastern North America. It is the most consistently dioecious species of *Sedum* in America. The common name, roseroot, alludes to the property of the dry rootstocks to have the scent of dry rose petals.

Description. Sample: 299 plants from 14 populations. Sources of plants, years of study at sites, mode of selection of plants from populations, and fraction studied from each population are indicated in the following table. (C = convenience, J = judgment, R = random, and St.L. = St. Lawrence).

Population	Year of study	N	n	n/N	Mode of selection	Stage when studied
B138, Shag Island, Gulf St.L.	1959	9,000	9	.001	R & J	Fl.
M411, Isl. N. of Forsyth Isl., Gulf St.L.	1959	>3,272	8	.002	R & J	Fl.
B203, Stevenson Isl., Gulf St.L.	1959	894	24	.03	R	Imm. fr.
N2, Gull Island, Gulf St.L.	1959	376	25	.07	R	Fr.
Natashquan, Que.	1959	—	5	—	J	Fl.
Little Moose Island, Me.	1967	1,127	4	.004	R & J	Fr.
St. Sauveur Mt., Me.	1967	34	2	.06	J	Fr.
Mt. Horrid, Vt.	1954	499	101	.2	J	Fl. & Fr.
Chittenango Falls, N.Y.	1946	15	1	.07	C	Fl.
Haines Falls, N.Y.	1947	~800	11	.014	J	Fl.
Esopus Creek, N.Y.	1954	2,075	71	.034	J	Fl. & Fr.
Millrift, Pa.	1954	358	30	.08	J	Fl. & Fr.
Nockamixon Rocks, Pa.	1954	151	7	.05	J	Fl. & Fr.
Roan Mountain, N.C.	1941	2	1	<.33	J	Fr.
Totals		>18,603	299			

In the following tables, descriptive data for the four populations in the Gulf of St. Lawrence are combined in a single expression, designated as Gulf St.L. Data for the two populations in Maine likewise are pooled and cited simply as Maine. Data for other populations are cited separately under names of localities.

Experiments of completely random design were in three situations: outdoors in full sun at the Cornell University Test Garden (TG), outdoors in partial shade at 1421 Slaterville Road, Ithaca, N.Y. (SR), and in a cold frame (cf.) in Ithaca. Numbers of plants in each experiment (n), numbers of plants which yielded data (n-d), and maximum number of years of survival (S-yrs.) are indicated in the following table.

Source of sample	TG			SR			Cf.		
	n	n-d	S-yrs. (n)	n	n-d	S-yrs. (n)	n	n-d	S-yrs. (n)
Shag Island	1	0	1	0	—	—	1	1	11
Island N. of Forsyth Isl.	0	—	—	0	—	—	4	4	12
Stevenson Island	1	0	1	6	1	3	6	6	12
Gull Island	3	0	1	7	7	3	5	5	12

Source of sample	TG			SR			Cf.		
	n	n-d	S-yrs. (n)	n	n-d	S-yrs. (n)	n	n-d	S-yrs. (n)
Natashquan	1	0	1	2	2	2	1	1	12
Machias Seal Island	1	1	6	0	—	—	0	—	—
Mt. Horrid	7	7	7	7	5	7	0	—	—
Esopus Creek	6	6	11	9	8	10	0	—	—
Millrift	3	3	11	5	2	10	0	—	—
Nockamixon Rocks	1	1	7	3	2	10	0	—	—
Totals	24	18		39	27		17	17	

Plants of *Sedum rosea* are dioecious, perennial herbs with stout rootstocks and pale brown, fibrous roots. The rootstocks are either erect or horizontal, pale brown, and .4–2.5 cm. in diameter. The floriferous stems arise in the axils of scalelike primary leaves and may not be completely elongated when plants are in flower. In summer, when leaves begin to turn yellow and wither, stems of male plants break or separate readily from the rhizome, but fructiferous stems at a similar stage remain firmly attached. Asterisks below indicate significant differences among units. Differences in height between sexes lack significance.

	n-pop.	n-pl.	n-rep.	n-obs.	$\bar{x}$	s
Maximum diam. (cm.)						
Gulf St.L.	2	43	—	43	7.8	34.6
Rhizomes—diam. (mm.)						
Gulf St.L.	2	45	—	45	10.9	59.7
Floriferous stems—n per plant						
Gulf St.L.	2	27	—	27	14	20.8
TG—1951	4	4	4	4	57	—
SR—1960	7	11*	19	19	5	3.7
Floriferous stems—length (cm.)						
Gulf St.L. ♂	3*	28	—	78	11	2.9
— ♀	2*	19	—	28	5	10.7
TG—1954 ♂	4	18	18	73	11	4.1
— ♀	4*	10	10	41	11	2.8
SR—1954 ♂ + ♀	4*	16	16	47	14	4.7
Height (cm.)						
Gulf St.L. ♂ + ♀	2*	46	—	46	5	3.3
Mt. Horrid ♂	1	40	—	40	18	5.8
— ♀	1	43	—	43	20	5.6
Esopus Creek ♂	1	30	—	30	20	7.7
— ♀	1	23	—	23	20	6.7
Millrift ♂	1	15	—	15	22	3.8
— ♀	1	13	—	13	22	4.9
Cf.—1972 ♂ + ♀	3	4	6	6	3	1.2

Leaves of the rhizomes are scalelike, ovate, acute or obtuse, 2–7 mm. long, 2–3 mm. wide, and reddish brown. Leaves of the floriferous stems are spirally arranged and vary from ovate or obovate in full light to oblanceolate-oblong in shaded situations. The apices are acute. Often the margins are dentate. Toothing may be coarse or remote. The blades are pale green and usually glaucous, sometimes red-tipped or even completely red in exposed situations. Dimensions below are for median leaves of floriferous stems at fruiting time. When plants are in anthesis, leaves may not be completely expanded. Differences among populations, based on data for plants in nature, are highly significant in all dimensions of leaves and significant in number of teeth per margin, but these differences are not confirmed when plants are grown under similar conditions in experiments. Shape of leaf is subject to great modification by the condition of light. One plant on Stevenson Island in the Gulf of St. Lawrence had the leaves consistently decussately opposite. Another plant on the same island had variegated leaves which were green with white margins.

	n-pop.	n-pl.	n-rep.	n-obs.	$\bar{x}$	s
Length (mm.)						
Gulf St.L. ♂	3**	28	—	114	15	2.1
— ♀	2**	19	—	41	9	2.3
Mt. Horrid ♂	1	40	—	200	20	4.1
— ♀	1	43	—	209	20	4.2
Esopus Creek ♂ + ♀	1	53	—	53	28	8
Millrift ♂ + ♀	1	29	—	106	25	5.5
Nockamixon Rocks ♂ + ♀	1	6	—	42	24	8.2
Roan Mt. ♂	1	1	—	1	26	—
TG—1954 ♂	5	18	18	74	20	5.1
— ♀	2	9	9	13	18	4.4
SR—1954 ♂ + ♀	4	17	17	48	22	4.8
Cf.—1972 ♂ + ♀	3	4	6	15	10	3.7
Width (mm.)						
Gulf St.L. ♂	3**	28	—	114	9.7	1.4
— ♀	2**	19	—	41	6	1.7
Mt. Horrid ♂	1	40	—	200	8	1.6
— ♀	1	43	—	209	8	1.5
Esopus Creek ♂	1	30	—	30	10	2.1
— ♀	1	23	—	23	8	1.5
Millrift ♂ + ♀	1	29	—	106	9	2.1
Nockamixon Rocks ♂ + ♀	1	6	—	42	8	2.2
Roan Mt. ♂	1	1	—	1	9	—
TG—1954 ♂	5	18	18	74	8	1.7
— ♀	2	9	9	13	7	1.2
SR—1954 ♂ + ♀	4	17	17	48	8	3.7
Cf.—1972 ♂ + ♀	3	4	6	15	6	1.1
Length/width						
Gulf St.L. ♂	3	28	—	114	1.5	.3
— ♀	2	19	—	41	1.4	.2
Mt. Horrid ♂	1	40	—	200	2.3	.3
— ♀	1	43	—	209	2.2	.3
Esopus Creek ♂ + ♀	1	53	—	53	3	.6
Millrift ♂ + ♀	1	29	—	106	2.7	.4
Nockamixon Rocks ♂ + ♀	1	6	—	42	2.9	.5
TG—1954 ♂	5*	18	18	74	2.5	.4
— ♀	2	9	9	13	2.5	.3
SR—1956 ♂ + ♀	4	12	12	60	2.2	.5
Cf.—1972 ♂ + ♀	3	4	6	15	1.8	.3
Teeth per margin (n)						
Gulf St.L. ♂ + ♀	2*	46	—	144	1	.8
SR—1960 ♂	7**	13*	26	67	.4	.2
Cf.—1972 ♂ + ♀	3	4	6	15	.04	.1

Inflorescences are terminal, corymbose cymes, .6–6.5 cm. in diameter. Number of flowers per cyme varies from 1 to 157. Floral bracts are elliptic-oblong or elliptic-spatulate and acute. Size of inflorescence, both diameter and number of flowers, is determined by the age and vigor of the plant.

	n-pop.	n-pl.	n-obs.	x̄	s
Flowers (n)					
Gulf St.L. ♀	2	19	41	6	35

Flowers are functionally either male or female and usually 4-merous, but sometimes 5-merous or rarely 3-merous. They are borne on slender, pale green pedicels. Male flowers at anthesis vary in diameter from 5–8 mm. and pistillate flowers with receptive stigmas from 3–6 mm. Staminate flowers persist on plants in withered condition for several weeks after pollen has been shed.

Sepals are linear-oblong or lanceolate, unequal, acute or obtuse, and pale green, yellow-green, or even pale canary yellow, sometimes red at tips. Sepals of pistillate flowers are longer than those of staminate flowers. The difference is highly significant. The observed width of sepals varies from .3–.8 mm.

	n-pop.	n-pl.	n-obs.	x̄ (mm.)	s (mm.)
Length					
Gulf St.L. ♀	1	1	1	2	—
Mt. Horrid ♂	1	11	11	1.3	.2
— ♀	1	13	13	1.9	.5
Esopus Creek ♂	1	12	12	1.6	.2
— ♀	1	20	20	2.1	.4
Millrift ♂	1	18	18	1.3	.3
— ♀	1	8	8	1.8	.3
Nockamixon Rocks ♂	1	3	3	1.4	.5
— ♀	1	4	4	1.7	.6
Width					
Gulf St.L. ♀	1	1	1	.7	—

Petals are oblong, canaliculate, hooded, rounded or acute at apex on ♂ plants, but flat and acute in ♀ plants. They are pale canary yellow, greenish yellow, or pale green, sometimes ruby red at apex. Red pigmentation appears primarily on petals of plants in exposed situations. Highly significant differences among populations in length of petals of staminate plants, on the basis of data for wild plants, are not confirmed by experiment. Differences between sexes are highly significant in length for plants along Esopus Creek and at Millrift. An experiment in the garden on the Slaterville Road confirms this difference. Likewise, an experiment in a cold frame confirms a difference between sexes in width of petals of plants from the Gulf of St. Lawrence. Rarely, stamens become petaloid. One plant exhibiting this condition on Shag Rock in the Gulf of St. Lawrence had 12 petals.

	n-pop.	n-pl.	n-rep.	n-obs.	$\bar{x}$	s
Number per flower						
Millrift ♂	1	18	—	18	4.2	.3
— ♀	1	8	—	8	4.5	.5
Length (mm.)						
Gulf St.L. ♂	1	2	—	2	1.4	.6
Mt. Horrid ♂ + ♀	1	24	—	24	2.2	.5
Esopus Creek ♂	1	12	—	12	3.1	.3
— ♀	1	18	—	18	2.2	.4
Millrift ♂	1	18	—	18	2.7	.4
— ♀	1	8	—	8	1.9	.4
Nockamixon Rocks ♂ + ♀	1	7	—	7	1.9	.5
TG—1955 ♂	4	12	12	80	2.6	.4
— ♀	2	5	5	34	1.9	.5
SR—1956 ♂	4	4	4	20	2.7	—
— ♀	4	6	6	66	2.1	.4
Gulf St.L.—cf. ♂	3	5*	8	13	2.5	1.7
— ♀	4	10	18	31	2.1	.2
Width (mm.)						
Gulf St.L. ♂	1	2	—	2	1	.2
Mt. Horrid ♂ + ♀	1	23	—	23	.7	.2
Esopus Creek ♂ + ♀	1	29	—	29	.6	.2
Millrift ♂	1	18	—	18	.7	.1
— ♀	1	8	—	8	.5	.2
TG—1955 ♂	4	12	12	80	1.1	.2
— ♀	2	5	5	34	.9	.2
SR—1956 ♂ + ♀	4	10	10	86	.9	.1
Gulf St.L.—cf. ♂	3	5	8	13	1.1	—
— ♀	4	10	18	31	.8	.1

Stamens have pale canary yellow filaments 2–4 mm. long. The anthers are Indian lake or yellow suffused with ruby red. One whorl of stamens is attached to the bases of the petals.

	n-pop.	n-pl.	n-rep.	n-obs.	$\bar{x}$ (mm.)	s (mm.)
Epipetalous filaments—length of adnation						
Millrift	1	18	—	18	0	0
Nockamixon Rocks	1	3	—	3	.1	.1
Anthers—length						
Mt. Horrid	1	7	—	7	.6	.1
Esopus Creek	1	4	—	4	.5	.2
Millrift	1	8	—	8	.5	.1
Nockamixon Rocks	1	3	—	3	.5	.1
TG—1955	4	9	9	50	.6	.1
SR—1961	4	5	7	31	.8	—
Gulf St.L.—cf.	3	4	4	6	.6	.2

Nectaries are quadrate or oblong, truncate and somewhat erose, undulate, or emarginate, and tangerine orange or deep yellow. Differences in length between sexes are highly significant for Nockamixon Rocks and significant for Millrift on the basis of data for wild plants, but experimental demonstration is lacking.

	n-pop.	n-pl.	n-rep.	n-obs.	x̄ (mm.)	s (mm.)
Length						
Millrift ♂	1	18	—	18	.7	.1
— ♀	1	8	—	8	.9	.3
Nockamixon Rocks ♂	1	3	—	3	.5	.2
— ♀	1	3	—	3	.9	.1
TG—1961 ♂	3	3	3	7	.6	—
SR—1961 ♂	5	12	15	50	1	.2
— ♀	2	3	3	6	1.1	.1
Gulf St.L.—cf. ♀	4*	12	21	36	.9	—
Width						
TG—1961 ♂	3	3	3	7	.7	—
SR—1961 ♂	5	12	15	50	.7	.1
—1964 ♀	2	3	3	6	1	.3
Gulf St.L.—cf. ♀	4	12	21	36	.6	.9

Carpels are erect and pale green, pink or red at apex, when stigmas are receptive. Differences in length among populations, based on data for wild plants, are highly significant, but dimensions change rapidly as carpels mature.

	n-pop.	n-pl.	n-obs.	x̄ (mm.)	s (mm.)
Carpels—length					
Mt. Horrid	1	13	13	6	1.6
Esopus Creek	1	21	21	7	2.6
Millrift	1	8	8	4	1.5
Nockamixon Rocks	1	4	4	3	1.8

Fruits are erect, brown follicles. Differences in length among populations are highly significant on the basis of wild plants, but are not confirmed in experiments. Follicles do not open all at once in an inflorescence, but over a period of days.

	n-pop.	n-pl.	n-rep.	n-obs.	x̄ (mm.)	s (mm.)
Length						
Gull Island	1	4	—	20	4.9	.6
Maine	1	3	—	15	5.7	.9
Mt. Horrid	1	41	—	103	6.1	.9
Esopus Creek	1	22	—	45	6.9	1.3
Millrift	1	12	—	22	7.7	1
TG—1954	4	7	7	14	5.4	0
SR—1954	2	3	3	7	5.4	0
Gulf St.L.—cf.						
1965	4	11	18	86	4.3	.2
1971	4	12	13	65	4.1	—

Seeds are pyriform, winged at both ends, finely ribbed, and brown or orange-brown. Differences among populations are significant in length and highly significant in diameter on the basis of data for wild plants. In experiments, differences from year to year surpass differences among populations, although some experiments suggest that the differences among populations in diameter are significant.

	n-pop.	n-pl.	n-rep.	n-obs.	x̄ (mm.)	s (mm.)
Length						
Gull Island	1	4	—	20	1.7	.1
Maine	1	2	—	10	1.9	.1
Mt. Horrid	1	40	—	187	1.9	.2
Esopus Creek	1	21	—	95	2	.8
Millrift	1	12	—	49	1.9	.1
TG—1954	4	7	7	27	1.8	.1
SR—1954	2	3	3	11	1.9	.3
Gulf St.L.—cf.						
1965	4	11	18	79	1.9	—
1968	4	8	13	26	2.2	—
Diameter						
Gull Island	1	4	—	20	.6	.06
Maine	1	2	—	10	.6	.05
Mt. Horrid	1	40	—	187	.4	.06
Esopus Creek	1	21	—	95	.4	.07
Millrift	1	12	—	49	.5	.06
TG—1954	4	7	7	27	.4	0
SR—1954	2	3	3	11	.5	.03
Gulf St.L.—cf.						
1965	4	11	18	79	.6	.02
1968	4**	9	14	28	.7	—

Chromosomes are small and remarkably similar, except for one pair which has a distinct constriction, as described by Banach-Pogan (1958:96–98, figs. 9–10) for plants from central Europe. Uhl (1952) has reported counts for several populations in eastern North America. His results are in agreement with those of Banach-Pogan and support the idea of a relationship between the North American and European populations. Banach-Pogan reported a triploid, s = 33, from Babia Góra in the western Carpathians. No triploids are known from North America. Only data applying to plants for which descriptive data are available above are cited here.

	n-pl.	g (n)	sp (2n)	Cytologist
Number				
Gull Island, Que.	1	11	—	Ilse Zandstra, 1969
Machias Seal Isl., N.B.	1	11	—	C. H. Uhl (1952)
Mt. Horrid, Vt.	1	11	—	C. H. Uhl (1952)
Chittenango Falls, N.Y.	1	11	—	C. H. Uhl (1952)
Haines Falls, N.Y.	1	—	22	C. H. Uhl (1952)
Esopus Creek, N.Y.	1	11	22	C. H. Uhl (1952)
Millrift, Pa.	1	11	22	C. H. Uhl (1952)

Variation. Many significant differences exist among populations of *Sedum rosea* in the latitudinal band extending from the Gulf of St. Lawrence to Roan Mountain in the southern Blue Ridge. Most of the morphological differences, however, are environmental modifications which lose significance when plants are grown under similar conditions in experiments. Data in the following table contrast differences among populations in the wild with results in experiments in gardens and cold frames at Ithaca, N.Y. (** = significant, P = .01; * = significant, P = .05; and n.s. = not significant).

Feature	Wild plants	Test garden	Garden on Slaterville Road	Cold frame
Staminate plants				
Height	**	n.s.	*	n.s.
Median leaves				
Length	**	n.s.	n.s.	n.s.
Width	**	n.s.	n.s.	n.s.
Length/width	**	n.s.	n.s.	n.s.
Teeth (n)	*	—	n.s.	n.s.
Sepals				
Length	*	—	—	—
Petals				
Length	**	n.s.	n.s.	n.s.
Width	n.s.	n.s.	n.s.	n.s.
Pistillate plants				
Height	**	*	n.s.	n.s.
Median leaves				
Length	**	n.s.	—	n.s.
Width	**	n.s.	—	n.s.
Length/width	**	n.s.	—	n.s.
Petals				
Length	n.s.	n.s.	n.s.	n.s.
Width	n.s.	n.s.	n.s.	n.s.
Nectaries				
Length	—	—	n.s.	*
Pistils				
Length	**	—	—	—
Follicles				
Length	**	n.s.	**	n.s.
Seeds				
Length	*	n.s.	n.s.	n.s.
Diameter	**	**	n.s.	**
Total significant differences	15	2	2	2

Along the northern shore of the Gulf of St. Lawrence, plants of *Sedum rosea* are common and often grow in places where they receive maximum sunlight. Southward, the species occurs in disjunct populations with exposure restricted to the north. Plants in such situations exhibit etiolation. They are taller and have longer, narrower leaves. Plants moved from shaded sites to full sunlight change drastically (fig. 152). Indication of this change is available from comparison of data for 13 ♀ plants on cliffs facing northward at Millrift, Pa., and 2 ♀ plants from the same cliffs, which had been cultivated for seven years in a garden receiving full sunlight at Ithaca, N.Y. All data are from the early summer 1954. The difference in height equals 3 standard deviations of the variation among plants in nature and the difference in length-width ratios of the leaves equals 2 standard deviations.

	n-pl.	Height (cm.)	Leaves—lgth./wdth.
Cliffs	13	22	2.5
Garden	2	7	1.5

NATIVE SPECIES 527

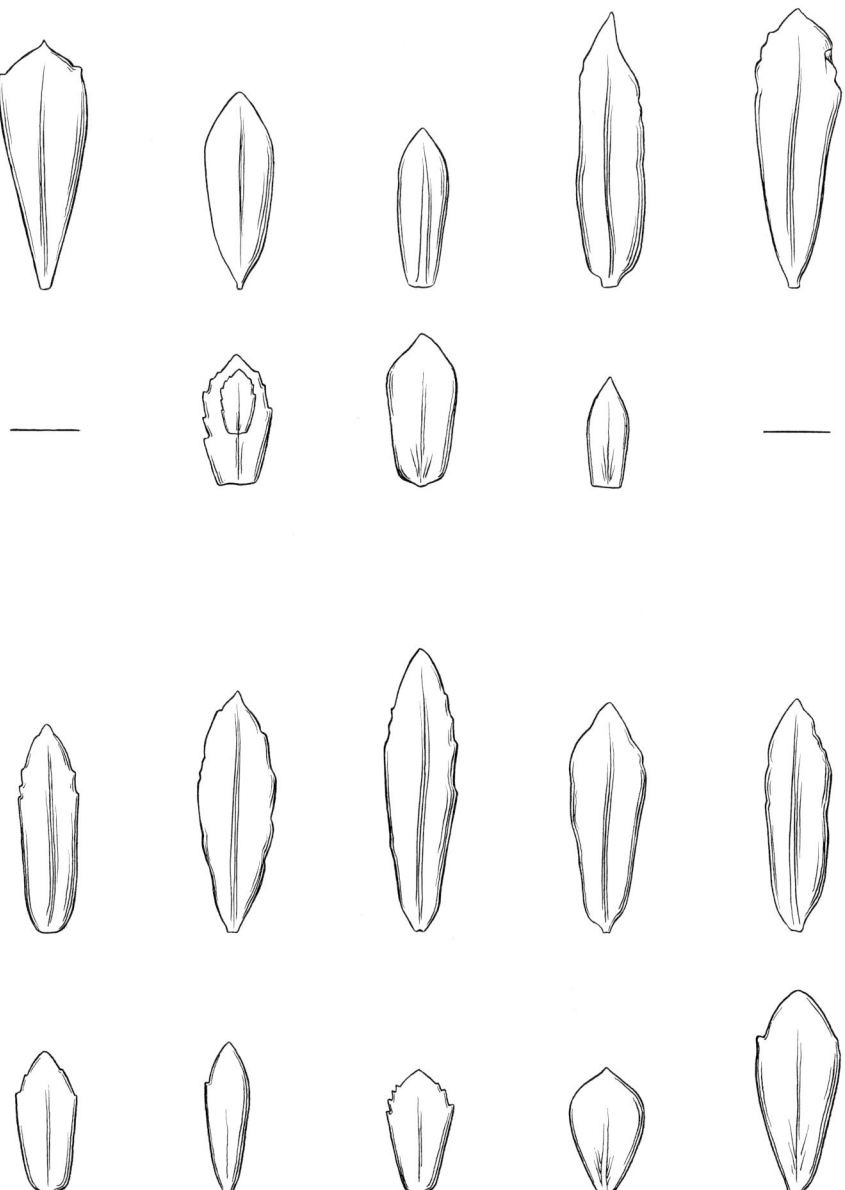

Fig. 152. Leaves of *Sedum rosea* (x 1). Top row: leaves of 5 plants in nature at Nockamixon Rocks, Bucks Co., Pa., from left to right—3 pistillate, 1 staminate, and 1 nonflowering. Next to top row: leaves of same 5 plants from Nockamixon Rocks after one year of cultivation in maximum potential sunlight in garden at Ithaca, N.Y. Next to bottom row: leaves of 5 plants in nature at Millrift, Pike Co., Pa., from left to right—2 staminate, 2 pistillate, and 1 staminate. Bottom row: leaves of same 5 plants from Millrift after one year of cultivation in maximum potential sunlight in garden at Ithaca, N.Y.

In this example, petals of the plants in cultivation exhibited more red pigment than those on the shaded cliffs.

Small differences in light on north-facing cliffs, resulting from shade from other species of plants, especially trees, have only a minor effect. As an illustration, 7 of the 13 ♀ plants on the cliffs at Millrift received maximum sunlight available for the northern exposure. The other 6 plants were variously shaded. The small differences between the two samples lack significance.

	n-pl.	Height (cm.)	Leaves—lgth./wdth.
Light	7	20	2.3
Shade	6	24	2.7

Plants of *Sedum rosea* grow poorly in cultivation at Ithaca, N.Y. They survive awhile, then die. For continued culture, natural conditions are necessary, including low temperatures in summer and adequate moisture. Survival of the plants from northern and southern populations has been similar in experiments. Evidence for ecotypes is lacking. Likewise lacking is a regional pattern of morphological variation which might indicate subspecies. Instead, the small differences which do exist among populations suggest local fluctuations in frequencies of genes. The data for eastern North America are in agreement with the idea of a recent arrival of the species here. For practical designation, citation of the binomial plus the locality of origin will identify a population.

Of all the North American species, *Sedum rosea* has advanced farthest toward the dioecious condition. Although most plants are functionally male or female, occasional plants have both male and pistillate flowers in the same cyme. Male plants outnumber female plants. Data in the following table indicate the situation in four populations and also provide a basis, in the case of Stevenson and Gull Islands, for evaluating the relative efficiency of small probability samples as compared with complete studies.

Population	Plants studied (n)	♂ (n)	♀ (n)	Sterile (n)	Ratio ♂:♀
Stevenson Island—3% sample	26	14	12	0	1.2:1
—complete study	894	475	377	42	1.3:1
Gull Island—5% sample	20	13	7	0	1.9:1
—complete study	376	229	146	1	1.6:1
Chittenango Falls—1955	23	13	8	2	1.6:1

Plants of the two sexes differ in various ways, but not consistently enough to make possible rigid generalizations. Data from several tests of the relationship reveal the situation.

Test 1: samples of 12♂ and 13♀ plants on Mount Horrid, Vt., 1954.

Feature	♂	♀
Height (cm.)*	19	23
Longest lvs. of floriferous stems—lgth. (mm.)	19	22
—wdth. (mm.)	9	10

Feature	♂	♀
Longest sepals—lgth. (mm.)**	1.3	1.9
Petals—lgth. (mm.)	2.3	2.1
—wdth. (mm.)	.7	.6

Test 2: samples of 20 ♂ and 22 ♀ plants in the gorge of Esopus Creek, N.Y., 1954.

Feature	♂	♀
Height (cm.)	25	24
Longest lvs. of floriferous stems—lgth. (mm.)*	33	26
—wdth. (mm.)*	11	9
Longest sepals—lgth. (mm.)**	1.6	2.1
Petals—lgth. (mm.)**	3.1	2.2
—wdth. (mm.)	.6	.5

Test 3: experiment in garden on Slaterville Road, Ithaca, N.Y., 1956, including 4 ♂ and 6 ♀ plants from Mount Horrid, Esopus Creek, Millrift, and Nockamixon Rocks.

Feature	♂	♀
Petals—lgth. (mm.)*	2.7	2.1
—wdth. (mm.)	1	.9

Test 4: experiment in a cold frame, Ithaca, N.Y., 1964, including 5 ♂ and 11 ♀ plants from three populations in the Gulf of St. Lawrence.

Feature	♂	♀
Height (cm.)	4	5
Petals—lgth. (mm.)	2.5	2.1
—wdth. (mm.)**	1.1	.8

The tests confirm that the differences between the sexes in *Sedum rosea* are not absolute. Petals of male plants tend to be longer and wider, but sepals are shorter.

Number of petals per flower usually is 4, with variation to 3 and 5. The following frequency distribution is for number of petals of 56 flowers on plants in populations at Esopus Creek, Millrift, and Nockamixon Rocks.

No. of petals	No. of flowers	%	Cumulative %
3	2	3.5	3.5
4	38	67.9	71.4
5	16	28.6	100
Total	56	100.0	

In 1950, a staminate plant from Mount Horrid, Vt., in the garden at Ithaca, produced leafy cymes with few flowers. The branches of the inflorescences developed as leafy shoots (fig. 153). In 1954, a pistillate plant from Nockamixon Rocks, Pa., in the garden, produced several unusual flowers. One of these had a stout green filament subtended by an orange, three-lobed, triangular nectary (fig. 154).

When leaves are dried, as for herbarium specimens, they shrink in both length and width, but in a differential way, with the result that they are relatively longer

Fig. 153. Floriferous stems with leafy cymes from staminate plant of *Sedum rosea* from Mount Horrid, Rochester Township, Windsor Co., Vt., cultivated in garden, Ithaca, N.Y., June 15, 1950 (x .8). Photo by W. R. Fisher.

Fig. 154. Three-lobed nectary and stout green filament of pistillate plant of *Sedum rosea* from Nockamixon Rocks, Bucks Co., Pa., cultivated in garden, Ithaca, N.Y., May 11, 1964 (x 50). Photo by Howard Lyon.

than wide after drying. The difference in length-width ratios, 3.1 for 5 fresh leaves and 3.4 for the same leaves in dried condition, barely misses significance, $P = .05$. The average loss of length of these 5 leaves was 3.8 mm. and the average loss of width was 2.2 mm. Such changes resulting from drying indicate the advantage of basing studies on living plants.

Chemistry. Principal attention to the biochemistry of *Sedum rosea* has been at Tomsk in the U.S.S.R. Two crystalline compounds, p-hydroxyphenyl-β-ethanol or p-thyrosol and its glucoside, p-hydroxyphenyl-[β-D-glucopyrancocyl]-ethanol, or rhodioloside, which occur in the rootstocks and roots, have high biological activity (Saratikov *et al.*, 1967, 1968). The leaves at the stage of development when the plants are in flower contain leucoanthocyanidin, abundant leucodelphinidin, and traces of leucocyanadin, as well as other flavonoids with brown fluorescence (Combier and Jay, 1967).

Medicinal properties. Roseroot has had a long history as a drug. The purified extract from the roots and rootstocks is rodozin. Work at Tomsk has demonstrated its stimulatory effect (Aksenova *et al.*, 1966). It increases excitation in the cerebral cortex and normalizes the pathological condition in patients with neuroses (Saratikov *et al.*, 1965; Kaliko and Tarasova, 1966). Other reported properties include improvement of patients with arterial hypotension (Fateeva, 1966), a tonic effect on diabetic patients (Kolmakova and Kutolina, 1966), improvement of hearing of workers in noisy situations (Oleinichenko, 1966), prevention of leukopenia in rabbits after intravenous injection of polyvaccine (Zotova, 1966), increase of the erythrocite sedimentation rate of rabbits (Sagaidak and Paznikova, 1966), inhibition of the development of the leucocyte reaction in castrated rabbits (Cherndyntsev and Zotova, 1966), and shortening barbital sodium sleep in mice (Marina and Prishchep, 1964, and Zotova, 1965). In contrast, Govorov and Lipskaya (1963) reported that an infusion of *Sedum rosea* had a pronounced sedative action on the central nervous system.

Pathology. Along the north shore of the Gulf of St. Lawrence, many plants appeared diseased in 1959. Symptoms were hypertrophy of leaves and flowers, with the leaves appearing pustulate, wrinkled, and chlorotic. Reduction in number of follicles with seeds resulted from the condition. The pathogen, which might be a virus, is unknown. This disease has not come to attention in the disjunct populations of *Sedum rosea* which occur southward in the Appalachian Highlands.

Few insects inhabit plants of roseroot, but Jacob (1964) described a species of aphid, *Thuleaphis sedi*, which feeds near the growing point of shoots, producing considerable distortion.

Nomenclature. *Sedum rosea* (L.) Scop., Flora Carniolica, ed. 2, 1:326 (1772), rendered as *Sedum Roseum*, but corrected by Sprague in Jour. Linn. Soc. Lond., Bot. 52:50, n. 181 (1939). Based on *Rhodiola Rosea* L. The gender of the specific epithet may be emended without changing the authority for the binary combination concerned.

Synonyms based on North American types or the basionym of Linnaeus are:

1753. *Rhodiola Rosea* L., Sp. Pl. 2:1,035. Type locality: Europe, mountains of Lapland. Type: C. Linnaeus (LINN). Fröderström (1930) published a photograph of the type sheet, pl. XI. Since Linnaeus collected the species in Lapland, one of his specimens from there is a reasonable choice for the holotype. The epithet *Rosea* is a former generic name used by Rivinus.

1903. *Sedum Roanense* Britton in Small, Fl. Southeastern U.S., p. 497. Type locality: North America—cliffs, Roan Mt., N.C. Type: J. K. Small and A. A. Heller (NY), July 16, 1891. The type is from Lyon's Bluff (= Roan High Bluff), at an elevation reported as 6,350 ft. The specimen at the right on the type sheet may serve as the holotype.

1903. *Rhodiola Roanensis* Britton, Bull. N.Y. Bot. Gard. 3:39. Based on *Sedum Roanense* Britton.

1930. *Sedum roseum* var. *roanense* (Britton) Berger, in Engler and Prantl, Nat. Pflanzenfam., ed. 2, 18a:440. Based on *Sedum roanense* Britton. Berger's publication lacks adequate bibliographical citation of the basionym.

American authors in the nineteenth century often used for the roseroot the name *Sedum rhodiola* DC., Plant. succ. hist. 2:143 (1804), and Fl. Franc., ed. 3, 4:386 (1805), based on *Rhodiola rosea* L. An earlier publication of *S. rhodiola* by Villars, Hist. Pl. Dauph. 1:278 (1786), is invalid because it lacks a description or citation of a previously published description, although the intent seems clear.

The common English name for *Sedum rosea* is roseroot. Other names are rosewort (Britton and Brown, 1913), Snowdon rose (Polunin, 1940), and scurvy grass (Wm. Dress in conversation, 1962).

Distribution. *Sedum rosea* inhabits cool temperate and subarctic situations (fig. 142). In North America, it ranges along the eastern coast from northern Labrador to eastern Maine and also occurs southward at eight disjunct locations in the Appalachian Highlands: Mount Horrid, Vt.; Haines Falls, N.Y.; Esopus Creek, N.Y.; Chittenango Falls, N.Y.; Millrift, Pa.; Nockamixon Rocks, Pa.; and Grandfather Mountain and Roan Mountain, N.C. Elsewhere, it occurs on Baffin Island, Greenland, and Iceland, and in Europe and Asia south to the Himalaya Mountains. Northward in North America, plants of *S. rosea* occur in crevices and in living mats of moss and other vegetation close to the shore. Southward, they grow on cliffs with northern exposure. Ecological data for 15 populations are summarized in the following tables. Dates after localities indicate the most recent year when a count of plants was made.

Population	Plants (n)	Area (m.^{2})	Density per m.^{2}	Exposure
Grenville Province				
Shag Rock—1959	8,960	7,000	1.3	E,W
Isl. N. of Forsyth Isl.—1959	1,900	5,000	.38	N,E,W
Stevenson Island—1959	894	10,000	.09	N,E,S,W
Isle Nadeau—1959	8	10,000	.001	S,W
Isl. W. of Hospital Isl.—1959	740	3,892	.19	N,E,S,W
Gull Island—1959	376	3,950	.095	N,E,S,W
New England Seaboard Lowland				
Schoodic Penin.—1967	87	—	—	~S
Little Moose Isl.—1967	1,127	—	—	SE–SSE
St. Sauveur Mt.—1967	34	—	—	E
Green Mts.				
Mt. Horrid—1954	499	7,200	.069	SSE–SSW
Catskill Mts.				
Esopus Creek—1954	2,075	14,000	.148	WNW–E
Glac. Allegheny Plat.				
Chittenango Falls—1973	23	~1,872	.006	NE,E
Millrift—1954	358	600	.597	N,NNE
Piedmont Plat.				
Nockamixon Rocks—1954	151	850	.178	N,NE,NW
Blue Ridge				
Roan Mt.—1972	2	~200	.01	N,NE

Population	Altitudinal range (m.)	pH, range	pH, tests (n)	Oldest record
Shag Rock	0–6	6.4–7	2	1959
Isl. N. of Forsyth Isl.	1–12	4.6–5	4	1959
Stevenson Island	5–15	4.4–6.4	28	1959
Isle Nadeau	1–12	5.2	1	1959
Isl. W. of Hospital Isl.	0–4	5.2–7	5	1959
Gull Island	1–4	5.0–7	25	1959
Schoodic Penin.	~6	—	0	1928
Little Moose Isl.	~20	—	0	<1970
St. Sauveur Mt.	50–60	—	0	1896
Mt. Horrid	732–793	4.6–5.4	4	1907
Esopus Creek	79–103	5.8–6.4	3	<1948
Chittenango Falls	220–256	—	—	1894
Millrift	143–158	4.8–5	2	1902
Nockamixon Rocks	85–148	5.0–5.2	2	1867
Roan Mt.	1,900	4.8	1	1841

Population	Latitude (° N.)	Rock	Drainage	Vitality
Shag Rock	50°57'30"	gneiss	moderate	vigorous
Isl. N. of Forsyth Isl.	50°44'	gneiss	mod.–good	vigorous
Stevenson Island	50°40'	gneiss	poor–mod.	vigorous
Isle Nadeau	50°38'	gneiss	good	vigorous
Isl. W. of Hospital Isl.	50°30'	gneiss	poor–excessive	vigorous
Gull Island	50°30'	gneiss	poor–mod.	vigorous
Schoodic Penin.	44°20'30"	granodiorite	moderate	vigorous
Little Moose Isl.	44°20'	granodiorite	moderate	vigorous
St. Sauveur Mt.	44°18'	granite	moderate	mod. vig.
Mt. Horrid	43°50'40"	mica-schist	poor	vigorous
Esopus Creek	41°54'–54'30"	blue sandstone	poor	vigorous
Chittenango Falls	42°58'40"	limestone	poor	persisting
Millrift	41°23'40"	shale	poor	mod. vig.
Nockamixon Rocks	40°33'40"	shale	very poor–mod.	weak
Roan Mt.	36°6'	gneiss	poor–mod.	persisting

The disjunct occurrence of *Sedum rosea* in the Appalachian Highlands, with outlying populations on the Piedmont Plateau in eastern Pennsylvania and in the southern Blue Ridge in western North Carolina, suggests that the species has been in North America since before the Pleistocene Epoch. On the other hand, populations along the northern coast, from Labrador to Maine, could be of recent origin since deglaciation and might be the result of colonizations from islands in the north Atlantic Ocean. The report of Fridrikkson (1964) of a stem and leaves of *S. rosea* on the volcanic island of Surtsey, six months after its origin, indicates how quickly colonization may occur. That *S. rosea* originally may have reached America from Europe, as suggested by Uhl (1952), is reasonable.

Species which compete with *Sedum rosea* are diverse. Along the north side of the Gulf of St. Lawrence, roseroot grows in association with tundra vegetation. Commonest competitors on the basis of random sampling from three populations are: *Ligusticum scothicum* .38, *Cochlearia officinalis* ssp. *groenlandica* .37, *Empetrum nigrum* .1, *Plantago maritima* ssp. *juncoides* .05, and *Betula pumila* .03. In the northern

Appalachian populations, species most often growing with *S. rosea* are miscellaneous, but *Arabis lyrata* is present in four out of six populations and *Physocarpus opulifolius* is present in two populations. On Roan Mountain, a sedge, *Carex misera*, and seedlings of *Saxifraga michauxii* are close to one plant of roseroot. The other plant lacks close competitors.

Wenner (1947) found a pollen grain of *Sedum rosea* at a depth of 15 cm. and an elevation of 14 m. above the sea on South Stag Island, off the coast of Cape Porcupine, in southern Labrador. The site is one where *Salix* heath has been replaced by *Empetrum* and *Ericaceae*.

Reproduction. Dispersal of *Sedum rosea* is by means of seeds and loose pieces of rhizomes. Most seeds probably germinate near where they are produced, but since they are small and also with wings, they are suitable for dissemination by the wind. On Mount Horrid, in 1954, seedlings occurred a little beyond old plants, suggesting that the population there might be expanding slightly.

Rhizomes break readily into pieces, and old parts of rhizomes rot. A result is that a large plant may break into separate pieces with the appearance of distinct plants. Water, ice, or strong wind can transport loose sections of rhizomes from one site to another. Where plants grow on cliffs, rhizomes sometimes drop from higher situations onto ledges or talus below and then take root and become established in the new habitat, while the original part of the same plant continues to survive in the old location.

Time of flowering varies with latitude, altitude, and the advance of the season: May in eastern Pennsylvania and New York, early June in Vermont, and July in the region of the Gulf of St. Lawrence. At 1,900 meters on Roan Mountain, plants are in flower in late June and July. Maturation of fruits takes six to eight weeks. In the garden at Ithaca, N.Y., plants from diverse localities have flowers from late April to June, with the height of bloom in May. Plants from sites in Pennsylvania and New York tend to be in flower a few weeks ahead of those from the Gulf of St. Lawrence, but the pattern is not absolute. Information on pollinators is lacking.

Relationships. If ancestral species of subgenus *Rhodiola* had large, perfect, 5-merous, pink flowers and occurred in temperate situations, then *Sedum rosea* with small, imperfect, 4-merous, yellow flowers and adaptation to subarctic conditions is highly specialized. *Sedum integrifolium*, with flowers (fig. 155) larger than those of *S. rosea* and sometimes perfect and with red petals, may be slightly less specialized. *Sedum rosea* possibly is the most advanced species of subgenus *Rhodiola*. This idea is wrong, however, if *S. integrifolium* is of allopolyploid origin and *S. rosea* is one of the parental species. Since experimental demonstration of an allopolyploid origin of *S. integrifolium* is lacking, the hypothesis that it is an autohexaploid derivative of some Asiatic species with 6 pairs of chromosomes is just as reasonable on the basis of available evidence.

Morphological data suggest that *Sedum rosea* and *S. integrifolium* are closely related. For discussion and comparative tables, see the section on relationships under *S. integrifolium*.

Fig. 155. Flowers of *Sedum rosea* (left) from Mount Horrid, Rochester Township, Windsor Co., Vt., plant in wild, and *S. integrifolium* (right) from Watkins Glen, Schuyler Co., N.Y., plant cultivated in garden, Ithaca, N.Y., June 9 and 3, 1947 (x 4). Photos by W. R. Fisher.

Other species, all Asiatic, to which *Sedum rosea* is related are *S. stephanii* Chamisso (including *S. crassipes* Wall.), with narrower, green leaves and longer petals; *S. algidum* Led., with larger flowers (petals >9 mm. long) and linear or linear-lanceolate, narrower leaves; and *S. kirilowii* Regel with green, linear or lanceolate leaves, usually broadest at the base and with flowers normally 5-merous. Except for *S. integrifolium*, *S. rosea* has no close relatives in North America. Neither does it anywhere make contact with *S. integrifolium*, although the population of the latter on Seneca Lake is only 100 kilometers from the population of *S. rosea* at Chittenango Falls. The arrival of *S. integrifolium* in eastern North America is a postglacial event, less than ten thousand years ago. If ever a genetic interaction occurred involving *S. rosea* and *S. integrifolium*, it must have occurred in Asia at a much earlier time.

References. Hultén, Eric (1949), and Uhl, C. H. (1952).

CHAPTER VI

The Species of *Sedum* Naturalized in North America North of the Mexican Plateau

Naturalized species have a different status from native species in any region because they are of recent origin. Their introduction usually is by man. In North America, this generally means since 1492 and often rather recently. If introduction has been only once, plants of a naturalized species may be remarkably similar, regardless how widespread or common they may be. If several introductions occur, the chance increases for different genotypes to become available. The principal potential effects of introductions are an increase of the variability of native species through hybridization involving either the introduction of new genes or the production of new types through allopolyploidy; competition with native species, which can involve their suppression or elimination from habitats; and a change in the direction of succession.

Only three exotic species of *Sedum* have become frequent in temperate North America. Others are scarcer and several appear ecologically insecure. Accounts of the three well-established species include concise descriptions and detailed discussions. The rest are merely listed, with brief distributional notes indicating the localities or areas where they occur.

1. *Sedum purpureum*** (figs. 156–157)

Deep pink petals, with the stamens equaling them or slightly shorter, are distinctive features of *Sedum purpureum* (fig. 156). The color and shape of the nectaries, yellow and longer than broad, are useful for separating *S. purpureum* from *S. telephioides*. The leaves differ from those of *S. telephioides* in being markedly reduced upward on the stems, one per node on large stems, but two per node on axillary shoots, and sessile, elliptic-oblong, and dull green.

English names are Live-forever, Garden Orpine, and Frogplant.

Description. Sample: 54 plants. Sources and numbers of plants from each locality are: New England–Acadian Highlands: E. Berkshire, Vt.—1; .8 km. N.N.E. of W. Berkshire, Vt.—5; and 1.8 km. N.N.E. of W. Berkshire, Vt.—2; Appalachian Plateau: Dryden Lake, N.Y.—1; and Connecticut Hill, Town of Newfield, N.Y.—44; and Grenville Province: Colton, N.Y.—1.

538 SEDUM OF NORTH AMERICA

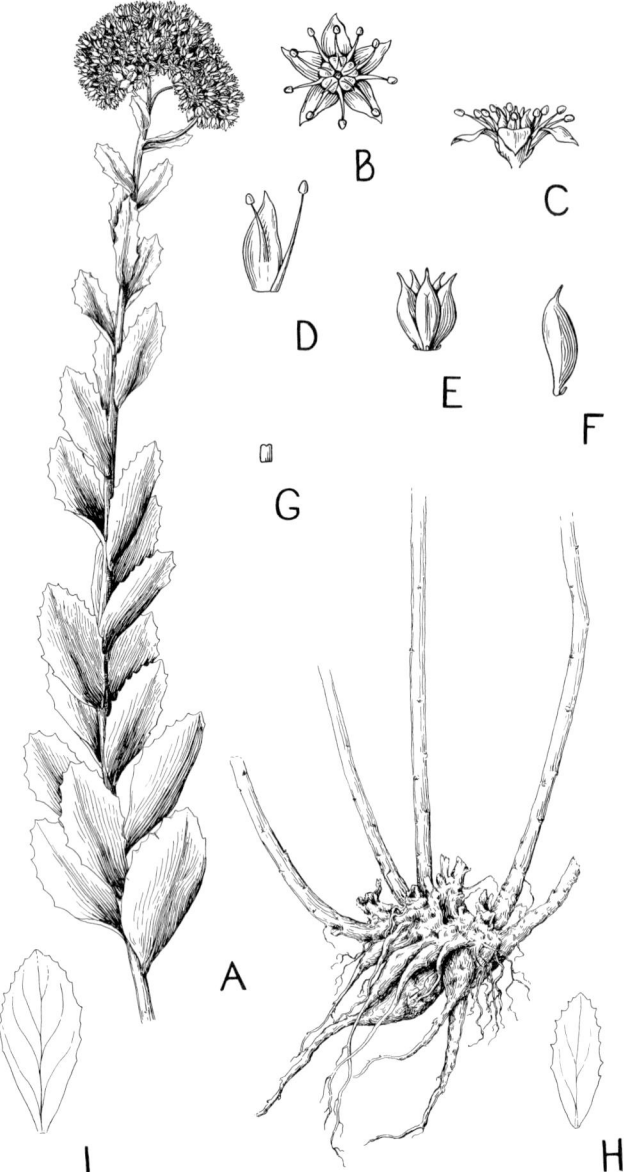

Fig. 156. Plant of *Sedum purpureum* from roadside south of Dryden Lake, Tompkins Co., N.Y., Aug. 23, 1945. A. Habit sketch (x .41). B. Flower from above (x 1.6). C. Flower from side (x 1.6). D. Petal and two stamens (x 2.5). E. Carpels (x 2.5). F. Carpel and nectary (x 3.3). G. Nectary (x 4.1). H and I. Leaves (x .41).

Although *Sedum purpureum* is a widespread weed, it seldom flowers in the Finger Lakes Region of New York and does poorly in cultivation at Ithaca. It thrives best in shade at 42° 30′ N., but rarely flowers under such conditions.

Plants of *Sedum purpureum* are erect, perennial herbs with clusters of tuberous, carrotlike, white roots, and glabrous, erect stems from short (.8–1 cm.), erect rootstocks. Stems are green below, but may be red upward. In shade, the stems may become reclining.

	n-pop.	n-pl.	n-obs.	$\bar{x}$	s
Stems per plant (n)	1	46	46	2	2
Lgth. flor. stems (cm.)	2	2	2	62	32

Leaves are spirally arranged, or decussately opposite on axillary shoots, flat, sessile, broadly cuneate at base, elliptic-oblong, dentate, especially above middle, obtuse, and dull green, subglaucous dorsally. They are markedly reduced in size upward on the stems.

	n-pop.	n-pl.	n-obs.	$\bar{x}$	s
Length (cm.)	6	54	93	4.8	.8
Width (cm.)	6	54	93	2.4	.5
Thickness (mm.)	6	54	93	.7	.2
Teeth per margin (n)	3	51	90	7.7	1.5

Inflorescences are dense, subglobose, convex cymes with elliptical or lanceolate, acute bracts.

	n-pop.	n-pl.	n-obs.	$\bar{x}$ (cm.)	s (cm.)
Diameter	3	3	3	5.8	1

Flowers are mostly 5-merous.

	n-pop.	n-pl.	n-obs.	$\bar{x}$ (mm.)	s (mm.)
Diameter	3	3	3	10.3	.6
Pedicels—lgth.	3	3	3	3.3	.6

Sepals are ovate or lanceolate, acute, and pale green or green streaked with pink. They are connate basally for about half their length.

	n-pop.	n-pl.	n-obs.	$\bar{x}$ (mm.)	s (mm.)
Length	3	3	3	6.1	.4
Width	3	3	3	2.8	.1

Petals are deep pink, erect below, then spreading widely and recurving, elliptic-oblong, and obtusely mucronate or acute.

	n-pop.	n-pl.	n-obs.	$\bar{x}$	s
Number	3	3	3	5	0
Length (mm.)	3	3	3	5.8	.5
Width (mm.)	3	3	3	2.4	.2

Stamens have pink filaments and dark red anthers.

	n-pop.	n-pl.	n-obs.	x̄ (mm.)	s (mm.)
Filaments—lgth.	3	3	3	3.2	.3
Adn. epipet. fil.—lgth.	2	2	2	.2	.3

Nectaries are quadrate or oblong, truncate, minutely emarginate, and yellow.

	n-pop.	n-pl.	n-obs.	x̄ (mm.)	s (mm.)
Length	3	3	3	.6	.3
Width	3	3	3	.4	.1

Carpels are erect and pink. The styles are bent outward.

	n-pop.	n-pl.	n-obs.	x̄ (mm.)	s (mm.)
Length	3	3	3	4.8	.3

Fruits are erect follicles, purple before dehiscence, eventually becoming brown.

	n-pop.	n-pl.	n-obs.	x̄ (mm.)	s (mm.)
Length	2	2	10	4	.1

Seeds usually are undeveloped. The cause of the infertility may be that plants are triploid hybrids.

Chromosomes are 2-armed with median or submedian constrictions (Baldwin, 1937).

	n-pl.	g (n)	sp (2n)	Cytologist
Number Virginia	3	—	36	Baldwin (1937)

Variation. Plants of *Sedum purpureum* at different localities in North America are remarkably similar. Most individuals appear like parts of one clone. No noteworthy variations have come to attention. These observations are compatible with the idea that *S. purpureum* is a sterile triploid which reproduces vegetatively.

Variations in phyllotaxy result from age or injury. Stems which develop early in the growing season have the leaves spirally arranged. Those which develop later in the year or from axillary buds after injury have decussately opposite leaves.

Another species, closely similar, but reproducing sexually and setting seeds, occurs in Europe. This is *Sedum fabaria* which is diploid and a separate species, not a variation of *S. purpureum*. It may be a rare introduction in North America.

Nomenclature. *Sedum purpureum* (L.) Schult., Oestr. Fl., ed. 2, 1:686 (1814). Based on *Sedum Telephium* β L., Sp. Pl., ed. 4, 2:760, which, as the same name in Sp. Pl., ed. 1, 1:430 (1753), is based on *Telephium purpureum majus* Bauh., Pin. 287.

Synonyms applied to plants in North America are:

1753. *Sedum Telephium β purpureum* L., Sp. Pl. 1:430. Type locality: Europe. Type: reference to Bauh., Pin., p. 287 (1623), which in turn refers back to Fuchs, *Telephium purpurascens*, in De Historia Stirpium, p. 801. An exact identification of Fuchs' plate is impossible, that is whether the plant was diploid, triploid, or tetraploid, but surely it belongs to the group of pink-flowered orpines common in Europe. The drawing of *Anacampseros purpurea* in Bauhin and Cherlerus, Historia Plantarum Universalis, p. 682 (1650–1651), cited by Linnaeus in Hort. Cliff., p. 176, under *Sedum foliis planiusculis patentibus serratis, corymbo terminatrici*, α, is a reduced copy of the illustration by Fuchs of *Telephium purpurascens*.

1821. *Sedum triphyllum* (Haw.) S. F. Gray, Nat. Arr. Br. Pl. 2:540. This applies to *S. purpureum* as to plant described, but possibly to *S. verticillatum* as to name bringing synonymy. The basionym, *Anacampseros triphylla* Haw., Syn. Plant. Succ., p. 111 (1812), is based on two different species: *S. purpureum* and *S. verticillatum*. The description of the leaves as in whorls of three suggests *S. verticillatum*. The rest of Haworth's account of the *Telephium* group of species also is confused.

1843. *Sedum purpurascens* Koch, Syn. Fl. Germ. et Helv., ed. 2, p. 284. Based on *Sedum Telephium β purpureum* L., Sp. Pl., ed. 2, 1:616 (1762), which has the same basis as the publication in 1753.

1909. *Sedum Telephium* L. ssp. *purpureum* (Schultes) Schinz and Keller, Flora der Schweiz, ed. 3, 1:255. Based on *S. purpurascens* Koch, which in turn is typified by *S. telephium β purpureum* L.

Distribution. *Sedum purpureum* is widely naturalized in the glaciated part of North America from Newfoundland to Washington. It requires a long light day for flowering and rarely flowers south of 42° N., although it blooms north of that latitude. Southernmost known occurrences in North America are at Highland Park, N.J. (CU, NY); Chester Co., Pa. (Stone, 1945); Bull Run Mountain, Va. (Allard, 1940); and several localities in Missouri (Steyermark, 1962). Usual habitats are situations disturbed by man: roadsides, old fields, and clearings in woods. Plants are established in North America at altitudes from ∼5 to 580 meters. They grow vigorously in soils with pH at 5–6.8. Elsewhere in the world, *S. purpureum* is widely distributed across northern Eurasia. The map (fig. 157) indicates definite records for the species in North America. Probably the species occurs in many other places. Also, some records from the literature may be based on faulty identifications.

Effect on native species. No hybrids of *Sedum telephioides* and *S. purpureum* are known. *Sedum telephioides* usually is diploid, rarely tetraploid, while *S. purpureum*, at least in North America, is triploid and does not flower in the area of potential overlap.

The ecological effect of the introduction of *Sedum purpureum* appears slight. It grows in places where native species of *Sedum* do not occur. At Chittenango Falls, N.Y., it grows only a few meters from cliffs with *S. rosea*, but does not invade the niche of the roseroot. Its main effect is on native herbaceous species of other genera which are excluded from sites where the *Sedum* is established. Also, it occupies space which might be inhabited by other introduced plants.

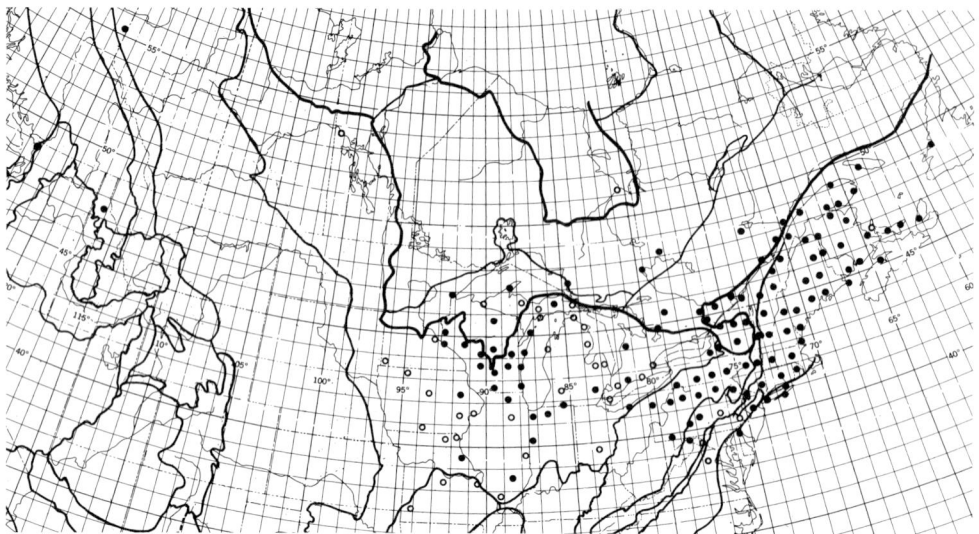

Fig. 157. Distribution of *Sedum purpureum* in North America as indicated by (●) personal observation or study of herbarium specimens, and (○) records from the literature.

Economic value. The young stems and leaves are edible and are useful as a salad (Fernald and Kinsey, 1958). Also, because the lower epidermis of the leaf is easy to remove, leaves have been used for studying the absorption of streptomycin (Lockwood, 1958).

Reproduction. The spread of *Sedum purpureum* in North America may be mostly by means of the tuberous roots. Plowing may break these apart and scatter them. Likewise, work on roads may move roots from place to place. In addition, broken pieces of stems may produce axillary rosettes which take root and become new plants. Lack of seeds makes unlikely any major dissemination by this means. *Sedum purpureum* surely reproduces only vegetatively in the southern part of its naturalized range, since it is a long-day plant and does not flower southward. Northward, where it commonly flowers, the production of seeds is a possibility, but evidence of viable seeds or seedlings is lacking. Although plants grow well in poorly drained soils in partial shade in the Finger Lakes Region of New York, they do poorly in the garden in Ithaca.

August is the principal month of flowering for *Sedum purpureum*. A few flowers may appear in late July and flowering may continue into September.

Relationships. The relationships of *Sedum purpureum* clearly are with *S. telephium*. Some authors include it within that species as a subspecies, namely ssp. *purpurascens* or ssp. *purpureum*. A reason for recognizing two species is that there seem to be two arrays of plants in Europe, one with white flowers and opposite leaves and the other with pink flowers and the leaves spirally arranged. Opinions of European botanists concerning the classification are conflicting. Although the two species are combined in *Flora Europaea*, the possibility exists that isolating mechanisms are effective enough to maintain species. Despite a long botanical history, many aspects of *Sedum purpureum* still require investigation.

References. Allard (1940), Becherer (1956), Fernald and Kinsey (1958), Lockwood (1958), and Webb (1961).

2. *Sedum sarmentosum*** (figs. 158–159)

Diagnostic features of *Sedum sarmentosum* (fig. 158) are the elongate, prostrate stems with pale green, elliptical leaves in whorls of three, and also the five-parted,

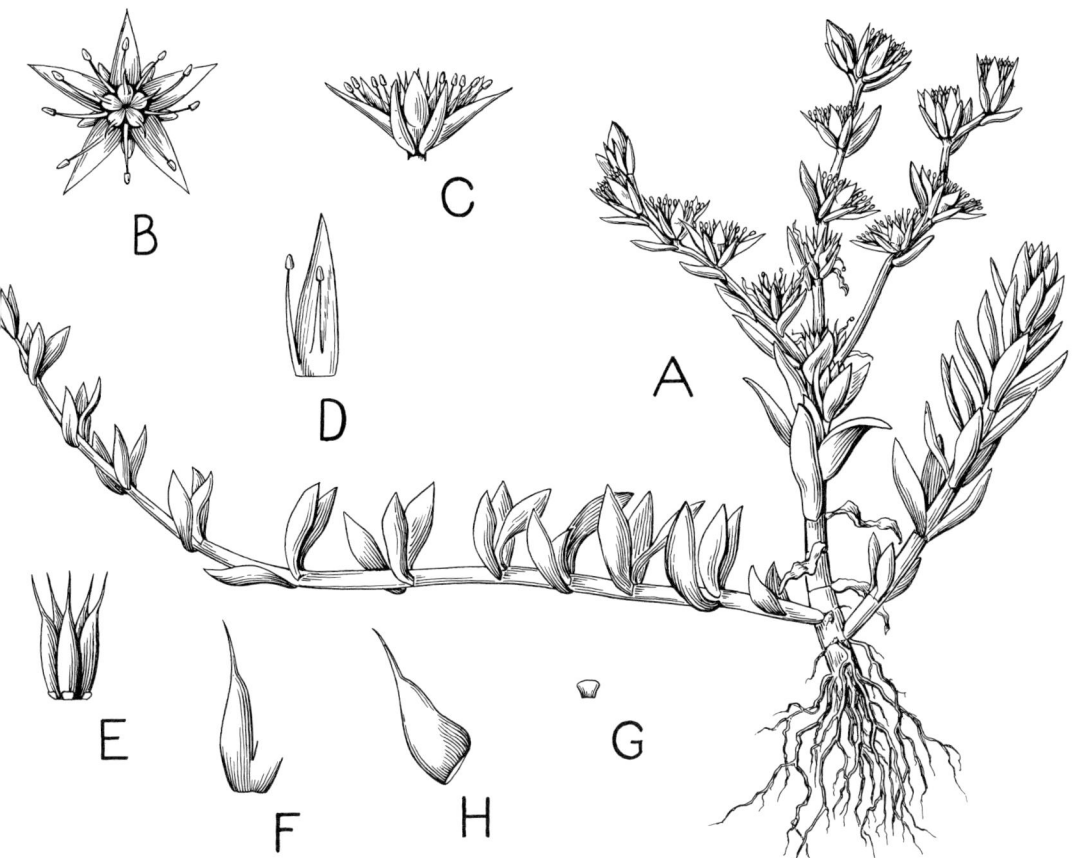

Fig. 158. Plant of *Sedum sarmentosum* from eastern side of Jermyn Creek, 1.6 km. north of Luray, Page Co., Va., cultivated in garden, Ithaca, N.Y., June 28, 1950. A Habit sketch (x 1.1). B. Flower from above (x 2.25). C. Flower from side (x 2.25). D. Petal and two stamens (x 3.3). E. Carpels (x 3.3). F. Single carpel and base of another (x 4.4). G. Nectary (x 5.5). H. Developing follicle (x 4.4).

yellow flowers. Since the stems readily take root at the nodes, the plants spread rapidly and often become a nuisance. *Sedum sarmentosum* is common in cultivation. Doubtless it has spread from gardens to wild situations. It endures temperatures much below freezing and has become naturalized in many places in temperate North America.

Description. Sample: 4 plants—1 on O'Shaughnessy Point, Monte Sano, Madison Co., Ala.; and 3 in garden in Ithaca—1 from Washington, D.C.; 1 from Jermyn Creek 1.6 km. N. of Luray, Page Co., Va.; and 1 from a canyon east of Chipari, Korea, R. V. Moran 4231.

Plants of *Sedum sarmentosum* are perennial herbs with fibrous, pale brown roots and prostrate, jointed, white stems which root at the nodes and give rise to prostrate, glabrous, pinkish white vegetative stems to 4 dm. long and 1.5–2 mm. in diameter, and pinkish white to green, decumbent, floriferous stems.

	n-pop.	n-pl.	n-obs.	x̄	s
Vegetative stems—lgth. (cm.)	1	1	3	27	—
Floriferous stems					
Number	1	1	1	3	—
Length (cm.)	2	2	13	19	9.4
Height (cm.)	3	3	3	8	2

Leaves are in whorls of three, short-spurred, short-petiolate, narrowly oblanceolate-elliptical, subacute, and yellow-green.

	n-pop.	n-pl.	n-obs.	x̄ (mm.)	s (mm.)
Length	4	4	21	18	7.5
Width	4	4	21	5	1.2
Thickness	3	3	10	.9	.2

Inflorescences are terminal cymes with few-flowered, zigzag cincinni. Floral bracts are similar to foliage leaves, but smaller.

	n-pop.	n-pl.	n-obs.	x̄	s
Cymes—diameter (cm.)	2	2	3	3.8	1.7
Cincinni—number	2	2	5	2.9	.6
Flowers—number	3	3	7	14.3	6.5

Flowers are sessile and 5-merous.

	n-pop.	n-pl.	n-obs.	x̄ (mm.)	s (mm.)
Diameter	3	3	5	11.8	.9

Sepals are short-spurred, divergent, lanceolate-oblong, obtuse, and green.

	n-pop.	n-pl.	n-obs.	x̄ (mm.)	s (mm.)
Length	4	4	15	4.4	.9
Width	4	4	16	1.1	.2

Petals are rotately spreading, slightly connate basally, lanceolate, acute, mucronate-appendaged, and sulphur yellow.

	n-pop.	n-pl.	n-obs.	$\bar{x}$	s
Number	3	3	5	5	0
Length (mm.)	4	4	15	6.2	.7
Length of cohesion (mm.)	2	2	4	.6	.05
Width (mm.)	4	4	15	1.7	.1

Stamens have yellow filaments which all are adnate to the tube of the corolla. Anthers are yellow flushed with dark reddish brown.

	n-pop.	n-pl.	n-obs.	$\bar{x}$ (mm.)	s (mm.)
Filaments—length	1	1	3	4.5	—
Adn. of filaments—lgth.	3	3	6	.8	.3
Anthers—length	1	1	2	.4	—

Nectaries are reniform-subquadrate, truncate, and some shade of orange.

	n-pop.	n-pl.	n-obs.	$\bar{x}$ (mm.)	s (mm.)
Length	3	3	5	.5	.1
Width	3	3	5	.5	.2

Carpels are erect in flower, connate basally, and yellow-green. Later, they become divergent, gibbous ventrally, and green.

	n-pop.	n-pl.	n-obs.	$\bar{x}$	s
Length (mm.)	3	3	5	4.9	.5
Length of cohesion (mm.)	2	2	4	1.1	.1
Ovules per ovary	1	1	3	11.3	—

Fruits and seeds do not develop on plants naturalized or cultivated in North America. The flowers wither after pollination.

Chromosomes are tiny and meiosis is irregular according to Uhl and Moran (1972).

Number	n-pl.	g (n)	sp (2n)	Cytologist
Luray, Va.	1	~36	—	C. H. Uhl, 1946
Ominato, Honshu, Japan	1	~36	—	C. H. Uhl (Uhl and Moran, 1972)

Variation. No important variations of *Sedum sarmentosum* have come to attention. Naturalized and cultivated plants are so much alike that all could be parts of a single clone. Whether the failure to produce seeds results because the plants are self-incompatible or because of the meiotic irregularity mentioned by Uhl and Moran (1972) is unknown.

Nomenclature. *Sedum sarmentosum* Bunge, Mém. Acad. Imp. Sci. St. Pétersbourg, par divers Savans 2:104 (1835). Type locality: near Peking, China. Type: a collection of A. Bunge in the herbarium of the Komarov Botanical Institute of the Academy of Sciences in Leningrad. Fröderström (1931) published a photograph, pl. 52, of what may be the type.

No other name has been used for plants of this species naturalized or cultivated in North America.

Distribution. *Sedum sarmentosum* (fig. 159) has spread from cultivation and become widely naturalized in temperate, eastern North America. Although it does not produce seeds, it spreads vegetatively and has reached many places remote from human habitations. Commonest habitats, however, are situations disturbed by man and close to places of cultivation. Records are available from the New England–Acadian Highlands, all Appalachian Provinces, the Adirondack Mountains, the Central Lowland, the Interior Highlands, and even the Coastal Plain. The plants grow on a variety of rocks. Sometimes they are a nuisance. The natural distribution of the species is in eastern Asia.

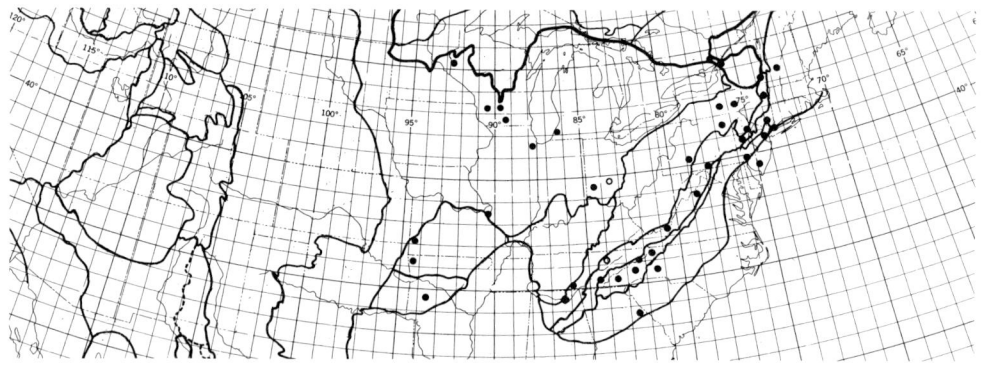

Fig. 159. Distribution of *Sedum sarmentosum* in North America as indicated by (●) personal observation or study of herbarium specimens, and (○) records from the literature.

Stems collected on Monte Sano near Huntsville, Ala., stored dry in a plastic bag for four months, from June 5 until Oct. 4, 1972, retained life and still had some fresh leaves at the end of the period.

Effect on native species. *Sedum sarmentosum* does not hybridize with any native North American species of *Sedum*. Neither does it hybridize with species cultivated in American gardens. Ecologically, however, it is more important because it spreads onto ledges and cliffs which might be occupied by native species. At Watkins Glen, N.Y., it has become established on cliffs on the north side of the ravine, at the entrance to the glen. It may have been a factor in the decline of the *S. integrifolium* there, because it now occupies some of the former habitat of the *S. integrifolium*.

Reproduction. *Sedum sarmentosum* reproduces vegetatively by means of runners which root at the nodes. Although the principal spread is for short distances, water or wind may carry pieces of stem for long distances. Further, stems, if transported with earth, easily may travel to new localities.

Flowering time in eastern North America is June, or southward and at low elevations, late May.

Relationships. The relationships of *Sedum sarmentosum* are with other yellow-flowered Asiatic species, notably *S. lineare*, which has narrower leaves and is less hardy, and *S. quaternatum*, if that is a species, with leaves in whorls of four.

Reference. R. T. Clausen (1939) and L. Marion and M. Chaput (1949).

3. Sedum acre** (figs. 160–161)

The combination of turgid, small, yellow-green, closely imbricate, ovate leaves; bright yellow, five-parted flowers; and a low creeping habit makes *Sedum acre* (fig. 160) an easy species to identify. The English name, Mossy Stonecrop, aptly describes the appearance. The plants spread mosslike over banks and walls. Although several North American species have yellow petals, they differ in important ways. For example, *S. divergens* has opposite turgid leaves, *S. nuttallianum* is an annual with elliptical or oblong leaves, *S. lanceolatum* has erect follicles, and *S. stenopetalum* has subterete, linear-lanceolate leaves with scarious bases.

Description. Sample: 4 plants—1 from Jamesville, N.Y., June 21, 1936, R. T. Clausen 2193; 1 from Cold Spring, Putnam Co., N.Y., Aug. 21, 1948, D. G. Huttleston, C48–48; and 2 from horticultural sources, one as var. *aureum*, 52–76, and the other a volunteer in the garden, 72–79.

Plants grow well outdoors at Ithaca, N.Y., and may develop weedy tendencies.

Plants of *Sedum acre* are smooth, evergreen, perennial herbs with fibrous roots arising from the nodes of the prostrate, branched stems. Sterile shoots are short, with crowded leaves. Floriferous shoots are erect and densely leafy, except at the base.

	n-pl.	n-obs.	$\bar{x}$ (cm.)	s (cm.)
Vegetative shoots—length	3	21	2.7	2.1
Floriferous stems				
Length	3	25	5.2	1.9
Height	1	1	8	—

Leaves are spirally arranged, densely crowded and imbricate, in many (>8) ranks. They are short-spurred, ovate, not much broader than thick, with entire margins, blunt, and yellow-green. Leaves of sterile and floriferous shoots are similar. In var. *aureum*, leaves at the ends of stems are yellow.

	n-pl.	n-obs.	$\bar{x}$ (mm.)	s (mm.)
Length	2	3	4.6	.7
Width	2	4	3.1	.8
Thickness	2	5	1.6	.6

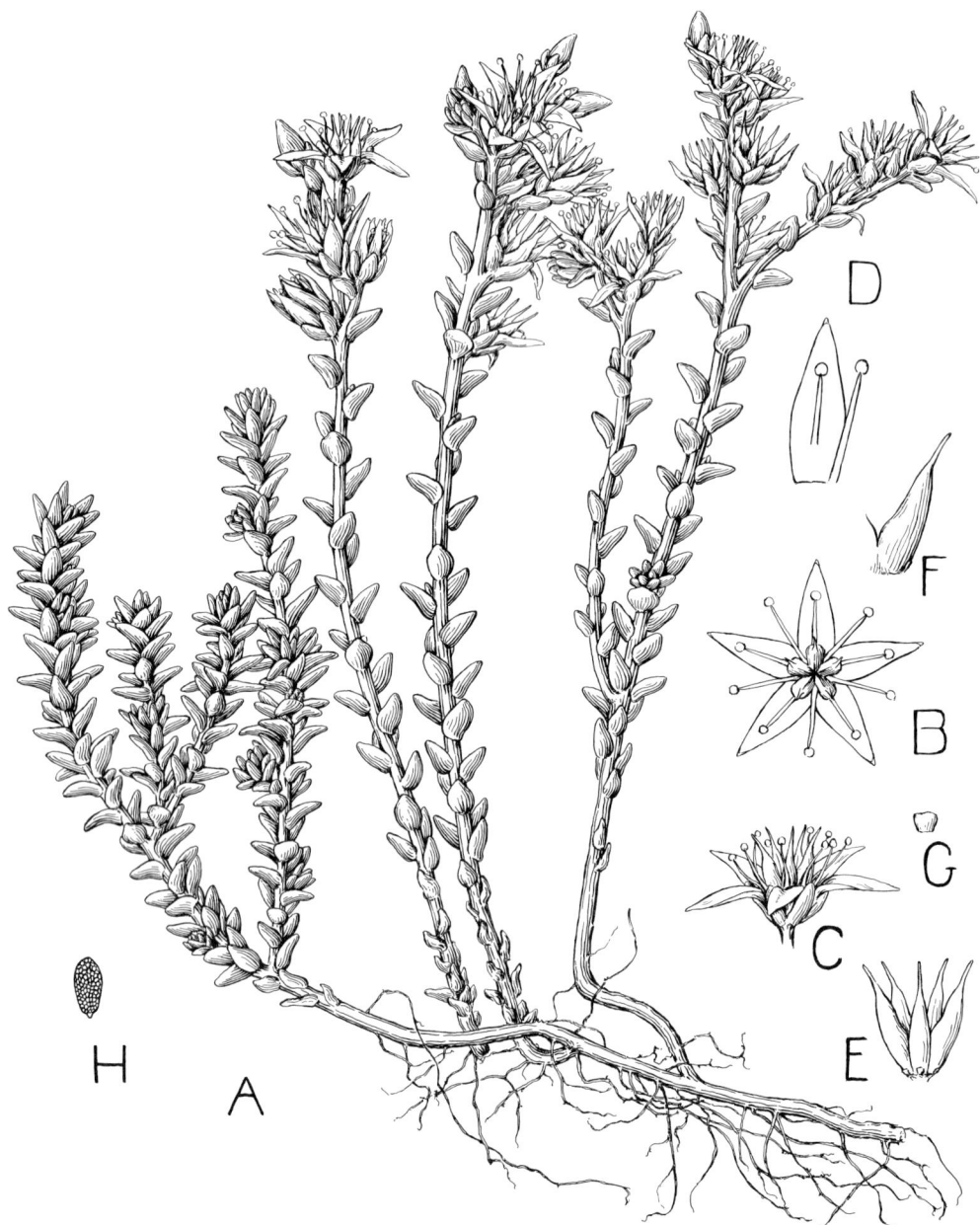

Fig. 160. Plant of Sedum acre cultivated in garden, Ithaca, N.Y., June 18, 1936. A. Habit sketch (x 1). B. Flower from above (x 2.6). C. Flower from side (x 2.6). D. Petal and two stamens (x 4.2). E. Carpels (x 4.2). F. Single carpel and part of base of another (x 5.2). G. Nectary (x 7.8). H. Seed (x 10).

Inflorescences are terminal cymes of usually 2 cincinni and 1 to 12 flowers.

	n-pl.	n-obs.	x̄	s
Cymes—diameter (cm.)	1	1	4	—
Cincinni—number	3	25	1.9	.1
Flowers—number	3	21	7.2	2.8

Flowers are sessile or subsessile and 5-merous.

	n-pl.	n-obs.	x̄ (mm.)	s (mm.)
Diameter	2	4	14	.4

Sepals are thick and fleshy, almost equal, spurred basally, ovate, blunt, and green.

	n-pl.	n-obs.	x̄ (mm.)	s (mm.)
Length	2	3	3	.8
Width	2	2	1.9	.6

Petals are slightly connate basally, lanceolate, acute, and yellow.

	n-pl.	n-obs.	x̄	s
Number	2	2	5	0
Length (mm.)	3	4	6.2	1.3
Length of connate part (mm.)	2	2	.4	0
Width (mm.)	2	2	2.4	.7

Stamens have yellow filaments and anthers. All are attached to the tiny tube of the corolla.

	n-pl.	n-obs.	x̄ (mm.)	s (mm.)
Filaments—length	1	1	4.8	1.3
Adn. epipet. fil.—lgth.	1	1	.2	—

Nectaries are quadrate or subquadrate, slightly emarginate, and yellowish green.

	n-pl.	n-obs.	x̄ (mm.)	s (mm.)
Length	3	4	.3	.2
Width	3	4	.4	.1

Carpels are slightly connate basally, erect in flower, then spreading in fruit and swelling on ventral side.

	n-pl.	n-obs.	x̄ (mm.)	s (mm.)
Length	2	3	5	.8
Length of cohesion	2	2	.9	.2

Fruits are widely divergent, gibbous ventrally, and pale brown, becoming almost white with age.

	n-pl.	n-obs.	x̄ (mm.)	s (mm.)
Length	2	19	2.8	.4

Seeds are pyriform or elliptic-oblong, with reticulate, yellow-brown integument.

	n-pl.	n-obs.	x̄ (mm.)	s (mm.)
Length	2	12	.6	.1
Diameter	2	12	.3	0

Chromosomes are in multiples of 20 ('T Hart, 1971). Other reported numbers need further confirmation.

	n-pl.	g (n)	sp (2n)	Cytologist
Number Ithaca, N.Y., 46–57	1	40	—	C. H. Uhl, 1946

Variation. *Sedum acre* is a polytypic species. Only ssp. *acre*, the common, northern European subspecies, is naturalized in temperate North America. That has ovate, imbricate leaves. 'T Hart (1971) discussed variation of this species in Europe. At least four subspecies occur. Besides having different geographic ranges, these have distinctive numbers of chromosomes: the northern European ssp. *acre* with ovate, imbricate leaves is octoploid, sp = 80; ssp. *neglectum* of southern Europe, with narrower, oblong leaves, is tetraploid; ssp. *majus*[1] of Morocco, with robust stems and leaves and pale yellow flowers, is hexaploid; and ssp. *glaciale* of the Alps and Pyrenees, with 3 to 4 large flowers, is either 12-ploid or decaploid. North American plants usually have more flowers per cyme than 'T Hart counted in European octoploids, but the leaves are characteristic of ssp. *acre*. The closeness of the correlation of number of flowers per cyme with number of chromosomes needs more attention. Several other numbers of chromosomes, not multiples of 10, have been reported, but 'T Hart did not encounter them.

A variation with the leaves at the tips of the stems yellow at the beginning of the growing season is occasional in gardens. This is var. *aureum* Masters, Gard. Chron., ser. 2, vol. 10:685 (1878). The subspecies *majus*, distinguished by being more robust, is rare in cultivation. Likewise rare is a tiny plant which is sold as var. *minus* or var. *minimum*.

Nomenclature. *Sedum acre* L., Sp. Pl. 1:432 (1753). Type locality: northern Europe, possibly Sweden. Type: a collection in the Linnaean Herbarium in the Riksmuseum, Stockholm. Fröderström (1932) published a photograph of this collection, pl. 39, fig. 1.

The following synonyms have been applied to North American plants:
1910. *Sedum Elrodi* M. E. Jones, Bull., Univ. of Montana, Biol. Ser. 15 (61):30, pl. 4.
Type locality: Somers, Mont., on loose soil and also on rocks. Type: a collection of Marcus E. Jones, no. 8,420 (POM), July 17, 1908.

---

[1] *Sedum acre* ssp. **majus** (Masters) comb. nov., fundatum super *Sedum acre* var. *majus* Masters, Gard. Chron., ser. 2, vol. 10:685 (1878).

NATURALIZED SPECIES 551

1915. *Sedum minimum* Nieuwland, Am. Midl. Natur. 4:56, not Rose (1903). Based on *Aizoon acre* Cordus, Hist. 98 (1561).

The common naturalized plants of *Sedum acre* in North America belong to ssp. *acre*. The var. *aureum* Masters and ssp. *majus* (Masters) Clausen occur only in cultivation.

Distribution. *Sedum acre* (fig. 161) is naturalized across the northern United States and southern Canada from Quebec to North Carolina in the east and from British Columbia to Oregon in the west. Records are available from the following geomorphic provinces: New England–Acadian, Piedmont Plateau, Blue Ridge, Ridge and Valley,

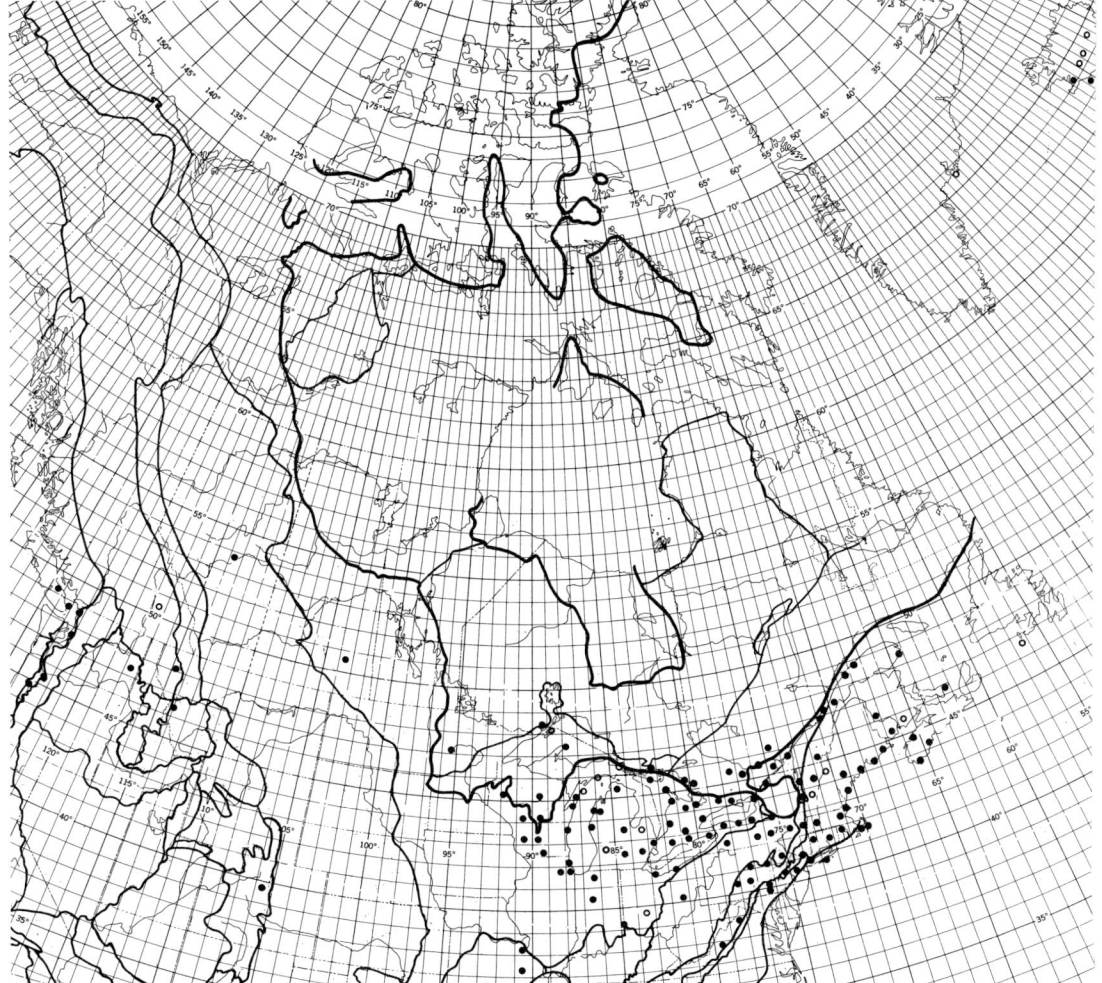

Fig. 161. Distribution of *Sedum acre* in North America as indicated by (●) personal observations or study of herbarium specimens, and (○) records from the literature.

Appalachian Plateau, Central Lowland, Ozark Plateaus, Great Plains, Grenville, Superior, Southern of the Canadian Shield, Southern Rocky Mountains, Basins and Ranges of Southwestern Montana and Idaho, Columbia Plateaus, Interior Plateaus and Ranges, Willamette–Puget Trough, and the Canadian Coastal Batholith. The plants grow on a variety of rocks, including limestone, shale, granite, and gneiss. Usual habitats are disturbed, rocky situations, along roads, near buildings, and sometimes in lawns. The plants take root in moss and in shallow depressions in rocks. The altitudinal range in North America is from ~5 to 2,298 meters. Because of the kind of habitat, populations may be unstable, continually starting and disappearing.

The natural range of *Sedum acre* is in Europe, northern Africa, and western Asia. Also it occurs in Iceland, eastern Greenland, and is naturalized in the Falkland Islands.

Effect on native species. No cases are known of hybridization between *Sedum acre* and any native North American species of *Sedum*. The genetic effects of its introduction therefore are zero.

Since plants of *Sedum acre* may persist on rocks long after introduction and initial disturbance of habitats, they must retard the re-establishment of native plants in such situations. With time, however, and increased shade from taller plants, especially trees and shrubs, the *Sedum* gradually dies out. Viewed from the standpoint of thousands of years, the ecological effect of its introduction may be negligible.

Reproduction. Vegetative reproduction by means of the rooting of loose pieces of stems, broken from parental plants, is an important mode of spread of *Sedum acre*. The occurrence of plants near buildings suggests recent introductions. In addition, seeds provide a possibility for long-distance as well as local dispersal. Production of seedlings is high.

Flowering time is June and July, rarely May.

Relationships. The two species most closely related to *Sedum acre* are *S. annuum* and *S. sexangulare*, both native in northern Europe. *Sedum annuum* usually is annual, with subterete, oblong or obovate-elliptical leaves and smaller flowers. *Sedum sexangulare* has linear, subterete leaves in six ranks and smaller flowers than *S. acre*. Hybrids of *S. sexangulare* with both *S. acre* and *S. annuum* are known.

References. Butruille *et al.* (1971), L. Marion (1945), L. Marion *et al.* (1951), and 'T Hart (1971).

Other Naturalized Species

Besides the three commonly naturalized species of *Sedum*, other species also need mention here. Most of these are escapes from cultivation. Any species in cultivation outdoors is a candidate for inclusion here. Some plants may persist for a short time and then disappear. Others may take hold and persist for many years. In reaching decisions about including species here, two criteria have been useful, namely persistence for at least five years and production of new individuals without human aid. The same criteria were used to decide whether to include weeds or escaped cultivated plants in a *Checklist of the Vascular Plants of the Cayuga Quadrangle* (Clausen 1949a:6).

Because all naturalized species listed below also are in cultivation, descriptive data are omitted here. For brief descriptions, see the chapter on cultivated species. Data provided here include the earliest known date of introduction, indication of localities where naturalized, and sometimes published reports of distribution.

1. *Sedum alboroseum* Boreau. Earliest record: 1895, Sept. 15, Marlborough, N.Y. (NY). Localities: 25. New England Seaboard Lowland: 1—Me.; Lake Champlain Valley: 1—Vt.; Coastal Plain: 4—N.J. to N.C.; Piedmont Plateau: 6—N.J. to N.C.; Blue Ridge: 1—N.C.; Ridge and Valley Province: 4—N.Y. to Pa.; Appalachian Plateau: 2—N.Y.; and Central Lowland: 6—Ind., Ill., Wisc., and Ia.

2. *Sedum praealtum* DC. ssp. *praealtum*.[2] Earliest and only record: 1948, Mar. 26, roadside near Pierpont Inn, Ventura, Calif.

3. *Sedum ternatum* Michx. Earliest record: 1808, June, Albion Cemetery, Albion, N.Y., G. W. Clinton (NYS). *Sedum ternatum* is naturalized north of its natural distributional range at many localities in New England, New York, northern Ohio, and southern Michigan. Sometimes it appears as though native in places where it undoubtedly has become established in recent time. See map (fig. 11).

4. *Sedum kamtschaticum* Fischer and Meyer ssp. *ellacombianum* (Praeger) R. T. Clausen. Earliest record: July 24, 1942, vicinity of Philadelphia, Pa., R. L. Schaeffer, Jr. (BH). According to Schaeffer, this *Sedum* was becoming naturalized in the region around Philadelphia. No similar reports are available for elsewhere. Fernald's (1950) report of the related *S. aizoon* may apply to *S. kamtschaticum* ssp. *ellacombianum*.

5. *Sedum spurium* M. B. Earliest record: 1900, July 16, Cumberland Center, Me. (GH, NY). Localities: 26. New England–Acadian Highlands: 14—Nfld. to Pa.; Ridge and Valley Province: 1—N.Y.; Appalachian Plateau: 4—N.Y. and Pa.; Coastal Plain: 1—Mass.; Great Plains: 1—Colo.; Southern Rocky Mountains: 2—Colo.; Columbia Plateau: 1—Ore.; and Canadian Shield: 2—N.Y. and Ontario. Wherry (1972) listed it from the Piedmont Plateau—Montgomery Co., Pa. This has been erroneously reported as *S. stoloniferum* (Rhodora 23:265. 1921.).

6. *Sedum album* L. Earliest record: 1934, on ledges of Rocky Fork Creek, Paint Tp., Highland Co., Ohio. Localities: 13. New England–Acadian Highlands: 2—Que. and Pa.; Piedmont Plateau: 2—Pa.; Ridge and Valley Province: 2—N.Y.; Appalachian Plateau: 1—N.Y.; Central Lowland: 4—Ont., N.Y., and Ohio; Columbia Plateau: 1—Ore.; and Klamath Mountains: 1—Ore. Also reported by Munz (1968) from near the coast in California.

7. *Sedum hispanicum* Jusl. Earliest record: 1880, June, cited from City Creek north of Salt Lake City, John Reading (GH), type of *Sedum Meehani* A. Gray, Proc. Am. Acad. 16:105 (1881). Otherwise, reported by Cody (1967) from near Ottawa, Ont., and Brompton, Que., and well established in a cedar glade .2 km. above north bank

---

[2] *Sedum praealtum* A. DC., ssp. *praealtum*, based on *S. praealtum* A. DC., Mém. Soc. Phys. et Hist. Nat. Genève 11:445–447 (1847–1848). For explanation of this change in interpretation, see the chapter on cultivated species.

of Lewis Creek, on the west side of the road to the State Fishing Access in the southwestern section of the Town of Ferrisburg, Vt. The plants grow at Ferrisburg in depressions and crevices of limestone among small trees of *Juniperus virginiana*. They cover an area of >10,600 m.² The late Leopold A. Charette found the population at Ferrisburg sometime prior to Mar. 26, 1968.

8. *Sedum sexangulare* L. Earliest record: 1942, June 28, grassy roadside abutting the Sawyer Estate near the Oyster River, Durham, N.H., A. R. Hodgdon (NEBC). Otherwise, L. A. Charette found it on June 28, 1969, on the western slope of the embankment of highway no. 5 in the southern section of the Town of Putney just north of the boundary with Dummerstown, Windham Co., Vt. The plants there probably spread from an adjacent nursery. Reference: A. R. Hodgdon (1959).

9. *Sedum ochroleucum* Chaix. Doubtfully established in North America. Two collections which might be this species are from Burden's Lake, Rensselaer Co., N.Y., Father Wibbe (NYS), and Lake Bonaparte, Lewis Co., N.Y., C. H. Peck (NYS). Another collection is from Weyauwega, Waupaca Co., Wisc., J. J. Davis (WIS). Fernald's (1950) reports of *S. anopetalum*, which is a synonym of *S. ochroleucum*, are based on plants of *S. reflexum*, as are his reports of *S. rupestre*. Likewise, a report of *S. rupestre* in Quebec by Cinq-Mars (1971) applies to *S. reflexum*.

10. *Sedum reflexum* L. Earliest record: 1876, Pigeon Cove, Essex Co., Mass., Mrs. Wheeler (GH). Localities: 5. New England–Acadian Highlands: 4—Que., Me., and Mass.; Coastal Plain—1. Hodgdon reported that plants still were persisting on a roadside at South Bristol, Me., after an interval of 46 years. References: A. Gray (1876), E. B. Chamberlain (1912), and A. R. Hodgdon (1959).

Summary

Thirteen species of *Sedum* are naturalized in North America north of the Mexican Plateau. Data concerning them are condensed in the following table where D = disturbed; f = fertile; R = of restricted distribution, known from fewer than ten populations; s = sterile; U = undisturbed; W = of wide distribution; and — = does not hybridize with native species.

Species	Continent of origin	Distrib. in N.A.	Oldest record in N.A.	Habitat	Hybrid. with nat. sp.	Genetic status
purpureum	Eu.	W	1858	D, U	—	s triploid
alboroseum	Asia	W	1895	D	—	s hybrid
praealtum	N.A.	R	1948	D	—	f tetraploid
ternatum	N.A.	W	1808	U	—	f tetraploid
kamtschaticum						
ssp. *ellacombianum*	Asia	R	1942	D	—	f tetraploid
sarmentosum	Asia	W	1889	D, U	—	s triploid
spurium	Eu.	W	1900	D	—	f tetraploid
album	Eu.	W	1934	D	—	f octoploid
hispanicum	Eu.	R	1880	D, U	—	f
sexangulare	Eu.	R	1942	D	—	f polyploid
acre	Eu.	W	1834	D, U	—	f octoploid
ochroleucum	Eu.	R	<1900	D	—	f diploid
reflexum	Eu.	R	1876	D	—	f hexaploid

Eight introduced species have come from Europe, three have come from Asia, and two are native elsewhere in North America. Seven of these species are now widely distributed across the North America continent, but none are common enough to have been found in any of my planned sampling surveys. In other words, all must be regarded as rare in terms of the total area surveyed. None of the introduced species has so far been of any importance in interacting genetically with native species. Likewise, since eight of the thirteen species are known only from disturbed situations, and four of the remaining five species similarly occur primarily in disturbed habitats, the ecological effects of these introductions are negligible. Ten of the thirteen species are polyploids and another is a sterile hybrid. Two of the polyploids, although sterile triploids, spred easily vegetatively and are aggressive weeds.

CHAPTER VII

The Species of *Sedum* Cultivated in North America North of the Mexican Plateau

Plants of many species of *Sedum* are attractive ornamentals. Various species are good for ground cover, borders, formal plantings, crevices in walls, rock gardens, and even hanging baskets. Nurseries offer more than a hundred species of *Sedum* for sale. Another important source of plants is transfer from one neighbor to another. Also, in areas where species of *Sedum* are native, people often take wild plants to cultivate by their houses.

Based on entries in a file at the Bailey Hortorium, Cornell University, 173 species of *Sedum* are in the North American horticultural trade, but many of the listed names are incorrect. Errors include misspellings, listings of subspecies and varieties as species, misidentifications, and invention of names in order to increase sales. Necessary corrections reduce the list to 108 species. More than one-third of the listed names, 65, are wrong. Further, of the species offered, only 14 are common in gardens. The 10 species most often seen in gardens, based on personal experience in traveling around the country, are: *S. spectabile*, *S. kamtschaticum*, *S. sarmentosum*, *S. spurium*, *S. album*, *S. hispanicum*, *S. sexangulare*, *S. acre*, *S. sediforme*, and *S. reflexum*.

My studies of *Sedum* have concentrated on native species. Attention to plants in cultivation has been secondary. However, cultivated plants are important in evolutionary studies from three standpoints. First, cultivated plants might hybridize with native species and thereby increase the variation of these species. So far, no evidence of this sort has come to attention for *Sedum*. Second, any cultivated plant might escape and become naturalized. This has happened several times. Information about the naturalized species is available in the preceding chapter. Third, cultivated plants may reveal aspects of variation which might be overlooked in the study of natural populations.

By way of indicating which species are in cultivation outdoors, I have organized in a table data for several localities. The listing for Ithaca, N.Y., includes the species which one sees in gardens and on walls when walking the streets. It omits my own large collection which is a special circumstance. The idea is to list only species which laymen cultivate. The listing for New York includes the species in the Thompson Memorial Rock Garden, as observed in June 1972. For Calgary, Alberta, the listing applies to the Reader Rock Garden, Sept. 1, 1965, and for Pasadena, it applies to the Huntington Botanical Garden, July 2, 1940.

Species	Ithaca	New York	Calgary	Pasadena
1. *S. sieboldii*	+			
2. *S. telephium*		+	+	
3. *S. purpureum*			+	
4. *S. spectabile*	+	+		+
5. *S. ewersii*		+	+	
6. *S. anacampseros*			+	
7. *S. praealtum*				+
(*S. praealtum monticola* × *Villadia batesii*)				(+)
8. *S. palmeri*				+
9. *S. treleasei*				+
10. *S. pachyphyllum*				+
11. *S. adolphii*				+
12. *S. rubrotinctum*				+
13. *S. stahlii*				+
14. *S. oaxacanum*				+
15. *S. moranense*				+
16. *S. aizoon*			+	
17. *S. kamtschaticum*		+	+	+
18. *S. sarmentosum*	+	+		
19. *S. spurium*	+	+	+	+
20. *S. multiceps*				+
21. *S. album*	+	+	+	
22. *S. brevifolium*				+
23. *S. dasyphyllum*				+
24. *S. anglicum*		+		
25. *S. bithynicum*		+		
26. *S. hispanicum*	+		+	
27. *S. sexangulare*	+	+		
28. *S. acre*	+	+	+	
29. *S. chrysanthum*				+
30. *S. sediforme*				+
31. *S. reflexum*	+	+		
Totals	9	12	10	17 + (1)

Annotated List of *Sedum* in Cultivation in North America North of the Mexican Plateau

Brief diagnoses are available for about half of the species of *Sedum* cultivated outdoors in North America north of the Mexican Plateau. Additional information about these species includes indication of the natural geographic distribution, type locality, date of earliest record in cultivation in North America, and months of flowering. Further, when species are listed in the horticultural trade under other specific names, these are indicated.

1. *Sedum sieboldii* Sweet ex Hooker, Botanical Magazine 89:pl. 5,358 (1863). Siebold's Stonecrop. Diagnosis (based on plants of horticultural origin, cultivated at Ithaca, N.Y.): perennial herbs with suberect, spreading stems; leaves in whorls of three, cuneate basally, obovate, and blue-green suffused with pink; inflorescences

corymbose cymes with pink petals. Distribution: Japan—Shikoku. Type locality: Japan. First record in cultivation: 1865 (Mich.). Flowers in September and October. Other names: none.

2. *Sedum telephium* L., Sp. Pl. 1:430 (1753). Diagnosis (based on a plant from Poland and also on plants of horticultural origin, cultivated at Ithaca, N.Y.): perennial herbs with tuberous, thickened roots; stems erect, about .5 m. tall; leaves opposite or subopposite, broad, dark green; and flowers in dense cymes with white or greenish white petals. Distribution: Europe. Type locality: Europe. First record in cultivation: <1902, Bailey (1900–1902). Flowers in August and September. Other names: *S. maximum*, *S. maximowiczii*.

3. *Sedum purpureum* (L.) Schult. For reference to original publication, diagnosis, statement of distribution, type locality, time of flowering, and other names, see pp. 537–543. First record in cultivation: 1846 (BH).

4. *Sedum spectabile* Boreau, Mém. Soc. Acad. Maine-et-Loire 20:116 (1866). Diagnosis (based on plants of horticultural origin, cultivated at Ithaca, N.Y.): perennial herbs (fig. 162) with erect stems about .5 m. tall; leaves opposite, in whorls of three, or alternate, elliptical, pale green, and glaucous; flowers in flat-topped clusters with pink petals and stamens much exceeding the petals. Distribution: eastern Asia. Type locality: unknown. First record in cultivation: <1902, Bailey (1900–1902). Flowers in August and September. Other names: several varieties are listed.

5. *Sedum alboroseum* Baker, in Saunders' Refugium Botanicum 1:pl. 33 (1868). Diagnosis (based on plants of horticultural origin, cultivated at Ithaca, N.Y.): perennial herbs (fig. 163) with erect stems to 6 dm. tall; leaves alternate or opposite, sessile or short-petiolate, elliptic, coarsely dentate toward apex, and light yellow-green; flowers in terminal corymbose cymes, about 10 mm. in diameter, with white petals which are green medianly, 0–10 stamens, and 0–5 pink carpels. Meiosis is irregular (Baldwin, 1937). Plants may be hybrids of *S. spectabile* and *S. viridescens*. Distribution: possibly in eastern Asia. Type locality: Japan. First record in cultivation: 1890 (GH). Flowers in August, September, and early October. Other names: none.

6. *Sedum populifolium* Pallas, Reise durch verschiedene Provinzen des Russischen Reiches 3(2):730, pl. O, fig. 2 (1776). Diagnosis (based on plants of horticultural origin, cultivated at Ithaca, N.Y.): subshrubs with alternate, petiolate, ovate, coarsely dentate leaves; flowers in cymes, with the petals white, pink at tips. Distribution: central and northern Asia. Type locality: near headwaters of Yenisei River in the Sajan Mountains of central Asia. First record in cultivation: <1902, Bailey (1900–1902). Flowers in August and September. Other names: none.

7. *Sedum ewersii* Ledebour, Flora Altaica 2:191 (1830). Diagnosis (based on plants of horticultural origin, cultivated at Ithaca, N.Y.): decumbent herbs with slender, horizontally spreading rootstocks; stems numerous, decumbent, and branched; leaves opposite, ovate-cordate, and bluish green; flowers 9 mm. in diameter, in corymbose cymes, with pink petals. Distribution: central Asia. Type locality: Altai

CULTIVATED SPECIES 559

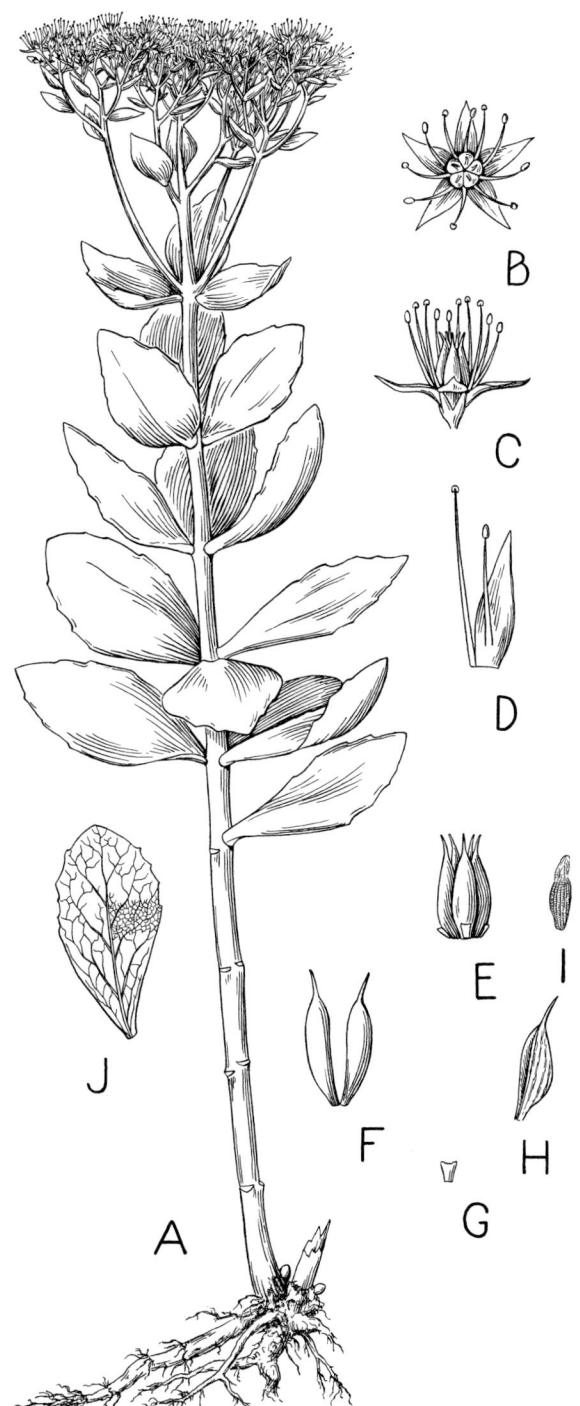

Fig. 162. Plant of *Sedum spectabile*, obtained from J. T. Baldwin, Jr., and studied by him for chromosomes (line 70), cultivated in garden, Ithaca, N.Y., Sept. 16, 1971. A. Habit sketch (x .45). B. Flower from above (x 1.8). C. Flower from side (x 1.8). D. Petal and two stamens (x 2.7). E. Carpels (x 2.7). F. Two carpels (x 3.6). G. Nectary (x 4.5). H. Follicle (x 3.6). I. Seed (x 9). J. Leaf (x .45).

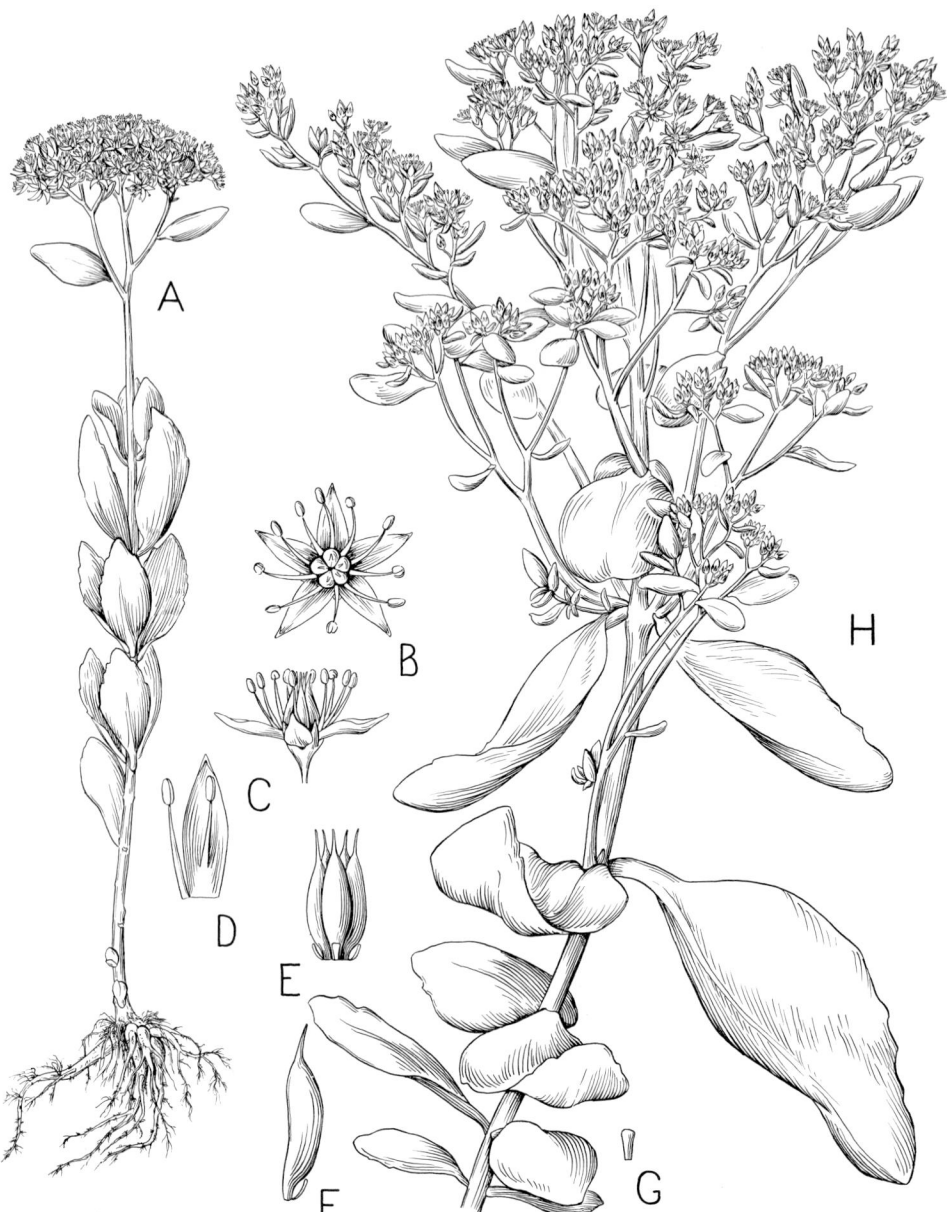

Fig. 163. Plant of *Sedum alboroseum*, obtained from cultivation in Kershaw Co., S.C., cultivated in garden, Ithaca, N.Y., Sept. 18, 1966. A. Habit sketch of plant in flower (x .43). B. Flower from above (x 1.7). C. Flower from side (x 1.7). D. Petal and two stamens (x 2.6). E. Carpels (x 2.6). F. Single carpel and nectary (x 3.4). G. Nectary (x 4.3). H. Habit sketch of floriferous stem with buds and a few flowers expanding from replicate of same plant cultivated in cold frame (x .43).

Mts. First record in cultivation: <1902, Bailey (1900–1902). Flowers in August and September. Other names: *S. hayesii*.

8. *Sedum anacampseros* L., Sp. Pl. 1:430 (1753). Diagnosis (based on plants of horticultural origin, cultivated at Ithaca, N.Y.): perennial herbs with alternate, entire, elliptical leaves; flowers with rose-purple petals, in terminal cymes. Distribution: Europe. Type locality: unknown, possibly in the mountains of southern France. First record in cultivation: <1902, Bailey (1900–1902). Flowers in July. Other names: none.

9. *Sedum praealtum* A. DC. ssp. *praealtum*, see p. 553. Diagnosis (based on plants from Río Blanco, Vera Cruz, cultivated at Ithaca, N.Y.): shrubs with sessile, oblong-elliptical leaves, 46–67 mm. long, 16–22 mm. wide, and 2–3 mm. thick; inflorescences are large, paniculate cymes; petals are yellow, less than 2.5 mm. wide. Distribution: eastern part of Trans-Mexican Volcanic Belt, alt. 1,330–1,370 m. Type locality: unknown (ssp. *praealtum*). First record in cultivation: 1930, listed by a nursery in California. Flowers in February, March, and April. Other names: *S. dendroideum* ssp. *praealtum*.

The change in interpretation of *Sedum praealtum*, once listed as a subspecies of *S. dendroideum* (Clausen, 1959), results from further study. Behavior of plants of the two species has been different. Further, additional features for separating them have come to attention. This change of status for ssp. *praealtum* requires the following two new combinations: *S. praealtum* ssp. **parvifolium** (R. T. Clausen), comb. nov., fundatum super *S. dendroideum* ssp. *parvifolium* R. T. Clausen, Sedum of the Trans-Mexican Volcanic Belt, p. 72 (1959), and *S. praealtum* ssp. **monticola** (T. S. Brandegee), comb. nov., fundatum super *S. monticola* T. S. Brandegee, Univ. Calif. Pub. Bot. 6:498 (1919). Hybrids of *S. praealtum* ssp. *monticola* X *Villadia batesii* are cultivated under the name of *Sedum amecamecanum*. These plants are small subshrubs with decumbent, repent, or even pendulous stems, elliptical leaves, 15–20 mm. long and 5 mm. wide, and white or pale yellow flowers.

10. *Sedum confusum* Hemsley, Diagnoses Plant. Nov. 1:10 (1878). Diagnosis (based on plants of horticultural origin, cultivated in a greenhouse at Ithaca, N.Y.): subshrubs with flat, spatulate leaves and dense cymes; flowers averaging 11 mm. in diameter, with sepals 1.2 mm. long and orange-yellow petals 5 mm. long. Distribution: Meyrán (1963), following my interpretation of *S. aoikon*, has cited as a locality cliffs between Zacapoaxtla and Huahuaxtla in the State of Puebla. Type locality: unknown. First record in cultivation: 1901. Flowers from December to April. Other names: *S. aoikon*. The plant described as *S. aoikon* in Sedum of the Trans-Mexican Volcanic Belt (Clausen 1959:82–87) is *S. confusum*. The plant interpreted there as *S. confusum* is *S. decumbens*. Additional study of plants in cultivation has led to this change in interpretation. The opinion now is that the types of both *S. confusum* and *S. aoikon* (see Cact. and Succ. Jour. 18:75, fig. 47, 1946; and 20:132, fig. 100, 1948); the figure by Praeger (1921, p. 212, fig. 121); and the illustration of Saunders, Refugium Botanicum 5:pl. 337 (1873), apply to the same species. For an earlier, similar opinion, see Rush (1941).

11. *Sedum decumbens* R. T. Clausen,[1] proposed here as a new species. Diagnosis (based on plants of horticultural origin, cultivated at Ithaca, N.Y.): subshrubs similar to *Sedum praealtum* ssp. *parvifolium* and *S. confusum*, but lower, less than 2.5 dm. tall, with smaller, emarginate, obovate or elliptic-spatulate leaves, 20 ± 5 mm. long and 12 ± 1 mm. wide; leaves forming rosettes at ends of sterile stems; inflorescences terminal, smaller, with three or more cincinni; sepals 3 mm. long, and petals lanceolate, yellow, 5 mm. long. Distribution: unknown, possibly on the central Mexican Plateau. Type locality: unknown, to be designated after the species is found in the wild and the pattern of intraspecific variation is understood. First record in cultivation: 1931, offered by a nurseryman in California. Flowers from March to June in greenhouse at Ithaca, N.Y. Other names: *S. confusum*.

12. *Sedum palmeri* S. Wats., Proc. Am. Acad. 17:355 (1882). Diagnosis (based on plants from five localities in the states of Nuevo León and Tamaulipas, Mexico, including the type localities of both *S. palmeri* and *S. compressum*): subshrubs, usually pendent, with glaucous, flat, spatulate leaves, aggregated in loose rosettes toward the ends of the stems; inflorescences terminal, but appearing lateral, 3–4-parted cymes; sepals linear-lanceolate, 4–7 mm. long; petals yellow, 4–7 mm. long; follicles divergent, but not gibbous ventrally. Distribution: Sierra Madre Oriental, Mexico. Type locality: Guajuco (Villa Santiago), 43 km. southeast of Monterrey, Mexico. First record in cultivation: 1932, offered by a nurseryman in New Jersey. Flowers from December until April. Other names: *S. compressum*.

13. *Sedum treleasei* Rose, Contr. U.S. Nat. Herb. 13(9):300 (1911). Diagnosis (based on plants of horticultural origin, cultivated in a greenhouse at Ithaca, N.Y.): succulent subshrubs with stems erect or procumbent, attaining a height of 4 dm.; leaves very fleshy, divergent and slightly upturned, oblong-obovate, blunt at apex, flattish ventrally, convex dorsally, and glaucous; flowers in corymbose cymes with glaucous, oblanceolate sepals and bright yellow petals. Distribution: vicinity of Tehuacán, State of Puebla, Mexico. Type locality: near El Riego, Tehuacán, Mexico. First record in cultivation: 1932, offered by a nurseryman in New Jersey. Flowers in January and February. Other names: none. Note: *Sedum treleasei* has been crossed with other species as *S. corynephyllum*, *S. pachyphyllum*, and *Graptopetalum amethystinum*.

14. *Sedum pachyphyllum* Rose, Contr. U.S. Nat. Herb. 13(9):299 (1911). Diagnosis (based on 5 plants of horticultural origin, cultivated in a greenhouse at Ithaca, N.Y.): succulent subshrubs with decumbent stems, attaining a height of 3 dm.; leaves very fleshy, terete, clavate, slightly upcurved, oblong-oblanceolate, rounded at apex, glaucous, and sometimes red-tipped; inflorescences corymbose cymes; flowers with

---

[1] *Sedum* **decumbens**, sp. nov., affine *Sedum praealtum* DC. and *S. confusum* Hemsley, sed humiliore, cum caulibus decumbentibus et foliis minoribus, viridibus-pallidis, obovatis vel spatulatis, et subcharactis, 20 ± 5 mm. longis et 12 ± 1 mm. latis; caulibus sterilibus cum rosulis ad extrema; cymis terminalibus, cum cincinnis 3 vel plus; floribus 10 mm. in linea media; sepalis 2.6 ± .4 mm. longis et petalis ovato-lanceolatis, luteis, 4.7 ± .2 mm. longis. Typus temporarius in Herbario Wiegand, Universitatis Cornellianae, originis hortensis, R. T. Clausen 44–119, 1952, Maius 25, est.

glaucous-green, clavately oblanceolate sepals and yellow petals. Distribution: hills in vicinity of San Luis, alt. 1,800–2,100 m., State of Oaxaca, Mexico. Type locality: hill near San Luis, Oaxaca, Mexico. First record in cultivation: 1931, offered by a nurseryman in California. Flowers in greenhouses in New York from February to May and outdoors in California in February and March. Other names: none. Note: *Sedum pachyphyllum* has been crossed with *S. praealtum*, *S. treleasei*, and *Echeveria derenbergii*.

15. *Sedum lucidum* R. T. Clausen, Cact. Succ. Jour. 23:125 (1951). Diagnosis (based on 20 plants from three localities in Vera Cruz, namely the Cerro Borrego, Río Blanco, and Maltrata): subshrubs with stems much branched, attaining a height of 4 dm.; leaves plano-convex, obovate to oblanceolate, upcurved, lustrous, and green; flowers on pedicels less than 7 mm. long, with ovate or oblong, green sepals, and white or pinkish white petals which are lanceolate or elliptic-oblong and obtuse. Distribution: on limestone in the northeastern part of the Sierra Madre del Sur, in an area extending from Maltrata to Orizaba, in the State of Vera Cruz. Type locality: probably the Cerro Borrego on the western side of the city of Orizaba. First record in cultivation: 1935, when taken to California by Eric Walther. Flowers in greenhouses at Ithaca, N.Y., from January to April. Other names: none.

16. *Sedum clavatum* R. T. Clausen.[2] Diagnosis (based on 5 plants from the gorge of the Tiscalatengo River in the vicinity of Villa Guerrero, Mexico, and cultivated at Ithaca, N.Y.): subshrubs with glaucous, thick, clavate, upcurved, short-spurred, elliptic-oblanceolate or obovate leaves; flower-bearing stems axillary; sepals clavate, 5–8 mm. long; petals elliptical and white; anthers red. For a fuller description, see Clausen (1959:102–107). Distribution: gorge of the Tiscalatengo River in the vicinity of Villa Guerrero, State of Mexico, Mexico. Type locality: cliff of andesite, alt. 2,180 m., exposed to southwest, 30 m. above Tiscalatengo River, in gorge about 4 km. west-southwest of Tenancingo and about 1 km. downstream from bridge of highway between Tenancingo and Villa Guerrero. First record in cultivation: 1947, at the New York Botanical Garden. Flowers in greenhouse at Ithaca, N.Y., from March to July. Other names: *Sedum* of the Tiscalatengo Gorge.

17. *Sedum nussbaumerianum* Bitter, Notizbl. Bot. Gart. u. Mus. Berlin-Dahlem 8(74):281 (1923). Diagnosis (based on 3 plants of horticultural origin, cultivated in greenhouse at Ithaca, N.Y.): subshrubs with decumbent, reddish brown, glabrous stems; leaves thick, acute, and yellow-green, averaging 30 mm. long, 13 mm. wide, and 7 mm. thick; cymes appearing umbellate; flowers white, on slender pedicels averaging 15 mm. in length. Distribution: known only from type locality, Zacuapan, Vera Cruz, Mexico. Type locality: sulphur spring, ravine at Zacuapan, Vera Cruz,

---

[2] *Sedum* **clavatum**, sp. nov., affine et simile *S. lucidum*, sed caulibus et foliis glaucis; sepalis longioribus, 5–8 mm. longis, et clavatis; et antheris rubris; aliter foliis crassis, clavatis, elliptico-oblanceolatis vel obovatis, brevicalcaratis, obtusis, et licinis; caulibus floriferentibus axillaribus; et floribus albis cum petalis ellipticis. Typus in Herbario Wiegand, Universitatis Cornellianae, ex cremno in Barranca de Tiscalatengo inter septentriones et orientem spectante de Villa Guerrero, Mexico, 1 km. secundo flumine de ponte viae de Tenancingo ad Villa Guerrero, cultus in Ithaca, N.Y., 1956, Martius 24, Robert T. Clausen TMV–T–Tis 1, est. Species dedicata ad Edward J. Alexander et amicum illium, Charles L. Gilly.

Mexico. First record in cultivation: 1944, listed by a nurseryman in California, but listings of *S. adolphii* from as far back as 1931 may apply to *S. nussbaumerianum*. Flowers in greenhouse at Ithaca, N.Y., from late January to early April. Other names: often confused with *S. adolphii* and may be identified as that species.

18. *Sedum adolphii* Raymond Hamet, Notizbl. Bot. Gart. u. Mus. Berlin-Dahlem 5:277 (1912). Diagnosis (based on 2 plants of horticultural origin, cultivated in greenhouse at Ithaca, N.Y.): subshrubs with decumbent or pendulous, pale brown stems; leaves thick, elliptic-oblong, almost plane ventrally, somewhat angulate, prominently carinate dorsally, averaging 42 mm. long, 18 mm. wide, and 8 mm. thick; cymes paniculate, with primary axis about 2.5 cm. long; flowers white, on pedicels 11–18 mm. long; petals ovate-lanceolate, 3–3.7 mm. broad. Distribution: unknown from any definite locality in wild. Type locality: somewhere in Mexico, possibly in the State of Vera Cruz. First record in cultivation: 1931, listed by a nurseryman in California, but listing may apply to *S. nussbaumerianum*. Flowers in greenhouse at Ithaca, N.Y., in January and February. Other names: often confused with *S. nussbaumerianum*. Note: Hybrids of *Sedum adolphii* and *Graptopetalum amethystinum* are available in cultivation.

19. *Sedum rubrotinctum* R. T. Clausen, Cact. Succ. Jour. 20:82 (1948). Diagnosis (based on plants of horticultural origin, cultivated in California and in greenhouse at Ithaca, N.Y.): subshrubs with decumbent, glabrous stems; leaves linear-oblong, clavate and terete, either lustrous green or more usually red or suffused with red, 4–21 mm. long, 2–8 mm. wide, and 2–7 mm. thick; cymes terminal; flowers 14–15 mm. in diameter, on pedicels 1 mm. long; petals lanceolate, acute, and yellow, 7 mm. long and 2 mm. wide. Distribution: unknown in wild. Type locality: unknown; possibly of horticultural origin. First record in cultivation: 1935, as *S. guatemalense*. Flowers in February, March, and September. Other names: has been confused with *S. guatemalense*.

20. *Sedum diffusum* S. Watson, Proc. Am. Acad. 25:148 (1890). Diagnosis (based on plants in the Sierra Madre Oriental and cultivated at Ithaca, N.Y.): perennial herbs with subterete, oblong-linear leaves, 3–12 mm. long, 1–3 mm. wide, 1–2 mm. thick; flowers with white, elliptic-lanceolate, acute petals, 4–4.8 mm. long, slightly connate basally; follicles widely divergent; seeds pyriform, dark brown, .6–.7 mm. long. Distribution: Sierra Madre Oriental of Mexico, in the states of Nuevo León and San Luis Potosi, and also either native or naturalized in Hidalgo. Type locality: Saddle Mountain near Monterrey, Nuevo León. First record in cultivation: 1932, listed by a nurseryman in Ohio. Flowers in greenhouse in Ithaca, N.Y., in May, June, and October, and in cold frame in June. Other names: has been offered as *S. potosinum*.

21. *Sedum morganianum* Walther, Cact. Succ. Jour. 10:35 (1938). Diagnosis (based on plants of horticultural origin, cultivated in Mexico, California, and in the greenhouse at Ithaca, N.Y.): perennial subshrubs with long, pendulous stems, attaining lengths of 9 dm. or more; leaves glaucous, falcate, oblong-lanceolate or elliptical and acute, subterete, 15–30 mm. long, 5–8 mm. thick; flowers with erect, orchid-purple petals 10–12 mm. long. Distribution: unknown. Type locality: unknown, the type

being a cultivated plant of unknown origin in the wild. First record in cultivation: 1940, listed by a nurseryman in southern California. Flowers irregularly, in June and December. Other names: none.

22. *Sedum stahlii* Solms, Sämereien bot. Gart. Univ. Strassburg 1900:4. Diagnosis (based on 7 plants from Maltrata and Acultzingo in Vera Cruz, cultivated in greenhouse at Ithaca, N.Y.): subshrubs with much-branched, spreading or trailing, puberulent stems; leaves opposite, thick, often globular, usually red or reddish, puberulent, averaging 12 mm. long, 5 mm. wide, and 5 mm. thick; cymes terminal; and petals elliptic-lanceolate and yellow. Distribution: in Mexico, in the northeastern part of the Sierra Madre del Sur and on the adjacent eastern slopes of the Trans-Mexican Volcanic Belt. Type locality: on limestone in the Cañada Istapam near Tehuacán, Puebla. First record in cultivation: 1921, listed by a nursery in Ohio. Flowers from February to May. Other names: none.

23. *Sedum hintonii* R. T. Clausen, Bull. Torr. Bot. Club 70:292 (1943). Diagnosis (based on type specimen and a plant of horticultural origin, cultivated in greenhouse, Ithaca, N.Y.): perennial herbs with dense rosettes of oblong or elliptical, densely pubescent leaves, 1.5–5 cm. long and .3–1 cm. wide; inflorescences paniculate cymes with broadly spreading branches and sessile flowers; petals oblong-lanceolate, recurved, white, and 4–6 mm. long. Distribution: Sierra Madre del Sur and Sierra Madre Occidental in states of Michoacán, Jalisco, and Durango. Type locality: Pinzan, alt. 830 m., District of Coalcomán, Michoacán, Mexico. First record in cultivation: 1947, offered by nursery in Mexico; available in California by 1960. Flowers in January and February in greenhouse at Ithaca, N.Y., and in April in Mexico. Other names: none.

24. *Sedum mexicanum* Britton, Bull. N.Y. Bot. Gard. 1:257 (1899). Diagnosis (based on plants cultivated in Mexico and in greenhouse at Ithaca, N.Y.): perennial herbs with decumbent stems and linear, subterete leaves, usually in whorls of four; cymes terminal; and flowers with yellow petals. Distribution: unclear, but possibly in central Mexico or Central America, or both. Rarely naturalized in Florida, 1930. Type locality: trap dike near the city of Mexico. First record in cultivation: 1930, listed by a nursery in California. Flowers from March to July. Other names: none.

25. *Sedum greggii* Hemsley, Diagn. Pl. Nov. 1:12 (1878). Diagnosis (based on 14 plants from four localities in Mexico and replicates of these cultivated in a greenhouse at Ithaca, N.Y.): perennial herbs with dense rosettes of obovate or elliptical, blunt, papillose leaves, 5–6 mm. long, and 2–4 mm. wide; floriferous stems erect, decumbent, or pendulous, with obovate or elliptical leaves; cymes terminal; flowers 10 mm. in diameter, with yellow petals; and nectaries .4 mm. wide. Distribution: Central Mexican Plateau and adjacent Trans-Mexican Volcanic Belt. Type locality: near Tultenango, Mexico. First record in cultivation: 1944, listed by a nursery in California. Flowers in greenhouse at Ithaca, N.Y., from January to May. Other names: none.

26. *Sedum moranense* H. B. K., Nov. Gen. et Sp. Plant. 6:37 (1823). Diagnosis (based on 36 plants from five localities in Mexico and replicates of these cultivated under glass at Ithaca, N.Y.): perennial herbs about 9 cm. tall, with divergent, ovate

leaves, closely set in five or six spiral ranks, about 4 mm. long and 1.2 mm. wide; cymes few-flowered; and petals white. Plants in cultivation are ssp. *moranense* with sepals averaging 2.4–2.6 mm. long and petals averaging 4.5–5.8 mm. long and 1.3–1.6 mm. wide. Distribution: eastern half of Trans-Mexican Volcanic Belt and southeastern part of Central Mexican Plateau. Type locality: near Real de Moran, alt. 2,534 m., Hidalgo, Mexico. First record in cultivation: 1931, listed by a nursery in California. Flowers at Ithaca, N.Y., from May to August. Other names: none.

27. *Sedum aizoon* L., Sp. Pl. 1:430 (1753). Diagnosis (based on plants from Manchuria and of horticultural origin, cultivated at Ithaca, N.Y.): large, perennial herbs (fig. 164) with thick rootstocks, tuberous, carrotlike roots, and erect floriferous stems, to 8 dm. tall; leaves elliptic-lanceolate, blunt, crenate-serrate except for the lower fourth of the margins, 4–8 cm. long; cymes flat-topped or even concave, many-flowered, to 11 cm. in diameter; petals yellow; and follicles divergent. Distribution: northern Asia, from China north to Siberia and west to the Ural Mts. Type locality: Siberia. First record in cultivation: 1901, Cambridge, Mass. (BH). Flowers in June and July. Other names: often offered by dealers under names of other species; also it may be listed as *S. maximowiczii* or *S. wallichianum*. Note: Hybrids of *S. aizoon* and *S. kamtschaticum* are in cultivation.

28. *Sedum kamtschaticum* Fischer, Index Seminum Hort. Petropol. 7:54 (1841) Diagnosis (based on plants from Japan and Korea and also of horticultural origin, cultivated at Ithaca, N.Y.): perennial herbs (fig. 165) with low, decumbent, non-creeping stems, not over 3 dm. high; leaves always broadest above the middle, spatulate or oblanceolate, 3–5 cm. long; cymes few-flowered, averaging less than 25 flowers; petals yellow; and follicles divergent.

Three subspecies are in cultivation: ssp. *kamtschaticum* with leaves spatulate, 7–25 mm. wide, remotely serrate and dark green, and with flowers 15–18 mm. in diameter; ssp. *ellacombianum* (Praeger) R. T. Clausen, with leaves spatulate, crenate, and light green, and with flowers 10–14 mm. in diameter; and ssp. *middendorffianum* (Max.) Fröd., with leaves narrowly linear or linear-oblanceolate, 2.5–10 mm. broad, and crenate-dentate toward apex, and with flowers about 15 mm. in diameter.

Distribution: eastern Asia; ssp. *kamtschaticum* in northeastern Asia from Kamchatka to northern Japan, Korea, and northern China, westward in Siberia to the Irtysh River; ssp. *ellacombianum* in Japan (Hokkaido) and Korea; and ssp. *middendorffianum* in Siberia, Manchuria, and Mongolia. Type locality: Kamchatka (ssp. *kamtschaticum*); Hakodote, Hokkaido, Japan (ssp. *ellacombianum*); and eastern Asia, probably Siberia (ssp. *middendorffianum*). First record in cultivation: 1901, Cambridge, Mass. (BH), ssp. *kamtschaticum*. Flowers in June and July. Other names: *S. ellacombianum*, *S. floriferum*, and *S. middendorffianum*. Note: Hybrids of *S. kamtschaticum* and *S. aizoon*, and perhaps of *S. kamtschaticum* and *S. hybridum*, are in cultivation.

29. *Sedum hybridum* L., Sp. Pl. 1:431 (1753). Diagnosis (based on plants of horticultural origin, cultivated at Ithaca, N.Y.): perennial herbs with creeping stems, rooting at the nodes, forming a dense mat; leaves obovate or spatulate, cuneate,

CULTIVATED SPECIES 567

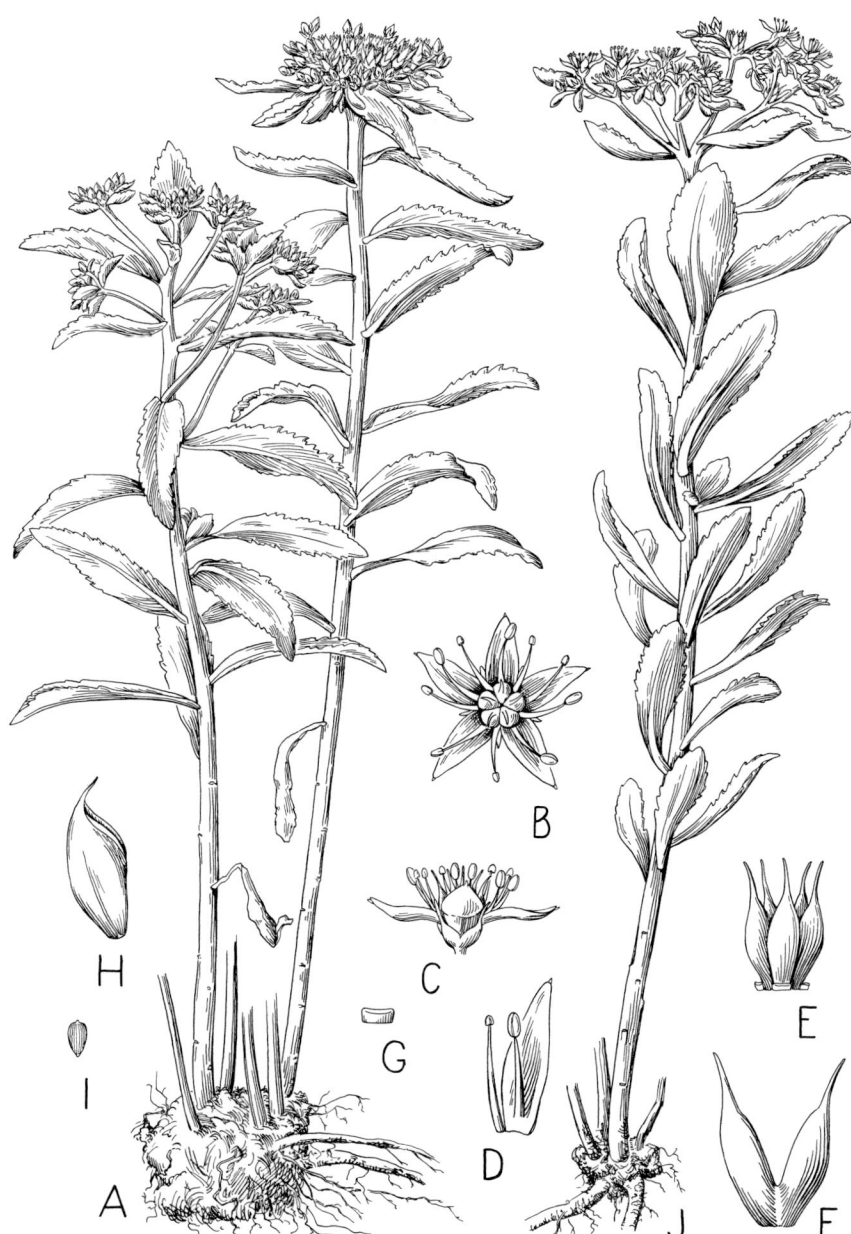

Fig. 164. Plants of *Sedum aizoon*, cultivated in garden, Ithaca, N.Y., June 25, 1973, A–I from Central Experimental Farm, Ottawa, Ont., and J from Reid Moran, collected at Yahiko on the island of Honshu, Japan. A. Habit sketch (x .41). B. Flower from above (x 1.6). C. Flower from side (x 1.6). D. Petal and two stamens (x 2.5). E. Carpels (x 2.5). F. Two carpels (x 3.3). G. Nectary (x 4.1). H. Follicle (x 3.3). I. Seed (x 4.1). J. Habit sketch (x .41).

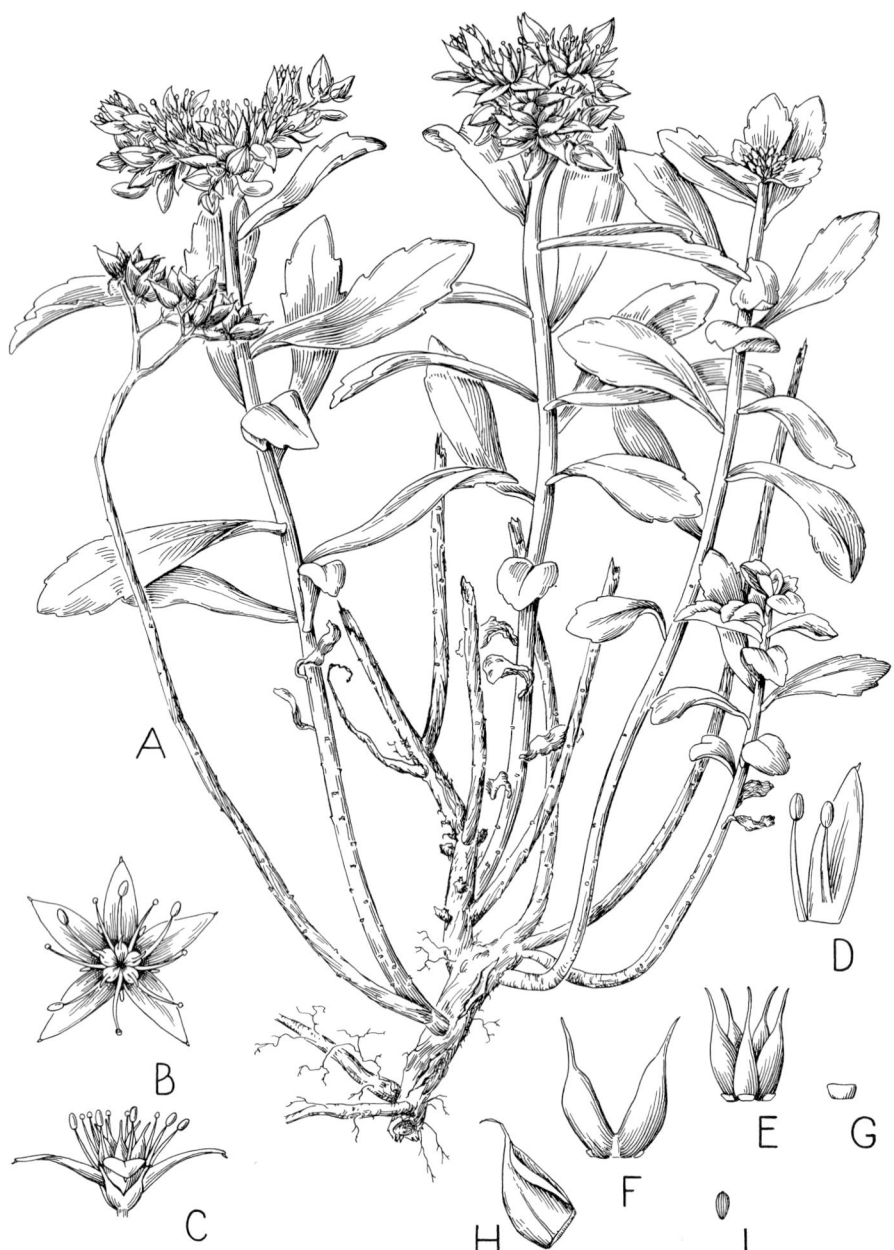

Fig. 165. Plant of *Sedum kamtschaticum* from horticultural source, cultivated in garden, Ithaca, N.Y., June 10, 1973. A. Habit sketch (x .82). B. Flower from above (x 1.6). C. Flower from side (x 1.6). D. Petal and two stamens (x 2.5). E. Carpels (x 2.5). F. Two carpels and nectaries (x 3.3). G. Nectary (x 4.1). H. Follicle (x 3.3). I. Seed (x 4.1).

remotely crenate-dentate at the rounded apices, 1–2.4 cm. long and .3–1.3 cm. wide; floriferous stems erect, with terminal cymes of 7–50 flowers, these 13–20 mm. in diameter; petals yellow, and follicles divergent. Distribution: north-central Asia from Kansu to the Abakan Steppe in Siberia. Type locality: the foot of the Ural Mountains. First record in cultivation: <1902, Bailey (1900–1902). Flowers in June and July. Other names: none, but sometimes confused with other species.

30. *Sedum lineare* Thunb., Flora Iaponica, p. 187 (1784). Diagnosis (based on plants of horticultural origin, cultivated in greenhouse and cold frame at Ithaca, N.Y.): perennial herbs, semihardy, with branched, decumbent stems and linear-lanceolate leaves, to 3 cm. long, and about 6 mm. wide, in whorls of three; flowers yellow, about 12 mm. in diameter, in terminal, few-flowered cymes. Distribution: Japan and eastern Asia. Type locality: Japan. First record in cultivation: 1925, greenhouse, Washington, D.C. (BH). Flowers in June. Other names: none.

31. *Sedum sarmentosum* Bunge. For reference to original publication, diagnosis, statement of distribution, type locality, time of flowering, and other names, see pp. 543–547. First record in cultivation: 1889, Norwich, Vt. (Vt.).

32. *Sedum spurium* Marschall von Bieberstein, Fl. taur. cauc. 1:352 (1808). Diagnosis (based on plants of horticultural origin, cultivated at Ithaca, N.Y.): perennial herbs (fig. 166) with branched creeping stems, rooting at the nodes; leaves obovate, cuneate basally, and coarsely crenate at apices, with prominently glandular margins, 1–3 cm. long, .5–2.2 cm. wide; flowers in terminal cymes, with petals erect below, then spreading, pink or white, 8–11 mm. long; and follicles erect or nearly so. Distribution: Caucasus or Turkestan. Type locality: rocks near hot springs in the Constantine Mountains in the Caucasus. First record in cultivation: 1901, Cambridge, Mass. (BH). Flowers in July and August. Other names: *S. oppositifolium*; also sometimes confused with *S. stoloniferum*. Listed as Creeping Sedum and Dragon's Blood.

33. *Sedum stoloniferum* S. G. Gmelin, Reise durch Russland 3:311, pl. 35, fig. 2 (1774). Diagnosis (based on plants of horticultural origin, cultivated at Ithaca, N.Y.): perennial herbs with prostrate stems, rooting at the nodes; leaves opposite, cuneate and subpetiolate basally, elliptic-spatulate, entire or crenate; flowers 9–11 mm. in diameter, with pink petals 6 mm. long; and follicles divergent, gibbous ventrally. Distribution: S.W. Asia. Type locality: wall, province of Gilan, Iran. First record in cultivation: 1921. Earlier offerings in the trade may apply to *S. spurium*. Flowers from June to September. Other names: sometimes confused with *S. spurium*.

34. *Sedum magellense* Tenore, Flora Napolitana 1:26 (1811). Diagnosis (based on plants of horticultural origin, cultivated at Ithaca, N.Y.): perennial herbs with decumbent stems; leaves alternate, oblong-elliptical or obovate, blunt or rounded at apices, yellow-green, glabrous, closely crowded, 2–6 mm. long and 2–4 mm. wide; inflorescences racemose cymes; and petals white. Distribution: Italian and Balkan Peninsulas and also western Asia Minor. Type locality: Abruzzes, Majella, Italy. First record in cultivation: 1931, Cronamere Nursery (BH). Flowers in May and June. Other names: none.

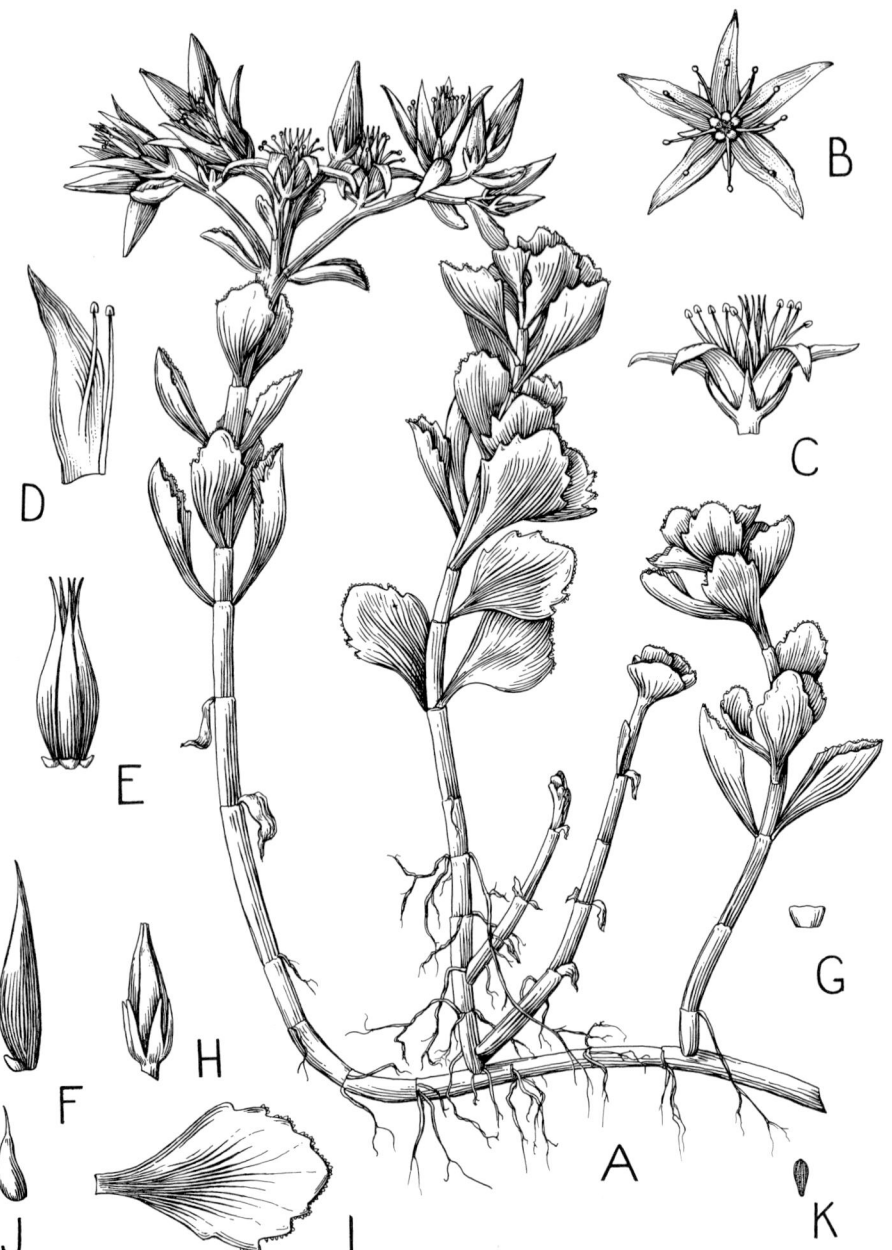

Fig. 166. Plant of *Sedum spurium* from horticultural source, cultivated in garden, Ithaca, N.Y., July 1942. A. Habit sketch (x .96). B. Flower from above (x 1.4). C. Flower from side (x 1.4). D. Petal and two stamens (x 2.4). E. Carpels (x 2.4). F. Single carpel and nectary (x 2.9). G. Nectary (x 4.8). H. Bud (x 1.4). I. Leaf (x 1.4). J. Follicle (x 1.9). K. Seed (x 4.8).

35. *Sedum monregalense* Balbis, Mém. Acad. Sc. Turin 7:339, pl. 6 (1802). Diagnosis (based on plants of horticultural origin, cultivated at Ithaca, N.Y.): perennial herbs with slender, creeping stems; leaves of sterile shoots spirally arranged, closely crowded, oblong-spatulate or elliptical, 2–8 mm. long, 1–3 mm. wide; leaves of floriferous stems in whorls of four, opposite, or alternate, oblong-spatulate, 2–6 mm. long, 1–3 mm. wide; flowers in paniculate cymes, nodding in bud, fragrant, on glandular-puberulent pedicels; petals white, lanceolate, acuminate, and glandular-puberulent dorsally; and follicles erect and glandular-hairy. Distribution: southern Europe, in Maritime Alps and Apennines, and also in Corsica. Type locality: on rocks in an alpine situation between Rastel and Blin, in the valley "Ellero Monregalensi de Blin," Piedmont Province, Italy. First record in cultivation: <1902, Bailey (1900–1902). Flowers in July and August. Other names: sometimes confused with *S. magellense*.

36. *Sedum album* L., Sp. Pl. 1:432 (1753). Diagnosis (based on plants of horticultural origin, cultivated at Ithaca, N.Y.): glabrous, perennial herbs (fig. 167) with creeping stems; leaves alternate, subterete to terete, linear-oblong to subglobose, 3–17 mm. long; flowers on slender pedicels in paniculate cymes; petals white and elliptic-lanceolate; follicles erect. Distribution: Europe, Asia, and northern Africa. Type locality: Europe. First record in cultivation: 1901, Cambridge, Mass. (BH). Flowers in June and July. Other names: none.

37. *Sedum dasyphyllum* L., Sp. Pl. 1:431 (1753). Diagnosis (based on plants of horticultural origin, cultivated at Ithaca, N.Y.): small, perennial herbs (fig. 168), to 12 cm. tall, with decumbent, creeping stems which are glandular-pubescent; leaves pale green or gray, sometimes glandular-pubescent, in dense, subglobose rosettes; flowers on glandular-pubescent pedicels, in terminal cymes; petals pinkish white. Plants in cultivation are ssp. *dasyphyllum* which is distinguished by mostly opposite, usually glandular-hairy, grayish leaves, 2–5 mm. long. Distribution: southern Europe (ssp. *dasyphyllum*) and northern Africa. Type locality: Europe. First record in cultivation: <1902, Bailey (1900–1902). Flowers in June and July. Other names: none.

38. *Sedum brevifolium* DC., Mém. Soc. Agr. Dépt. Seine 11:79 (1808). Diagnosis (based on plants of horticultural origin, cultivated at Ithaca, N.Y.): small, perennial herbs with slender, procumbent or creeping stems; leaves globular, gray, or reddish, and densely pruinose, in four ranks, easily detachable, 2–3 mm. long, and 1–2 mm. in diameter; cymes terminal, of 3–6 flowers; and petals white with pink midribs, broadly elliptical, and 3–4 mm. long. Distribution: southern Europe, in mountains. Type locality: on rocks, Hautes-Pyrénées, France. First record in cultivation: <1902, Bailey (1900–1902). Flowers in spring, from April to July under glass at Ithaca, N.Y. Other names: none.

39. *Sedum lydium* Boissier, Diagn. Plant. Orient. Nov. 1 (fasc. 3):17 (1843). Diagnosis (based on plants of horticultural origin, cultivated at Ithaca and Erin, N.Y.): perennial herbs with decumbent stems, densely leafy at tips; leaves linear, terete, 3–9 mm. long; flowers in dense, terminal cymes; petals white, elliptic-lanceolate, hooded, 4 mm. long; pistils erect. Distribution: southwestern Asia Minor. Milton

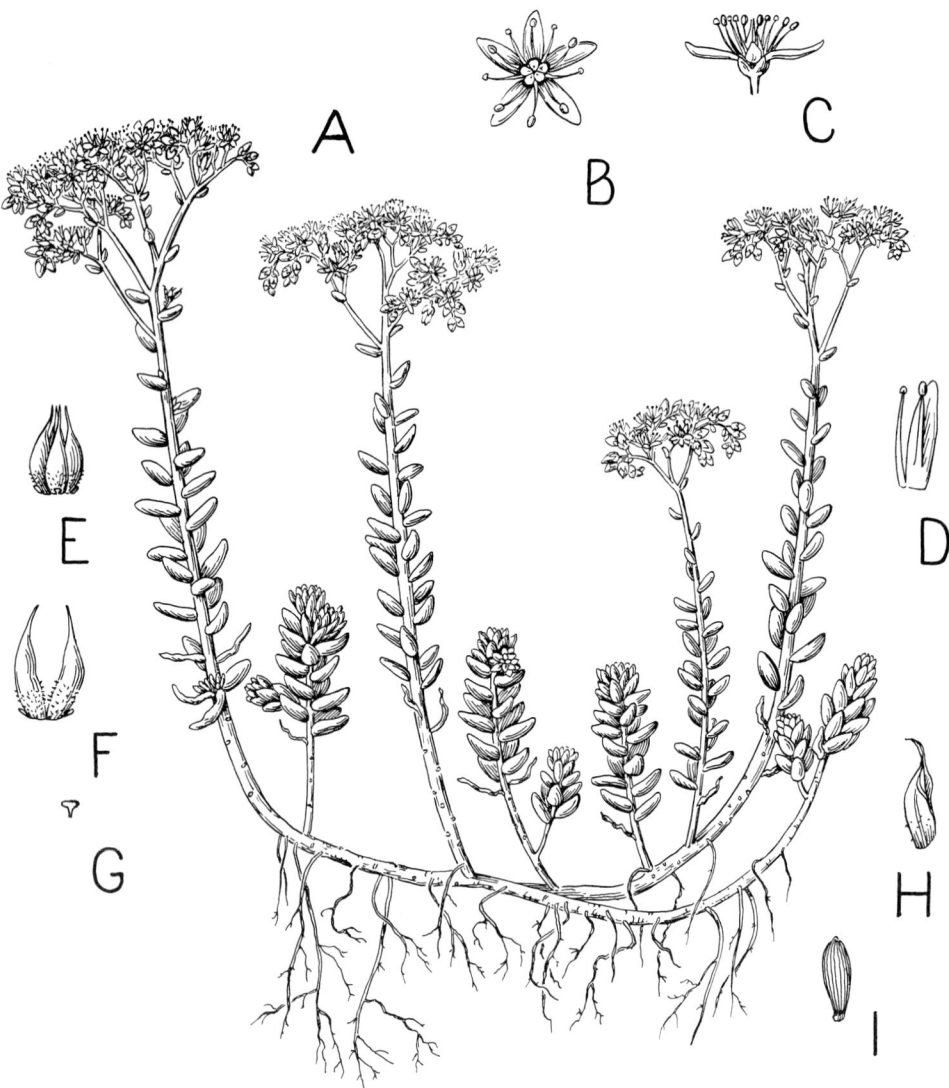

Fig. 167. Plant of *Sedum album* from bank of Beaver Creek Road, Jackson Co., Ore., cultivated in cold frame, Ithaca, N.Y., June 29, 1971. A. Habit sketch (x .6). B. Flower from above (x 2.4). C. Flower from side (x 2.4). D. Petal and two stamens (x 3.6). E. Carpels (x 3.6). F. Two carpels and nectaries (x 4.8). G. Nectary (x 6). H. Follicle (x 4.8). I. Seed (x 12).

Jack found plants naturalized in the vicinity of Hurricane Ridge, Port Angeles region of Olympic Peninsula, Wash., in 1944. Type locality: rocks of Mount Tmolus in the Boz Dagh southeast of Sardis, Turkey, Asia Minor. First record in cultivation: <1902, Bailey (1900–1902). Flowers in June. Other names: none.

40. *Sedum anglicum* Hudson, Flora Anglica, ed. 2, 1:196 (1778). Diagnosis (based

CULTIVATED SPECIES 573

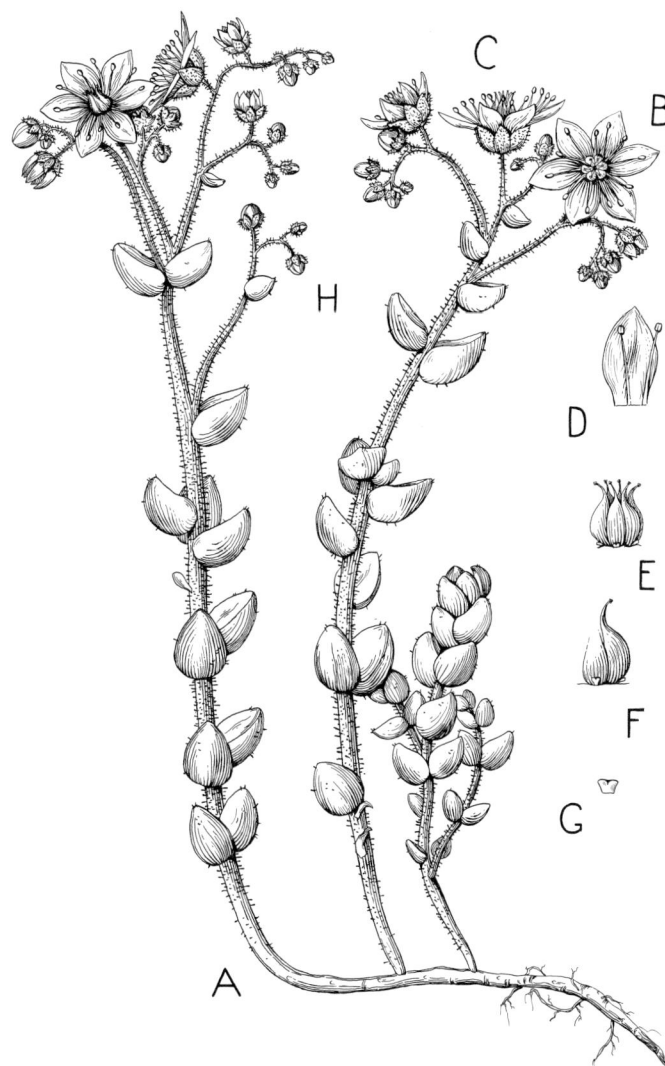

Fig. 168. Plant of *Sedum dasyphyllum* cultivated in garden, Ithaca, N.Y., June 1942. A. Habit sketch (x 1.9). B. Flower from above (x 1.9). C. Flower from side (x 1.9). D. Petal and two stamens (x 3). E. Carpels (x 3). F. Two carpels and nectaries (x 3.7). G. Nectary (x 5.6). H. Bud (x 1.9).

on plants of horticultural origin, cultivated at Ithaca, N.Y.): small perennial herbs with elliptical or ovate-elliptical leaves, broader than thick, 3–5 mm. long and 1–2 mm. wide; flowers in one- or two-parted cymes; petals pinkish white, lanceolate, 3–5 mm. long; pistils kyphocarpic. Distribution: western Europe and N.W. Africa. Type locality: England. First record in cultivation: 1931, Brooklyn Botanic Garden (BH). Flowers in spring, from March to May in greenhouse at Ithaca, N.Y., and in June outdoors at New York, N.Y. Other names: none.

41. *Sedum gracile* C. A. Meyer, Enum. Pl. Cauc., p. 151 (1831). Diagnosis (based on plants of horticultural origin, cultivated at Ithaca, N.Y.): small perennial herbs

with procumbent stems; leaves spirally arranged, linear, blunt, minutely papillose, densely crowded, 2–5 mm. long, 1 mm. wide; cymes terminal, with 2 or 3 cincinni; flowers 4–7 mm. in diameter; petals white, lanceolate, 2–3 mm. long; follicles divergent, gibbous ventrally. Distribution: western Asia, from the Caucasus to Iran. Type locality: rocks of subalpine region of Mount Gutgora, alt. 2,012 m., Caucasus. First record in cultivation: 1932, listed by two nurseries. Flowers in June. Other names: none.

42. *Sedum hispanicum* Juslenius in Linnaeus, Centuria I. Plantarum, p. 12 (1755). Diagnosis (based on plants of horticultural origin, cultivated at Ithaca, N.Y.): biennial herbs (fig. 169), 4–17 cm. tall; stems glabrous below, glandular-pubescent upward; leaves alternate, pale glaucous-green, oblong-linear, and terete, 4–15 mm. long; cymes glandular-pubescent; flowers 4–8-merous; petals white, with pink or green dorsal keels, 4–6 mm. long; follicles divergent, gibbous ventrally, glandular-hirsute, 2–4 mm. long. Distribution: southern Europe, Asia Minor, the Caucasus, and western Himalaya Mountains. Type locality: cited as Spain, but most likely somewhere else. First record in cultivation: <1902, Bailey (1900–1902). Flowers from May to July. Other names: *S. glaucum*.

43. *Sedum multiceps* Cosson et Durieu, Bull. Soc. Bot. France 9:171 (1862). Diagnosis (based on plants of horticultural origin, cultivated at Ithaca, N.Y.): much-branched, tiny subshrubs to 1 dm. tall; leaves alternate, closely crowded in dense clusters toward ends of branches, linear, 5–9 mm. long, with enlarged, scarious, sheathing spurs; flowers 12–14 mm. in diameter, in terminal cymes; petals oblong-lanceolate, acuminate, yellow, 6 mm. long; pistils kyphocarpic. Distribution: northern Africa, in mountains of Algeria. Type locality: Algeria, between Collo and Bougie in eastern Kabylia, up to 1,200–1,300 m. First record in cultivation: 1931, nursery in California. Flowers in June in greenhouse at Ithaca, N.Y. Other names: none.

44. *Sedum sexangulare* L., Sp. Pl. 1:432 (1753). Diagnosis (based on plants of horticultural origin, cultivated at Ithaca, N.Y.): evergreen perennial herbs with decumbent, creeping stems; leaves linear, blunt, subterete, 3–6 mm. long, in six ranks; cymes terminal, 3-parted; flowers 7–9 mm. in diameter; petals lanceolate, acute, canary yellow, 3–4 mm. long; follicles divergent, gibbous ventrally, pale brown, 2–4 mm. long. Distribution: Eurasia, from Scandinavia to the Caucasus. Type locality: Europe—Scandinavia. First record in cultivation: 1883 (BH). Flowers in June and July. Other names: *S. boloniense* and *S. hildebrandtii*. Note: See also p. 554.

45. *Sedum acre* L. For reference to original publication, diagnosis, statement of distribution, type locality, time of flowering, and other names, see pp. 547–552. First record in cultivation: <1902, Bailey (1900–1902).

46. *Sedum caeruleum* L., Mant. Plant. Alt. 2:241 (1771). Diagnosis (based on plants of horticultural origin, cultivated at Ithaca, N.Y.): annual herbs; stems erect or ascending, diffusely branched, 1–2 dm. tall; leaves alternate, sessile, oblong, more or less terete, 3–8 mm. long; cymes lax; flowers 5–9-merous, 6 mm. in diameter, on pedicels 1–2 mm. long; petals blue or white, elliptic-oblong, 3–4 mm. long; and follicles bearing 1–3 seeds. Distribution: islands of the Mediterranean Sea, the Italian

CULTIVATED SPECIES 575

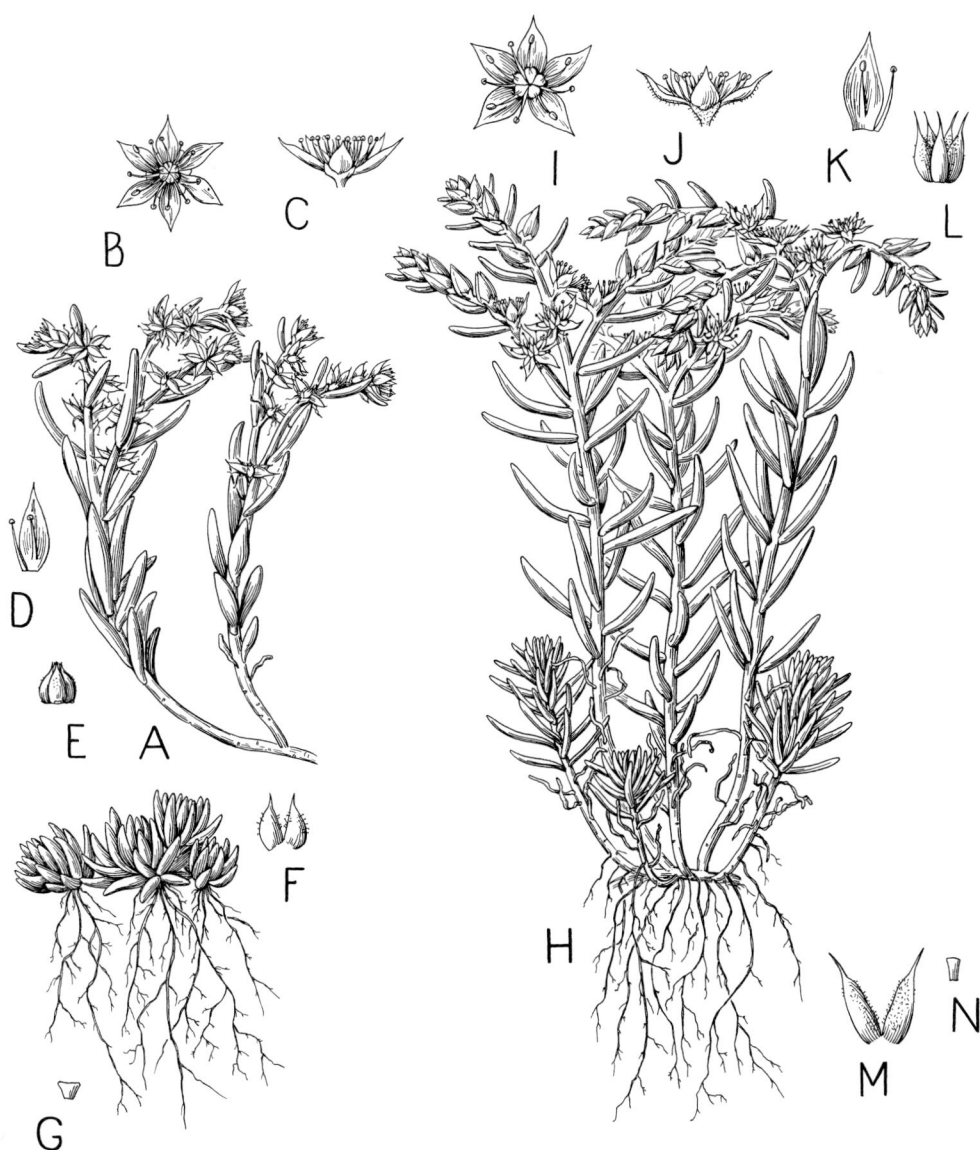

Fig. 169. Plants of *Sedum hispanicum* and *S. bithynicum* cultivated in garden, Ithaca, N.Y., June 19, 1960, A–G, biennial plant of *S. hispanicum* from Ferrisburg, Vt., and H–N, perennial plant of *S. bithynicum* from horticultural source. A and H. Habit sketches (x .9). B and I. Flowers from above (x 1.8). C and J. Flowers from side (x 1.8). D and K. Petals, each with two stamens (x 2.7). E and L. Carpels (x 2.7). F and M. Two carpels (x 3.6). G and N. Nectaries (x 4.5).

Peninsula, and northern Africa. Type locality: unknown, but erroneously cited as Cape of Good Hope, South Africa. First record in cultivation: <1902, Bailey (1900–1902). Flowers in garden at Ithaca, N.Y., from June to September. Other names: none

47. *Sedum hirsutum* Allioni, Flora Pedemontana 2:122, pl. 65, fig. 5 (1785). Diagnosis (based on plants of horticultural origin, cultivated at Ithaca, N.Y.): glandular-hairy, spicy-aromatic, perennial herbs, to 18 cm. tall; stems pale salmon, creeping or ascending, producing numerous sterile shoots; leaves aggregated at ends of stems, oblanceolate or elliptic-oblanceolate, blunt, densely pubescent with white, gland-tipped hairs; cymes 3-parted; flowers 1–2 cm. in diameter, on pedicels 1–8 mm. long; sepals glandular-hairy; petals connate basally for .5–2 mm., then spreading, ovate or oblong-lanceolate, white except for green or red median dorsal keel; pistils erect. Plants in cultivation are either ssp. *hirsutum* with the main stems creeping, 1–1.5 mm. in diameter, and with rosulate leaves lustrous, veronese green, 2–10 mm. long and 1–3 mm. wide; or ssp. *baeticum* Rouy, with main stems erect, 2–5 mm. in diameter, and with rosulate leaves pale green, not lustrous, 5–15 mm. long and 2–5 mm. wide. Distribution: southern Europe and northwestern Africa. Type locality: Italy—on rocks around Zuasse (ssp. *hirsutum*) and southern Spain—rocks in the Sierra de Palma (ssp. *baeticum*). First record in cultivation: 1933, listed by a nurseryman in Independence, Ohio. Ssp. *hirsutum* flowers in winter in the greenhouse at Ithaca, N.Y., but ssp. *baeticum* does not flower under similar circumstances. Other names: *S. winkleri* (= ssp. *baeticum*). Reference: Clausen (1943).

48. *Sedum pilosum* Marschall von Bieberstein, Fl. taur. cauc. 1:352 (1808). Diagnosis (based on a plant from the Georgian Caucasus and on another plant of horticultural origin, cultivated at Ithaca, N.Y.): biennial herbs with short rootstocks terminated upward in dense rosettes of glandular-hirsute, oblanceolate-oblong leaves, 5–10 mm. long and 2–4 mm. wide; cymes terminal, bearing 9–15 flowers on glandular-hirsute pedicels 1–6 mm. long; flowers 5-merous, 6–8 mm. in diameter; petals elliptic-spatulate, erect for about two-thirds of length, then widely spreading, pink, 6–8 mm. long; and pistils erect. Distribution: the Caucasus and adjacent Asia Minor; also reported from northern Iran (Fröderström, 1931). Type locality: central Caucasus. First record in cultivation: 1930, offered by nurseryman in California. Flowers in June in garden at Ithaca, N.Y. Other names: none.

49. *Sedum chrysanthum* (Boissier et Heldreich) Raymond Hamet, Candollea 4:27 (1929). Diagnosis (based on plants of horticultural origin, cultivated at Ithaca, N.Y.): glandular-hairy, perennial herbs with short, branched rootstocks; leaves in dense rosettes, oblong-spatulate, blunt, glandular-pubescent and glandular-ciliate, 10–15 mm. long, 3–6 mm. wide; floriferous stems to 2 dm. tall, bearing elongate cymes with ascending branches; flowers 4–5-merous, on glandular-pubescent pedicels to 10 mm. long; petals erect, united for one-third to one-half their length, elliptic-oblong, cuspidate, 8–15 mm. long, yellowish white or golden yellow; and follicles erect, pale brown, 6–7 mm. long. Plants in cultivation are ssp. *chrysanthum* with yellowish white petals, 12–15 mm. long, and floriferous stems 5–20 cm. tall. Distribution: Taurus Mountains of Asia Minor; also in Armenia. Type locality: western summit of Mount

Ghei Dagh, alt. 1,830 m., in the Isaurian Taurus above Marla (ssp. *chrysanthum*), and high mountains of the western Taurus Range (ssp. *aizoon*[3]). First record in cultivation: 1936 (BH). Flowers from mid-June to early July in garden at Ithaca, N.Y. Other names: none.

50. *Sedum sediforme* (Jacq.) Pau, Acta y Memorias Prim. Cong. Nat. Esp., Zaragozo, p. 246 (1909). Diagnosis (based on a plant from the Isle of Capri and on other plants of horticultural origin, cultivated in garden at Ithaca, N.Y.): perennial herbs (fig. 170) with decumbent stems and glaucous or green leaves which are narrowly elliptic-lanceolate and plane above, and taper to soft, spinescent tips, 5–20 mm. long and 1.8–4 mm. wide; floriferous stems erect from beginning, not nodding in bud, to 4.2 dm. tall; flowers usually 6 (5–7)-merous, 9–13 mm. in diameter; sepals ovoid, 1.5–3 mm. long; petals elliptic-oblong or oblanceolate, blunt, white, 5–6 mm. long and 1.8–2 mm. wide; and follicles erect. Distribution: southern Europe, Asia Minor, and northern Africa. Type locality: unknown. Jacquin described the species from a cultivated plant. First record in cultivation: 1932, Poughkeepsie, N.Y. (BH). Flowers in garden at Ithaca, N.Y. in June, July, and early August. Other names: *S. altissimum* and *S. nicaeense*.

51. *Sedum ochroleucum* Chaix in Villars, Histoire des Plantes de Dauphiné 1:325 (1786). Diagnosis (based on a plant from the Island of Vis in the Adriatic Sea and on other plants of horticultural origin, cultivated in garden at Ithaca, N.Y.): perennial herbs with creeping, decumbent stems with subterete, narrowly linear-lanceolate or linear leaves, 5–12 mm. long and 1–2.5 mm. wide; floriferous stems erect in development and at maturity, 1.6–2.4 dm. tall; flowers 5–7-merous, sessile or on short pedicels 1 mm. long; sepals lanceolate, acute, concave dorsally, and 3–6 mm. long; petals white or yellowish white, erect or suberect, narrowly oblong-lanceolate, acute, 6–8 mm. long, and 1.5–2 mm. wide; and follicles erect, stramineous or pale brown. Distribution: southern Europe from Iberian to Balkan Peninsulas; also reported from Asia Minor. Doubtfully established in eastern North America. Type locality: hills of Baux, France. First record in cultivation: 1931 (BH). Flowers in garden at Ithaca, N.Y., in late June and early July. Other names: *S. anopetalum*. Note: *Sedum ochroleucum* hybridizes with both *S. sediforme* and *S. reflexum*.

52. *Sedum rupestre* L., Sp. Pl. 1:431 (1753). Diagnosis (based on plants of horticultural origin, cultivated in garden at Ithaca, N.Y.): perennial herbs with prostrate, branched stems which give rise to numerous sterile shoots with usually blue-green, linear or linear-oblanceolate leaves, somewhat flattened ventrally toward tips, and closely crowded in dense, subglobose rosettes; leaves are 6–12 mm. long and 1–1.6 mm. wide; floriferous stems reflexed and with inflorescences globose in bud, later erect, 1.4–2.7 dm. tall; flowers 5–8-merous, 11–13 mm. in diameter; sepals ovate, acute, 2.5–3.5 mm. long; petals spreading, aureolin, elliptic-oblong, blunt, 5–6 mm. long; and follicles erect. Distribution: western Europe and northern Africa. Naturalized in England. Type locality: Europe, wherever Dillenius obtained his "*Sedum*

---

[3] *Sedum chrysanthum* ssp. **aizoon** (Fenzl), comb. nov., fundatum super *Umbilicus Aizoon* Fenzl, Pugillus Plantarum Novarum Syriae et Tauri Occidentalis Primus, p. 15 (1842).

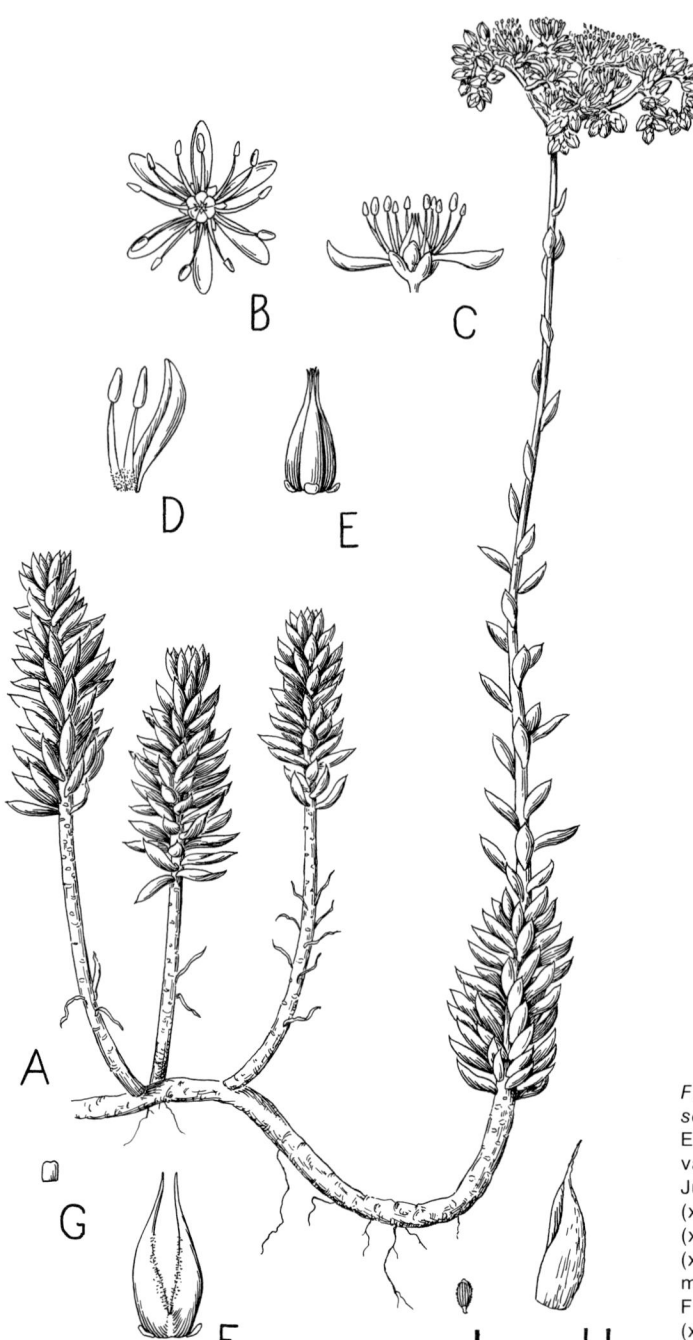

Fig. 170. Plant of *Sedum sediforme* from cultivation in El Dorado Co., Calif., cultivated in garden, Ithaca, N.Y., June 27, 1973. A. Habit sketch (x .47). B. Flower from above (x 1.9). C. Flower from side (x 1.9). D. Petal and two stamens (x 2.8). E. Carpels (x 2.8). F. Two carpels and nectaries (x 3.8). G. Nectary (x 4.7). H. Follicle (x 3.8). I. Seed (x 4.7).

rupestre repens." First record in cultivation: 1919, Cambridge, Mass, (BH). Flowers in garden at Ithaca, N.Y., in June and July. Other names: *S. forsterianum, S. elegans,* and *S. aureum.*

53. *Sedum reflexum* L., Sp. Pl., ed. 2, 1:618 (1762). Diagnosis (based on plant of horticultural origin, cultivated at Ithaca, N.Y.): perennial herbs (fig. 171) with much-branched, prostrate stems, forming loose mats, giving rise to many sterile shoots with linear-lanceolate, terete or subterete leaves, 5–12 mm. long and 1–2.5 mm. wide; floriferous stems reflexed before anthesis, later erect, to 3.3 dm. tall; flowers 5–7-merous, 8–16 mm. in diameter, sessile or on pedicels to 2 mm. long; sepals ovate or ovate-lanceolate, acute, 2–3.3 mm. long; petals spreading, canary-yellow, oblong-lanceolate, 5–6.5 mm. long, and 2 mm. wide; and follicles erect. Distribution: western Europe and northern Africa. Also naturalized in eastern North America. Type locality: somewhere in Europe. First record in cultivation: about 1816, Gray (1876). Flowers in garden at Ithaca, N.Y., from June to August. Other names: *S. cristatum.* Often misidentified and confused with both *S. ochroleucum* and *S. rupestre.*

Other Cultivated Species

Species in the following list, with a few exceptions, are less common in cultivation than those described in the preceding part of the chapter. Fewer nurseries offer them. Likewise, they are less frequent in gardens. Many are rare in cultivation. Although almost every species of *Sedum* is a candidate for horticultural use, many require special conditions for culture and others have not been introduced into American gardens. Altogether, about a third of the total number of known species in the genus is in cultivation in North America.

Entries in the present list are brief, simply the scientific name with citation of the publishing author, indication of continent and area of nativity, and a reference to a published description of the species. For sources of descriptive data, the species fall into four categories: those which are described elsewhere in this book—22; those which are described in *Sedum of the Trans-Mexican Volcanic Belt,* Clausen (1959)—6; those described by Praeger (1921)—16; and those described by the authors of the names or by subsequent monographers—11.

1. *Sedum telephioides* Michx. N.A.—Appalachian Highlands and Interior Low Plateaus. See pp. 70–91.

2. *Sedum verticillatum* L. Asia—Eastern Highlands and Japanese Islands. See Praeger (1921), pp. 94–96.

3. *Sedum tatarinowii* Maxim. Asia—Eastern Highlands. See Praeger (1921), pp. 101–104.

4. *Sedum cyaneum* Rudolph, incl. *S. pluricaule* (Maxim.) Kudo. Asia—Eastern Siberian Highlands. See Praeger (1921), pp. 106–107.

5. *Sedum tortuosum* Hemsley. N.A.—Sierra Madre del Sur and Trans-Mexican Volcanic Belt. See Clausen (1959). pp. 48–60.

6. *Sedum dendroideum* A. P. DC. N.A.—Central American Volcanic Upland and Sierra Madre del Sur. See Clausen (1959), pp. 69–70.

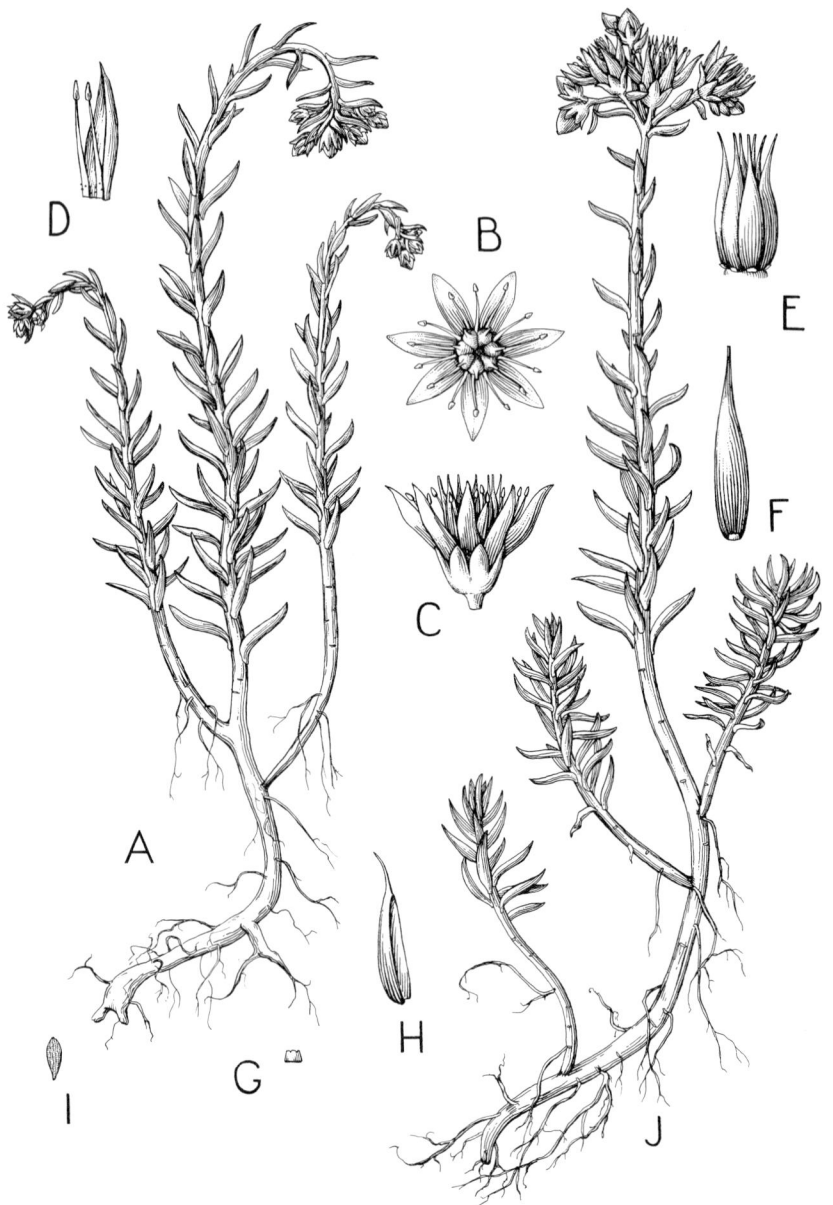

Fig. 171. Plant of *Sedum reflexum* from horticultural source, cultivated in garden, Ithaca, N.Y., July 1942. A. Habit sketch showing floriferous stems with floral buds (x .84). B. Flower from above (x 2.1). C. Flower from side (x 2.1). D. Petal and two stamens (x 2.5). E. Carpels (x 2.5). F. Single carpel and nectary (x 3.4). G. Nectary (x 4.2). H. Follicle (x 3.4). I. Seed (x 4.2). J. Habit sketch showing floriferous stem at stage when flowers are expanding (x .84).

7. *Sedum allantoides* Rose. N.A.—Sierra Madre del Sur. See Praeger (1921), p. 153.

8. *Sedum platyphyllum* E. J. Alexander. N.A.—Sierra Madre del Sur. See Cact. and Succ. Jour. 14:76–78 (1942).

9. *Sedum craigii* Clausen. N.A.—Sierra Madre Occidental. See Cact. and Succ. Jour. 15:167–169 (1943).

10. *Sedum oxypetalum* H. B. K. N.A.—Trans-Mexican Volcanic Belt. See Clausen (1959), pp. 113–126.

11. *Sedum frutescens* Rose. N.A.—Trans-Mexican Volcanic Belt. See Clausen (1959), pp. 126–134.

12. *Sedum torulosum* Clausen. N.A.—Sierra Madre del Sur. See Cact. Succ. Jour. 18:151–152 (1946).

13. *Sedum retusum* Hemsley. N.A.—Sierra Madre Oriental and eastern Mexican Plateau. See Praeger (1921), pp. 147–149.

14. *Sedum griseum* Praeger. N.A.—Trans-Mexican Volcanic Belt and central Mexican Plateau. See Clausen (1959), pp. 159–170.

15. *Sedum pulvinatum* Clausen. N.A.—Sierra Madre del Sur. See Cact. and Succ. Jour. 20:36–37 (1948).

16. *Sedum wrightii* A. Gray. N.A.—Sierra Madre Oriental, eastern Mexican Plateau, and southern Edwards Plateau. See pp. 202–211.

17. *Sedum hemsleyanum* Rose. N.A.—Sierra Madre del Sur and Trans-Mexican Volcanic Belt. See Clausen (1959), pp. 227–236.

18. *Sedum humifusum* Rose. N.A.—Mexican Plateau. See Praeger (1921), p. 244.

19. *Sedum furfuraceum* Reid Moran. N.A.—Mexican Plateau. See Cact. and Succ. Jour. 33:103–105 (1961).

20. *Sedum cupressoides* Hemsley. N.A.—Sierra Madre del Sur. See Praeger (1921), pp. 244–245.

21. *Sedum ternatum* Michx. N.A.—Appalachian Highlands, Interior Low Plateaus, Central Lowland, inner Coastal Plain, Ozark Plateau, and Ouachita Mountains. See pp. 92–108.

22. *Sedum nevii* A. Gray. N.A.—southern Appalachian Highlands. See pp. 109–121.

23. *Sedum glaucophyllum* Clausen. N.A.—Appalachian Highlands. See pp. 122–142.

24. *Sedum pulchellum* Michx. N.A.—Appalachian Ridge and Valley Province, Cumberland Plateau, Interior Low Plateaus, Till Plains, Ozark Plateaus, Ouachita Mountains, and Osage Plains. See pp. 142–161.

25. *Sedum niveum* Davidson. N.A.—San Bernardino and Santa Rosa Mountains in California, and Baja California. See pp. 178–189.

26. *Sedum lanceolatum* Torrey. N.A.—western Great Plains to Pacific Mountain System, and from 35°–61° N. See pp. 211–236.

27. *Sedum stenopetalum* Pursh. N.A.—Central and Northern Rocky Mountains westward through the interior plateaus and ranges to the Cascade Mountains, and southward to the Modoc Plateau in California. See pp. 264–281.

28. *Sedum leibergii* Britton. N.A.—Columbia Plateau and Idaho Batholith. See pp. 281–290.

29. *Sedum divergens* Watson. N.A.—middle and northern Cascade Mountains, Olympic Mountains, and Coast Mountains of British Columbia north to the Queen Charlotte Islands. See pp. 303–312.

30. *Sedum parvum* Hemsley (listed as *S. nanifolium*). N.A.—Northern Mexican Plateau and Sierra Madre Oriental. See Fröderström (1930–1935), pt. 4, pp. 65 and 96.

31. *Sedum nuttallianum* Rafinesque. N.A.—western Interior Highlands, southern Osage Plains, inner West Gulf Coastal Plain, and eastern edge of Great Plains in Texas. See pp. 324–334.

32. *Sedum makinoi* Maxim. Asia—Japanese Volcanic Chain. See Fröderström (1930–1935), pt. 2, pp. 95–96.

33. *Sedum japonicum* Siebold. Asia—Eastern Highlands and Japanese Volcanic Chain. See Praeger (1921), pp. 254–256.

34. *Sedum involucratum* M. Bieb. Asia—Caucasus. See Fröderström (1930–1935), pt. 3, p. 11.

35. *Sedum villosum* L. Europe and northern Africa, and also in northeastern N.A. See pp. 334–344.

36. *Sedum stribrnyi* Velenovsky. Europe—Balkan Range and Rhodope Massive. See Praeger (1921), pp. 248–251.

37. *Sedum alpestre* Villars. Europe and Asia Minor. See Praeger (1921), pp. 256–257.

38. *Sedum stellatum* L. Southern Europe and northern Africa. See Praeger (1921), pp. 293–295.

39. *Sedum nevadense* Coss. Southwestern Europe and northern Africa. See Fröderström (1930–1935), pt. 3, pp. 91–93.

40. *Sedum bithynicum* Boissier. Southwestern Europe and Asia Minor. See Fröderström (1930–1935), pt. 3, pp. 82–84.

41. *Sedum oreganum* Nuttall ssp. *oreganum*. N.A.—Coast Ranges of Oregon and Washington, Puget Trough, and Coast Batholith of British Columbia and southern Alaska. See pp. 345–358.

42. *Sedum obtusatum* A. Gray ssp. *obtusatum*. N.A.—Sierra Nevada. See pp. 358–379.

43. *Sedum laxum* (Britton) Berger. N.A.—Klamath Mts. and northern California Coast Ranges. See pp. 379–403. All subspecies are in cultivation.

44. *Sedum oregonense* (S. Watson) M. E. Peck. N.A.—Cascade and Klamath Mts. See pp. 410–424.

45. *Sedum moranii* Clausen. N.A.—Klamath Mountains, in canyon of Rogue River. See pp. 433–439.

46. *Sedum spathulifolium* Hooker. N.A.—Pacific Mountain System, from Transverse Ranges of California northward to Vancouver Island. See pp. 439–473. All subspecies are in cultivation.

47. *Sedum sempervivoides* Fischer ex M. Bieb. Asia—Asia Minor and Caucasus. See Praeger (1921), pp. 281–283.

48. *Sedum tenuifolium* (Sibth. and Smith) Strobl. Southern Europe, northern Africa, and Asia Minor. See Praeger (1921), p. 279.

49. *Sedum pruinatum* Brotero. Europe—western Spanish Meseta, in the Sierra de Gerez of Portugal. See Praeger (1921), pp. 277–278.

50. *Sedum dumulosum* Franchet. Asia—Eastern Highlands to northeastern Tibetan Highlands. See Praeger (1921), pp. 61–63.

51. *Sedum stephanii* Chamisso. Asia—northeastern and south-central parts. See Praeger (1921), pp. 58–61.

52. *Sedum himalense* Don. Alpine regions of south-central Asia. See Fröderström (1930–1935), pt. 1, p. 31.

53. *Sedum rhodanthum* A. Gray. N.A.—Southern and Central Rocky Mountains and also higher parts of Colorado Plateau. See pp. 475–487.

54. *Sedum integrifolium* Raf. N.A.—western mountains from 33° N. northward to the Brooks Range in Alaska, and also eastward at disjunct localities. See pp. 487–517. Subspecies *leedyi* and *procerum* are most likely in cultivation.

55. *Sedum rosea* (L.) Scop. N.A.—Appalachian Highlands, mostly northward, and also northeastern Canadian Shield. Elsewhere in Greenland, Iceland, Europe, and Asia. See pp. 517–536.

Summary

About 108 species of *Sedum* are in cultivation outdoors in North America north of the Mexican Plateau. None of these species hybridizes with native species. Thirteen of them have become naturalized. Sources of the cultivated species by continents are as follows:

North America	51
South America	0
Asia	26
Europe	26
Africa	1
Unknown	4
Total	108

CHAPTER VIII

Species of Genera of *Crassulaceae*, Related to *Sedum*, Native in North America North of the Mexican Plateau

Genera of North American *Crassulaceae*, related to *Sedum*, are *Lenophyllum*, *Parvisedum*, and *Diamorpha*.

Lenophyllum comprises a group of six perennial species with distinctive aspect, characterized by having the flowers in racemose or spicate cymes and at least some of the leaves opposite or whorled. Petals tend to be recurved and leaves usually are concave ventrally. In addition, the placentas are weakly or moderately lobed. The species of *Lenophyllum* occur in the Sierra Madre Oriental, on the adjacent eastern Mexican Plateau, and on the Mexican and East Gulf Coastal Plains.

Parvisedum is a genus of four annual species, native on the eastern slopes of the Sierra Nevada, in the Great Valley of California, and in the eastern part of the Coast Ranges. These species have in common small yellow flowers with the petals minutely connate basally and only one ovule per carpel. Their distinction from species of *Sedum* which occur in California is well marked.

Diamorpha differs from *Sedum* in important respects, as has been indicated by Sherwin and Wilbur (1971). The fruits have a circumscissile type of dehiscence which involves the breaking away of a dorsal valve. The mature carpels lack ventral suture lines and the dorsal bundle is much reduced. Further, the ventral bundles are homocarpous and the sepals receive only a single vascular bundle. In addition, the petals are hooded and the carpels are connate basally for about a quarter of their length. The one species, *D. cymosa*, occurs in depressions on flat rocks in the southern Appalachian Highlands.

No crucial test is available to confirm the validity of the separation of any of these three genera from *Sedum*. Instead, personal judgment, based on experience, must be the guide. Two kinds of considerations are important: first, the value of the diagnostic generic characteristics, and second, the presence or absence of intermediate species. Considering both criteria, *Lenophyllum*, *Parvisedum*, and *Diamorpha* seem reasonable as genera. Intermediate species are lacking, and the diagnostic features are substantial.

My experience in the field has been with three species of *Lenophyllum*, including *L. texanum*, all species of *Parvisedum*, and the one species of *Diamorpha* at eight localities.

Other genera of *Crassulaceae* native in North America north of the Mexican Plateau are *Dudleya* with axillary floriferous stems; *Hasseanthus*, with corms and also with axillary floriferous stems; and *Crassula*, with opposite leaves and flowers with a single whorl of stamens. In addition, the following genera are naturalized: *Sempervivum* with flowers with 6 to 16 petals and with prominent rosettes which give rise to offsets; and *Kalanchoe*, with 4-merous flowers with connate sepals and petals.

Descriptive data are available here for each of the species of the three genera related to *Sedum*, together with brief consideration of their evolutionary relationships. Then follows a list of species in the other genera, with pertinent notes when new data are available. Citation of references provides a guide to sources of fuller information.

1. *Lenophyllum**** (figs. 172–173)

The elongate cymes of *Lenophyllum*, which appear racemose or spicate, suggest a relationship with *Villadia* or *Echeveria*. Several species of these two genera have elongate inflorescences, but the condition is rare in *Sedum* and does not occur in any species which might be related. Another peculiarity of *Lenophyllum* is the lobed placenta. Mauritzon (1933) noted this condition in *L. weinbergii*. It is known elsewhere in the *Crassulaceae* only in other species of *Lenophyllum*, *Echeveria*, and *Pachyphytum*. On the basis of floral anatomy, Quimby (1971) put *Lenophyllum* in his group 3 of the *Crassulaceae*, to which belong *Echeveria*, *Pachyphytum*, and a majority of the species of *Sedum* which he studied. The plant of *Lenophyllum* examined by Quimby, C67 of horticultural origin and labeled *L. texanum*, probably is *L. acutifolium*. Quimby's group 3 of the *Crassulaceae* exhibits considerable fusion of vascular bundles and surely is phyletically advanced. This evidence is in agreement with the concept that *Lenophyllum* is an advanced genus and perhaps derived from the same phyletic line which gave rise to *Villadia*, *Echeveria*, and *Pachyphytum*. Evidence about chromosomes, resulting from studies by Uhl and suggesting a base number 11 for *Lenophyllum*, does not preclude such a relationship.

Lenophyllum texanum (fig. 172) is the only species of the genus which occurs in the part of North America covered by this book. Its range is on the East Gulf Coastal Plain in the lower valley of the Rio Grande and eastward near the Gulf of Mexico as far as Corpus Christi Bay. Relationships are with *L. acutifolium*, which it resembles. Besides being smaller and slenderer than *L. acutifolium*, *L. texanum* differs in having the leaves and sepals simply acute, not spinescent, and the nectaries oblong rather than reniform.

Description. Sample: 4 plants, a judgment sample—1 from Loma Alta, alt. 6 m., 11 km. northeast of Brownsville, Tex., and 3 from bluffs, alt. 10 m., overlooking bay east of Corpus Christi, Tex. All 4 plants were cultivated in a greenhouse at Ithaca, N.Y., and flowered there. Quantitative data in the following tables apply to the 3 plants from Corpus Christi, abbreviated C. C.

Plants of *Lenophyllum texanum* are cespitose, perennial herbs with copious, pale brown, capillary roots and white rootstocks. Stems are tufted, either erect or somewhat reclining, and pink and glabrous. Secondary shoots arise in axils of lowest

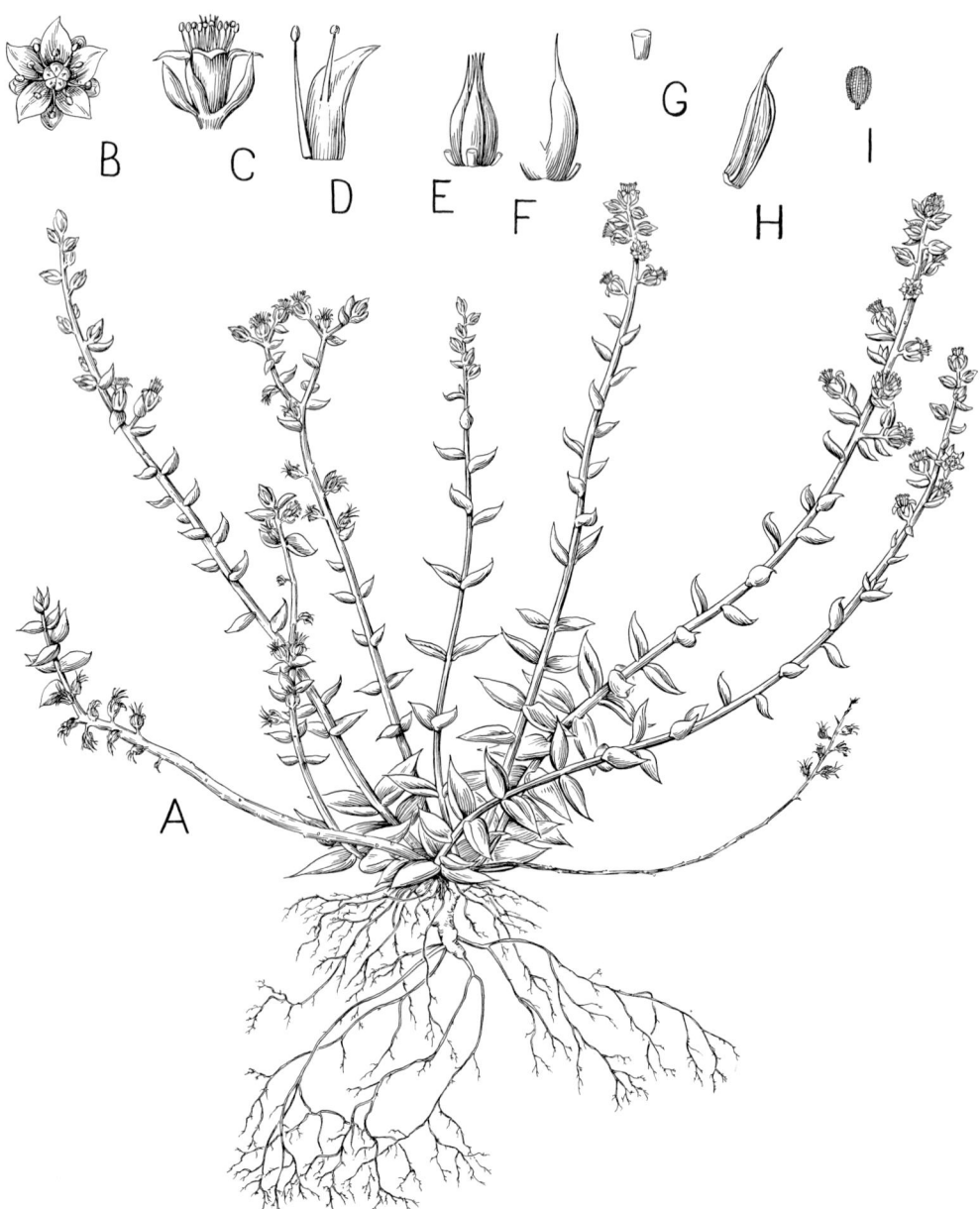

Fig. 172. Plant of *Lenophyllum texanum* from Corpus Christi, Tex., cultivated in greenhouse, Ithaca, N.Y., Oct. 1966. A. Habit sketch (x .46). B. Flower from above (x 1.8). C. Flower from side (x 1.8). D. Petal and two stamens (x 2.8). E. Carpels (x 2.8). F. Single carpel and base of another (x 3.7). G. Nectary (x 4.6). H. Follicle (x 3.7). I. Seed (x 4.6).

leaves. Stems take a year to develop and then lengthen to produce elongate, terminal inflorescences. After bearing fruits, the old stems die.

	n-pl.	n-obs.	$\bar{x}$	s
Diameter (cm.)				
C. C.	3	3	8	1.8
C. C.—gh.	1	1	51	—
Height—veg. stems (cm.)				
C. C.	3	3	11	3.5
Floriferous stems				
Length (cm.)—C. C.—gh.	2	10	35	18
Number—C. C.—gh.	2	2	15	4

Leaves are opposite, at least below, but may be alternate upward. They are plano-convex, concave ventrally, ovate or elliptic-lanceolate, abruptly acute or cuspidate, and green when young, but lavender-green or reddish with age. They are caducous and readily detach from the stems on the slightest pressure.

	n-pl.	n-obs.	$\bar{x}$ (mm.)	s (mm.)
Length				
C. C.	3	15	21	4
C. C.—gh.	2	10	21	19
Width				
C. C.	3	15	10	1
C. C.—gh.	2	10	10	3
Thickness				
C. C.	3	15	3	1
C. C.—gh.	2	10	3	1

Inflorescences are terminal, elongate, racemose cymes. Occasionally, the cymes are branched. Floral bracts are similar to other leaves, but smaller.

	n-pl.	n-obs.	$\bar{x}$ (n)	s (n)
Branches per cyme				
C. C.—gh.	2	11	2.4	4.3
Flowers per cyme				
C. C.—gh.	2	11	20	22

Flowers are on short pedicels, but appear almost sessile.

	n-pl.	n-obs.	$\bar{x}$ (mm.)	s (mm.)
Pedicels—length				
C. C.—gh.	2	8	1	2
Flowers—diameter				
C. C.—gh.	1	3	7	—

Sepals are divergent, connate basally for .1–.5 mm., ovate-oblong, acute, and green.

588 SEDUM OF NORTH AMERICA

	n-pl.	n-obs.	x̄ (mm.)	s (mm.)
Length				
C. C.—gh.	2	8	3.2	1.2
Width				
C. C.—gh.	2	8	1.5	.02

Petals are erect below, then divergent or recurved above middle, slightly connate basally, oblanceolate-oblong, acute and apiculate, and pale primrose yellow with dark red blotch at apex and on mucro.

	n-pl.	n-obs.	x̄	s
Number				
C. C.—gh.	2	8	5	.2
Length (mm.)				
C. C.—gh.	2	8	5.5	1.8
Length of cohesion (mm.)				
C. C.—gh.	2	8	.3	0
Width (mm.)				
C. C.—gh.	2	8	1.7	.04

Stamens have deep yellow filaments and yellow anthers.

	n-pl.	n-obs.	x̄ (mm.)	s (mm.)
Filaments—length				
C. C.—gh.	1	5	3.6	—
Epipetalous filaments, adnation—length				
C. C.—gh.	2	8	2.3	.8

Nectaries are oblong or obovately oblong, truncate and erose, and either orange or deep yellow.

	n-pl.	n-obs.	x̄ (mm.)	s (mm.)
Length				
C. C.—gh.	2	8	.8	.2
Width (mm.)				
C. C.—gh.	2	8	.5	.03

Carpels are erect, connate basally, and green except for white stigmas.

	n-pl.	n-obs.	x̄	s
Length (mm.)				
C. C.—gh.	2	8	5	.6
Length of cohesion (mm.)				
C. C.—gh.	2	8	1.2	.3
Ovules per ovary (n)				
C. C.—gh.	2	8	29	17

Follicles are divergent, with subdivergent beaks, and brown. They are connate basally for about a millimeter. Sepals, petals, and stamens persist in withered condition around follicles.

	n-pl.	n-obs.	x̄ (mm.)	s (mm.)
Length				
C. C.	2	11	4.4	.2
Length of beaks				
C. C.	2	11	1.7	.9

Seeds are oblong-elliptical, verrucose, and brown. They have very short tails .1 mm. long.

	n-pl.	n-obs.	x̄ (mm.)	s (mm.)
Length				
C. C.	2	11	.6	.2
Diameter				
C. C.	2	11	.3	.03

Chromosomes are tiny. C. H. Uhl has reported a gametophytic number of 44 for my collection from near Brownsville, Tex.

Variation. No important differences between plants from Corpus Christi and Brownsville have come to attention. Whether or not different levels of polyploidy occur in populations identified as *Lenophyllum texanum* awaits further study.

Nomenclature. *Lenophyllum texanum* (J. G. Smith) Rose, Smithsonian Misc. Coll. 47:162 (1904). Based on *Sedum texanum* J. G. Smith.

Synonyms are:

1895. *Sedum Texanum* J. G. Smith, Rep. Mo. Bot. Gard. 6:114, pl. 50. Type locality: chapparal near Corpus Christi, Tex. Type: G. C. Nealley (isotype at US), Oct. 1894. Type is at the Missouri Botanical Garden. Drawing of flower in plate 50 does not properly show the orientation of the petals.

1903. *Villadia Texana* (J. G. Smith) Rose, Bull. N.Y. Bot. Gard. 3:3. Based on *Sedum texanum* J. G. Smith.

Distribution. The known distribution of *L. texanum* (fig. 173) is on the East Gulf Coastal Plain from Corpus Christi to the Rio Grande and inland as far as Laredo. At least nine populations are known. The plants grow in sandy soil in desert shrub. At Corpus Christi, on Feb. 12, 1963, plants were growing in the shade of shrubs, back about 5 meters from the edge of bluffs overlooking the bay. Exposure was to the north-northeast. The pH of the soil about the roots ranged from 6.6–7.4. Two problems affect the survival of the population by Corpus Christi Bay. The bluffs are caving in and falling down onto the shore. Despite this situation, the land has mostly been cleared and developed as a residential area. The source of my sample was a vacant lot which by now may have been cleared.

Reproduction. The ease with which leaves detach from the stems and sprout roots is a factor of importance in the perpetuation of the species. In the greenhouse, this

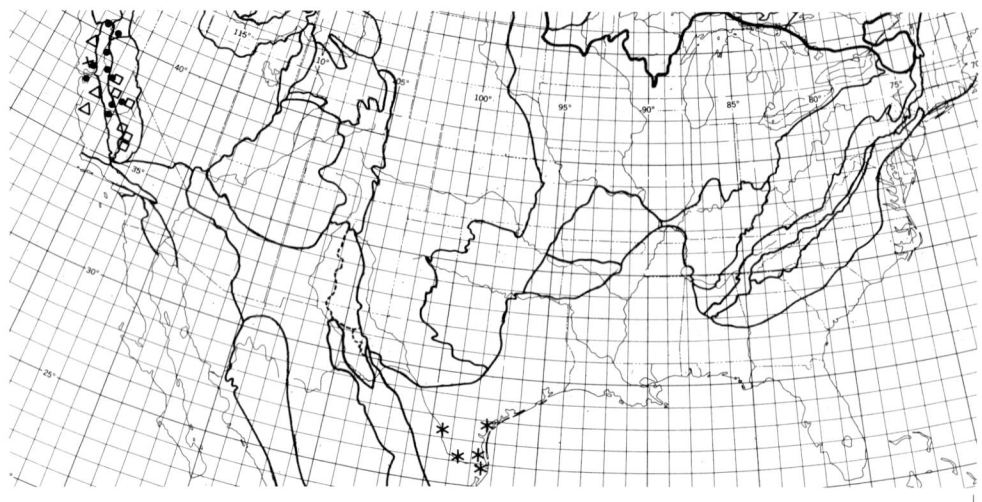

Fig. 173. Known distribution of *Lenophyllum texanum* (∗), *Parvisedum pumilum* (●), *P. congdonii* (□), *P. leiocarpum* (x), and *P. pentandrum* (△).

propensity makes *Lenophyllum texanum* a nuisance. In nature, sprouting of detached leaves may be more important than germination of seeds in determining local distribution. For more distant dispersal, the tiny seeds are well adapted for dissemination by the wind.

Flowering time in nature is from October to March, and in the greenhouse at Ithaca, N.Y., from October to December.

Relationships. The species most closely related to *Lenophyllum texanum* is *L. acutifolium*. That has reniform nectaries, broader than long, and is more robust, with larger leaves and flowers. Sometimes the two species are confused.

Sedum wrightii resembles *Lenophyllum texanum* in the easily detachable leaves, but differences are many. The flowers of the *Sedum* are in cymes of one or two cincinni, the leaves are smaller and yellow-green, the petals are white and separate basally, the nectaries are stipitate-reniform, and the flowers are musky-scented. The ranges of the two species do not overlap. A close relationship is unlikely. The caducous condition of the leaves probably is an adaptive response by two unrelated species to similar environmental stress. In a period of drought, a sprouted leaf would have a larger reservoir of water than a tiny seedling.

2–5. *Parvisedum**** (figs. 173 and 174–177)

Plants of *Parvisedum* are small, annual herbs which bloom in the spring. The flowers are yellow, with carpels each bearing a single ovule. The solitary ovule is diagnostic and separates *Parvisedum* from all other *Crassulaceae*. Because only four species are known and these are similar, data in this account are organized as a table. This arrangement makes comparisons easy.

Description. Sample: 77 plants. In the following table, names of species are abbreviated as follows: *P. pum.* = *P. pumilum* (fig. 174), *P. c.* = *Parvisedum congdonii* (fig. 175), *P. l.* = *P. leiocarpum* (fig. 176), and *P. pent.* = *P. pentandrum* (fig. 177). Data from plants in the greenhouse at Ithaca, N.Y., are prefixed by gh. Under *P. congdonii*, data for a plant from Woods Creek, cultivated in the greenhouse in 1965, are kept separate and designated WC. Arithmetical means, plus or minus the standard deviation, are cited for all quantitative features. Differences among populations and

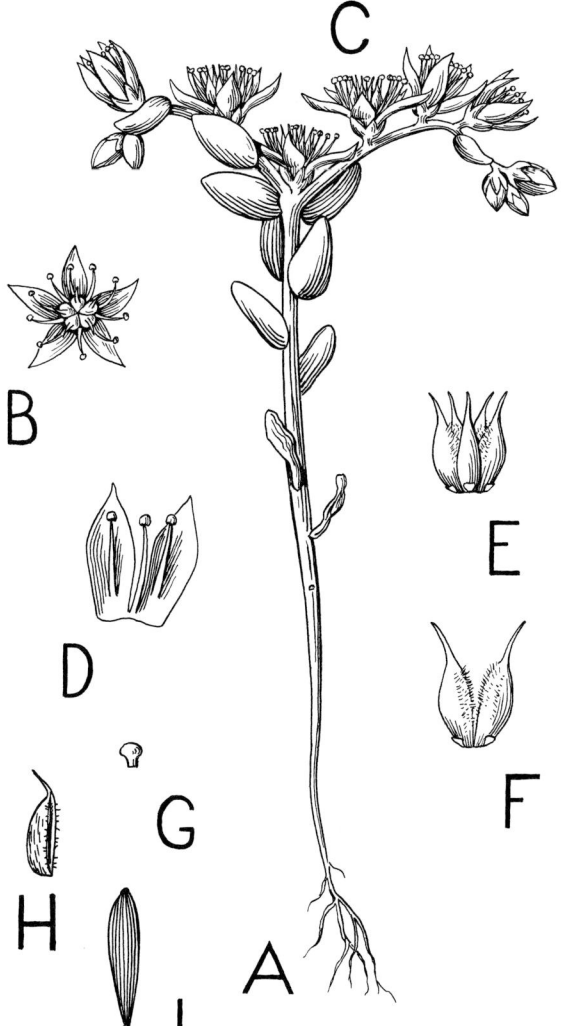

Fig. 174. Plant of *Parvisedum pumilum* from east of Mutual Ditch southeast of Chico, Butte Co., Calif. A. Habit sketch (x 2.8). B. Flower from above (x 2.8). C. Flower from side (x 2.8). D. Two petals and stamens (x 5.6). E. Carpels (x 5.6). F. Two carpels and nectaries (x 7). G. Nectary (x 14). H. Follicle (x 5.6). I. Seed (x 14).

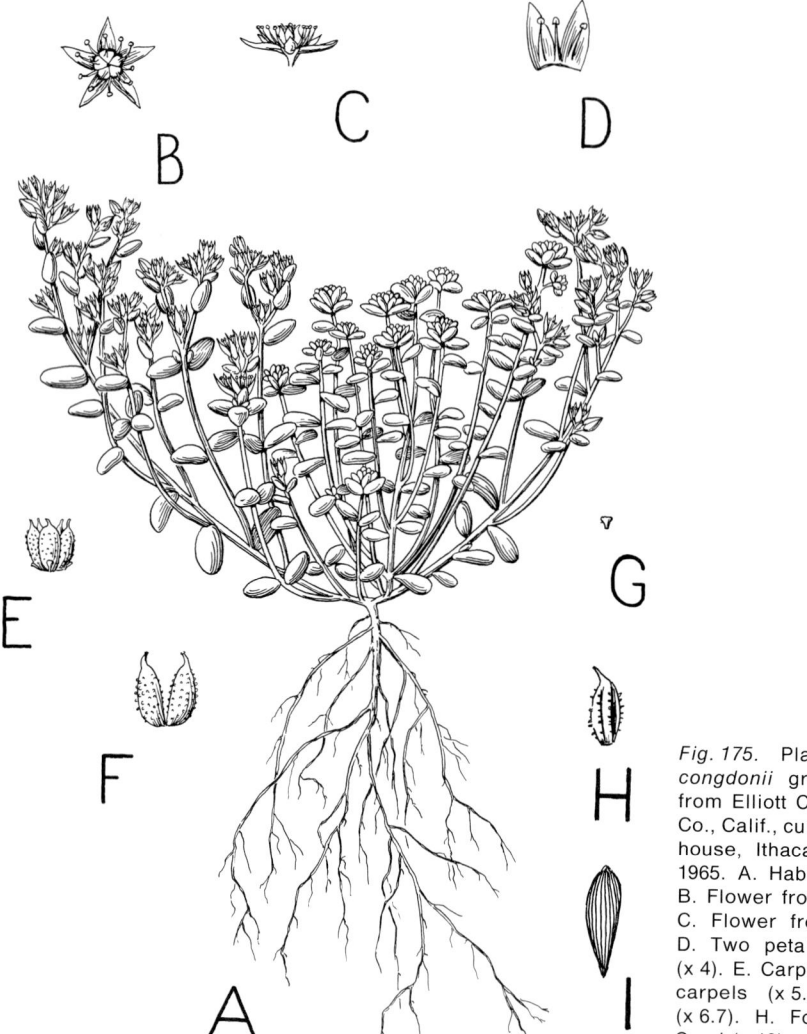

Fig. 175. Plant of *Parvisedum congdonii* grown from seed from Elliott Corner, Mariposa Co., Calif., cultivated in greenhouse, Ithaca, N.Y., June 1, 1965. A. Habit sketch (x 1.3). B. Flower from above (x 2.7). C. Flower from side (x 2.7). D. Two petals and stamens (x 4). E. Carpels (x 4). F. Two carpels (x 5.4). G. Nectary (x 6.7). H. Follicle (x 5.4). I. Seed (x 13).

among species were tested by analysis of variance for five features, indicated by underlining the name of the feature. Asterisks after the names indicate significant differences among species and asterisks after means indicate significant differences among populations within species. Underlining beneath means indicates lack of significant differences. Detailed information about locations of populations and ecology is available in the section on distribution. Discovery of the populations at Chico, Woods Creek, and both Lower and Upper Forbes Creek was by means of sampling surveys with unrestricted randomization. Selection of other populations was on the basis of

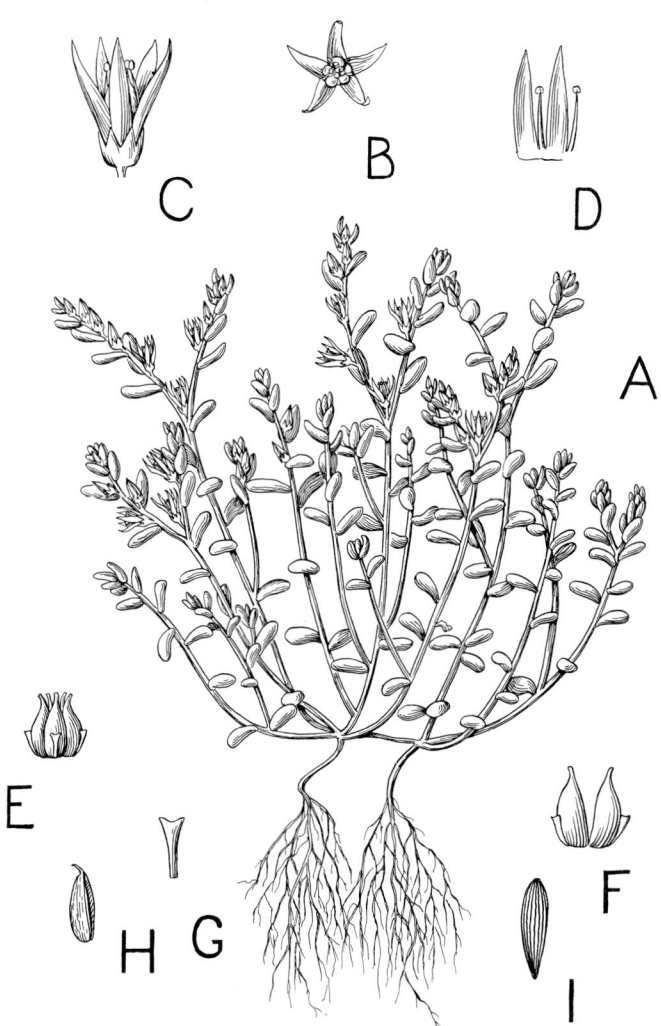

Fig. 176. Plant of *Parvisedum leiocarpum* grown from seed from Manning Flat, Lake Co., Calif., cultivated in greenhouse, Ithaca, N.Y., Apr. 21, 1964. A. Habit sketch (x 1.1). B. Flower from above (x 4.5). C. Flower from side (x 4.5). D. Two petals and stamens (x 4.5). E. Carpels (x 4.5). F. Two carpels and nectaries (x 5.6). G. Nectary (x 11). H. Follicle (x 5.6). I. Seed (x 11).

personal judgment. Sampling within populations was random from randomly selected clusters.

	P. pum.	*P. c.*	*P. l.*	*P. pent.*
Sample—n-pops.	3	3	1	3
Pops. and n-pl.	Chico—5	Ione—3	Manning Flat—27	Lower Forbes Cr.—12
	fl.—1	Elliott Corner—5	fl.—20	fl.—10
	fr.—2	fl.—gh.—2	fr.—6	fr.—2
	gh.—2	fr.—3	gh.—1	Upper Forbes Cr.—12
	Rescue—2	Woods Creek—5		fl.—1
	Pentz—2	fl.—gh.—1		fr.—11
		fr.—4		Pinnacles—4
				fl.—2
				fr.—2

594   SEDUM OF NORTH AMERICA

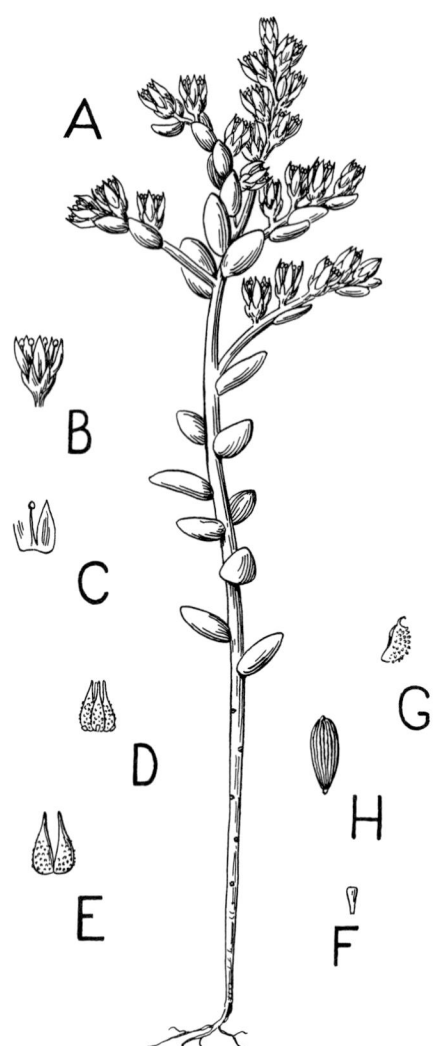

Fig. 177. Plant of *Parvisedum pentandrum* from below Bear Gulch Reservoir, Pinnacles National Monument, San Benito Co., Calif. A. Habit sketch (x 2.3). B. Flower from side (x 3.5). C. Two petals and stamen (x 3.5). D. Carpels (x 4.6). E. Two carpels (x 4.6). F. Nectary (x 5.8). G. Follicle (x 5.8). H. Seed (x 12).

	P. pum.	*P. c.*	*P. l.*	*P. pent.*
Height (cm.)	4 ± .002 gh. 5 ± 0	4 ± .6 gh. 5 ± .7 WC—gh. 3	2.9 ± .7 gh. 7.5	4.2 ± 1.3 —
Branches (*n*)	1 ± .9 gh. 11 ± 2.7	4 ± 2.4 gh. 12 ± 5.3 WC—gh. 10	1 ± 0 gh. 5	3.8 ± 1 —
Median leaves Length (mm.)	4.1 ± 1.2 gh. 3.4	4.3 gh. 4.4 ± 2.1 WC—gh. 4.3	3.3 ± .6 gh. 3.8	4.6 ± 1.2 —

	P. pum.	P. c.	P. l.	P. pent.
Width (mm.)	1.8 ± .04 gh. 1.1	2.1 gh. 2.6 ± 1.6 WC—gh. 2.2	1.8 ± .4 gh. 2	3 ± .8 —
Thickness (mm.)	1.3 ± .1 gh. .7	1.5 gh. 1.8 ± .5 WC—gh. 1.7	.9 ± .2 gh. 1.5	1.4 ± .5 —
Color	yellow-green speckled with red	glaucous	green streaked with red	pale green streaked with red
Inflorescences				
Cincinni (n)	3 ± .9 gh. 3.4 ± .6	2.4 ± .4 gh. 13 WC—gh. 1.5	1 ± .2 gh. 1.8	9 ± 1.7 —
Flowers (n)	10 ± 19 gh. 12 ± 6	7 ± 2 gh. 5 ± 1.3 WC—gh. 9.7	4.3 ± 2.5 gh. 4.9	7 ± 1.9 —
Flowers				
Diameter (mm.)*	4.9 ± 1.3 gh. 6.3 ± .4	5.4 ± .7 gh. 4.1 ± .3 WC—gh. 7.8	2.1 ± .6 gh. 2.8	2.4 ± .5 —
Sepals				
Length (mm.)	1.1 ± .3 gh. .9 ± .1	.9 ± .04 gh. .6 ± .1 WC—gh. .6	1.1 ± .2 gh. .8	1.8* ± .2 —
Width (mm.)	.7 ± .2 gh. .7 ± .1	.6 ± .04 gh. .7 ± 0 WC—gh. .5	.7 ± .1 gh. .5	.8 ± .1 —
Petals				
Number	5 ± .4 gh. 5 ± 0	5 ± .2 gh. 5 ± .4 WC—gh. 5.2	4.4 ± .5 gh. 4.6	4.7 ± .2 —
Length (mm.)*	2.8 ± .7 gh. 3.3 ± .2	2.2 ± .04 gh. 2.2 ± .3 WC—gh. 4	2.6 ± .3 gh. 2.9	1.9 ± .2 —
L. of coh. (mm.)	.2 gh. .3 ± .02	— gh. .1 ± .1 WC—gh. .4	.3 ± .08 gh. .2	.2 —
Width (mm.)	1.0 ± .1 gh. 1.2 ± .04	.9 ± .04 gh. .9 ± .04 WC—gh. .8	.7 ± .2 gh. .7	.7 ± .2 —
Color	sulphur yellow	sulphur or canary yellow	Naples yellow or chartreuse green, streaked with red dorsally	barium yellow
Orientation				
—In fl.	spreading	spreading	divergent	erect
—In fr.	erect	divergent	erect	erect or subdivergent
Stamens				
Number	10 ± .9 gh. 10 ± 0	10 ± .6 gh. 10 ± .7 WC—gh. 10.2	4 ± .9 gh. 4.4	4.6 ± .2 —
Length (mm.)	— gh. 1.3 ± .4	2.1 ± .2 —	1.3 ± .3 gh. 1.6	1.6 ± .2 —
Anthers—l. (mm.)	.6 ± .1 —	.3 gh. .2 ± .3 WC—gh. .5	.3	.3
—color	yellow	yellow	yellow suffused with red	yellow
Adn. ep. fil. (mm.)	—	— gh. .3 ± .2 WC—gh. .4	—	.2
Nectaries				
Length (mm.)	.4 ± .1 gh. .3 ± .04	.3 ± .04 gh. .3 ± .05 WC—gh. .3	.5 ± .1 gh. .5	.3 ± .02 —

	P. pum.	P. c.	P. l.	P. pent.
Width (mm.)	$.3 \pm .2$	$.3 \pm .05$	$.3 \pm .05$	$.2 \pm .04$
	gh. $.3 \pm .03$	gh. $.2 \pm .1$	gh. .3	
		WC—gh. .3		
Color	yellow	yellow	yellow or pale red	orange or yellow
Carpels				
Number	—	—	$4.2 \pm .9$	—
	gh. $5.1 \pm .2$	gh. $5 \pm .7$	gh. 4.5	—
		WC—gh. 5		
Length (mm.)	1.9 ± 1.1	$1.6 \pm .2$	$1.7 \pm .2$	$1.5 \pm .1$
	gh. $2.3 \pm .1$	gh. $1.7 \pm .1$	gh. 1.4	—
		WC—gh. 2.4		
Styles—l. (mm.)	$.8 \pm .2$	$.8 \pm .2$	$.2 \pm .05$	$.2 \pm .02$
	gh. $.8 \pm .1$	gh. $.3 \pm .1$	gh. .3	
		WC—gh. .7		
Ovules per ovary (n)	1 ± 0	1 ± 0	1 ± 0	1 ± 0
	gh. 1 ± 0	1 ± 0	—	—
Ovaries—vesture	papillose puberulent	puberulent and ciliate on ventral suture and at apex	smooth	strongly hirtellous-papillose
Follicles				
Length (mm.)**	$1.6 \pm .02$	$1.5 \pm .1*$	$1.7 \pm .05$	$1.1 \pm .07$
		WC $1.4 \pm .4$		
Orientation	erect	divergent	erect or subdivergent	erect
Seeds				
Length (mm.)**	$1.4 \pm .04$	$1.1 \pm .1*$	$1.2 \pm .05$	$.9 \pm .04*$
		WC $1.1 \pm .2$		
Diameter (mm.)	$.3 \pm .02$	$.3 \pm .05$	$.3 \pm .01$	$.3 \pm .02$
		WC $.3 \pm .1$		
Chromosomes				
Baldwin (1940)	s = 18	s = 18	—	s = 18

Variation. The four species of *Parvisedum* fall into two groups, one with the flowers large and with 10 stamens, *P. congdonii* and *P. pumilum*, and the other with smaller flowers and only one whorl of stamens, *P. leiocarpum* and *P. pentandrum*. The first two species are less advanced, but which of these is least advanced is difficult to decide. *Parvisedum pumilum* has larger seeds, but occurs in a more rigorous environment. *Parvisedum pentandrum*, with tiny fruits and seeds, short petals, and adaptation to a habitat which is extremely dry and hot in summer, is most advanced of the four species.

The large population of *Parvisedum* at Woods Creek, near Jamestown, Calif., was interesting because the follicles were either divergent or erect. Further, a single plant, which flowered in the greenhouse at Ithaca in 1965, had larger flowers than normal, even surpassing in diameter flowers of *P. pumilum*. This evidence caused doubt about identification. Yet, the dimensions of fruits and seeds agree with expectations for *P. congdonii*. The population at Woods Creek provides interesting data relevant to the problem of relationships of *P. pumilum* and *P. congdonii*.

Parvisedum leiocarpum may be the most variable of the four species. The plants are tiny and reddish, with the appearance of a small annual *Crassula*. The number of petals per flower is more often 4 than 5. Sometimes one carpel is undeveloped. Also, the pedicels are enlarged upward to the calyx. In length of petals and seeds, *P. leiocarpum* resembles *P. pumilum* and *P. congdonii*. Although my data clearly classify the plants of *Parvisedum* into four categories, the proper taxonomic rank for them remains a question. If they are inbred local subspecies which are interfertile, revision of the classification may be desirable.

For practical separation of species, length of seeds is useful because it is conservative within populations and it permits interpopulational distinctions. Small samples are adequate. For example, samples of one seed per plant and one plant per cluster are adequate to detect a difference of .1 mm. in length. In contrast, sepals are so variable that they do not aid in separating species. Yet, significant differences in length of sepals between populations within species suggest the potential value of genetic and physiological studies of this feature.

Branching, a variable feature, may be a function of the age and vigor of the plants. Number of flowers per plant likewise is highly variable and may be similarly explained.

Nomenclature. *Parvisedum* R. T. Clausen, Cact. and Succ. Jour. 18:58 (1946). Based on *Sedella* Britton and Rose.

A synonym is:

1903. *Sedella* Britton and Rose, Bull. N.Y. Bot. Gard. 3:45 (not *Sedella* Fourreau, 1868). Type species: *Sedum pumilum* Benth.

Nomenclatural information for the four known species follows:

1. *Parvisedum pumilum* (Benth.) R. T. Clausen, Cact. and Succ. Jour. 18:58 (1946). Based on *Sedum pumilum* Benth.

Synonyms are:

1849. *Sedum pumilum* Benth., Plantae Hartwegianae, p. 310. Type locality: gravelly soil in the Sacramento Valley, probably near Chico, Butte Co., Calif. Type: R. Hartweg 1735 (BM).

1903. *Sedella pumila* (Benth.) Britton and Rose, Bull. N.Y. Bot. Gard. 3:45. Based on *Sedum pumilum* Benth.

2. *Parvisedum congdonii* (Eastwood) R. T. Clausen, Cact. Succ. Jour. 18:58 (1946). Based on *Sedum Congdonii* Eastwood.

Synonyms are:

1898. *Sedum Congdoni* Eastwood, Proc. Calif. Acad. III. 1:135, pl. 11, fig. 5. Type locality: Grant's Springs, Mariposa Co., Calif. Type: J. W. Congdon (CAS), Apr. 9, 1898.

1903. *Sedella Congdoni* (Eastwood) Britton and Rose, Bull. N.Y. Bot. Gard. 3:45. Based on *Sedum Congdoni* Eastwood.

1925. *Sedella pumilum* var. *congdonii* Jepson, Man. Fl. Pl. Calif., p. 450. Based on *Sedella congdoni* Britton and Rose.

3. *Parvisedum leiocarpum* (H. K. Sharsmith) R. T. Clausen, Cact. Succ. Jour. 18:58 (1946). Based on *Sedella leiocarpa* H. K. Sharsmith.

A synonym is:

1940. *Sedella leiocarpa* H. K. Sharsmith, Madroño 5:192. Type locality: in bed of vernal pool, 6.5 miles north of Lower Lake, on Kelseyville Highway, Lake Co., Calif. Type: Milo S. Baker 8971 (UC), May 3, 1938.

4. *Parvisedum pentandrum* (H. K. Sharsmith) R. T. Clausen, Cact. Succ. Jour. 18:58 (1946). Based on *Sedella pentandra* H. K. Sharsmith.

A synonym is:

1936. *Sedella pentandra* H. K. Sharsmith, Madroño 3:240, pl. 12, figs. 4 and 5. Type locality: in moss on surface of large sandstone boulder near stream bed, alt. 259 m., Arroyo del Puerto, about 10 km. above mouth of canyon, Mt. Hamilton Range, Stanislaus Co., Calif. Type: C. W. and H. K. Sharsmith 1831 (UC), Apr. 21, 1935.

Distribution. The general distribution of *Parvisedum* (fig. 173) is in the Pacific Mountain System of California: the Sierra Nevada, Great Valley, Coast Ranges, and Cascade Mountains. The most widespread species is *P. pumilum*. Also, it is the species with the fewest advanced characteristics. It occurs at altitudes from 30 to 1,220 meters in the lower foothills of the Sierra Nevada, the extreme southern Cascade Mountains, the Great Valley, and a few places in the Coast Ranges on either side of San Francisco Bay. The plants grow in gravel on a variety of rocks—andesite, tuff, basalt, granite, and serpentine, and also sometimes in depressions in the mud of vernal pools. A collection from bluffs of the Crooked River in Crook Co., Ore. (GH), on the Columbia Plateau, is from outside the expected range and needs confirmation. Populations in the area of overlap of range with *P. congdonii* are of special interest in understanding interspecific relationships. So far, I have been able to separate populations to my satisfaction, but many populations remain for further study. As presently understood, *P. congdonii* is endemic in the western foothills of the Sierra Nevada, at altitudes from 45 to 1,525 meters. It occurs in shallow soil or in mats of moss on a variety of rocks: granite, schist, and serpentine. *Parvisedum leiocarpum* is the most localized of the four species. It is known only from the northern California Coast Ranges, in the vicinity of Clear Lake, where it occurs in the beds of vernal pools and in the clay of flats at several localities west and south of the lake, at an altitude of about 570 meters. The most advanced species, *P. pentandrum*, occurs in the inner California Coast Ranges, both south and north of San Francisco Bay. The plants grow on rhyolite, shale, sandstone, and serpentine, at altitudes from 259 to 701 meters.

In the following table, data are organized for several populations of *Parvisedum* which I have studied in some detail.

Species and location	n-plants	Area (m.²)	Alt. range (m.)	pH	Drainage
P. pumilum					
Chico	32,545	~8,640	50– 54	6.2–6.4	poor
Sugarloaf Peak	~3,600,000	~500,000	115–135	5.4	good
N. of Pentz	40	110	375	5.4	excessive
P. congdonii					
Elliott Corner	8,689	59	915	5.2–5.6	good
Woods Creek	12,910	16,540	400–410	6.4–6.6	good-excessive
P. leiocarpum					
Manning Flat	59,360	1,600	570	6.2–6.4	poor
P. pentandrum					
Pinnacles	67,789	33	550	5.6	excessive
Lower Forbes Creek	573	<450,000	470–525	6.2	excessive
Upper Forbes Creek	1,045	326	500–550	5.4–6.4	excessive

The population at Chico is along the old Oroville–Chico Highway east of the Mutual Ditch and north of the Mesa Road. The plants of *Parvisedum pumilum* grow in slight depressions at the western end of a low ridge of andesite and also on an adjacent lower tract with boulders and gravel of andesite. This site, chosen without restriction of randomization, by coincidence is near to the type locality of the species and could even be the place where Hartweg originally collected *P. pumilum*. The population on Sugarloaf Peak is on the western slope below the summit. Selection of this locality was because it is the highest point in the part of the Sacramento Valley which I was surveying. Studies at Chico were in March and June 1963 and on Sugarloaf Peak on Mar. 22, 1963. The population north of Pentz is on the eastern side of the Pentz–Magalia Road 1.1 km. north of the Nelson Bar Road. Plants there grow in thin soil and in a mat of moss over basalt. Selection of this population was deliberate to compare with the population at Chico and also with populations of *P. congdonii* at localities farther south. My visit to Pentz was on August 8, 1964.

The population of *Parvisedum congdonii* at Elliott Corner was the nearest that I could find to Grant's Springs, the type locality. The site is a flat exposure of granite on the eastern side of the road .2 km. south of Elliott Corner and 5.6 km. northwest of the Sulphur Springs which formerly were known as Grant's Springs. This could be the place where J. W. Congdon originally found the species in 1898. If not, it is nearby. My visit there was on July 11, 1964. The site along the eastern side of Woods Creek is a low ridge of serpentine southwest of Quartz Mountain and 4 km. SSW of Jamestown. This was a randomly selected sampling unit in the survey of the Sierra Nevada. I was there on Aug. 12, 1964.

The site for *Parvisedum leiocarpum* at Manning Flat is on the northern side of the Lower Lake Road which goes to Kelseyville. It must be within a few kilometers of the type locality. Manning Flat is about 8 km. WNW of Lower Lake. I was there in May and August 1963.

The population of *Parvisedum pentandrum* studied at the Pinnacles National Monument was below the Bear Gulch Reservoir. Plants there grow in dense patches in shallow soil on rhyolitic tuff. My studies there were on May 5 and 6 and Aug. 7, 1963. The populations at Lower and Upper Forbes Creek are in randomly selected sampling units about 4 km. SW of Lakeport. The plants there occur in open areas, in gravelly and rocky situations which are dry and hot in summer. The rock is rhyolite. Observations were on May 15 and 16 and Aug. 10, 1963.

Other vegetation is sparse at sites where *Parvisedum* occurs. The habitats are rigorous, hot and dry in summer, and with shallow soil. Few vascular plants can endure such conditions. As a result, plants of *Parvisedum* occur in virtually pure stands. Competitors are few and miscellaneous.

Reproduction. All reproduction of *Parvisedum* is sexual and by means of seeds. Flowering time for each of the species is spring. Extreme dates for flowers in anthesis are indicated in the following table:

Species	Dates for flowers
P. pumilum	Mar. 20–May 30
P. congdonii	Mar. 18–May 25
P. leiocarpum	Apr. 15–May 18
P. pentandrum	Mar. 29–May 19

The occurrence of plants of *Parvisedum* in dense patches in nearly pure stands is a good condition for outbreeding of plants, but since the populations are of local distribution and do not overlap, inbreeding of species and populations results.

Relationships. The four species of *Parvisedum* are closely related. The two species with twice as many stamens as petals probably are the most primitive. *P. pentandrum*, with stamens as many as petals, hairy fruits, and adaptation of seeds to high temperatures and prolonged summer drought, appears the most advanced.

The relationships of *Parvisedum* are with *Sedum*. Probably it is derived from the same ancestral stock which gave rise to species such as *S. nuttallianum* and *S. radiatum*. Evolution has involved basal cohesion of petals, reduction of ovules per carpel to one, and loss of the perennial habit.

References. Baldwin (1940) and Sharsmith (1936, 1940).

6. *Diamorpha**** (figs. 178–179)

Lateral dehiscence of the fruits by means of a dorsal valve is the diagnostic feature of *Diamorpha*. This condition, together with hooded petals, 4-merous flowers, basally connate carpels, annual habit, and small red leaves, distinguishes the genus from all other *Crassulaceae*. *Diamorpha cymosa* (fig. 178) is the only species.

Description. Sample: 94 plants—44 from Flat Rock north of Camden, S.C., the type locality; 7 from Kelly Rock, 5 km. northwest of Flat Rock; and 43 from Forty-acre Rock, 3 km. southwest of Taxahaw, Lancaster Co., S.C. Samples of plants were simple random from randomly selected clusters. In estimating means and variances for populations, weighting was according to the sizes of the selected clusters. For data on fruits and seeds, plants were brought to Ithaca to mature. Dehiscence of fruits occurred about six weeks after anthesis.

Plants of *Diamorpha cymosa* are annual herbs with slender, fibrous, white roots and erect, branched stems. The seeds germinate in winter. The main roots may be brownish white and up to .5 mm. in diameter. In shallow soil, the roots develop horizontally. The stems are finely ribbed, appearing striate and red. Hypocotyls are about 6 mm. long. Rarely a rootstock develops from which rootlets arise.

	n-pl.	n-obs.	$\bar{x}$ (cm.)	s (cm.)
Roots—length				
Flat Rock	19	19	1	.8
Kelly Rock	2	2	1	.4
Forty-acre Rock	12	12	1.2	.4
Height				
Flat Rock	28	28	3.2	.3
Kelly Rock	2	2	4.7	1.5
Forty-acre Rock	18	18	1.4	.1

Leaves are divergent or appressed, one per node, short-spurred, oblong-elliptical or elliptical, subpetiolate, rounded, smooth or minutely papillose on margins, and red. The first leaves on the stems appear whorled. They are petiolate and spatulate.

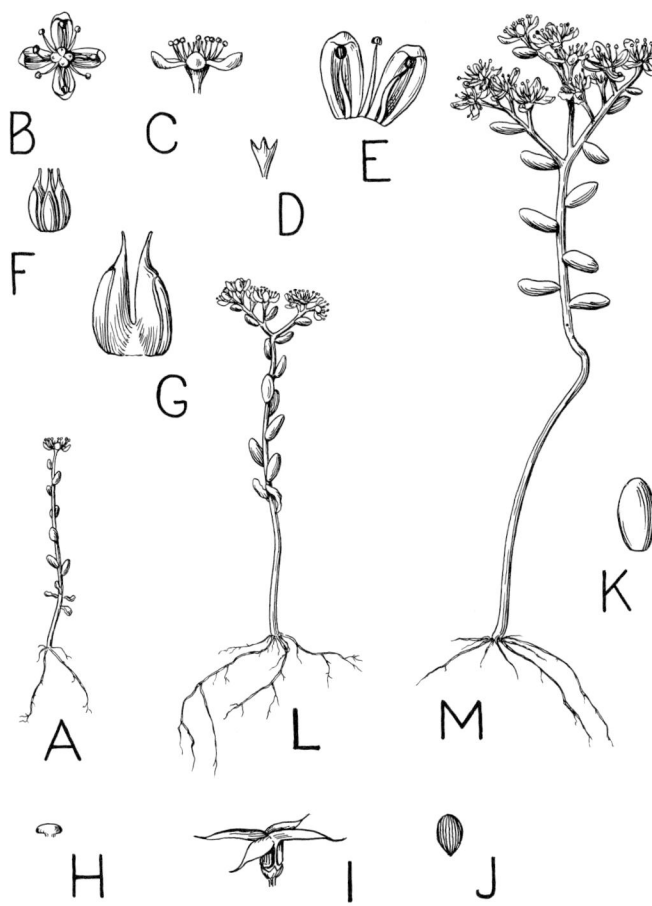

Fig. 178. Plants of *Diamorpha cymosa* from Flat Rock above Camden, Kershaw Co., S.C., the type locality, A–L, and Forty-acre Rock, Lancaster Co., S.C., M. A, L, and M. Habit sketches (x 1.1). B. Flower from above (x 2.8). C. Flower from side (x 2.8). D. Calyx from side (x 3.4). E. Two petals and stamens (x 5.6). F. Carpels (x 5.6). G. Two carpels (x 11). H. Nectary (x 11). I. Follicle (x 5.6). J. Seed (x 11). K. Single leaf (x 2.8).

	n-pl.	n-obs.	$\bar{x}$ (mm.)	s (mm.)
Length				
Flat Rock	28	28	2.8	1
Kelly Rock	2	2	2	1
Forty-acre Rock	18	18	2.8	.6
Width				
Flat Rock	28	28	1.7	.4
Kelly Rock	2	2	1	.5
Forty-acre Rock	18	18	1.8	.1
Thickness				
Flat Rock	28	28	1.2	.2
Kelly Rock	2	2	.7	.4
Forty-acre Rock	18	18	.4	.3

Inflorescences are terminal cymes with floral bracts similar to foliage leaves, but smaller.

	n-pl.	n-obs.	$\bar{x}$	s
Diameter (cm.)				
Flat Rock	28	28	.9	.1
Kelly Rock	2	2	1.3	.3
Forty-acre Rock	18	18	.7	.1
Cincinni—number				
Flat Rock	28	28	1.8	.7
Kelly Rock	2	2	1	1.7
Forty-acre Rock	18	18	1.8	.5
Flowers—number				
Flat Rock	28	28	5	1.1
Kelly Rock	2	2	3	3.5
Forty-acre Rock	18	18	5	.9

Flowers are pedicellate, with pedicels enlarged upward and continuous with tube of calyx.

	n-pl.	n-obs.	$\bar{x}$ (mm.)	s (mm.)
Pedicels—length				
Flat Rock	26	34	1.3	.4
Kelly Rock	2	3	1.9	.1
Forty-acre Rock	6	7	2.1	.6
Diameter				
Flat Rock	28	40	4.7	.4
Kelly Rock	2	3	4.7	.4
Forty-acre Rock	18	23	4.3	.7

Sepals are connate basally for .2 mm., ovate, acute, and green below, but otherwise pink. Viewed from below, they form a perfect square with pink corners and rim and a green circular center.

	n-pl.	n-obs.	$\bar{x}$ (mm.)	s (mm.)
Length				
Flat Rock	28	37	.6	.04
Kelly Rock	2	3	.6	0
Forty-acre Rock	18	23	.5	.07
Width				
Flat Rock	28	37	.7	.1
Kelly Rock	2	3	.7	.1
Forty-acre Rock	18	23	.5	.05

Petals are widely divergent, elliptical, channeled, and hooded. They are white, slightly flushed with pink at the rounded tips. Usually they are 4 per flower, rarely 5 or 3.

	n-pl.	n-obs.	$\bar{x}$	s
Number				
Flat Rock	28	37	4	0
Kelly Rock	2	3	4	0
Forty-acre Rock	18	23	4	.1
Length (mm.)				
Flat Rock	28	37	2.4	.3
Kelly Rock	2	3	2.4	.2
Forty-acre Rock	18	23	2.4	.2

	n-pl.	n-obs.	x̄	s
Width (mm.)				
Flat Rock	28	37	1.4	.2
Kelly Rock	2	3	1.3	.1
Forty-acre Rock	18	23	1.2	.2

Stamens are twice as many as the petals and shorter than them. The filaments are white except deep pink just below the dark red anthers. Half of the stamens are epipetalous. The others are attached to the tube of the calyx. Before anthesis, anthers of the epipetalous stamens fit into the hoods of the petals.

	n-pl.	n-obs.	x̄ (mm.)	s (mm.)
Epipetalous filaments, adnation—length				
Flat Rock	28	37	.1	.04
Kelly Rock	2	3	.2	.1
Forty-acre Rock	18	23	.1	.03
Anthers—length				
Flat Rock	16	17	.6	.1
Kelly Rock	1	1	.5	—
Forty-acre Rock	8	11	.5	.01

Nectaries are short-stipitate, reniform, broadly rounded, and translucent, but appearing white.

	n-pl.	n-obs.	x̄ (mm.)	s (mm.)
Length				
Flat Rock	28	37	.3	.04
Kelly Rock	2	3	.2	.04
Forty-acre Rock	18	23	.2	.004
Width				
Flat Rock	28	37	.3	.04
Kelly Rock	2	3	.3	0
Forty-acre Rock	18	23	.3	.04

Carpels are connate basally for about one-quarter of length, erect, with slender styles, at first white, then pink, and eventually red.

	n-pl.	n-obs.	x̄	s
Length (mm.)				
Flat Rock	28	37	1.4	.2
Kelly Rock	3	5	1.5	.1
Forty-acre Rock	18	23	1.6	.3
Styles—length (mm.)				
Flat Rock	28	37	.4	.1
Kelly Rock	3	5	.5	.1
Forty-acre Rock	18	23	.4	.08
Ovules per ovary (n)				
Flat Rock	28	37	2.3	.8
Kelly Rock	2	3	3.5	.9
Forty-acre Rock	18	23	3.8	1.1

Fruits are widely divergent and brown. They split on the sides in such a way that the dorsal section falls away, exposing the seeds.

	n-pl.	n-obs.	x̄ (mm.)	s (mm.)
Length				
Flat Rock	13	28	1.5	.2
Kelly Rock	5	10	1.6	.1
Forty-acre Rock	8	14	1.2	.2

Seeds are subglobose-ovoid, rounded, finely longitudinally ribbed, and yellow-brown.

	n-pl.	n-obs.	x̄ (mm.)	s (mm.)
Length				
Flat Rock	12	26	.5	.05
Kelly Rock	5	10	.5	0
Forty-acre Rock	8	14	.5	.05
Diameter				
Flat Rock	12	26	.3	.03
Kelly Rock	5	10	.3	0
Forty-acre Rock	8	14	.3	.03

Chromosomes are nearly metacentric. Baldwin (1940) reported the number as $g = 9$ and $sp = 18$. Other studies confirm this result.

Variation. The millions of little plants of *Diamorpha cymosa* which occur on granitic flat rocks of the southern Piedmont Upland are remarkably similar. A notable exception is one cluster of about 13,000 much larger plants in a depression at Forty-acre Rock. Although these plants are taller than those in other clusters, differences among groups within this cluster are highly significant, and the intragroup variation is greater than the variation among clusters. The circumstance may not be of any taxonomic importance.

Number of petals per flower normally is 4. Exceptions are rare. Out of a sample of 67 flowers, only 3 had 5 petals and 1 had 3 petals.

Nomenclature. *Diamorpha cymosa* (Nutt.) Britton ex Small, Fl. Southeastern U.S., ed. 1, p. 498 (1903). Based on *Tillaea? cymosa* Nutt., with *D. pusilla* cited as a synonym. Wilbur (1964) designated *D. cymosa* as illegitimate because, according to him, it was nomenclaturally superfluous when published. Article 63, which Wilbur cited, states that a name is superfluous if, as circumscribed by its author, it included the type of a name or epithet which ought to have been adopted under the rules. *Tillaea? cymosa*, as circumscribed by Nuttall, did not include *Sedum pusillum* Michx., even if Nuttall mistakenly listed it as a synonym. No name was available which ought to have been adopted. A new name was necessary. Nuttall's description of the capsules of *Tillaea? cymosa* as opening externally precludes the possibility of including *Sedum pusillum* Michx. within the circumscription of his species.

Synonyms are:

1818. *Tillaea? cymosa* Nutt., Gen. N. Am. Pl. 1:110. Type locality: Flat Rock above Camden, S.C., listed by Nuttall as N.C. Type: Nuttall (Ph), 1816.

1818. *Diamorpha pusilla* Nutt., Gen. N. Am. Pl. 1:293. Based on *Tillaea? cymosa* Nutt., not on *Sedum pusillum* Michx., which is wrongly listed as a synonym.

1905. *Diamorpha Smallii* Britton, N.A. Flora 22:56. Type locality: falls of the Yadkin River, Stanly Co., N.C. Type: J. K. Small (NY), April 20–24, 1896. Construction of a dam destroyed the type locality. Probably *D. smallii* is no more than a slender, developmental stage of *D. cymosa*.
1936. *Sedum cymosum* (Nutt.) Fröd., Act. Hort. Goth. 10 (App.):137. Based on *Tillaea*? *cymosa* Nutt.
1936. *Sedum cymosum* var. *Smallii* (Britton) Fröd., Act. Hort. Goth. 10 (App.):138. Based on *Diamorpha Smallii* Britton.
1964. *Sedum smallii* (Britton) Ahles, Jour. Elisha Mitchell Sci. Soc. 80:172. Based on *Diamorpha smallii* Britton.

Distribution. *Diamorpha cymosa* (fig. 179) occurs on many granitic flat rocks in the southern Piedmont Upland, from North Carolina to Alabama, and also on flat areas of sandstone in the southern parts of the Ridge and Valley Province and the Cumberland Plateau. The plants often grow in vast numbers in vernal pools in depressions in the rocks. Gravel or sand in such depressions may be very shallow, only 3 mm. in depth, in which case the roots of the *Diamorpha* grow horizontally. Because of the shallowness of the gravel and sand in which plants grow, the roots must endure extreme conditions from freezing in winter to very hot at times in spring. Diurnal fluctuations in temperatures are enormous. The altitudinal range is from 120 to 579 meters.

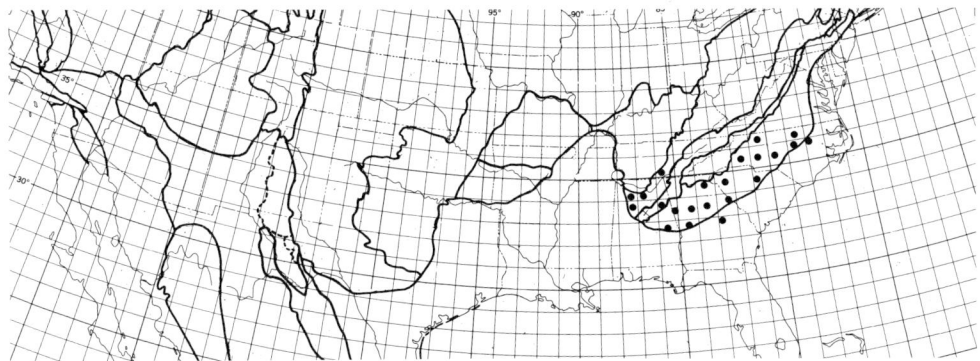

Fig. 179. Known distribution of *Diamorpha cymosa* (●).

Of eight populations visited in the field, data follow for three which received most attention. Further information about these localities is available in Chapter X on geography, in the discussion of the Piedmont Plateau.

Location	n-plants	Area (m.²)	Alt. range (m.)	pH	Drainage
Flat Rock	427,750	1,139	120–160	4.4–5.4	none-excessive
Kelly Rock	755,577	2,043	177	4.4	poor
Forty-acre Rock	1,644,334	1,054	140–150	4.6–5.4	poor

Plants of *Diamorpha* usually grow in pure stands or with rare individuals of other flowering plants. Sometimes they are rooted in living mats of moss and lichens. Mosses include *Grimmia ciliata* and *Polytrichum*. At Flat Rock, the two seed-bearing plants most often associated with *Diamorpha* were a species of *Cyperus* and *Bulbostylis capillaris*. At Forty-acre Rock, besides *Cyperus*, *Arenaria brevifolia* sometimes constituted 10% of the vegetation in a depression with *Diamorpha*.

Kelly Rock sometimes is called Red Rock because of the redness of the leaves of the great masses of plants of *Diamorpha* there.

Baskin and Baskin (1972) have demonstrated that the seeds of *Diamorpha* are almost completely dormant at the time that they are shed and also that maximum germination temperature increases as the seeds age.

Reproduction. Perpetuation of *Diamorpha* from year to year is by means of seeds. Insects, especially small bees and flies, cross-pollinate the flowers. The tiny seeds are ideal for dispersal by the wind. Transport from one flat rock to another probably is by this means. Germination of seeds is in autumn or winter.

Flowering time of *Diamorpha* is April and May, sometimes in March.

Relationships. *Diamorpha* has no close relatives. Although unique in the mode of dehiscence of the fruits, it probably is derived from *Sedum*. Sherwin and Wilbur (1971) provided an excellent discussion of the anatomy of *Diamorpha* and also of its generic relationships. Its distinctive features include sepals, each with one vascular bundle, cucullate petals, carpels fused for about a quarter of their length, homocarpous ventral traces, a lack of ventral suture lines on carpels at maturity, great reduction of the dorsal carpellary bundle, and lateral dehiscence of the fruits. So many important distinctions justify generic status. The work of Sherwin and Wilbur demonstrates the value of careful anatomical studies.

Plants of *Diamorpha* often occur in pure stands. Occasionally, however, as at Forty-acre Rock, plants of *Diamorpha* may grow intermixed with *Sedum pusillum*. When this happens, the *Diamorpha* is far behind the *Sedum*. Its floral buds may be unopened when the *Sedum* has immature fruits.

Hybridization does not occur between *Diamorpha* and *Sedum pusillum*. Sherwin and Wilbur (1971) have discussed this possibility and dismissed it, as do I.

References: Baskin, J. M., and Baskin, C. C. (1972); Lammers, W. T. (1959); McCormick, J. F., and Platt, R. B. (1962 and 1964); Murdy, W. H. (1968); O'Connell, J. E. (1949); Sherwin, P. A., and Wilbur, R. L. (1971); Wiggs, D. N., and Platt, R. B. (1962); and Wilbur, R. L. (1964).

Other Genera and Species

Dudleya***

Thirteen species of *Dudleya* occur in the part of North America covered by this book. These are listed below, with the general areas of distribution. Of the thirteen species, four were found in randomly selected areas in a survey of the Pacific Border region and two others were encountered on trips between units. The remaining seven species are of limited distribution and were not found. For descriptions of all thirteen species and a key for identification, see Munz and Keck (1959) and Munz (1968).

1. *Dudleya cymosa* (Lem.) Britton and Rose. Coast Ranges and Sierra Nevada, below 2,745 m. Ssp. *cymosa* was in 2 out of 21 sampling units, each about a kilometer square, in the Sierra Nevada. Plants occurred from 410 to 1,850 m. Also, I found ssp. *cymosa* in a randomly selected unit at Chalk Point in the Mayacamas Mountains, at 655 m. Ssp. *minor* (Rose) Moran was growing near *Sedum spathulifolium* ssp. *yosemitense*, on quartz monzonite, in a canyon in the San Bernardino Mountains, at an altitude of 1,550 m. Ssp. *setchellii* (Jeps.) Moran was on cliffs at the Pinnacles, San Benito Co., Calif.

2. *Dudleya bettinae* Hoover. Southern Coast Ranges—San Luis Obispo Co., Calif. Not found.

3. *Dudleya parva* Rose and Davidson. Local in Transverse Ranges. Not found.

4. *Dudleya farinosa* (Lindl.) Britton and Rose. Bluffs along the coast from southern Oregon to the region of Los Angeles. The northernmost population which came to attention was on Table Rock at Bandon, Ore., latitude 43° 5′ N. Plants growing in sand on top of bluffs by the ocean between Pacifica and Pescadero Beach, San Mateo Co., Calif., had lustrous green leaves, sepals 2.6 mm. long, and pedicels 2–11 mm. long. These appeared different from typical *D. farinosa* and either constitute a distinctive race or might be the result of hybridization between that and some other species.

5. *Dudleya palmeri* (Wats.) Britton and Rose. Southern Coast Ranges. Not found.

6. *Dudleya caespitosa* (Haw.) Britton and Rose. Plants on a slope east of Avila, San Luis Potosi Co., Calif., R. T. C. 4,730, had lanceolate, glaucous leaves.

7. *Dudleya candelabrum* Rose. Channel Islands. Not visited.

8. *Dudleya greenei* Rose. Channel Islands. Not visited.

9. *Dudleya abramsii* Rose ssp. *murina* (Eastwood) Moran. An interesting result of a survey of the Gabilan Mountains in 1963 was discovery of a population of 336 plants of this subspecies, growing on quartzite on the northwestern slope of the ridge southeast of the Llano Grande Canyon, Monterey Co., Calif. Plants in this population had median cauline leaves 14–21 mm. long, floriferous stems 5–24 cm. tall, and petals connate for 1.4–2.8 mm. They also had short caudices, .8–1.7 cm. in diameter, with 1–5 rosettes. The petals were Naples yellow and the reniform nectaries were orange. Seeds were tiny, elliptical, ribbed, and brown, .4 mm. long and .2 mm. in diameter. Otherwise, ssp. *murina* is known only from serpentine rocks near Chorro Creek, San Luis Obispo Co., Calif.

10. *Dudleya lanceolata* (Nutt.) Britton and Rose. Coast Ranges, Transverse Ranges, and southward. In my survey in 1963, this species occurred in 4 out of 8 randomly selected sampling units, each about a kilometer square, in the Santa Monica Mountains, and in 4 out of 4 similar units in the Perris Upland. A plant from the northern Simi Hills, which flowered in a greenhouse at Ithaca, N.Y., had the floriferous stems 7.7 dm. tall as in *D. lanceolata*, but the petals connate for 2.8 mm., as in *D. saxosa* ssp. *aloides* (Rose) Moran, Madroño 14:108 (1957).

11. *Dudleya saxosa* (Jones) Britton and Rose. Panamint Mountains and Transverse Ranges. Not found.

12. *Dudleya pulverulenta* (Nutt.) Britton and Rose. Coast Ranges from San Luis Obispo Co., Calif., southward. Present in 1 sampling unit out of 8 in the Santa Monica Mountains, namely in the western part of the Twentieth-Century-Fox ranch.

13. *Dudleya densiflora* (Rose) Moran. Southern base of San Gabriel Mountains. Not found.

Hasseanthus***

Three species of *Hasseanthus* are known. A fourth is putative. Two occur in the geographic area covered by this book. Arguments for and against regarding them as constituting a genus are cited in the discussion of *Sedum blochmanae* in the introduction to Chapter V. All species are restricted to the coastal region of California. For descriptions and keys, see Clausen, Moran, and Uhl (1945); and Moran (1951 a and b).

1. *Hasseanthus elongatus* Rose. Transverse Ranges of California, Los Angeles Basin, and coastal part of the Peninsular Ranges. Seen near Pomona.

2. *Hasseanthus blochmaniae* (Eastwood) Rose. Coastal part of the southern California Coast Ranges from San Simeon, north of Morro Bay, to La Jolla. Seen on bluffs between Casmailia Beach and Point Sal, Santa Barbara Co., Calif., and on bluffs east of Estero Bay, San Luis Obispo Co.

Crassula***

Three species of *Crassula* occur in North America north of the Mexican Plateau. These are described and distinguished, as *Tillaea*, by Munz (1959).

1. *Crassula erecta* (H. & A.) Berger. Sierra Nevada, Great Valley of California, and Coast and Peninsular Ranges. Found in 2 out of 4 randomly selected, kilometric sampling units in the Chico region of the Sacramento Valley, 4 out of 7 units in the Santa Monica Mountains, 2 out of 2 units in the Gabilan Range, 4 out of 4 units on the Perris Upland, and 3 out of 4 units in the Mayacamas Mountains. The plants often occur in vast numbers in pure stands in open places. The largest population seen was where a fire had burned the year before, on the northern slope of the ridge southeast of Winchester, Calif. The estimated number of plants there was 92,000. Flowering plants may be only a little over a centimeter in height, with flowers 1 mm. in diameter. Ovules are 1 or 2 per carpel.

2. *Crassula muscosa* (L.) Roth. This European species now is naturalized and common in parts of the Great Valley of California and in the southern part of the northern Coast Ranges. Found in 3 out of 4 randomly selected, kilometric sampling units in the Chico region of the Great Valley and in 2 out of 4 units in the Mayacamas Mountains. The largest population, 10,000,000 plants, is along Moore Creek southwest of Kirkwood. Sepals are 3 per flower and are scarious-apiculate. Seeds are .4 mm. long and .2 mm. in diameter.

3. *Crassula aquatica* (L.) Schönl. Widespread across North America, but found by me only on tidal mud flats along the St. Lawrence River in Quebec and in a vernal

pool in a pasture in the Sacramento Valley, Calif. In the survey of the Pacific Border region, it was in only 1 out of 40 randomly selected, kilometric sampling areas.

Sempervivum**

Hens-and-Chickens are frequent in gardens. Two species are naturalized.

1. *Sempervivum tectorum* L. Native in Europe, naturalized at a few places in New England and also reported as naturalized at one locality in Michigan.
2. *Sempervivum heuffelii* Schott. Native in southeastern Europe. Naturalized in an area ~ 25 m.² on Eagles Peak, S.E. of Mellen, Ashland Co., Wisc. Found by R. R. Kowal and Michael Nee on Aug. 7, 1971, and collected by Nee, no. 4,441 (WIS).

Kalanchoe**

Many species of *Kalanchoe* are in cultivation in North America. Three are naturalized in the southern part of the Florida Peninsula.

1. *Kalanchoe crenata* Haw. Native in tropical Africa.
2. *Kalanchoe pinnata* (Lam.) Pers. Nativity uncertain.
3. *Kalanchoe verticillata* Elliot (*K. tubiflora* Hamet, not *Bryophyllum tubiflorum* Harvey). Native in Madagascar.

Summary

Eight genera of *Crassulaceae*, other than *Sedum*, occur in North America north of the Mexican Plateau. Of these, *Lenophyllum*, *Parvisedum*, and *Diamorpha* are indigenous and related to *Sedum*. Probably each of these is derived from *Sedum*. *Dudleya*, *Hasseanthus*, and two species of *Crassula* likewise are native in North America. Of the remaining two genera, *Sempervivum* is native in Europe and northern Africa, and *Kalanchoe* is native in Africa, Madagascar, and tropical Asia. Species of these two Old World genera have escaped from cultivation in North America and become naturalized.

Numbers of species of the genera of *Crassulaceae*, other than *Sedum*, native in North America north of the Mexican Plateau are:

Lenophyllum	1
Parvisedum	4
Diamorpha	1
Dudleya	11
Hasseanthus	2
Crassula	2
Total	21 species

Lenophyllum, with racemose or spicate cymes and some leaves opposite, appears more advanced than *Sedum* and allied to the phyletic line which gave rise to *Villadia*. *Parvisedum* likewise is more advanced than *Sedum* and may be descended from yellow-flowered species which are perennial. The condition with one ovule per ovary appears to be at the end of a reduction series. *Diamorpha* is most advanced of all and seemingly remote from the other genera. It has a special mode of dehiscence and other specializations of the carpels. The derivation of *Dudleya* and *Hasseanthus*, with axillary floriferous stems, may be from the same ancestral stock which gave rise to *Sedum*. The two native species of *Crassula*, on the other hand, are advanced members of a genus which primarily is African.

Because species of *Lenophyllum*, *Parvisedum*, and *Diamorpha* can be confused with *Sedum*, their description here is essential for clarity.

CHAPTER IX

Keys to Species

Three keys are available. These provide a means for identifying all native and naturalized species of *Sedum* in North America, north of the Mexican Plateau, and also all other genera of *Crassulaceae* occurring naturally or naturalized in this area. The first key, covering 49 species of *Sedum, Lenophyllum, Parvisedum*, and *Diamorpha*, and also five other genera, is for plants in flowering condition. To use the key, a living plant with fresh flowers is necessary, and also a hand lens and metric ruler. For questions about terminology of floral parts, see fig. 6 (p. 77), which shows an enlarged flower of *Sedum telephioides*, with parts labeled. In interpreting a plant, use average conditions, not extremes. The second key is for plants in fruiting condition. The third key makes possible identification of 53 species of *Sedum* which are common in cultivation. This requires plants with flowers. If the source of the plant for identification is known and is in North America, north of the Mexican Plateau, use either the first or the second key, depending on the condition of the specimen. If the plant originated in the Trans-Mexican Volcanic Belt, see the keys in *Sedum of the Trans-Mexican Volcanic Belt* (Clausen, 1959).

In identification of cultivated specimens, the problem will arise that a plant may belong to a species not included in the key. No easy solution is available because unnamed hybrids and newly described species, not included in any general treatment, are possibilities in cultivation. Some aid is available in the second half of the chapter on cultivated species, where, under the heading of other cultivated species, 55 additional species, not included in the key, are listed, together with references to sources of information.

Although keys may serve as a practical aid in identification, other steps are necessary. After a name has been found, compare the plant with the description of the species, including the accompanying quantitative data. The tables of data form a sort of continuous key. For subspecies, see the special keys in the sections on variation under the various species.

KEY TO NATIVE AND NATURALIZED SPECIES:
PLANTS IN FLOWERING CONDITION

A. Stamens twice as many as petals .... B
   B. Petals completely separate or not connate basally for more than 1 mm. C
      C. Petals not at the same time hooded, 4 per flower, and white; sepals with three vascular bundles .... D
         D. Petals averaging 4 or 5 per flower, if more, then margins not ciliate .... E

E. Rootstocks lacking or not with scalelike leaves and axillary annual, flower-bearing stems .... F
F. Petals white, pink, or purple .... G
G. Rootstocks short, bearing several or many annual flower-bearing stems from crown; leaves broad, flat, elliptic-ovate or oblong, 3–12 cm. long, 1–6 cm. broad; ovaries stipitate .... H
H. Petals white; sepals less than 4 mm. long; leaves not markedly reduced in size upward on stems I
I. Ovaries white, pale green, or pink; nectaries white or pale yellow; carpels 5 .... *S. telephioides*, p. 70
II. Ovaries pink; nectaries yellow; carpels 0–5
.... *S. alboroseum*, p. 558
HH. Petals deep pink; sepals >5 mm. long; leaves markedly reduced in size upward on stems .... *S. purpureum*, p. 537
GG. Rootstocks lacking or elongate; leaves of flower-bearing stems not at same time broad, flat, and over 3 cm. long; ovaries not stipitate
.... H
H. Adnation of epipetalous stamens slight, ⩽.3 mm.; flowers mostly 4-parted; plants glabrous I
I. Principal leaves mainly in whorls of 3, sometimes opposite, or rarely alternate or in whorls of 4; nectaries yellow, rarely white or orange .... *S. ternatum*, p. 92
II. Principal leaves alternate; nectaries white or pale yellow ... J
J. Leaves of flower-bearing stems linear, sagittate-spurred at base, glabrous; petals pink or white
.... *S. pulchellum*, p. 142
JJ. Leaves of flower-bearing stems narrowly elliptical, oblanceolate-spatulate, or elliptic-oblong, short-spurred, but not sagittate basally, papillose; petals white .... K
K. Plants perennial; sepals 2–9 mm. long; petals 4–9 mm. long L
L. Ratio of width to the thickness of leaves of flower-bearing stems <1.7; leaves usually green
.... *S. nevii*, p. 109
LL. Ratio of width to thickness of leaves of flower-bearing stems >2; leaves sometimes glaucous
.... *S. glaucophyllum*, p. 122
KK. Plants annual; sepals .4–1 mm. long; petals 1.4–4.2 mm. long .... *S. pusillum*, p. 162
HH. Adnation of epipetalous stamens >.3 mm.; flowers mostly 5-parted; plants glabrous or hairy I
I. Leaves of floriferous stems alternate, spirally arranged ... J
J. Stems glabrous .... K
K. Flowers with pungent scent; leaves thick, fleshy, and caducous; petals erect for more than half their length, then curved outward .... *S. wrightii*, p. 202
KK. Flowers not with pungent scent; leaves not caducous; petals widely spreading from near bases or erect in one species with linear leaves L
L. Sepals 2–12 mm. long; petals 5–10 mm. long
.... M
M. Leaves of flower-bearing stems obovate, oblanceolate, elliptical, or even suborbicular, 2–14 mm. wide .... N
N. Plants with tuberous or fibrous roots, lacking rootstocks; basal rosettes small at time of flowering .... O

KEY TO SPECIES 613

       O. Leaves of rosettes averaging about 2 mm. in thickness; stems elongate and creeping; flowers 2–3 per stem; petals slightly connate basally
.... *S. niveum*, p. 178

       OO. Leaves of rosettes averaging about 1 mm. in thickness; stems short; flowers 5 or more per stem; petals separate .... *S. cockerellii*, p. 189

     NN. Plants with stout rootstocks; basal rosettes prominent at time of flowering
.... O

       O. Petals widely spreading; cymes usually 3-parted
.... *S. spathulifolium*, p. 439

       OO. Petals erect; cymes paniculate
.... P

         P. Sepals usually >3.5 mm. long; petals white or pink .... Q

           Q. Petals white, <7.5 mm. long; anthers yellow ......
S. obtusatum ssp. *boreale*,
.... p. 373

           QQ. Petals white or pink, >7.5 mm. long; anthers red or yellow
.... *S. laxum*, p. 379

         PP. Sepals usually <3.5 mm. long; petals white or yellowish white
.... *S. oregonense*, p. 410

   MM. Leaves of flower-bearing stems linear-lanceolate or lanceolate, 1–4 mm. wide
.... N

     N. Sepals separate, 1.7–3.7 mm. long; petals widely spreading .... O

       O. Plants annual; leaves lanceolate with scarious, truncate spurs
.... *S. radiatum*, p. 247

       OO. Plants perennial; leaves lanceolate, with scarious, 3-lobed spurs
.... *S. stenopetalum*, p. 264

     NN. Sepals connate basally, 3–6 mm. long; petals erect or suberect
.... *S. ochroleucum*, p. 577

  LL. Sepals .5–1.5 mm. long; petals 2–4.5 mm. long; flowers on slender pedicels in paniculate cymes
.... *S. album*, p. 571

JJ. Stems pubescent, at least upward; plants annual or biennial .... K

  K. Stems and leaves glandular-hirtellous; petals pink; sepals 2–2.8 mm. long; nectaries .2 mm. long and red
.... *S. villosum*, p. 334

  KK. Stems glandular-pubescent upward; leaves glabrous; petals white with pink or green dorsal keels; sepals 1–1.2 mm. long; nectaries .5 mm. long, yellow
.... *S. hispanicum*, p. 574

II. Leaves of floriferous stems opposite, obovate, cuneate basally,

coarsely crenate at apex with margin glandular; petals pink or white .... *S. spurium*, p. 569
FF. Petals yellow .... G
G. Carpels each with several ovules .... H
H. Cymes 3-parted or paniculate, not racemose or spicate; leaves flat or convex ventrally I
I. Leaves alternate, 1 per node .... J
J. Cymes with central flower, basically 3-parted, although number of cincinni varies .... K
K. Plants with prominent basal rosettes; rosulate leaves spatulate or oblanceolate L
L. Plants biennial with fibrous roots; sepals separate; epipetalous stamens adnate to petals for .3–.6 mm.; nectaries .3 mm. long ... *S. leibergii*, p. 281
LL. Plants perennial with rootstocks; sepals connate basally; epipetalous stamens adnate to petals for .8–1.8 mm.; nectaries .4–.7 mm. long .... *S. spathulifolium*, p. 439
KK. Plants either lacking prominent basal rosettes or rosulate leaves neither spatulate nor oblanceolate L
L. Leaves spatulate to linear; anthers red; rootstocks present .... *S. kamtschaticum*, p. 566
LL. Leaves lanceolate, elliptical, elliptic-oblong, obovate, ovate, linear-lanceolate, or linear; anthers yellow; stems creeping or plants annual ..... M
M. Petals separate from base; leaves various, but not ovate .... N
N. Plants perennial .... O
O. Leaves elliptic-lanceolate, elliptic-oblong, lanceolate, or elliptical ... P
P. Sepals separate from base; carpels becoming divergent with papillose ovaries .... Q
Q. Leaves of rosettes elliptical or obovate, 2.6–3.4 mm. wide; axillary rosettes restricted to lower parts of stems; ovaries yellow .... *S. borschii*, p. 290
QQ. Leaves or rosettes oblong-elliptical with scarious bases, 2.2–2.7 mm. wide; vegetative offsets common in axils of leaves of floriferous stems and even in place of flowers in cymes; ovaries pale green or yellow-green ... *S. stenopetalum*, p. 264
PP. Sepals connate basally; carpels erect with glabrous ovaries .... Q
Q. Leaves lanceolate, 1.5–2.8 mm. wide, firmly attached; petals yellow ..... *S. lanceolatum*, p. 211

KEY TO SPECIES

                    QQ. Leaves ovate, 1.9–3.2 mm. wide, readily detachable; petals orange-yellow ..... *S. rupicolum*, p. 236
              OO. Leaves linear or linear-lanceolate .... P
                    P. Vegetative offsets common in axils of upper leaves of floriferous stems and even in place of flowers in cymes; bases of leaves scarious and often 3-lobed ... *S. stenopetalum*, p. 264
                    PP. Vegetative offsets usually lacking from floriferous stems and cymes; bases of leaves not scarious and 3-lobed ...... Q
                          Q. Leaves linear, obtuse, in 6 ranks; petals 3–4 mm. long ..... *S. sexangulare*, p. 574
                          QQ. Leaves linear-lanceolate, acute or apiculate, spirally arranged in many ranks; petals 5–8 mm. long ... R
                                R. Sepals lanceolate, 3–6 mm. long, concave dorsally; petals erect or suberect; flower-bearing stems erect in development .... *S. ochroleucum*, p. 577
                                RR. Sepals ovate, 2–3.3 mm. long, plane dorsally; petals spreading; flower-bearing stems reflexed in development ... *S. reflexum*, p. 579
              NN. Plants annual or biennial .... O
                    O. Leaves acuminate or acute; petals 6–10 mm. long ... *S. radiatum*, p. 247
                    OO. Leaves obtuse; petals 1.5–4 mm. long .... *S. nuttallianum*, p. 324
        MM. Petals connate basally for .4 mm.; leaves ovate .... *S. acre*, p. 547
  JJ. Cymes paniculate .... K
        K. Plants large subshrubs with leaves elliptic-oblanceolate, 4–8 cm. long and 13–25 mm. wide; flowers with petals widely spreading .... *S. praealtum*, p. 561
        KK. Plants perennial herbs with stout rootstocks and basal rosettes of leaves; leaves of floriferous stems .3–3 cm. long and 4–13 mm. wide; flowers with petals erect, then spreading above middle .... *S. obtusatum*, p. 358
II. Leaves in whorls of three or opposite .... J
  J. Leaves in whorls of 3, oblanceolate-elliptical; stems prostrate and jointed .... *S. sarmentosum*, p. 543
  JJ. Leaves opposite and elliptical, suborbicular, or obovate;

616 SEDUM OF NORTH AMERICA

                stems decumbent .... K
             K. Leaves pale glaucous green, speckled with pink, ellip-
                tical and clasping; petals connate basally; carpels
                erect .... S. debile, p. 312
             KK. Leaves green or reddish green, suborbicular to ob-
                ovate, thick and turgid, slightly subcordate clasping;
                petals separate basally; carpels becoming divergent
                in age .... S. divergens, p. 303
          HH. Cymes racemose or spicate, sometimes branched; leaves con-
             cave ventrally, lavender-green or reddish with age, and caducous
             .... Lenophyllum texanum, p. 585
       GG. Carpels with only 1 ovule; plants annual .... H
          H. Leaves yellow-green speckled with red; anthers .6 mm. long;
             carpels remaining erect .... Parvisedum pumilum, p. 591
          HH. Leaves glaucous; anthers .3 mm. long; carpels becoming diver-
             gent in age .... Parvisedum congdonii, p. 591
    EE. Rootstocks stout, with scalelike leaves and axillary, annual, flower-bearing
       stems .... F
       F. Petals pink, 8–13 mm. long; flowers perfect and slightly perigynous; in-
          florescences elongate and racemelike .... S. rhodanthum, p. 475
       FF. Petals red or yellow, 1–5 mm. long; flowers imperfect or perfect, hypog-
          ynous; inflorescences corymbose cymes .... G
          G. Petals of staminate plants usually dark red, 1.3–1.5 mm. wide
             .... S. integrifolium, p. 487
          GG. Petals of staminate plants usually yellow or greenish yellow, .7–1.1
             mm. wide .... S. rosea, p. 517
 DD. Petals 6–16, with margins prominently ciliate; plants with rootstocks bearing large
    rosettes of oblong-lanceolate or oblong-obovate leaves ..... Sempervivum, p. 609
 CC. Petals hooded, 4 per flower, and white; sepals with a single vascular bundle; leaves
    elliptical, rounded, and red; plants small annuals .... Diamorpha cymosa, p. 600
BB. Petals connate basally for >1 mm.; flower-bearing branches either terminal or axillary
 C
 C. Flowers usually 5-parted; leaves alternate, except sometimes opposite on seedlings and
    shoots of offsets .... D
    D. Plants either with or without thick rootstocks, but not with corms; leaves of rosettes
       persisting throughout year .... E
       E. Flowers 1–9 per cyme; petals only slightly connate, seldom for more than a
          mm.; plants lacking rootstocks .... F
          F. Petals white streaked with pink; anthers red; leaves alternate, obovate or
             oblanceolate, and green .... S. niveum, p. 178
          FF. Petals yellow streaked with red; anthers yellow; leaves opposite, ellip-
             tical, and glaucous .... S. debile, p. 312
       EE. Flowers 4–121 per cyme; petals mostly strongly connate, usually for more
          than a mm.; plants with rootstocks .... F
          F. Flower-bearing stems mostly terminal .... G
             G. Flower-bearing stems glabrous, erect before expansion of flowers
                .... H
                H. Cymes 3-parted; petals narrowly lanceolate, aristate; leaves spat-
                   ulate and clavate, shiny green often suffused with red
                   .... S. oreganum, p. 345
                HH. Cymes paniculate; petals oblong, often oblanceolate-oblong, not
                   aristate; leaves not lustrous I
                   I. Petals yellow, rarely white .... J
                      J. Leaves of rosettes not white-margined; leaves of flower-
                         bearing stems mostly <20 mm. long
                         .... S. obtusatum, p. 358
                      JJ. Leaves of rosettes white-margined; leaves of flower-
                         bearing stems ≥20 mm. long .

.... *S. albomarginatum*, p. 424
II. Petals white, creamy white, or pink .... J
  J. Sepals usually >3.5 mm. long; petals white or pink .... K
    K. Stems green or glaucous, but not pruinose upward; leaves of floriferous stems obovate, spatulate, or oblanceolate.. L
      L. Petals white, <7.5 mm. long; anthers yellow .... *S. obtusatum* ssp. *boreale*, p. 373
      LL. Petals white or pink, >7.5 mm. long; anthers red or yellow .... *S. laxum*, p. 379
    KK. Stems pruinose upward; leaves of flower-bearing stems oblanceolate-oblong .... *S. oblanceolatum*, p. 403
  JJ. Sepals usually <3.5 mm. long; petals white or yellowish white .... *S. oregonense*, p. 410
  GG. Flower-bearing stems glandular-pubescent upward, recurved before time of flowering; cymes 3-parted; petals yellow ... *S. moranii*, p. 433
 FF. Flower-bearing stems always axillary .... *Dudleya*, p. 606
DD. Plants with corms; leaves of rosettes not persisting, vernal ... *Hasseanthus*, p. 608
CC. Flowers usually 4-parted; leaves always opposite .... *Kalanchoe*, p. 609
AA. Stamens same number as petals; plants small annuals .... B
 B. Petals yellow, 4–5 per flower, 1.7–2.9 mm. long; leaves 1 per node C
  C. Petals 2.3–2.9 mm. long, divergent in flower; nectaries .4–.6 mm. long; ovaries glabrous .... *Parvisedum leiocarpum*, p. 591
  CC. Petals 1.7–2.1 mm. long, erect in flower; nectaries .3 mm. long; ovaries strongly hirtellous-papillose .... *Parvisedum pentandrum*, p. 591
 BB. Petals white, sometimes pinkish, 3–4 per flower, .4–1.1 mm. long; leaves opposite, 2 per node, connate-perfoliate .... *Crassula*, p. 608

KEY TO NATIVE AND NATURALIZED SPECIES: PLANTS IN VEGETATIVE OR FRUITING CONDITION

A. Plants perennial, with green leaves at time of dehiscence of fruits .... B
 B. Principal leaves alternate, spirally arranged C
  C. Leaves of rosettes broadest above middle .... D
   D. Fruits divergent .... E
    E. Fruit-bearing stems glabrous .... F
     F. Plants herbs, chamaephytes .... G
      G. Leaves 3–5 cm. long; fruits 4–6 mm. long ... *S. kamtschaticum*, p. 566
      GG. Leaves <2 cm. long; fruits 2–5 mm. long .... H
       H. Plants with stout, branched rhizomes; seeds .9–1.2 mm. long .... *S. spathulifolium*, p. 439
       HH. Plants lacking rhizomes; seeds .5–.9 mm. long I
        I. Leaves green or gray-green, not glaucous; seeds averaging .7 mm. long .... *S. nevii*, p. 109
        II. Leaves pale green or blue-green, sometimes glaucous; seeds averaging .8 mm. long .... *S. glaucophyllum*, p. 122
     FF. Plants shrubs with paniculate clusters of fruits on axillary branches .... *S. praealtum*, p. 561
    EE. Fruit-bearing stems glandular-puberulent upward .... *S. moranii*, p. 433
   DD. Fruits erect .... E
    E. Plants either lacking rhizomes or with rhizomes without scalelike leaves; seeds <1.8 mm. long .... F
     F. Rosettes of large basal leaves not prominent at time of flowering; rootstocks lacking or short; seeds reticulate .... G
      G. Leaves of rosettes about 2 mm. thick; seeds obscurely reticulate, lustrous, and glabrous .... *S. niveum*, p. 178

GG. Leaves of rosettes about 1 mm. thick; seeds reticulate and hairy
.... *S. cockerellii*, p. 189
FF. Rosettes of large basal leaves prominent at time of flowering; rootstocks present; seeds finely longitudinally ribbed .... G
G. Clusters of fruits in 3-parted cymes; withered petals erect and aristate
.... *S. oreganum*, p. 345
GG. Clusters of fruits in paniculate cymes; withered petals not aristate
.... H
H. Leaves of rosettes not white-margined I
I. Fruit-bearing stems glabrous or glaucous, but not pruinose upward; leaves of rosettes oblanceolate, spatulate, or obovate, green or glaucous .... J
J. Persistent sepals >3.5 mm. long; seeds brown .... K
K. Leaves of rosettes 1–4 mm. thick, mostly less than 2.6 mm. .... *S. obtusatum*, p. 358
KK. Leaves of rosettes 2.8–3.7 mm. thick
.... *S. laxum*, p. 379
JJ. Persistent sepals <3.5 mm. long; seeds pale brown .
.... *S. oregonense*, p. 410
II. Fruit-bearing stems pruinose upward; leaves of rosettes oblanceolate-oblong and pruinose ... *S. oblanceolatum*, p. 403
HH. Leaves of rosettes white-margined ... *S. albomarginatum*, p. 424
EE. Plants with rhizomes with scalelike leaves; seeds oblong, 1.8–2 mm. long
.... *S. rhodanthum*, p. 475
CC. Leaves of rosettes broadest at or below middle .... D
D. Leaves appearing flat, >5 times broader than thick .... E
E. Follicles erect; plants with or without rhizomes, never with corms .... F
F. Leaves not ciliate; rosettes prominent or absent at time of fruiting ...... G
G. Rosettes with well-developed leaves present at time of fruiting
.... *Dudleya*, p. 606
GG. Rosettes with well-developed leaves lacking at time of fruiting ... H
H. Fruit-bearing stems arising from crown of short rootstock; follicles stipitate I
I. Leaves markedly reduced upward on stems; roots carrotlike
.... *S. purpureum*, p. 537
II. Leaves not markedly reduced upward on stems; roots stout, but not carrotlike .... J
J. Follicles erect, 2–5 mm. long; seeds pyriform, regularly developing; leaves green or yellow-green, sometimes glaucous .... *S. telephioides*, p. 70
JJ. Follicles subdivergent, 4–6 mm. long, empty, not producing seeds; leaves pale yellow-green, often glaucous
.... *S. alboroseum*, p. 558
HH. Fruit-bearing stems arising from axils of scalelike leaves on stout rhizomes; follicles not stipitate I
I. Fruits in elongate, racemelike clusters; seeds oblong and pale brown; midveins of leaves depressed ... *S. rhodanthum*, p. 475
II. Fruits in corymbose clusters; seeds pyriform or oblanceolate and brown; midveins of leaves not depressed .... J
J. Leaves spinach green or glaucous (postpone identification until flowers are available) ...... *S. integrifolium*, p. 487
JJ. Leaves usually glaucous (wait for flowers)
.... *S. rosea*, p. 517
FF. Leaves ciliate; prominent rosettes present at time of fruiting
.... *Sempervivum*, p. 609
EE. Follicles spreading; plants with corms or rhizomes .... F
F. Plants with corms; leaves of rosettes not persisting, vernal

.... *Hasseanthus*, p. 608
 FF. Plants with rhizomes; leaves of rosettes usually persisting
.... *Dudleya*, p. 606
 DD. Leaves not appearing flat, <5 times broader than thick .... E
  E. Fruits erect .... F
   F. Leaves elliptical, elliptic-lanceolate, oblong, or subglobose .... G
    G. Seeds papillose or mammillate, .6–1 mm. long; leaves broadly rounded or obtuse .... H
     H. Leaves detaching on slight pressure, 3.8–5.4 mm. wide; fruits 4–5 mm. long .... *S. wrightii*, p. 202
     HH. Leaves firmly attached, 2–3.5 mm. wide; fruits 3–4 mm. long
.... *S. album*, p. 571
    GG. Seeds longitudinally ribbed, .9–1.3 mm. long; leaves obtusely apiculate or obtuse .... H
     H. Leaves firmly attached, lanceolate, 1.5–2.8 mm. wide
.... *S. lanceolatum*, p. 211
     HH. Leaves readily detachable, ovate, 1.9–3.2 mm. wide
.... *S. rupicolum*, p. 236
   FF. Leaves linear-lanceolate, acute or apiculate .... G
    G. Dry sepals lanceolate, concave dorsally, 3–6 mm. long; clusters of fruits flat-topped .... *S. ochroleucum*, p. 577
    GG. Dry sepals ovate, plane dorsally, 2–3.3 mm. long; clusters of fruits concave .... *S. reflexum*, p. 579
  EE. Fruits spreading and gibbous ventrally .... F
   F. Seeds longitudinally ribbed, brown to pale yellow, .8–1.3 mm. long .... G
    G. Leaves elliptic-oblong, 4–9 mm. long .... *S. borschii*, p. 290
    GG. Leaves linear or linear-lanceolate, 4–36 mm. long .... H
     H. Leaves sagittate-spurred; basal rosettes, if present, with oblanceolate leaves .... *S. pulchellum*, p. 142
     HH. Leaves with scarious spurs which often are 3-lobed; leaves of rosettes linear-lanceolate .... *S. stenopetalum*, p. 264
   FF. Seeds reticulate, brown, .5–.7 mm. long .... G
    G. Leaves ovate, spirally arranged in many ranks .... *S. acre*, p. 547
    GG. Leaves linear, in 6 ranks .... *S. sexangulare*, p. 574
BB. Principal leaves in whorls of 3 or opposite C
 C. Leaves mostly in whorls of 3 .... D
  D. Leaves obovate to elliptical, rounded at apex; plants producing fruits and seeds
.... *S. ternatum*, p. 92
  DD. Leaves oblanceolate-elliptical, subacute; plants with repent stems, not producing fruits and seeds .... *S. sarmentosum*, p. 543
 CC. Leaves mostly opposite .... D
  D. Stems erect, >2 dm. tall .... E
   E. Follicles stipitate, separate from each other; withered sepals separate; fruits erect in corymbose clusters .... F
    F. Follicles erect, 2–5 mm. long; seeds pyriform ...... *S. telephioides*, p. 70
    FF. Follicles subdivergent, 4–6 mm. long, empty, not producing seeds
.... *S. alboroseum*, p. 558
   EE. Follicles not stipitate, connate basally; withered sepals united, at least basally, and sometimes bladdery inflated; fruits either erect or pendulous in paniculate cymes .... *Kalanchoe*, p. 609
  DD. Stems low, <2 dm. tall, or creeping .... E
   E. Fruits erect .... F
    F. Leaves elliptical, clasping basally, broadly rounded, 4–6 mm. long
.... *S. debile*, p. 312
    FF. Leaves obovate, cuneate basally, and coarsely crenate apically, 10–30 mm. long .... *S. spurium*, p. 569
   EE. Fruits divergent .... F

620   SEDUM OF NORTH AMERICA

    F. Fruits in 2- or 3-parted cymes; leaves suborbicular or obovate, thick and turgid; seeds pyriform, longitudinally ribbed, .9–1.1 mm. long .... *S. divergens*, p. 303
    FF. Fruits in racemose or spicate cymes; leaves ovate, abruptly acute or cuspidate, 3 times wider than thick; seeds oblong-elliptical, verrucose, .4–.8 mm. long .... *Lenophyllum*, p. 584
AA. Plants annual or biennial, rarely perennial, without green leaves at time of dehiscence of fruits .... B
 B. Fruits dehiscing along ventral suture C
  C. Leaf scars or withered leaves alternate .... D
   D. Stems glandular-hairy upward; fruits hairy .... E
    E. Fruits suberect; seeds lustrous, amber-yellow, longitudinally ribbed and verrucose .... *S. villosum*, p. 334
    EE. Fruits divergent and slightly gibbous; seeds brown and longitudinally ribbed .... *S. hispanicum*, p. 574
   DD. Stems glabrous; fruits glabrous or papillose-puberulent; fruits divergent or erect; seeds brown .... E
    E. Fruits divergent; seeds pyriform or ovoid, seldom more than twice as long as wide, usually several per follicle .... F
     F. Fruits divergent and gibbous ventrally .... G
      G. Seeds finely longitudinally ribbed; fruits 1.9–6 mm. long .... H
       H. Plants lacking subterranean offsets; seeds .9–1.3 mm. long, brown I
        I. Leaves linear with sagittate spurs at base; follicles 3.2–4.3 mm. long .... *S. pulchellum*, p. 142
        II. Leaves lanceolate with short, scarious spurs, either 3-lobed or unlobed; follicles 3–3.1 mm. long .... *S. radiatum*, p. 247
       HH. Plants with either subterranean offsets or corms; seeds .7–.9 mm. long, yellow-brown or brown I
        I. Plants with subterranean offsets; seeds yellow-brown .... *S. leibergii*, p. 281
        II. Plants with corms; seeds brown .... *Hasseanthus*, p. 608
      GG. Seeds obscurely alveolate; fruits 1.3–2.6 mm. long .... *S. nuttallianum*, p. 324
     FF. Fruits widely divergent, but not gibbous ventrally .... G
      G. Seeds pyriform, .6–.9 mm. long, 0–10 per follicle .... *S. pusillum*, p. 162
      GG. Seeds narrowly oblanceolate-oblong, 1–1.2 mm. long, 1 per follicle .... *Parvisedum congdonii*, p. 591
    EE. Fruits erect; seeds oblanceolate or narrowly elliptic-oblong, 3–4 times longer than broad, 1 per follicle .... F
     F. Fruits papillose-puberulent .... G
      G. Fruits averaging 1.6 mm. long; seeds 1.2–1.5 mm. long .... *Parvisedum pumilum*, p. 591
      GG. Fruits 1–1.2 mm. long; seeds averaging .9 mm. long .... *Parvisedum pentandrum*, p. 591
     FF. Fruits smooth, 1.7–1.8 mm. long .... *Parvisedum leiocarpum*, p. 591
  CC. Leaves or leaf scars opposite; fruits tiny, in clusters of 3 or 4 .... *Crassula*, p. 608
 BB. Fruits dehiscing laterally by means of a dorsal valve; fruits divergent, 1–1.7 mm. long, with tiny seeds, .4–.6 mm. long .... *Diamorpha cymosa*, p. 600

KEY TO COMMONLY CULTIVATED SPECIES:
PLANTS IN FLOWERING CONDITION

A. Leaves alternate, 1 per node .... B
 B. Both stems and leaves glabrous C

C. Leaves appearing flat, >5 times broader than thick .... D
   D. Margins of leaves toothed .... E
      E. Flowers pink, purple, or white .... F
         F. Plants shrubby, with leaves ovate, long-petiolate, cordate or truncate basally, and coarsely dentate .... *S. populifolium*, p. 558
         FF. Plants herbaceous, with leaves elliptical, sessile or short-petiolate, and variously toothed .... G
            G. Stamens about same length as petals .... H
               H. Leaves markedly reduced in size upward on stems; inflorescences dense, subglobose; petals deep pink .... *S. purpureum*, p. 537
               HH. Leaves not markedly reduced in size upward on stems; inflorescences corymbose; petals white .... *S. alboroseum*, p. 558
            GG. Stamens much longer than petals .... *S. spectabile*, p. 558
      EE. Flowers yellow .... F
         F. Stems decumbent or erect; numerous sterile shoots not present at time of flowering .... G
            G. Stems erect and stout, 1.5–8 dm. tall, 4–8 mm. in diam. basally; cymes broad, flat-topped or concave; leaves broadest at or below middle, toothed for two-thirds of length .... *S. aizoon*, p. 566
            GG. Stems erect or decumbent and slender, 1–2.8 dm. tall, 1–4 mm. in diam. basally; cymes few-flowered, averaging about 25 flowers; leaves broadest above middle, toothed only toward apex .... *S. kamtschaticum*, p. 566
         FF. Stems extensively creeping and branched; numerous sterile shoots present at time of flowering .... *S. hybridum*, p. 566
   DD. Margins of leaves entire .... E
      E. Flowers pink, purple, or white .... F
         F. Stems erect; stems with flowers >3 dm. tall; stamens much longer than petals .... *S. spectabile*, p. 558
         FF. Stems decumbent or creeping, those with flowers <2 dm. tall; stamens about same length as petals .... G
            G. Leaves 12–32 mm. long; inflorescences dense cymes; petals rose purple .... *S. anacampseros*, p. 561
            GG. Leaves 2–6 mm. long; inflorescences racemose; petals white .... *S. magellense*, p. 569
      EE. Flowers yellow .... F
         F. Cymes paniculate on axillary branches .... G
            G. Leaves averaging 2 or more mm. in thickness; sepals 1.5–6.1 mm. long; anthers .9–1.3 mm. long .... *S. praealtum*, p. 561
            GG. Leaves averaging 1.8 mm. in thickness; sepals .8–2.1 mm. long; anthers .6–.9 mm. long .... *S. confusum*, p. 561
         FF. Cymes 3–5-parted, terminal, but sometimes apparently lateral .... G
            G. Low, decumbent plants with obovate or elliptic-spatulate leaves, emarginate at apex and pale green, 15–20 mm. long; sepals 2.2–3.4 mm. long; petals yellow, 4.2–5.2 mm. long .... *S. decumbens*, p. 562
            GG. Subshrubs, often pendulous in nature, with spatulate or elliptic-oblong leaves, acute or obtuse or sometimes apiculate at apex and glaucous, 11–54 mm. long; sepals 4–7 mm. long; petals buttercup yellow, 4–7 mm. long .... *S. palmeri*, p. 562
CC. Leaves not appearing flat, often as thick as broad, and always less than 5 times broader than thick .... D
   D. Plants perennial herbs or subshrubs; petals white or yellow, but never blue ..... E
      E. Leaves broadest above middle .... F
         F. Plants subshrubs with thick, fleshy leaves, >3 mm. thick .... G
            G. Petals white .... H
               H. Leaves green or yellow, not glaucous; sepals ovate or oblong, .6–5 mm. long I

I. Inflorescences paniculate; pedicels 1–7 mm. long; anthers yellow; leaves lustrous, green .... S. lucidum, p. 563
II. Inflorescences umbellate; pedicels 8–20 mm. long; anthers salmon-pink; leaves dull green or yellow .... J
    J. Flowers appearing all in same plane; petals lanceolate, 2.8–3.2 mm. wide; principal leaves 5–8 mm. thick, convex and obscurely carinate dorsally
.... S. nussbaumerianum, p. 563
    JJ. Flowers at different levels in cymes, not all in one plane; petals ovate-lanceolate, 3–3.5 mm. wide; principal leaves 8–10 mm. thick, strongly carinate dorsally .
.... S. adolphii, p. 564
  HH. Leaves glaucous; sepals clavate, 5–8 mm. long .
.... S. clavatum, p. 563
GG. Petals yellow .... H
  H. Leaves glaucous, markedly upcurved; flower-bearing branches lateral in axils of leaves I
    I. Leaves nearly flat on upper surface, pruinose .
.... S. treleasei, p. 562
    II. Leaves nearly round in cross-section, not flattened on upper surface, glaucous .... S. pachyphyllum, p. 562
  HH. Leaves lustrous green strongly tinged with red, ascending, but not markedly upcurved; inflorescences terminal on main stems
.... S. rubrotinctum, p. 564
FF. Plants herbs with the leaves moderately fleshy, less than 3 mm. thick
.... G
  G. Petals white; inflorescences racemose ...... S. magellense, p. 569
  GG. Petals yellow; inflorescences cymes of 1–4, mostly 3, cincinni .
.... S. greggii, p. 565
EE. Leaves broadest below middle or of same width throughout length .... F
  F. Petals white or purple .... G
    G. Petals white .... H
      H. Leaves ovate, subglobose, elliptical, or oblong I
        I. Inflorescences cymes of 1 to 3 cincinni .... J
          J. Plants herbs; petals 3–5 mm. long .... K
            K. Leaves elliptical, often tinged with red, in several ranks on stems; petals often pinkish .
.... S. anglicum, p. 572
            KK. Leaves ovoid or subglobose, mealy, in 4 or 5 ranks
.... S. brevifolium, p. 571
          JJ. Plants subshrubs with the leaves closely set in several spiral ranks; petals 4–8 mm. long .
.... S. moranense, p. 565
        II. Inflorescences paniculate, with flowers on slender pedicels; leaves subglobose to linear-oblong, green or suffused with red .... S. album, p. 571
      HH. Leaves linear I
        I. Petals 2–5 mm. long; flowers mostly 5-parted; sepals separate or nearly so .... J
          J. Leaves oblong-linear, subterete, and glaucous; petals 4–7 mm. long; plants with long, creeping stems .
.... S. diffusum, p. 564
          JJ. Leaves linear, somewhat flattish or subterete, green and often spotted or suffused with red; petals 2–4 mm. long; stems decumbent or creeping .... K
            K. Inflorescences dense; carpels becoming pink or red; stems prostrate, forming a mat ..... S. lydium, p. 571

                    KK. Inflorescences mostly 2-parted, open cymes; carpels
                         pale green; stems procumbent, not creeping, with a
                         tufted habit .... S. gracile, p. 573
                  II. Petals 5–8 mm. long; flowers 5–7-parted; sepals connate
                      basally .... J
                      J. Sepals ovate, flat on back, 1.5–3 mm. long; leaves elliptic-
                         lanceolate, flattened on upper surface .
                         .... S. sediforme, p. 577
                      JJ. Sepals lanceolate, erect, concave on back, 3–6 mm. long;
                          leaves linear-lanceolate .... S. ochroleucum, p. 577
            GG. Petals purple, erect, 10–12 mm. long; leaves falcate, glaucous, 15–30
                mm. long; stems pendulous .... S. morganianum, p. 564
        FF. Petals yellow .... G
            G. Leaves ovate, closely crowded and imbricate .... S. acre, p. 547
            GG. Leaves linear .... H
                H. Leaves 3–6 mm. long, in 6 ranks; petals 3–4 mm. long .
                   .... S. sexangulare, p. 574
                HH. Leaves 5–12 mm. long, spirally arranged in many ranks; petals
                    5–8 mm. long I
                    I. Miniature subshrubs; leaves with spurs enlarged, scarious,
                       and sheathing, and with apices obtuse .... S. multiceps, p. 574
                    II. Herbs with decumbent, much-branched stems; leaves short-
                        spurred, with spurs not enlarged, not scarious, and not
                        sheathing, and with apices acute, apiculate, or spinulose
                        .... J
                        J. Sepals lanceolate, concave dorsally, 3–6 mm. long .
                           .... S. ochroleucum, p. 577
                        JJ. Sepals ovate, plane dorsally, 2–3.5 mm. long .... K
                            K. Leaves linear-oblanceolate to subulate, plane ven-
                               trally, incurved, in dense subglobose to oblong rosettes
                               at ends of secondary shoots ...... S. rupestre, p. 577
                            KK. Leaves linear-lanceolate or lanceolate-linear, sub-
                                terete or terete, densely crowded toward ends of
                                secondary shoots, but not incurved or forming sub-
                                globose rosettes .... S. reflexum, p. 579
        DD. Plants annual; petals blue .... S. caeruleum, p. 574
BB. Stems or leaves hairy C
    C. Leaves broadest above middle .... D
       D. Leaves glabrous; inflorescences paniculate; petals separate, widely spreading;
          stems slender, creeping .... S. monregalense, p. 571
       DD. Leaves glandular-hairy; inflorescences 2- or 3-parted cymes or elongate; petals
           connate basally for .5–4 mm.; plants with prominent rosettes of basal leaves at
           time of flowering .... E
           E. Petals connate for less than a quarter of their length .... F
              F. Sepals separate; petals white; plants perennial ..... S. hirsutum, p. 576
              FF. Sepals connate basally; petals pink; plants biennial .... S. pilosum, p. 576
           EE. Petals connate for a quarter or more of their length; inflorescences elongate;
               petals white or yellow .... S. chrysanthum, p. 576
    CC. Leaves broadest at or below middle .... D
        D. Leaves glabrous; inflorescences 2–3-parted cymes or paniculate .... E
           E. Plants perennial; leaves ovate, suborbicular, or elliptical .... F
              F. Leaves yellow-green; cymes paniculate .... S. monregalense. p. 571
              FF. Leaves bluish green or grayish, glaucous; cymes usually 2-parted .
                  .... S. dasyphyllum, p. 571
           EE. Plants biennial or annual; leaves oblong-linear or oblong .... F
               F. Petals white, 4–6 mm. long; leaves glaucous; inflorescences usually 3-
                  parted .... S. hispanicum, p. 574

                    FF. Petals blue, 3–4 mm. long; leaves green, not glaucous; inflorescences paniculate .... *S. caeruleum*, p. 574
               DD. Leaves densely pubescent; inflorescences paniculate; plants perennial with prostrate rhizomes, 5 mm. in diameter, and prominent rosettes of elliptic-oblong leaves .... *S. hintonii*, p. 565
AA. Leaves opposite or in whorls .... B
  B. Both stems and leaves glabrous C
    C. Leaves in whorls of 3 or 4 .... D
      D. Leaves obovate or elliptical, toothed or entire; plants with short rootstocks; petals pink or white; inflorescences corymbose .... E
        E. Leaves obovate and blue-green; stamens about same length as petals; nectaries orange .... *S. sieboldii*, p. 557
        EE. Leaves broadly elliptical to elliptic-spatulate, pale green, and glaucous; stamens longer than petals .... *S. spectabile*, p. 558
      DD. Leaves linear, linear-lanceolate, or oblanceolate-elliptical, with margins untoothed; plants lacking rootstocks; stems creeping or decumbent; petals yellow; cymes 2–5-parted .... E
        E. Stems long-creeping; leaves oblanceolate-elliptical ... *S. sarmentosum*, p. 543
        EE. Stems decumbent; leaves linear-lanceolate .... F
          F. Leaves flattish on upper surface, usually in whorls of 3 .... *S. lineare*, p. 569
          FF. Leaves subterete, usually in whorls of 4 .... *S. mexicanum*, p. 565
    CC. Leaves opposite .... D
      D. Flower-bearing stems erect; cymes corymbose; leaves elliptical, elliptic-ovate, or oblong, usually toothed on margins .... E
        E. Petals white; stamens about same length as petals .... F
          F. Carpels greenish white or pale green; flowers 6–8 mm. in diameter . ... *S. telephium*, p. 558
          FF. Carpels pink or red; flowers 9–13 mm. in diameter . ... *S. alboroseum*, p. 558
        EE. Petals pink; stamens longer than petals .... *S. spectabile*, p. 558
      DD. Flower-bearing stems decumbent or creeping; cymes corymbose, umbellate, or 3-parted; leaves obovate, elliptic-spatulate, or ovate-cordate, entire or toothed at apex .... E
        E. Stems decumbent; leaves ovate-cordate, entire; cymes dense, corymbose .... *S. ewersii*, p. 558
        EE. Stems creeping; leaves obovate or elliptic-spatulate, toothed at apex; cymes umbellate or 3-parted .... F
          F. Leaves coarsely crenate at apex, with prominently glandular margins, dark green; petals erect below, then spreading, 8–11 mm. long, pink or white .... *S. spurium*, p. 569
          FF. Leaves entire or crenate toward apex, light green; petals widely spreading from base, 6 mm. long, pink .... *S. stoloniferum*, p. 569
  BB. Stems hairy; leaves hairy or glabrous C
    C. Stems and also leaves puberulent; leaves elliptical or globular, about as thick as wide, usually suffused with red; petals yellow; subshrubs .... *S. stahlii*, p. 565
    CC. Stems glandular-puberulent; leaves oblong-spatulate, obovate, elliptical, or suborbicular, wider than thick, glabrous or sometimes glandular-puberulent; petals white; herbs .... D
      D. Leaves yellow-green; cymes paniculate .... *S. monregalense*, p. 571
      DD. Leaves bluish green or grayish, glaucous; cymes usually 2-parted . ... *S. dasyphyllum*, p. 571

CHAPTER X

Geography of *Sedum* in North America North of the Mexican Plateau

Sedum is widely distributed in North America. The plants characteristically grow on rocks in disjunct populations. Although species occur along both the Atlantic Coast northward from Maine to Baffin Island and the Pacific Coast from California to Alaska, *Sedum* is lacking from vast areas of the continent. It seldom is a dominant feature of the landscape.

Sampling surveys have been useful to get information about the geographical distribution of *Sedum* and to gain an unprejudiced perspective. Since 1959, I have surveyed one or more regions in each growing season. Resulting information indicates the environmental conditions where *Sedum* grows and also the situations where it does not occur. Information is organized in tables with explanatory descriptive text. Data of the tables are the principal basis for conclusions about distribution, and they provide descriptions of the sites which were the source of plants cultivated in experiments.

Methods of sampling have varied in different surveys, but the general plan always has been the same. Differences in size and shape of listing units and in method of subsampling have been consequences of problems presented by different species. No available model is entirely satisfactory for the survey of *Sedum*. For discussion of sampling procedures, see the chapter on sampling.

The first table includes a list of all forty-three geomorphic provinces of North America north of the Mexican Plateau, together with information about years of survey, time spent in each for studying *Sedum*, details about kind and number of sampling units, and number of species of *Sedum* found by me in each province. A second table includes information about total area of each province, area surveyed, altitudinal range, and total number of species of *Sedum* which are native. The main part of the chapter is a detailed inventory of the species of *Sedum* occurring in each geomorphic province. Accounts of provinces are brief, without detailed description of sampling units. Exceptions are the Piedmont Plateau, Ridge and Valley Province, Central Rocky Mountains, Sierra Nevada, and Grenville Province. Accounts of these provinces are fuller and serve to demonstrate how surveys were conducted and also the kinds of data which resulted. Final sections include a summary of all geographic data and conclusions.

In the following lists, geomorphic provinces are in a sequence similar to that in the chapter on geology. This arrangement keeps together data for related provinces.

Abbreviations for kinds of sampling units are: p = populations of plants, q = quadrats, and t = topographic units. In the column, *Sedum*, only native species are included.

Greenland, otherwise omitted from this treatment, is included in the tables of this chapter because it is related geographically to North America, even though the affinities of the species of *Sedum* occurring there are with Europe.

The record of surveys follows:

Geomorphic province	Years of surveys	Time spent (n-weeks)	Kind of unit	Sedum (n-species)
New England–Acad. High.	1954, 1967	1	p	1
Ridge and Valley	1966, 1969	6	p,t	4
Blue Ridge	1960, 1970	13	t,p	5
Piedmont Plateau	1946, 1966	6	p,t	4
Appalachian Plateau	1969, 1972	8	p	5
Ouachita Mt. System	1949	1	t	0
Central Rocky Mts.	1966	9	t	5
Southern Rocky Mts.	1949, 1962	11	t,p	4
Colorado Plateau	1940	1	p	1
Wyoming Basin	1966	.5	t	0
Idaho Batholith	1961	.3	p	2
Wyomide Ranges	1940	.3	p	1
Great Basin	1940, 1964	2	t	0
Mts. and Basins of S.W. Mont. and Ida.	1961	.5	p	2
Sierra Nevada	1964, 1970	16	t,p	7
Klamath Mts.	1963, 1965	2	t,p	7
Great Valley of Calif.	1963	2	t	0
Calif. Coast Ranges	1963, 1970	12	t	4
Transverse Ranges of Calif.	1968	9	p,t	2
Columbia Plateau	1961	7	p,t	5
Cascade Mts.	1961, 1965	15	p,t	8
Willamette–Puget Trough	1963	1	t	2
Coast Ranges of Ore. and Wash.	1963	9	t	2
Olympic Peninsula	—	0	—	—
Northern Rocky Mts.	1961	.7	p	4
Int. Plat. and Ranges	1965	.5	p	1
Coast Batholith	—	0	—	—
Alaskan Peninsula	—	0	—	—
Superior	1965	.4	—	0
Slave	—	0	—	—
Bear	—	0	—	—
Churchill	—	0	—	—
Hudson Bay Lowland	—	0	—	—
Southern Province of Can. Shield	1965, 1966	1	t	0
Grenville	1959	8	q,p	2
Arctic Island	—	0	—	—
Greenland	—	0	—	—
St. Lawrence Lowland	1937	.5	t	0
Central Lowland	1949	.5	p	1
Great Plains	1952, 1962, 1970	6	p	3
Int. Low Plateaus	1949, 1970	1	p	2
Ozark Plateaus	1940, 1968	2	t,p	2
Coastal Plain	1938, 1940	4	t	0

In the second table, figures for areas are approximate. Original measurements were necessary because no source is available which provides the necessary information. Since interest for *Sedum* is in terrestrial environment, areas of major bodies of water are omitted from dimensions of geomorphic provinces. Figures for areas surveyed include only the units carefully investigated on foot for *Sedum*. In almost all provinces, much larger areas were visited, but not as part of a planned sampling survey for *Sedum*. Under altitudinal range, citation is of extremes in elevation above the level of the sea for each province.

Geomorphic province	Area (km.²)	Area surveyed (km.²)	Alt. range (m.)	Sedum native (n-species)
New England–Acad. High.	475,000	.4	0–1,916	1
Ridge and Valley	120,000	6.2	0–1,363	5
Blue Ridge	42,500	8.8	64–2,047	5
Piedmont Plateau	197,500	2.1	0– 885	6
Appalachian Plateau	270,000	17.5	116–1,481	6
Ouachita Mt. System	57,500	~7	90– 839	2
Central Rocky Mts.	97,400	8.3	2,034–4,176	5
Southern Rocky Mts.	125,000	10	1,524–4,359	4
Colorado Plateau	327,500	~1	1,525–3,871	5
Wyoming Basin	105,000	<1	1,981–2,286	1
Idaho Batholith	45,000	~1	915–3,856	5
Wyomide Ranges	30,000	<1	1,525–3,581	2
Great Basin	497,500	~12	−85–4,341	3
Mts. and Basins of S.W. Mont. and Ida.	77,500	<1	978–3,442	4
Sierra Nevada	75,000	20	50–4,421	7
Klamath Mts.	37,500	~3	0–2,744	10
Great Valley of Calif.	37,500	4.5	0– 518	0
Calif. Coast Ranges	80,000	4.3	0–2,468	5
Transverse Ranges of Calif.	15,000	4.1	0–3,500	2
Columbia Plateau	500,000	5.9	30–2,861	7
Cascade Mts.	105,000	17	150–4,392	9
Willamette–Puget Trough	20,000	7	0– 735	4
Coast Ranges of Ore. and Wash.	35,000	10.1	0–1,249	2
Olympic Peninsula	12,500	0	0–2,413	5
Northern Rocky Mts.	532,500	<1	0–3,948	4
Int. Plat. and Ranges	1,655,000	.1	0–3,390	4
Coast Batholith	352,500	0	0–4,041	6
Alaskan Peninsula	415,000	0	0–6,187	1
Superior	1,545,000	0	0– 653	0
Slave	182,500	0	158– 654	0
Bear	112,500	0	0– 518	0
Churchill	2,207,500	0	0–1,622	1
Hudson Bay Lowland	307,500	0	0– 152	0
Southern Province of Can. Shield	165,000	<1	177– 680	0
Grenville	652,500	7.9	0–1,629	2
Arctic Island	752,500	0	0–2,604	1
Greenland	2,175,082	0	0–3,700	4
St. Lawrence Lowland	47,500	<1	0– 120	0
Central Lowland	2,245,000	<1	75– 762	4
Great Plains	2,210,000	~8	457–4,152	5
Interior Low Plateaus	130,000	<1	92– 427	3
Ozark Plateaus	142,500	~1	122– 686	2
Coastal Plain	1,247,500	<1	0– 305	2

Tables for geomorphic provinces, except for the five for which accounts are detailed, include number of known populations of each native species. Such information comes from both surveys and herbarium records. For provinces not visited, records from herbaria are the main source of information. These records may be incomplete. Also, data about distribution change as conditions change. An asterisk after the name of a species indicates that it is endemic in the province for which it is listed. Estimates of number of populations, although possible for surveyed provinces, are omitted because they require larger samples than usually were possible.

My concept of population is broad. Sometimes populations extend for many kilometers. To distinguish a group of plants as a population, some disjunction is necessary. Here, a problem exists as to how much disjunction is requisite, and figures for number of populations vary accordingly. No rule is possible.

Sizes of populations, in both number of plants and area, are indicated in the listings under distribution in the accounts of species. Altitudinal range and range of pH are observed extremes. Letters to designate pattern of distribution are explained in the chapter on sampling. When data about pattern of distribution are in parentheses, they apply to populations not randomly chosen. Otherwise, selection of populations or areas containing them was random. Subspecies of some species, for example *Sedum lanceolatum*, are difficult to determine from dried specimens. These require experimental culture for positive identification. For this reason, assignment of some populations to subspecies may be faulty when experimental culture was impossible.

New England–Acadian Highlands—1

Species	Known pops. (n)	Alt. range (m.)	pH, range	Pattern of distrib., A–D
S. rosea	60	20–793	4.6–5.4	(B–.75, D–.25)

Ridge and Valley Province—5

Five species of *Sedum* occur naturally in the Ridge and Valley Province. Of these, *S. pulchellum* reaches its eastern limits in the province in Tennessee and Georgia. *Sedum glaucophyllum* is in Maryland, Virginia, and West Virginia; *S. nevii* is in Alabama; *S. telephioides* is in southern Pennsylvania, Maryland, Virginia, and West Virginia; and *S. ternatum* is widespread from Pennsylvania southward.

My principal trips in the Ridge and Valley Province have been in 1942, with Reid Moran, Robert Thorne, and my wife, to the valley of the Cahaba River in Alabama to study *Sedum nevii*; in 1946, with L. J. Kezer, to Pigeon Mountain, Georgia, to study *S. pulchellum*; in 1947, with D. G. Huttleston and L. J. Kezer, to the vicinity of Warm Springs, Virginia, to study *S. glaucophyllum*; in 1966, May 19–June 13, with my daughter Joanna, to the drainage of the New River, to explore the area, with special attention to *S. glaucophyllum*; and in 1969, with my son Tom, to the part of the province draining into and north of the Potomac River between the mouth of Licking Creek and the eastern base of Wills Mountain, to study *S. telephioides*. Only the trips in 1966 and 1969 were sampling surveys, with an attempt to get random samples from the various populations.

Choice of the drainage of the New River for particular attention was for two reasons. It includes the type locality for *Sedum glaucophyllum* and also is at the southwestern limits of the known distribution of the species. This drainage comprises .033 of the Ridge and Valley Province, which has a total area of 120,000 km.² Because *S. glaucophyllum* is strongly aggregated in its distribution, a general survey for it is inappropriate. It is too rare a species. Instead, a survey based on its distribution is necessary. For this, I listed all sixteen populations known from the drainage of the New River and selected six of these for survey. Three populations are a simple random sample. Another is the supposed type population, near the Cascades of Little Stony Creek, Mountain Lake, Virginia. The other two populations include the one farthest upstream and also southernmost, near Allisonia, and the one nearest Mountain Lake, on the western slope of Bald Knob. Finally, I surveyed one randomly selected, local area at Hicksville, Virginia.

The three randomly selected populations of *Sedum glaucophyllum* which I studied in the drainage of the New River have an area of about .0008 km.² Using this figure as the basis for an estimate, the probable total area of *S. glaucophyllum* in the drainage of the New River is .006 km.², or .000001 of the drainage. This tiny fraction is ample reason for the kind of survey which I made. Had I employed a general survey, my expectation of finding *S. glaucophyllum* in a randomly selected unit was one in a million. I might have explored units for several years before finding one with the species. The results of a survey of that kind would be interesting, but attention to other problems has seemed more important.

Within areas chosen for study, the fraction of land occupied by *Sedum glaucophyllum* likewise was tiny. For that reason, I first determined the total area occupied by the species by means of pacing. I then subdivided the occupied area into quadrats 10^2 m.² Quadrats of so big a size were desirable because finding and laying them out in the rough terrain, and also moving from one to the other, took a large amount of time. The task of locating many small quadrats would have taken too much time. Further, density of the perennial species of *Sedum* present was low. Even with large quadrats, numbers of included plants were small. The goal was to study 20% of the quadrats and 10% of the plants within them. The large quadrats have the additional advantage that, for tests of association, they exceed the size of the biggest plants of *Sedum* encountered, thus eliminating the condition of negative association which results when individual plants are so large that they occupy an entire quadrat, leaving no room for other plants.

Data for the areas surveyed in the New River drainage are in the following table and paragraphs. The areas are grouped in three strata: R = stratum with unrestricted randomization; GR = stratum comprising all populations of *Sedum glaucophyllum* known in the primary sampling unit and from which selection was random; and GO = an optimum stratum, with all included areas surveyed.

Sampling area	31–182	4–48	4–66	27–1	15–114	7–93	Sums	Estimates primary un
Status	R1	GR1	GR2	GR3	GO1	GO2	—	—
Area (km.^{2})	1.2	.32	.4	.4	.4	.44	3.16	3,951
Fraction of primary unit	3×10^{-4}	8×10^{-5}	10^{-4}	10^{-4}	10^{-4}	10^{-4}	8×10^{-4}	—
Lowest altitude (m.)	648	494	899	495	573	1,067	494	457
Highest altitude (m.)	792	594	1,036	576	670	1,330	1,330	1,402
pH, range	5.4	7.4–7.6	5.8–6.6	7–7.2	5.8–6.6	4.4–5.8	4.4–7.6	4.4–7.6
pH, tests (n)	1	3	7	3	13	5	32	32
Sedum—species (n)	0	1	1	2	2	1	2	3
S. ternatum—plts. (n)	0	0	0	>1,000e	>5,000e	0	>6,000e	7.5×10^6e
S. glaucophyllum								
Plts. (n)	0	68e	26	26	391e	76e	587e	1,027e
Area (m.^{2})	0	300	100	400	800	1,700	3,300	6,224e
Fraction of sampling area	0	9×10^{-4}	2×10^{-4}	10^{-3}	2×10^{-3}	4×10^{-3}	10^{-3}	10^{-6}

Hicksville area no. 31–182 (Bland, 1:62,500) is situated in the drainage of Wolf Creek at Hicksville, Bland Co., Va. The area includes part of the valley of Wolf Creek and the hollow southeast of Hicksville. Boundaries are N.—Wolf Creek; E.—the crest of a hill and southward a ridge east of Hicksville; S.—the crest of the high ridge south of Hicksville; and W.—the crest of the ridge west of Hicksville and then Wolf Creek. The rock is Kimberling shale. The valley parts of the area, about 60% of the total, are largely cleared and used for agricultural purposes. The steep slopes and crests of the ridges, and also the hill, are wooded. Common trees include species of *Pinus*, *Acer*, and *Quercus*. Small gravelly areas near the base of the northern slope of the ridge and minor outcrops of shale on the ridge southwest of Hicksville lack *Sedum*. Date of survey was June 11, 1966.

Klotz area no. 4–48 (Pearisburg 1:62,500) is part of the ridge northwest of Klotz, Giles Co., Va. Boundaries are N.—a tributary of Stony Creek; E.—a low place on the ridge above the northwestern end of the railroad yard; S.—the New River; and W.—a line crossing the ridge west of the highest point and passing down a gully to the northwest. The rock is limestone and forms high cliffs on the side toward the New River. Deciduous forest occurs on less steep parts of the slope to the west of the quarry. Trees include *Celtis occidentalis*, *Ulmus rubra*, and *Quercus muhlenbergii*. *Sedum glaucophyllum* occurs sparsely on shaded cliffs on the lower part of the slope. Date of survey was June 3, 1966.

Cascades area no. 4–66 (Pearisburg 1:62,500) is part of the lower, eastern slope of Butt Mountain, Giles Co., Va. Boundaries are: N.—the crest of the first ridge south of the junction of the roads south of Laurel Creek; E.—Little Stony Creek; S.—a tributary of Little Stony Creek with mouth at the Cascades; and W.—the 1,036 m. (= 3,400 ft.) contour line. The rock is sandstone, exposed in small outcrops. The slope is wooded, with *Tsuga canadensis*, *Pinus strobus*, species of *Quercus*, and *Cornus florida* common. *Sedum glaucophyllum* occurs on rocks on the western bank of the road between the little and big Cascades, just south of a point where the road rounds a rocky ridge and 450 meters north of the place where the road curves downward to the big Cascades. I interpret this occurrence as the type population. The only indication of locality on the label accompanying the type specimen is "near Mountain Lake." Another specimen, collected on the same day, June 17, 1939, by the collector of the type, has the additional information, "in woods along trail near Cascades of Little Stony Creek." The site is near Mountain Lake, only 3.8 km. to the west. It fits the available data. Further, an alternate proposition is lacking. Dates of my survey were June 4 and 8, 1966.

Pembroke area no. 27–1 (Pearisburg 1:62,500) is the ridge around which the New River flows on the southern side of Pembroke. Boundaries are: N. and E.—the New River; S.—gullies to the south of the ridge; and W.—Stone Quarry Brook. The rock is limestone and forms cliffs of varying heights. Where not too steep or cleared, the slopes are wooded. *Sedum glaucophyllum* occurs on low cliffs of limestone, south of the bridge, between the tunnel of the railroad and the river. *Sedum ternatum* occurs at various places along both the road and the New River below the bridge. Date of survey was June 6, 1966.

Allisonia area no. 15–114 (Macks Mountain and Max Meadows, 1:62,500) is the northern end of the ridge southwest of Reed Island Creek, across from Rich Hill School and south-southwest of Allisonia, Pulaski Co., Va. Boundaries are: N.—the road which crosses Reed Island Creek on a low bridge near Rich Hill School; E.—Reed Island Creek; S.—the crest of the ridge and of a spur oriented northwestward; and W.—the road from Allisonia to Kayoulah. The rock is limestone and sandstone. The highest cliffs in the vicinity of Allisonia are along Reed Island Creek within area no. 114. Pine and oak forest occurs above the cliffs, with *Pinus virginiana*, *Quercus prinus*, *Q. borealis*, and *Ostrya virginiana* the commonest trees. *Thuja occidentalis*, *Acer nigrum*, and *Corydalis aurea* are on the talus at the base of the cliffs. *Sedum glaucophyllum*, discovered there by Howard Shriver in 1877, occurs on ledges at two places on the cliffs and at four places on or near their crest. *Sedum ternatum* is widespread on the talus at the base of the cliffs and also along their crest. Dates of survey were May 24, 26, 28, 29, 30, and June 9, 1966.

Bald Knob area no. 7–93 (Pearisburg 1:62,500) is the western slope of Bald Knob, just south of Mountain Lake, Giles Co., Va. Boundaries are: N.—a line extending down the slope westward from the summit of the knob; E.—the crest of the ridge of which the knob is the highest point; S.—a gully 1 km. south of the knob; and W.—the main road to Mountain Lake. The rock is sandstone, occurring as large boulders and forming cliffs near the summit of the slope. The slope is wooded, with oaks predominant, especially *Quercus prinus* and *Q. borealis*. Trees are stunted on the windswept summit of the knob. *Sedum glaucophyllum* occurs on rocks in the woods along the road. Dates of survey were May 20 and June 2, 5, and 12, 1966.

My survey in the vicinity of Mountain Lake included, in addition to the areas described in detail, the vicinity of the Biological Station of the University of Virginia, the valley of the Pond Drain for a distance of 2 km. below Mountain Lake, and the northern slope of Bald Knob. I did not find *Sedum* in these places.

The survey in the drainage of the Potomac River in Maryland and adjacent Pennsylvania was primarily for the study of *Sedum telephioides*, but also provided some information on *S. ternatum*. The work in the period from Aug. 13–18 and Aug. 31–Sept. 8, 1969, was designed to answer questions about the northernmost populations and limits of *S. telephioides* and *S. glaucophyllum*. Results for *S. telephioides* are included in the account of that species. No plants of *S. glaucophyllum* were found at the localities which were surveyed.

The trips to the valley of the Cahaba River, Pigeon Mountain, and Warm Springs, Virginia, although resulting in the collection of plants for study, were too brief for detailed investigation of the populations and the sites in which they grew.

In summary, five species of *Sedum* are native in the Ridge and Valley Province. Data for these are summarized in the following table. I interpret all occurrences of *S. telephioides* and *S. ternatum* north of southern or central Pennsylvania respectively as introductions from farther south. Further, I have been unable to confirm reports of *S. pulchellum* from Virginia.

Species	Latitudinal range (N.)	Altitudinal range (m.)	pH, range
S. telephioides	37° 17′–39° 46′	237–1,482	4.4–6.8
S. ternatum	33° 1′–40° 45′	75–1,220	5.8–6.6
S. nevii	33° 1′	75	—
S. glaucophyllum	36° 55′–39° 32′	244–1,152	4.4–7.6
S. pulchellum	34° 43′–36° 5′	241–305	—

Blue Ridge—5

Species	Known pops. (n)	Alt. range (m.)	pH, range	Pattern of distrib., A–D
S. telephioides	44	600–1,790	3.8–4.8	A–1 (B–.25, D–.75)
S. ternatum	~40	265–1,690	5.2–6.8	A–1 (D–1)
S. nevii	1	260–307	5.6–6.8	A–1 (D–1)
S. glaucophyllum	>15	152–1,067	4.8–6.8	A–1 (D–1)
S. rosea	2	1,900	4.8	A–1 (B–1)

Piedmont Plateau—6

Six species of *Sedum* are native on the Piedmont Plateau. *Sedum pusillum* is endemic and confined to granitic flat rocks of the Carolinas and Georgia. *Sedum rosea* occurs in a small population along the Delaware River in Pennsylvania, where it survives as a relic from the ice ages. *Sedum nevii*, *S. glaucophyllum*, and *S. telephioides* are at the eastern limits of their distribution. Only *S. ternatum* is widespread, but even that is rare southward.

I lived on the Piedmont Plateau for eleven years and made many trips there. Trips for *Sedum* or other *Crassulaceae* were to Heggie's Rock, Columbia Co., Ga., for *S. pusillum* and *Diamorpha cymosa*, April 10, 1938, with Rogers McVaugh and others; the granite exposure at Liberty, S.C., for *D. cymosa*, Sept. 15, 1938, with Harold Trapido; the Smith River above Leaksville–Spray, N.C., for *S. glaucophyllum*, April 13, 1946, with my wife Edna; the Roanoke River above Goode's Ferry bridge, Va., for *S. glaucophyllum*, May 11, 1946, with Charles Uhl; the Nockamixon Rocks, Bucks Co., Pa., for *Sedum rosea*, May 25, 1946, with J. L. Edwards, and also May 15 and June 30, 1954; the vicinity of Pine Mountain, Warm Springs, Ga., to search for *S. glaucophyllum* or *S. nevii*, but resulting in finding *D. cymosa*, April 7–10, 1947, with

D. G. Huttleston and L. J. Kezer; and the vicinity of Flat Rock, Kershaw and Lancaster Counties, S.C., for survey of *Crassulaceae*, with special attention to *S. pusillum* and *D. cymosa*, March 20 to April 19, 1966. On trips in the Flat Rock region, I was accompanied on several occasions by James Sweet, assessor for Kershaw County. He helped me in many ways, especially in gaining access to privately owned land. In the work at Forty-acre Rock, I was assisted on four trips by Reuben Harris of Kershaw.

Short trips on the Piedmont Plateau yielded specimens for study, but did not involve detailed survey of populations. My intensive studies of populations were in the region of Flat Rock, S.C., and at Nockamixon Rocks, Pa.

I chose Flat Rock, S.C., as a special area for study because it is the type locality for both *Sedum pusillum* and *Diamorpha cymosa*. Furthermore, other populations of both species are in the vicinity. Although populations of these species comprise numerous plants, the sites where they occur are both disjunct and relatively infrequent. For that reason, sampling was restricted to reasonable sites. I recognized three strata:

1. Optimum, in which *Sedum pusillum* is known to occur.
2. Reasonable, comprising all known outcrops of granite and also stony areas.
3. Doubtfully reasonable, including the remainder of the region not included in either numbers 1 or 2.

For purposes of survey, I defined the Flat Rock region as a primary sampling unit. It is the part of the Piedmont Plateau between the Wateree (Catawba) River and the Lynches River, extending from the Fall Line upstream to Waxhaw Creek. I sampled completely the optimum stratum, visited two areas in the reasonable stratum, but neglected the third stratum. The fractions in the following table indicate what a tiny part of the whole primary sampling unit is occupied by *Crassulaceae*. Only a special survey, based on the distribution of the species, is practical. O = optimum stratum for *Sedum pusillum* and R = reasonable stratum for *S. pusillum* and *Diamorpha cymosa*.

Sampling area	20	9	17	1	21	23	Sums	Estimates for primary unit
Status	R1	R2	R4	R5	O1	O2	—	—
Area (km.^{2})	.47	.28	.44	.32	.01	.52	2.04	1,434
Fraction of primary unit	3×10^{-4}	2×10^{-4}	3×10^{-4}	2×10^{-4}	7×10^{-6}	36×10^{-5}	14×10^{-4}	—
Lowest alt. (m.)	120	104	122	91	134	113	91	43
Highest alt. (m.)	168	165	155	152	150	177	177	223
pH, range	4.6–6	—	—	4.6–5.4	4.6–5.8	4.6–5.6	4.6–6	4.6–6.6
pH, tests (n)	23	0	0	29	20	4	76	77
Crassulaceae species (n)	2	0	0	1	1	2	2	3
S. pusillum								
Plts. (n)	968,562e	0	0	0	130,935e	11,400e	1,110,897e	5,227,285e
Area (m.^{2})	17,750	0	0	0	375	113	18,238	93,675e
Fraction of sampling area	378×10^{-4}	0	0	0	375×10^{-4}	22×10^{-5}	89×10^{-4}	6×10^{-5}
Diamorpha cymosa								
Plts. (n)	1,644,334e	0	0	427,750e	0	755,577e	2,827,661e	22,512,459e
Area (m.^{2})	1,054	0	0	1,139	0	2,043	4,236	11,513e
Fraction of sampling area	2×10^{-4}	0	0	35×10^{-4}	0	39×10^{-4}	21×10^{-4}	77×10^{-7}

Forty-acre Rock area no. 20 (Monroe 1:125,000) is the first randomly selected site in the stratum reasonable for *Sedum pusillum* and *Diamorpha cymosa*. It is part of the drainage of a tributary of Flat Creek, including an extensive exposure of granite across which the stream flows, 3 km. southwest of Taxahaw, Lancaster Co., S.C. Boundaries are: N.—the northern edge of the pine woods north of the main exposure of granite; E.—the eastern edge of the wooded area, southwest of cultivated land, and a line projected both northwestward and southeastward through lobes of woods; S.—a prominent dike which runs from southwest to northeast across the valley of the brook; and W.—a line on the ridge to the west of the rocky area, but east of cultivated fields. In the north, the road of access is the western boundary. The surface of the granite, besides being eroded by the main stream and lesser rivulets, is further marked with many, often circular depressions. *Diamorpha cymosa* is abundant in these depressions and *S. pusillum* grows in thin mats of moss about the edges of the open areas. Trees grow where soil has developed, on flat surfaces as well as on slopes. *Juniperus virginiana* is the commonest species in rocky areas. The pine of the vicinity is *Pinus taeda*. Huntley (1939) has made a study of Forty-acre Rock. Dates of my survey were March 30 and April 2, 4, 8, 9, and 16, 1966.

Mt. Calvary Church area no. 9 (Camden, 1:62,500) is the southern slope of a ridge on the northern side of a tributary of Beaver Creek 1.3 km. west of Mt. Calvary Church, Kershaw Co., S.C. Boundaries are: N.—a road starting just north of Mt. Calvary Church and going to Stoneboro; E.—a gully .8 km. northwest of Mt. Calvary Church; S.—the tributary of Beaver Creek; and W.—a gully just east of a bench mark at 403 feet. A few large boulders of granite occur on the slope, but no outcrop. The site is wooded. Lack of suitable exposures of rock probably explains the absence of *Crassulaceae*. Date of survey was April 11, 1966.

Liberty Hill area no. 17 (map for Soil Survey of Kershaw Co., S.C., 1:62,500) is a slope on the northwestern side of Liberty Hill, S.C. Boundaries are: N.—a branch of a tributary of Singleton's Creek, north of Liberty Hill; E.—the highway from Chester to Camden; S.—a road going west from Liberty Hill toward the Wateree River; and W.—the tributary of Singleton's Creek. The rock, muscovite granite, occurs as small outcrops and scattered boulders on a slope wooded with *Juniperus virginiana*, *Pinus*, and *Quercus*. Sophie Richards owns the part of the area with the exposures of rock. Although the site appears reasonable for *Crassulaceae*, no plants are there. Date of survey was April 11, 1966.

Flat Rock area no. 1 (Camden, 1:62,500) is the site known as Flat Rock in Kershaw Co., S.C., a ridge of granite between Flat Rock and Little Flat Rock Creeks. Boundaries are: N.—a road leading to a pond on a branch of Little Flat Rock Creek on the east, and a hedgerow and minor gully leading to Flat Rock Creek on the west; E.—a branch of Little Flat Rock Creek; S.—a secondary road south of a quarry on the east, and a minor gully leading to Flat Rock Creek on the west; and W.—Flat Rock Creek. Michaux found *Sedum pusillum* at this site in 1795 and Nuttall collected *Diamorpha cymosa* there in 1816–1817. The Georgia Granite Company presently has an active quarry on the eastern side of the road at Flat Rock. *Sedum pusillum* no longer occurs there, but *Diamorpha* continues to thrive in depressions northwest of the quarry and in gravel and mats of moss about the edges of the rock and also on small exposures along the road northward. About half of the area around the exposure of rock is cultivated. The rest is wooded, with *Pinus taeda, Juniperus virginiana*, and species of *Quercus* dominant. Dates of survey were March 21, 22, 23, 25, 26, and 31, and April 5 and 15, 1966.

Midway area no. 21 (Monroe, 1:125,000) is a small outcrop of granite on both sides of South Carolina Highway 265 in Lancaster Co., 1.3 km. W.S.W. of Midway Crossroads and 10.6 km. N.E. of Kershaw. Boundaries are: N.—the edge of a field north of the rocky area; E.—a brook; S.—the upper edge of the rocky area; and W.—the western edge of the rocky area and a road north of the highway. According to Frank Small, who lives in the house on the south side of the highway, the Georgia Granite Company once did some quarrying here, but stopped work because of rust in the rock. About twenty-five years ago, the highway department scraped soil from the granite north of the road. *Sedum pusillum* now grows in gravel and sand about the edges of the scraped area, where the granite is exposed. Mr. Small thinks that the *Sedum* appeared since the scraping. In addition to the main population, a few plants occur on rocks to the west, beyond a brook, in an area totaling about 7 m.^{2} Shrubs and trees about the edges of the exposure of granite include *Ulmus alata, Rhus copallina, Quercus nigra*, and *Prunus umbellata*. Dates of survey were March 27-29 and April 1, 6, 13, and 14, 1966.

Kelly Rock area no. 23 (Camden, 1:62,500; fig. 180) is part of a low ridge northwest of Calvary Church and 5 km. N.W. of Flat Rock, Kershaw Co., S.C. The Rev. Mr. E. C. Leonard, of the Thornhill Baptist Church, brought this site to my attention. Boundaries are: N.—a tributary of Beaver Creek; E.—a gully northwest of the Mt. Calvary Church; S.—highway 13 and the road going from near Calvary Church to Stoneboro; and W.—minor gullies to the west of the nearly circular exposure of granite. Besides the almost flat surface of granite in the northern part of the area, the slope toward the branch of Beaver Creek is rocky in places. Species of *Quercus* and *Pinus taeda* predominate in woods around the rocky sites. *Sedum pusillum* is in gravel at the edge of sloping granite and in open places in the woods west and northwest of the main exposure of rock. *Diamorpha cymosa* is in depressions on the nearly flat surface. Dates of survey were March 23 and 27, and April 7 and 18, 1966.

In addition to the studies on the flat rocks, I visited a site for *Sedum ternatum* near where J. W. Hardin and W. H. Duncan found it in 1953, along the Catawba River N.N.W. of Lancaster, S.C. There, the plants grow close to *Arundinaria tecta*, in alluvial soil with pH 6.6, on the crest of the steep eastern bank of the river, at an altitude of about 150 m. The site is in woods of *Carya* and *Ulmus*, just below the bridge of the Seaboard Railroad. In a short inspection on April 12, 1966, I found seven plants of *S. ternatum*. My impression was that it is restricted to the bank of the river and probably had washed down from upstream.

My only detailed study elsewhere on the Piedmont Plateau was at the Nockamixon Rocks in the Delaware River primary sampling unit (area 2,389 km.^{2}). This locality

Fig. 180. Kelly Rock, 5 km. northwest of Flat Rock, Kershaw Co., S.C., Mar. 27, 1966, with extensive patches of *Diamorpha cymosa*, and Janice and Grady Coffey at site of *Isoetes melanopoda*.

had interest because of the occurrence there of *Sedum rosea*. The location is within 30 kilometers of the southern edge of the last great glacier which covered northeastern North America. The Delaware River makes a bend there and briefly flows eastward around a more resistant ridge of shale and sandstone. The *S. rosea* could have come down the river from farther north and survived at this site during the Wisconsin stage of glaciation. The extent of the population in 1954 is indicated by the following data.

Area (m.^{2})	850
Fraction of primary unit	.0000000003
Plants (*n*)	151
Lowest altitude (m.)	85
Highest altitude (m.)	148
pH range	5.2
pH tests (*n*)	2

The Nockamixon Rocks (Easton, 1:62,500) are on the south side of the Delaware River 1.5 km. east of Kintnersville, Bucks Co., Pa. The site for *Sedum rosea* is between two power lines, opposite a factory on the northern bank of the river, and about 30 meters above the highway at the base of the cliffs. The plants grow both on moist ledges of shale on the northern side of a small brook in a big ravine and on north-facing cliffs to the east of the ravine. *Arabis lyrata, Heuchera americana, Saxifraga penn-*

sylvanica, and *Achillea borealis* are associated with the *Sedum* on the ledges. *Hydrangea arborescens* is common in the ravine. Dates of survey were May 25, 1946, and May 15 and June 30, 1954.

In summary, the known distributional status on the Piedmont Plateau is indicated in the following table for the six native species of *Sedum*, and also for *Diamorpha cymosa* and four naturalized species. An exclamation point indicates a personal observation.

Species	Latitudinal range (N.)	Altitudinal range (m.)	pH, range
Native			
S. telephioides	35°54′–39°55′	~60–533	—
S. ternatum	33°57′–40°46′	~60–~460	6.6!
S. nevii	32°30′	~120	—
S. glaucophyllum	~36°–36°35′!	60!–165!	—
S. pusillum	33°25′–34°55′	140!–366	4.6!–6!
S. rosea	40°34′!	85!–148!	5.2!
Diamorpha cymosa	32°45′–36°8′	120!–366	4.4!–5.4!
Naturalized			
S. purpureum	38°45′–40°30′	—	—
S. alboroseum	35°30′–36°32′	—	—
S. sarmentosum	33°31′–40°50′	30–300	—
S. acre	40°15′– 33′	—	—

Three of the native species, *Sedum nevii, S. glaucophyllum*, and *S. rosea*, are confined to bluffs along major rivers. The other indigenous species are of wider distribution.

Appalachian Plateau—6

Species	Known pops. (n)	Alt. range (m.)	pH, range	Pattern of distrib., A–D
S. ternatum	>20	235–320	—	A–1(D–1)
S. nevii	2	55	—	A–1
S. glaucophyllum	1	—	—	A–1
S. pulchellum	4	261–466	7.4–8.2	A–1(D–1)
S. integrifolium ssp. leedyi	2	136–157	6.8–7.6	A–1(D–1)
S. rosea	3	143–610	4.8–5	A–1(B–1)

Ouachita Mountain System—2

Species	Known pops. (n)	Alt. range (m.)	pH, range	Pattern of distrib., A–D
S. ternatum	1	—	—	—
S. pulchellum	2	—	—	—

Central Rocky Mountains—5

Five species of *Sedum* are known from the Central Rocky Mountain region. *Sedum debile* reaches its northeasternmost limits there, and *S. stenopetalum* ssp. *stenopetalum* its southeasternmost limits.

The area of the Central Rocky Mountain region is 97,400 km.^{2} For sampling, subdivision was into 35 natural, primary units. Ranges of mountains, basins, or parts of these, make the primary units. The average size of three units, two chosen randomly and one chosen deliberately to ensure experience with *Sedum debile*, was 1,951 km.² Survey was between July 4 and September 4, 1966. Statistics follow.

Primary unit	Area (km.²)	Fraction of stratum	Randomly surveyed areas (km.²)	Fraction of primary unit randomly surveyed
Random stratum				
N.E. Big Horn Basin	3,122	.033	4.65	.0015
Pinyon Peak Highlands	1,175	.012	1.51	.0013
Optimum stratum				
Teton Mountains	1,555	.016	2.12	.0014

Because of the large size of the primary units, simple random samples of sampling areas would have been impractical. Not only would a prohibitive amount of time have been necessary to enumerate areas for whole primary units, but these would have entailed excessive travel. Instead, random selection involved a stage of secondary units and then, within the chosen secondary units, the ultimate units or local sampling areas. Data for secondary units follow.

Primary unit	N	n	Area of n (km.²)	Average area (km.²)	Area of n / Area of N
N.E. Big Horn Basin	35	3	238.52	79.51	.076
Pinyon Peak Highlands	16	2	154.54	77.27	.131
Teton Mountains	13	3	585.07	195.02	.376

Northeastern Big Horn Basin

The Big Horn Basin is an extensive area of Mesozoic and Cenozoic fill, with some Paleozoic sediments near the borders. Drainage is into the Big Horn River. Almost surrounded by mountains, the basin is semiarid. Conditions for growth of plants are rigorous because of deficient rainfall, high summer temperatures, and low temperatures in winter.

The northeastern Big Horn Basin, comprising the part of Big Horn County, Wyoming, north of the Greybull River, is the first randomly selected primary sampling unit in the Central Rocky Mountain region. Data for the secondary units surveyed there follow.

Unit no.	Name	Area (km.²)	Sampling areas (N)	(n)	Area of n (km.²)	Fraction of area covered by n
23	Foster Gulch	93.24	33	1	2.5	.0268
31	Tillett	50.38	59	2	1.09	.0216
13	Bear Creek	94.9	165	2	1.06	.0112
	Totals	238.52	257	5	4.65	.0195

Data for the local areas are in the following table.

Sampling areas	23-9	31-23	31-31	13-92	13-91	Sums	Estimates for primary unit
Area (km.^{2})	2.5	.55	.54	.37	.69	4.65	3,122
Fraction of secondary unit	.027	.01	.01	.0038	.0072	—	—
Lowest alt. (m.)	1,311	1,140	1,128	1,414	1,432	1,140	1,115
Highest alt. (m.)	1,402	1,170	1,170	1,520	1,506	1,520	1,919
pH, range	7.4–7.6	7.2	7	7.2	8–8.2	7–8.2	7–8.2
pH, tests (n)	2	1	1	1	2	7	7
Sedum—species (n)	0	0	0	0	0	0	0

Foster Gulch area no. 23–9 (Cody, 1:250,000) is a lateral ridge south of a tributary of Foster Gulch, on the western slope of the main ridge east of the gulch and 13 km. south of Lovell. Boundaries are: N.E. and N.W.—gullies tributary to a larger gully which enters Foster Gulch 12 km. S.S.W. of Lovell; S.E.—crest of main ridge; and S.W.—crest of ridge between two gullies. Rock, Paleocene sandstone, occurs as small exposures and as gravel on the crests of the ridges. Soil varies from loamy fine sand to clay. Drainage is excessive. The sparse vegetation is of the desert and basin type, with *Opuntia polyacantha* and *Grayia spinosa* the commonest species. The largest plants in the area are two bushes of *Juniperus osteosperma*, part of the northeasternmost population known to me. Date of survey was August 21, 1966.

Tillett area no. 31–23 (Kane NW, 1:24,000) is a low ridge with steep eastern slope, west of the headquarters of the Wyoming Game and Fish Commission, in T. 57 N., R. 94 W., sections 30 and 31. The site is about 19 km. east-northeast of Lovell. Boundaries are: N.—an obscure roadway; E.—the bottom of a gully; S.—the road from Lovell to the Game and Fish Headquarters; and W.—the western boundaries of sections 30 and 31. Cretaceous shale forms small cliffs near the eastern border. The soil is sand and gravel. Drainage is excessive. Vegetation is of the desert and basin type, with *Opuntia polyacantha*, *Grayia spinosa*, and *Artemisia arbuscula* ssp. *nova*, the commonest species. Date of survey was August 22, 1966.

Tillett area no. 31–31 (Kane NW, 1:24,000) is a hill northeast of the headquarters of the Wyoming Game and Fish Commission. The location is about 5 km. northeast of area no. 31–23, in the northern half of section 29, T. 57 N., R. 94 W. Boundaries are: N. and E.—gullies; and S. and W.—roads. Lower Cretaceous sandstone is extensively exposed. The soil is gravel and sand and is excessively drained. Desert and basin vegetation occupies about 20% of the area. Commonest plants are *Opuntia polyacantha* and species of *Eriogonum* and *Chrysothamnus*. The northeastern part of the area is an alkaline waste, almost devoid of plants. Date of survey was August 22, 1966.

Bear Creek area no. 13–92 (Bear Creek Ranch, 1:24,000) is the western slope of the highest ridge between Bear Creek and its big western tributary, in T. 54 N., R. 92 W., sections 17 and 18, about 25 km. northeast of Greybull. Boundaries are: N.—the saddle between the highest ridge and the next ridge to the west; E.—the crest of the highest ridge; S.—a gully west of a southernmost high point; and W.—the bottom of the gully west of the highest ridge. Lower Cretaceous shale is exposed along gullies. The soil, either clay or gravel, is excessively drained. Desert and basin vegetation covers about half of the land. *Artemisia arbuscula* and *Opuntia polyacantha* are the commonest species. Date of survey was August 24, 1966.

Bear Creek area no. 13–91 (Bear Creek Ranch, 1:24,000) is a portion of the ridge between two tributaries of Bear Creek, 3 km. south-southwest of Black Butte, and adjacent to area 92 on the east, in T. 54 N., R. 92 W., section 18. Boundaries are: N.—a gully leading northwestward in the northeastern corner of section 18; E.—the bottom of the big tributary of Bear Creek west of the highest ridge; S.—gullies oriented southeastward and southwestward; and W.—the large tributary of Bear Creek, east of the road. Lower Cretaceous shale is exposed on the banks of gullies. The soil is gravel or clay and is excessively drained. Desert and basin vegetation covers about half of the area. The three commonest plants are *Artemisia arbuscula* ssp. *nova*, *Opuntia polyacantha*, and a species of *Stipa*. Date of survey was August 24, 1966.

In summary, present conditions in the northeastern Big Horn Basin appear unsuitable for *Sedum*. The genus is absent there.

Pinyon Peak Highlands

The Pinyon Peak Highlands, northeast of Jackson Hole and south of the Yellowstone Plateau, are erosion remnants of Tertiary sediments. They comprise ridges of sandstone and conglomerate, largely wooded, but with extensive meadows along the larger streams. Fire has produced open areas or parks on many slopes. Although the Pinyon Peak Highlands are close to the Teton Range, where four species of *Sedum* occur, I found only *S. lanceolatum* there. The lack of species such as *S. debile* may be due to the chemical composition of the rocks and soil, sparsity of large exposures of rocks for long periods of time, competition, and fire.

Data for the secondary units surveyed there follow.

Unit no.	Name	Area (km.^{2})	Sampling areas (N)	(n)	Area of n (km.^{2})	Fraction of area comprised by n
4	Small lakes	57.75	76	1	.75	.0129
15	Whetstone Mountain	96.79	96	2	.76	.0078
Totals		154.54	172	3	1.51	.0097

Next follow data for the three local sampling areas.

Sampling area	4–63	15–45	15–63	Sums	Estimates for primary unit
Area (km.^{2})	.75	.44	.32	1.51	1,175
Fraction of secondary unit	.012	.0045	.0033	—	—
Lowest alt. (m.)	2,591	2,332	2,554	2,332	2,103
Highest alt. (m.)	2,865	2,572	2,792	2,865	3,006
pH, range	6.6–7.2	5.6	5.6	5.6–7.2	5.6–7.2
pH, tests (n)	2	1	1	4	4
Sedum—species (n)	1	0	0	1	1
S. lanceolatum (n plts.)	576e	0	0	576e	332,705e

North Buffalo Fork area no. 4–63 (Mount. Leidy, 1:125,000) is the western slope of a peak with elevation 2,865 m., west of the North Buffalo Fork and about midway between Soda and Joy Creeks. It is in the Teton Wilderness Area, in T. 46 N., Ranges

111 and 112 W. Boundaries are: N. and S.—ridges extending northwestward and southwestward respectively from the main peak; E.—the summit of the peak; and W.—the 2,591 m. (= 8,500 ft.) contour line. Limestone is exposed at the summit. The soil near the summit is gravelly and well drained. The area is about 75% forest, with *Pinus flexilis* and *Picea engelmannii* common, about 24% meadow, and 1% rock and gravel. *Sedum lanceolatum* is on rocks and in gravel at the summit. Date of survey was August 8, 1966.

Upper Pilgrim Peak area no. 15–45 (Huckleberry Mountain and Mount Hancock, 1:62,500) comprises the western and northern slopes of a grassy knoll in the upper drainage of Pilgrim Creek, 3 km. west of Bobcat Ridge, in the Teton Wilderness Area of the Teton National Forest. Boundaries are: N.W.—Pilgrim Creek; E. and N.—a tributary flowing northwestward; and S.—a line from the point where the tributary turns northward, extending across the summit of the knoll and then down a minor depression on the western side of the ridge to Pilgrim Creek. The only rock exposed is some conglomerate on a steep bank facing northward. The soil is gravelly loam developed on a glacial deposit containing some sandstone boulders. About 75% of the area is forest, with *Picea engelmannii*, *Pinus flexilis*, *Abies lasiocarpa*, and *Pinus contorta*. The remaining 25% of the area is open park, the result of an old burn, with *Carex*, *Lupinus*, and grasses common plants. Date of survey was August 11, 1966.

Upper Pilgrim Creek area no. 15–63 (Mount Hancock 1:62,500) is the northwestern slope of the high ridge 2.5 km. west of Bobcat Ridge, in the Teton Wilderness Area. Boundaries are: N.—the stream which flows around the grassy knoll in area 45; E.—the crest of the ridge; and S. and W.—the crest of a lateral ridge which leads toward the grassy knoll. Rocks are not exposed. The soil is rocky or gravelly loam of glacial origin. About 95% of the area is forest, with *Pinus albicaulis*, *Abies lasiocarpa*, and *Picea engelmannii*, and about 5% near the crest of the main ridge is park. The part which is park is more windswept and drier than the open portion of area no. 45. Date of survey was August 11, 1966.

Teton Mountains

The core of the Teton Range (fig. 181) is a block of crystalline rocks, uplifted and tilted westward, raising the covering layers of sedimentary rocks and causing them to slope to the west. The sedimentary rocks persist on some summits and are the prevailing rocks on the western side of the mountains. Following uplift and erosion, the whole range was extensively glaciated. As a result, sites for plants are comparatively young. All populations of *Sedum* probably have colonized the Teton Mountains since the period of maximum glaciation in the Pleistocene epoch. For a detailed description of the geology of the Teton Mountains, see Love and Reed (1968).

The three randomly selected secondary units, surveyed in 1966, included both crystalline and sedimentary rocks and a wide range of altitudes. Four species of *Sedum* occur in these units. The experiences in the sampling areas, together with observations made along the routes to these, are the primary basis for my information about the status of each species. In addition, various friends contributed specimens and reports which enhanced understanding.

Fig. 181. Jackson Lake and Teton Mountains, Grand Teton and Cascade Canyon in left center, July 12, 1966.

Data for the secondary units surveyed in the Teton Range follow.

Unit no.	Name	Area (km.^{2})	Sampling areas (N)	(n)	Area of n (km.^{2})	Fraction of area comprised by n
9	Mt. Bannon	183.2	148	1	.5	.0027
8	Baldy Knoll	294.4	192	2	.82	.0028
3	Grand Teton	107.4	161	1	.8	.0074
Totals		585.0	501	4	2.12	.0036

The following table has data for the four randomly selected sampling areas.

Sampling areas	9–127	8–104	8–152	3–21	Sums	Estimates for primary unit
Area (km.^{2})	.5	.2	.62	.8	2.12	1,555
Fraction of secondary unit	.0027	.00067	.0021	.0074	.0036	.0014
Lowest alt. (m.)	3,050	2,194	2,347	2,066	2,066	2,034
Highest alt. (m.)	3,322	2,404	2,697	2,438	3,322	4,176
pH, range	6.6–7	6–6.6	5.8	5.4–6.6	5.4–7	5.4–7
pH, tests (n)	2	2	1	42	47	47
Sedum—species (n)	2	0	0	2	3	4

Sampling areas	9-127	8-104	8-152	3-21	Sums	Estimates for primary unit
S. lanceolatum—plts. (n)	111	0	0	19,425e	19,536e	7,039,967e
S. debile—plts. (n)	0	0	0	43,925e	43,925e	15,674,742e
S. rhodanthum—plts. (n)	0	0	0	0	0	217e
S. integrifolium—plts. (n)	34	0	0	0	34	33,114e

Descriptions of sampling areas

Alaska Basin area no. 9–127 (Grand Teton National Park 1:62,500) is the southern slope of the southwestern spur of The Wall. Boundaries are: N.—the crest of the ridge; E.—a draw southwest of The Wall; and S. and W.—the 3,050 meter contour line. The rock is primarily a bluish limestone, with abundant marine fossils, overlying a foundation of gneiss exposed at the base of the slope. Evidences of glaciation are abundant. Slides of gravel make up about 40% of the area. Alpine vegetation covers perhaps 80% of the area. The remainder is either bare rock or unvegetated gravel. *Sedum lanceolatum* occurs primarily on flat surfaces of gneiss at the base of the declivity, but a few plants are at a less well-drained site on a ridge toward the eastern end of the slope. *Sedum integrifolium* occurs only in turf near a stream on the flat lower surface. Date of survey was July 27, 1966.

Upper Nordwall Canyon area no. 8–104 (Driggs, 1:62,500) is the southwestern section of the canyon. Boundaries are: N.—a minor gully arising near the center of the highest point on the ridge to the west; E.—the bottom of the canyon; and S. and W.—the crest of the bordering ridge. Little rock is exposed. Some sandstone boulders and gravel occur near the summit of the ridge to the west. Eastward, the rock is limestone. The area is about 95% forested, with *Abies lasiocarpa* and *Picea engelmannii* the commonest trees. Shrubs occur mostly along the top of the ridge, where common species are *Ceanothus velutinus* and dwarf *Populus tremuloides*. The absence of *Sedum* may be due to lack of large outcroppings of rock and too great competition. *Sedum lanceolatum* occurs only a few meters beyond the boundary, on the southern crest of the ridge, on limestone. Date of survey was August 1, 1966.

Headwaters of North Fork of Game Creek area no. 8–152 (Driggs, 1:62,500; Grand Teton, 1:125,000) is the slope of a ridge at the head and between two branches of the North Fork of Game Creek, southeast of Baldy Knoll. Boundaries are: N.—the crest of the ridge east of Baldy Knoll, which is the divide between Fox and Game Creeks; E. and S.—the upper North Fork of Game Creek; and W. and N.W.—the gully of a tributary. Limestone occurs as small outcroppings in the upper part of the area. Vegetation is about 90% grassland and forbs, and 5% forest, with *Pinus flexilis*, *Abies lasiocarpa*, and *Populus tremuloides*. The remaining 5% of the area is rocky. Original destruction of the forest on the ridge probably was by fire. The Forest Service contour-trenched and seeded the area with grasses in 1956 and 1957. Absence of *Sedum* probably results from sparsity of suitable habitat, fire, and grazing. *Sedum lanceolatum* occurs sparingly on the Fox Creek Ridge, only 16 meters westward from the sampling area. Date of survey was August 2, 1966.

Eastern Teewinot Mountain area no. 3–21 (Grand Teton National Park 1:62,500; fig. 182) is a part of the eastern slope of Teewinot Mountain, below 2,440 m., in Grand

Fig. 182. Teewinot Mountain, showing the site of sampling area no. 3–21, July 15, 1966.

Teton National Park. Boundaries are: N.—a gully .7 km. south of Cascade Creek; E.—the western shore of Jenny Lake; S.—a gully leading toward Moose Pond; and W.—the 2,440 m. contour line. Thirty-five exposures of granite and gneiss, in the middle and upper portions of the tract, comprise about a quarter of the total area. The lower slope and also some sites upward, about 40% of the area, are forest, with *Abies lasiocarpa*, *Pseudotsuga menziesii*, *Pinus contorta*, and *Picea engelmannii*. The remaining 35% of the area is montane meadow. Two species of *Sedum* are common in the rocky areas, sparse in the meadows, and lacking in the forest. This area was my principal place for study of both *S. debile* and *S. lanceolatum* in 1966. Time of survey was on 14 days between July 15 and August 31, 1966.

Northeastern base of Teewinot Mountain area no. 3–1 (Grand Teton National Park 1:62,500) is the lower, northeastern slope of the mountain, in Grand Teton National Park. Selection was deliberate because of the known occurrence of *Sedum debile* there (Louis Williams—1934). Boundaries are: N.—Cascade Creek; E.—Jenny Lake; S.—the gully .7 km. south of Cascade Creek; and W.—the highest part of the ridge on the northeastern slope. Altitudinal range is 2,066–2,774 meters. The rock is gneiss, exposed in many places. Vegetation is sparse in the rocky areas, but, toward the base of the slope, is forest of *Pinus contorta*, *Abies lasiocarpa*, and *Pseudotsuga menziesii*. *Sedum debile*, an estimated 142 plants, and *S. lanceolatum*, 24 plants, are on rocks below the upper trail for horses. Because of both insufficient time and problems in getting around on steep terrain, coverage was only about 10% and in the lower part. Dates of survey were July 7 and 12, and August 29, 1966.

South Fork of Cascade Canyon area no. 3–19 (Grand Teton National Park 1:62,500) is a part of the southeastern slope of the ridge northeast of Table Mountain and northwest of the South Fork of Cascade Creek, in Grand Teton National Park. Selection was deliberate because of the occurrence of *Sedum debile* there (Clausen—1966). Boundaries are: N.E.—a gully; S.E.—the South Fork of Cascade Creek; S.W.—a gully; and N.W.—the crest of the high ridge northeast of Table Mountain. Altitudinal range is 2,515–3,025 meters. The rock, granite, is abundantly exposed. Vegetation is forest on the lower slopes, with *Abies lasiocarpa* and *Pinus albicaulis*; meadow, with shrubby willows, in flat places along the stream; and a sparse growth of herbs and small shrubs on the rocks. *Sedum debile* (545e plants) is in a living mat of moss on a nearly flat exposure of granite; *S. rhodanthum* (15 plants) is in moss on ledges; and *S. lanceolatum* (>29e plants) is with the *S. debile*, as well as in gravel on benches and elsewhere in crevices in the rock. Dates of survey were July 7 and 9, and August 28, 1966.

Treasure Mountain Camp area no. 9–141 (Grand Teton 1:125,000) is the part of the southern slope of Teton Canyon, in the Targhee National Forest, in which the Treasure Mountain Camp of the Boy Scouts of America is located. Selection was deliberate because of the known occurrence of *Sedum debile* there (Anderson—1956). Boundaries are: N.—Teton Creek; E.—a gully oriented northeastward; S.—the summit of the first high knob south of the camp; and W.—the crest of the ridge south of the camp. Altitudinal range is 2,042–2,835 meters. Limestone is exposed as high cliffs, comprising about a quarter of the area, and as rock slides, comprising perhaps another quarter. The lower slope, along the creek, is forest, with *Pseudotsuga menziesii*, *Pinus contorta*, and *Abies lasiocarpa* dominant. Where these trees extend upward, they are more scattered. *Populus tremuloides* is frequent on midslope. *Sedum debile* (1,344e plants) and *S. lanceolatum* (14e plants) are on slides and cliffs. Dates of survey were July 11 and 26, and August 30, 1966.

Jackson Hole

Jackson Hole was not part of the planned sample in the Central Rocky Mountain region. Yet, because of its proximity to the Teton Mountains, I made many observations there. Also, I lived there for more than five weeks. The land is largely glacial outwash from the adjacent mountains.

Sedum lanceolatum is a common and widely distributed species on the sagebrush flats. *Sedum stenopetalum* occurs at four or more places near Moose. The northernmost of these is on the outwash plain along the trail to Taggart and Bradley Lakes. Where the two species occur together, the *S. stenopetalum* is in situations more protected from the north. From the four sites for *S. stenopetalum* known to me, I chose one randomly for study. This is at the edge of a grove of *Populus tremuloides* along the Wilson Road. Seven clusters of plants are there. Of these clusters, I chose two randomly. Both selected clusters are in open gravelly areas. In general, *S. lanceolatum* is more widely distributed, but it is the less frequent species in the two clusters surveyed, as indicated in the following table.

Cluster no.	4	1
Area (m.^{2})	91	88
Altitude (m.)	1,970	1,970
pH, range	6.4	6.6
pH, tests (*n*)	1	1
Sedum lanceolatum	10	2
S. stenopetalum	18	43
S. lanceolatum × stenopetalum	4	0

Although I made no counts of *Sedum* in Jackson Hole, other than the two clusters described above, my guess is that the number of plants of *S. lanceolatum* must be in the millions. On the basis of an expansion of my data for the two clusters, the estimate for *S. stenopetalum* is 854 plants, but more likely the true number is several thousand. The population along the trail to Bradley Lake is the largest of the four. I neither made counts there nor determined the total extent of plants at that site.

Beartooth Mountains

The Beartooth Mountains, astride the boundary between Wyoming and Montana, north of the Clark's Fork of the Yellowstone River, have an altitudinal range from about 1,341 to 3,901 meters. Principal rocks are granites and gneiss. Three species of *Sedum* occur: *S. lanceolatum*, *S. rhodanthum*, and *S. integrifolium*.

My first visit to the Beartooth Range was on August 31, 1963, when I collected *Sedum lanceolatum* and *S. integrifolium* at 3,250 m., north of Beartooth Summit. On July 5, 1966, I stopped at a meadow, elevation 2,850 m., southeast of Beartooth Lake, to collect fruits of *S. rhodanthum*, but was unsuccessful. Apparently, an early freeze in the late summer of 1965 had frozen flowers still in anthesis. Numerous follicles were available at the same site on September 3, 1966. These were already dehisced, but with seeds still in position. North of Frozen Lake, many flowers of *S. rhodanthum* were in anthesis on September 3. Likewise, *S. lanceolatum* was in flower near there and at Beartooth Summit. At lower elevations, as at 1,640 m., on the outwash flats along Rock Creek 5 km. north of Red Lodge, Mont., plants had mature fruits, with seeds ready to drop.

Summary for Central Rocky Mountain region

Information about *Sedum* in the Central Rocky Mountain region is summarized in the following table. Figures in parentheses indicate personal judgments when these differ greatly from the results of the sampling survey.

Species	Plants (estimates of N)	Altitudinal range (m.)	pH, range
S. lanceolatum	14,140,840	1,700–4,115	5.4–7.2
S. debile	15,674,742	2,100–3,170	5.4–7.2
S. stenopetalum	840 (3,000)	1,970–2,020	6.4–6.6
S. rhodanthum	240 (2,500)	2,285–3,660	5
S. integrifolium	33,117	2,975–3,250	6.6

Before 1966, my impression was that *Sedum debile* is an uncommon species, at the periphery of its distributional range in the Teton Mountains. Instead, the survey there revealed that it occurs in large populations and over a wide elevational range. In addition, detailed information about the variation, relationships, distribution, and pollination of *S. debile* resulted from the work in 1966. Other consequences of the studies in the Central Rocky Mountains were discovery on Teewinot Mountain of plants of *S. lanceolatum* with extra carpels and no functional stamens in the flowers, of hybrids of *S. lanceolatum* and *S. stenopetalum* in Jackson Hole, and that large bees cross-pollinate *S. rhodanthum* in the Beartooth Mountains.

Southern Rocky Mountains—4

Species	Known pops. (n)	Alt. range (m.)	pH, range	Pattern of distrib., A–D
S. cockerellii	31	2,330– 2,715	6.2–7.6	A–1(A–.3, B–.3, D–.3)
S. lanceolatum				
ssp. lanceolatum	103	2,049– >3,000	5.6–8.4	A–.5, D–.5(D–1)
ssp. subalpinum	>6	<3,375– 4,048	5.6–6.8	A–.75, B–.25 (B–.67, D–.33)
S. rhodanthum	98	2,785– 3,645	4.8–6.4	A–1(B–.25, D–.75)
S. integrifolium				
ssp. procerum	34	2,532– 4,050	5.2–6.6	A–.75, B–.25(D–1)
ssp. integrifolium	>24	2,897– 4,000	4.8–6.2	A–1(B–.33, D–.67)

Colorado Plateau—5

Species	Known pops. (n)	Alt. range (m.)	pH, range	Pattern of distrib., A–D
S. cockerellii	1	1,680	6.8	(B–1)
S. lanceolatum				
ssp. lanceolatum	23	1,770–3,300	—	—
S. debile	4	1,576–3,508	—	—
S. rhodanthum	6	2,700–3,508	—	(D–1)
S. integrifolium				
ssp. procerum	1	2,800–3,400	—	—

Wyoming Basin—1

Species	Known pops. (n)	Alt. range (m.)	pH, range	Pattern of distrib., A–D
S. lanceolatum				
ssp. lanceolatum	2	1,830	—	—

Idaho Batholith—5 (6)

Species	Known pops. (n)	Alt. range (m.)	pH, range	Pattern of distrib., A–D
S. lanceolatum				
ssp. lanceolatum	12	2,135–3,050	—	—
ssp. subalpinum	1	1,940–1,955	4.6–4.8	(B–1)
S. debile	6	2,440–3,172	—	—
S. stenopetalum				
ssp. stenopetalum	13	915–2,073	—	(B–.5, D–.5)
S. leibergii	1	—	—	—
S. borschii	6	2,042–2,120	4.8–5	(B–1)
S. integrifolium	?	—	—	—

Wyomide Ranges—2 (3)

Species	Known pops. (n)	Alt. range (m.)	pH, range	Pattern of distrib., A–D
S. lanceolatum				
ssp. lanceolatum	9	1,830–3,330	—	(D–1)
S. debile	23	1,616–3,203	—	(D–1)
S. integrifolium	?	—	—	—

Great Basin—3

Species	Known pops. (n)	Alt. range (m.)	pH, range	Pattern of distrib., A–D
S. lanceolatum				
ssp. lanceolatum	18	1,586–3,292	—	—
S. debile	26	1,800–2,958	—	—
S. integrifolium				
ssp. integrifolium	12	2,745–3,050	—	—

Mountains and Basins of Southwestern Montana and Idaho—4

Species	Known pops. (n)	Alt. range (m.)	pH, range	Pattern of distrib., A–D
S. lanceolatum				
ssp. lanceolatum	11	610–1,982	—	—
ssp. subalpinum	1	3,355	—	—
S. stenopetalum				
ssp. stenopetalum	>17	1,270–3,050	—	(B–.5, D–.5)
S. borschii	2	1,270–1,570	4.8–5.6	(D–1)
S. integrifolium				
ssp. integrifolium	5	2,745–2,897	—	—

Sierra Nevada—7

Seven species of *Sedum* occur in the Sierra Nevada. If *S. pinetorum* is synonymous with *S. niveum* and really occurs in the Sierra Nevada, then the number of species is eight. In addition, two species of *Parvisedum*, two species of *Dudleya*, and three species of *Crassula* occur on the western and southern slopes, mostly at low elevations. *Sedum albomarginatum* is endemic. *Sedum lanceolatum*, *S. stenopetalum* ssp. *monanthum*, *S. obtusatum*, and *S. integrifolium* are at their southwestern limits. And *S. spathulifolium* and *S. radiatum* are at their eastern limits on the western slope.

The area of the Sierra Nevada is about 75,000 km.² For sampling, subdivision of the total area was into 24 large, primary sampling units. Boundaries for these included the main crest of the range and the principal rivers. A list of the species of *Sedum* in each of the primary units, compiled from personal experience (July 3–14, 1940, and 6 days in 1963) and records in herbaria and the literature, made possible definition of an optimum stratum, after the manner of Dalenius (1953), including units which together contain all species of *Sedum* surely known from the Sierra Nevada. Only two units, South Yosemite and Feather River, were necessary to meet these requirements. Two-thirds of the time of the survey in 1964 was allotted to these two units because of the large number of species in them. Within each unit, half of the time was devoted to an optimum substratum composed of listing units known or expected to contain species of *Sedum*. Survey of this substratum was complete. The other half of the time was spent in visiting listing units in a random substratum in which selection of units was entirely random. The random substratum served as a check on the optimum substratum.

Attention in the remaining third of the time in the Sierra Nevada in 1964 was to two primary units chosen randomly from the list of 24 from which the two optimum units had been selected. Samples of listing units within these primary units were simple random.

Information about the survey with respect to the distribution of time and the discovery of *Sedum* is indicated in the following summary. The period of study in the Sierra Nevada in 1964 was from June 17 to September 10, and in 1970 from July 20 to 30, when the effort was devoted to search for *S. pinetorum*.

	Random stratum	Optimum stratum	
		Random substratum	Optimum substratum
Time (weeks)	4	4	4
Listing units surveyed (*n*)	10	11	10
Species of *Sedum* found (*n*)	4	1	6

Statistics for the surveyed primary sampling units are as follows:

Primary unit	Area (km.²)	Fraction of stratum	Randomly selected units surveyed (km.²)	Fraction of primary unit randomly surveyed
Random stratum				
North Yosemite	4,021	.06	3.39	.0008
Mammoth Lakes	1,315	.02	2.42	.0018
Optimum stratum				
South Yosemite	4,501	.70	8.08	.0018
Feather River	1,967	.30	6.15	.0031

Climatic differences in the Sierra Nevada are great. Data for the years 1960–1964, available from the reports of the U.S. Weather Bureau (1960–1964), indicate the conditions at several altitudes on both sides of the range.

Station	Alt. (m.)	Annual precip. (cm.)	Precip. in July (cm.)	Max. summer temp. °C	Min. winter temp. °C	Av. temp. °C
Eastern slope						
Bishop W B Airport	1,252	5.6– 9.6	.0– .5	39–42	−17– −12	12–13
Bishop Union Carbide	2,862	21.8– 53.6	.0–1.6	28–29	−20– −18	6– 7
Western slope						
Cathay Bull Run Ranch	434	33.4– 64.9	.0– .4	39–41	−8– − 7	14–15
Yosemite Park, S. entrance	1,561	63.4–166.1	.0– .7	33–37	−17– −11	9–10
Sonora	558	52.3– 93.2	.0– .5	40–45	−8– − 4	14–15
Hetch Hetchy	1,179	57.9–113.5	.0–1.3	34–39	−12– − 8	11–12
Oroville	52	54.8– 82.8	.0– .2	40–46	−5– − 4	16–17
Sierra City	1,275	109.7–176.5	T– .1	38–39	−15– − 8	11–12

North Yosemite Primary Sampling Unit

The first randomly selected primary sampling unit in the Sierra Nevada was no. 14, the North Yosemite unit. It is part of the western slope of the Sierra Nevada between the Tuolumne and the Middle Fork of the Stanislaus Rivers. To select a simple random sample of listing units, each approximately a kilometer square, within this area, I delimited 34 natural regions, measured the size of the largest of these, and then assigned consecutive runs of numbers to all regions on the basis of the maximum number of kilometric units—200—to be expected in the largest. The result was an estimated 6,800 units in the entire primary sampling unit. A simple random sample then was selected from the whole series. In selection, four sets of numbers were mismatched because the regions in which they applied were too small to have such high numbers. McCarthy (1957:140–142) has described this method of drawing a random sample.

Data for the chosen listing units, comprising .096% of the total area of the primary unit, are summarized in the following table and subsequent paragraphs. The survey was from August 10–26, 1964.

Sampling unit (no.) (name)	5,639 Hull Creek	4,726 Reed Creek	2,208 Rancheria Mtn.	5,657 Camp Lee Price	896 Woods Creek	Sums	Estimates for primary unit
Area (km.²)	.92	.4	.8	.68	.59	3.39	4,021
Lowest alt. (m.)	1,707	1,074	1,158	1,597	312	312	152
Highest alt. (m.)	1,935	1,403	1,951	1,731	411	1,951	3,978
pH, range	5.2	5.2	4.8–5.4	4.8–5.4	6.4–6.6	4.8–6.6	4.8–6.6
pH, tests (n)	2	1	8	2	7	20	20
Crassulaceae—species (n)	0	0	3	0	2	4	6
S. lanceolatum—plts. (n)	0	0	0	0	0	0	>50
S. stenopetalum							
ssp. monanthum—plts. (n)	0	0	41	0	0	41	38,883
S. obtusatum							
ssp. obtusatum—plts. (n)	0	0	18	0	0	18	16,888
S. integrifolium—plts. (n)	0	0	0	0	0	0	>2
Parvisedum congdonii—plts. (n)	0	0	0	0	12,910e	12,910e	9,034,664
Dudleya cymosa							
ssp. cymosa—plts. (n)	0	0	14	0	40	54	41,255

Hull Creek unit no. 5,639 (Long Barn 15′) is the eastern slope of the ridge between Hull and Wrights Creeks, in sections 18 and 19, T. 3 N., R. 18 E., and sections 13 and 24, T. 3 N., R. 17 E. The southern boundary is the fourth gully north of road 3 NO 1 of the Miwok District, Stanislaus National Forest. The northern boundary is the seventh gully north of the road. The land along Hull Creek and on the slope is forested with *Pinus lambertiana*, *P. jeffreyi*, *Libocedrus decurrens*, and *Abies concolor*. The summit of the ridge is either open and gravelly or covered with chaparral, especially *Arctostaphylos patula*, *Ceanothus cordulatus*, and *Castanopsis pumila*. The rocks are andesite, occurring as outcroppings on the summit of the ridge. The top of the ridge appears too dry for *Sedum*, and suitable habitats are lacking on the slope. My survey was on August 24, 1964.

Reed Creek unit no. 4,726 (Tuolumne 15′) is the steep southern slope of a ridge on the northern side of lower Reed Creek, about 3 km. above its junction with the Clavey River, in sections 5 and 6, T. 1 N., R. 18 E., and section 31, T. 2 N., R. 18 E. The eastern and western boundaries are minor gullies. Reed Creek is the southern boundary. The spur of the ridge north of the creek, at an elevation of 1,403 meters, is the northern boundary. The slope is largely wooded with *Libocedrus decurrens*, *Pinus ponderosa*, *Pseudotsuga menziesii*, *Quercus kelloggii*, and some *Pinus lambertiana*. A few openings with grass and shrubs occur. The rocks are metavolcanic. They form low cliffs along Reed Creek. Lack of suitable habitat in most of the unit and the southern exposure may explain the absence of *Sedum*. Logging long ago probably destroyed the original vegetation. My survey was on August 22, 1964.

Rancheria Mountain unit no. 2,208 (Hetch Hetchy Reservoir 15′) is the steep southern slope of the southwestern spur of Rancheria Mountain, east of LeConte Point and north of Hetch Hetchy Reservoir in Yosemite National Park. The reservoir is the southern boundary. The northern boundary is the crest of the ridge of the spur, which is south of the trail. The eastern boundary is a minor ridge west of the first high point on the ridge, and the western boundary is the crest of the ridge east of the brook east of LeConte Point. The slope is both steep and rocky. *Pinus ponderosa* is the commonest tree. *Pinus lambertiana* occurs in the upper part of the unit, and *P. sabiniana* is on the lower slope. The rock is granite. *Sedum* occurs on the crest of the ridge in places where rocks shield the plants from the rays of sun from the south. My survey was on August 20, 1964.

Rancheria Creek unit no. 3,824 (Tower Peak 15′), alt. 2,097–2,268 m., is the fourth in the random sequence. It is a portion of the ridge on the eastern side of Rancheria Creek, west of a small lake which is southwest of the chain of lakes on Breeze Creek, in Yosemite National Park. The southern boundary is the outlet of the small lake. Rancheria Creek and a tributary which flows northward form the northern boundary. Both the north-flowing tributary and a gully which drains into the lake make the eastern boundary. A gully tributary to the outlet from the lake and a gully on the ridge west of the lake make the western boundary. Attempts to reach this unit on foot from Bear Valley on August 15 and 16, 1964 were unsuccessful, although I was within .5 km. of my goal. A ridge with steep sides still intervened. The land is rocky and difficult to traverse. The rock is granite. Probably *Sedum obtusatum* occurs there and possibly also *S. lanceolatum*.

Lee Price Camp unit no. 5,657 (Long Barn 15′) is the western slope of a ridge on the eastern side of Hull Creek south of Hulls Meadows, in section 35, T. 3 N., R. 17 E., and sections 1 and 2, T. 2 N., R. 17 E. The southern boundary is the second spur of the ridge south of Lee Price Camp. The northern boundary is the first spur which projects into Hulls Meadows north of the camp. The crest of the ridge and Hull Creek are respectively the eastern and western boundaries. The slope of the ridge is about 85% wooded. *Abies concolor* is the commonest tree. Other trees are *Pinus jeffreyi*, *P. lambertiana*, *P. contorta*, *Libocedrus decurrens*, and *Populus tremuloides*. The nonwooded area has been burned and now is grown up to shrubs, with *Ceanothus cordulatus* the commonest species. The rock is schist. Habitats suitable for *Sedum* are few and no plants occur. My survey was on August 23, 1964.

Woods Creek unit no. 896 (Sonora 7.5′) is a low ridge of serpentine along the eastern side of Woods Creek southwest of Quartz Mountain, in section 27, T. 1 N., R. 14 E. Boundaries are a brook southwest of Stent and a gully oriented northwestward into Woods Creek, on the south; a gully west of the southern end of Quartz Mountain, on the north; a low area leading into the brook at the south and the gully at the north, on the east; and Woods Creek on the west. The crest of the ridge is open grassland with sparse vegetation in the serpentine gravel and scattered trees of *Quercus douglasii* and *Pinus sabiniana*. This area is used as pasture. The slopes to the west and in the gullies have a dense growth of *Adenostoma fasciculatum* and occasional trees of *Pinus sabiniana*. The rock of the western slope appears to be schist. *Parvisedum congdonii* occurs in the open areas of serpentine gravel. *Dudleya cymosa* occurs on outcroppings of serpentine, both on the crest of the ridge and on cliffs of the small canyon on the southwestern side of the unit. My survey was on August 12, 1964.

Mammoth Lakes Primary Sampling Unit

The second randomly selected primary sampling unit in the Sierra Nevada was no. 10, the Mammoth Lakes unit. Its area is the eastern slope of the Sierra Nevada between Bishop and Lee Vining Creeks. Selection of sampling units was according to the method described for the North Yosemite unit. Seventeen natural regions with an estimated maximum number of 150 sampling units per region resulted in a potential 2,550 units. From these, a simple random sample of units was drawn.

Data for the chosen sampling units, comprising .18% of the total area of the primary unit, are summarized in the following table and paragraphs. My work in the Mammoth Lakes unit was from August 26 to September 10, 1964, and from July 20–30, 1970.

Sampling unit (no.) (name)	1,977 Reversed Peak	2,285 Bohler Canyon	2,156 Gem Pass	966 Hilton Creek	461 Round Valley	Sums	Estimates for primary unit
Area (km.^{2})	.8	.3	.36	.68	.28	2.42	1,315
Lowest alt. (m.)	2,286	3,048	3,145	3,017	1,524	1,524	1,524

Sampling unit (no.) (name)	1,977 Reversed Peak	2,285 Bohler Canyon	2,156 Gem Pass	966 Hilton Creek	461 Round Valley	Sums	Estimates for primary unit
Highest alt. (m.)	2,515	3,706	3,365	3,817	1,585	3,817	4,258
pH, range	5.4	4.8	5–6.4	5.2	5.6	4.8–6.4	4.8–6.4
pH, tests (n)	2	1	8	2	1	14	14
Crassulaceae—species (n)	0	0	3	1	0	3	3
S. lanceolatum—plts. (n)	0	0	12e	0	0	12e	2,347
S. obtusatum ssp. obtusatum—plts. (n)	0	0	252e	0	0	252e	49,296
S. integrifolium—plts. (n)	0	0	104e	130	0	234e	68,380

Reversed Peak unit no. 1,977 (Mono Craters 15′) is the eastern slope of the long, northeastern ridge of Reversed Peak, in sections 26, 34, and 35, T. 1 S., R. 26 E. Boundaries are a minor gully on the north, a secondary road on the east, minor ridges and gullies on the south, and the crest of the northeastern ridge of Reversed Peak on the west. The ridge and adjacent hills are morainic, with mixtures of granite and basalt boulders. The vegetation is about 85% sagebrush scrub, with *Artemisia tridentata* and *Purshia tridentata* dominant; 10% woodland, with *Pinus jeffreyi* dominant; and 5% chaparral, with *Ceanothus velutinus* dominant. The land is part of the Inyo National Forest. Absence of *Sedum* probably is due to both dry climate and recency of habitat. My survey was on August 28, 1964.

Bohler Canyon unit no. 2,285 (Mono Craters 15′) is the southern slope of a ridge between Kidney Lake and the north fork of Bohler Creek, in the Mt. Dana Minarets Wild Area of the Inyo National Forest. Boundaries are the crest of the ridge south of Kidney Lake on the north, the 3,048 meter contour line on the east, the north fork of Bohler Creek on the south, and the summit of the ridge south of Kidney Lake on the west. The ridge is morainic, composed of glacial debris: sand, gravel, and granitic boulders. Enormous slides of granitic boulders are a distinctive feature. The lower part of the slope is wooded with *Pinus albicaulis*. The trees are scrubby upward. Vegetation is sparse both at the higher altitudes and in the boulder areas. The site is dry, without either streams or springs. Drainage is excessive. *Sedum*, mosses, and ferns probably are lacking because of the dryness. My survey was on September 3, 1964, when I was accompanied by Jack Reveal.

Gem Pass unit no. 2,156 (Mono Craters 15′) is the northeastern slope of the ridge northwest of Gem Pass, in the Mt. Dana Minarets Wild Area of the Inyo National Forest. Boundaries are a minor spur of a ridge north of a pond northwest of Gem Pass on the north, the trail north of the pass on the east, Gem Pass and the trail northward on the south, and the crest of the ridge on the west. The unit is rocky, with cliffs of metavolcanic rocks comprising about half of the area. A small pond near the northeastern corner of the unit is surrounded by alpine meadow. Vegetation is sparse, consisting of herbs and small shrubs. Small trees of *Pinus albicaulis* are on a bluff south of the pond. The three species of *Sedum* are in the middle and northern part of the unit. *Sedum obtusatum* is in dry situations which receive maximum sunlight. The other two species are mostly in shadier, moister sites and are intermixed. My survey was on August 30, 1964.

Hilton Creek unit no. 966 (Mt. Abbot 15′) is a granitic ridge extending from the second Hilton Lake to a high peak east of Stanford Lake, near the crest of the Sierra Nevada, in the High Sierra Wilderness Area of the Inyo National Forest. Boundaries are a tributary brook of the second Hilton Lake on the north; the second lake and Hilton Creek on the east; the outlet of the fourth Hilton Lake, the lake itself, its inlet for 400 meters upstream, and then the crest of the ridge, on the south; and the summit of the high peak east of Stanford Lake on the west. The crest of the ridge and the northern slope mostly are bare granite. Intrusions of andesite appear in places between the beds of granite. The next to the highest summit on the ridge has sheer cliffs. Below these is an enormous slide of boulders. About half of the unit is forested. *Pinus contorta* is dominant on the lower slopes. *Pinus albicaulis* replaces it upward and becomes stunted and shrubby in the more exposed situations. Small meadows occur at the upper end of the fourth Hilton Lake and also along the brook on the northern boundary of the unit. *Sedum integrifolium* grows on the northern slope of the ridge, on gravelly benches, in crevices in the granite, and in the turf along the stream at the base. My survey was on September 5, 1964.

Round Valley unit no. 461 (Mt. Tom 15′) is a gently sloping area of glacial outwash—sand, gravel, and granitic boulders—in the southwestern part of the Round Valley, 1.2 km. southwest of the Vanadium Ranch in northern Inyo Co., Calif. Boundaries are the point where a road southwest of the Vanadium Ranch crosses the 1,524 meter contour on the north; the 1,524 meter contour on the east; a minor gully on the south; and the road southwest of the Vanadium Ranch, between the 1,524 and 1,585 meter contours, on the west. The vegetation is sagebrush scrub, with shrubs of *Artemisia tridentata*, *Coelogyne ramosissima*, *Purshia glandulosa*, and *Ephedra viridis* dominant. The absence of *Sedum* probably results from the dryness of the site and the lack of rocky habitats. My survey was on September 1, 1964.

South Yosemite Primary Sampling Unit

Selection of the South Yosemite primary sampling unit, no. 11, for inclusion in an optimum stratum within the Sierra Nevada, was for two reasons. First, available information indicated a maximum number of species, namely six, occurring within the unit. Second, type localities of four binomials are included: Vernal Falls—*Sedum obtusatum*; Short Trail, Yosemite Valley—*S. rubroglaucum*; between Vernal and Nevada Falls—*S. yosemitense*; and Grant's Springs, Mariposa Co., Calif.—*S. congdonii*. The area of the primary sampling unit is the western slope of the Sierra Nevada between the Middle Fork of the San Joaquin and Merced Rivers. Subdivision of the unit was into an optimum substratum comprising four sampling units, each known to contain one or more populations of interest, and a random substratum. Survey of the optimum substratum was complete. The random substratum included all sampling units not part of the optimum substratum. Within the random substratum, selection of sampling units was according to the method described for the North Yosemite unit. Fifty natural regions, with an estimated maximum number of 200 sampling units per region, resulted in a potential 10,000 units. From these, a simple random sample was drawn.

Data for the chosen sampling units are summarized in the following two tables, one table for each substratum, and the following paragraphs. By combining the estimates for the two substrata, an estimate is available for the whole primary unit. My work in the South Yosemite primary sampling unit was from June 17 to July 15, 1964. The following table provides data for the random substratum.

Sampling unit (no.) (name)	7,276 Bishop Creek	5,409 Cathey's Valley	5,217 Drunken Gulch	7,237 Mosquito Creek	7,807 Little Yosemite	Sums	Estimates for substratum
Area (km.^{2})	1.84	.52	.83	.72	.45	4.36	4,497.28
Lowest alt. (m.)	1,847	390	609	1,457	1,859	390	152
Highest alt. (m.)	2,268	470	991	1,719	2,073	2,268	4,010
pH, range	5.2–5.6	6.2–6.4	5.4	6.4	5–6.2	5–6.4	5–6.4
pH, tests (n)	3	2	1	2	3	11	11
Crassulaceae—species (n)	0	0	0	0	1	1	6
S. obtusatum ssp. obtusatum—plts. (n)	0	0	0	0	520e	520e	536,392

Bishop Creek unit no. 7,276 (Yosemite 15′) is the western slope of the ridge at the head of Bishop Creek and southwest of Westfall Meadows in Yosemite National Park, in sections 27, 33, and 34, T. 3 S., R. 21 E. Boundaries are Bishop Creek and the gully leading to it on the north, the crest of the ridge southwest of Westfall Meadows on the east, the south fork of Bishop Creek on the south, and the forks of Bishop Creek on the west. Several large open areas occur on the upper slope and summit of the ridge. Granite is exposed at these sites. Otherwise, the vegetation is largely chaparral. Dominant species are *Ceanothus cordulatus*, *Arctostaphylos mariposa*, and *Castanopsis sempervirens*. The area had been logged before it became part of the national park. Also, fire has ravaged the slope. In a few places, large trees of *Abies concolor* and *Pinus jeffreyi* still stand. Competition and lack of suitable habitat, plus the logging and burning, probably explain the absence of *Sedum*. My survey was on July 7, 1964.

Cathey's Valley unit no. 5,409 (Catheys Valley 7.5′) is part of the drainage area of a tributary of Mariposa Creek, in sections 23 and 24, T. 6 S., R. 17 E. Boundaries are the crest of the ridge south of Stonehouse on the north, the gully of a branch of a tributary of Mariposa Creek on the east, the gully of a branch of a tributary of Mariposa Creek on the south, and a tributary of Mariposa Creek on the west. Many outcroppings of metasedimentary rocks occur on the low ridges within the unit. The land is about 90% grassland and is used as pasture. Trees of *Querus douglasii* are occasional. Also, a few trees of *Salix* occur along the brook. The lack of *Sedum* may be due to the dryness of the site. My survey was on July 2, 1964.

Drunken Gulch unit no. 5,217 (Bear Valley 7.5′) is part of the eastern slope of the northwestern spur of Bullion Mountain, in sections 15, 16, and 22, T. 4 S., R. 17 E. Boundaries are the crest of the ridge north of Drunken Gulch on the north, the 609 m. contour line on the east, the stream bed of Drunken Gulch on the south, and an abandoned road on the west. Metasedimentary rocks are exposed at many places. The soil is gravelly. The upper part of the ridge is open woodland with *Quercus douglasii* and *Pinus sabiniana*. This area is used for pasture. A dense growth of chaparral, with abundant *Adenostoma fasciculatum*, covers the northern spur of the ridge and the slopes of the southern spur below 750 m. The site appears too dry for *Sedum*. My survey was on July 4, 1964.

Mosquito Creek unit no. 7,237 (Yosemite 15′) is the wooded northern slope of the western end of Turner Ridge, Yosemite National Park, in sections 20, 21, and 28, T. 4 S., R. 21 E. Boundaries are the Wawona Road on the north; Mosquito Creek, which is a tributary of Alder Creek, and a gully leading to Mosquito Creek, on the east; and the crest of the western prong of Turner Ridge on the south and west. The rock is granite, but outcrops are few. Dominant trees are *Pinus lambertiana*, *P. ponderosa*, *Abies concolor*, *Pseudotsuga menziesii*, and *Libocedrus decurrens*. Habitats suitable for *Sedum* are sparse. My survey was on July 10, 1964.

Little Yosemite unit no. 7,807 (Yosemite Valley, 1:24,000) is a part of the Little Yosemite Valley, on the south side of the Merced River. It includes the alluvial plain of the river and also the steep, rocky slope north of Starr King Lake. Boundaries are the Merced River on the north, the outlet of Starr King Lake on the east, the 2,073 m. contour line on the south, and a gully leading to an oxbow north of Starr King Lake on the west. Much of the unit is a steep, granitic slope which is nearly smooth and almost devoid of vegetation, except for lichens and mosses. The alluvial plain is forested with *Pinus contorta* and *Abies concolor*. *Pseudotsuga menziesii* and *Pinus jeffreyi* occur in cracks and small gullies in the granite of the slope. *Sedum obtusatum* is in similar situations. My survey was on June 29, 1964.

Data for sampling units in the optimum substratum of the South Yosemite primary sampling unit follow.

Sampling unit (no.) (name)	7,802 Vernal Falls	6,810 Inspiration Point	4,201 Elliott Corner	7,401 Chilnualna Falls	Sums
Area (km.^{2})	.79	1.14	.67	1.12	3.72
Lowest alt. (m.)	1,378	1,195	826	1,853	826
Highest alt. (m.)	2,012	2,012	960	2,219	2,219
pH, range	5–5.8	5.2–5.8	5.2–5.6	5–5.2	5–5.8
pH, tests (*n*)	13	11	4	6	34
Crassulaceae—species (*n*)	2	1	1	2	5
S. radiatum—plts. (*n*)	0	1,722e	0	0	1,722e
S. stenopetalum					
ssp. *monanthum*—plts. (*n*)	0	0	0	440	440
S. obtusatum					
ssp. *obtusatum*—plts. (*n*)	2,896e	0	0	374e	3,270e
S. spathulifolium					
ssp. *yosemitense*—plts. (*n*)	24,233e	0	0	0	24,233e
Parvisedum congdonii—plts. (*n*)	0	0	8,689e	0	8,689e

Vernal Falls unit no. 7,802 (Yosemite Valley, 1:24,000) is the part of the slope on the south side of the Merced River from the junction of the Mist and Nevada Falls Trails to Nevada Falls, in Yosemite National Park. Boundaries are the Merced River on the north, the trail up the slope south of Nevada Falls on the east, the Glacier Point–Panorama Cliff Trail on the south, and the Nevada Falls Trail and its projection up the slope to Panorama Cliff on the west. The rock is granite which is exposed throughout, in places forming high cliffs, and in other places a nearly smooth slope with gradient of about 45°. The cliffs below Vernal Falls are perpetually bathed in spray. These cliffs make the wettest, coolest part of the unit in summer. The base of the slope along the river below Vernal Falls has an extensive area of huge boulders, mostly bare or with tiny patches of moss. Wherever cracks are in the granite, trees occur. Common species are *Pseudotsuga menziesii, Abies concolor, Libocedrus decurrens*, and *Pinus ponderosa-jeffreyi*. *Sedum* occurs almost throughout the unit, except that it is rare on the big boulders along the river and on the crest of the cliff east of Panorama Cliff, absent from the cliffs and talus bathed in spray, and lacking from most smooth surfaces of granite, unless mosses and lichens have first formed a mat in which it can take root. My survey was on June 19, 20, 22, 23, 24, and 27, 1974.

Inspiration Point unit no. 6,810 (Yosemite Valley, 1:24,000) is part of the slope on the southern side of the Yosemite Valley above the Wawona Tunnel in Yosemite National Park. Boundaries are the Wawona Tunnel on the north, Artist Creek on the east, the ridge leading to the western end of the Wawona Tunnel on the west, and the Pohono Trail east of Fort Monroe on the south. The rock is granite, exposed in many places. In summer the site is dry. Commonest trees are *Pinus ponderosa, Libocedrus decurrens, Pseudotsuga menziesii*, and *Quercus kelloggii*. *Sedum radiatum* grows in open situations which are partially shaded and exposed to the north. It is particularly common along a small brook which is dry in summer. My survey was on July 5, 8, and 13, 1964.

Elliott Corner unit no. 4,201 (Stumpfield Mtn. 7.5′) is a low ridge of granite north of Paloni Mountain in sections 27 and 34, T. 5 S., R. 20 E. Boundaries are the junction of the road from Elliott Corner with the Humby Road on the north, a tributary of the East Fork of the Chowchilla River on the east, the Bootjack Road on the south, and the road from Elliott Corner to the Humby Road on the west. The rock is granite, with flat exposures near Elliott Corner. *Parvisedum congdonii* occurs in mats of moss and thin soil on a flat area of granite. My survey was on July 11, 1964.

Chilnualna Falls unit no. 7,401 (Yosemite 15′) is a ridge on the northern side of Wawona Dome, above Chilnualna Falls, in section 25, T. 4 S., R. 21 E., and section 30, T. 4 S., R. 22 E. Boundaries are Chilnualna Creek on the north, a gully east of the northern spur of Wawona Dome on the east, the crest of the ridge including Wawona Dome on the south, and the crest of Chilnualna Falls and the escarpment southward on the west. The rock is granite. *Sedum stenopetalum* ssp. *monanthum* is in granite gravel on ledges above the waterfall. *Sedum obtusatum* is intermixed with the *S. stenopetalum* and also occurs more widely on granite exposed to the north. My survey was on July 12, 1964.

The following table is a summary of data for the two substrata and also for the entire South Yosemite primary sampling unit. Figures after names of species are estimates of sizes of populations within the designated areas. Plus marks indicate species not included in the selected sample of areal sampling units, but observed elsewhere within the random substratum.

Unit	Random substratum	Optimum substratum	Entire p.s.u.
Area (km.²)	4,497.28	3.72	4,501
Listing units (N)	4,498	4	4,502
Listing units (n)	5	4	9
Area of n	4.36	3.72	8.08
Area of n/area of N	.001	1	.0018
Crassulaceae—species (n)	1(+2)	5	5
S. radiatum	+	1,722	>1,722
S. stenopetalum ssp. monanthum	0	440	440
S. obtusatum ssp. obtusatum	536,392	3,270	539,662
S. spathulifolium ssp. yosemitense	+	24,233	>24,233
Parvisedum congdonii	0	8,689	8,689

In addition to the species of *Crassulaceae* studied in the survey, *Sedum integrifolium* and *Dudleya cymosa* also occur in the South Yosemite primary sampling unit.

Feather River Primary Sampling Unit

Inclusion of the Feather River primary sampling unit, no. 24, in the optimum stratum was because one species and two subspecies of *Sedum*, not known elsewhere in the Sierra Nevada, occur there, and also because its complex geological structure makes it interesting for *Sedum*. The unit comprises the western slope of the Sierra Nevada from the North Yuba River to the northern end of the range south of Lassen Peak and north of the Feather River. Mesozoic and Paleozoic Rocks there meet the Tertiary lavas of the Cascade Mountains. Subdivision of the unit was into an optimum substratum comprising six sampling units, each known to contain one or more populations of interest, and a random substratum, comprising the rest of the unit. Survey of the optimum substratum was complete. Selection of a sample in the random substratum was according to the method described for the North Yosemite unit. Sixty natural regions, with an estimated maximum number of 300 sampling units per region, resulted in a potential 18,000 units. From these, a simple random sample was drawn.

Data for the chosen sampling units are summarized in the following paragraphs and in the next two tables. My work in the Feather River primary sampling unit was on March 28, June 6, and August 16, 1963, and from July 15 to August 10, 1964. Data for sampling units in the random substratum of the Feather River primary sampling unit are in the following table.

Sampling unit (no.) (name)	5,490 Oroville	10,576 Bloomer Hill	3,328 Sierra Buttes	7,910 Feather Falls	7,847 Crystal Hill	19,872 Long Lake	Sums	Estimates for whole substratum
Area (km.²)	.36	.75	.56	.38	.29	.48	2.82	1,963.67
Lowest alt. (m.)	120	567	1,585	774	457	1,798	120	61
Highest alt. (m.)	150	792	2,103	915	615	1,878	2,103	2,615
pH, range	5.4	5.6	4.8–5.2	5.6	5.2	4.8	4.8–5.6	4.8–5.6
pH, tests (n)	1	1	5	1	1	1	10	10
Crassulaceae—species (n)	0	0	1	0	0	0	1	3

Sampling unit (no.) (name)	5,490 Oroville	10,576 Bloomer Hill	3,328 Sierra Buttes	7,910 Feather Falls	7,847 Crystal Hill	19,872 Long Lake	Sums	Estimates for whole substratum
S. obtusatum								
ssp. boreale	0	0	407e	0	0	0	407e	448,319
S. albomarginatum	0	0	0	0	0	0	0	>1
S. spathulifolium								
ssp. purdyi	0	0	0	0	0	0	0	>348

Oroville unit no. 5,490 (Oroville 7.5′) is part of a ridge, now a residential area, on the eastern side of Oroville, in section 15, T. 19 N., R. 4 E. Boundaries are the old road from Oroville to Forbestown on the north, a gully on the east, Hilldale Avenue on the south, and Arbol Avenue on the west. The rock appears to be andesite. The soil is mostly sandy loam. The area is largely fields, but the slopes of the two gullies are wooded, with oaks—*Quercus douglasii*, *Q. wislizenii*, and *Q. kelloggii*, and *Pinus sabiniana*. The site appears too dry for *Sedum* and without suitable habitat for *Parvisedum*. My survey was on August 8, 1964.

Bloomer Hill unit no. 10,576 (Las Plumas 7.5′) is part of the southeastern slope of Bloomer Hill in sections 30 and 31, T. 21 N., R. 5 E. Boundaries are a small gully on the north, a brook tributary to Canyon Creek on the east, a gully and Ponderosa Way on the south, and the crest of the ridge south of Bloomer Hill and east of Ponderosa Way on the west. Large exposures of rock are lacking, but metavolcanic boulders are exposed at several places. The unit is about 80% chaparral, with *Arctostaphylos mariposa*, *A. patula*, *Ceanothus*, and *Quercus*. Scattered trees of *Pinus ponderosa*, *P. sabiniana*, *Arbutus menziesii*, *Quercus chrysolepis*, *Q. douglasii*, and *Q. kelloggii* occur. The unit appears too dry for *Sedum* and also lacks suitable habitat. My survey was on August 4, 1964.

Sierra Buttes unit no. 3,328 (Sierra City 15′) is part of the eastern slope of the Sierra Buttes southwest of Mountain Mine in sections 16 and 21, T. 20 N., R. 12 E. Boundaries are the crest of a spur and a gully southwest of Mountain Mine on the north, a gully and creek below Mountain Mine on the east, the fourth gully southwest of Mountain Mine on the south, and the crest of the northwestern spur of the Sierra Buttes on the west. The unit is about 80% bare rock—siliceous argillite and quartz porphyry, with chaparral—two species of *Arctostaphylos* and *Holodiscus*—on rock slides and in gullies. Isolated trees of *Pinus jeffreyi* occur on small benches. *Sedum obtusatum* ssp. *boreale* occurs on rock slides, in crevices, and on ledges. My survey was on July 29, 1964.

Feather Falls unit no. 7,910 (Forbestown 7.5′) is the eastern slope of a ridge south of Feather Falls, in sections 14 and 23, T. 20 N., R. 6 E. Boundaries are a hill southwest of the sawmill in Feather Falls on the north, a brook tributary to Sucker Run on the east, a gully tributary to this brook on the south, and the crest of the ridge southwest of Feather Falls on the west. The unit lacks outcroppings of rocks. It is largely forested, with *Pinus ponderosa* the commonest species of tree. Other trees include *Pseudotsuga menziesii*, *Libocedrus decurrens*, *Abies concolor*, *Torreya californica*, *Arbutus menziesii*, and *Quercus kelloggii*. Lack of suitable habitat seems to explain the absence of *Sedum*. My survey was on August 5, 1964.

Crystal Hill unit no. 7,847 (Forbestown 7.5′) is the central part of the southern slope of Crystal Hill, in sections 25 and 26, T. 20 N., R. 5 E. Boundaries are the summit of Crystal Hill on the north, a gully and brook tributary to Oregon Gulch on the east, the 457 m. contour line on the south, and the crest of the ridge which is the southwestern spur of Crystal Hill on the west. Granite rich in quartz is exposed in a few places. The vegetation is largely chaparral, the common species being *Arctostaphylos mariposa*, *Quercus wislizenii*, and *Heteromeles arbutifolia*. Occasional trees occur, especially *Pinus ponderosa*, *P. sabiniana*, and *Quercus kelloggii*. The site appears too dry and warm in summer for *Sedum*. My survey was on August 6, 1964.

Long Lake unit no. 19,872 (Pulga 15′) is the eastern slope of a ridge of volcanic rocks and granite west of Long Lake and its outlet in sections 30 and 31, T. 25 N., R. 6 E. Boundaries are a gully leading to the northwestern corner of Long Lake on the north, Long Lake and its outlet on the east, a line passing through the crest of a hill south of the lake and another hill south-southwest of the lake on the south, and the crest of the ridge west of the lake on the west. The unit is on the boundary between the formations of the Sierra Nevada and Southern Cascade Mountains. The southern end of the ridge is granite overrun by volcanic rocks from the north. The slope has been cleared of large trees and now is covered with shrubs: *Amelanchier*, *Salix scouleriana*, *Arctostaphylos patula*, and *Quercus vaccinifolia*. Trees which remain are *Pinus contorta*, *P. jeffreyi*, *P. monticola*, *Abies concolor*, and *Populus tremuloides*. Conditions appear suitable for *Sedum*, but clearing of the land and possible ancient fires may explain its absence. My survey was on July 22, 1964.

The following table gives data for sampling units in the optimum substratum of the Feather River primary sampling unit.

Sampling unit (no.) (name)	18,601 Nelson Bar	19,501 Rock Creek	20,408 Serpentine Canyon	18,001 Keddie	14,101 Spanish Creek	16,801 Eureka Peak	Sums
Area (km.²)	.29	.84	1.28	.36	.12	.44	3.33
Lowest alt. (m.)	297	500	811	984	1,067	1,585	297
Highest alt. (m.)	402	1,073	1,634	1,225	1,103	1,824	1,824
pH, range	5.4	5–5.6	6.4–7.2	5.2–5.4	5.2–6.2	4.6–5.2	4.6–7.2
pH, tests (*n*)	2	9	10	2	4	10	37
Crassulaceae—species (*n*)	1	1	1	1	1	1	4
S. obtusatum ssp. boreale	0	0	0	0	0	781e	781e
S. albomarginatum	0	0	68	0	0	0	68
S. spathulifolium ssp. purdyi	0	197	0	1,252	462e	0	1,911e
Parvisedum pumilum	40	0	0	0	0	0	40

Nelson Bar unit no. 18,601 (Cherokee 7.5′) is the eastern slope of a ridge of lava west of the Nelson Bar Road in sections 6 and 7, T. 21 N., R. 4 E. Boundaries are the Penstock Ditch on the north, the stream between the power house and Lime Saddle on the east, the first gully north of Lime Saddle on the south, and the Pentz-Magalia Road on the west. The unit is on the boundary between the formations of the Southern Cascade Mountains and the Sierra Nevada. The ridge is Pliocene andesite. Vegetation on the lava is sparse and consists largely of *Selaginella hansenii* and annual herbs. *Parvisedum pumilum* occurs in thin soil in nearly bare areas where competition is slight. My survey was on August 8, 1964.

Rock Creek unit no. 19,501 (Pulga 15′) is the eastern, steep, rocky slope of Sugar Loaf Mountain in sections 35 and 36, T. 24 N., R. 5 E., and sections 30 and 31, T. 24 N., R. 6 E. Boundaries are a gully on the north side of Sugar Loaf Mountain and Rock Creek on the north, the North Fork of the Feather River on the east, the southeastern ridge of Sugar Loaf Mountain on the south, and the summit of Sugar Loaf Mountain on the west. The rocks are granite. Trees on the slope are *Pinus ponderosa*, *Libocedrus decurrens*, *Umbellularia californica*, *Acer macrophyllum*, *Quercus chrysolepis*, and *Q. kelloggii*. *Sedum spathulifolium* ssp. *purdyi* occurs in moss and in gravel on granite exposed to the north and northeast. My survey was on July 20, 1964.

Serpentine Canyon unit no. 20,408 (Almanor 15′) is part of the steep, southern slope of the eastern end of Red Hill in sections 11, 12, and 13, T. 25 N., R. 7 E. Boundaries are the crest of the eastern end of Red Hill on the north, the third gully east of the mouth of Rattlesnake Gulch on the east, the East Branch of the North Fork of the Feather River on the south, and the first gully west of Rattlesnake Gulch on the west. The rock is serpentine, forming cliffs on the lower part of the slope. Above the cliffs, the vegetation is woodland, with *Quercus chrysolepis* predominating, and also *Pinus ponderosa*, *Pseudotsuga menziesii*, and *Libocedrus decurrens*. *Garrya fremontii* is common on the slopes of the gully on the eastern boundary. *Sedum albomarginatum* occurs in crevices and on ledges of serpentine, where few or no other plants occur, and mostly in situations which are shaded for part of the day. My survey was on June 6 and August 16, 1963, and July 18, 1964.

Keddie unit no. 18,001 (Quincy 7.5′) is the northern slope of the hill between Big and Little Blackhawk Creeks in sections 26, 27, 34, and 35, T. 25 N., R. 9 W. Boundaries are Little Blackhawk Creek on the north and east, a ridge which reaches the creek where that is crossed by the road on the south, and a gully tributary to Big Blackhawk Creek on the west. The slope is steep, rocky, and wooded. *Pseudotsuga* is the dominant species of tree. *Sedum spathulifolium* ssp. *purdyi* grows in partial shade in moss in small open areas. My survey was on August 2, 1964.

Spanish Creek unit no. 14,101 (Bucks Lake 15′) is the northern point of the ridge along Spanish Creek just below Slate Creek in section 18, T. 24 N., R. 9 W. Boundaries are Spanish Creek on the north and east, the Bucks Lake Road on the south, and Rock and Spanish Creeks on the west. The rock of the northern end of the point is schist. The ridge is wooded, with *Pseudotsuga menziesii* the dominant tree. *Sedum spathulifolium* ssp. *purdyi* occurs only on the northern side of the ridge, in partial shade, on the steeper, rocky parts of the slope. My survey was on July 24, 1964.

Eureka Peak unit no. 16,801 (Sierra City 15′) is a part of the lower southeastern slope of Eureka Peak, opposite the mouth of Little Jamison Creek, around and above the campground in the Plumas–Eureka State Park. Boundaries are the first ridge on Eureka Peak north of the mouth of Little Jamison Creek on the north, Jamison Creek on the east, the first ridge on Eureka Peak south of the mouth of Little Jamison Creek on the south, and the 1,824 m. contour line on the west. The rocks of the unit appear to be granite and andesite, with glacial deposits along the stream. Some trees of *Pinus ponderosa* and *Libocedrus decurrens* occur in the flatter areas opposite the mouth of Little Jamison Creek. Chaparral is abundant on the rocky slopes, especially *Quercus vaccinifolia* and *Arctostaphylos patula*. *Sedum obtusatum* occurs in shallow soil on open areas of rock. My survey was on July 27 and 31, 1964.

Summary for the Feather River primary sampling unit is as follows:

	Random substratum	Optimum substratum	Entire p.s.u.
Area (km.^{2})	1,963.67	3.33	1,967
N sampling units	1,964	6	1,970
n sampling units	6	6	12
Area of n (millions m.^{2})	2.82	3.33	6.15
Area of n/area of N	.0014	1	.003
Crassulaceae—species (n)	3	4	4
Sedum obtusatum			
ssp. *boreale*	448,319e	781e	449,100e
S. albomarginatum	>1	68	>69
S. spathulifolium			
ssp. *purdyi*	>348	1,911e	>2,259e
Parvisedum pumilum	0	40	>40

Summary for the Sierra Nevada is as follows:

Species	Plants (estimates of N)	Altitudinal range (m.)	pH, range
S. lanceolatum	30,771	2,135–3,660	5.2–5.4
S. radiatum	>1,722	1,489–1,572(915–2,287)	5.4–5.8
S. stenopetalum			
ssp. *monanthum*	499,764	1,865–1,970(1,798)	5.0–5.2
S. obtusatum			
ssp. *obtusatum*	850,439,187	1,370–3,660	4.8–5.6
ssp. *boreale*	449,100	1,570–2,033	4.6–5.2
S. albomarginatum	>69	365–858	6.4–7.2
S. spathulifolium			
ssp. *yosemitense*	539,662	430–1,970	5.0–6.4
ssp. *purdyi*	>2,259	400–1,180(1,432)	5.0–6.2
S. integrifolium	878,237	2,776–3,660(2,135–4,034)	5.2–6.4
Parvisedum pumilum	>40	350–375 (1,220)	5.4
P. congdonii	116,043,782	480–915 (45)	5.2–6.6
Dudleya cymosa			
ssp. *cymosa*	529,752	465–1,850	5–6
D. lanceolata	—	—	—

Figures for the numbers of plants of each species in the Sierra Nevada are derived from the survey in 1964. They include several sources of variation, namely among clusters, among listing units, and among primary units. For this reason, they are only approximations. Estimates are by means of the following formula, in which A = area in square kilometers, D = density per square kilometer, O = optimum, R = random, and S = stratum.

$$\text{Est. } N = A_{S_R} \frac{(A_{p.s.u._{R1}} \times D_{p.s.u._{R1}}) + (A_{p.s.u._{R2}} \times D_{p.s.u._{R2}})}{A_{p.s.u._{R1}} + A_{p.s.u._{R2}}} +$$

$$A_{S_O} \frac{(A_{p.s.u._{O1}} \times D_{p.s.u._{O1}}) + (A_{p.s.u._{O2}} \times D_{p.s.u._{O2}})}{A_{p.s.u._{O1}} + A_{p.s.u._{O2}}}$$

Density per square kilometer equals the number of plants in any unit divided by the area of the unit in square kilometers.

Figures for altitudinal range and pH in the above table likewise are derived from the survey in 1964, but when data from records in the literature or from labels in herbaria exceed my experience, these are indicated in parentheses after my figures.

Besides the sources of variation resulting from subsampling, rugged terrain and environmental diversity presented problems in getting accurate data. Sometimes subsampling and also substratification were necessary. As a result, numbers of plants indicated for listing units often are estimates and not actual counts. Throughout the tables, an e after a number indicates an estimate.

The Vernal Falls unit no. 7,802 in the optimum substratum of the South Yosemite primary unit is an example of the problem of getting a count within a unit. It possesses both high cliffs and steep slopes. Three substrata are recognizable: 1—base of slope along river, 2—intermediate slope, and 3—crest of slope. The size of each of these substrata is determinable on maps. For sampling, a subunit in each was chosen randomly. From the number of plants of *S. obtusatum* found within each chosen subunit, the total number for the substratum was calculated by simple expansion. Because only one subunit per substratum was surveyed, an estimate of variance is impossible. This is a weakness of the data, but time was insufficient to permit further attention to the Vernal Falls unit. An additional difficulty was a condition in substratum 1, not understood at first, that four subunits, none of which were chosen for study, were within range of spray from the fall. Lack of *Sedum* in these units, determined by walking through them, probably resulted from excessive wetness of the site. Properly, these subunits should have comprised another substratum.

Summary for *S. obtusatum* in the Vernal Falls unit 7,802 is as follows:

Substratum	Area of surveyed subunit (m.²)	Plants (N)	Plants (n per m.²)	Area of substratum (m.²)	Plants (N)
1. Base of slope	10,000	0	0	70,000	0
2. Intermediate slope	5,450	29	.0053	484,000	2,565.2
3. Crest of slope	18,186	25	.0014	236,000	330.4
Totals	33,636	54		790,000	2,895

The data for N for the whole Vernal Falls unit, added to the data for the unit by Chilnualna Falls, provide the figure for N for the optimum substratum of the South Yosemite primary sampling unit.

The data for the random substratum of the South Yosemite unit were handled similarly. In a sample of five units, *Sedum obtusatum* occurs in only one, the Little Yosemite unit. This unit consists of two natural areas or substrata:

Substratum	Subunits (N)	Area (m.^{2})
1. Valley	12	190,000
2. Slope	16	260,000
Totals	28	450,000

For survey, I selected randomly one subunit in each substratum.

Substratum	Subunit no.	Area (m.^{2})	Sedum obtusatum (n plants)
1. Valley	28	30,000	0
2. Slope	12	12,000	24

The number of plants per m.^{2} in the slope substratum, on the basis of the one count, is .002, and the estimated total for that substratum is .002 × 260,000 or 520. This also is the estimate for the whole unit.

Because samples of units are small, I have not used ratio estimates. According to Cochran (1963:157, 166), formulas for the variance of the estimate are valid only for large samples, when n exceeds 30. For this reason, my estimates are simple expansions of the mean per unit. Variances are large and confidence intervals correspondingly wide, but this must be expected in such circumstances. At least the data are better than a guess and indicate the gap between precise understanding and present knowledge.

The small number of areal sampling units surveyed in each primary unit in the Sierra Nevada is explained by the following facts. Units were expensive to reach in both time and money. The average time in going to and coming from listing units was 5.88 hours. In addition, time devoted to making arrangements to visit units and in gaining permission for access is a cost of the survey. The average time per unit for this purpose was 1.6 hours. The size of the unit makes little difference with respect to the time necessary for either gaining access or reaching the unit. This view is supported by a test of correlation, plotting the time for these purposes per unit against the sizes of the units. The correlation coefficient, $r = +.23$, is not significant. This kind of evidence has encouraged me to use a few large areal units or clusters, rather than many small ones. Were the size of the sampling unit considerably reduced, I still might survey in a season no more units than I did in 1964.

Plotless sampling had the advantage of permitting prompt attention to *Sedum*, rather than an elongate procedure of laying out quadrats. The efficiency of the method is indicated by subunit 19 in the Vernal Falls listing unit. There, *Sedum spathulifolium* ssp. *yosemitense* attains maximum density. The subunit comprises 5,450 square meters. By count, the number of plants of *S. spathulifolium* in this area is 267 and the total space occupied by them is 363 square meters. This space is only 6.7% of the total area. Randomly selected quadrats of a size similar to the average size of the plants, namely 1 square meter, would be expected to include plants at the rate of 1 quadrat in 14, provided that the plants were randomly distributed. The rate of inclusion would decrease with smaller size of quadrat and increase as the quadrat increases. The plants are not randomly distributed, however, but are concentrated in certain parts of the area. As a result, most quadrats, unless large, for example 100 m.^{2}, would be in areas lacking plants.

Stratification was valuable in the survey of the Sierra Nevada. Although units in the random stratum included four basic species of *Sedum*, the optimum stratum was essential for others. Had there been no optimum stratum, three species and three subspecies would have been missed. The distribution of species by strata follows. An asterisk indicates that the species is found only in the optimum stratum.

Random stratum: *Sedum lanceolatum, S. obtusatum* ssp. *obtusatum, S. integrifolium, S. stenopetalum* ssp. *monanthum,* and *Parvisedum congdonii.*

Optimum stratum:

Random substratum: *S. obtusatum* ssp. *obtusatum* and ssp. *boreale** (and *S. spathulifolium* ssp. *yosemitense**).

Optimum substratum: *S. albomarginatum*, S. obtusatum* ssp. *boreale*, S. spathulifolium* ssp. *yosemitense** and ssp. *purdyi*, S. radiatum*, S. stenopetalum* ssp. *monanthum, Parvisedum congdonii,* and *P. pumilum*.*

Finally, finding species at the same site, without evidence of hybridization, is one of the best tests of specific status. The following combinations of species met this test.

Pairs of species	No. of units in which together
S. lanceolatum, S. obtusatum ssp. obtusatum	1(+ 2 other sites)
S. lanceolatum, S. integrifolium	1
S. obtusatum ssp. obtusatum, S. integrifolium	1
S. obtusatum ssp. obtusatum, S. stenopetalum ssp. monanthum	2
S. obtusatum ssp. obtusatum, S. spathulifolium ssp. yosemitense	1

Klamath Mountains—10

Species	Known pops. (n)	Alt. range (m.)	pH, range	Pattern of distrib., A–D
S. lanceolatum				
ssp. lanceolatum	6	1,754–2,791	—	—
S. radiatum				
ssp. radiatum	14	305–1,646	—	—
ssp. ciliosum	13	220– >940	6.4	(D–1)
ssp. depauperatum	2	1,586–1,617	6–6.6	(B–1)
S. stenopetalum				
ssp. monanthum	8	1,585–2,013	5.6	(B–1)
S. divergens	2	1,707	—	—
S. spathulifolium				
ssp. spathulifolium	14	~10–1,296	—	—
ssp. purdyi	31	420–935	—	—
S. obtusatum				
ssp. retusum	30	465–2,165	—	—
S. laxum				
ssp. laxum	20	0–1,220	6.4–6.6	(D–1)
ssp. latifolium	9	61–549	6.6	(D–1)
ssp. heckneri	22	152–1,769	6.2–6.6	(B–1)

Species	Known pops. (n)	Alt. range (m.)	pH, range	Pattern of distrib., A–D
S. oblanceolatum*	3	457–1,582	5.0–5.8	(B–1)
S. oregonense	11	1,798–2,681	—	—
S. moranii*	1	210–220	6.4–7.2	(B–1)

California Coast Ranges—5

Species	Known pops. (n)	Alt. range (m.)	pH, range	Pattern of distrib., A–D
S. radiatum				
ssp. radiatum	33	244–2,145	—	(D–1)
S. stenopetalum				
ssp. monanthum	1	1,951	—	—
S. spathulifolium				
ssp. spathulifolium	52	405–450	5.2–5.4	A–1(D–1)
ssp. pruinosum	6	3–9	—	A–1(D–1)
S. obtusatum				
ssp. retusum	33	1,480–2,287	5.4–5.6	A–.75, B–.25
S. laxum				
ssp. eastwoodiae	2	1,065–1,240	5.6–6.8	(D–1)

Transverse Ranges of California—2

Species	Known pops. (n)	Alt. range (m.)	pH, range	Pattern of distrib., A–D
S. niveum	8	2,740–2,980	5.0–6.6	(B–.5, D–.5)
S. spathulifolium				
ssp. yosemitense	15	1,065–2,286	6.4–6.8	A–1(D–1)

Columbia Plateau—7

Species	Known pops. (n)	Alt. range (m.)	pH, range	Pattern of distrib., A–D
S. lanceolatum				
ssp. lanceolatum	27	2,120–2,380	5.4–5.6	A–1(D–1)
S. debile	3	2,440	—	—
S. stenopetalum				
ssp. stenopetalum	64	1,065–2,065	5.0–6.8	A–.7, D–.3 (A–.8, D–.2)
S. leibergii	71	52–1,005	7.2–7.8	A–.7, D–.3 (A–.8, D–.2)
S. borschii	2	1,670–1,775	6.4–7.6	A–1(A–.8, D–.2)
S. spathulifolium				
ssp. spathulifolium	1	—	—	—
S. integrifolium				
ssp. integrifolium	8	2,250–2,715	4.8–5.6	A–1(A–.8, B–.2)

Cascade Mountains—9 (fig. 183)

Species	Known pops. (n)	Alt. range (m.)	pH, range	Pattern of distrib., A–D
S. lanceolatum				
ssp. lanceolatum	30	305–2,135	4.6–6.6	A–.5, D–.5(A–.75, D–.25)
S. rupicolum*	12	860–2,144	5.0–6.4	A–1(A–.3, B–.3, C–.3, D–.1)

Fig. 183. Summit of O'Leary Mountain, Cascade Mountains, Lane Co., Ore., July 9, 1965.

Species	Known pops. (n)	Alt. range (m.)	pH, range	Pattern of distrib., A–D
S. stenopetalum				
ssp. *stenopetalum*	22	1,214–1,220	4.9–6.4	A–1(A–.7, C–.3)
ssp. *monanthum*	6	1,490–1,790	5.6	A–1(A–.9, B–.1)
S. *divergens*	40	1,095–1,380	4.8–6.2	A–.5, D–.5(A–.7, D–.3)
S. oreganum				
ssp. *tenue**	29	825–1,550	5.0–6.4	A–.5, D–.5(A–.9, B–.1)
S. obtusatum				
ssp. *boreale*	1	1,707	—	(B–1)
S. *oregonense*	25	1,190–1,685	5.2–6.2	A–1(A–.9, D–.1)
S. spathulifolium				
ssp. *spathulifolium*	18	630–1,590	5.2–6	A–1(A–.9, D–.1)
S. integrifolium				
ssp. *integrifolium*	8	1,830–2,440	—	—

Willamette–Puget Trough—4

Species	Known pops. (n)	Alt. range (m.)	pH, range	Pattern of distrib., A–D
S. lanceolatum				
ssp. *nesioticum**	13	4–20	6–6.6	(D–1)
S. stenopetalum				
ssp. *stenopetalum*	3	—	—	—

Species	Known pops. (n)	Alt. range (m.)	pH, range	Pattern of distrib., A–D
S. oreganum				
ssp. oreganum	1	—	—	—
S. spathulifolium				
ssp. spathulifolium	7	25–92	6	A–1 (D–1)
ssp. pruinosum	9	—	—	—

Coast Ranges of Oregon and Washington—2

Species	Known pops. (n)	Alt. range (m.)	pH, range	Pattern of distrib., A–D
S. oreganum				
ssp. oreganum	25	0–460	6.4–6.6	(B–.6, D–.4)
S. spathulifolium				
ssp. spathulifolium	19	30–965	5.4–6.4	C–1 (D–1)
ssp. pruinosum	3	3–6	6.4–6.6	(D–1)

Olympic Peninsula—5

Species	Known pops. (n)	Alt. range (m.)	pH, range	Pattern of distrib., A–D
S. lanceolatum				
ssp. lanceolatum	3	1,830–1,981	—	—
S. stenopetalum				
ssp. stenopetalum	1	1,220	—	—
ssp. monanthum	1	—	—	—
S. divergens	9	1,219–1,676	—	—
S. oreganum				
ssp. oreganum	4	—	—	—
S. spathulifolium				
ssp. spathulifolium	5	1,372±	—	—

Northern Rocky Mountains—4

Species	Known pops. (n)	Alt. range (m.)	pH, range	Pattern of distrib., A–D
S. lanceolatum				
ssp. lanceolatum	32	1,220–2,896	—	(D–1)
ssp. subalpinum	>3	2,190–2,440	—	(D–1)
S. stenopetalum				
ssp. stenopetalum	30	1,062–1,830	—	(D–1)
S. divergens	1	650	—	—
S. integrifolium				
ssp. integrifolium	34	1,450–2,499	—	(D–1)

Interior Plateaus and Ranges of British Columbia, Yukon, and Alaska—4

Species	Known pops. (n)	Alt. range (m.)	pH, range	Pattern of distrib., A–D
S. lanceolatum				
ssp. lanceolatum	43	746–2,135	—	—
S. stenopetalum				
ssp. stenopetalum	14	393–2,050	—	—
S. divergens	7	1,678–1,830	—	—
S. integrifolium				
ssp. integrifolium	61	1,525–2,195	—	—

Coast Batholith—6

Species	Known pops. (n)	Alt. range (m.)	pH, range	Pattern of distrib., A–D
S. lanceolatum				
ssp. lanceolatum	5	—	—	—
ssp. nesioticum	6	6–2,135	—	—
S. stenopetalum				
ssp. stenopetalum	2	—	—	—
S. divergens	29	3–2,012	—	—
S. oreganum				
ssp. oreganum	12	3–457	—	—
S. spathulifolium				
ssp. spathulifolium	8	—	—	—
ssp. pruinosum	6	—	—	—
S. integrifolium				
ssp. integrifolium	34	0–?	—	—

Alaskan Peninsula—1

Species	Known pops. (n)	Alt. range (m.)	pH, range	Pattern of distrib., A–D
S. integrifolium				
ssp. integrifolium	32	0–1,300	—	—

Canadian Shield

The Canadian Shield comprises eight provinces: Superior, Slave, Bear, Churchill, Southern, Grenville, Artic Island, and Greenland. The only part of the Canadian Shield which I have surveyed is a tiny portion along the northern coast of the Gulf of St. Lawrence. The purpose of my survey there in 1959 was to study *Sedum villosum* in the field. For that reason, attention was concentrated on two regions where the species was known to occur, namely the Harrington Harbour region, defined as a quadrangle, 50°–51° N., 59°–60° W., and the Great Mecatina region, 50°–51° N., 58°–59° W. The selected quadrangles were subdivided into blocks 10′ N.–S. and E.–W. Selection of blocks was random within the part of each quadrangle containing islands, 4 out of 15 in the Harrington Harbour quadrangle and 3 out of 4 in the Great Mecatina quadrangle. The areas of islands in each 10′ block were measured on hydrographic charts by means of a compensating polar planimeter. From the information about areas, the number of quadrats possible, each $(100 \text{ m.})^2$, was determined for each island. Numbers were assigned to these potential quadrats in a single sequence for all islands within a block. Simple random samples of quadrats then were chosen for study. Because of loss of time due to bad weather and other circumstances, three randomly selected blocks were not surveyed. These were number 3 in the Harrington Harbour quadrangle and numbers 1 and 2, in order of random selection, in the Great Mecatina quadrangle. Otherwise, the survey proceeded as planned. Selected quadrats were subsampled by means of simple random samples of meter squares. In total, I surveyed 9 large quadrats, each $(100 \text{ m.})^2$, and 89 included meter quadrats. The randomly selected units were situated on 9 islands. In addition, I studied 5 populations of *Sedum villosum* using meter quadrats for selection of plants and environmental conditions. These squares were within the areas actually inhabited by *S. villosum* and were selected randomly after the populations were found. This procedure was necessary because *S. villosum* is of such limited distribution that it did not turn up in any of the units which were part of the general sampling scheme. Discovery of the species resulted from rapid cruising of whole islands. The populations studied were on 5 islands. In addition, I searched for the species on 6 other islands, bringing to 18 the total number of islands which I visited.

The principal results of the survey in the Gulf of St. Lawrence were data about the variation and distribution of *Sedum villosum* and *S. rosea*. Equally important, the experience there demonstrated the advantage of large listing units and the superiority of natural over artificial boundaries.

Information about the occurrence of *Sedum* on the Canadian Shield is available from the following publications: Böcher, Holmen, and Jakobsen (1957), *Grønlands Flora*; Lewis (1931–1932), *An Annotated List of Vascular Plants Collected on the North Shore of the Gulf of St. Lawrence, 1927–1930*; Polunin (1959), *Circumpolar Arctic Flora*; and Porsild (1957), *Illustrated Flora of the Canadian Arctic Archipelago*. A total of four native species is known. Data are summarized in the following table where names of authors are abbreviated as B = Böcher, Holmen, and Jakobsen; L = Lewis; Pol = Polunin; and Por = Porsild. In the column at the right, numbers indicate the observed altitudinal range.

Species	Province	Survey—1959, Alt. (m.)
Sedum acre	Greenland (B)	—
S. annuum	Greenland (B)	—
S. villosum	Grenville (L)	1–20
	Greenland (B, Pol)	—
S. rosea	Grenville (L, Por)	1–40
	Churchill (Pol, Por)	—
	Greenland (B, Pol, Por)	—

Trips in the Adirondack Mountains and in the Southern Province in Ontario, Michigan, Wisconsin, and Minnesota have not resulted in the discovery of any native populations of *Sedum*.

Data for the primary sampling units in the Gulf of St. Lawrence in 1959 are organized in two charts, one for the Harrington Harbour quadrangle and the other for the Great Mecatina quadrangle. Each chart is followed by brief descriptions of the sampling units.

Sampling unit	B203	B85	N2	N13	M411	Sum
Area (m.²)	10,000	10,000	10,000	3,892	5,000	38,892
Block	Bald Islands (30′–40′ N., 0–10′ W.)	Bald Islands	Netagamu River (30′–40′ N., 30′–40′ W.)	Netagamu River	Cape Mecatina (40′–50′ N., 0–10′ W.)	3

Harrington Harbour Quadrangle, 50°–51° N., 59°–60° W.

	Harrington Harbour Quadrangle, 50°–51° N., 59°–60° W.					
Sampling unit	B203	B85	N2	N13	M411	Sum
Island	Stevenson	Nadeau	Gull	W. of ctr. of Hospital	N. of Forsyth	5
Meter quadrats surveyed (n)	10	10	7	0	21	48
Lowest alt. (m.)	3	1	0	0	0	0
Highest alt. (m.)	20	12	4	4	12	20
pH, range	4.4–6.4	5.2	5–7	5.2–7	4.2–7	4–7
pH, tests (n)	28	1	25	5	15	74
Crassulaceae—species (n)	1	1	1	1	1	1
Sedum rosea						
n	894	8	376	740e	1,900e	3,918e
frequency (% of 1 m.²)	0	0	0	—	9	—

The sums of the areas of the islands within the selected blocks of the Harrington Harbour quadrangle and the fractions which were surveyed are as follows:

Block	Total area of islands (km.²)	Area of surveyed units (km.²)	Fraction surveyed (%)
Bald Islands	2.04	.020	.98
Netagamu River	1.90	.014	.73
Cape Mecatina	6.23	.005	.08

Sampling of the Cape Mecatina block is so disproportionate from the other two blocks that the data for the unit in that are not included in the estimates of $\bar{x}$ and s for *Sedum rosea* for the whole quadrangle.

Stevenson Island unit no. B203 (Chart 4469, Flat I. to Little Mecatina I., 1:76,200) is at the northern end of the island. The north-central point is the northernmost point near the northern center of the island. Rock, which is gneiss, is exposed in many places. The soil is organic matter. The vegetation is heath, with *Empetrum nigrum* and *Betula pumila* the commonest species. *Sedum rosea* mostly is at the edge of the vegetation around areas of bare rock and in crevices. My survey was on August 4, 5, 7, and 10, 1959.

Ile Nadeau unit no. B85 (Chart 4469, Flat I. to Little Mecatina I., 1:76,200) is on the southern peninsula of the island. The south-central point is 50 meters E.S.E. of the southeastern corner of the cove on the western side of the southern peninsula. The rock is gneiss and is exposed in many places. A soil of organic matter has developed in depressions on and among the rocks. Lichens are common on the rocks. Otherwise, the vegetation is heath, with *Empetrum nigrum* predominating. My survey was on August 11, 1959.

Gull Island unit no. N2 (Chart 4468, Little Mecatina I. to St. Mary Is., 1:75,000) is at the northern end of a small island which is on a line east from the falls, 6 m. high, of the Netagamu River. The northwestern corner of the quadrat is the northwestern-most point on the island. The rock is gneiss and is exposed in many places. Organic matter on and among the rocks supports a heath vegetation and meadow. *Empetrum nigrum*, *Myrica gale*, and a species of *Festuca* are common in different parts of the unit. My survey was on August 14, 15, 16, and 17, 1959.

Island west of center of Hospital Island, no. N13 (Chart 4468, Little Mecatina I. to St. Mary Is., 1:75,000) is the southern part of a nameless islet west from the center of Hospital Island. The northwestern corner of the unit is the corner of the cove on the northern side of the island. The western boundary is a line S.S.W. from this point to the southern shore. The rock is gneiss and the soil is organic matter. The vegetation is mostly heath, with *Empetrum nigrum* and *Myrica gale* the commonest species. My survey was on August 18, 1959.

Island north of Forsyth Island, unit no. M411 (Chart 4469, Flat I. to Little Mecatina I., 1:76,200) is in the southeastern part of the island, just north of the island with Forsyth Point. It is east of Crooked Passage and Deadman's Island. The south-central point of the unit is 50 m. east from the innermost part of the cove in the forked peninsula on the western side of the island. The rock, exposed in many places, is gneiss. The soil typically is organic matter, but in a few places is black sand. The vegetation is tundra, with the three commonest species *Empetrum nigrum*, *Vaccinium uliginosum*, and *Myrica gale*. My survey was on July 13, 14, and 17, 1959.

Great Mecatina Quadrangle (50°–51° N., 58°–59° W.)

Sampling unit	B276	B138	B76	B233	Sum
Area (m.^{2})	10,000	7,000	10,000	10,000	37,000
Block	Bun Islands 50'–60' N., 40'–50' W.	B.I.	B.I.	B.I.	1
Island	Kécarpoui	Shag Rock	W. Affligée	N. Query	4
Meter quadrats surveyed (n)	10	10	10	11	41
Lowest alt. (m.)	15	0	5	9	0
Highest alt. (m.)	25	8	15	30	30
pH, range	4–4.4	6.4–7	4.4–5.8	4–4.4	4–7
pH, tests (n)	10	2	9	10	31
Crassulaceae—species (n)	0	1	0	0	1
Sedum rosea					
n	0	8,960	0	0	8,960
frequency (% of 1 m.^{2})	0	28	0	0	5

The islands of the Bun Islands block have a combined area of 3.68 km.^{2} Of these, I surveyed .04 km., or a fraction of about 1%.

Kécarpoui Island unit no. B276 (Chart 4474, Bun Islands to Mutton Bay, 1:36,500) is the northeastern summit of the hill on Kécarpoui Island. The southwestern corner is the northeasternmost point of the depression on the southwestern side of the hill. Rock is exposed in many places. Where developed, the soil is organic matter. The vegetation is tundra, with *Empetrum nigrum* the commonest species. *Sedum* may be lacking because of the acid conditions. My survey was on July 21 and 22, 1959. Fig. 184 shows the southwestern part of Kécarpoui Island as seen from the summit of the hill.

Shag Rock unit no. B138 (Chart 4474, Bun Islands to Mutton Bay, 1:36,500) is the southern portion of Shag Rock, a tiny island to the southeast of the Iles Affligées, with a maximum elevation of 7.6 m. The southwestern corner of the unit is a point

Fig. 184. Pond and southwestern part of Kécarpoui Island as seen from the highest point, July 22, 1959.

40 m. east of the western shore on the southernmost part of the western half of the island. About half of the unit is bare rock, some of which is covered by water at high tide. Where soil is in crevices and depressions in the higher places, it is organic matter, sand, or clay. Vegetation is sparse, composed of littoral herbs. *Sedum rosea* is common on the western, higher half of the island. In the past, cormorants nested there, giving the island its name. My survey was on July 28, 1959.

Western Ile Affligée unit no. B76 (Chart 4474, Bun Islands to Mutton Bay, 1:36,500) is the southwestern part of the Western Ile Affligée. The southwestern corner of the unit is also the southwesternmost point on the island. The unit is rocky along the shore. Eastward from the rocks, the soil is organic matter. The vegetation is tundra, with *Empetrum nigrum* dominant in 80% of the area. My survey was on July 21, 23, and 26, 1959.

Northern Query Island unit no. B233 (Chart 4474, Bun Islands to Mutton Bay, 1:36,500) is at the southern end of the island, on the eastern side, south of a small bay. The central point on the northern boundary is the southwestern corner of the small bay. A rock covered with barnacles marks this point. The soil is organic matter and the vegetation is tundra, with *Empetrum nigrum* the commonest species, occupying about 60% of the area. My survey was on July 29 and 30, 1959.

The eighteen islands which I cruised rapidly to find *Sedum villosum* have a combined area of 7.8 km.² Pertinent data about the five islands on which *S. villosum* occurs are in the following table:

Name of island	Area (m.²)	Area of S. villosum (m.²)	Fraction of island inhabited by S. villosum
Middle Kécarpoui	332,000	25	.00007
Eastern Affligée	226,000	52	.00023
Western Affligée	374,000	31	.00008
Big Gull	435,483	80	.00018
Stevenson	580,644	6	.00001

Middle Kécarpoui Island is labeled Isle du Milieu on hydrographic chart 4474, Bun Islands to Mutton Bay. Frank Jones found *Sedum villosum* there while walking on the hill on the northwestern side of the island. On the Eastern Isle Affligée, *S. villosum* grows along the western side. On the Western Isle Affligée, site of sampling unit B76, the *Sedum* is in disjunct groups near the southern end, especially on the eastern side. The population on Big Gull Island is in a compact area near shore on the outer, northern side of the island. The location is 2.3 km. west of Forsyth Point. Big Gull Island is crescent-shaped and is shown without a name on hydrographic chart 4469, Flat Island to Little Mecatina. The population on Stevenson Island is on the west-central side of the island, near where we landed to visit unit no. B203. In the cases of both Stevenson Island and the Western Isle Affligée, finding *S. villosum* followed landing to survey randomly selected quadrats. Landing on the other islands, both where *S. villosum* occurs and where it is lacking, was part of a deliberate search for the species. The method of searching large natural areas was adopted as a regular policy in surveys after 1959. It has the advantage of increasing the chance of finding rare items because it provides a large space in which to look.

Greenland—4 (following Böcher, 1938)

Species	Known pops. (n)	Alt. range (m.)	pH, range	Pattern of distrib., A–D
S. annuum	57	—	—	—
S. acre				
ssp. acre	5	~5	—	—
S. villosum	49	—	—	—
S. rosea	173	—	—	—

St. Lawrence Lowland—0/Central Lowland—4

Species	Known pops. (n)	Alt. range (m.)	pH, range	Pattern of distrib., A–D
S. ternatum	17	—	—	—
S. pulchellum	19	—	—	—
S. nuttallianum	12	—	—	—
S. integrifolium				
ssp. leedyi	1	347–384	—	(D–1)
ssp. integrifolium	7	61–305	—	—

Great Plains—5

Species	Known pops. (n)	Alt. range (m.)	pH, range	Pattern of distrib., A–D
S. wrightii	2	335–350	7.2–8.2	(D–1)
S. lanceolatum				
ssp. lanceolatum	43	1,372–3,470	6.4–6.6	(D–1)
S. stenopetalum				
ssp. stenopetalum	1	—	—	—
S. nuttallianum	7	—	—	—
S. integrifolium				
ssp. integrifolium	2	3,240–3,570	5.6–6.6	(C–1)

Interior Low Plateaus—3

Species	Known pops. (n)	Alt. range (m.)	pH, range	Pattern of distrib., A–D
S. telephioides	9	240	4.6–5.2	(C–1)
S. ternatum	27	—	—	—
S. pulchellum	46	135–255	—	(B–.16, D–.84)

Ozark Plateaus—2

Species	Known pops. (n)	Alt. range (m.)	pH, range	Pattern of distrib., A–D
S. ternatum	3	—	—	—
S. pulchellum	49	92–385	—	(D–1)

Coastal Plain—2

Species	Known pops. (n)	Alt. range (m.)	pH, range	Pattern of distrib., A–D
S. ternatum	4	—	—	—
S. nuttallianum	10	—	—	—

Extent of area occupied by a species is one measure of its distribution. Another measure is the number of geomorphic provinces in which it occurs. No species occurs in all forty-three geomorphic provinces in North America, just as no species occurs in all eight sections of the Trans-Mexican Volcanic Belt. *Sedum lanceolatum* occurs in the maximum number of provinces—17. It is followed by *S. integrifolium* in 15 provinces and *S. stenopetalum* in 14 provinces. Eight species occur in only one province in North America, but three of these occur elsewhere—*S. villosum* in Europe, *S. wrightii* on the Mexican Plateau, and *S. niveum* in Lower California. The remaining five species are endemic to the provinces in which they occur: *S. rupicolum* in the Cascade Mountains, *S. pusillum* on the Piedmont Plateau, *S. oblanceolatum* and *S. moranii* in the Klamath Mountains, and *S. albomarginatum* in the Sierra Nevada. In the following list, numbers of provinces are indicated in the column on the left, and in the right column is indicated the number of species which occur in that number of provinces.

No. of provinces	No. of species
19	0
18	0
17	1
16	0
15	1
14	1
13	0
12	0
11	0
10	1
9	1
8	0
7	1
6	2
5	1
4	7
3	2
2	4
1	8
Total	30

Size of geomorphic province is not correlated with number of species occurring there. For demonstration, see the second long table in this chapter. Ten species of *Sedum*, the maximum number found in any province, are native in the Klamath Mountains, although that province is thirty-eighth in size. Explanation for the high number of species in the Klamath Mountains appears to be the length of time that the area has been available for occupation by plants and also the great environmental diversity prevailing there. The Cascade Mountains, with nine native species of *Sedum*, comprise a province which is twenty-sixth in order of size. While some parts of the Cascade Mountains are young and were covered with glaciers in the Pleistocene, the outer foothills, especially in Oregon, are older and escaped recent glaciation. These older regions could have served as refuges in which species survived for long periods of time. The Sierra Nevada, thirty-third in size, has the third largest number of species, seven. Explanation is the same as for the Klamath and Cascade Mountains. In the East, the glaciated New England–Acadian Highlands, which together comprise the largest part of the Appalachian region, have only one native species of *Sedum*. In contrast, the smallest Appalachian province, the Blue Ridge, which virtually escaped Pleistocene glaciation, has five native species of *Sedum*.

The Churchill Province, of the Canadian Shield, is the third largest geomorphic province in North America. Pleistocene ice completely covered it, and only one native *Sedum* occurs. The Central Lowland and Great Plains, respectively the largest and second largest geomorphic provinces in North America, include both unglaciated portions in the south and glaciated parts in the north. Four species of *Sedum* are native on the Central Lowland, three on the unglaciated part and one on the glaciated northern part. Five species of *Sedum* are native on the Great Plains, three on unglaciated and two on glaciated parts.

The conclusion is evident that past history of a region, particularly length of time that the region has been available for growth of plants, is of primary importance in explaining the species which are there. This historical factor is more important than extent of area or even contemporary physical factors.

Major drainage slopes of North America are the Atlantic, Gulf, and Pacific. Interior drainages, small or large, are related here to the adjacent major slopes. Only *Sedum integrifolium* occurs on the Arctic slope. It is the sole species which occurs on all slopes. Of thirty species of *Sedum* native in North America, eight occur on two or more drainage slopes. The others are each limited to a single slope. The greatest diversity of species, two-thirds, is on the Pacific slope. Of the twenty species found there, only four (one of these on all slopes) occur on both the Pacific and Gulf slopes. The situation illustrates the point that highlands, making divides between drainages, may be effective barriers to dispersal.

Slope	No. of species
Atlantic	2
Gulf	4
Pacific	16
Both Atlantic and Gulf	4
Both Gulf and Pacific	3
All slopes	1
Total	30

Twenty-four of the species of *Sedum* native in North America north of the Mexican Plateau, that is 80%, occur at altitudes between sea level and 1,000 meters. In contrast, only two species occur at altitudes above 4,000 meters. Both of these are phylogenetically advanced. In the Trans-Mexican Volcanic Belt, *Sedum* is lacking from the altitudinal zone below 1,000 meters. The majority of species there occur in the zone from 2,001 to 3,000 meters. The situation in temperate North America is shown in the following chart.

Altitudinal zone	No. of species
0–1,000 m.	24
1,001–2,000 m.	21
2,001–3,000 m.	14
3,001–4,000 m.	6
4,001–5,000 m.	2

Most North American species of *Sedum* occur on a variety of rocks. Nine species occur on both limestone and granite. In contrast, nine of the thirty species seem limited to just one kind of rock. Of this set of nine, three are endemic on serpentine, namely *S. laxum*, *S. albomarginatum*, and *S. moranii*. Half of the North American species occur in regions which were recently glaciated. Eleven of these fifteen species occur only in glaciated regions. This circumstance indicates invasion by half of the species of areas which have been available for occupation by plants for a time no longer than ten thousand years. Of course, this fact does not require that these species evolved in the last ten thousand years. Instead, they may have moved into glaciated territory from some adjacent refuge from which they subsequently disappeared. The following table provides a summary of information about numbers of species on different kinds of rocks and in glaciated and unglaciated regions, and also concerning the numbers of kinds of interspecific hybrids in these different situations. As in the Trans-Mexican Volcanic Belt, areas recently available for colonization by plants do not have a higher incidence of hybridization of species of *Sedum* than older areas.

Kinds of rocks	No. of species	No. of interspecific and intergeneric hybrids
Limestone	16	1
Sandstone and shale	15	0
Granite	17	0
Gneiss	9	0
Basalt and andesite	10	1
Serpentine	7	1
Unglaciated areas	19	2
Glaciated areas	15	2

Twenty-four of the thirty species of *Sedum* known to occur in North America north of the Mexican Plateau are endemic. The remaining six species, 20%, occur in other regions, some in two other regions. Species which occur naturally elsewhere are *S. cockerellii*, *S. niveum*, *S. wrightii*, *S. villosum*, *S. integrifolium*, and *S. rosea*. Other regions in which these species occur are indicated in the following list.

No. of species	Other regions
3	Mexican Plateau
1	Sierra Madre Oriental
1	Peninsular Range Province
2	Asia
2	Europe
1	Africa

Conclusions

Thirty species of *Sedum* occur in North America north of the Mexican Plateau. The largest number of species in any geomorphic province is 10 in the Klamath Mountains. The next largest number of species is 9 in the Cascade Mountains. The geomorphic province in the East with the greatest number of native species of *Sedum* is the Appalachian Plateau, with 6. The 3 species with the widest distribution, as determined by number of geomorphic provinces in which they occur, are *S. lanceolatum*—17, *S. integrifolium*—15, and *S. stenopetalum*—14. Each of these 3 species is advanced from an evolutionary standpoint and either exhibits polyploidy or may be the product of polyploidy. Eight species occur in only one geomorphic province. Of these, 5 are endemic to the provinces in which they occur.

Twenty species of *Sedum* occur on the Pacific slope of North America, compared with 6 species on the Atlantic slope. More species, 17, grow on granite or granitic rocks than on any other kind of rocks. Sixteen species occur on limestone. About two-thirds of the species occur only in unglaciated areas. Three species also occur on the Mexican Plateau, but only 2 species are common to both North America and Europe. Although 24 species are found at elevations below 1,000 meters above the sea, only 2 species occur above 4,000 meters.

The high degree of endemism of the North American species of *Sedum* suggests a long, independent history. *Sedum telephioides*, with closest relatives in Europe and Asia, is a primitive species. Likewise, *S. ternatum*, another primitive species, either is derived from species of the Mexican highlands or from ancestors in Eurasia.

CHAPTER XI

Relationships of Species

The levels of relationships among plants are many. Individuals of a local population have varying degrees of relationship with each other, then local populations of a subspecies, subspecies of a species, species of a genus, and on up through the taxonomic hierarchy.

Subgenera and Sections

Species of *Sedum* of North America, north of the Mexican Plateau, are not equally related to each other. Some species or groups of species are so distinct that a question arises whether they should be regarded as constituting separate genera. Other species clearly have affinity with each other and appear closely related. The North American species of *Sedum* comprise no less than eleven groups on the basis of natural relationships. Six of these groups probably originated elsewhere and invaded North America independently: subgenus *Telephium* from Europe or Asia, the section *Cockerellia* from the Mexican Plateau, the *S. lanceolatum* group either from the Mexican Plateau or Asia, *S. villosum* from Europe, *Clementsia* from Asia, and *Rhodiola* from Asia, Europe, or both. The remaining groups—*S. ternatum* and related species, *S. pusillum* (*Tetrorum*), and the three sections of subgenus *Gormania*—may be of American origin.

No single criterion is available as an infallible means of assessing natural relationships. Instead, many criteria are necessary. These include gross morphology, internal anatomy, chromosomes, genetic compatibility, physiological properties, geographic distribution, and chemical composition. When data from all sources provide a consistent result, an interpretation may be reasonably clear. When evidence from different sources conflicts, the interpretation may be unclear. In the present study, evidence is best for gross morphology and geographic distribution. Some evidence is available for most species about chromosomes, internal anatomy, and physiological properties. Least evidence is available concerning chemical composition. Rightfully, information about chemistry must come last, after other details have been learned, otherwise a frame of reference would be lacking. Someone must first work out the pattern of morphological variation and distribution. That has been a primary objective of my study.

The subgenus *Telephium* comprises about fifteen species, distributed as follows: Asia—11, Europe—2, both Europe and Asia—1, and North America—1. An Asiatic origin seems likely. Further, the species with the most primitive floral anatomy, *Sedum sieboldii*, occurs in Japan (see Quimby, 1971). Since the North American *S. telephioides* appears close to the European *S. telephium* in both morphology and chromosomes, a European origin for the American species is a possibility. Such an interpretation envisions a migration across the northern Atlantic Ocean in the pre-Pleistocene, possibly when the distribution of land and water was different from the present condition. Otherwise, an origin from an ancestral species in Asia and a migration across the Bering Strait could explain the Appalachian species.

Flowers of subgenus *Telephium* exhibit the least fusion of vascular bundles of any group of species of *Sedum* (Quimby, 1971). For this reason, subgenus *Telephium* comes first in the sequence of groups of species. The occurrence of *S. telephioides* in North America appears to be relict. The twelve pairs of subequal chromosomes do not detract from the idea that *S. telephioides* is an old species, but the cytological data require direction from the morphological and geographic evidence. Among North American species, *S. telephioides* is distinct. It does not interbreed with any other species. Relationships are with species of subgenus *Telephium* in Eurasia, not with other North American species of *Sedum*. If *S. telephioides* evolved in North America, it has descended from ancestors which occurred elsewhere and migrated to this continent independently of other North American species.

A group of four species with filaments flattened basally and the epipetalous filaments only slightly adnate, mostly less than .3 mm., has its distributional center in the southern Appalachian highlands. Here belong *Sedum ternatum*, *S. nevii*, *S. glaucophyllum*, and *S. pulchellum*. Other features of these species, besides the distinctive stamens, are the usually 4-merous flowers, white or pinkish white petals, and kyphocarpic carpels which develop into divergent follicles with prominent ventral lips. The origin of this group is unclear, but the species are primitive with respect to other members of subgenus *Sedum*. For thoughts about origin, see the discussion of relationships under *S. ternatum*. That appears to be a primary species of the group, although Quimby's (1971) study of floral anatomy places it in his group 3 with moderate vertical compression and fusion of vascular bundles. In this respect, it is more advanced than *S. rhodocarpum*, but that in turn is more advanced in having greater adnation of the epipetalous filaments. Information about morphology, distribution, and chromosomes supports the idea of a close relationship among the four Appalachian species considered in this paragraph.

Sedum pusillum, an annual species of the southern Piedmont Upland, also with slight adnation of the epipetalous filaments, is so distinctive that Rose proposed to make it a separate genus, *Tetrorum*. The carpels are deeply set in the receptacle and become widely spreading, but not gibbous ventrally. Sherwin and Wilbur (1971) considered the vascular pattern unique, but derived from Quimby's group 3. The flowers of *S. pusillum* are slightly perigynous and the lateral bundles arise close to the ventrals. Anatomical evidence supports the idea that *S. pusillum* is different from other species. Likewise, it is distinctive in having four pairs of chromosomes, the lowest number known in the *Crassulaceae*. From an evolutionary standpoint, *S. pusillum* is advanced. It could have evolved from the *S. ternatum* group. Surely, it must stand by itself among North American species of subgenus *Sedum*, although with some doubt whether it deserves subgeneric status.

Section *Cockerellia* of subgenus *Sedum*, described in 1943 for several white-flowered, perennial species, now appears to comprise a group with distribution centered on the Mexican Plateau. These species prevailingly have 5-merous flowers, erect follicles which are not gibbous ventrally, and tiny, reticulate or papillose seeds. Species of the present treatment belonging to this group are *S. niveum*, *S. cockerellii*, and *S. wrightii*. The species of the section *Cockerellia* differ from the *S. ternatum* group in the condition of the carpels, the greater adnation of the epipetalous filaments, number of floral parts, and adaptation to more rigorous conditions for growth. They may be more advanced than the Appalachian species.

The largest group of species of subgenus *Sedum* occurs in the western Cordillera. The nine species of this group have in common several characters which distinguish them: yellow petals, anthers, nectaries, and ovaries; separate petals; 3-parted cymes; narrow, elongate leaves; a lack of rootstocks; and either eight pairs of chromosomes or some multiple of eight. Exceptions to these conditions are white petals in some subspecies of *S. radiatum* and *S. stenopetalum*, occasional greenish ovaries, petals basally connate in *S. debile*, variation in the number of cincinni per cyme, suborbicular or obovate leaves in *S. divergens*, and ten pairs of chromosomes in *S. nuttallianum*. The yellow or green ovaries are a good practical means of distinguishing this group of species from the preceding four groups which have white ovaries.

The Cordilleran species of subgenus *Sedum* comprise at least five subgroups. *Sedum lanceolatum* and *S. rupicolum* are close and hybridize naturally. Likewise, *S. radiatum* and *S. stenopetalum* are close. Further, the latter hybridizes with *S. lanceolatum*, indicating a genetic relationship between these two subgroups. Also, *S. stenopetalum* hybridizes with *S. borschii*, which is close to *S. leibergii*. The two species with opposite leaves are advanced in different ways, *S. divergens* with kyphocarpic carpels and *S. debile* with connate petals. A supposed hybrid of *S. divergens* and *S. lanceolatum* provides a further clue of genetic relationship. *Sedum nuttallianum* is most different from the other species, but possibly nearer to them than to any other group. If the Cordilleran species are derived from the ancestral stock of *Sedum* in the Mexican highlands, then *S. nuttallianum* may be similarly derived. Both *S. stenopetalum* and *S. nuttallianum* are moderately advanced anatomically (Subramanyam, 1955, and Quimby, 1971). Except for *S. nuttallianum*, the other eight species appear to be interrelated. Together, they constitute a reticulate complex.

Sedum villosum is a European species which has no close relatives in North America. It has only slightly invaded North America in the region of the Gulf of St. Lawrence and stands by itself in America. Its relationships are with other European species with orthocarpic carpels and short styles, such as *S. pedicellatum* and *S. dasyphyllum*.

Distinctive characteristics of subgenus *Gormania* are prominent rosettes of spatulate or oblanceolate leaves, perennial habit, and basal cohesion of petals in all except one species. In addition, the plants often have rootstocks or stout decumbent stems. *Gormania* comprises eight species of the Pacific Mountain System. Quimby (1971) studied the floral anatomy of two of those species and found a moderately advanced condition which he designated as group 3. The usual number of chromosomes in *Gormania* is fifteen pairs or some multiple of fifteen. An exception is *S. oreganum* with twelve pairs.

The species of subgenus *Gormania* fall into three groups or sections: *Oreganica*—*S. oreganum*; *Gormania*—*S. obtusatum, S. laxum, S. oblanceolatum, S. oregonense, S. albomarginatum,* and *S. moranii*; and *Rosulata*—*S. spathulifolium*. The origin of *Gormania* must have been long ago. Further, the sections imply the possibility of three independent origins. The least advanced of the species still are more advanced than the most primitive members of subgenus *Sedum*. That is why *Gormania* follows *Sedum* in the list of natural groups.

The most distinctive subgenus of North American *Sedum* and the one most often segregated as a separate genus is *Rhodiola*. It comprises three species in North America: *S. rhodanthum* of section *Clementsia*, and *S. integrifolium* and *S. rosea* of section *Rhodiola*. Diagnostic features are stout rootstocks with scalelike primary leaves; axillary, annual floriferous stems; and maximum fusion of the vascular bundles of the flowers, Quimby's group 4. In floral anatomy, *Rhodiola* is the most advanced of the eight natural groups of species of *Sedum* in North America. Haploid chromosome numbers are 7, 11, and 18.

Lenophyllum, Parvisedum, and *Diamorpha* are three small genera related to *Sedum* and probably derived from it. All occur only in North America. Their origin on this continent appears likely. *Lenophyllum*, with its spicate inflorescences, may have evolved from the same ancestral stock which gave rise to the advanced genus *Villadia*. The yellow-flowered *Parvisedum*, with one ovule per ovary, may be descended from the phyletic line which gave rise to the section *Lanceolata*, and *Diamorpha* may be a remote derivative of the line from which originated the sections *Ternata* and *Tetrorum*.

Floral anatomy, gross morphology, and geographic distribution provide the basis for the arrangement of the natural groups of *Sedum* in North America north of the Mexican Plateau. Other kinds of information harmonize with the sequence which is adopted. The preferred classification and arrangement of the native species is indicated in the following list.

Subgenus *Telephium*
  S. telephioides
Subgenus *Sedum*
  Section *Ternata*
    S. ternatum, S. nevii, S. glaucophyllum, S. pulchellum
  Section *Tetrorum*
    S. pusillum
  Section *Cockerellia*
    S. niveum, S. cockerellii, S. wrightii
  Section *Lanceolata*
    S. lanceolatum, S. rupicolum, S. radiatum, S. stenopetalum, S. leibergii, S. borschii,
    S. divergens, S. debile, S. nuttallianum

Section *Villosa*
  S. villosum
Subgenus *Gormania*
  Section *Oreganica*
    S. oreganum
  Section *Gormania*
    S. obtusatum, S. laxum, S. oblanceolatum, S. oregonense, S. albomarginatum, S. moranii
  Section *Rosulata*
    S. spathulifolium
Subgenus *Rhodiola*
  Section *Clementsia*
    S. rhodanthum
  Section *Rhodiola*
    S. integrifolium, S. rosea

Morphological Relationships

Morphological evidence is the primary basis for arrangement of the species in phylogenetic sequence. For this purpose, the work of Quimby (1971) and of Sherwin and Wilbur (1971) has been particularly helpful. Also, information about chromosomes has been useful to indicate which species are allopolyploids and which are basic diploids. Other kinds of evidence have helped either to confirm relationships or to resolve troublesome problems.

Eleven features of gross morphology have served for rating the species with respect to phylogenetic advancement. Features included are ones which were noted for each species. Other features, such as number of vascular traces per sepal or degree of fusion of vascular bundles in flowers, although important, are not known for all species.

A list of advanced conditions follows:

1. Roots capillary, annual or biennial, not from rhizomes or other perennial stems
2. Inflorescences cincinnal cymes
3. Flowers small, <5 mm. in diameter
4. Flowers imperfect, unisexual
5. Sepals perigynous
6. Sepals connate basally
7. Petals connate basally
8. Carpels connate basally
9. Ovules few, <5 per carpel
10. Follicles with corky lips
11. Seeds winged

Despite limitations of method, the rating of species on the basis of relative advancement in gross morphology provides a general idea of relationships. Further, it seems to appraise correctly what is the most primitive species and which are the most advanced. Exceptions to the results of the ratings are discussed after the list below.

Advanced conditions (n)	Number and names of species
0	1: *S. telephioides*.
1	0:—
2	2: *S. wrightii* and *S. albomarginatum*.
3	16: *S. ternatum, S. nevii, S. glaucophyllum, S. niveum, S. cockerellii, S. lanceolatum, S. rupicolum, S. stenopetalum, S. borschii, S. villosum, S. obtusatum, S. laxum, S. oregonense, S. rhodanthum, S. integrifolium,* and *S. rosea.*
4	8: *S. pulchellum, S. radiatum, S. debile, S. leibergii, S. oreganum, S. oblanceolatum, S. moranii,* and *S. spathulifolium.*
5	3: *S. divergens, S. pusillum,* and *S. nuttallianum.*

The above list needs adjustment to make ratings agree with other kinds of evidence. The two species listed as with only two advanced conditions probably deserve higher ratings than indicated. *Sedum wrightii* has thick, caducous leaves and is adapted to a hot, arid environment. *Sedum albomarginatum* is adapted to growth on serpentine. On the other hand, the carpels of *S. albomarginatum* are only loosely closed and can be opened all the way to the tips of the styles. Two allopolyploids, *S. stenopetalum* and *S. borschii*, need higher ratings than their diploid parents, respectively *S. radiatum* and *S. leibergii*. The three species of subgenus *Rhodiola*, despite both the advanced anatomical condition, demonstrated for *S. integrifolium* by Quimby (1971), and the tendency toward unisexual flowers, are not so advanced in other respects. *Sedum divergens* and the two annual species, one, *S. pusillum*, sometimes segregated as a separate genus, appear to be the most advanced in the North American series. The results suggest what is obvious, namely that advances have occurred in different directions in each group of related species and that advancement sometimes has been great in groups which still retain relatively primitive species.

Physiological Relationships

All species of North American *Sedum* can endure some freezing. Perennial species least able to survive after either prolonged freezing or temperatures much below the freezing point are *S. wrightii, S. nevii,* and some of the species of subgenus *Gormania*. Adaptation to drought varies. One adjustment is the annual habit, permitting the species to survive the adverse period in the form of seeds. Another response is the thick, succulent leaf, as in *S. wrightii* and *S. laxum*.

Few insects feed on *Sedum*. Possibly most species have developed distinctive chemical defenses. We are only at the beginning of accumulating knowledge on this subject.

Although physiological adaptation to the environment is of crucial importance for the life of an individual plant, the use of physiological characters for classification is limited because of the ecotypic differentiation which occurs within species. Grouping together all plants which can endure some particularly extreme environmental condition will only result in putting together members of diverse phyletic lines, for example *Sedum villosum* and *S. rosea* on islands in the Gulf of St. Lawrence, *S. lanceolatum* and *S. integrifolium* in the Seven Devils Mountains, or *S. stenopetalum* and *S. oregonense* on O'Leary Mountain. Such groupings are unreasonable from an evolutionary standpoint and serve to explain why students of classification and relationships give more attention to form than to function.

Ecological Relationships

The thirty species of *Sedum* of North America north of the Mexcian Plateau exhibit four life forms. Unlike the situation in the Trans-Mexican Volcanic Belt, where 35% of the species are phanerophytes, none are in this category north of the Mexican Plateau. As in the Volcanic Belt, chamaephytes predominate. Two-thirds of the species are in this category. Interestingly, the least advanced species, *S. telephioides*, and the three species of subgenus *Rhodiola*, including one with the most advanced condition of floral anatomy, have in common that they are hemicryptophytes. *Sedum leibergii*, with subterranean offsets, qualifies as a geophyte. The five therophytes are advanced species of subgenus *Sedum*, but belong to three different sections. One species, *S. pulchellum*, listed here as a therophyte, occasionally is a chamaephyte.

Life form	No. of species
Chamaephytes	20
Hemicryptophytes	4
Geophytes	1
Therophytes	5

Adaptation to diverse edaphic conditions makes possible wide geographic distribution. One aspect of such adaptation is adjustment to pH. North American species of *Sedum* are able to grow in soils with a wide range of pH, from 3.8 to 8.4. In cultivation, where competition is lacking, plants will grow at levels of pH different from those in which they occur in nature. Most species grow in soils in the range of 5.1–7. Only three species grow primarily where the pH is above 7.1 and no species grows primarily where the pH is below 5. To some extent, pH is a factor in keeping the distributions of species separate in nature.

		3.1–4	4.1–5	5.1–6	6.1–7	7.1–8	8.1–9
1.	S. telephioides	+	+	+	+		
2.	S. ternatum			+	+		
3.	S. nevii			+	+		
4.	S. glaucophyllum		+	+	+	+	
5.	S. pulchellum					+	+
6.	S. pusillum		+	+			
7.	S. niveum			+	+		
8.	S. cockerellii				+	+	
9.	S. wrightii					+	+
10.	S. lanceolatum		+	+	+	+	+
11.	S. rupicolum			+	+		
12.	S. radiatum			+	+		
13.	S. stenopetalum		+	+	+		
14.	S. leibergii					+	
15.	S. borschii			+	+	+	
16.	S. divergens			+	+		

		3.1–4	4.1–5	5.1–6	6.1–7	7.1–8	8.1–9
17.	S. debile			+	+	+	
18.	S. nuttallianum			+			
19.	S. villosum				+	+	
20.	S. oreganum			+	+		
21.	S. obtusatum		+	+			
22.	S. laxum			+	+		
23.	S. oblanceolàtum			+			
24.	S. oregonense			+	+		
25.	S. albomarginatum				+	+	
26.	S. moranii				+	+	
27.	S. spathulifolium			+	+		
28.	S. rhodanthum		+	+	+		
29.	S. integrifolium		+	+	+	+	
30.	S. rosea		+	+	+		
	Total no. of species	1	11	23	23	12	3

Two species regarded as of possible allopolyploid origin, namely *S. glaucophyllum* and *S. borschii*, occur over a wide band of pH values, from 4.1 to 8. *Sedum lanceolatum*, the species which occurs in the largest number of geomorphic provinces, also occurs over the widest band of pH. Viewed from the standpoint of evolutionary relationships, closely related species, as *S. nevii* and *S. pulchellum* or *S. stenopetalum* and *S. leibergii*, may adapt in different directions. The adjustment to level of pH cannot serve as a basis for determining natural relationships.

Time of flowering is important in reducing or preventing the interbreeding of species. In a general way, groups of North American species of *Sedum* exhibit seasonal patterns. *Sedum telephioides*, of subgenus *Telephium*, flowers in summer. The four species of section *Ternata* and *S. pusillum* of section *Tetrorum* flower in spring. The three species of section *Cockerellia* flower in summer and autumn. Species of the remaining groups, section *Lanceolata* of subgenus *Sedum* and all sections of subgenera *Gormania* and *Rhodiola*, flower in spring and summer, depending on latitude and altitude. From the standpoint of effective isolation, the important condition is the relative time of flowering at any particular site. For example, the earlier flowering of *S. rupicolum* on the ridge above Scotty Creek in the Wenatchee Mountains of Washington limits the possibility of hybridization with *S. lanceolatum* where plants of the two species grow adjacent to each other. Such details are concealed in the following chart which serves only to indicate general patterns of flowering. Plus marks indicate months when a species may be in flower anywhere within its range.

		Mar.	Apr.	May	June	July	Aug.	Sept.	Oct.	Nov.
1.	S. telephioides					+	+	+		
2.	S. ternatum		+	+	+					
3.	S. nevii		+	+						
4.	S. glaucophyllum			+	+					
5.	S. pulchellum	+	+	+	+					
6.	S. pusillum	+	+							
7.	S. niveum					+	+			
8.	S. cockerellii						+	+		

		Mar.	Apr.	May	June	July	Aug.	Sept.	Oct.	Nov.
9.	S. wrightii						+	+	+	+
10.	S. lanceolatum			+	+	+	+	+		
11.	S. rupicolum				+	+	+			
12.	S. radiatum			+	+	+	+			
13.	S. stenopetalum			+	+	+				
14.	S. leibergii		+	+	+					
15.	S. borschii			+	+	+				
16.	S. divergens				+	+	+	+	+	
17.	S. debile				+	+	+			
18.	S. nuttallianum		+	+	+					
19.	S. villosum					+	+			
20.	S. oreganum				+	+	+	+		
21.	S. obtusatum				+	+	+			
22.	S. laxum			+	+	+				
23.	S. oblanceolatum					+				
24.	S. oregonense				+	+	+			
25.	S. albomarginatum				+					
26.	S. moranii			+	+					
27.	S. spathulifolium		+	+	+	+	+			
28.	S. rhodanthum				+	+	+	+		
29.	S. integrifolium			+	+	+	+			
30.	S. rosea			+	+	+				
	Total no. of species	2	7	15	22	19	16	7	2	1

Sizes of populations of *Sedum* vary enormously. Altogether, I have studied in some detail 153 populations. These range in size from a low of 1 plant to a high of 968,562 plants. Three populations consist of only one individual: *S. oreganum* on O'Leary Mountain in Oregon, *S. spathulifolium* ssp. *yosemitense* along the Merced River at El Portal in California, and *S. integrifolium* at Watkins Glen, N.Y. The population at Watkins Glen has been larger in the time of my experience, 7 plants in 1961. It is an example of a population which is dying out. The populations on O'Leary Mountain and at El Portal may be in the same predicament. The population of the annual *S. pusillum* at Forty-acre Rock, S.C., is the largest which has come to attention. Twelve populations of four species in sampling units chosen without any restriction of randomization had an average size of 13,612 plants and a median size of 169 plants. Similarly, 141 populations of 30 species, either in optimum strata or selected specially, had an average size of 13,588 plants and a median size of 332 plants. The major differences between means and medians indicate asymmetric distributions which are skewed to the right. For comparison, the average size of 75 populations in the Trans-Mexican Volcanic Belt was 611 plants with a median value of 73 plants. The inference is that populations of *Sedum* in temperate North America are larger than those in the Trans-Mexican Volcanic Belt, which lies entirely within the tropical zone.

Potential for Hybridization

One measure of potential for hybridization is the number of species which occur together at a site. Other measures are the number of species which flower at the same time in a region or have the same number of chromosomes. The number of interspecific hybrids which occur is an indication of the actual extent of present or past hybridization.

Thirty species of *Sedum* are native in North America north of the Mexican Plateau. Possible gaps to be bridged by intermediates are

$$\frac{n(n-1)}{2} = \frac{30 \times 29}{2}$$

Statistics for potential for hybridization follow.

No. of species	30	
No. of interspecific gaps	435	
No. of interspecific gaps with species on either side in the same square kilometer	34	8%
No. of interspecific gaps with species on either side in same square meter	15	3%
No. of interspecific gaps bridged by natural hybrids	5	1%
No. of interspecific gaps involving species of other genera, bridged by natural hybrids	0	0%

The low potential for hybridization suggests that processes other than crossing of species are more important in the evolution of North American *Sedum* at the present time.

To investigate the reason for the low number of interspecific hybrids, I gave attention to ten species, each of which occur within one meter of some other species. Such proximity is adequate to provide a good chance for interbreeding. The present inventory covers fifteen interspecific gaps. After the name of each species is indication of flowering time, the names of other species which occur within one meter, and also their flowering times. Asterisks indicate that natural hybrids occur.

1. *S. ternatum*—Apr.–June: *S. glaucophyllum*—May–June.
2. *S. pulchellum*—Mar.–June: *S. nuttallianum*—Apr.–June.
3. *S. lanceolatum*—May–Sept.: *S. rupicolum**—June–Aug.; *S. stenopetalum**—May–July; *S. debile*—June–Aug.; *S. integrifolium*—May–Aug.
4. *S. stenopetalum*—May–July: *S. borschii**—May–July; *S. obtusatum*—June–Aug.; *S. oregonense*—June–Aug.
5. *S. divergens*—June–Oct.: *S. oreganum*—June–Sept.
6. *S. villosum*—July–Aug.: *S. rosea*—May–July.
7. *S. oreganum*—June–Sept.: *S. oregonense*—June–Aug.
8. *S. obtusatum*—June–Aug.: *S. spathulifolium*—Apr.–Aug.
9. *S. oregonense*—June–Aug.: *S. spathulifolium*—Apr.–Aug.
10. *S. rhodanthum*—June–Sept.: *S. integrifolium*—May–Aug.

In addition, Mr. and Mrs. R. N. Payne have found *S. laxum* ssp. *heckneri* and *S. spathulifolium* ssp. *spathulifolium* growing near to each other and hybridizing.

Where two species occur close together, flowering time usually overlaps. When hybrids are lacking, the species either have different numbers of chromosomes or have developed genetic incompatibilities. Out of 15 observed situations optimum for hybridization, only 20% yielded hybrids.

Pollinators of *Sedum* are insects: bees of several kinds, especially sweat bees and bumblebees; flies, including bombyliids, syrphus flies, and muscid flies; at least six species of butterflies; and beetles. These visit the flowers and carry pollen from one plant to another. Since many of the pollinators are strong fliers, they can transfer pollen for many meters.

Plants occurring within as small an area as a square kilometer may receive pollen from other plants in the vicinity, whether of the same or different species. Since few of the pollinators appear to restrict their visits to just one kind of flower, chances for transfer from one species to another are great. Differential behavior of pollinators seems to be of little consequence in isolating species of *Sedum*. Genetic isolation, especially that resulting from differences in number of chromosomes, is more important in keeping species distinct at a site.

With few exceptions, competitors of species of *Sedum* in North America are miscellaneous. The lack of other vegetation often marks a site as suitable for *Sedum*. Plants of the genus abound where other vegetation is sparse. Any attempt to define habitats of *Sedum* in terms of prevailing other vegetation is sure to encounter difficulty. Exceptions are situations prevailing in certain local regions, for example on Sugarloaf Peak in the San Bernardino Mountains, where *Selaginella watsonii* is a common associate of *Sedum niveum*; the region along Jordan Creek in southwestern Missouri, where *Festuca octoflora* grows with *S. nuttallianum*; islands of the Gulf of St. Lawrence, where *Epilobium palustre* is a common associate of *S. villosum*; and the canyon of the Rogue River where *S. moranii* and *Selaginella wallacei* are associated. When larger plants, as grasses and shrubs, take over a site, *Sedum* disappears. It is a plant of the open places and does not endure strong competition.

Genetic Relationships

Occurrence of natural hybrids is a good demonstration of genetic relationships. Five cases of interspecific hybrids have come to attention in the study of North American species of *Sedum*: *S. lanceolatum* ssp. *lanceolatum* × *rupicolum*, *S. lanceolatum* ssp. *lanceolatum* × *stenopetalum*, *S. lanceolatum* ssp. *lanceolatum* × *divergens*, *S. stenopetalum* ssp. *stenopetalum* × *borschii*, and *S. laxum* ssp. *heckneri* × *spathulifolium* ssp. *spathulifolium*. Although these hybrids may not be fertile, they can endure for long periods of time by means of vegetative propagation. They indicate that the chromosomes of the parental species are sufficiently similar to work together in individuals which are functionally efficient. Finally, four species may themselves be the products of ancient hybridization. These are *S. glaucophyllum*, *S. stenopetalum*, *S. borschii*, and *S. integrifolium*.

Chromosomes are another indication of genetic relationships. Both their number and shape are important. Some information is available for most species of *Sedum* in North America. Listing in the following table includes first the gametophytic number of chromosomes and then the names of species which have that number. Altogether, 29 species have no less than 20 numbers of chromosomes.

Number	Species
4	S. pusillum
5	—
6	S. nevii
7	S. rhodanthum
8	S. ternatum, S. lanceolatum, S. radiatum, S. leibergii, S. divergens, S. debile
9	—
10	S. nuttallianum
11	S. pulchellum, S. rosea
12	S. telephioides, S. wrightii, S. oreganum
13	—
14	S. glaucophyllum, S. cockerellii
15	S. obtusatum, S. laxum, S. oblanceolatum, S. albomarginatum, S. moranii, S. spathulifolium
16	S. ternatum, S. niveum, S. cockerellii, S. lanceolatum, S. rupicolum, S. borschii
17	S. cockerellii
18	S. integrifolium
19	—
20	—
22	S. glaucophyllum, S. pulchellum
24	S. telephioides, S. ternatum, S. wrightii, S. lanceolatum, (?) S. borschii
28	S. glaucophyllum
32	S. stenopetalum
33	S. pulchellum
36	S. wrightii
45	S. oregonense
64	S. niveum

The genetic variation of species, especially within local populations, reveals what morphological and physiological conditions are available for the work of natural selection. Data derived from plants in the wild are an imperfect measure of heritable variation because they include an unassessed amount of environmental effect. Data resulting from planned experiments, in which clones are replicated, can provide a basis for noting real differences among plants within local populations. The following table includes species of which two or more plants, each with two or more replicates, were grown in planned experiments at Ithaca, N.Y. Experiments were of completely random design. Two measures of genetic differences within species are available: the number of significant differences noted among populations and the number of populations within which significant differences were detected among plants.

	Species	Populations cultivated	Differences among populations	Populations with possible genetic variation
1.	S. telephioides	6	5	3
2.	S. ternatum	7	3	0
3.	S. nevii	3	5	1
4.	S. glaucophyllum	14	5	2
5.	S. pulchellum	7	5	2
6.	S. niveum	4	6	0
7.	S. cockerellii	3	1	—
8.	S. wrightii	1	—	1
9.	S. lanceolatum	32	10	0
10.	S. rupicolum	5	0	0
11.	S. stenopetalum	5	3	0

Species	Populations cultivated	Differences among populations	Populations with possible genetic variation
12. S. leibergii	2	0	0
13. S. borschii	3	3	4
14. S. divergens	3	0	0
15. S. debile	3	0	0
16. S. oreganum	4	1	1
17. S. obtusatum	12	6	2
18. S. laxum	5	8	2
19. S. oregonense	4	2	3
20. S. albomarginatum	1	—	0
21. S. moranii	1	—	1
22. S. spathulifolium	25	18	4
23. S. integrifolium	6	5	1
24. S. rosea	8	4	1

The above information on genetic variation is encumbered by various problems. Sometimes the most distinctive clones did not survive to yield data in experiments. For some species, larger replication might have detected significant differences. Yet, the results may be correct that some populations lack marked genetic variation. Since plants often reproduce easily vegetatively, all in some populations could have the same genetic constitution, in which case a whole population might comprise parts of a single clone. Further, even seedlings could have similar genotypes because of inbreeding in isolated populations, with the result that members of a population may belong to one or a few biotypes.

A discovery of the present century is that species of plants are of different kinds. Camp and Gilly (1943) provided a useful terminology for the kinds of species. Babcock (1947), in his monograph of *Crepis*, demonstrated the application of this terminology and indicated how it can help to make clear the genetic structure of a large and complex genus. The following table is an attempt to indicate the situation in North American *Sedum*. For definition of terms, see the paper by Camp and Gilly.

Kind of species	No. of species	% of species	Species
Homogeneon	7	23	S. pusillum, S. leibergii, S. divergens, S. nuttallianum, S. oblanceolatum, S. rhodanthum, and S. rosea.
Parageneon	5	17	S. nevii, S. debile, S. villosum, S. moranii, and S. albomarginatum.
Rheogameon	6	20	S. radiatum, S. oreganum, S. obtusatum, S. laxum, S. spathulifolium, and S. integrifolium.
Dysploidion	1	3	S. cockerellii.
Euploidion	6	20	S. telephioides, S. ternatum, S. pulchellum, S. wrightii, S. lanceolatum, and S. niveum.
Alloploidion	5	17	S. glaucophyllum (?), S. rupicolum (?), S. stenopetalum (?), S. borschii, and S. oregonense (?).
Totals	30	100	

For comparison, percentages for kinds of species in the Trans-Mexican Volcanic Belt are: homogeneon—17, parageneon—55, rheogameon—14, dysploidion—3.5, euploidion—7, and alloploidion—3.5. Estimates of percentages for the Trans-Mexican Volcanic Belt take into consideration changes of interpretation of species since 1959. Information about chromosomes of species of *Sedum* of the Trans-Mexican Volcanic Belt still does not match the data for species of North America north of the Mexican Plateau. Additional cytological information may move species from the category of parageneon into the categories of euploidion and alloploidion. However, such changes will not alter the interesting conclusion that, in both North America north of the Mexican Plateau and in the Trans-Mexican Volcanic Belt, not more than one-fifth of the species are rheogameons, a type of species common among vertebrate animals.

Summary

The thirty species of *Sedum* native in North America north of the Mexican Plateau comprise no less than eleven groups on the basis of natural relationships. These groups are not of equal rank and are variously related to each other.

Sedum telephioides, of the predominantly Eurasian subgenus *Telephium*, is the most primitive species of *Sedum* in temperate North America. In contrast, the two most advanced species are *S. pusillum* and *S. nuttallianum*, both annuals adapted to rigorous habitats. Although one of the three species of subgenus *Rhodiola*, *S. integrifolium*, has the most advanced pattern of floral anatomy known in the genus, it is less advanced in other respects.

Morphology is the primary basis for ideas about phylogeny of species, but other kinds of evidence also are important.

Most North American species of *Sedum* are chamaephytes. The optimum range of pH of soils in which they grow is 5.1–7. Peak months for flowering are June and July. Populations vary in number of plants from 1 to 968,562.

The potential for hybridization of species of *Sedum* in North America north of the Mexican Plateau is low. Of 435 interspecific gaps, only 5 are bridged by hybrids. The percentage of interspecific gaps with species on either side in the same square kilometer is 8 and in the same square meter is 3. Differences in number of chromosomes of species which occur together, and sometimes genetic incompatibilities when species have the same number of chromosomes, explain the small amount of interspecific hybridization.

Species of *Sedum* of North America north of the Mexican Plateau are of six kinds. Six species, 20%, are euploidions, the same number that are rheogameons. Seven species, 23%, are homogeneons and lack either subspecies or marked varieties.

The comparative study of relationships makes possible a phylogenetic arrangement of the species.

CHAPTER XII

Conclusions

1. Thirty species of *Sedum* occur naturally in North America north of the Mexican Plateau, an area of 21,000,000 sq. km. In comparison, 29 species occur naturally in the Trans-Mexican Volcanic Belt, which, with an area of about 97,000 sq. km., is about 216 times smaller.

Britton and Rose (1905) listed 38 species of *Sedum, Tetrorum, Gormania, Clementsia,* and *Rhodiola* in North America north of the Mexican Plateau. For the same area and taxonomic groups, Berger (1930) listed 34 species and Fröderström (1930–1935) listed 33 species. A comparison of the present interpretation of species with the treatments of Berger and Fröderström follows:

Species in this book	Species according to Berger	Species according to Fröderström
1. *S. telephioides*	*S. telephium*	*S. telephioides*
2. *S. ternatum*	*S. ternatum*	*S. ternatum*
3. *S. nevii*	*S. nevii*	*S. nevii*
4. *S. glaucophyllum*	*S. nevii*	*S. nevii*
5. *S. pulchellum*	*S. pulchellum*	*S. pulchellum*
		S. vigilmontis
6. *S. pusillum*	*S. pusillum*	*S. pusillum*
7. *S. niveum*	not listed	*S. niveum*
	S. pinetorum	*S. pinetorum*
8. *S. cockerellii*	not listed	*S. cockerellii*
	S. wootonii	*S. wootoni*
		S. anomiosepalum
9. *S. wrightii*	*S. wrightii*	*S. wrightii*
10. *S. lanceolatum*	*S. stenopetalum*	*S. stenopetalum*
	S. shastense	*S. shastense*
11. *S. rupicolum*	not listed	not listed
12. *S. radiatum*	*S. radiatum*	*S. radiatum*
13. *S. stenopetalum*	*S. douglasii*	*S. douglasii*
14. *S. leibergii*	*S. leibergii*	*S. leibergii*
15. *S. borschii*	not listed	not listed
16. *S. divergens*	*S. divergens*	*S. divergens*
17. *S. debile*	*S. debile*	*S. debile*
18. *S. nuttallianum*	*S. nuttallianum*	*S. nuttallianum*
19. *S. villosum*	*S. villosum*	*S. villosum*
20. *S. oreganum*	*S. oreganum*	*Gormania oregana*
21. *S. obtsatum*	*S. obtusatum*	*Gormania obtusata*
	S. hallii	
	S. burhamii	
	S. rubroglaucum	*Gormania rubroglauca*
	S. sanhedrinum	
22. *S. laxum*	*S. laxum*	not listed
	S. eastwoodiae	

Species in this book	Species according to Berger	Species according to Fröderström
23. S. oblanceolatum	not listed	not listed
24. S. oregonense	S. watsonii	Gormania watsonii
25. S. albomarginatum	not listed	not listed
26. S. moranii	not listed	not listed
27. S. spathulifolium	S. spathulifolium	S. spathulifolium
	S. californicum	S. californicum
	S. woodii	S. woodii
	S. pruinosum	S. pruinosum
	S. anomalum	Gormania anomala
	S. yosemitense	S. yosemitense
28. S. rhodanthum	S. rhodanthum	S. rhodanthum
29. S. integrifolium	S. roseum	S. roseum
30. S. rosea	S. roseum	S. roseum
Totals 30	34	33

2. The thirty species of *Sedum* of North America north of the Mexican Plateau are of six kinds in genetic structure: homogeneon—7, parageneon—5, rheogameon—6, dysploidion—1, euploidion—6, and alloploidion—5.

3. The most primitive species of *Sedum* of North America north of the Mexican Plateau is *S. telephioides*, a broad-leaved perennial herb with closest relatives in Europe and Asia. The most advanced species are *S. divergens*, *S. pusillum*, and *S. nuttallianum*. Two of these species are small, annual herbs of probable American origin. *Sedum integrifolium*, of subgenus *Rhodiola*, has the most advanced pattern of vascular anatomy of the flowers, with maximum fusion of bundles.

4. On the basis of natural relationships, the North American species of *Sedum* comprise no less than eleven groups. These groups are variously interrelated and unequal in rank. Morphological evidence is the primary basis for arranging the groups in phylogenetic sequence.

5. Six natural groups of species of *Sedum* in North America north of the Mexican Plateau probably originated elsewhere: subgenus *Telephium* in Europe or Asia, section *Cockerellia* on the Mexican Plateau, section *Lanceolata* either on the Mexican Plateau or in Asia, *S. villosum* in Europe, section *Clementsia* in Asia, and section *Rhodiola* in Asia or Europe. Five groups—sections *Ternata*, *Tetrorum*, *Oreganica*, *Gormania*, and *Rosulata*—are endemic and probably originated in North America. Of these, the section *Ternata* is least advanced.

6. The most widespread species of *Sedum* in North America north of the Mexican Plateau is *S. lanceolatum*. That occurs in 17 geomorphic provinces, from the Great Plains to the Pacific coastal provinces. Next most widespread are *S. integrifolium* and *S. stenopetalum*. In contrast, eight species occur in only one geomorphic province, and five of these are each endemic in the province in which it occurs.

7. Geomorphic provinces in North America with the maximum number of species of *Sedum* are the Klamath Mountains with ten species and Cascade Mountains with nine species.

8. Twenty-four of the thirty species of *Sedum* native in North America north of the Mexican Plateau, 80%, are endemic.

9. Sizes of populations of North American *Sedum* vary from a low of one plant in

three cases, each a different perennial species, to a high of 968,562 plants in a population of the annual *S. pusillum*. Two estimates of the median size of populations are 169 and 332 plants.

10. Current geographic distribution provides a low potential for interspecific hybridization. Of 435 interspecific gaps, only 5 are bridged by hybrids in nature. The frequency of interspecific gaps with species on either side growing in the same square kilometer is only 8%, and in the same square meter is 3%.

11. Twenty different diploid numbers of chromosomes occur in species of *Sedum* native in North America north of the Mexican Plateau. These numbers range from a low of 4 pairs in *S. pusillum* to a high of 45 pairs in *S. oregonense*. A report of a haploid number of 64 chromosomes for a plant of *S. niveum* from Lower California is the maximum for any species with range extending into the area covered.

12. Genetic differences among populations of a species tend to be more frequent than heritable differences among plants within populations. The ease with which plants of perennial species reproduce vegetatively may explain the comparative uniformity of populations. Likewise, the high degree of geographic isolation of populations, restricting chances for hybridization, leads to inbreeding and homogeneity.

13. Morphology is the primary basis for ideas about the phylogeny of the species. Advanced conditions include fusion of floral parts and vascular bundles, reduction in size of flowers and number of ovules, the annual habit, perigyny, dioecism, and development of corky lips on follicles and of wings on seeds.

14. Species of *Sedum* are remarkably free from diseases and predators. Exceptions are hypertrophy of leaves and flowers of *S. rosea* in the region of the Gulf of St. Lawrence, possibly caused by a virus; a geometrid larva feeding on leaves of *S. telephioides* in the Blue Ridge; and larvae of *Parnassius* feeding on stems of *S. lanceolatum* in the Southern and Central Rocky Mountains. In the greenhouse, both nematodes and mealy bugs are problems. Also, stems are subject to rot.

15. The principal human use of North American species of *Sedum* is cultivation for ornament. Twenty-two of the native species are cultivated. In the north, *S. rosea* is used as a salad and potherb. Rodozin, found in the roots of *S. rosea*, has medicinal properties.

16. Two species, both of subgenus *Gormania*, *Sedum oblanceolatum* in Oregon and *S. albomarginatum* in California, are described here for the first time. Also, *S. borschii* is elevated to specific status. New subspecies are three and new subspecific combinations are nine. In addition, *S. decumbens* is described from cultivation as new, *S. clavatum* is the name supplied for the *Sedum* of the Tiscalatengo Gorge, and three new subspecific combinations are made under names of species in cultivation. A peripheral new combination is *Acer nigrum* ssp. *saccharophorum*.

17. Sampling surveys provided representative data for a vast area of land and made possible estimates of sizes of populations, even when numbers of plants were enormous.

18. Two-thirds of the North American species of *Sedum* occur in the western part of the continent, suggesting that evolution of this genus has been more active there.

This, like the Trans-Mexican Volcanic Belt, is an area of diverse and disjunct environmental conditions.

Consideration of the Ten Propositions of Chapter I

1. The arrangement of species in groups, the arrangement of these groups in a sequence, and interpretations of the relative importance of variations within species have been on the assumption of an evolutionary process. The test of the validity of the classification will depend on whether additional evidence confirms or disproves it.

2. Unevenness has been evident at all levels in the hierarchy of variation of North American *Sedum*. This unevenness has involved local populations, subspecies, species, and groups of species. No level has seemed more natural than another. Each phyletic line possesses its own cohesiveness and is as natural as others.

3. The last part of proposition 3, namely that difficulties in definition and application exist at all levels in the taxonomic hierarchy, is the more important part. Decisions concerning the subspecies of *Sedum radiatum* and *S. stenopetalum* have been difficult. Likewise, the problem of species and subspecies in subgenus *Gormania* has been troublesome, no less so than problems at the generic level. Proposition 3 may be deleted or the last part combined with proposition 2.

4. Most local populations of *Sedum* in North America north of the Mexican Plateau are remarkably distinct. This fact has been a continuing surprise.

5. Only eight of the thirty species of *Sedum* in North America north of the Mexican Plateau comprise subspecies. Five species most likely arose through allopolyploidy. Others probably resulted from adjustment in number of chromosomes and not through geographic variation. The data do not contradict proposition 5, but suggest that subspecies may be less important in evolution of *Sedum* than in groups less prone to polyploidy and structural changes of chromosomes.

6. The concept of species is no more objective than the concepts of local population, subspecies, genus, or other taxonomic groups. The working hypothesis of proposition 6 that species should differ from each other in several correlated characteristics is essential for a useful classification. Abandonment of emphasis on differences would result in breaking up all dysploid and euploid species. The resulting microspecies would have as their primary means of recognition number of chromosomes. Such a classification of *Sedum* would be impractical and would separate plants which have in common many morphological and physiological features. The kinds of genes which are present are more important than the way that these are packaged or the number of times that they are replicated.

7. Differences between populations of species of *Sedum* with the same number of chromosomes, as in *S. spathulifolium*, or among individuals within populations, as in *S. obtusatum*, confirm the importance of gene mutation in the evolution of *Sedum*. Further, the existence of six species with eight pairs of chromosomes and a similar number of species with fifteen pairs of chromosomes, with some of these species occurring together in the same habitat without interbreeding, suggests the existence of incompatibilities at the genic level.

8. The study of the potential for interbreeding revealed that only 8% of the species on either side of interspecific gaps occur in the same square kilometer. This is testimony that most species occur in different environments. Many are ecospecies. Yet, only 20% of the species are rheogameons which have developed geographic subspecies with ecotypic status. Data from *Sedum* of North America north of the Mexican Plateau favor rewording the proposition to delete universal, but to retain important.

9. The twenty different numbers of chromosomes found in North American species of *Sedum* indicate that differences in numbers of chromosomes are important in isolating populations and can be significant in the evolution of species. Similarly, the evident spatial isolation of the majority of species is a key to understanding the patterns of variation which prevail. The ideas of proposition 9 are in harmony with the evidence for *Sedum* in North America north of the Mexican Plateau.

10. The five species listed as alloploidions, namely *S. glaucophyllum*, *S. rupicolum*, *S. stenopetalum*, *S. borschii*, and *S. oregonense*, if properly interpreted, are products of interspecific hybridization. Any evolutionary process which accounts for as many as one-sixth of the species is important.

General Conclusions

1. The data for *Sedum* of North America north of the Mexican Plateau suggest that the species have come into being in different ways. Only 20% of the species exhibit the pattern of geographic variation which is widely accepted as a usual way in which evolution proceeds. Almost as many species, 17%, may have originated through allopolyploidy. Although only one species, 3% of the total, is dysploid, six of the seven species which are homogeneons are dysploid with respect to each other, and probably originated in this way rather than as geographic subspecies. The same comment applies to three of the five parageneons. The remaining 20% of the species exhibit autoployploidy. The data seem more to support the critique of the neo-Darwinist position by Goldschmidt (1940) than to confirm commonly accepted views about geographic variation and the gradual accumulation of small genic differences.

2. Most North American species of *Sedum* are noncommunity plants. They thrive where other plants do not grow, in open rocky situations. They have few associates or competitors belonging to other species. These other species are miscellaneous and do not provide a pattern of association with *Sedum*. Often the plants of *Sedum* form nearly pure stands.

3. Planned sampling surveys were useful in the present study to obtain representative data. A lesson learned from them is to adopt a good plan in the beginning and then to make as few changes as possible after the work has started.

4. Projects such as this study of *Sedum* require prolonged attention, longer than most sponsors consider necessary. Yet, the amount of detail which is essential for clarity of understanding requires many years of investigation and unhurried thought. No quick and magic route exists to explain the mysteries of the universe, of which *Sedum* is a part.

Gazetteer

Entries in this gazetteer include all localities at which I have made detailed studies of populations of *Sedum*, and in addition, all type localities of species, subspecies, and varieties. Three items of information are available for each locality: the geomorphic province, designated by an abbreviation, using the designations provided for the map in the chapter on geology, see p. 43; the location as a distance in kilometers and compass direction from a locality shown in the School Atlas of Goode (1943); and the appropriate topographic map of the U.S. Geological Survey or Canadian Hydrographic Chart, designated by name of map or chart, state or province, and scale. Sometimes, when both large- and small-scale maps are now available, I have cited the small-scale map when this was the only one accessible at the time when I visited the locality. Consultation of the cited maps will provide information about coordinates and altitudes.

Affligées—see Isles Affligées.
Alaska Basin—CRM, high, broad valley on western slope of Teton Mts. 6 km. SSW of Grand Teton; Grand Teton Natl. Park, Wyo., 1:62,500.
Allisonia—FA, town on New R. 13 km. SSE of Pulaski, Va.; Macks Mountain, Va., 1:62,500.
Almeda—KM, mine 3 km. N of Galice in valley of Rogue R. 30 km. NW of Grants Pass, Ore.; Galice, Ore., 1:62,500.
Arroyo del Puerto—CCR, creek on W side of Diablo Range 40 km. SW of Modesto, Calif.; Mt. Boardman, Calif., 1:62,500.
Bald Knob—FA, mountain, S of Mountain Lake and 24 km. NNW of Radford, Va.; Pearisburg, Va., 1:62,500.
Bald Mountain—CM, mountain, 23 km. NW of Mt. Baker, Wash.; Mt. Baker, Wash., 1:62,500.
Bandon—CR, city on Pacific Coast of Oregon, 43° 5′ N, 124° 25′ W; Bandon, Ore., 1:62,500.
Barnes Gap—FA, gap in Town Hill, 33 km. SE of Bedford, Pa.; Bellegrove, Md.–Pa., 1:24,000.
Bear Gap—CM, gap in ridge of Cascade Mts., 22 km. ENE of Mt. Rainier, Wash.; Mt. Aix, Wash., 1:125,000.
Bear Mountain (Calif.)—CCR, mountain, 11 km. NE of Healdsburg, Calif.; Jimtown, Calif., 1:24,000.
Bear Mountain (Ore.)—CR, mountain, 26 km. W of Roseburg, Ore.; Camas Valley, Ore., 1:62,500.
Beard's Hollow—CR, cove on W side of Cape Disappointment, 2 km. W of Ilwaco, Wash.; Cape Disappointment, Wash., 1:62,500.
Beartooth Lake—CRM, lake draining into Clarks Fork of Yellowstone River, 63 km. NW of Cody, Wyo.; Beartooth Butte, Wyo., 1:62,500.
Beartooth Summit—CRM, crest of ridge of Beartooth Mts. on Cooke City Highway, 59 km. NW of Cody, Wyo.; Deep Lake, Wyo., 1:62,500.
Beaver Creek Road—KM, road up valley of Beaver Creek and NW ridge of Dutchman's Peak, 27 km. SW of Medford, Ore.; Talent, Ore.–Calif., 1:62,500.

Berkshire (Vt.)—NEA, town 34 km. NE of St. Albans, Vt.; Enosburg Falls and Jay Peak, Vt., 1:62,500.

Berthoud Pass—SR, pass in Colorado Front Range, 27 km. WNW of Idaho Springs, Colo.; Berthoud Pass, Colo., 1:24,000.

Big Gull Island—Gr P, island, 9.1 km. SW of Mutton Bay, Que.; Can. Hydrographic Chart 4469, Flat I. to Little Mecatina I., 1:76,200.

Bluffs Mountain—BR, mountain, 25 km. NNW of North Wilkesboro, N.C.; Wilkesboro, N.C., 1:125,000.

Bonita Peak—SR, peak in San Juan Mts., 10 km. NE of Silverton, Colo.; Handies Peak, Colo., 1:24,000.

Breeze Creek—SN, tributary of Rancheria Creek, Yosemite Natl. Park, Calif.; Hetch Hetchy Reservoir and Tower Peak, Calif., 1:62,500.

Briery Branch Gap—FA, gap in Shenandoah Mountain, 32 km. W of Harrisonburg, Va.; McDowell, Va.–W. Va., 1:62,500.

Brownsville—CP, city in southern Texas, 25° 55′ N, 97° 30′ W; West Brownsville, Tex., 1:24,000.

Buck Spring Gap—BR, gap on Blue Ridge Parkway, 25 km. SW of Asheville, N.C.; Dunsmore Mountain, N.C., 1:24,000.

Buckhorn—GP, place in Black Hills, 36 km. SE of Sundance, Wyo.; Sundance, Wyo.–S.D., 1:125,000.

Burnt Mills—PP, town on N. Branch of Raritan R., 23 km. W of Plainfield, N.J.; Gladstone, N.J., 1:24,000.

Burnt Mountain—CR, mountain, not named on map, drained by forks of Brummit Creek and 30 km. E of Coquille, Ore.; Sitkum, Ore., 1:62,500.

Cabins—AP, place 8 km. W of Petersburg, W. Va.; Petersburg, W. Va.–Va., 1:62,500.

Cameron Pass—SR, pass between Medicine Bow Mts. and Colo. Front Range, 7 km. W of NW corner of Rocky Mt. Natl. Park; Home, Colo.–Wyo., 1:125,000.

Carberry Creek—KM, tributary of Applegate R., 45 km. SSE of Grants Pass, Ore.; Ruch, Ore.–Calif., 1:62,500.

Carthage—ILP, city on Cumberland R., 47 km. ESE of Gallatin, Tenn.; Carthage, Tenn., 1:62,500.

Cascade Canyon—CRM, canyon of Cascade Creek, 2.5 km. north of the Grand Teton; Grand Teton Natl. Park, Wyo., 1:62,500.

Cascades—FA, rocky part of valley of Little Stony Creek, west of Mountain Lake and 25 km. N of Radford, Va.; Pearisburg, Va., 1:62,500.

Casmaila Beach—CCR, beach on Pacific Coast, 28 km. NW of Lompoc, Calif.; Point Sal, Calif., 1:62,500.

Catawba River—PP, name for upper Wateree R. in Carolinas; site in South Carolina is 18 km. NNW of Lancaster; Monroe, N.C.–S.C., 1:125,000.

Cedar Cliff Mountain—BR, small mountain, 5 km. east of Tuckasegee and 26 km. SSW of Waynesville, N.C.; Tuckasegee, N.C., 1:24,000.

Chaneysville—FA, town in valley of Town Creek, 21 km. S of Bedford, Pa.; Clearville, Pa., 1:62,500.

Charlton Peak—TR, mountain, 3 km. NW of San Gorgonio Mt., highest peak in San Bernardino Mts., Calif.; San Gorgonio Mt., Calif., 1:62,500.

Chico—GV, city in valley of Sacramento R., Calif., 39° 43′ N, 121° 51′ W; Chico, Calif., 1:24,000.

Chilnualna Falls—SN, waterfalls of Chilnualna Creek, 5 km. NE of Wawona and 17 km. S of Yosemite Valley, Yosemite Natl. Park, Calif.; Yosemite, Calif., 1:62,500.

Chinese Wall—MP, rocky ridge in Chisos Mts., Big Bend Natl. Park, 130 km. SSE of Alpine, Texas; Emory Peak, Texas–Mexico, 1:250,000.

Chinook—CR, town on N side of Columbia R., 8 km. ESE of Ilwaco, Wash.; Chinook, Wash., 1:24,000.

Chittenango Falls—AP, falls of Chittenango Creek, 4 km. N of Cazenovia, N.Y.; Cazenovia, N.Y., 1:24,000.

Chopaka Mountain—CM, mountain, 23 km. WNW of Oroville, Wash.; Loomis, Wash., 1:62,500.

City Creek—WR, creek on north side of Salt Lake City, Utah, arising in mountains to northeast; Ft. Douglas, Utah, 1:125,000, and Salt Lake City, Utah, 1:250,000.

Clear Creek—SR, creek entering South Platte R. just N of Denver, Colo.; Empire, Colo., 1:24,000, and Denver, Colo., 1:250,000.

Clements Mountain—NR, mountain on Continental Divide and west of Logan Pass, Glacier Natl. Park, 56 km. WNW of Browning, Mont.; Chief Mountain, Mont., 1:125,000.

Cleveland (Va.)—FA, town on Clinch R., 30 km. NW of Abingdon, Va.; Carbo, Va., 1:24,000.

Clifton Forge—FA, city in Virginia, 37° 50′ N, 79° 50′ W; Clifton Forge, Va., 1:24,000.

Cold Mountain—BR, mountain between Little East Fork and East Fork of Pigeon R., 15 km. ESE of Waynesville, N.C.; Cruso, N.C., 1:24,000.

Cold Spring—FA, town, 9 km. S of Beacon, N.Y.; West Point, N.Y., 1:24,000.

Colton—Gr P, town on Raquette R., 13 km. SSE of Potsdam, N.Y.; Colton, N.Y., 1:24,000.

Columbia River—CR, river forming boundary between states of Oregon and Washington; Hoquiam, Wash.–Ore., 1:250,000.

Cook—CM, town on N bank of Columbia R., 61 km. ENE of Camas, Wash.; Hood River, Ore.–Wash., 1:62,500.

Coos Ridge—CR, ridge N of S Fork of Coos R., 30 km. E of Coos Bay, Ore.; Ivers Peak, Ore., 1:62,500.

Corpus Christi—CP, city on Gulf Coast of Texas, 27° 48′ N, 97° 24′ W; Corpus Christi, Tex., 1:24,000.

Cottonwood Creek—KM, creek entering Sacramento R. below Anderson, Calif.; Chanchelulla Peak and Ono, Calif., 1:62,500.

Craggy Pinnacle—BR, mountain, 11 km. SW of Mt. Mitchell, N.C.; Craggy Pinnacle, N.C., 1:24,000.

Crawford Creek—BR, tributary of East Fork of Pigeon R., 18 km. ESE of Waynesville, N.C.; Cruso, N.C., 1:24,000.

Crescent City—CCR, city on Pacific Coast in NW California, 41° 45′ N, 124° 13′ W; Crescent City and Sister Rocks, Calif., 1:24,000.

Crows—FA, town, 20 km. SW of Covington, Va.; New Castle, Va., 1:62,500.

Cuba (N.M.)—SR, town in valley of Rio Puerco, 100 km. WNW of Santa Fe, N.M.; Cuba, N.M., 1:62,500.

Custer—GP, city in Black Hills of South Dakota, 43° 46′ N, 103° 36′ W; Harney Peak, S.D., 1:125,000.

Dalton Canyon—SR, canyon on eastern slope of Sangre de Cristo Mts., 20 km. ESE of Santa Fe, N.M., McClure Reservoir, N.M.; 1:24,000, and Santa Fe, N.M., 1:125,000.

Dartmouth Creek—CM, tributary of North Fork of Middle Fork of Willamette R., 48 km. SE of Springfield, Ore.; Sardine Butte, Ore., 1:62,500.

Deadman's Point—KM, NW end of ridge NW of Dutchman's Peak, 29 km. S of Medford, Ore.; Talent, Ore.–Calif., 1:62,500.

Deadwood—GP, city in Black Hills of South Dakota, 44° 23′ N, 103° 42′ W; Deadwood, S.D., 1:125,000.
Deer Butte—CM, knob, 55 km. SSW of Mt. Jefferson, Ore.; Three Sisters, Ore., 1:62,500.
Deerfield—FA, town, 30 km. WNW of Staunton, Va.; Craigsville, Va., 1:62,500.
Devils River—GP, tributary of Rio Grande, entering R. above Del Rio, Tex.; Devils River and Del Rio, Tex., 1:62,500.
Dry Gulch—SR, valley on western side of Laramie Range, 17 km. ESE of Laramie, Wyo.; Sherman Mountains, Wyo., 1:62,500.
Dryden Lake—AP, lake south of Dryden and 18 km. E of Ithaca, N.Y.; Dryden, N.Y., 1:24,000.
Dutchman's Peak—KM, mountain, 30 km. S of Medford, Ore.; Talent, Ore.–Calif., 1:62,500.
Earl Peak—CM, peak on ridge S of Ingalls Creek, 25 km. SW of Leavenworth, Wash.; Mt. Stuart, Wash., 1:125,000.
East Humboldt Mountains—GB, range of mountains SW of Wells, Nev.; Wells, Nev., 1:250,000.
El Paso—MP, city in western Texas, 31° 46′ N, 106° 28′ W; El Paso, Tex., 1:24,000.
El Portal—SN, town in valley of Merced R. at W entrance to Yosemite Natl. Park, Calif.; El Portal, Calif., 1:62,500.
Elliott Corner—SN, crossroad in Mariposa Co., Calif., 32 km. SW of Yosemite Valley; Stumpfield Mt., Calif., 1:24,000.
Elwood Pass—SR, pass in San Juan Mts., 8 km. NW of Summit Peak, Colo.; Summitville, Colo., 1:125,000.
Esopus Creek—AP, creek entering Hudson R. N of Kingston, N.Y.; Ashokan, N.Y., 1:24,000.
Etna Mills—KM, town in northern California, 37 km. SW of Yreka; Etna, Calif., 1:62,500.
Eureka Gulch—SR, valley, 10 km. NE of Silverton, Colo.; Howardsville, Colo., 1:24,000.
Eureka Peak—SN, mountain, 22 km. SW of Portola, Calif.; Sierra City, Calif., 1:62,500.
Eureka Springs—OP, city in Arkansas, 36° 24′ N, 93° 44′ W; Eureka Springs, Ark., 1:125,000.
Feather River—SN, tributary of Sacramento R., Calif.; Chico and Susanville, Calif., 1:24,000.
Fern Canyon—MP, canyon in Davis Mts., 19 km. N of Alpine, Tex.; Fort Davis, Tex., 1:125,000.
Fincastle—FA, town, 22 km. N of Roanoke, Va.; Oriskany, Va., 1:24,000.
Flat Rock—PP, a ridge of granite, 24 km. N of Camden, S.C.; Camden, S.C., 1:62,500.
Forbes Creek—CCR, creek entering Clear Lake on south side of Lakeport, Calif.; Lakeport, Calif., 1:24,000.
Forsee Creek—TR, creek draining NW slope of San Bernardino Mt., 25 km. NE of Redlands, Calif.; San Gorgonio Mt., Calif., 1:62,500.
Forsyth Island—Gr P, island, 6.1 km. SW of Mutton Bay, Que.; Can. Hydrographic Chart 4469, Flat I. to Little Mecatina I., 1:76,200.
Forty-acre Rock—PP, a granitic flat rock, 22 km. ESE of Lancaster, S.C.; Monroe, N.C.–S.C., 1:125,000.
Fremont Peak—Co P, mountain, 14 km. N of Flagstaff, Ariz.; Flagstaff, Ariz., 1:125,000.
Front Royal—FA, city in Va., 38° 56′ N, 78° 11′ W; Front Royal, Va., 1:62,500.
Frozen Lake—CRM, small lake near headwaters of Canyon Creek, a tributary of the Clarks Fork of the Yellowstone R., 60 km. NW of Cody, Wyo.; Deep Lake, Wyo., 1:62,500.
Galice—KM, place in valley of Rogue R., 27 km. NW of Grants Pass, Ore.; Galice, Ore., 1:62,500.
Garden of the Gods—ILP, bluffs in Shawnee Natl. Forest, SE Saline Co., 19 km. SE of Harrisburg, Ill.; Herod, Ill., 1:24,000.
Gavillan Peak—CCR, also rendered as Gabilan Peak, mountain, 16 km. NE of Salinas, Calif.; San Juan Bautista, Calif., 1:62,500.

Gem Pass—SN, pass in ridge E of Blacktop Peak, 33 km. SSW of Mono Lake, Calif.; Mono Craters, Calif., 1:62,500.

Gennargentu Mountains—Europe, range of mountains in central Sardinia, Italy; Touring Club Italiano, Carta Automobilistica d'Italia, sheet 29, 1:200,000.

Gibbs Peak—SR, peak in Sangre del Cristo Range, western Custer Co., Colo., 48 km. SW of Cañon City; Pueblo, Colo., 1:250,000.

Gilpin—FA, town, 17 km. ENE of Cumberland, Md.; Flint Stone, Md.–Pa., 1:24,000.

Glacier Way—CM, valley of Glacier Creek, 54 km. S of Mt. Jefferson, Ore.; Three Sisters, Ore., 1:62,500.

Glassy Mountain—BR, mountain, 31 km. NE of Greenville, S.C.; Tigerville, S.C.–N.C., 1:62,500.

Glenora—AP, point on western side of Seneca Lake, N.Y., 12 km. N of the SW corner of the lake; Reading Center, N.Y., 1:24,000.

Goat Rock Dam—PP, dam on Chattahoochee R., Harris Co., Ga., 20 km. N of Columbus; Smiths, Ala.–Ga., 1:24,000.

Golconda—ILP, city on Ohio R. in southern Illinois, 37° 22′ N, 88° 29′ W; Golconda, Ky.–Ill., 1:62,500.

Gold Lake—SN, lake, 20 km. NE of Downieville, Calif.; Sierra City, Calif., 1:62,500.

Goode's Ferry—PP, site of bridge across Roanoke R., 69 km. ESE of South Boston, Va.; South Hill, N.C.–Va., 1:62,500.

Grant's Springs—SN, Sulphur Springs, NE of Miami Mt., 5.6 km. SE of Elliott Corner and 30 km. SSW of Yosemite Valley, Calif.; Bass Lake, Calif., 1:62,500.

Grave Creek—KM, tributary of Rogue R., 21 km. N of Grants Pass, Ore.; Galice and Glendale, Ore., 1:62,500.

Groghan Hole—KM, hollow, 1 km. S of North Trinity Mt. and W of Trinity Summit Guard Station, at confluence of Oregon and Tish Tang A Tang Creeks, about 50 km. ENE of Fieldbrook, Calif.; Salmon Mt., Calif., 1:62,500.

Guerrero (Chih.)—MP, town, 140 km. W of city of Chihuahua, Chih.; Chihuahua, Mexico, 1:1,000,000.

Gulf of Georgia—WPT, large body of water, N of San Juan Islands, between NW Wash. and southern Vancouver Island, B.C.; Victoria, Wash.–Canada, 1:250,000.

Gull Island—Gr P, small island about 3 km. W of Hospital I. and 22 km. NNE of St. Mary Is., Que.; Can. Hydrographic Chart 4468, Little Mecatina I. to St. Mary Is., 1:75,000.

Haines Falls—AP, town, 18 km. WSW of Catskill, N.Y.; Kaaterskill, N.Y., 1:24,000.

Harney Peak—GP, highest mountain in Black Hills of South Dakota, 43° 52′ N, 103° 30′ W; Harney Peak, S.D., 1:125,000.

Hart's Pass—CM, pass in crest of Cascade Mts., 85 km. E of Mt. Baker, Wash.; Concrete, U.S.–Canada, 1:250,000.

Heavens Gate—Col P, mountain in Seven Devils Mts., 64 km. SSW of Grangeville, Ida.; Riggins, Ida., 1:125,000.

Heggie's Rock—PP, granitic flatrock, 25 km. NW of Augusta, Ga.; not named, but 5.5 km. NW of Grovetown and W of Oak Lake, on Grovetown, Ga., 1:24,000.

Hetch Hetchy—SN, reservoir and canyon of the Tuolumne R. in Yosemite Natl. Park, Calif.; Lake Eleanor and Hetch Hetchy Reservoir, Calif., 1:62,500.

Hilton Lake 4—SN, lake NE of Mt. Huntington and 36 km. NW of Bishop, Calif.; Mt. Abbot, Calif., 1:62,500.

Holt—AP, town, 9 km. ENE of Tuscaloosa, Ala.; Cottondale, Ala., 1:62,500.

Hopland Grade—CCR, grade in Mayacamas Mts. of highway from Hopland to Lakeport, Calif.; Highland Springs, Calif., 1:24,000.

Hospital Island—Gr P, island with Grenfell Hospital and Harrington Harbour, 22 km. NNE of St. Mary Is., Que.; Can. Hydrographic Chart 4468, Little Mecatina I. to St. Mary Is., 1:75,000.

Ingalls Creek—CM, tributary of Peshastin Creek, S of Mt. Stuart and 17 km. S of Leavenworth, Wash.; Mt. Stuart, Wash., 1:125,000.

Ingleside—FA, town, 17 km. ENE of Bluefield, W. Va.; Princeton, W. Va., 1:24,000.

Ione—SN, town, 14 km. W of Jackson, Calif.; Ione, Calif., 1:24,000.

Iron Mountain—SR, mountain, SE of Cameron Pass and 8 km. W of NW corner of Rocky Mt. Natl. Park, Colo.; Home, Colo.–Wyo., 1:125,000.

Isle Nadeau—Gr P, island, 17 km. SW of Mutton Bay, Que.; Can. Hydrographic Chart 4469, Flat I. to Little Mecatina I., 1:76,200.

Isles Affligées—Gr P, islands, S of Grand Rigolet, 16 km. NNE of Mecatina I., Que.; Can. Hydrographic Chart 4474, Bun Is. to Mutton Bay, 1:36,500.

Ithaca—AP, city in New York, 42° 27′ N, 76° 31′ W; Ithaca East, N.Y., 1:24,000.

Jamesville—AP, town, 8 km. SE of Syracuse, N.Y.; Jamesville, N.Y., 1:24,000.

Joe Wright Creek—SR, tributary of Fall R., N of Cameron Pass and 10 km. W of NW corner of Rocky Mt. Natl. Park, Colo.; Home, Colo.–Wyo., 1:125,000.

Joplin—OP, city in Missouri, 37° 5′ N, 94° 31′ W; Joplin East, Mo., 1:24,000.

Jordan Creek—OP, tributary of Sinking Creek, 15 km. SE of Greenfield, Mo.; South Greenfield and Everton, Mo., 1:24,000.

Kécarpoui Island—Gr P, = Isle du Milieu, island near mouth of Kécarpoui R., 18 km. NNE of Mecatina I., Que.; Can. Hydrographic Chart 4474, Bun Is. to Mutton Bay, 1:36,500.

Keddie—SN, place on railroad, 35 km. S of Westwood, Calif.; Greenville, Calif., 1:62,500.

Kelly Rock—PP, a granitic flatrock, 5 km. NW of Flat Rock and 27 km. NNW of Camden, S.C.; Camden, S.C., 1:62,500.

Kinney Point—Col P, peak on western side of Seven Devils Mts., 95 km. SSW of Grangeville, Ida.; Cuprum, Ida.–Ore., 1:62,500.

Klotz—FA, town on New River, 27 km. NNW of Pearisburg, Va.; Pearisburg, Va., 1:62,500.

Knoxville—FA, city in Tennessee, 35° 58′ N, 83° 57′ W; Knoxville, Tenn., 1:24,000.

La Cueva—MP, place on W slope of Organ Mts., 17 km. E of Las Cruces, N.M.; Las Cruces, N.M., 1:125,000, and Organ Peak, N.M., 1:62,500.

Lake Eleanor—SN, reservoir, 6 km. NW of Hetch Hetchy Reservoir, Yosemite Natl. Park, Calif.; Lake Eleanor, Calif., 1:62,500.

Lake Emma—SR, lake at elev. of 3,730 m. on NE side of Bonita Peak, in western San Juan Mts., 7 km. NNE of Silverton, Colo.; Silverton, Colo., 1:62,500.

Lake Peak—SR, mountain, 18 km. NE of Santa Fe, N.M.; Aspen Basin, N.M., 1:24,000.

Lake Tahoe—SN, lake on border between California and Nevada, 39° 2′ N, 120° 0′ W; Tahoe, Calif.–Nev., 1:62,500, and Fallen Leaf Lake, Calif., 1:62,500.

Leaksville-Spray—PP, two towns on Dan R. in Rockingham Co., N.C., now called Eden, 36° 31′ N, 79° 45′ W; Spray, N.C.–Va., 1:24,000.

Lebanon—ILP, city in Tennessee, 36° 12′ N, 86° 17′ W; Lebanon, Tenn., 1:24,000.

Lithonia—PP, town in DeKalb Co., Ga., 10 km. NW of Conyers, Ga.; Conyers, Ga., 1:24,000.

Little Elk Creek—GP, tributary of Elk Creek in Black Hills, 20 km. NE of Rapid City, S.D.; Rapid, S.D., 1:125,000.

Little Moose Island—NEA, small island, 16 km. ESE of Bar Harbor, Me.; Acadia Natl. Park, Me., 1:24,000.

Little Yosemite Valley—SN, valley of Merced R. between Nevada Fall and Bunnel Cascade, Yosemite Natl. Park, Calif.; Merced Peak and Yosemite, Calif., 1:62,500.

Llano Grande Canyon—CCR, canyon on western slope of Gabilan Range, 14 km. NW of King City, Calif.; Greenfield, Calif., 1:62,500.

Lolo Creek—IB, tributary of Bitterroot R., SW of Missoula, Mont.; Missoula, Mont.–Ida., 1:125,000.

Long Island—CR, island in southeastern part of Willapa Bay, 15 km. NE of Ilwaco, Wash.; Long Island, Wash., 1:24,000.

Lookout Mountain—FA, SW of Chattanooga, Tenn.; Fort Oglethorpe, Ga.–Tenn., 1:24,000.

Lower Lake—CCR, town at SE end of Clear Lake, Calif.; Lower Lake, Calif., 1:24,000.

Lulu Pass—SR, high pass, also called Thunder Pass, in Colorado Front Range, on NW boundary of Rocky Mt. Natl. Park; Fall River Pass, Colo., 1:24,000.

Lummi Island—WPT, island in Puget Sound, 16 km. W of Bellingham, Wash.; Anacortes, Wash., 1:62,500.

Lummi Rocks—WPT, rocks off western shore of Lummi Island, 17 km. WSW of Bellingham, Wash.; Anacortes, Wash., 1:62,500.

Luray—FA, city in Virginia, 38° 39′ N, 78° 28′ W; Stony Man, Va., 1:62,500.

Macomber Peak—SR, mountain, 6 km. NE of Silverton, Colo.; Howardsville, Colo., 1:24,000.

Makanda—ILP, town, 13 km. S of Carbondale, Ill.; Carbondale, Ill., 1:24,000.

Manning Creek—CCR, creek draining eastern slope of Mayacamas Mts. and entering Clear Lake S of Lakeport, Calif.; Highland Springs and Lakeport, Calif., 1:24,000.

Manning Flat—CCR, flat area, 5 km. S of Wheeler Point on S side of Clear Lake, Calif.; Clearlake Highlands, Calif., 1:24,000.

Marks Butte—NR, mountain, 35 km. NNE of Elk R., Ida.; Spokane, Wash.–Ida., 1:250,000.

Middle Brummit Creek—CR, Middle Fork of East Fork of Brummit Creek, a tributary of East Fork of Coquille R., 30 km. E of Coquille, Ore.; Sitkum, Ore., 1:62,500.

Midway—PP, a small granitic flat rock along highway 265 about 27 km. ESE of Lancaster, S.C.; Monroe, N.C.–S.C., 1:125,000.

Millboro Spring—FA, town, 27 km. NE of Clifton Forge, Va.; Millboro, Va., 1:62,500.

Millicoma River—CR, tributary of Coos R., with confluence 9 km. E of Coos Bay, Ore.; Coos Bay and Ivers Peak, Ore., 1:62,500.

Millrift—AP, locality on Delaware R., 6 km. NW of Port Jervis, N.Y.; Port Jervis North, N.Y., 1:24,000.

Milwaukee (Ore.)—WPT, suburb of Portland, Ore., between Portland and Oregon City; Oregon City, Ore., 1:62,500.

Mission Peak—CM, also called Old Baldy, mountain on ridge of Wenatchee Mts., 17 km. SW of Wenatchee, Wash.; Wenatchee, Wash., 1:62,500.

Monroe Canyon—GP, canyon of a tributary of Hat Creek, 13 km. N of Harrison, Neb.; Alliance, Neb., 1:250,000.

Monte Sano—AP, mountain, E of Huntsville, Ala.; Huntsville and Meridianville, Ala., 1:24,000.

Moore Creek—GV, tributary of Sour Grass Creek, 35 km. N of Willows, Calif.; Kirkwood, Calif., 1:24,000.

Moose—CRM, post office, 12 km. SE of the Grand Teton; Grand Teton Natl. Park, Wyo., 1:62,500.

Morrison Ridge—Col P, ridge, extending eastward from Heavens Gate between South Fork of Rapid R. and Shingle Creek, in Seven Devils Mts., 80 km. SSW of Grangeville, Ida.; Riggins, Ida., 1:125,000.

Morro Bay—CCR, bay, 17 km. WNW of San Luis Obispo, Calif.; Cayucos, 1:62,500.

Mount Adams—CM, mountain in southern Washington; Steamboat Mt. and Mt. Adams, Wash., 1:125,000.

Mount Eddy—KM, mountain, 14 km. W of town of Mt. Shasta, Calif.; Weed, Calif., 1:62,500.

Mount Flora—SR, mountain, 22 km. WNW of Idaho Springs, Colo.; Empire, Colo., 1:24,000.

Mount Hoffmann—SN, mountain, 11 km. NNE of E end of Yosemite Valley, Yosemite Natl. Park, Calif.; Hetch Hetchy Reservoir, Calif., 1:62,500.

Mount Horrid—NEA, mountain in Green Mts., 11 km. ENE of Brandon, Vt.; Rochester, Vt., 1:62,500.

Mount Linn—CCR, highest peak in South Yolla Bolly Mts., 56 km. WSW of Red Bluff, Calif.; Yolla Bolly, Calif., 1:62,500.

Mount Morgongiori—Europe, mountain in western Sardinia, Italy: Touring Club Italiano, Carta Automobilistica d'Italia, sheet 29, 1:200,000. (Map shows Morgongiori, but not the mountain).

Mount Shasta—CM, mountain in northern California, 41° 25′ N, 122° 12′ W; Mt. Shasta, Calif., 1:62,500.

Mount Stuart—CM, highest peak in Wenatchee Mts., 22 km. SW of Leavenworth, Wash.; Mt. Stuart, Wash., 1:125,000.

Mountain Home Creek—TR, tributary of Mill Creek, in San Bernardino Mts., 20 km. NE of Redlands, Calif.; Redlands and San Gorgonio, Calif., 1:62,500.

Mountain Lake—FA, lake, 25 km. NNW of Radford, Va.; Pearisburg, Va., 1:62,500.

Mud Lake (Ore.)—CM, small lake, 6 km. SSW of Horse Mt., 84 km. ESE of Springfield, Ore.; Maiden Peak, Ore., 1:125,000.

Nagai Island—Al P, longest island of Shumagin Is., S of Alaska Peninsula; Port Moller, Simeonof I., and Stepovak Bay, Alaska, 1:250,000.

Nashville—ILP, city in Tennessee, 36° 10′ N, 86° 49′ W; Nashville West, Tenn., 1:24,000.

Natashquan—Gr P, town on north shore of Gulf of St. Lawrence, Que., 50° 10′ N, 61° 47′ W; Mingan–Cape Whittle, Que., 1:506,880.

Nockamixon Rocks—PP, north-facing cliffs overlooking Delaware R., 15 km. SSE of Easton, Pa.; Easton, Pa., 1:62,500.

Noel—OP, town, 37 km. SSW of Neosho, Mo.; Noel, Mo., 1:62,500.

North Buffalo Fork—CRM, tributary of Snake R., 35 km. E of Jackson Lake, Wyo.; Mt. Leidy, Wyo., 1:125,000.

North Mountain (Collierstown, Va.)—FA, mountain west of Collierstown, 20 km. W of Lexington, Va.; Millboro, Va., 1:62,500.

Oak Creek Canyon—Co P, canyon of Oak Creek, 26 km. SSW of Flagstaff, Ariz.; Camp Verde, Ariz., 1:125,000, and Mountainaire, Ariz., 1:24,000.

Ocoee River—BR, tributary of Hiwassee R., SE Tenn., W of Ducktown, Tenn.; Caney Creek, Tenn., 1:24,000.

Old Hollowtop—NR, mountain, WSW of Pony and 65 km. SE of Butte, Mont.; Harrison, Mont., 1:62,500.

O'Leary Mountain—CM, mountain, 8 km. SSW of McKenzie Bridge and 80 km. E of Eugene, Ore.; McKenzie Bridge, Ore., 1:62,500.

Oregon City—WPT, city of Oregon in valley of Willamette R., 45° 21′ N, 122° 37′ W; Oregon City, Ore., 1:62,500 and 1:24,000.

Papoose Creek—Col P, tributary of Squaw Creek on eastern slope of Seven Devils Mts., 68 km. SSW of Grangeville, Ida.; Riggins, Ida., 1:125,000.

Peking—Asia, capital city of China, 39° 55′ N, 116° 25′ E; U.S. Army Map Service Series L–500, Peiping, China, 1:250,000.

Pembroke—FA, town in valley of New R., 22 km. NNW of Radford, Va.; Pearisburg, Va., 1:62,500.

Pennsboro—OP, town, 10 km. SSE of Greenfield, Mo.; South Greenfield, Mo., 1:24,000.

Pentz—SN, place, 15 km. N of Oroville, Calif.; Cherokee, Calif., 1:24,000.

Peshastin Creek—CM, creek entering Wenatchee R. between Leavenworth and Cashmere, Wash.; Mt. Stuart and Chiwaukum, Wash., 1:125,000.

Pigeon Mountain—FA, mountain, 7 km. W of La Fayette, Ga.; Estelle, Ga., 1:24,000.

Pikes Peak—SR, mountain, W of Colorado Springs, Colo., 38° 50′ N, 105° 3′ W; Pikes Peak, 1:24,000.

Pine City—SN, deserted mining town above Old Mammoth and 60 km. NW of Bishop, Calif.; Mt. Morrison and Devils Postpile, Calif., 1:62,500.

Pinnacles—CCR, National Monument, N of King City, Calif.; Greenfield, Calif., 1:62,500.

Piute Peak—SN, mountain, S of Kerrick Canyon, Yosemite Natl. Park, Calif.; Tower Peak, Calif., 1:62,500.

Platoro—SR, town in valley of Conejos R., 15 km. E of Summit Peak, Colo.; Summitville, Colo., 1:125,000.

Plum Creek—SR, creek entering South Platte R. at Denver, Colo.; Larkspur, Colo., 1:24,000, and Denver, Colo., 1:250,000.

Pluma—GP, locality, 1.8 km. S of Deadwood, S.D.; Deadwood, S.D., 1:125,000.

Pohono Trail—SN, trail on S side of Yosemite Valley, starting at E end of Wawona Tunnel, Yosemite Natl. Park; Yosemite, Calif., 1:62,500.

Pole Mountain—SR, mountain in Laramie Range, 19 km. ESE of Laramie, Wyo.; Sherman Mountains, Wyo., 1:62,500.

Pores Knob—PP, knob in Brushy Mts., 14 km. S of North Wilkesboro, N.C.; Moravian Falls, N.C., 1:24,000.

Post Creek—KM, tributary of Rattlesnake Creek, E of Forest Glen and 88 km. WNW of Red Bluff, Calif.; Dubakella Mt., Calif., 1:62,500.

Pratt's Ferry—FA, site of bridge on Cahaba R., 2 km. NW of Piper, Ala.; Blocton, Ala., 1:62,500.

Pulga—SN, town in valley of N Fork of Feather R., 32 km. NE of Oroville, Calif.; Pulga, Calif., 1:62,500.

Rancheria Mountain—SN, mountain, 5 km. NE of Hetch Hetchy Reservoir, Yosemite Natl. Park, Calif.; Hetch Hetchy Reservoir, Calif., 1:62,500.

Rattlesnake Creek—MB, tributary of Blackfoot R., NE of Missoula, Mont.; Bonner, Mont., 1:125,000.

Rawlings—FA, town in valley of N Branch of Potomac R., 17 km. SW of Cumberland, Md.; Lonaconing, Md.–W. Va., 1:24,000.

Red Mountain—CCR, mountain, 67 km. NNW of Willits, Calif.; Leggett, Calif., 1:62,500.

Red River—CP, river making boundary between Texas and states of Oklahoma and Arkansas; Texarkana, Ark.–Tex., 1:250,000.

Rescue—SN, town, 14 km. W of Placerville, Calif.; Shingle Springs, Calif., 1:24,000.

Rich Bar—SN, bar on E Branch of N Fork of Feather R., 20 km. S of Lake Almanor, Calif.; Almanor, Calif., 1:62,500.

Rileyville—FA, town in valley of S Fork of Shenandoah R., 24 km. SW of Front Royal, Va.; Strasburg, Va., 1:62,500.

Rio Hondo Canyon—SR, canyon on eastern slope of Sangre de Cristo Mts., 18 km. NNE of Taos, N.M.; Taos and vicinity, N.M., 1:125,000.

Roan Mountain—BR, mountain, 43 km. SSE of Elizabethton, Tenn.; Carvers Gap and Bakersville, N.C., 1:24,000.

Rock Creek—SN, tributary of N Fork of Feather R., 50 km. NNE of Oroville, Calif.; Pulga, Calif., 1:62,500.
Rock Point—BR, on crest of Blue Ridge about 18 km. SSW of Waynesboro, Va.; Lovingston, Va., 1:62,500.
Rocky Knob Spring—BR, along Blue Ridge Parkway, 54 km. ESE of Galax, Va.; Floyd, Va., 1:62,500.
Rogue River—KM, river in southwestern Oregon; Medford, Ore., 1:250,000.
Root River—CL, tributary of Mississippi R., 13 km. SSE of Rochester, Minn.; Stewartville, Minn., 1:62,500.
Roseburg—CR, city of Oregon in valley of South Umpqua R., 43° 11′ N, 123° 20′ W; Roseburg, Ore., 1:62,500 and 1:125,000.
Roubaix—GP, locality in Black Hills, 12 km. SE of Deadwood, S.D.; Deadwood, S.D., 1:125,000.
Rutledge—FA, town, 22 km. WNW of Morristown, Tenn.; Dutch Valley, Tenn., 1:24,000.
St. Sauveur Mountain—NEA, small mountain in south-central part of Mt. Desert Island, Me.; Acadia Natl. Park, Me., 1:24,000.
San Diego—LC, city in southwestern California, 32° 44′ N, 117° 10′ W; Point Loma, Calif., 1:24,000.
San Luis Obispo County—CCR, county including San Luis Obispo, Calif.; San Luis Obispo, Calif., 1:62,500.
Sandia Mountains—MP, mountains, 25 km. NE of Albuquerque, N.M.; Sandia Crest, N.M., 1:24,000.
Sanhedrin Mountains—CCR, mountains, 25 km. ENE of Willits, Calif.; Potter Valley and Eden Valley, Calif., 1:62,500.
Schoodic Peninsula—NEA, peninsula, 12 km. ESE of Bar Harbor, Me.; Acadia Natl. Park, Me., 1:24,000.
Scotty Creek—CM, tributary near headwaters of Peshastin Creek, 25 km. S of Leavenworth, Wash.; Mt. Stuart, Wash., 1:125,000.
Seaview—CR, village, 3 km. NW of Ilwaco, Wash.; Cape Disappointment, Wash., 1:62,500.
Serpentine Canyon—SN, canyon of E Branch of N Fork of Feather R. between Rich Bar and Rich Gulch, 17 km. S of Lake Almanor, Calif.; Almanor, Calif., 1:62,500.
Seven Devils Lake—Col P, small lake, NE of He Devil, in Seven Devils Mts., 70 km. SSW of Grangeville, Ida.; He Devil, Ore.–Ida., 1:62,500.
Shag Island—Gr P, tiny island, SE of Isles Affligées and 16 km. NNE of Mecatina I., Que.; Can. Hydrographic Chart 4474, Bun Is. to Mutton Bay, 1:36,500.
Shenandoah Mountain—FA, mountain, 48 km. W of Harrisonburg, Va.; Fort Seybert, Va.–W. Va., 1:62,500.
Shingle Creek—Col P, tributary of Rapid R. on E side of Seven Devils Mts., 80 km. SSW of Grangeville, Ida.; Riggins, Ida., 1:125,000.
Sierra Blanca—MP, high peak in south-central New Mexico, E of Rio Grande; Sierra Blanca Peak, 1:62,500.
Sierra Buttes—SN, mountains, N of Sierra City and 15 km. ENE of Downieville, Calif.; Sierra City, Calif., 1:62,500.
Silver Lake—SN, lake on E side of Sierra Nevada, 19 km. S of Mono Lake; Mono Craters, Calif., 1:62,500.
Silver Mills—FA, locality on Sideling Hill Creek, 31 km. SE of Bedford, Pa.; Clearville, Pa., 1:62,500.

Six Mile Creek—AP, large creek entering Cayuga Inlet on south side of Ithaca, N.Y.; Ithaca East and Ithaca West, N.Y., 1:24,000.

Smith River—KM, river in NW California, NE of Crescent City; Crescent City, Calif., 1:62,500.

Snake River—Col P, river making part of boundary between Idaho and Oregon; He Devil, Ore.–Ida., 1:62,500.

Snow Camp Lookout—KM, mountain, S of Rogue R. and 28 km. SE of Cape Blanco, Ore.; Collier Butte, Ore., 1:62,500.

Somers—IPR, town at northwest corner of Flathead Lake, Mont.; Kalispell, Mont., 1:250,000.

Sonora Pass—SN, pass in crest of Sierra Nevada, 60 km. NW of Mono Lake, Calif.; Sonora Pass, Calif., 1:62,500.

Spanish Creek—SN, tributary of E Branch of N Fork of Feather R., Calif.; Bucks Lake and Quincy, Calif., 1:62,500.

Sperry Glacier—NR, glacier on slope of Gunsight Mountain east of Lake McDonald, Glacier Natl. Park, 64 km. W of Browning, Mont.; Chief Mountain, Mont., 1:125,000.

Spray—PP, town in Rockingham Co., N.C., now included in Eden, 36° 31′ N, 79° 45′ W; Spray, N.C.–Va., 1:24,000.

Stevenson Island—Gr P, island, 15 km. SW of Mutton Bay, Que.; Can. Hydrographic Chart 4469, Flat I. to Little Mecatina I., 1:76,200.

Stringtown—SN, bar on S fork of Feather R. and also mountain, 15 km. E of Oroville, Calif.; Bidwell Bar, Calif., 1:24,000.

Sugarloaf Mountain—TR, mountain in San Bernardino Mts., 40 km. NE of Redlands, Calif.; San Gorgonio Mt., Calif., 1:62,500.

Sugarloaf Peak (in Sacramento Valley)—GV, highest point in Campbell Hills, 5 km. NW of Oroville, Calif.; Oroville, Calif., 1:24,000.

Summit Peak—SR, peak in San Juan Mts. of Colorado, 37° 18′ N, 106° 42′ W; Summitville, Colo., 1:125,000.

Tazewell—FA, town, 31 km. N of Marion, Va.; Pounding Mill, Va., 1:62,500.

Teewinot Mountain—CRM, mountain, 2 km. NE of Grand Teton, Wyo.; Grand Teton Natl. Park, Wyo., 1:62,500.

Terry Peak—GP, mountain in Black Hills, 7 km. SW of Lead, S.D.; Deadwood, S.D., 1:125,000.

Tesuque Peak—SR, peak of ridge SW of Lake Peak, 18 km. NE of Santa Fe, N.M.; Aspen Basin, N.M., 1:24,000.

Teton Canyon—CRM, canyon on western slope of Teton Mts., 12 km. W of Grand Teton; Grand Teton, Wyo., 1:125,000.

Teton Mountains—CRM, range of mountains in western Wyoming; Grand Teton Natl. Park, Wyo., 1:62,500.

Tuerto Mountain—SR, small mountain, east of Santa Fe R. and 15 km. E of Santa Fe, N.M.; McClure Reservoir, N.M., 1:24,000.

Tunnel Hill—ILP, town, 13 km. NNE of Vienna, Ill.: Creal Springs, Ill., 1:24,000.

Tuolumne Meadows—SN, subalpine meadow along upper Tuolumne R., Yosemite Natl. Park, Calif.; Yosemite Natl. Park, Calif., 1:125,000, and Tuolumne Meadows, 1:62,500.

Tuscaloosa—FA, city in Alabama, 33° 11′ N, 87° 35′ W; Tuscaloosa, Ala., 1:62,500.

Tye River Gap—BR, gap in crest of Blue Ridge, 32 km. SW of Waynesboro, Va.; Vesuvius, Va., 1:62,500.

Unaka Mountain—BR, mountain, 6 km. SE of Erwin, Tenn.; Unicoi, Tenn.–N.C., 1:24,000.

Unimak Island—Aleutian Islands, island, SW of Alaska Peninsula, 54° 40′ N, 160° W; Unimak (S) and False Pass (S), Alaska, 1:250,000.

Valentine Ridge—KM, eastern ridge of Yolla Bolly Mts., 50 km. WSW of Red Bluff, Calif.; Yolla Bolly, Calif., 1:62,500.

Vernal Falls—SN, waterfalls of Merced R., SSW of Half Dome, Yosemite Natl. Park, Calif.; Yosemite, Calif., 1:62,500.

Waldo—KM, locality in drainage of E Fork of Illinois R. just NW of Takilma and 50 km. SSW of Grants Pass, Ore.; Cave Junction, Ore., 1:62,500.

Warm Springs—FA, town, 42 km. NW of Lexington, Va.; Warm Springs Run, Va., 1:24,000.

Washington (D.C.)—PP, city, capital of U.S., 38° 55′ N, 77° W; Washington West, D.C.–Md.–Va., 1:24,000.

Watkins Glen—AP, gorge, SW of Seneca Lake, N.Y.; Beaver Dams, N.Y., 1:24,000.

Wedge Mountain—CM, mountain, 11 km. SSW of Leavenworth, Wash.; Chiwaukum, Wash., 1:125,000.

West Hill (Ithaca, N.Y.)—AP, hill on west side of Ithaca, N.Y.; Ithaca West, N.Y., 1:24,000.

West Spanish Peak—GP, mountain, 30 km. SW of Walsenburg, Colo.; Spanish Peaks, Colo., 1:125,000.

White Mountain Peak—MP, see Sierra Blanca.

Wildwood—KM, town, 38 km. SSW of Weaverville, Calif.; Dubakella Mt., Calif., 1:62,500.

Willow Creek—SR, tributary of Pecos R., draining western slope of Rincon Peak, 30 km. ENE of Santa Fe, N.M.; Santa Fe, N.M., 1:125,000 and 1:250,000.

Windy Ridge—IB, ridge, W of Papoose Creek and S of Lolo Trail, in northern Idaho Co., 136 km. NE of Grangeville, Ida.; Lolo, Ida.–Mont., 1:250,000.

Woods Creek—SN, tributary of Tuolumne R., SW of Sonora, Calif.; Chinese Camp and Sonora, Calif., 1:24,000.

Yadkin River—PP, river in North Carolina; Albemarle, N.C., 1:62,500, and Mount Gilead West, N.C., 1:24,000.

Yellville—OP, town, 37 km. E of Harrison, Ark.; Yellville, Ark., 1:24,000.

Yosemite Gorge—SN, gorge of the Merced R. from Turtleback Dome to Nevada Falls, Yosemite Natl. Park, Calif.; Yosemite, Calif., 1:62,500.

Yosemite Valley—SN, broad gorge of Merced R. between Turtleback Dome and Half Dome, Yosemite Natl. Park, Calif.; Yosemite, Calif., 1:62,500.

Bibliography

Aksenova, R. A., M. I. Zotova, M. F. Nechoda, and S. G. Tscherdinzeff. 1966. [Russian title] The stimulating and adaptogenic action of purified preparation of roseroot *Sedum*, rodozin. In: Stimuliatory tsentral'noi nervnoi sistemy. Tomsk, Tomsk Univ. pp. 77–79.

Alexander, E. J. 1943. *Sedum stenopetalum*. Addisonia 22 (1):5–6, 1 pl.

Allard, H. A. 1940. *Sedum telephium* L. in the Bull Run Mountain area, Virginia. Castanea 5:17–19.

Allard, H. A., and W. W. Garner. 1940. Further observations on the response of various species of plants to length of day. U.S. Dept. Agric. Tech. Bull. 727. 64 p., 1 pl., 26 figs.

Anderson, Alfred L. 1941. Physiographic subdivisions of the Columbia Plateau in Idaho. Jour. of Geomorphology 4 (3):206–222.

Atwood, Howland. 1953. *Sedum glaucophyllum*. Cact. Succ. Jour. 25:148–149.

Babcock, Ernest B. 1947. The genus *Crepis*. Part one. University of California Publications in Botany 21:1–198, illus.

Bailey, Edgar H. 1966. Geology of northern California. San Francisco, California Division of Mines and Geology. 520 p., illus.

Bailey, L. H. 1900–1902. Cyclopedia of American horticulture. New York, Macmillan. 4 vols.

Bailey, L. H. 1914. The standard cyclopedia of horticulture. New York, Macmillan. 5 vols.

Bailey, L. H., and Ethel Zoe Bailey. 1941. Hortus second. New York, Macmillan. 778 p.

Bailey, Thomas L., and Richard H. Jahns. 1954. Geology of the Transverse Range Province, southern California. California Division of Mines, Bull. 170 (11):83–106.

Baldwin, Ewart M. 1959. Geology of Oregon. Eugene, University of Oregon Cooperative Book Store. 136 p.

Baldwin, J. T., Jr. 1935. Somatic chromosome numbers in the genus *Sedum*. Bot. Gaz. 96:558–564, figs. 1–14.

Baldwin, J. T., Jr. 1936. Polyploidy in *Sedum ternatum*. Jour. of Heredity 27:241–248, figs. 8–13.

Baldwin, J. T., Jr. 1937. The cyto-taxonomy of the *Telephium* section of *Sedum*. Am. Jour. Bot. 24:126–132, figs. 1–41.

Baldwin, J. T., Jr. 1939. Certain cytophyletic relations of *Crassulaceae*. Chron. Bot. 5:415–417.

Baldwin, J. T., Jr. 1940. Cytophyletic analysis of certain annual and biennial *Crassulaceae*. Madroño 5:184–192, pl. 17.

Baldwin, J. T., Jr. 1942a. Cytological basis for specific segregation in the *Sedum Nevii* complex. Rhodora 44:11–14, figs. 1–4.

Baldwin, J. T., Jr. 1942b. Polyploidy in *Sedum ternatum* Michx. II. Cytogeography. Am. Jour. Bot. 29:283–286, figs. 1–5.

Baldwin, J. T., Jr. 1943. Polyploidy in *Sedum pulchellum*—I. Cytogeography. Bull. Torrey Bot. Club 70:26–33, figs. 1–10.

Baldwin, J. T., Jr. 1944. Affinities of *Sedum Nevii*. Rhodora 46:450–451.
Baldwin, J. T., Jr. 1945. Chromosomes of *Cruciferae*. II. Cytogeography of *Leavenworthia*. Bull. Torrey Bot. Club 72:367–378, 9 figs.
Banach-Pogan, Eugenia. 1958. Cytological studies in three species of the genus *Sedum* L. Acta Biol. Cracoviensia, Ser. Bot. 1 (2):91–101, pl. 10 and 11.
Baskin, Jerry M., and Carol C. Baskin. 1972. Germination characteristics of *Diamorpha cymosa* seeds and an ecological interpretation. Oecologia (Berl.) 10:17–28, 3 figs., 1 pl.
Becherer, A. 1956. Bemerkungen zur Nomenklatur der Farn-und Blütenpflanzen der Schweiz. Ber. Schweiz. bot. Ges. 66:224–236.
Berger, Alwin. 1930. *Crassulaceae*. In:Engler, Die naturlichen Pflanzenfamilien. 2d. ed. 18a: 352–485.
Birman, Joseph P. 1964. Glacial geology across the crest of the Sierra Nevada, California. Geol. Soc. Am. Spec. Paper no. 75:1–80, figs. 1–5, pls. 1–5, tables 1–11.
Böcher, Tyge W. 1938. Biological distributional types in the flora of Greenland. Medd. om Grønl. 106, no. 2:1–339, pl. 1–2.
Böcher, Tyge W., Kjeld Holmen, and Knud Jakobsen. 1957. Grønlands Flora. Copenhagen, P. Haase and Sons. 313 p.
British Colour Council. 1938–1942. Horticultural colour chart. Banbury and London, Henry Stone and Son Ltd. 2 vols.
British Columbia, Department of Agriculture. 1928—. Climate of British Columbia. 1—.
Britton, Nathaniel Lord, and Addison Brown. 1913. An illustrated flora of the northern United States, Canada, and the British Possessions. New York, Charles Scribner's Sons. 3 vols.
Britton, Nathaniel L., and Joseph N. Rose. 1903. New or noteworthy *Crassulaceae*. Bull. N.Y. Bot. Gard. 3:1–45.
Britton, Nathaniel L., and Joseph N. Rose. 1905. *Crassulaceae*. In North American Flora 22:7–74.
Brown, Richard M. 1970. Notes on the larva and habitat of *Callophrys fotis bayensis* (*Lycaenidae*). Jour. Res. Lepidopt. 8:49–50.
Brown, W. L., Jr., and E. O. Wilson. 1956. Character displacement. System. Zool. 5:49–64.
Butruille, D., G. Fodor, C. S. Huber, and F. Letourneau. 1971. Absolute configuration of the alkaloid (+) sedridine and of (−) allosedridine. Tetrahedron 27 (11):2055–2067.
Buwalda, John P. 1954. Geology of the Tehachapi Mountains, California. Calif., Div. of Mines, Bull. 170 (Chapt. 2):131–142, figs. 1–6.
Camp, W. H., and C. L. Gilly. 1943. The structure and origin of species. Brittonia 4:323–385.
Canada, Department of Mines and Technical Surveys. 1965. Principal mineral areas of Canada. Map no. 900A. 15th ed. Scale 1/7,603,200.
Canada, Department of Transport, Meteorological Branch. 1959—. Arctic Summary 1—.
Canada, Department of Transport, Meteorological Branch. 1916—. Monthly record. 1—.
Candolle, Augustin P. de. 1828. *Crassulaceae*, in Prodromus Systematis Naturalis Vegetabilis. Vol. 3:381–414.
Caudle, Carol, and Jerry M. Baskin. 1968. The germination pattern of three winter annuals. Bull. Torrey Bot. Club 95:331–335.
Chamberlain, E. B. 1912. A *Sedum* new to North America. Rhodora 14:227–228.
Chapman, Alvan W. 1860. Flora of the southern United States. New York, Ivison, Phinney. 621 p.
Cherdyntsev, S. G., and M. I. Zotova. 1966. [Russian title] The mechanism of the inhibition of the leucocyte reaction by an extract of roseroot *Sedum*. In: Stimuliatory tsentral'noi nervnoi sistemy. Tomsk, Tomsk. Univ. pp. 80–82.

Cinq-Mars, L., R. Van den Hende, C. Rousseau, J. P. Bernard, C. Leduc, and J. G. Perras. 1971. Notes sur la flore du Quebec: Additions. Natur. Can. 98:194–197, illus.

Clausen, Jens, David D. Keck, and William M. Hiesey. 1940. Experimental studies on the nature of species. I. Effect of varied environments on western North American plants. Carn. Inst. Wash. Pub. 520:i–vii, 1–452.

Clausen, Robert T. 1939. Some plants of New York. Torreya 39:1–9.

Clausen, Robert T. 1942. Studies in the *Crassulaceae*. *Sedum*, subgenus *Gormania*, section *Eugormania*. Bull. Torrey Bot. Club 69 (1):27–40, figs. 1–3.

Clausen, Robert T. 1943. *Sedum hirsutum*. Desert Plant Life 15:90–94, illus.

Clausen, Robert T. 1944. A description and name for *Sedum XYZ*. Cact. Succ. Jour. 16:7–8, figs. 5–6.

Clausen, Robert T. 1946. Nomenclatural changes and innovations in the *Crassulaceae*. Cact. Succ. Jour. 18:58–61, 74–77, figs. 39–40, 47–48.

Clausen, Robert T. 1948. A reinterpretation of *Sedum stenopetalum* and *Sedum lanceolatum*. Cact. Succ. Jour. 20:143–146, figs. 105–107.

Clausen, Robert T. 1949a. Checklist of the vascular plants of the Cayuga Quadrangle, 42°–43° N., 76°–77° W. Cornell Univ. Exp. Sta. Mem. 291, pp. 3–87.

Clausen, Robert T. 1949b. The distribution and variation of *Sedum nevii*. Cact. Succ. Jour. 21:180–185.

Clausen, Robert T. 1950. Description of a *Sedum* from the Sierra Madre Oriental of Mexico. Cact. Succ. Jour. 22:86–89, figs. 34–37.

Clausen, Robert T. 1959. *Sedum* of the Trans-Mexican Volcanic Belt: An exposition of taxonomic methods. Ithaca, Cornell University Press. 380 p.

Clausen, Robert T., Reid V. Moran, and Charles H. Uhl. 1945. The taxonomy and cytology of *Hasseanthus*. Desert Plant Life 17:69–83.

Clausen, Robert T., and Charles H. Uhl. 1943. Revision of *Sedum Cockerellii* and related species. Brittonia 5:33–46, figs. 1–20.

Clausen, Robert T., and Charles H. Uhl. 1944. The taxonomy and cytology of the subgenus *Gormania* of *Sedum*. Madroño 7:161–180, fig. 1, pl. 22.

Clements, Frederic E., Emmett V. Martin, and Frances L. Long. 1950. Adaptation and origin in the plant world. Waltham, Chronica Botanica. 332 p.

Cochran, William G. 1963. Sampling techniques. 2d ed. New York, John Wiley and Sons. 413 p.

Cody, W. J. 1967. *Sedum* in the Ottawa district. Can. Field Natur. 81:273–274.

Combier, Henri, and Maurice Jay. 1967. Recherches chimiotaxinomiques sur les plantes vasculaires. 11. Distribution des flavonoides chez les Crassulacées. Plantes Médicinales Phytothérapie 1 (4):165–170.

Cook, Earl Ferguson. 1954. Mining geology of the Seven Devils region. Idaho Bureau of Mines and Geology. Pamphlet No. 97. 22 p., 6 figs., 1 pl.

Cooke, William Bridge. 1941. First supplement to the flora of Mount Shasta. Am. Midl. Nat. 26:74–84.

Dalenius, Tore. 1953. The problem of optimum stratification in a special type of design. Skandinavisk Aktuarietidskrift 35:61–70.

Davis, Helen Burns. 1936. Life and work of Cyrus Guernsey Pringle. Burlington, Free Press. 756 p.

Davis, Ray J. 1952. Flora of Idaho. Dubuque, Wm. C. Brown. 828 p.

Denffer, Dietrich von. 1941. Über die photoperiodische Beeinflussbarkeit von Habitus und Sukkulenz bei einigen Crassulaceen-Arten. Jahr. wiss. Bot. 89 (4):543–573, figs. 1–21.

Dixon, Wilfrid J., and Frank J. Massey. 1951. Introduction to statistical analysis. New York, McGraw-Hill. 370 p.

Eames, Arthur J. 1931. The vascular anatomy of the flower with refutation of the theory of carpel polymorphism. Am. Jour. Bot. 31:147–188, figs. 1–29.

Eardley, A. J. 1962. Structural geology of North America. 2d ed. New York, Harper and Row. 743 p.

Emiliani, Cesare. 1958. Ancient temperatures. Scientific American 198 (2):54–63, illus.

Ericson, David B., Maurice Ewing, and Goesta Wollin. 1964. The Pleistocene epoch in deep-sea sediments. Science 146 (3645):723–732, figs. 1–5.

Ewan, Joseph A. 1950. Rocky Mountain naturalists. Denver, University of Denver Press. 358 p.

Fateeva, A. P. 1966. [Russian title] The use of an extract of roseroot *Sedum* in arterial hypertension. In: Stimuliatory tsentral'noi nervnoi sistemy. Tomsk, Tomsk Univ. pp. 121–123.

Fenneman, Nevin M. 1931. Physiography of western United States. New York, McGraw-Hill. 534 p., 173 figs., map.

Fenneman, Nevin M. 1938. Physiography of eastern United States. New York, McGraw-Hill. 714 p., 197 figs., 7 pl.

Fernald, Merritt Lyndon. 1950. Gray's Manual of Botany. 8th ed. New York, American Book. 1632 p.

Fernald, Merritt L., and Alfred C. Kinsey. 1958. Edible wild plants of eastern North America. Rev. ed. by Reed C. Rollins. New York, Harper. 452 p.

Flint, Richard Foster. 1957. Glacial and Pleistocene geology. New York, John Wiley and Sons. 553 p., illus.

Flynn, Nellie F. 1909. Plants new to Vermont. Rhodora 11:198–199.

Frémont, John Charles. 1887. Memoirs of my life. Chicago, Belford, Clarke. 655 p.

Fridriksson, Sturla. 1964. [Danish title] The colonization of the dryland biota on the island of Surtsey off the coast of Iceland. Natturfraedingurinn 34 (2):83–89.

Fröderström, Harald. 1930–1935. The genus *Sedum* L. Act. Hort. Goth. 5, 6, 7, and 10 (App.).

Funke, G. L. 1943. Observations on the flowering photoperiodicity. Rec. Trav. Bot. Néerland. 40:393–412, figs. 1–4.

Gleason, H. A. 1922. The vegetational history of the Middle West. Ann. Assoc. Amer. Geog. 12:39–85.

Goldschmidt, Richard. 1940. The material basis of evolution. New Haven, Yale University Press. 436 p.

Goode, J. Paul. 1943. Goode's School Atlas. New York, Rand McNally. xvi, 286 p.

Govorov, V. P., and N. A. Lipskaya. 1963. [Russian title] Pharmacological properties of *Rhodiola rosea*. Trudy Omskogo Med. Inst. 45:15–22.

Graustein, Jeanette E. 1967. Thomas Nuttall, naturalist; explorations in America, 1808–1841. Cambridge, Harvard University Press. 481 p.

Gray, Asa. 1868. Manual of the botany of the northern United States. New York, Ivison, Blakeman, Taylor. 703 p., 20 pl.

Gray, Asa. 1876. *Sedum reflexum* L. Am. Nat. 10:553.

Gray, Asa. 1889. Scientific papers of Asa Gray, selected by Charles S. Sargent. Boston and New York, Houghton, Mifflin. 2 vols.

Grieg-Smith, P. 1964. Quantitative plant ecology. 2d ed. London, Butterworths. 256 p.

Gronovius, Joh. 1762. Flora Virginica. 2d ed. Leyden, 1762. 176 p.

Günthart, A. 1902. Beiträge zur Blüthenbiologie der Cruciferen, Crassulaceen und der Gattung *Saxifraga*. Bibl. Bot., Stuttgart, H. 58. ix, 97 p., 11 pl.

Guilday, John E., Paul S. Martin, and Allen D. McCrady. 1964. New Paris No. 4: A late Pleistocene cave deposit in Bedford County, Pennsylvania. Bull. Natl. Speleological Soc. 26 (4): 121–194.

Hansen, Morris H., William N. Hurwitz, and William G. Madow. 1953. Sample survey methods and theory. New York, John Wiley and Sons. 2 vols.

Hara, Hiroshi. 1952. Contributions to the study of variations in the Japanese plants closely related to those of Europe or North America. Part 1. Jour. Fac. Sci., Univ. Tokyo, sect. 3, Bot. 6:29–96.

Henslow, G. 1891. On the vascular systems of floral organs, and their importance in the interpretation of the morphology of flowers. Jour. Linn. Soc. London, Bot. 28:151–197.

Heusser, Calvin. 1954. Nunatak flora of the Juneau Ice Field, Alaska. Bull. Torrey Bot. Club 81:236–250.

Hodgdon, A. R. 1959. *Sedum sexangulare* in New Hampshire. Rhodora 61:247.

Hollingshead, Lillian. 1942. Chromosome studies in *Sedum*, subgenus *Gormania*, section *Eugormania*. Bull. Torrey Bot. Club 69:41–43.

Hooker, J. D. 1876. *Sedum puchellum*. Bot. Mag. ser. 3, vol. 32: pl. no. 6223.

Hooker, William J. 1829–1840. Flora Boreali-Americana. London, Treuttel and Würtz. 2 vols.

Hultén, Eric. 1949. On the races in the Scandinavian flora. Svensk. Bot. Tidskr. 43 (2/3):383–406, illus.

Hultén, Eric. 1950. Atlas över växternas utbredning i Norden; fanerogamer och ormbungsväxter. Stockholm, Generalstabens litografiska anstalts förlag. 512 p.

Huntley, Dorothy. 1939. A survey of the vegetation of Forty-acre Rock, Lancaster County, South Carolina. M. A. thesis, Duke University. 33 p., 12 figs. (typewritten).

Hylander, Nils. 1945. Nomenklatorische und Systematische Studien über nordische Gefasspflanzen. Uppsala Universitets Årsskrift 1945:7:1–337.

Irwin, William P. 1960. Geologic reconnaissance of the northern Coast Ranges and Klamath Mountains, California, with a summary of the mineral resources. California, Division of Mines, Bull. 179:3–80, illus., map.

Ives, Sumner A. 1944. The vascular plants of Greenville County, South Carolina. Bull. Furman University 27 (4):5–28.

Jacob, F. H. 1964. A new species of *Thuleaphis* from Wales, Scotland, and Iceland (*Thuleaphis sedi* n. sp. on *Sedum rosea*). Proc. Roy. Ent. Soc., London, Ser. B, Taxonomy 33 (5/6):111–116.

Jacobsen, Hermann. 1960. A handbook of succulent plants. London, Blandford Press. 3 vols.

Jahandiez, Emile, and René Maire. 1931–1941. Catalogue des Plantes du Maroc. Alger, Imprimerie Minerva. 4 vols.

Jalas, Jaakko. 1954. Populationsstudien an *Sedum telephium* L. in Finnland. Ann. Bot. Soc. Zool. Bot. Fennicae "Vanamo" 26 (3):1–47, illus.

James, W. O. 1958. Succulent plants. Endeavor 17 (66):90–95.

Jensen, Lawrence C. W. 1968. Primary stem vascular patterns in three subfamilies of the *Crassulaceae*. Am. Jour. Bot. 55:553–563.

Jessen, Raymond J., *et al.* 1947. On a population sample for Greece. Jour. Am. Stat. Assoc. 42 (239):357–384.

Jones, G. Neville. 1941. How many species of plants are there? Science 94:234.

Kaliko, I. M., and A. A. Tarasova. 1966. [Russian title] The action of extracts of *Leuzea* and roseroot *Sedum* [*Sedum rosea*] on the dynamic features of higher nervous activity. In:Stimuliatory tsentral'noi nervnoi sistemy. Tomsk, Tomsk Univ. pp. 115–120.

Kendrew, Wilfrid G., and Donald P. Kerr. 1955. The climate of British Columbia and the Yukon Territory. Ottawa, E. Cloutier. ix, 222 p.

Kennedy, Partick B. 1912. Alpine plants—XI. *Rhodiola integrifolia* Raf. Muhlenbergia 8:95, fig. 14.

Kerner, Anton. 1894. The natural history of plants. Translation by F. W. Oliver. London, Blackie and Son. 2 vols.

King, Philip B. 1959. The evolution of North America. Princeton, Princeton University Press. 189 p., illus.

Knowlton, Clarence H. 1917. Preliminary lists of New England plants—XXV. Rhodora 19:217–219.

Kolmakova, L. F., and N. I. Kutolina. 1966. [Russian title] Clinical observations on the action of extracts of *Leuzea, Eleutherococcus*, and roseroot *Sedum* in patients with diabetes mellitus and, other diseases. In: Stimuliatory tsentral'noi nervnoi sistemy. Tomsk, Tomsk Univ. pp. 131–132.

Lammers, William Tuthill. 1959. A study of certain environmental and physiological factors influencing the adaptation of three granite outcrop endemics: *Amphianthus pusillus* Torr., *Isoetes melanospora* Engelm., and *Diamorpha cymosa* (Nutt.) Britton. Diss. Abst. 19:3096.

Lanjouw, J., and F. A. Stafleu. 1956. Index Herbariorum. Part 1. The herbaria of the world. 3d ed. Reg. Veg. 6:1–224.

Lewis, Harrison F. 1931–1932. An annotated list of vascular plants collected on the north shore of the Gulf of St. Lawrence, 1927–1930. The Canadian Field-Naturalist 45:129–135, 174–179, 199–204, 225–228; 46:12–18, 36–40, 64–66, 89–95.

Linnaeus, Carolus. 1753. Species Plantarum. Stockholm, L. Salvius. 2 vols.

Lockwood, John L. 1958. A method for studying absorption of streptomycin by using leaf disks of *Sedum purpureum*. Phytopath. 48 (3): 150–155.

Love, J. D., and John C. Reed, Jr. 1968. Creation of the Teton landscape, the geologic story of Grand Teton National Park. Grand Teton Natural History Association, Moose, Wyo. 120 p.

Marina, T. F., and T. P. Prishchep. 1964. [Russian title] Pharmacology of *Rhodiola rosea*. Izv. Sibirsk Otd. Akad. Nauk S.S.R. Ser. Biol. Med. Nauk 4 (1):49–55.

Marion, Léo. 1945. The alkaloids of *Sedum acre* L. Canadian Jour. Res., Sect. B, Chem. Sci. 23 (5):165–166.

Marion, Léo, and Marcel Chaput. 1949. A new occurrence of dl–methylisopelletierine. Canadian Jour. Res., Sect. B, 27 (4):215–217.

Marion, Léo, Robert Lavigne, and Lionel Lamay. 1951. The structure of sedamine. Canadian Jour. Chem. 29 (5):347–351.

Martin, Paul S. 1958. Taiga-tundra and the full-glacial period in Chester County, Pennsylvania. Am. Jour. Sci. 256:470–502.

Masters, M. T. 1874. *Sedum pulchellum*, Michx. The Gardeners' Chronicle. N.S. 2:552, 623, 657, fig. 111.

Masters, M. T. 1878. Hardy stonecrops: Sedums. The Gardeners' Chronicle, N.S. 10:266–268, 302–303, 336–337, 376–377, 463, 590–591, 626, 658, 684–685, 716–717, 750–751, and 784.

Matthes, Francois E. 1950. The incomparable valley. Berkeley, University of California Press. 160 p., 50 pl., 11 figs.

Mauritzon, Johan. 1933. Studien über die Embryologie der Familien *Crassulaceae* und *Saxifragaceae*. Lund, Gleerupska Univ.-Bokhandela. 152 p., 44 fig.

McCarthy, Philip J. 1957. Introduction to statistical reasoning. New York, McGraw-Hill. 402 p.

McCormick, J. Frank, and Robert B. Platt. 1962. Effects of ionizing radiation on a natural plant community. Radiation Botany 2:161–188.

McCormick, J. Frank, and Robert B. Platt. 1964. Ecotypic differentiation in *Diamorpha cymosa*. Bot. Gaz. 125 (4):271–279.

McCormick, J. Frank, and William N. Rushing. 1964. Differential radiation sensitivities of races of *Sedum pulchellum* Michx. Radiation Botany 4:247–251, figs. 1–2.

McKelvey, Susan Delano. 1955. Botanical exploration of the Trans-Mississippi West, 1790–1850. Jamaica Plain, Mass., Arnold Arboretum of Harvard University. 1,144 p.

McVaugh, Rogers. 1943. The vegetation of the granitic flat-rocks of the southeastern United States. Ecological Monographs 13:119–166.

Meyrán, Jorge. 1963. *Sedum Aoikon*. Cactaceas y Succulentas Mexicanas 8 (2):51–53.

Michaux, André. 1803. Flora Boreali-Americana. Paris, Levrault. 2 vols.

Mitra, Jyotirmay. 1965. An improved method for plant chromosome preparation. The Nucleus 8 (2):179–182.

Moran, Reid. 1950. Whence *Sedum pinetorum* Brandegee? Leaflets of Western Botany 6:62–63.

Moran, Reid. 1951a. Notes on *Hasseanthus*—II. Desert Plant Life 22:99–105, figs. 5–8.

Moran, Reid. 1951b. Natural hybrids between *Dudleya* and *Hasseanthus*. So. Calif. Acad. Sci. 50:57–67, pls. 19–23.

Moran, Reid V. 1951c. A revision of *Dudleya* (*Crassulaceae*). Summary of the Dissertation, Univ. Calif., Grad. Div., No. Sect. 4 p.

Moran, Reid. 1969. *Sedum* in Baja California. Cact. Succ. Jour. 41:20–25, figs. 1–7.

Mori, Antonio. 1879. Saggio monografico sulla struttura istologica della *Crassulaceae*. Nuov. Giorn. Bot. Ital. 11:161–187, pl. 1–3.

Munk, Walter H., and Gordon J. F. Macdonald. 1960. The rotation of the earth. A geophysical discussion. Cambridge, The University Press. 323 p.

Munz, Philip A. 1968. A supplement to a California flora. Berkeley, University of California Press. 224 p.

Munz, Philip A., and David D. Keck. 1959. A California flora. Berkeley, University of California Press. 1681 p.

Murdy, W. H. 1968. Plant speciation associated with granite outcrop communities of the southeastern Piedmont. Rhodora 70:394–407.

Murray, Grover E. 1961. Geology of the Atlantic and Gulf Coastal Province of North America. New York, Harper and Brothers. 692 p.

O'Connell, J. E. 1949. Cytology, morphology, and taxonomy of *Diamorpha cymosa*. Jour. Elisha Mitchell Sci. Soc. 65:194.

Oleinichenko, V. F. 1966. [Russian title] The effect of extracts of *Eleutherococcus* and roseroot *Sedum* [*Sedum rosea*] on the functional condition of the ear in workers in noisy departments of the Tomsk Electromechanical Factory and of pilots at the Tomsk airport. In: Stimuliatory tsentral'noi nervnoi sistemy. Tomsk, Tomsk Univ. pp. 124–127.

Pearson, E. S., and H. O. Hartley. 1954. Biometrika Tables for Statisticians. Vol. 1. Cambridge, The University Press. 238 p.

Peech, Michael. 1945. Determination of exchangeable cations and exchange capacity of soils—rapid micromethods utilizing centrifuge and spectrophotometer. Soil Science 59:25–38.

Penck, Walther. 1953. Morphological analysis of land forms. Translated by Hella Czech and Katharine Cumming Boswell. London, Macmillan. 429 p., illus.

Piaget, Jean. 1966. Observations sur le mode d'insertion et la chute des rameaux secondaires chez les *Sedum*. Candollea 21:137–239, illus.

Plukenet, Leonard. 1696. Almagestum botanicum, sive Phytographiae Plukenetianae Onomasticon methodo synthetica digestum. London, Plukenet. 402 p.
Polunin, Nicholas. 1940. Botany of the Canadian eastern Arctic. Part I. *Pteridophyta* and *Spermatophyta*. National Museum of Canada. Bull. 92. 408 p.
Polunin, Nicholas. 1959. Circumpolar Arctic flora. Oxford, Clarendon Press. 514 p.
Porsild, A. E. 1957. Illustrated flora of the Canadian Arctic Archipelago. National Museum of Canada. Bull. 146:iii, 209 p., 70 figs., 332 maps.
Praeger, R. Lloyd. 1921. An account of the genus *Sedum* as found in cultivation. Jour. Roy. Hort. Soc. 46:1–314, figs. 1–185.
Putnam, William C. 1954. Marine terraces of the Ventura Region and the Santa Monica Mountains, California. California Division of Mines, Bull. 170 (v):45–48.
Quimby, Maynard Ward. 1940. The floral morphology of the *Crassulaceae*. In: Abstracts of Theses, Cornell University, 1939. pp. 376–377.
Quimby, Maynard Ward. 1971. The floral morphology of the *Crassulaceae*. University, Mississippi, M. W. Quimby. 42 p., XI pl.
Radford, Albert E., Harry E. Ahles, and C. Ritchie Bell. 1964. Guide to the vascular flora of the Carolinas. Chapel Hill, The Book Exchange. 383 p.
Radford, Albert E., Harry E. Ahles, and C. Ritchie Bell. 1965. Atlas of the vascular flora of the Carolinas. North Carolina Tech. Bull. 165. 208 p.
Ray, Louis L. 1940. Glacial chronology of the Southern Rocky Mountains. Bull. Geol. Soc. America 51:1,851–1,917, 6 pls., 12 figs.
Riedel, Spencer M., and William T. Wilson. 1967. Pollen collection and behavioral characteristics of the honey bee at high altitudes. Am. Bee Jour. 107 (1):10–12, illus.
Rinehart, Dean. 1964. Geologic story. In *The Mammoth Lakes Sierra* by Genny Schumacher. San Francisco, The Sierra Club. Pp. 73–84.
Robinson, Benjamin L., and Merritt L. Fernald. 1908. Gray's New Manual of Botany. 7th ed. New York, American Book Company. 926 p.
Robinson, B. L., and M. L. Fernald. 1909. Emendations of the seventh edition of Gray's Manual—I. Rhodora 11:33–61.
Rocky Mountain Association of Geologists. 1948. Guide to the geology of central Colorado. Quarterly of the Colorado School of Mines 43 (2):1–176, 14 pls., 38 figs.
Rose, Joseph N. 1903. See Britton and Rose.
Runcorn, S. K., ed. 1962. Continental drift. New York, Academic Press. 338 p.
Rush, Ethel. 1941. *Sedum confusum* Hemsley, rediscovered. Cact. Succ. Jour. 13 (7):146–148, figs. 86–88.
Sagaidak, L. P., and O. G. Paznikova. 1966. [Russian title] The effect of the roseroot *Sedum* on the reactiveness of the body and elaboration of tetanus antitoxin in rabbits. In: Stimuliatory tsentral'noi nervnoi sistemy. Tomsk, Tomsk Univ. pp. 97–98.
Saint-Lager, Jean Baptiste, and Victor Vivian-Morel. 1876. Discussion sur la valeur specifique de la glaucescence des *Sedum*. Ann. Soc. Bot. Lyon 3 (2):107–108.
Saratikov, A. S., E. A. Krasnov, L. A. Khnykina, and L. N. Dubidzon. 1967. [Russian title] Extraction of and study of individual biologically active substances from *Rhodiola rosea* [*Sedum*] and *Rhodiola quadrifida*. Izv. Sib. Otd. Akad. S.S.R. Ser. Biol. Med. Nauk 5:54–60.
Saratikov, A. S., E. A. Krasnov, L. A. Khnykina, L. N. Dubidzon, M. I. Zotova, T. F. Marina, M. F. Nechoda, R. A. Aksenova, and S. G. Tscherdinzeff. 1968. Rhodiolosid, ein neues Glykosid aus *Rhodiola rosea* und seine pharmakologischen Eigenschaften. Pharmazie 23 (7):392–395.
Saratikov, A. S., T. F. Marina, and I. M. Kaliko. 1965. [Russian title] Stimulating effect of

Rhodiola rosea on the higher sections on the brain. Iz. Sibirsk. Otd. Akad. Nauk Ser. Biol. Med. Nauk 8 (2):120–125. Illus.

Schmidt, Karl P. 1938. Herpetological evidence for the postglacial eastward extension of the steppe in North America. Ecology 19 (3):396–407.

Shapley, Harlow, ed. 1953. Climatic change. Cambridge, Harvard University Press. 318 p.

Sharpe, C. F. Stewart. 1960. Landslides and related phenomena. Paterson, N.J., Pageant Books. 137 p.

Sharsmith, Helen K. 1936. The genus *Sedella*. Madroño 3:1–8, fig. 1, pl. 12.

Sharsmith, Helen K. 1940. Further notes on the genus *Sedella*. Madroño 5:192–196, fig. 1.

Sherwin, Priscilla A., and Robert L. Wilbur. 1971. The contributions of floral anatomy to the generic placement of *Diamorpha smallii* and *Sedum pusillum*. Jour. Elisha Mitchell Sci. Soc. 87 (3):103–114.

Shriver, Howard. 1877a. Some notes on *Nepeta glechoma* and other plants. Bot. Gaz. 2:118–119.

Shriver, Howard. 1877b. Notes from southwestern Virginia. Bot. Gaz. 2:134–135.

Simonson, Roy W. 1957. What soils are. In Soil, Yearbook of Agriculture, 1957. Washington, United States Department of Agriculture. pp. 16–31.

Simpson, G. G. 1960. Types and name-bearers. Science 131 (3414):1684.

Smith, Harriet E. 1943. Polyploidy in *Sedum pulchellum*—II. Stomatal size and frequency. Bull. Torrey Bot. Club 70:261–264.

Smith, Harriet E. 1946. *Sedum pulchellum*: A physiological and morphological comparison of diploid, tetraploid, and hexaploid races. Bull. Torrey Bot. Club 73 (6):495–541, 42 figs.

Snedecor, George W. 1956. Statistical methods applied to experiments in agriculture and biology. 5th ed. Ames, Iowa State College Press. xiii, 534 p.

Soil Survey Staff, Soil Conservation Service, United States Department of Agriculture. 1960. Soil classification, a comprehensive system, 7th approximation. 265 p.

Steel, Robert G. D., and James H. Torrie. 1960. Principles and procedures of statistics. New York, McGraw-Hill. 481 p.

Stevenson, John S. 1962. The tectonics of the Canadian Shield. University of Toronto Press, 1962. 180 p., illus.

Steyermark, Julian A. 1934. Some features of the flora of the Ozark region in Missouri. Rhodora 36:214–233.

Steyermark, Julian A. 1942. Rediscovering *Sedum pulchellum* at its northeastern limit. Rhodora 44:73–76.

Steyermark, Julian A. 1962. Flora of Missouri. Ames, Iowa State University Press. 1,725 p.

Stockwell, C. H. 1961. Structural provinces, orogenies, and time classification of rocks of the Canadian Precambrian Shield. In: J. A. Lowdon, Age determinations by the Geological Survey of Canada. Report 2—Isotopic ages. Geol. Survey of Canada, Paper 61–17:108–118.

Stockwell, C. H. 1963. Second report on structural provinces, orogenies, and time classification of rocks of the Canadian Precambrian Shield. Geol. Surv. of Canada, Paper 62–17:123–133.

Stone, Hugh E. 1945. A flora of Chester County, Pennsylvania, with especial reference to the Flora Cestrica of Dr. William Darlington. 2 vols.

Stoutamire, W. P., and J. H. Beaman. 1960. Chromosome studies of Mexican alpine plants. Brittonia 12:226–230.

Strasburger, E. 1866–1867. Ein Beitrag zur Entwicklungsgeschichte der Spaltöffnungen. Jahrbücher für wissenschaftliche Botanik 5:297–342, pl. 35–42.

Subramanyam, Krishna. 1955. Morphological studies of some species of *Sedum*. I. Floral anatomy. Am. Jour. Bot. 42:850–855.

Subramanyam, Krishna. 1963. Embryology of *Sedum ternatum*. Jour. Ind. Bot. Soc. 42A:259–275.
Taylor, Roy L., and Gerald L. Mulligan. 1968. Flora of the Queen Charlotte Islands. Part 2. Cytological aspects of the vascular plants. Ottawa-Research Branch-Canada Dept. of Agriculture. Monograph no. 4, part 2. 148 p.
'T Hart, H. 1971. Cytological and morphological variation in *Sedum acre* L. in western Europe. Acta Bot. Neerl. 20 (3):282–290. Illus., maps.
Thornbury, William D. 1965. Regional geomorphology of the United States. New York, John Wiley and Sons, 1965. 609 p.
Torrey, John, and Asa Gray. 1838–1840. A flora of North America. Vol. 1. New York, Wiley and Putnam. 711 p.
Transeau, E. N. 1935. The prairie peninsula. Ecology 16:423–437.
Turesson, Göte. 1938. Chromosome stability in Linnean species. Annals Agric. Coll. Sweden 5:405–416.
Uhl, Charles H. 1952. Heteroploidy in *Sedum rosea* (L.) Scop. Evolution 6:81–86, figs. 1–5.
Uhl, Charles H. 1961. Some cytotaxonomic problems in the *Crassulaceae*. Evolution 15:375–377.
Uhl, Charles H. 1962. Chromosomes of *Sedum* in the western United States. Am. Jour. Bot. 49 (no. 6, pt. 2):664.
Uhl, Charles H. 1970. Heteroploidy in *Sedum glaucophyllum*. Rhodora 72:460–479.
Uhl, Charles H. 1972. Intraspecific variation in chromosomes of *Sedum* in the southwestern United States. Rhodora 74:301–320.
Uhl, Charles H., and Reid Moran. 1953. The cytotaxonomy of *Dudleya* and *Hasseanthus*. Am. Jour. Bot. 40:492–502, figs. 1–25.
Uhl, Charles H., and Reid Moran. 1972. Chromosomes of *Crassulaceae* from Japan and South Korea. Cytologia 37:59–81, figs. 1–55.
United States Department of Agriculture. 1941. Climate and man, yearbook of agriculture. Washington, U.S. Govt. print. off. xii, 1248 p.
United States Geological Survey. 1935. Geologic map of Colorado.
United States Geological Survey. 1955. Geologic map of Wyoming.
United States Weather Bureau. 1885—. Climatological data for the various states.
United States Weather Bureau. 1948—. Monthly climatic data for the world. Vol. 1—.
United States Weather Bureau. 1950—. Climatological data, national summary. Vol. 1—.
United States Weather Bureau. 1960–1964. Climatological data, California. Vols. 64–68.
von Engeln, O. D. 1940. Symposium: Walter Penck's contribution to geomorphology. Annals of the Association of American Geographers 30:219–284, figs. 1–4.
Webb, D. A. 1961. What is the type of *Sedum telephium* L.? Fedde's Repert. Specierum Nov. Reg. Veg. 64 (1):18–19.
Wenner, Carl-Gösta. 1947. Pollen diagrams from Labrador, a contribution to the Quaternary geology of Newfoundland–Labrador, with comparisons between North America and Europe. Geografiska Annaler 29:1–241, illus.
Wherry, Edgar T. 1934. The Sedums of the eastern United States. Gard. Chron. Amer. 38:264–266, illus.
Wherry, Edgar T. 1936. The ranges of our eastern Parnassias and Sedums. Bartonia 17:17–20.
Wherry, Edgar T. 1950. Saxiflora. Plateau stonecrop: *Sedum pulchellum*. Am. Rock Gard. Soc. Bull. 8:85–86, pl. 27.
Wherry, Edgar T. 1972. A checklist of the flora of Montgomery County, Pennsylvania. Bartonia No. 41:71–84.

Whitmore, Frank C., Jr., K. O. Emery, H. B. S. Cooke, and D. J. P. Swift. 1967. Elephant teeth from the Atlantic Continental Shelf. Science 156:1477–1481.

Wiens, Del, and Dianne Halleck. 1962. Chromosome numbers in Rocky Mountain plants. I. Botaniska Notiser 115 (4):455–464.

Wiggs, Deems N., and Robert B. Platt. 1962. Ecology of *Diamorpha cymosa*. Ecology 43:654–670.

Wilbur, Robert L. 1964. Notes on the genus *Diamorpha* (*Crassulaceae*). Rhodora 66:87–92.

Wilcoxon, Frank, and Roberta A. Wilcox. 1964. Some rapid approximate statistical procedures. Pearl River, Lederle Laboratories. 60 p.

Williams, Howel. 1957. A geologic map of the Bend quadrangle and a reconnaissance map of the central portion of the high Cascade Mountains. Ore. Dept. Geol. and Mineral Industries.

Williams, Howel, ed. 1958. Landscapes of Alaska, their geologic evolution. Berkeley, University of California Press. 148 p., 23 pl., 6 maps, 3 figs.

Wilson, M. E. 1958. Precambrian classification and correlation in the Canadian Shield. Bull. Geol. Soc. Am. 69:757–774, figs. 1–2.

Winterringer, Glen S., and Arthur G. Vestal. 1956. Rock-ledge vegetation in southern Illinois. Ecol. Monogr. 26 (2):105–130, illus.

Wooton, Elmer Otis. 1906. Southwestern localities visited by Charles Wright. Bull. Torrey Bot. Club 33:561–567.

"Zotova, M. I. 1965. [Russian title] The production of a preparation from *Rhodiola rosea* which has stimulating and antihypnotic action. In: Materialy teoreticheskoi i klinicheskoi, meditsiny. Tomsk Univ.: Tomsk Gosud. medits. inst. 5:86–89."

Zotova, M. I. 1966. [Russian title] A comparative characterization of the stimulating and adaptogenic action of extracts of roseroot *Sedum* and *Eleutherococcus*. In: Stimuliatory tsentral 'noi nervnoi sistemy. Tomsk, Tomsk Univ. pp. 67–71.

Abbreviations and Symbols

*	denotes a significant difference, the probability of no difference between populations being $<.05 > .01$
**	denotes a highly significant difference, the probability of no difference between populations being $<.01$
μ	arithmetical mean (average) of a total population
σ	standard deviation of a population
χ^2	chi-square
cf.	cold frame
D	minimal significant difference between means
e	estimated
F	variance ratio $= t^2$
fl.	flowering
fr.	fruiting
g	gametophytic or haploid number of chromosomes
gd.	garden
gh.	greenhouse
N	number, the total number of items in a sampled population
n	number of items in a sample
N_c	total number of plants in sampled clusters
n-obs.	number of observations of a characteristic in a whole sample
n-pl.	number of plants studied in a sample of plants
n-pop.	number of populations studied in a sample of populations
n-rep.	number of replicates, that is duplicates, propagated vegetatively and with identical genotypes
n.s.	not significant
N_u	total number of plants in a sampling unit
P	probability, as $P = .05$, probability of event is 5 in 100 or 1 in 20
pH	logarithm of the reciprocal of the hydrogen ion concentration
s	standard deviation of a sample; unless indicated otherwise, the standard deviation among plants of a sample
$s_{\bar{x}}$	standard error of the mean
SG	garden, 1421 Slaterville Road, Ithaca, N.Y.
sp	sporophytic or unreduced number of chromosomes
ssp.	subspecies
t	a statistic with a symmetric distribution and involving the relationship of $\bar{x}$ and $s_{\bar{x}}$

TG   Test Garden of Department of Floriculture, Cornell University, located in valley of Fall Creek between Forest Home and Varna, east of Ithaca, N.Y.
$\bar{x}$ arithmetical mean (average) of a sample
w   observed range of variation of a sample, the absolute extreme dimensions being cited

(For abbreviations of names of herbaria, see page 15; for abbreviations of names of geomorphic provinces, see p. 43; for designations of patterns of distribution, see p. 24; and for abbreviations of localities used in descriptions, see the accounts of the species.)

Supplementary Notes

Sedum stelliforme Watson, a white-flowered species of the mountains of the northern part of the Mexican Plateau, with gibbous, spreading follicles and elliptic-linear, short-spurred leaves, possibly should be added to the list of native species known from north of the Mexican Plateau. It probably occurs in the Zuni Mountains near the southern edge of the Colorado Plateau, although no recent collections are available since Dr. W. Matthews obtained it near Fort Wingate, N.M., in 1883 (NY). *Sedum stelliforme* is frequent in mountains of the adjacent Mexican Plateau: Magdalena Mts., 2438 m.; Black Range, 2073–2797 m,; Pinos Altos Mts., ~2134 m.; Mogollon Mts., 2329–3263 m.; and San Francisco Mts., 2140–2200 m.

Plants of *Sedum wrightii* at altitudes of 2515–2560 m. in the Sacramento Mts., N.M., are small and slender, with few-flowered cymes. Despite smallness and the cooler, moister habitat, these plants remarkably resemble plants at lower altitudes.

Additional records for maps are as follows:
Fig. 26, *Sedum pulchellum*: 34°–35° N, 97°–98° W and 35°–36° N, 95°–96° W
Fig. 32, *S. wrightii*: 32°–33° N, 105°–106° W
Fig. 76, *S. nuttallianum*:
    31°–32° N: 96°–97° W and 99°–100° W
    34°–35° N: 95°–96° W and 96°–97° W
    35°–36° N: 94°–95° W, 95°–96° W, and 97°–98° W
    36°–37° N: 95°–96° W, 97°–98° W, and 102°–103° W, the last, if the data are correct, on the High Plains of the Great Plains
Fig. 142, *S. integrifolium* ssp. *procerum*: 34°–35° N, 107°–108° W and 36°–37° N, 108°–109° W

Index

Principal entries for species and subspecies are indicated by numerals in italic type. New names of plants appear in boldface type.

Abbe, Elfriede, 89
Abbreviations, 722–723
Abnormalities, 529–531
Abruzzes, 569
Acer nigrum ssp. **saccharophorum**, 106, 696
Achillea borealis, 503, 637
Adaptation, 687
Aectyson sagittatum, 158
Affligée Islands, *see* Isles Affligées
Ahles, Harry E., 175
Aizoon acre, 551
Aksenova, R. A., 532
Alaska Basin, 232, 513, 643
Alaska Range, 53–54
Alaskan Peninsula, 46, 53, 626–627, 669
Albion, 553
Alexander, Edward J., 563
Algeria, 574
Allard, H. A., 37, 40, 89, 140, 541
Allisonia, 106, 121, 138–140, 629–631
Allopolyploidy, 684–685, 687, 698
Almeda, 438
Altai Mountains, 558, 561
Altitudes, 677
American Harbour Island, 342
Amphidiploidy, 141, 161
Anacampseros, 70
Anacampseros purpurea, 541
Anacampseros telephioides, 86
Anacampseros ternata, 105
Anacampseros triphylla, 541
Analysis of variance, 66, 136
Anatomy, 19, 35–37, 40, 91, 108, 235, 281, 344, 606, 680, 683–684, 693
Anderson, Alfred L., 51
Anderson, Loran C., 645
Aphid, 532
Apomixis, 227, 233, 471–472
Appalachian Mountains, 41–45, 56, 681

Appalachian Plateau, 43, 45, 58, 626–627, 637, 678
Appalachian Ridge and Valley Province, *see* Ridge and Valley Province
Applegate River, 391, 410
Arctic Island Province, 55, 626–627, 669
Arroyo del Puerto, 598
Atlantic continental shelf, 142
Aurora, 373
Auroral Plotting Map, 6
Autoploidy, 37, 104, 142, 161, 280, 698
Axenova, R. A., *see* Aksenova, R. A.
Azóuzetta Lake, 308

Babcock, Ernest B., 371, 692
Bailey, Edgar H., 49
Bailey, Ethel Zoe, 39–40
Bailey, L. H., 39–40, 558–559, 561, 569, 571–572, 574, 576
Bailey, Thomas L., 51
Bailey Hortorium, 556
Baker, Kenneth, 32
Baker, Milo S., 597
Bald Islands, 670–671
Bald Knob, 122, 139, 629–631
Bald Mountain, 310–311, 357
Baldwin, Ewart M., 57
Baldwin, J. T., Jr., 37, 40, 79, 102, 104, 108, 115, 120, 133, 151–152, 156, 161, 171, 329, 540, 604
Baldy Knoll, 642
Banach-Pogan, Eugenia, 525
Bandon, 467, 607
Banister, John, 31–32
Banks, Joseph, 35
Bard, Gily E., 138
Barnes Gap, 87–88
Basin and Range Province, 41, 46, 49
Baskin, Carol C., 606
Baskin, Jerry M., 161, 606
Bauhin, C., 541

INDEX

Baux, 577
Bayleaf, 138
Bear Creek, 638–640
Bear Gap, 245
Bear Mountain, 465
Bear Province, 54–55, 626–627, 669
Beard's Hollow, 355
Beartooth Lake, 485
Beartooth Mountains, 646
Beartooth Summit, 513
Beaver Creek Road, 257, 261
Bees, 89, 140, 161, 177, 188, 233–234, 246, 263, 289, 300, 311, 324, 333, 409, 423, 472, 486, 514, 647, 690
Beetles, 141, 280, 690
Belden, 432
Bell, C. Ritchie, 175
Benjamin, L. and R., 80, 115, 196, 202, 307, 497
Berg, N. K., 33
Bergen Park, 224
Berger, Alwin, 34, 177, 694
Berkshire, 537
Berthoud Pass, 484–485
Beyrich, Heinrich C., 117, 137
Bibliography, 711–721
Big Gull Island, 341–344, 674
Big Horn Basin, 46, 638–640
Big Spruce Knob, 45
Bigelow, J. M., 208
Bigness, 81–83
Bird's-claw Sedum, 143
Birman, Joseph P., 56
Bishop Creek, 655
Black Hills, 230
Blackfoot River, 298
Blankinship, J. W., 231
Blin, 571
Blochman, Ida M., 67
Bloomer Hill, 658–659
Blue Ridge, 14–15, 43–44, 626–627, 632, 676
Bluff Moss, 143
Bluffs Mountain, 87–88
Böcher, Tyge W., 14, 670, 674
Bohler Canyon, 652–653
Bonita Peak, 506–508
Borsch, Fred J., 299
Brandegee, Katherine, 69
Brandegee, T. S., 244
Breeze Creek, 230
Brewer, William H., 32–33, 262, 371
British Colour Council, 65
British Isles, 39
Britton, Nathaniel L., 34–35, 40, 324, 345, 533, 694
Brown, H. E., 33, 228, 463
Brown, N. E., 138
Brown, W. L., Jr., 360
Brownsville, 585

Bryan, Kirk, 61
Bryophyllum tubiflorum, 609
Buckhorn, 229
Buffalo Fork, 230
Bunge, A., 546
Burden's Lake, 554
Burnham, S. H., 371
Burnt Mountain, 465
Butterfield, H. M., 468
Butterflies, *see Lepidoptera*
Buwalda, John P., 49

Cabins, 138
Cahaba River, 628, 632
Calcium, 138
Calgary, 556–557
California Coast Ranges, 15, 23, 46, 50–51, 626–627, 666
Cameron Pass, 231–232
Camp, W. H., 692
Canada, 39, 59
Canadian Shield, 41, 54–55, 57–58, 669–674, 676
Canby, W. M., 137
Candolle, Augustin de, 34–35, 105
Canyon Creek, 470
Carolina, 104, 158
Carpels, 36, 108, 120, 177, 226, 334, 681; *see also* under Description in accounts of species
Carthage, 156, 158–159
Cascade Canyon, 224, 322, 645
Cascade Mountains, 15, 46, 51, 57–58, 310, 420, 626–627, 666–667, 676, 678, 695
Cascades, 122, 137, 139, 629–631
Casmaila Beach, 67, 608
Castle Rock, 464
Catawba River, 106, 635
Cathedral Bluff, 89
Cathey's Valley, 655
Caucasus, 569, 576
Caudle, Carol, 161
Cedar Cliff Mountain, 138–139
Central Lowland, 55, 626–627, 674–676
Central Rocky Mountains, 15, 46, 625–627, 637–647
Cerro Borrego, 563
Chamberlain, E. B., 554
Chandler, Harvey P., 393
Chapman, Alvan W., 86
Charlton Peak, 187
Chattahoochee River, 118
Chemistry, 531, 680, 685
Cherdyntsev, S. G., 532
Chickering, John, 32
Chico, 597–598
Chilnualna Falls, 279, 373, 656–657
Chimaera, 210
Chinook, 353–354

INDEX 727

Chittenango Falls, 533–534, 541
Chopaka Mountain, 224, 230
Chromosomes, 13, 31, 37, 40, 65, 104, 108, 119–120
  122, 124, 135, 140–142, 152, 157, 161, 196–197
  201–202, 209–211, 224, 227, 263, 281, 290, 298,
  302, 311–312, 324, 334, 358, 370, 461, 550, 585,
  681, 684, 690–691, 696–698; *see also* under Description in accounts of species
Churchill Province, 54–55, 626–627, 669, 676
Cinq-Mars, L., 554
Citations, 16, 19
City Creek, 68
Clark, William, 32
Clausen, Edna R., 209, 628, 632
Clausen, Eric, 310, 355
Clausen, Heidi, 423
Clausen, Jens, 19
Clausen, Joanna M., 628
Clausen, Robert T., 12, 18, 19, 33, 35, 37, 60, 65,
  68–69, 89, 184, 196, 198, 201, 207, 211, 224, 228,
  258, 299, 319, 345, 352, 355, 373, 388, 394, 401,
  405, 408, 432, 453, 506, 547, 553, 561, 576, 579,
  581, 608, 611, 645
Clausen, Thomas P., 209, 628
Clayton, John, 31–32, 68, 105
Clear Creek, 484
Clements, Frederic E., 498
Clements Mountain, 230, 513
Clementsia, 484, 486, 694
Clementsia rhodantha, 484
Cleveland, D., 69
Climate, 41, 60–62, 649–650
Clinton, G. W., 553
Clove, 86
Clusters, 115, 172–174, 176
Coast Batholith, 46, 53, 626–627, 669
Coast Mountains of British Columbia, 53
Coast Ranges of Oregon and Washington, 23, 46,
  53, 626–627, 668
Coastal Plain, 41, 55–56, 626–627, 675
Cochran, William G., 26, 65, 664
Cockerell, Theodore, 33, 198, 231, 233
Cody, W. J., 553
Coefficient of variation, 85, 184, 197, 256
Coffey, G. and J., 636
Colchester, 86
Cold Mountain, 106
Colorado, 48
Colorado Plateau, 46, 48, 323, 626–627
Colors, 65, 353, 370, 460, 499–500
Colton, 537
Columbia Plateau, 46, 48, 51, 57, 626–627, 666
Columbia River, 275, 354, 462–463
Combier, Henri, 531
Compatibility, 472
Competitors, 690, 698; *see also* under Distribution in accounts of species

Conclusions, 678
Congdon, J. W., 597
Congdonia pinetora, 67
Connecticut Hill, 537
Convergence, 358
Cook, Earl F., 57
Cooke, William B., 228, 373, 463
Cooks, 277
Coos Ridge, 355, 465
Copeland, E. B., 422
Copper Butte, 261
Cordillera, 41, 46, 56, 58, 60
Corpus Christi, 589
Cottonwood Creek, 470
Cotyledon anomala, 463
Cotyledon brittoniana, 396
Cotyledon burnhamii, 371
Cotyledon debilis, 321
Cotyledon glandulifera, 438
Cotyledon mendocinoana, 398
Cotyledon obtusata, 371
Cotyledon oregana, 354
Cotyledon oregonensis, 345, 419–420
Cotyledon retusa, 375
Cotyledon yosemitensis, 371
Craggy Pinnacle, 87
Crassula, 36, 585, 608–610, 617, 620, 648
Crassula aquatica, 177, 608–609
Crassula erecta, 608
Crassula muscosa, 608
Crassulaceae, 11, 13, 34, 36–38, 63, 177, 344, 611,
  681
Crater Lake, 412
Creep, 61
Creeping Sedum, 569
Crepis, 692
Crescent City, 467
Cristate condition, 146, 273–274
Crooked River, 598
Crystal Hill, 658, 660
Cuba, 190, 196
Cultivated species, 39–40, 556–583
Cultivation, 17, 39–40, 557, 583, 611, 696
Cumberland Plateau, 15, 152
Curtiss, A. H., 137, 158
Cusick, W. C., 287
Custer, 229, 233

Dalenius, Tore, 23, 649
Dalton Canyon, 200
Dartmouth Creek, 465
Data, 11, 21–22, 28, 30, 611
Davidson, Anstruther, 35
Davis, H. A., 140
Davis, Helen Burns, 198
Davis, Ray J., 244
De Chalmot, 140

Deadman's Point, 257–258, 261, 405, 408
Deadwood, 229
Deer Butte, 422
Deer Butte Trail, 465
Demaree, Delzie, 190
Dempster, Lawrence, 33, 261
Denffer, Dietrich von, 38
Density, 16, 21–23, 25–26, 30, 173
Description, 33, 40, 64–65; *see also* Description in accounts of species
Desiccation, 184, 376, 423
Devils River, 207–209
Diagnosis, 64, 66
Diamorpha, 13, 37, 40, 63, 163, 174, 177, 334, 584, 600–606, 609–611, 635, 683
Diamorpha cymosa, 177, 334, 584, *600–606*, 616, 620, 632–637
Diamorpha pusilla, 604
Diamorpha smallii, 605
Dickson, 33, 257
Dillenius, J. J., 577
Dimensions, 26, 85, 144–145, 172, 185
Dioecious condition, 528
Diptera, 188, 300, 376, 472, 486
Discovery, 31–33, 35, 40, 67
Diseases, *see* Pathology
Dispersal, 41, 535, 542
Distribution, 16, 18, 21, 28, 31, 35, 41, 59, 61–62, 64, 66–67, 92, 141, 235, 675, 696; *see also* Distribution in accounts of species
Dixon, Wilfrid J., 65
Don, George, 35
Donn, James, 35
Donner Pass, 372
Douglas, David, 32, 275, 462–463
Draba arabisans, 503
Dragon's Blood, 569
Drainage slopes, 677
Drake, 33, 257
Dress, William, 533
Drought, 685
Drunken Gulch, 655–656
Dry Gulch, 230
Dryden Lake, 537
Drying, 85, 185, 484, 529, 531
Dudleya, 63, 67–68, 379, 585, 606–610, 617–619, 648
Dudleya abramsii ssp. *murina*, 607
Dudleya bettinae, 607
Dudleya caespitosa, 68, 607
Dudleya candelabrum, 607
Dudleya cymosa, 607, 652, 658
Dudleya cymosa ssp. *cymosa*, 379, 607, 650, 662
Dudleya cymosa ssp. *minor*, 607
Dudleya cymosa ssp. *setchellii*, 607
Dudleya densiflora, 608
Dudleya edulis, 68

Dudleya farinosa, 607
Dudleya greenei, 607
Dudleya hallii, 372
Dudleya lanceolata, 372, 607, 662
Dudleya palmeri, 607
Dudleya parva, 607
Dudleya pulverulenta, 608
Dudleya saxosa, 608
Dudleya saxosa ssp. *aloides*, 607
Durham, 554
Dutchman's Peak, 263, 279–280, 405, 408, 410
Dyal, S. C., 137

Eames, Arthur J., 35
Eardley, A. J., 53
Earl Peak, 245
East Berkshire, 537
East Humboldt Mountains, 321
Eastern Isle Affligée, 341, 674
Eastwood, Alice, 33, 67–68, 398, 465
Echeveria, 372, 379, 585
Echeveria brittonii, 371
Echeveria debilis, 321
Echeveria derenbergii, 563
Echeveria gormania, 396
Echeveria hallii, 372
Echeveria obtusata, 371
Echeveria oregana, 354
Echeveria watsonii, 420
Ecology, 18, 31, 35, 40, 686–688; *see also* under Distribution in accounts of species
Ecotypes, 12
Edwards, J. L., 632
Edwards Plateau, 15
Eggleston, W. W., 198
El Paso, 208
El Portal, 688
El Salto, 202
Ellerd, 342
Elliott Corner, 598–599, 656–657
Elmer, Adolf, 32
Elwood Pass, 485–486, 506, 508
Embryology, 36, 40, 108
Emiliani, Cesare, 56
Endemism, 678–679, 695
England, 573
Epeteium americanum, 334
Ericson, David B., 56
Esopus Creek, 533–534
Estero Bay, 608
Etiolation, 185, 526
Etna Mills, 468
Eureka Gulch, 506, 508
Eureka Peak, 375, 660–662
Eureka Springs, 159
Europe, 340, 550, 558, 571, 577, 579, 679

INDEX 729

Evermann, B. W., 323
Evolution, 11–12, 60, 62, 235, 378, 403, 424, 535, 685, 693, 695–698; *see also* under Relationships in accounts of species
Ewan, Joseph, 69, 228
Experiments, 13, 16, 19–20, 26, 37, 64–66, 72–73, 80–81, 83, 89, 94, 103, 109, 116, 124, 134–136, 140, 142, 153, 185, 209–210, 214, 224–226, 238, 249, 264, 272, 280–281, 283, 287, 292, 297–298, 301, 308, 314, 347, 353, 360, 369–370, 390, 409, 412, 418, 423, 426, 435, 438, 443, 455–459, 461, 472, 477, 489, 516, 519–520, 522, 525–526, 691–692
Eyerdam, W. J., 511

False Pass, 511
Fateeva, A. P., 532
Feather Falls, 658–659
Feather River, 424, 432, 658–662
Fenneman, Nevin M., 44, 49
Fernald, Merritt L., 86, 542, 554
Fincastle, 138
Fire, 119, 608, 641, 643, 655
Fisher's Landing, 86
Flat Rock, 174, 605–606, 633–634
Flavonoids, 534
Flies, 177, 201, 233, 246, 263, 280, 289, 311, 324, 333, 343–344, 376, 400, 472, 514, 690
Flint, Richard Foster, 56–57
Flowering, 37–39, 154–155, 157–158, 208, 226, 459, 611, 687–688, 693; *see also* under Reproduction in accounts of species
Flowers, 135, 156, 184, 208, 226, 273, 298, 320, 330; *see also* under Description in accounts of species
Flyn, Nellie F., 86
Follicles, *see* under Description in accounts of species
Fontana-cungiada, 341
Forbes Creek, 598–599
Forsee Creek, 468
Forsyth Island, 533–534, 671–672
Forty-acre Rock, 172–173, 175–177, 604–606, 633–634, 688
Foster Gulch, 638–639
France, 561
Freer, Jay, 89, 109, 281
Freezing, 685
Fremont, John C., 32, 371
Fremont Peak, 262
Frequency, 16, 21–22
Fridriksson, Sturla, 534
Fröderström, Harald, 34–35, 69, 81, 86, 158, 177, 198, 231, 287, 308, 324, 334, 339, 372, 465, 532, 546, 550, 576, 582–583, 694
Frogplant, 537
Frozen Lake, 485

Fruits, *see* under Description in accounts of species
Funke, G. L., 38

Galice, 257, 397
Garden of the Gods, 87
Garden Orpine, 537
Garner, W. W., 37, 40, 89, 140
Gattinger, A., 118
Gavillan Peak, 262
Gazeteer, 699–710
Gem Pass, 230, 234, 373, 513, 652–653
Genera, 13, 63, 67, 583, 611
Genetics, 19, 37, 40, 472, 690
Genotypes, 16
Geography, 67, 625–679
Geology, 41–62
Geomorphic provinces, 15, 23, 625–628, 675–676, 678, 695, 699
Georgia, 158
Germination, 156, 161
Gibbs Peak, 231
Gilan, 569
Gilly, Charles L., 563
Gilpin, 87
Glaciation, 56–59, 641, 676
Glaciers, 56–60
Glassy Mountain, 81, 86–87
Glaucescence, 37
Gleason, H. A., 506
Glenora, 503
Golconda, 155, 159
Gold Beach, 392
Gold Lake, 375
Goldschmidt, Richard, 698
Goode, J. Paul, 699
Goode's Bridge, 138
Goode's Ferry, 632
Gorman, M. W., 345
Gormania, 63, 324, 345, 694
Gormania anomala, 463, 695
Gormania burnhami, 371
Gormania debilis, 321
Gormania eastwoodiae, 398
Gormania glandulifera, 438
Gormania hallii, 371
Gormania laxa, 396
Gormania obtusata, 371, 694
Gormania oregana, 354, 694
Gormania retusa, 375
Gormania rubroglauca, 372, 694
Gormania watsoni, 420, 695
Govorov, V. P., 532
Grand Teton, 47, 642
Grandfather Mountain, 85
Grant's Springs, 597, 654
Graptopetalum amethystinum, 562, 564

Graustein, Jeanette E., 354
Grave Creek, 257
Gray, Asa, 34–35, 40, 137, 158, 174
Grazing, 643
Great Basin, 41, 49, 323, 626–627, 648
Great Plains, 55, 60, 230, 626–627, 675–676
Great Valley of California, 23, 46, 50, 626–627
Greenland, 14, 55, 58, 626–627, 669, 674
Greig-Smith, P., 30
Grenville Province, 55, 625–627, 669
Grinnell, Hilda and Joseph, 32
Groghan Hole, 308
Gronovius, Joh., 31, 105
Groups of plants, 24–26
Guajuco, 562
Günthart, A., 344
Guerrero, 198
Guilday, John E., 506
Gulf of Georgia, 232
Gulf of St. Lawrence, 14, 16, 22, 28–30, 339–343, 669–674, 690
Gull Island, 533–534, 671

Habitats, 21, 61; *see also* under Distribution in accounts of species
Haines Falls, 519
Hall, H. M., 371, 467
Hansen, Morris H., 65
Hara, Hiroshi, 511
Hardin, James W., 635
Harney Peak, 229
Harper, Roland M., 118, 152
Harper's Ferry, 138
Harrington Harbour, 30, 669–672
Harris, Reuben, 633
Hartley, H. O., 29
Hartweg, R., 597
Hasseanthus, 67, 585, 608–610, 617, 619–620
Hasseanthus blochmaniae, 67–68, 608
Hasseanthus blochmaniae ssp. *brevifolius*, 68
Hasseanthus elongatus, 68–69, 608
Hasseanthus multicaulis, 68–69
Hasseanthus nesioticus, 67
Hasseanthus variegatus, 69
Hautes-Pyrénées, 571
Haworth, A. H., 541
Heaven's Gate, 230
Heggie's Rock, 632
Heller, Amos A., 33, 372–373, 375, 463
Hemiptera, 333
Henderson, Louis F., 33, 35, 232, 420
Henslow, G., 35
Herbaria, 15–16
Hess, W., 485
Hetch Hetchy, 468
Heusser, Calvin, 35

Hexaploidy, 156, 378, 424
Hicksville, 629–630
Hiesey, William M., 19
Hilton Creek, 652, 654
Hilton Lake, 4, 513, 652–653
Historical factor, 677
History, 31–40
Hodgdon, A. R., 554
Hokkaido, 566
Hollingshead, Lillian, 37, 152, 307, 319, 368, 388, 418, 438, 453
Holt, 109
Hooker, William J., 35, 228, 275, 462
Hopland Grade, 455
Horn, Charles, 80, 102, 133, 171, 223–224, 243, 255, 297, 319, 408, 453
Hornibrook, Murray, 67
Horse Mountain, 388
Hospital Island, 533–534, 671–672
Howell, Joseph, 32, 289, 420, 422
Howell, Thomas J., 32–33, 35, 257, 276, 396–397, 420, 422
Hudson Bay Lowland, 626–627
Hull Creek, 651
Hultén, Eric, 90
Hummingbird, 423
Huntley, Dorothy, 634
Hurwitz, William N., 65
Hutchison, Paul, 202, 388, 390–391, 408, 432–433
Huttleston, D. G., 547, 628, 633
Hybridization, 13, 21, 25, 58, 67, 91, 103–104, 108–109, 120, 132, 140–142, 177, 224, 228, 234, 243, 247, 273, 280–281, 290, 300, 311–312, 333, 357, 379, 424, 497–498, 552, 556, 577, 583, 606, 611, 665, 687–690, 693, 696, 698
Hybrids, 37, 67, 272, 280–281, 292, 301–303, 311–312, 400–403, 541, 552, 564, 566, 646, 689–690, 693
Hylander, Nils, 90
Hymenoptera, 188, 472
Hypertrophy, 532

Idaho, 299
Idaho Batholith, 46, 48, 323, 626–627, 647
Independence, 28
Inflorescences, 273; *see also* under Description in accounts of species
Ingalls Creek, 230, 234, 245, 310–311
Insecticides, 17
Insects, 685
Inspiration Point, 656–657
Interior Low Plateaus, 55, 626–627, 675
Interior Lowland, 41, 55, 58
Interior Plateaus and Ranges of British Columbia, Yukon, and Alaska, 53, 626–627, 668
Iron Mountain, 232
Irwin, William P., 50

INDEX   731

Isle Nadeau, 533–534, 671
Isles Affligées, 342, 673–674
Isoetes melanopoda, 636
Isolation, 13, 235, 696, 698
Ithaca, 16–17, 556–557
Ives, Sumner A., 118

Jack, Milton, 571–572
Jackson Hole, 46, 224, 234, 281, 645–647
Jackson Lake, 642
Jacob, F. H., 532
Jacquin, N. J., 577
Jahandiez, Emile, 341
Jalas, Jaakko, 90
James, Edwin P., 32, 227–228, 331, 510
James, W. O., 39
Jamesville, 547
Japan, 558, 569
Jemez Creek, 190
Jensen, Lawrence, 37
Jepson, Willis L., 35, 394, 468
Jermyn Creek, 544
Jessen, Raymond J., 24
Joe Wright Creek, 485–486
Jones, Frank, 674
Jones, George N., 19, 35, 277
Jones, Marcus E., 35, 276, 323, 550
Joplin, 159–160
Jordan Creek, 331–333, 690
Joyce, Barbara, 80, 108, 115, 352, 453
Juniperus osteosperma, 639

Kalanchoe, 585, 609, 617, 619
Kalanchoe crenata, 609
Kalanchoe pinnata, 609
Kalanchoe tubiflora, 609
Kalanchoe verticillata, 609
Kaliko, I. M., 532
Kamchatka, 566
Kécarpoui Island, 341, 343, 672–674
Keck, David D., 19
Keddie, 470, 660–661
Kelly Rock, 175–177, 605–606, 635–636
Kendrew, Wilfrid G., 61
Kerner, Anton, 336
Kessler, R., 186
Kezer, L. J., 628, 633
Keys, 462, 611–624; cultivated species, flowering condition, 620–624; native species, flowering condition, 611–617; native species, fruiting condition, 617–620
Kildale, Doris, 33
Kimnach, Myron, 202
King, Philip B., 41, 48
King's Crown, 487
Kinney Point, 275–276

Klamath, 465
Klamath Mountains, 23, 46, 48–50, 389, 626–627, 665–666, 676, 678, 695
Klickitat, 289
Klotz, 139, 629
Knowlton, Clarence H., 86
Knoxville, 31, 158
Kohlmeister, Benjamin, 32
Kolmakova, L. F., 532
Kowal, Robert R., 609
Krathwohl, Kathy, 152, 209

Lake Bonaparte, 554
Lake Eleanor, 371
Lake Emma, 224, 485–486
Lake Peak, 308, 506, 508
Lake Vernon, 371
Lapland, 532
Larkspur, 228
Lawrenceburg, 156
Leach, John R. and Lilla, 32, 257, 438, 463
Leaves, 28, 38, 134, 136, 155–156, 209–210, 226–227, 245, 320, 499, 517; *see also* under Description in accounts of species
Lebanon, 159
Lee, Robert, 121
Lee Price Camp, 652
Leedy, John L., 502
Lembert Dome, 371
Lenophyllum, 13, 63, 584–590, 609–611, 616, 620, 683
Lenophyllum acutifolium, 585, 590
Lenophyllum texanum, 584, *585–590*, 616
Lenophyllum weinbergii, 585
Leonard, E. C., 635
Lepidoptera, 177, 233, 324, 423, 690, 696
Lewis, Harrison, F., 342–343, 670
Lewis, Meriwether, 32
Lewis and Clark Expedition, 274
Liberty, 632
Liberty Hill, 634
Life forms, 686
Light, 17, 38, 89, 208, 521, 526, 528
Linnaeus, Carolus, 34–35, 532
Listing units, 23–25, 625
Literature, 19–20
Little Elk Creek, 229
Little Fishery Island, 342
Little Gull Island, 30
Little Moose Island, 533–534
Little Stony Creek, 137, 629–631
Little Yosemite Valley, 373, 655–656, 664
Live-forever, 537
Llano Grande Canyon, 607
Localities, *see* Gazetteer, 699–710
Lockwood, John L., 542
Lolo, 274

732   INDEX

Lolo Creek, 274–276
Loma Alta, 585
Long Island, 355
Long Lake, 658, 660
Lookout Mountain, 158, 244
Love, J. D., 641
Lower Lake, 597
Lulu Pass, 232, 512–513
Lummi Island, 465
Lummi Rocks, 224, 233–234, 465
Luray, 544

Macomber Peak, 224, 230
Macoun, John, 33, 467
Madow, William G., 65
Magazine Mountain, 45
Makanda, 159
Mammoth Lakes, 652–654
Manitou Springs, 228
Manning Canyon, 262
Manning Flat, 598–599
Maps, 6, 16–18, 21, 25, 42, 66, 699
Marina, T. F., 532
Marks Butte, 231–232
Marla, 577
Marlborough, 553
Martin, Paul S., 59
Massey, Frank J., 65
Masters, M. T., 39, 117, 137, 143
Mastrangelo, Iris, 80, 152, 207, 209, 223–224, 272, 297, 302, 319
Matthes, François E., 57
Mauritzon, Johan, 36, 40, 474, 585
Mayacamas Mountains, 464
McCarthy, Philip J., 25, 65, 650
McCormick, J. Frank, 156, 177
McDonald, J. V., 57
McKelvey, Susan D., 228
McVaugh, Rogers, 35, 632
Mean, 26–27, 65–66, 145, 443
Measurements, 18, 25, 28, 65
Mecatina Sanctuary, 342
Medicinal properties, 532, 696
Meehan, Thomas, 68
Methods, 13, 19, 21–22, 26
Mexican Plateau, 14, 679
Mexico, 565
Meyrán, Jorge, 561
Michaux, André, 31–32, 35, 85, 104, 158, 174, 177, 634
Middle Brummit Creek, 465
Middle Kécarpoui Island, 341, 343, 674
Middle Rocky Mountains, 46
Midway, 172–173, 175–177, 635
Miller, Faith, 89
Millicoma River, 355
Millrift, 533–534

Milwaukee, 276
Mission Peak, 234, 243, 245
Missouri Plateau, 58
Misty Harbor, 510
Mitchell, Elisha, 140
Mitra, Jyotirmay, 223
Modifications, 25, 82–84, 103, 115, 135, 152, 154, 157, 174, 184, 208, 243, 287, 320, 330, 369, 461, 521, 525
Mohr, C., 118
Monroe Canyon, 229
Monte Sano, 152–155, 158–159
Monument, 228
Moore Creek, 608
Moose, 275–276
Moran, Reid, 67–69, 179, 184, 234, 608, 628
Morgan, Avanell, 423
Mori, Antonio, 35
Moricetown, 234
Moris, G. G., 341
Morphology, 18, 38, 40, 64, 201, 227, 684–685, 693, 696
Morrison Ridge, 275–276
Morro Bay, 68
Mosquito Creek, 655–656
Mossy Stonecrop, 547
Moth, 433
Mount Adams, 308
Mount Baker, 244
Mount Bannon, 642
Mount Calvary Church, 634
Mount Eddy, 228, 373, 463
Mount Evans, 486
Mount Flora, 224, 232, 512–513
Mount Ghei Dagh, 576–577
Mount Gutgora, 574
Mount Hamilton, 598
Mount Hoffman, 371
Mount Hood, 420
Mount Horrid, 533–535
Mount Linn, 361
Mount Morgongiori, 341
Mount Rainier, 52, 57
Mount Shasta, 228, 373, 463
Mount Stuart, 51, 244–245
Mount Tmolus, 572
Mount Washington, 44
Mountain Home Creek, 468
Mountain Lake, 122, 137, 630–631
Mountains and Basins of southwestern Montana and adjacent Idaho, 49, 626–627, 648
Mud Creek Canyon, 373
Mud Lake, 422
Muhlenberg, Gotthilf H. E., 104
Mulligan, Gerald L., 307
Munk, Walter H., 60
Munz, Philip A., 607–608

INDEX   733

Murray, Grover E., 56
Mutations, 12, 461, 697

Nagai Island, 510
Nashville, 159
Natashquan, 520
Naturalized species, 537–555, 556, 583, 611, 617
Nealley, G. C., 589
Nectary, 36, 196–197; *see also* under Description in accounts of species
Nee, Michael, 609
Nelson, Elias, 244
Nelson Bar, 660–661
Nevada Falls, 467, 656
Nevius, Reuben, 32, 117–118
New England-Acadian Higlands, 41–44, 626–628, 676
New River, 138–139, 628–631
New York, 556–557
New York Mountains, 186
Newcomer, E. J., 244
Nieuwland, Julius A., 35
Niimoto, Dorothy, 102, 132–133, 224, 272, 319, 418
Nockamixon Rocks, 533–534, 632–633, 635–637
Noel, 159
Nomenclature, 64, 66; *see also* under Nomenclature in accounts of species
Nordwall Canyon, 643
North America, 11, 14, 21–22, 31–32, 34, 41–62, 69, 105, 625–679
North Buffalo Fork, 640–641
Northern Rocky Mountains, 46, 53, 626–627, 668
Number of plants, 21–22, 25–26, 30, 65, 245, 279, 310, 355–356, 628, 662–663, 688, 693
Numbering, 18, 25
Numbers, 18
Nutrition, 336
Nuttall, Thomas, 32, 35, 86, 158, 331, 353–354, 604, 634
Nye, Clarice, 308

Oak Creek Canyon, 199–200
Observatory Peak, 185
Ocoee Gorge, 118–119
Ocoee River, 106–107, 118–119
Offsets, 357, 376
Old Hollowtop, 510
O'Leary Mountain, 279, 357, 422, 465, 667, 688
Oleinichenko, V. F., 532
Olympic Mountains, 23, 53
Olympic Peninsula, 626–627, 668
Optimum stratum, 23, 649
Oregon City, 463
Oregon Coast Ranges, 49, 53, 668
Organ Mountains, 198

Origin, 234, 280, 312, 324, 358, 424, 433, 680–681, 683
Origin of species, 698
Ornamentals, 556
Orobanche fasciculata, 432
Orobanche uniflora, 468
Oroville, 658–659
Osborne, Charles, 342
O'Shaughnessy Point, 544
Ouachita Mountains, 41, 45, 626-627, 637
Ozark Plateaus, 55, 331, 626–627, 675

Pachyphytum, 585
Pacific Coast Ranges, 46
Pacific Mountain System, 15
Pacifica, 607
Papoose Creek, 289, 299–300
Parry, C. C., 484
Parvisedum, 13, 58, 63, 161, 324, 334, 590–600, 609–611, 648, 683
Parvisedum congdonii, 590, *591–600*, 616, 620, 650, 652, 656–658, 662, 665
Parvisedum leiocarpum, 590, *591–600*, 617, 620
Parvisedum pentandrum, 263, 590, *591–600*, 617, 620
Parvisedum pumilum, 590, *591–600*, 616, 620, 660–662, 665
Pasadena, 557
Pathology, 513, 532, 696
Patrick Creek, 394
Pattern of distribution, 24–25, 30, 62, 628
Paxton, Joseph, 39
Payne, Helen, 401–402, 473, 689
Payson, E. B. and L. B., 323
Peaks of Otter, 138
Pearson, E. S., 29
Peck, C. H., 554
Peck, M. E., 391, 396, 438
Peech, Michael, 138
Peirson, F. W., 187
Peking, 545
Pembroke, 139, 631
Penck, Walther, 61
Pennsboro, 159
Pennsylvania, 104
Penthorum sedoides, 68
Pentz, 598–599
Performance, 23, 26, 28
Perry, James B., 204
Peshastin Creek, 244–245
Petals, 84–85, 104, 116, 135, 158, 174, 184, 197, 207, 226, 273, 287, 298, 320, 339, 353, 370, 459–460, 484, 500–501, 517, 529, 604; *see also* under Description in accounts of species
pH, 17–18, 21, 66, 686–687, 693; *see also* under Distribution in accounts of species

Phillips, Arthur III, 186, 326, 329, 333
Photoperiod, 37–38, 40, 89, 140, 279
Phyletic lines, 12
Phyllotaxy, 37, 540
Physiological changes, 12
Physiology, 38, 40, 201, 458–459, 685–686
Piaget, Jean, 36
Piedmont Plateau, 43–45, 59, 625–627, 632–637
Pigeon Cove, 554
Pigeon Mountain, 157, 159, 628, 632
Pigmentation, 84, 185, 208
Pikes Peak, 233, 512–513
Pilgrim Creek, 641
Pine City, 69
Pine Mountain, 632
Pine Ridge Escarpment, 212
Pinnacles, 465, 598–599
Pinyon Peak Highlands, 638, 640–641
Pinzan, 565
Pistils, *see* under Description in accounts of species
Piute Peak, 506, 508
Plaskett Ranger Station, 370
Platoro, 229, 234, 506, 508
Pleistocene, 56–59, 142, 177, 247, 379, 389, 534
Plukenet, Leonard, 31
Plum Creek, 223, 227–229
Pluma, 229
Pohono Trail, 248
Pole Mountain, 224, 230, 233
Pollen, 58–59
Pollination, 67, 89, 140, 161, 233–234, 376, 400
Pollinators, 141, 188, 246, 263, 280, 289, 311, 324, 343, 400, 409, 423, 433, 472, 486, 514, 690
Polunin, Nicholas, 533, 670
Polyploidy, 13, 37–39, 103, 120, 135–136, 140, 142, 156, 224, 244, 555, 678, 697
Pomona, 608
Population, 14, 16, 21–22, 26, 61, 65, 80–81, 135, 140, 276, 280, 297, 339–340, 462, 628, 688, 695, 697; *see also* under Distribution in accounts of species
Populations, local, 12, 18
Porsild, A. E., 512, 670
Post Creek, 393
Potentilla, 19
Potomac River, 628
Powers, Margaret, 223, 247
Praeger, R. Lloyd, 39, 138, 464, 561, 579, 581–583
Pratt's Ferry, 118
Pringle, C. G., 198
Pronsky, Zaneta, 19
Propositions, 12, 697–698
Puget Trough, 52, 667–668
Pulga, 432
Purdy, Carl, 468
Pursh, Frederick, 32, 35, 105, 140, 274
Putnam, William C., 51

Quadrats, 22–23, 29–30, 172, 174, 176, 186–188, 626, 629, 669–670
Query Island, 672–673
Quimby, Maynard W., 36, 40, 69–70, 91, 108, 235, 344, 474, 585, 681–685

Races, 133, 136–138, 142
Radford, Albert E., 175
Radiation, 156–157
Rafinesque, 35, 331
Rancheria Creek, 373, 651
Rancheria Mountain, 279, 651
Range, 64
Rank sum test, 81, 120, 172, 410
Rate of sampling, 26
Ratio estimates, 664
Raton, 321
Rattlesnake Creek, 299–300
Rawlings, 138
Ray, Louis L., 56
Reading, John, 68
Real de Moran, 566
Red Lodge, 223
Red Mountain, 53, 398
Red River, 331
Red Rock, 606
Redding, 470
Reduction, 85, 273, 298
Reed Creek, 651
References, 19, 67, 711–721
Relationships, 18, 67, 680–693; *see also* Relationships in accounts of species
Replicates, 65, 136, 140
Reproduction, 61–62, 67, 85, 279; *see also* Reproduction in accounts of species
Rescue, 593
Reveal, Jack, 653
Reversed Peak, 652–653
Rhodiola, 63, 474, 694
Rhodiola alaskana, 506, 510
Rhodiola integrifolia, 509–510
Rhodiola neomexicana, 509
Rhodiola polygama, 510
Rhodiola quadrifida, 718
Rhodiola roanensis, 532
Rhodiola rosea, 510, 532–533
Rhodiola rosea ssp. *integrifolia*, 511
Rhodiola rosea var. *integrifolia*, 510
Rhodioloside, 531
Rich Bar, 432
Rich Gulch, 432
Ridge and Valley Province, 15, 44, 625–632
Riedel, Spencer M., 486
Rinehart, Dean, 49
Rio Hondo Canyon, 506, 508
Rivinus, 532
Roan Mountain, 532–534

Robinson, Benjamin L., 118
Rock Creek, 470, 660–661
Rock Point, 87–88
Rocks, 41, 59, 61, 677–679; *see also* under Distribution in accounts of species
Rocky Mountains, 510
Roderick, W., 261
Rodia, 474
Rodozin, 532, 696
Rogers Pass, 273
Rogue River, 257, 396, 438, 690
Root River, 502–503
Rose, Frank H., 298–299
Rose, Joseph N., 34–35, 40, 486, 681
Rosea, 532
Roseburg, 257
Roseroot, 487, 533
Rosewort, 533
Roubaix, 229
Round Valley, 652, 654
Ruidoso Creek, 190
Runcorn, S. K., 60
Rush, Ethel, 561
Rushing, William N., 156
Russell Springs, 158
Rydberg, Per A., 35, 323, 510

Sacramento Valley, 597
Saddle Mountain, 564
Sagaidak, L. P., 532
St. Augustin Square, 342
St. John, Harold, 32, 341
Saint-Lager, 37
St. Lawrence Lowland, 626–627, 674
St. Saveur Mountain, 533–534
Sajan Mountains, 558
Samples, 23, 25–28, 30, 64, 81, 186, 307, 431, 439–441, 650; *see also* under Description in accounts of species
Sampling, 13–14, 19, 21–30, 41, 64, 163, 336, 625, 664, 696, 698
Sampling units, 18, 21–26, 29–30, 66, 625–626, 664
San Bernardino Mountains, 15, 51, 185–188
San Diego, 69
San Juan Mountains, 47
San Luis, 563
San Luis Obispo, 464
San Luis Obispo County, 463
San Pedro River, 208
Sandia Mountains, 190, 512–513
Sangre de Cristo Mountains, 46–47, 60
Sanhedrin Mountains, 375
Santa Rosa Mountains, 186
Saratikov, A. S., 531–532
Scandinavia, 574
Schaeffer, R. L., 553
Schmidt, Karl P., 506

Schoodic Peninsula, 533–534
Scotty Creek, 224, 230, 234, 245, 275–276
Scurvy grass, 533
Seaview, 355, 420
Secondary sexual characteristics, 500, 522, 528–529
Sections, 680
Sedella, 597
Sedella congdoni, 597
Sedella leiocarpa, 597
Sedella pentandra, 597
Sedella pumila, 597
Sedella pumila var. *congdonii*, 597
Sedum, 5, 11, 13–14, 16, 18–25, 31–40, 58–62, 63, 67, 70, 626–627, 694
Sedum, group *Rosulata*, 345
Sedum, section *Eusedum*, 92
Sedum, section *Seda genuina*, 92
Sedum, section *Telephium*, 70
Sedum, subgenus *Gormania*, 35, 37, 312, 344, 357–358, 403, 426, 433, 435, 439, 473, 680, 682–685 687, 696–697
Sedum, subgenus *Gormania*, section *Gormania*, 35, 345, 358, 683–684, 695
Sedum, subgenus *Gormania*, section *Oreganica*, 345, 357–358, 683–684, 695
Sedum, subgenus *Gormania*, section *Rosulata*, 345, 358, 683–684, 695
Sedum, subgenus **Rhodiola**, 473–475, 486, 519, 535, 683, 686–687, 693, 695
Sedum, subgenus *Rhodiola*, section *Clementsia*, 474, 680, 683–684, 695
Sedum, subgenus *Rhodiola*, section *Rhodiola*, 474–475, 680, 683–684, 695
Sedum, subgenus *Sedum*, 91, 92, 344, 682–683
Sedum, subgenus *Sedum*, section *Cockerellia*, 92, 680, 682–683, 687, 695
Sedum, subgenus *Sedum*, section **Lanceolata**, 92, 333, 683, 687, 695
Sedum, subgenus *Sedum*, section *Ternata*, 92, 683, 687, 695
Sedum, subgenus *Sedum*, section **Tetrorum**, 92, 680, 683, 687, 695
Sedum, subgenus *Sedum*, section **Villosa**, 92, 684
Sedum, subgenus *Telephium*, 69–70, 91, 94, 474, 680–681, 683, 687, 693, 695
Sedum acre, 14, 36, *547–552*, 554, 556–557, 574, 615, 619, 623, 637, 670
Sedum acre ssp. *acre*, 550–551, 674
Sedum acre ssp. *glaciale*, 550
Sedum acre ssp. **majus**, 550–551
Sedum acre ssp. *neglectum*, 550
Sedum acre var. *aureum*, 550–551
Sedum acre var. *majus*, 550
Sedum acre var. *minimum*, 550
Sedum acre var. *minus*, 550
Sedum adenotrichum, 67
Sedum adolphii, 557, *564*, 622

Sedum aizoon, 38, 557, *566–567*, 621
Sedum alaskanum, 510
Sedum albidum, 67
Sedum **albomarginatum**, 32, 34, 40, 345, 359, 378–379, 394, *424–433*, 439, 617–618, 648, 659–662, 665, 675, 677, 683–685, 687–688, 691–692, 695–696
Sedum alboroseum, 553–554, *558*, *560*, 612, 618–619, 621, 624, 637
Sedum album, 67, 92, 553–554, 556–557, *571–572*, 613, 619, 622
Sedum algidum, 517, 536
Sedum allantoides, 581
Sedum alpestre, 582
Sedum alsinefolium, 344
Sedum altissimum, 577
Sedum ambiflorum, 473
Sedum amecamecanum, 561
Sedum americanum, 105
Sedum anacampseros, 70, 557, *561*, 621
Sedum anglicum, 557, *572–573*, 622
Sedum annuum, 14, 334, 552, 670, 674
Sedum annuum caule compresso, 31, 105
Sedum anoicum, 67
Sedum anomalum, 463, 467, 695
Sedum anomiosepalum, 198, 694
Sedum anopetalum, 554, 577
Sedum aoikon, 561
Sedum atropurpureum, 511, 517
Sedum aureum, 579
Sedum beyrichianum, 117, 137–138
Sedum bithynicum, 557, 575, 582
Sedum blochmanae, 67, 608
Sedum boloniense, 574
Sedum **borschii**, 32, 34, 40, 92, 244, 280–281, 283, 289, *290–303*, 614, 619, 648, 666, 683, 685–692, 694, 696, 698
Sedum botterii, 60
Sedum brevifolium, 557, *571*, 622
Sedum burhamii, 371, 694
Sedum caducum, 201–202, 211
Sedum caerulescens, 228
Sedum caeruleum, *574*, *576*, 623–624
Sedum californicum, 463, 695
Sedum chrysanthum, 410, 557, *576–577*, 623
Sedum chrysanthum ssp. **aizoon**, *577*
Sedum chrysanthum ssp. *chrysanthum*, *576–577*
Sedum ciliosum, 257
Sedum **clavatum**, *563*, 622
Sedum cockerellii, 32, 34–35, 92, 122, 175, 178 188, *189–202*, 208, 211, 235, 263, 613, 618, 647, 678, 682–683, 685–687, 691–692, 694
Sedum coeruleum, 36
Sedum compactum, 69
Sedum compressum, 562
Sedum confusum, *561*, 562, 621
Sedum congdonii, 597, 654
Sedum corynephyllum, 562

Sedum cotyledon, 68
Sedum craigii, 581
Sedum crassipes, 536
Sedum cristatum, 579
Sedum cupressoides, 581
Sedum cyaneum, 70, 579
Sedum cymosum, 334, 605
Sedum cymosum var. *smallii*, 605
Sedum dasyphyllum, 344, 557, *571*, 573, 623–624, 682
Sedum dasyphyllum ssp. *dasyphyllum*, *571*, *573*
Sedum debile, 32, 34, 92, 230, 233–234, 264, 304, 308–309, 311, *312–324*, 616, 619, 637–638, 643–645, 647–648, 666, 682–683, 685, 687–689, 691–692, 694
Sedum **decumbens**, 561, *562*, 621, 696
Sedum deficiens, 105
Sedum dendroideum, 561, 579
Sedum dendroideum ssp. *parvifolium*, 561
Sedum dendroideum ssp. *praealtum*, 561
Sedum diffusum, *564*, 622
Sedum divaricatum, 287
Sedum divergens, 32, 34, 37, 57, 92, 224, 233–234, 247, 264, 281, 290, *303–312*, 324, 357–358, 547, 582, 616, 620, 665–669, 682–683, 685–686, 688–692, 694, 695
Sedum douglasii, 274–275, 277, 694
Sedum douglasii f. *uniflorum*, 277
Sedum douglasii ssp. *ciliosum*, 257
Sedum douglasii ssp. *radiatum*, 262
Sedum douglasii var. *monanthum*, 277
Sedum douglasii var. *uniflora*, 276
Sedum dumulosum, 487, 583
Sedum eastwoodiae, 398, 694
Sedum ebracteatum, 210–211
Sedum elegans, 579
Sedum ellacombianum, 566
Sedum elongatum, 68
Sedum elrodi, 550
Sedum engleri, 107, 473
Sedum ewersii, 70 557, *558*, *561*, 624
Sedum fabaria, 69, 540
Sedum flaccidum, 188
Sedum floriferum, 566
Sedum formosanum, 473
Sedum forsterianum, 579
Sedum frigidum, 510
Sedum frutescens, 581
Sedum furfuraceum, 581
Sedum gertrudianum, 68
Sedum glanduliferum, 438
Sedum glandulosum, 341
Sedum glandulosum var. *minus*, 341
Sedum glaucophyllum, 32–33, 37, 40, 58, 92, 94, 106–109, 117–121, *122–142*, 161, 210, 439, 473, 581, 612, 617, 628–632, 637, 681, 683, 685–687, 689–692, 694, 698
Sedum glaucum, 574

INDEX 737

Sedum gracile, *573–574*, 623
Sedum grandipetalum, 235
Sedum greggii, *565*, 622
Sedum griffithsii, 178, 188, 190, 197–199, 201
Sedum griseum, 581
Sedum guatemalense, 564
Sedum hallii, 371, 694
Sedum hayesii, 561
Sedum heckneri, 391
Sedum hemsleyanum, 581
Sedum hesperium, 60, 89
Sedum hildebrandtii, 574
Sedum himalense, 583
Sedum hintonii, *565*, 624
Sedum hirsutum, *576*, 623
Sedum hirsutum ssp. *baeticum*, *576*
Sedum hirsutum ssp. *hirsutum*, *576*
Sedum hispanicum, 68, 198, 553–554, 556–557, *574–575*, 613, 620, 623
Sedum humifusum, 581
Sedum hybridum, *566*, *569*, 621
Sedum insulare, 341
Sedum integrifolium, 26–28, 32, 34–35, 37, 57–58, 86, 230, 234–235, 474–475, 486, *487–517*, 519, 535–536, 546, 583, 616, 618, 643, 646–648, 650, 653–654, 658, 662, 665–669, 675, 678, 683–685, 687–693, 695
Sedum integrifolium ssp. *integrifolium*, 502, 504, 506, *509–513*, 517, 647–648, 666–669, 674–675
Sedum integrifolium ssp. **leedyi**, 32, 488, 501, *502–506*, 514, 583, 637, 674
Sedum integrifolium ssp. **neomexicanum**, 33, 502, 504, 506, *509*
Sedum integrifolium ssp. **procerum**, 33, 502–504, *506–508*, 509–510, 515, 583, 647, 723
Sedum involucratum, 582
Sedum japonicum, 582
Sedum jepsonii, 396
Sedum kamtschaticum, 556–557, *566*, 568, 614, 617, 621
Sedum kamtschaticum ssp. *ellacombianum*, 553–554, *566*
Sedum kamtschaticum ssp. *kamtschaticum*, *566*, 568
Sedum kamtschaticum ssp. *middendorffianum*, *566*
Sedum kirilowii, 517, 536
Sedum lanceolatum, 32–33, 37, 92, 202, *211–236*, 236–238, 243–244, 246–248, 263–264, 272, 274, 280–281, 290, 302, 312, 324, 331, 379, 547, 581, 614, 619, 628, 640–641, 643–648, 650, 652–653, 662, 665–669, 675, 678, 680, 683, 685–692, 694–696
Sedum lanceolatum ssp. *lanceolatum*, 215–219, *227–230*, 231, 246, 513, 647–648, 665–666, 668–669, 675, 690
Sedum lanceolatum ssp. *nesioticum*, 33, 215–217, 227, 229, *232–233*, 667, 669
Sedum lanceolatum ssp. **subalpinum**, 33, 215–217, 227–229, *231–232*, 647–648, 668

Sedum lanceolatum var. *nesioticum* 232
Sedum lanceolatum var. *rupicolum*, 244
Sedum laxum, 32, 34, 335, 345, 358, 360, 378, *379–403*, 408–410, 425, 433, 439, 475, 582, 613, 617–618, 677, 683–685, 687–689, 691–692, 694
Sedum laxum ssp. **eastwoodiae**, 33, 379, 385, 390, 394, *398–399*, 400, 666
Sedum laxum ssp. *heckneri*, 33, 379, 384, 389–390, *391–394*, 399–403, 410, 421, 424, 439, 473, 665, 689–690
Sedum laxum ssp. *latifolium*, 33, 379, 384, 389–390, 392, *394–396*, 665
Sedum laxum ssp. *laxum*, 379–380, 389–390, 392, 394, *396–398*, 399–401, 410, 665
Sedum laxum ssp. *perplexum*, 396
Sedum laxum ssp. *retusum*, 375, 398
Sedum leibergii, 32, 34, 58, 92, 264, *281–290*, 300–302, 311–312, 473, 582, 614, 620, 648, 666, 683 685–688, 691–692, 694
Sedum leibergii var. *borschii*, 298
Sedum lenophylloides, 211
Sedum lineare, 547, *569*, 624
Sedum lucidum, *563*, 622
Sedum lumholtzii, 201
Sedum luteviride, 68
Sedum lydium, 36, *571–572*, 622
Sedum magellense, 344, *569*, 571, 621–622
Sedum makinoi, 582
Sedum maximowiczii, 558, 566
Sedum maximum, 558
Sedum meehani, 68
Sedum mexicanum, *565*, 624
Sedum middendorffianum, 566
Sedum minimum, 69, 178, 188, 551
Sedum minimum ssp. *minimum*, 189
Sedum monanthum, 276–277
Sedum monregalense, *571*, 623–624
Sedum monticola, 561
Sedum moranense, 557, *565–566*, 622
Sedum moranii, 32, 34, 40, 345, 359, 379, 394, 400, 425, *433–439*, 472–473, 582, 617, 666, 675, 677, 683–685, 687–688, 691–692, 695
Sedum morganianum, *564–565*, 623
Sedum multicaule, 68
Sedum multiceps, 557, *574*, 623
Sedum napiferum, 188
Sedum nesioticum, 232
Sedum nevadense, 177, 582
Sedum nevii, 32, 34–35, 37, 58, 92, 103, 106–108, *109–121*, 122, 136–138, 141, 152, 154, 157, 161, 177, 190, 334, 439, 473, 581, 612, 617, 628, 632, 637, 681, 683, 685–687, 691–692, 694
Sedum nevii var. *beyrichianum*, 117, 138
Sedum nicaeense, 577
Sedum niveum, 32, 34, 40, 69, 92, 175, *178–189*, 190, 201, 581, 613, 616–617, 648, 666, 675, 678, 682–683, 685–687, 690–692, 694, 696
Sedum nussbaumerianum, *563–564*, 622

Sedum nuttallianum, 32–33, 58, 92, 144, 160–161, 177, 309, *324–334*, 547, 582, 600, 615, 620, 674–675, 682–683, 685, 687–695, 723
Sedum nuttallii, 331
Sedum oaxacanum, 557
Sedum **oblanceolatum**, 32, 34, 40, 345, 360, 394, 400–401, *403–410*, 617–618, 666, 675, 683–685, 687–688, 691–692, 695–696
Sedum oblongorhizum, 68
Sedum obtusatum, 32, 34, 230, 234–235, 279, 345, *358–379*, 398, 401, 410, 426, 432, 439, 615–616, 618, 648, 652–658, 662–667, 683–685, 687–689, 691–692, 694, 697
Sedum obtusatum ssp. *boreale*, 33, 369, 371–372, *373–375*, 376, 379, 401, 404, 409–410, 421, 424, 613, 617, 659–660, 662, 665, 667
Sedum obtusatum ssp. *obtusatum*, 281, 359, 362, 365, 369, *371–373*, 376, 379, 582, 650, 653, 656, 658, 662
Sedum obtusatum ssp. **retusum**, 33, 370–372, *375–378*, 665–666
Sedum obtusatum var. *hallii*, 371
Sedum ochroleucum, 36, 235, 554, *577*, 579, 613, 615, 619, 623
Sedum octogonum, 105
Sedum oppositifolium, 569
Sedum oreganum, 32, 34, 36, 57, 281, 311–312, 344, *345–358*, 420, 422, 616, 618, 683–685, 687–689, 691–692, 694
Sedum oreganum ssp. *oreganum*, 349–350, *354–355*, 356–357, 582, 668–669
Sedum oreganum ssp. **tenue**, 33, 354, *355–357*, 667
Sedum oregonense, 32, 34, 37, 281, 345, 357–358, 360, 378–379, 400–401, 403, 408–409, *410–424*, 465, 582, 613, 617–618, 666–667, 683–685, 687–689, 691–692, 695–696, 698
Sedum oxypetalum, 581
Sedum pachyphyllum, 36, 557, *562–563*, 622
Sedum palmeri, 473, 557, *562*, 621
Sedum parvum, 333–334, 582
Sedum pedicellatum, 177, 344, 682
Sedum penthorum, 68
Sedum pilosum, *576*, 623
Sedum pinetorum, 67–69, 186, 648–649, 694
Sedum platyphyllum, 581
Sedum pluricaule, 579
Sedum polygamum, 510
Sedum populifolium, 36, *558*, 621
Sedum portulacoides, 104
Sedum potosinum, 564
Sedum praealtum, 557, *561*, 563, 615, 617, 621
Sedum praealtum ssp. **monticola**, 557, 561
Sedum praealtum ssp. **parvifolium**, 561
Sedum praealtum ssp. **praealtum**, 553–554, *561*
Sedum pruinatum, 583
Sedum pruinosum, 465, 695
Sedum puberulum, 198

Sedum pulchellum, 31–33, 37–40, 92, 107–109, 118–119, 122, *142–161*, 163, 177, 324, 334, 475, 581, 612, 619–620, 628, 632, 637, 674–675, 681, 683, 685–687, 689, 691–692, 694, 723
Sedum pulchrum, 158
Sedum pulvinatum, 581
Sedum pumilum, 597
Sedum purdyi, 468
Sedum purpurascens, 541
Sedum purpureum, 36, 70–71, 87, 89, 91, *537–543*, 554, 557–558, 612, 618, 621, 637
Sedum pusillum, 32–33, 35, 37, 40, 58, 92, *162–178*, 324, 334, 604, 606, 612, 620, 632–635, 637, 675, 680–681, 683, 685–688, 691–696
Sedum quaternatum, 547
Sedum radiatum, 32, 34, 92, *247–263*, 264, 272, 280, 290, 600, 613, 615, 620, 648, 656–658, 662, 665–666, 682–683, 685–686, 688, 691–692, 694, 697
Sedum radiatum ssp. **ciliosum**, 33, 256, *257–259*, 260, 263, 400, 665
Sedum radiatum ssp. **depauperatum**, 33, 256–258, 260–262, 263, 279, 408–409, 665
Sedum radiatum ssp. *radiatum*, 252, 257, 260, *262–263*, 665–666
Sedum reflexum, 36, 211, 235, 554, 556–557, 577, *579–580*, 615, 619, 623
Sedum retusum, 581
Sedum rhodanthum, 32, 34, 334, 474, *475–487*, 519, 583, 616, 618, 643, 645, 646–647, 683–685, 687–688, 691–692, 695
Sedum rhodiola, 509
Sedum rhodocarpum, 94, 107–108, 681
Sedum roanense, 532
Sedum rosea, 16, 18, 28–30, 32–33, 36–37, 59, 343, 474–475, 486–487, 504, 514–516, *517–536*, 541, 583, 616, 618, 628, 632, 636–637, 670–674, 678, 683–685, 687–689, 691–692, 695–696
Sedum rosea var. *leedyi*, 502
Sedum roseum, 474, 532
Sedum roseum ssp. *integrifolium*, 511
Sedum roseum ssp. *integrifolium* var. *frigidum*, 511
Sedum roseum var. *alaskanum*, 510
Sedum roseum var. *aleuticum*, 511
Sedum roseum var. *integrifolium*, 511
Sedum roseum var. *neo-mexicanum*, 509
Sedum roseum var. *polygamum*, 510
Sedum roseum var. *roanense*, 532
Sedum rubroglaucum, 371, 654, 694
Sedum rubrotinctum, 557, *564*, 622
Sedum rupestre, 211, 235, 554, *577*, 579, 623
Sedum rupicolum, 32, 34, 92, 211–212, 224, 228–230, 234, *236–247*, 311, 615, 619, 666, 675, 683, 685–692, 694, 698
Sedum sanctae monicae, 69
Sedum sanhedrinum, 375, 694
Sedum sarmentosum, *543–547*, 554, 556–557, 569, 615, 619, 624, 637

Sedum sediforme, 235, 474, 556–557, *577–578*, 623
Sedum semenovii, 475, 486–487
Sedum sempervivoides, 582
Sedum sexangulare, 36, 552, 554, 556–557, *574*, 615, 619, 623
Sedum shastense, 228, 694
Sedum sieboldii, 69, *557–558*, 624, 681
Sedum Silvermoon, 401–403
Sedum smallii, 605
Sedum sparsiflorum, 331
Sedum spathulifolium, 32, 34, 36, 58, 232, 234–235, 290, 344–345, 357–358, 402–403, 422–423, *439–473*, 582, 613–614, 617, 648, 683–685, 687–692, 695, 697
Sedum spathulifolium ssp. *anomalum*, 463
Sedum spathulifolium ssp. *pruinosum*, 33, 456, 462, 464, *465–467*, 666, 668–669
Sedum spathulifolium ssp. **purdyi**, 33, 400, 451–452, 458, 462, 464, *468–470*, 471, 659–662, 665
Sedum spathulifolium ssp. *spathulifolium*, 400–402, 440, 451, 460, *462–465*, 467, 665–669, 689–690
Sedum spathulifolium ssp, **yosemitense**, 32, 379, 457, 462, 464, *467–468*, 607, 656, 658, 662, 664–666, 688
Sedum spathulifolium var. *majus*, 467
Sedum spathulifolium var. *minus*, 463
Sedum spathulifolium var. *pruinosum*, 466
Sedum spathulifolium var. *purpureum*, 464
Sedum spectabile, 38, 69, 86–87, 556–557, *558–559*, 621, 624
Sedum spurium, 35, 553–554, 556–557, *569–570*, 614, 619, 624
Sedum stahlii, 304, 312, 557, *565*, 624
Sedum stellatum, 582
Sedum stelliforme, 723
Sedum stenopetalum, 32–33, 36, 40, 58, 92, 198, 211, 224, 227–228, 230, 234, 246–248, 263, *264–281*, 290, 292, 300–303, 547, 581, 613–615, 619, 645–647, 675, 678, 682–683, 685–692, 694–695, 697–698
Sedum stenopetalum f. *rubrolineatum*, 231
Sedum stenopetalum ssp. *ciliosum*, 257
Sedum stenopetalum ssp. **monanthum**, 32, 260, 274, *276–279*, 280, 379, 422, 648, 650, 656–658, 662, 665–668
Sedum stenopetalum ssp. *nesioticum*, 232
Sedum stenopetalum ssp. *radiatum*, 262
Sedum stenopetalum ssp. *stenopetalum*, 234, 260, 273, *274–276*, 280–281, 302, 357, 637, 648, 666–669, 675
Sedum stenopetalum var. *subalpinum*, 231
Sedum stephanii, 517, 536, 583
Sedum stoloniferum, *569*, 624
Sedum stribryni, 582
Sedum subalpinum, 231
Sedum subclavatum, 69, 228
Sedum tatarinowii, 69, 579

Sedum telephioides, 31–33, 37, 60, 69, *70–91*, 94, 107, 537, 541, 579, 611–612, 618–619, 628, 632, 637, 675, 679, 681, 683, 685–687, 691–696
Sedum telephium, 35, 38, 60, 70–71, 90–91, 540, 542, 557, *558*, 624, 681, 694
Sedum telephium ssp. *fabaria*, 70
Sedum telephium ssp. *purpurascens*, 542
Sedum telephium ssp. *purpureum*, 70, 541–542
Sedum telephium var. *album*, 90
Sedum telephium var. *purpureum*, 540
Sedum tenuifolium, 583
Sedum ternatum, 31–33, 35–37, 39–40, 58, *92–108*, 109, 119–120, 122, 136, 139–140, 141–142, 161, 177, 312, 334, 439, 473, 553–554, 581, 612, 619, 628–629, 631–632, 635, 637, 674–675, 679–683, 685–687, 689, 691–692, 694
Sedum ternatum var. *minus*, 105
Sedum texanum, 589
Sedum torreyi, 331
Sedum tortuosum, 579
Sedum torulosum, 581
Sedum treleasei, 557, *562*, 563, 622
Sedum triphyllum, 541
Sedum umbellatum, 308
Sedum uniflorum, 276–277
Sedum variegatum, 69
Sedum verticillatum, 541, 579
Sedum vigilmontis, 158, 694
Sedum villosum, 22–23, 28–29, 32, 34, 92, 309, *334–344*, 475, 582, 613, 620, 669–670, 674–675, 678, 680, 682, 684–685, 687–690, 692, 694–695
Sedum villosum var. *arcticum*, 339
Sedum villosum var. *aristatum*, 341
Sedum villosum var. *glabrum*, 339
Sedum vinicolor, 188
Sedum viridescens, 60, 91
Sedum wallichianum, 566
Sedum watsoni, 420
Sedum winkleri, 576
Sedum woodii, 463, 695
Sedum wootoni, 198, 694
Sedum wrightii, 32, 34, 37, 92, 175, 190, 198–199, 201, *202–211*, 581, 590, 612, 619, 675, 678, 682–683, 685–686, 688, 691–692, 694, 723
Sedum yosemitense, 467, 654, 695
Seedlings, 84, 140, 154, 157, 184, 201, 233, 261, 323
Seeds, 140, 155–156, 184, 201, 209–211, 233, 245, 289, 300, 323, 333, 342, 423; *see also* under Description in accounts of species
Selese Mountain, 356
Sempervivum, 585, 609, 616, 618
Sempervivum heuffellii, 609
Sempervivum tectorum, 609
Seneca Lake, 26, 86, 505
Sepals, *see* under Description in accounts of species
Serpentine, 244, 378, 677
Serpentine Canyon, 432–433, 660–661

Seven Devils Lake, 230, 234, 513
Seven Devils Mountains, 14, 51, 57–58, 244, 288, 300
Sex ratios, 500, 528
Shag Rock, 533–534, 672–673
Shapley, Harlow, 60
Sharpe, C. F. Stewart, 61
Sharsmith, C. W. and H. K., 598
Shawnee Hills, 86
Sheldon, E. P., 355
Shenandoah Mountain, 122
Sherwin, Priscilla, 37, 40, 177, 334, 584, 606, 681, 684
Shikoku, 558
Shingle Creek, 289
Shoal Creek, 333
Short Trail, 654
Siberia, 566
Siebold's Stonecrop, 557
Sierra Blanca, 509
Sierra Buttes, 375, 658–659
Sierra City, 650
Sierra de Palma, 576
Sierra del Carmen, 208
Sierra Madre Occidental, 199
Sierra Madre Oriental, 678
Sierra Nevada, 15, 46, 48–49, 56–57, 60, 69, 186, 279, 625–627, 648–665, 676
Sierra San Pedro Mártir, 185
Silver Lake, 361
Silver Mills, 87–88
Simmers, R. W., Jr., 109
Simonson, Roy W., 59
Simpson, G. G., 66
Siskiyou Mountains, 50
Sizes of populations, 30, 688, 695
Skeena River, 464
Slave Province, 54, 626–627, 669
Slopes, 61
Small, Frank, 635
Small, J. K., 532, 605
Smith, Harriet, 38, 40, 155–156, 161
Smith, S. J., 109, 118
Smith River, 394, 632
Snake River, 287
Snedecor, George W., 16, 65, 73, 136, 443
Snow Camp Lookout, 463
Snowdon rose, 533
Soil, 17, 21, 59
Somers, 550
Sonora, 201, 650
Sonora Pass, 361
Sonoran Desert, 41
South Carolina, 163
South Stag Island, 535
Southern Blue Ridge, 14

Southern Province, 54–55, 626–627, 669
Southern Rocky Mountains, 15, 46–48, 56, 230, 626–627, 647
Spain, 574
Spanish Creek, 470, 660–661
Sparta, 86
Species, 12, 19, 33, 40, 64, 66, 692, 697
Sperry, Omer, 204
Sperry Glacier, 231
Spray, 138
Stamens, *see* under Description in accounts of species
Standard deviation, 30, 65, 145, 443
Standley, Paul C., 86
Statistics, 13, 65
Staunton, F. W., 33, 470
Steel, Robert G. D., 65, 73, 81, 116, 453, 483
Stein's test, 26–27, 360
Stevenson, John S., 54
Stevenson Island, 341–342, 533–534, 671, 674
Steyermark, Julian A., 118, 541
Stockwell, C. H., 54
Stomates, 35
Stone, Hugh E., 541
Stoutamire, W. P., 189
Strait of Georgia, 233
Strasburger, E., 35
Strata, 23, 25, 28, 41, 65, 441–442, 629, 633, 665
Stratton, Thomas, 244
Stringtown, 470
Subgenera, 680
Subramanyam, Krishna, 36, 40, 108, 281, 682
Subsampling, 24, 115
Subspecies, 12, 32–33, 64, 66, 227, 257, 273–274, 353, 370, 390, 457–458, 501–502, 611, 628, 697–698
Succulence, 39
Sugarloaf Mountain, 184–188
Sugarloaf Peak, 598–599, 690
Suksdorf, Wilhelm N., 32, 277, 308
Summers, R. W., 463
Summit Peak, 231–232, 234, 506, 508
Sunlight, 83, 188, 526–527
Superior Province, 54, 626–627, 669
Surtsey, 534
Survey, 13–15, 21–23, 41, 625–626, 629, 649, 665, 696, 698
Sweet, James, 633
Sweetser, Albert R., 464
Syczawinski, A. F., 308
Sydney, 228
Symbols, 722–723

Taurus Range, 576–577
Taxonomy, 13, 19, 21

INDEX 741

Taylor, Roy L., 307–308
Techniques, 18
Teewinot Mountain, 47, 224, 226, 230, 233–234, 322, 643–645, 647
Tehuacán, 562, 565
Telephium, 70, 474
Telephium album, 90
Telephium purpurascens, 541
Telephium purpureum majus, 540
Telephium Virginianum petraeum album, 31
Telephium vulgare, 70, 90
Temperature, 17, 188
Terrell, E. E., 72, 83
Terrero, 196
Terry Peak, 229
Tesuque Peak, 232
Teton Canyon, 322, 645
Teton Mountains, 46–47, 322–323, 485–486, 638, 641–645, 647
Tetraploidy, 83, 104, 143, 155–156
Tetrorum, 92, 174, 681, 694
Tetrorum pusillum, 174
'T Hart, H., 550
Thornbury, William D., 48–49
Thorne, Robert, 628
Thrips, 263
Thuleaphis sedi, 532
Tillaea, 36, 608
Tillaea cymosa, 604–605
Tillett, 638–639
Tiscalatengo River, 563
Tony Lake, 319
Topography, 23, 26, 41, 62
Torrey, John, 32–35, 86, 158, 331, 372, 510
Townsend, C. H., 510
Tracy, J. P., 308
Tracy, S. M., 321
Trans-Mexican Volcanic Belt, 13, 18–19, 65, 611, 677–678, 686, 688, 692–693, 694
Transeau, E. N., 504, 506
Transverse Ranges of California, 23, 46, 51, 626–627, 666
Trapido, Harold, 373, 632
Traveller's Rest Creek, 274
Treasure Mountain Camp, 645
Trinity River, 393
Trinity Summit, 308
Trips, 14
Tuerto Mountain, 198, 200–201
Tukey, J. W., 340
Tultenango, 565
Tunnel Hill, 159
Tuolumne Meadows, 371
Turesson, Göte, 90
Turkey Mountain, 86
Tuscaloosa, 117–118

Type localities, 66–67, 699

Uhl, Charles H., 35, 37, 40, 67, 80, 83, 102, 108–109, 115, 118, 122, 132–133, 152, 184, 196, 207, 223–224, 243, 255, 272, 287, 352, 370, 388, 431, 453, 482–483, 487, 497, 525, 534, 545, 550, 585, 632
Unimak Island, 511
United States, 39, 59
United States South Pacific Exploring Expedition, 289
Upper Marias Pass, 276
Ural Mountains, 569
Urbana, 158

Valentine Ridge, 376
Vancouver Island, 233, 467
Variance, 26–27, 65
Variation, 24–28, 38, 40, 65–66, 115, 556, 611, 663, 691–692, 698; *see also* under Variation in accounts of species
Ventura, 553
Vernal Falls, 654, 656–657, 663
Villa García, 207–209
Villa Guerrero, 563
Villa Santiago, 562
Villadia, 585, 609, 683
Villadia batesii, 557, 561
Villadia Texana, 589
Virginia, 104–105, 158
Vivian–Morel, Victor, 37
Vocabulary, 19
von Engeln, O. D., 61

Wadesboro, 174
Waldo, 396–397
Wallowa Mountains, 57
Walther, Eric, 563
Warm Springs, 140, 628, 632
Warnock, Barton H., 204
Warrior River, 109, 118–119
Watering, 17
Watkins Glen, 503, 546, 688
Watson, Sereno, 32, 35, 40, 198, 308, 321, 323, 420
Wax, 459
Webb, D. A., 90
Wedge Mountain, 355–357
Wenatchee Mountains, 15, 224, 245, 281, 310–311
Wenner, Carl-Gösta, 59, 535
West Berkshire, 537
West Indies, 14
West Spanish Peak, 229, 510, 512–513
Western Isle Affligée, 341–342, 672–674
Wheeler, Mrs. Alonzo, 554
Wherry, Edgar T., 35, 143, 156, 158
White Mountain Peak, 509

Whitehorse, 228
Whitmore, Frank C., Jr., 142
Whittaker, Robert H., 32, 408
Wibbe, John H., 554
Widow's Cross, 143
Wiens, Del, 483
Wilbur, Robert, 37, 40, 177, 334, 604
Wilcox, Roberta A., 81, 120, 172
Wilcoxon, Frank, 81, 120, 172, 281
Wildwood, 393
Wilkes Expedition, 289, 308
Willamette-Puget Trough, 23, 52–53, 626–627, 667–668
Willdenow, Carl, 35, 104
Williams, Howel, 52, 54
Williams, M., 261
Willow Creek, 198
Wilson, M. E., 54
Winchester, 608
Windy Ridge, 299–300
Winsor's Ranch, 509
Wisconsin glaciation, 56, 58
Wood, A., 463

Woods Creek, 596, 598–599, 652
Wooton, Elmer O., 32–33, 198, 208, 509
Wright, Charles, 32, 198, 208–209
Wright, John D., 373
Wyomide Ranges, 46, 48, 323, 626–627, 648
Wyoming, 48
Wyoming Basin, 46, 48, 626–627, 647

Yadkin River, 605
Yellowknife Province, 54
Yellville, 159
Yosemite Gorge, 262, 373, 468
Yosemite National Park, 263, 280, 467
Yosemite Valley, 371, 654
Youngs, George B., 32, 432
Yukon Plateau, 53

Zacapoaxtla, 561
Zacuapan, 563
Zandstra, Ilse, 80, 102, 132, 184, 196, 223–224, 272, 368, 418, 453, 497, 525
Zotova, M. I., 532
Zuasse, 576